Fate and Effects of Petroleum Hydrocarbons
in
Marine Ecosystems and Organisms

PERGAMON TITLES OF RELATED INTEREST

AMAVIS & SMEETS — Principles and Methods for Determining Ecological Criteria on Hydrobiocenoses

BADCOCK & MERRETT — Midwater Fishes in the Eastern North Atlantic — Vertical Distribution and Associated Biology in 30°N, 23°W, With Developmental Notes on Certain Myctophids

KRAUS — Modelling and Prediction of the Upper Layers of the Ocean

PEARSON & FRANGIPANE — Marine Pollution and Marine Waste Disposal

Fate and Effects of Petroleum Hydrocarbons in Marine Ecosystems and Organisms

Proceedings of a Symposium
November 10-12, 1976, Olympic Hotel, Seattle, Washington

Edited by
Douglas A. Wolfe

with assistance from
J. W. Anderson, D. K. Button, D. C. Malins
T. Roubal and U. Varanasi

Sponsored by
National Oceanographic and Atmospheric Administration
Environmental Protection Agency

PERGAMON PRESS
Oxford / New York / Toronto / Sydney / Paris / Frankfurt

Pergamon Press Offices:

U.S.A.	Pergamon Press Inc., Maxwell House, Fairview Park, Elmsford, New York 10523, U.S.A.
U.K.	Pergamon Press Ltd., Headington Hill Hall, Oxford OX3, OBW, England
CANADA	Pergamon of Canada, Ltd., 207 Queen's Quay West, Toronto 1, Canada
AUSTRALIA	Pergamon Press (Aust) Pty. Ltd., 19a Boundary Street, Rushcutters Bay, N.S.W. 2011, Australia
FRANCE	Pergamon Press SARL, 24 rue des Ecoles, 75240 Paris, Cedex 05, France
WEST GERMANY	Pergamon Press GmbH, 6242 Kronberg/Taunus, Frankfurt-am-Main, West Germany

Copyright © 1977 Pergamon Press Inc.

Library of Congress Catalog Card No. 77-76464

All Rights Reserved. No part of this publication may be reproduced, stored in a retrieval system or transmitted in any form or by any means: electronic, electrostatic, magnetic tape, mechanical, photocopying, recording or otherwise, without permission in writing from the publishers.

ISBN 0-08-021613-7

Printed in the United States of America

CONTENTS

List of Illustrations   ix

Preface   xix

Part I. Invited Papers

1. Oil Spills in the Alaskan Coastal Zone - The Statistical Picture. J. R. Harrald, B. D. Boyd, and C. C. Bates. . . .   1

2. Hydrocarbons in the Water Column, D. G. Shaw   8

3. Dispersal and Alteration of Oil Discharged on a Water Surface. Clayton D. McAuliffe.   19

4. Biodegradation of Aromatic Petroleum Hydrocarbons. David T. Gibson   36

5. Biotransformation of Petroleum Hydrocarbons in Marine Organisms Indigenous to the Arctic and Subarctic. Donald C. Malins   47

6. Accumulation and Turnover of Petroleum Hydrocarbons in Marine Organisms. Richard F. Lee   60

7. Food Chain Transfer of Hydrocarbons. John M. Teal.   71

8. Comparative Oil Toxicity and Comparative Animal Sensitivity. Stanley D. Rice, Jeffrey W. Short, and John F. Karinen   78

9. Responses to Sublethal Levels of Petroleum Hydrocarbons: Are they Sensitive Indicators and do they Correlate with Tissue Contamination? J. W. Anderson   95

10. The Effects of Petroleum Hydrocarbon Exposure on the Structure of Fish Tissues. Joyce W. Hawkes   115

11. The Effects of Petroleum Hydrocarbons on Marine Populations and Communities. A. D. Michael   129

Part II. Contributed Papers

A. Biological Effects

12. Effects of Certain Petroleum Products on Reproduction and Growth of Zygotes and Juvenile Stages of the Alga *Fucus Edentatus* De La Pyl. Richard Lewis Steele.   138

13. Effect of Crude Oil on Trout Reproduction. H. O. Hodgins, W. D. Gronlund, J. L. Mighell, J. W. Hawkes, and P. A. Robisch   143

14. Thermal Conductance of Immersed Prinniped and Sea Otter Pelts before and after Oiling with Prudhoe Bay Crude. G. L. Kooyman, R. W. Davis and M. A. Castellini.   151

15. Effects of External Applications of Fuel Oil on Hatchability of Mallard Eggs. Peter H. Albers.   158

16. Effects of External Applications of No. 2 Fuel Oil on Common Eider Eggs. Robert C. Szaro and Peter H. Albers . .   164

17. The Effect of Petroleum Hydrocarbons on the Survival and Life History of Polychaetous Annelids. Robert Scott Carr and Donald J. Reish. 168

18. Cytological Damage in Mercenaria mercenaria Exposed to Phenol. C. R. Fries and M. R. Tripp. 174

19. Interactive Effects of Temperature, Salinity Shock and Chronic Exposure to No. 2 Fuel Oil on Survival, Development Rate and Respiration of the Horseshoe Crab, Limulus Polyphemus. R. B. Laughlin, Jr. and J. M. Neff 182

20. Effects of Dispersed Crude Oil upon the Respiratory Metabolism of an Arctic Marine Amphipod, Onisimus (Boekisimus) Affinis. J. A. Percy. 192

21. Accumulation of Naphthalenes by Grass Shrimp: Effects on Respiration, Hatching, and Larval Growth. H. E. Tatem. . 201

22. Effects of a Seawater-Soluble Fraction of Cook Inlet Crude Oil and its Major Aromatic Components on Larval Stages of the Dungeness Crab, Cancer magister Dana R. S. Caldwell, E. M. Caldarone and M. H. Mallon. 210

23. Molting and Survival of King Crab (Paralithodes Camtschatica) and Coonstripe Shrimp (Pandalus Hypsinotus) Larvae Exposed to Cook Inlet Crude Oil Water-Soluble Fraction. T. Anthony Mecklenburg, Stanley D. Rice, and John F. Karinen 221

24. Response of the Clam, Macoma balthica (Linnaeus), Exposed to Prudhoe Bay Crude Oil as Unmixed Oil, Water-Soluble Fraction, and Oil-Contaminated Sediment in the Laboratory. Tamra L. Taylor and John F. Karinen. 229

25. Long Term Biological Effects of Bunker C Oil in the Intertidal Zone. Martin L. H. Thomas 238

26. Biological Survey of Intertidal Areas in the Straits of Magellen in January, 1975, Five Months after the Metula Oil Spill. Dale Straughan. 247

B. Bioaccumulation and Metabolism

27. Studies on Petroleum Biodegradation in the Arctic. Ronald M. Atlas 261

28. Arctic Hydrocarbon Biodegradation. S. D. Arhelger, B. R. Robertson, and D. K. Button 270

29. Bioavailability of Sediment-Sorbed Naphthalenes to the Sipunculid Worm, Phascolosoma agassizii. J. W. Anderson, L. J. Moore, J. W. Blaylock, D. L. Woodruff, and S. L. Kiesser 276

30. Factors Affecting the Retention of a Petroleum Hydrocarbon by Marine Planktonic Copepods. R. P. Harris, V. Berdugo, E. D. S. Corner, C. C. Kilvington and S. C. M. O'Hara. 286

31. Effects of Temperature and Salinity of Naphthalenes Uptake in the Temperature Clam, Rangia cuneata and the Boreal Clam, Protothaca staminea. Kenneth W. Fucik and Jerry M. Neff 305

32. The Accumulation and Depuration of No. 2 Fuel Oil by the Soft Shell Clam, Mya arenaria L. Dennis Stainken. 313

33. Effects of Chlorinated Biphenyls and Petroleum Hydrocarbons on the Activity of Hepatic Aryl Hydrocarbon Hydroxylase of Coho Aalmon (Oncorhyunchus kisutch) and Chinook Salmon (O. tshawytscha). Edward H. Gruger, Jr., Marleen M. Wekell, and Paul A. Robisch. 323

34. The Fate of Petroleum Hydrocarbons from a No. 2 Fuel Oil Spill in a Seminatural Estuarine Environment. Rudolf H. Bieri and Vassilios C. Stamoudis 332

C. Distribution and Movement of Hydrocarbons

35. Interlaboratory Calibration for the Analysis of Petroleum Levels in Sediment. S. A. Wise, S. N. Chesler, B. H. Gump, H. S. Hertz and W. E. May. 345

36. Determination of the Leeway of Oil Slicks. Craig L. Smith . 351

37. Evaporation and Solution of C_2 to C_{10} Hydrocarbons from Crude Oils on the Sea Surface. Clayton D. McAuliffe 363

38. Input of Low-Molecular-Weight Hydrocarbons from Petroleum Operations into the Gulf of Mexico. James M. Brooks, Bernie B. Bernard and William M. Sackett. 373

39. Intertidal Sediment Hydrocarbon Levels at Two Sites on the Strait of Juan De Fuca. W. D. MacLeod, Jr., D. W. Brown, R. G. Jenkins, and L. S. Ramos 385

40. Characterization of Volatile Hydrocarbons in Flowing Seawater Suspensions of Number 2 Fuel Oil. R. M. Bean and J. W. Blaylock 397

41. Sediment Hydrocarbons as Environmental Indicators in the Northeast Gulf of Mexico. Julia S. Lytle and Thomas F. Lytle. 404

42. The Stability of Emulsified Crude Oils as Affected by Suspended Particles. C. P. Huang and H. A. Elliott. 413

43. Chemical Carcinogens in the Marine Environment. Benzo(a)Pyrene in Economically-Important Bivalve Mollusks from Oregon Estuaries. Michael C. Mix, Ronald T. Riley, Keith I. King, Steven R. Trenholm and Randy L. Schaffer 421

44. Seasonal Variations of Hydrocarbons in the Water Column of the MAFLA Lease Area. John A. Calder. 432

45. Distribution of Petroleum Hydrocarbons in Westernport Bay (Australia): Results of Chronic Low Level Inputs. Kathryn A. Burns Ph.D. and Jonathan L. Smith 442

Part III. Panel Discussion on Research Needs

46. Discussion 454

List of Symposium Papers Presented and Authors. 465

Index 467

LIST OF ILLUSTRATIONS

Fig. 2.1 Tetrahedral structure of ice due to hydrogen bonding. Distortions in the liquid state lead to closer packing and a denser structure.

Fig. 2.2 Structure of chlorine hydrate, $Cl_2 \cdot 8H_2O$. The water structure only is shown. A chlorine molecule resides in the center of each larger cavity.

Fig. 2.3 Schematic representation of a spherical micelle composed of anions of fatty acids.

Fig. 2.4 Schematic representation of a portion of a bilayer of amphiphiles.

Fig. 3.1 Percent of low-boiling hydrocarbons remaining in South Louisiana crude oil slick.

Fig. 3.2 Size distributions (Coulter counter) of o/w emulsions prepared from Richfield-Kraemer crude oil, saline water, and Chevron dispersant NI-W.

Fig. 4.1 Cycle of hydrocarbons in the environment.

Fig. 4.2 Formation of 7,8-dihydro-7,8-dihydroxybenzo[a]pyrene-9,10-oxide by mammalian microsomes.

Fig. 4.3 Co-oxidation of aromatic hydrocarbons.

Fig. 4.4 Initial reactions in the degradation of benzene by *Pseudomonas putida*.

Fig. 4.5 Bacterial oxidation of catechol.

Fig. 4.6 Oxidation of benzo[a]pyrene by a *Beijerinckia* species.

Fig. 4.7 Differences between the reactions used by eucaryotic and procaryotic organisms to initiate the oxidation of aromatic hydrocarbons.

Fig. 5.1 Pathway of electron flow for hydroxylations in liver microsomes [From White et al. 1973].

Fig. 5.2 Pathways involved in the metabolism of benzo[a]anthracene. [From Swaisland et al. 1973].

Fig. 5.3 Conjugating reactions Sulfation of 1-naphthol.

Fig. 8.1 Concentration of total aromatic hydrocarbons from water-soluble fractions of Cook Inlet crude oil measured by gas chromatography at 24-hr intervals. Solutions were nonaerated, and held at either $5°$, $8°$, or $12°C$ (from Cheatham et al. In prep.).

Fig. 9.1 Concentrations of WSFs or specific hydrocarbons producing 50% mortality of *Palaemonetes pugio* life stages in 96 hr. Vertical bars represent standard deviations. (Redrawn from Tatem, 1975).

Fig. 9.2 Concentrations of WSFs or specific hydrocarbons producing 50% mortality of *Penaeus aztecus* or *P. setiferus* life stages in 96 hr. Vertical bars represent standard deviations. (Redrawn from Cox, 1974).

Fig. 9.3 Concentrations of No. 2 Fuel oil and So. Louisiana Crude oil WSFs producing 50% mortality of *Neanthes arenaceodentata* life stages in 96 hr. Vertical lines indicate standard deviations and represents gravid females. (From Rossi and Anderson, 1976).

Fig. 9.4 Relationship between respiratory rate of mysids, *Mysidopsis almyra*, and the hydrocarbon content of exposure water. Vertical bars represent standard errors. (From Anderson, 1975). A. Measured during exposure to WSFs of No. 2 Fuel oil (5 animals/container and 5 containers/concentration. B. Measured in clean seawater after a 2 hr exposure to OWDs of No. 2 Fuel oil (10 animals/container and 3 containers/concentration).

Fig. 9.5   Relationship between the respiratory rate of postlarval *Penaeus aztecus* and the rates of accumulation and release of naphthalenes. Determinations were made during a 4 hr exposure to a 30% WSF of No. 2 Fuel oil (0.85 ppm TN), and after 6 and 26 hr in clean seawater. (Redrawn from Cox, 1974). A. Vertical bars represent standard deviations, when n-12 for both exposed (e) and control (C) animals. B. Each data point represents the analysis of a pooled sample of 15 to 20 postlarvae. After 4 hr of exposure (as in respiratory phase, A), a large group os postlarvae were transferred to clean seawater.

Fig. 9.6   Relationship between the respiratory rate of *Palaemonetes pugio* and the accumulation and release of naphthalenes. Shrimp were exposed fcr 5 hr to an OWD of No. 2 Fuel oil containing 3 ppm TH and 0.1 ppm TN, during which respiration was measured. (Redrawn from Anderson, 1975). A. The same 8 groups of 12-14 shrimp were used in all respiratory rate determinations over a period of 10 days. Vertical bars represent standard deviations. B. A large group of shrimp were exposed to the same hydrocarbon solution for 5 hr before transfer to clean seawater for 72 hr. Analyses of tissues at both 5 and 72 hr were from 2 sets of 10 animals each.

Fig. 9.7   Relationship between the rate of blood chloride regulation of *Penaeus aztecus* and the uptake and release of naphthalenes. Shrimp were exposed to a 20% WSF of No. 2 Fuel oil (1.3 ppm TH and 0.4 ppm TN) for 10 hr in 20 $^0/_{00}$S seawater before transfer to clean water at either 10 $^0/_{00}$S or 30 $^0/_{00}$S. Each blood chloride value represents the 3 replicates, each from the pooled blood of 3 shrimp. Vertical bars represent standard deviations and the dashed line indicates the final acclimation level. (Redrawn from Anderson et al., 1974a, and Cox, 1974). A. Blood chloride concentrations of control (solid line) and exposed (broken line) transferred to a salinity of 30 $^0/_{00}$. B. Blood chloride concentrations of shrimp transferred to 10 $^0/_{00}$. C. After blood samples were taken, the tissues of the exposed shrimp (9 per time interval) were analyzed for content of total naphthalenes. Analyses were conducted after the 10 hr exposure and at 1, 6, 24 and 96 hr in clean water.

Fig. 9.8. Relationship between the adjustment of pericardial fluid chloride concentration by *Crassostrea virginica*, and the accumulation and release of petroleum hydrocarbons. Oysters were exposed to 1% (v/v) OWDs of No. 2 Fuel oil or South Louisiana Crude oil for 96 hr before transfer from 20 $^0/_{00}$S to either 10 or 30 $^0/_{00}$ salinities. (Redrawn from Anderson, R.D., 1975, and Anderson and Anderson, 1975). A. Each point represents the mean pericardial chloride concentration of 3 oysters. The solid line indicates the isochloride line, and vertical bars are shown for standard deviations greater than 10%. B. Content of petroleum hydrocarbons in oyster tissues exposed to a 1% OWD of No. 2 Fuel oil for 96 hr and then held in clean water for 52 days.

Fig. 9.9   Relationship between the behavioral stage of *Fundulus similus* and the content of naphthalenes in 4 organs. Five fish at each stage were selected from a large group exposed to an acute concentration of No. 2 Fuel oil WSF (6.3 ppm TH and 1.9 ppm TN). Organs from all 5 were pooled for analyses at each behavioral stage during exposure (about 4 hr) and at 3 intervals of depuration in clean water. The schematic of the stages of stress proceeding from "normal" (I) to death (VI) and the corresponding content of naphthalenes are shown. Stage V illustrates slight opercular movements but not swimming, and fish transferred at this stage to clean water were held for 3.5, 23 and 366 hr before analyses. (Redrawn from Dixit and Anderson, 1977).

Fig. 9.10 Viability of *Rithropanopeus harrisii* larvae exposed to WSFs of No. 2 Fuel oil. (From Neff et al., 1976).

Fig. 9.11 Growth of *Neanthes arenaceodentata* larvae under control and exposed conditions. (From Rossi, 1976).

Fig. 9.12 Growth of *Neanthes arenaceodentata* juveniles under control and exposed conditions. (From Anderson, 1976).

Fig. 10.1 (Upper) SEM of gill from an untreated coho salmon. X 370.

Fig. 10.2 (Lower) SEM of gill from an WSF (10 ppb) petroleum-exposed coho salmon. One filament appears undamaged and the two on the left show severe cellular disruption. Many surface cells have sloughed and there is an abundance of exuded mucus (arrow). X 370.

Fig. 10.3.   (Upper) Surface cells of the gill of starry founder after exposure to 10 ppb WSF with SEM. The cells with microridges appear normal but a group of cells have sloughed. The surface of the underlying cell layer is evident. X 2,000.

Fig. 10.4   (Lower left) The trematode, *Gyrodactylus* sp. on the gill of coho salmon with SEM. X320.

Fig. 10.5   (Lower right) Higher magnification of the opishaptor of *Gyrodactylus* sp. The attachment to the skin does not appear to harm the surface cells of the salmon gill. X 1,400.

Fig. 10.6   (Upper) SEM of the skin surface of an untreated English sole. The mucous glands (arrows) are liberally scattered throughout the normal skin between the filament-containing cells which are sculptured on their outer surface by microridges. X 3,500.

Fig. 10.7   (Lower) English sole, five days after exposure to a 13% solution of WSF. The mucous glands (arrow) are numerous and conspicuously open. These fish felt more slimy than the untreated. X 3,500.

Fig. 10.8   (Upper) Light micrograph of sections of the liver of untreated rainbow trout. This and the accompanying micrograph were photographed from sections prepared with a polychrome stain adapted for plastic sections. The glycogen deposits are bright red against the blue background of the cytoplasm. Parts of the larger glycogen reserves were lost during preparation. Lipid droplets (circled) are abundant and red blood cells (R) are often apparent in the small hepatic vessels or sinusoids. X 620.

Fig. 10.9   (Lower) Companion micrograph to Fig. 10.8 of petroleum-fed rainbow trout (see text). No glycogen or lipid deposits are seen in this section and the cytoplasm is an overall blue tone. X 620.

Fig. 10.10   (Upper) TEM of thin sections of untreated rainbow trout liver. The hepatocytes are rich in glycogen (G) and lipid (L). X 6,400.

Fig. 10.11   (Lower) Thin section of liver from a rainbow trout fed petroleum for 75 days (see text). In addition to a lack of glycogen in the hepatocytes, the endoplasmic reticulum (ER) has proliferated to an extreme degree. X 7,500.

Fig. 10.12   (Upper) Thin section of liver from rainbow trout fed petroleum as above. There was an abundance of cochlear ribosomes (arrow) in the petroleum-fed fish. The mitochondria appear unchanged by petroleum exposure. X 26,000.

Fig. 10.13   (Lower) TEM of liver from rainbow trout fed petroleum as above. Small pools of glycogen (G) were found after considerable searching with the electron microscope. The deposits were small enough that they did not show in the thick sections used for light microscopy. In addition, there are unusual crystalline inclusions (arrow) of unknown significance in some of the hepatocytes. X 16,000.

Fig. 10.14   (Upper) SEM of fibers from the eye lens of an untreated rainbow trout. The fiber surface, exclusive of connecting projections and pits, is smooth. The lateral interdigitations are quite regular. X 3,100.

Fig. 10.15   (Lower) Lens fibers from a rainbow trout fed 17 mg crude oil/kg body weight per weekday for about one year (see text). These fibers are from an enlarged lens and appear undulated and irregular. X 3,100.

Fig. 13.1   Timing of maturation for petroleum-exposed and non-petroleum exposed rainbow trout.

Fig. 13.2   Typical spectrophotofluorometric curves from the analyses of standard Prudhoe Bay crude oil and extracts of fish eggs. Excitation at 262 nm. Solvent: methylene chloride in petroleum ether, 20:80 (v/v), Curve No. 1 - Prudhoe Bay crude oil (10 µg/ml) standard, Curve No. 2 - Oil-fed fish egg extract, Curve No. 3 - Control fish egg extract.

Fig. 14.1   Thermal conductance of sea otter, fur seal and sea lion pelts during normal immersion, after oiling, and after cleaning.

Fig. 18.1   Normal Gill. Light Micrograph (LM; 750X). Filament with ciliated epithelial (E) cells surrounding a blood sinus (BS).

Fig. 18.2   Treated Gill: 1000 ppb. (LM; 750x). Epithelial cells at the tips of the filaments are sloughed off. Chitinous rods (CR) are apparent.

Fig. 18.3   Treated Gill: 10000 ppb. (LM; 140x). Damaged demibranch. Hemolymph vessels (HV) contain precipitate.

Fig. 18.4   Treated Gill: 25000 ppb. (LM; 140x). General disintegration of the demibranch.

Fig. 18.5   Treated Gill: 50000 ppb. (LM; 750x). Chitinous rods (CR) and little connective tissue remain.

Fig. 18.6   Normal Gill. Electron Micrograph (EM; 9000x). Epithelial cells from the base of a filament. The nucleus of the epithelial cell contains mostly peripheral chromatin. A nucleolus is evident in most cells. The cytoplasm contains mitochondria (M), electron dense bodies (DB), Golgi (G) apparatus, numerous vesicles (Ve) as well as basal bodies (BB), rootlets (R), microvilli (MV) and other parts of the cilia (C).

Fig. 18.7   Treated gill: 1000 ppb. (EM; 17500x). Small epithelial cell at the base of a filament. Nucleus (N) and plasma membranes (PM) remain intact. Cytoplasm essentially devoid of components; remnants of what were probably mitochrondria can be seen.

Fig. 18.8   Treated Gill: 100 ppb. (EM: 4500x). Cytoplasm of the epithelial cells at the tip of the filament shows a general loss of cytoplasmic components; mitochrondria and cell membranes remain intact.

Fig. 18.9   Treated Gill: 10000 ppb. (EM; 4500x). Disintegration of epithelial cells at the tip of the filament. Cell membranes have lysed; few cytoplasmic components are seen.

Fig. 18.10 Normal Digestive Gland. (LM; 750x). Typical tubules consisting of epithelial cells.

Fig. 18.11 Treated Digestive Gland. 10000 ppb. (LM; 750x). Disorganization of the epithelium of a digestive tubule.

Fig. 18.12 Normal Digestive Gland. (EM; 4500x). The edge of normal digestive tubule. Epithelial cells contain great quantities of endoplasmic reticulum (vesicular form predominantly) and Golgi.

Fig. 18.13 Treated Digestive Gland. 100ppb. (EM; 17500x). Two tubule (epithelial) cells. The endoplasmic reticulum (ER) is altered as is the Golgi (G) in the adjacent cell. Only remnants of the cell membrane can be seen. The mitochondria (M) appear normal.

Fig. 18.14 Normal Hindgut. (LM; 750x). Typical ciliated (C) columnar epithelium lining the lumen of the gut.

Fig. 18.15 Treated Hindgut. 10000 ppb. (LM; 750x). Separation (sloughing) of epithelial cells from the adjacent connective tissue.

Fig. 18.16 Normal Hemocytes. (LM; 1850x). Cells have phagocytosed yeast (Y). Nucleus (N), cytoplasm (CY) and phagocytic vacuoles (PV) surrounding the yeast are apparent.

Fig. 18.17 Treated Hemocytes. 50000 ppb. (Lm; 1850x). Disintegrating hemocytes. Many have lysed; only the nuclei (N) remain.

Fig. 18.18 Normal Hemocytes. (EM; 17500x). Hemocyte has phagocytosed a yeast (Y) particle. Note numerous electron dense bodies (DB). See Cheng and Goley (1975).

Fig. 18.19 Treated Hemocytes. 100 ppb. (EM; 4500x). Granulocytes from heart hemolymph. Most cytoplasmic organelles have disintegrated; electron-dense bodies (DB) remain.

Fig. 18.20 Treated Hemocytes. 10000 ppb. (EM; 17500x). Granulocyte in gill tissue. Few cytoplasmic organelles remain. Electron-dense bodies (DB) remain.

Fig. 19.1 Rate of loss of total naphthalenes from three concentration of the WSF of No. 2 fuel oil at 20, 25 and 30°C. Conditions were the same as those in the exposure of three determinations. For values at 5 hrs, 25°C, $F = 2.89$ df = 2,5 ns.

Fig. 19.2 Survival of Limulus polyphemus during chronic exposure to WSFs of No. 2 fuel oil at 20, 25 and 30°C and 32 ppt salinity. Points represent the percentage of animals which survived the indicated stage and molted to the succeeding one.

Fig. 19.3 Mean intermolt periods (development rates) of Limulus polyphemus exposed to 5 concentrations of the WSF of No. 2 fuel oil at 3 temperatures at 32 ppt salinity. No data are available for 25°C, 50° WSF. Vertical lines represent standard deviations.

Fig. 19.4 Respiratory rates of first-tailed stage larvae of Limulus polyphemus exposed to 0, 5, 10 and 25% WSF of No. 2 fuel oil at 20, 25 and 30°C at 32 ppt salinity (A) or immediately after transfer to 20 (B) or 10 ppt (C) salinity sea water. Animals exposed to 25% WSF were available in sufficient numbers for respiration measurements at only 20°C.

Fig. 20.1 Effect of 24 hr exposure to light, medium and heavy dispersions of crude oils upon the respiratory metabolism of Onisimus. Asterisk indicates difference significant at $p < 0.05$ level. (N = 8 in each group.)

Fig. 20.2 Effect of 24 hr exposure to light, medium and heavy concentrations of Corexit, alone and in combination with crude oils, upon the respiratory metabolism of Onisimus. Asterisk indicates difference significant at $p < 0.05$ level. (N = 8 or 9 in each group.)

Fig. 20.3 Effect of exposure to light, medium and heavy dispersions of Norman Wells crude upon the respiration of cell free homogenates of < 0.05 level. (N = 7-9 in each group.)

Fig. 20.4 Effect of exposure to various concentrations of seawater extracts of Norman Wells and Atkinson Point crude oils upon the respiration of Onisimus. Asterisk indicates difference significant at $p < 0.05$ level. (N = 13-22 for each group.)

Fig. 20.5 Effect of exposure to light dispersions of Norman Wells crude for various periods upon the respiration of Onisimus. Animals not fed between 0 hr and 96 hr, but refed after 96 hr. Vertical lines indicate standard errors of means. (N = 8-14 in each group.)

Fig. 21.1 Accumulation and release of naphthalenes by grass shrimp exposed for 24 hr to PH from a No. 2 fuel oil WSF.

Fig. 21.2 Respiratory rates of eight groups of grass shrimp subjected to a 48-hr starvation period and to 3.0-3.6 ppm PH from a NO. 2 fuel oil OWD. Standard errors and 95% confidence intervals of the data are shown. Tissue analyses of a separate group of shrimp are also shown. These shrimp were exposed to 3.0-3.6 ppm PH for 5 hr and allowed to depurate in clean seawater for 72 hr.

Fig. 21.3 Hatching success of Palaemonetes larvae after gravid female grass shrimp were exposed to PH from a NO. 2 fuel oil WSF for 72 hr. Standard errors and 95% confidence intervals are shown. Larvae hatched in uncontaminated seawater.

Fig. 21.4 Growth of grass shrimp larvae exposed for a total of 14 days to PH from a No. 2 fuel oil WSF. Each data point represents the mean weight of 10-12 larvae. Standard deviations of the data are shown.

Fig. 22.1 Survival of C. magister zoeae exposed continuously to different concentrations of crude oil WSF, benzene and naphthalene. Results are from the first long-term experiment which employed the progeny of an Oregon female crab.

Fig. 22.2 Survival of C. magister zoeae exposed continuously to different concentrations of crude oil WSF, benzene and naphthalene. Results are from the second long-term experiment which employed the progeny of an Alaskan female crab.

Fig. 23.1   Molting success of larvae of coonstripe shrimp, *Pandalus hypsinotus*, exposed for increased lengths of time to increased concentrations of the WSF of Cook Inlet crude oil. After the exposure time listed, the exposure water was replaced with clean seawater to make a total of 144 hr of observation for each test. Molting success at 144 hr = stage II larvae/initial total of stage I larvae.

Fig. 23.2   Molting success of larvae of king crab, *Paralithodes camtschatica*, exposed for increased lengths of time to increased concentrations of the WSF of Cook Inlet crude oil. After the exposure time listed, the exposure water was replaced with clean seawater to make a total of 120 hr of observation for each test. Molting success at 120 hr = stage II larvae/initial total of stage I larvae.

Fig. 24.1   Response of buried <u>Macoma balthica</u> exposed to concentrations of the WSF of Prudhoe Bay crude oil (given in ppm naphthalene equivalents) in a recirculating water-sediment system. The percentage of clams that responded by coming to the sediment surface is graphed. Control clams made no response throughout. Solid arrows = days WSF added or changed; open arrow = the day clean seawater was added.

Fig. 24.2   Response of unburied clams to concentrations of the Prudhoe Bay crude oil WSF (given in ppm naphthalene equivalents) in a recirculating water-sediment system. The percentage of clams that remained on the surface is graphed.

Fig. 24.3   Percentage of clams that responded to 24-hr exposure to oil-treated sediment by coming to the surface vs. grams of sediment added cm^{-2} squared. Broken lines for oil-treated sediment = $\pm$ standard deviation.

Fig. 25.1   The Chedabucto Bay area of Nova Scotia, Canada, showing the general study area and the location of stations. The inset shows the location of Chedabucto Bay in the Maritime Provinces of Canada.

Fig. 25.2   Surface oil cover at two tidal levels at Chedabucto Bay stations from 1970-1975. Abbreviations are: MHW, mean high water; MLW, mean low water.

Fig. 26.1   Map of Straits of Magellan to show survey sites. Site A is at Porvenir - see inset.

Fig. 26.2   Normal and Inverse Classification Dendrograms with Resultant Two-way Table. Petroleum content of sediments was determined on the basis of % total $C Cl_4$ extractables and gas chromatography to determine if the <u>Metula</u> oil was present.

Fig. 27.1   Map showing sampling locations.

Fig. 28.1   Study area locations.

Fig. 28.2   Production of CO_2 from <u>in situ</u> incubation of an amino acid mixture and from dodecane. Initial substrate concentrations are also shown. Open symbols show 1 µM $HgCl_2$ poisoned control rates. Each datum point represents total $^{14}CO_2$ collection from a 1 m dark bottle after incubation for the time shown. Sample and incubation conditions were: for Port Valdez, 10 m depth, 3 C, salinity 30°/oo, April 1972; Point Barrow, 4 m depth through the shore ice, 2 C, salinity 28.3°/oo, April 1973; T-3 Ice Island, 3 m depth through the ice, -1.7 C, salinity 30.1°/oo, January 1973.

Fig. 29.1   Total hydrocarbons and total naphthalenes in oil-in-sediment exposure system over the 2-week period. Number of samples analyzed, means, and standard deviations >10% (vertical bars) are shown for each time interval.

Fig. 29.2   Content of methyl- and dimethylnaphthalenes in sipunculids (*Phascolosoma*) exposed to oil-in-sediments. Means, standard deviations (>10%), and the number of samples are shown for exposed (squares), depurated (triangles), and control (circles) tissues. Solid symbols designate GC analyses, while all others were determined by u.v. Depuration of the 40-hr exposed animals was for 48 hr, while that of the 2-week exposed groups was either 2 weeks (4 samples) or undetermined (5 samples).

Fig. 30.1   Retention of radioactivity by different species of marine copepod immersed in seawater containing ^{14}C-1-naphthalene at different concentrations for 24 hr (no food present). Regression equations are given in Table 1.

Fig. 30.2 Relationship between 24 hr retention of radioactivity (calculated for a concentration of 1 µg/l. naphthalene from the regression equations in Table 1) and dry weight (●—●), ash-free dry weight (○—○) and total lipid (▲—▲) content for different species of marine copepod. Details of regression equations: a) Log y = 1.2426 log x + 0.0705, r = 0.9125, n = 9; b) Log y = 1.2814 log x - 0.0763, r = 0.9106, n = 9; c) Log y = 1.1230 log x - 0.7692, r = 0.9503, n = 9.

Fig. 30.3 The retention of radioactivity at different temperatures by female *Calanus helgolandicus* immersed in sea water solutions naphthalene (concentration 1 µg/l.) for 24 hr in the absence of food. Each point is the mean of four observations; vertical bars represent ± standard deviation.

Fig. 30.4 Levels of radioactivity in female *Calanus helgolandicus* immersed in sea water containing ^{14}C-1-naphthalene at different concentrations, when starving (○—○) and feeding on *Biddulphia sinensis* cells (●—●),: regression line for starving animals is log y = 0.832 log x + 1.712, r = 0.993, n = 12; for feeding animals log y = 0.854 log x + 2.04, r = 0.986, n = 12. Each point represents the mean of three determinations at the same concentration. Other details as in text.

Fig. 30.5 Retention of radioactivity by female *Calanus helgolandicus* immersed in sea water containing ^{14}C-1-naphthalene at different concentrations under various conditions. 1) (———) Starving, 24 hr exposure, log y = 0.9794 log x + 1.7584, r = 0.9888, n = 26; 2) (— - — -) Feeding on the dinoflagellate, *Peridinium trochoideum*, at a food concentration of about 100 µg algal carbon/l., log y - 1.003 log x + 1.7698, r = 0.9453, n = 26; 3) (— — —) Starving, exposed for 30 seconds only, log y = 0.8833 log x + 0.4778, r = 0.9975, n = 12; 4) (--------) Dead animals (heat treated at 37°C for 5 mins), log y = 0.9916 log x + 1.5315, r = 0.9971, n = 12). For further details see text.

Fig. 30.6 Levels of radioactivity in *Eurytemora affinis* transferred to uncontaminated conditions following 24 hr exposure to ^{14}C-1-naphthalene at various concentrations.

Fig. 31.1 Uptake of naphthalenes by clams exposed to a 25% WSF of So. Louisiana crude oil after a 24 hr. acclimation period to the different salinity-temperature regimes.

Fig. 32.2 Uptake and depuration of naphthalenes by clams exposed to a 25% WSF of So. La. crude oil. These clams were acclimated to the varying conditions over a 14 day period.

Fig. 32.3 Filtration rates of *Rangia cuneata* at various temperatures and salinities. Each point is the average rate of three clams measured over 19 hours. No measurements were made at 20°C, 30°/oo because of mortalities.

Fig. 32.4 Naphthalenes uptake and depuration by the clam, *P. staminea*, exposed for three days to a 25% WSF of So. La. crude oil. The depuration period lasted for 3 days.

Fig. 33.1 Gas chromatograms.

Fig. 33.2 Control animal gas chromatograms.

Fig. 33.3 Gas chromatograms clams exposed to 10 ppm #3 Fuel Oil.

Fig. 33.4 Gas chromatograms clams exposed to 50 ppm #2 Fuel Oil.

Fig. 33.5 Gas chromatograms clams exposed to 100 ppm #2 Fuel Oil.

Fig. 34.1 Aerial view of the two spill-areas at Penniman Spit, Cheatham Annex, Yorktown, VA. At the time this picture was taken, the booms had not yet been emplaced.

Fig. 34.2 Aliphatic fractions of post-spill extracts from oyster, demonstrating the dramatic changes in composition taking place between +6 and +25 hours. In the lower chromatogram, the approximate position of the n-alkanes is indicated by vertical lines.

Fig. 35.1 Concentration vs time histograms of (A) Soxhlet extracted and (B) headspace-sampled Katalla River sediment. Peak heights from the respective gas chromatograms have been reduced to the internal standard peak heights (positions noted by vertical lines) and plotted as single species concentrations based on wet weight.

Fig. 35.2   GC-MS analysis of Katalla River sediment: (A) composite m/e 142, 156 and 170 single ion records indicating presence of C_1, C_2, and C_3-naphthalenes, respectively, (B) m/e 43 single ion record, (C) total ion chromatogram. C_x=alkane containing x carbon atoms. $C_x\text{-}\emptyset$=benzene substituted with x carbon atoms (e.g. $C_3\text{-}\emptyset$ could be trimethyl-, propyl-, isopropylbenzene, etc.) Peaks labeled 1,2,3,4 are the internal standards methyl-C_{11}, methyl-C_{14}, methyl-C_{16}, and methyl-C_{18}, respectively. Identifications followed by "?" are not definite due to incompletely resolved spectra.

Fig. 36.1   Experimental Layout.

Fig. 36.2   Oil slick leeway as a function of wind speed.

Fig. 36.3   Tailing phenomenon.

Fig. 36.4   Wind factor as a function of wind speed.

Fig. 37.1   Percent of aromatic hydrocarbons remaining in surface oil slick - first La Rosa spill.

Fig. 38.1   Relative light hydrocarbon concentrations on the Louisiana Shelf --- Cruise 74-G-8 (May 1974).

Fig. 38.2   Relative light hydrocarbon concentrations on the Louisiana Shelf --- Cruise 74-G-8 (May 1974).

Fig. 38.3   Surface methane concentrations on South Texas Shelf (units = nl/L).

Fig. 38.4   Mass spectrum of gas collected from underwater vent.

Fig. 38.5   Gas chromatogram of gas collected from an underwater vent ---same gas as Fig. 4.

Fig. 38.6   Gas chromatogram of hydrocarbons in a produced brine in the Gulf of Mexico.

Fig. 38.7   Station locations.

Fig. 38.8   Discharge locations of oil field brines in 1973 on the Louisiana Shelf outside of the three-mile limit (U.S.G.S., personal communications).

Fig. 39.1   The Strait of Juan de Fuca and the Olympic Peninsula, Washington.

Fig. 39.2   Peabody Creek intertidal sampling area at Port Angeles, Washington.

Fig. 39.3   Dungeness Bay intertidal sampling area at Dungeness, Washington.

Fig. 39.4   Schematic details of the GC sample train: injector, column, and detector.

Fig. 39.5   Schematic of sediment analysis.

Fig. 39.6   Glass capillary gas chromatograms of unsaturated hydrocarbons from (a) Port Angeles sediments and (b) Dungeness sediments.

Fig. 39.7   Glass capillary gas chromatograms of saturated hydrocarbons extracted from (a) Port Angeles sediments and (b) Dungeness sediments.

Fig. 40.1   Gas sampling loop.

Fig. 40.2   n-Hexane calibration using 25 ml sample loop. Instrument conditions as in text.

Fig. 40.3   Gas chromatograms of seawater suspension of No. 2 fuel oil.

Fig. 40.4   Partitioning of monocyclic aromatics from a seawater suspension of No. 2 fuel oil (0.15 ppm by IR).

Fig. 41.1   Sample stations for 1975-1976. Sation Nos. at all Transects run consecutively seaward except Transect 4 stations which run shoreward. The Dotted line represents the 50 fathom contour.

Fig. 41.2   Aliphatic hydrocarbons of Florida shelf sediments in 1974 and 1975. The chromatograms are from the station designated Area II-46 in 1974 and Transect III-16 in 1975. The 'x' refers to a group of phytadiene peaks.

Fig. 41.3   Mass spectrum of a Kovats Index 2070-2075 component of Florida shelf aliphatics.

Fig. 41.4   Aliphatic hydrocarbons of Mississippi shelf sediments in 1974 and 1975. The chromatograms are from the station designated Area V-2 in 1974 and Transect VI-38 in 1975.

Fig. 41.5   Aliphatic hydrocarbons of outer continental shelf of northeastern Gulf of Mexico. Period 1 represents sampling of June, 1975 and period 3, January, 1976.

Fig. 42.1   The electrophoretic character of crude oil emulsions.

Fig. 42.2   The electrophoretic character of inorganic solid particles.

Fig. 42.3   The effect of silica on the stability of Nigerian crude oil droplets. Time (in min.) shown for various periods of quiescence following system agitation.

Fig. 42.4   The distribution of Nigerian oil droplets as affected by silica particles.

Fig. 42.5   The stability of crude oil emulsions as affected by silica particles.

Fig. 42.6   The effect of relative particle size on the stability of Nigerian crude oil emulsions.

Fig. 42.7   Schematic representation of the stability of oil emulsion as affected by similarly-charged inorganic solid particles (Type I behavior).

Fig. 42.8   The effect of $\alpha\text{-}Al_2O_3$ on the stability of Nigerian crude oil emulsions (Type II behavior).

Fig. 42.9   Schematic representation of the stability of oil emulsions as affected by counter-charged inorganic solid particles.

Fig. 42.10 The effect of ionic strength on the electrokinetics of Nigerian crude oil emulsions.

Fig. 42.11 The effect of ionic strength on the stability of the Nigerian - SiO_2 (Cabosil) system.

Fig. 43.1   Oregon bays and estuaries.

Fig. 43.2   Coos Bay, with a total surface of 10,000 acres of which 50% are tidelands, drains a basin of 605 square miles with a very high freshwater yield. Coos Bay is the most heavily industrialized of all Oregon bays. Timber, fish resources and agricultural activities are of major economic importance. Many major lumber manufacturers are located in the bay and there is a large amount of shipping traffic concentrated around Coos Bay. Two commercial oyster growers operate in South Slough and there is moderate to heavy exploitation of the clam populations.

Fig. 43.3   Tillamook Bay, with a total surface area of 8,660 acres of which 50-60% are tidelands, drains a basin of 540 square miles with a high freshwater yield. Major industries: timber, agricultural products, fish and seafoods, tourism. Not considered to be highly industrialized. Three commercial oyster companies are in the bay and there is a moderate amount of recreational clamming.

Fig. 43.4   Alsea Bay, with a total surface area of 2,140 acres of which 45-50% are tidelands, drains a basin of 474 square miles with a high freshwater yield. Lumber-related activities, tourism and agriculture are of major economic importance. Little industrial use of the bay. Clams are moderately exploited.

Fig. 43.5   Netarts Bay, with a total surface area of 2,200 acres of which 65-90% are tidelands, drains a basin of 14 square miles with a very low freshwater yield. Manufacturing companies are lacking completely; clam digging very popular. The bay is considered to be relatively pristine.

Fig. 43.6   Yaquina Bay, with a total surface area of 4,000 acres of which 35-61% are tidelands, drains a basin of 253 square miles with a medium freshwater yield. A major industrial estuary, the bay is a center for lumbering and commercial fishing activities. Toledo is the focal point of the forest industry processing facilities for the entire Mid-Coast Basin. Newport is the center for commercial fishing activities and there are numerous fish processing plants along the bayfront. Numerous marinas are scattered throughout the bay. Four commercial oyster growers are in the bay and clams are heavily dug.

Fig. 44.1 Dissolved hydrocarbon distribution during summer (upper), fall (middle) and winter (lower) sampling seasons. Station locations are indicated by closed circles and are numbered sequentially nearshore to offshore beginning with transect 1.

Fig. 44.2 Gas chromatograms of representative dissolved hydrocarbon samples.

Fig. 44.3 Particulate hydrocarbon distribution during summer (upper), fall (middle) and winter (lower) sampling seasons.

Fig. 44.4 Gas chromatograms of representative particulate hydrocarbon samples.

Fig. 45.1 Map of Westernport Bay (Victoria, Australia) showing mussel (numbers) and sediment (letters) sampling stations. Areas bounded by solid lines are subject to high chronic input of petroleum hydrocarbons (>450 ppm found in mussels). Areas bounded by dashed lines show approximately 100 to 200 ppm in mussels.

Fig. 45.2 Sample chromatograms of total hydrocarbon extracts of mussels collected in Westernport Bay showing various types of oil discharged into the system. Numbers are the positions of n-alkanes of the indicated carbon chain length. Dashed lines are column bleed signals a) fresh fuel oil (3220 ppm) from boat dock at Sta. 3; b) degraded crude oil (570 ppm) from refinery wharf Sta. 4; c) biogenics (no petroleum) from clean area Sta. 14; d) degraded diesel oil (100 ppm) from small boat dock Sta. 12; e) lube oil (126 ppm) from Sta 12; f) mixture of degraded diesel and lube oils (160 ppm) from Sta. 12.

Fig. 45.3 Coupled excitation-emission fluorescence spectra of 20% benzene/hexane column fractions of : A. source oils and B. environmental samples (about 50 µg hydrocarbon/ml hexane). Samples scanned 260 to 540 nm emission with excitation monochromaters set 23 nm lower. Band pass width 5 nm; sensitivity 30. *Based on Gordon and Keizer, 1974. a) hexane blank. b) diesel oil; c) lube oil; d) Gippsland crude oil; e) mussels from a refinery wharf; f) clean mussels; g) mussels from small boat wharf; h) sediments near refinery; i) clean sediments.

PREFACE

This volume presents most of the papers which were presented November 10-12, 1976 at an international symposium by the same title, held at the Olympic Hotel in Seattle, Washington. The Symposium, attended by approximately 350 persons, consisted of three half-day plenary sessions of invited papers and three half days of concurrent sessions for contributed papers describing original research results. On the final afternoon, a Panel was convened to discuss future research priorities in this field. This Symposium was sponsored by the National Oceanic and Atmospheric Administration and the Environmental Protection Agency to provide a vehicle for documenting the current status of research on the fates and effects of petroleum in marine environments, and for identifying areas still in need of future research.

During the plenary sessions twelve invited scientists, each a recognized expert in his field, summarized various aspects of the following topics:

a) <u>Inputs and Physical Transport Processes</u> influencing the distribution and composition of petroleum hydrocarbons in marine systems;

b) <u>Bioaccumulation and Metabolism</u> of hydrocarbons by marine organisms; and

c) <u>Biological and Ecological Effects</u> of petroleum exposure in marine systems.

These summary review presentations appear in Part I of this volume; Part II consists of those contributed papers which were accepted for publication; and Part III is the transcript of the Panel Discussion.

Not all the papers presented at the meeting appear in this volume; a complete list appears on pages 465-466. The Editorial Committee carefully selected from the submitted manuscripts those that were most scientifically sound and most representative of the scope of the Symposium. A few accepted manuscripts suffered attrition from failure to meet submission deadlines for publication. Each of the contributed papers was technically reviewed by outside experts and the Editorial Committee made its selections on the basis of the technical reviews. The draft manuscripts were then revised by the authors in accordance with the comments of the outside referees and the editors. Beyond that, the chapters have received only minor editing. For those errors which may have crept into the volume during retyping and have gone so far undetected, I assume responsibility and offer my apologies.

I thank all those who contributed to the success of the Symposium and to the preparation of these Proceedings. They include the Editorial Committee, the Session Chairmen and the technical referees who evaluated the scientific quality of the submitted papers. I especially wish to thank Mr. George Snyder, of the NMFS Northwest and Alaska Fisheries Center, Seattle, for overseeing the registration and meeting facilities for the Symposium. The authors cooperated fully with a rigorous production schedule which helped bring these proceedings out quickly after the Symposium. I also thank Pam Schulz who helped with the final preparation of the manuscripts for publication.

CHAPTER 1

OIL SPILLS IN THE ALASKAN COASTAL ZONE
THE STATISTICAL PICTURE

J. R. Harrald, B. D. Boyd, and C. C. Bates

U. S. Coast Guard
400 Seventh Street
Washington, D. C. 20590

Introduction

As noted by the authors in another recent paper (Boyd et al., 1976), two collateral methods are frequently referred to when generating an estimate of how much petroleum hydrocarbon (PHC) enters the sea annually. The gross global budget approach developed during a National Academy of Science (1975) study of the problem was sponsored in part by U. S. Coast Guard funding. In addition, the U. S. Coast Guard, under authority contained in the Water Quality Improvement Act of 1970, has developed a nationwide "Pollution Incident Reporting System (PIRS)" limited to domestic waters and pertaining only to non-continuous discharges creating a visible signature (sheen) on the water's surface.

If one omits the numerous and comprehensive studies of oil spill statistics incorporated into the evolving gray literature of environmental impact analyses, it can be said that there have been four definitive studies published in the past six years regarding the influx of petroleum hydrocarbons into the ocean as a whole. These were generated by a study of the Massachusetts Institute of Technology (1970); the University of Oklahome technological assessment of outer continental shelf oil and gas operations (Kash, 1973); a U.S. Coast Guard (1973) environmental impact statement for Prevention of Pollution from Ships, and the National Academy of Science (1975) study. Bates and Pearson (1975) have analyzed these four global budgets and concluded that the study accomplished between 1972 and 1974 for the National Academy of Science (1975) is probably the most reliable. At first glance, it might be concluded that the Coast Guard's PIRS system and the National Academy of Science's (NAS) statistical compilation of PHC inputs might be easily correlated one with the other. After all, much of the background data used in the NAS effort was derived from United States data and multiplied by a factor of 3.3 to give the global picture in view of the fact that the United States uses about 30 percent of the world's output of petroleum. Unfortunately, the two statistical data bases do not mesh directly for they have been assembled for separate reasons and differ in scope and type of reporting. Accordingly, for the purposes of the present conference, the remaining portion of this paper is restricted to delineating the nature of these data bases in order that workers in the broad field of petroleum pollution may be familiar with what these statistical assemblages bases do--and do not--provide.

The NAS Budget of Petroleum in the Marine Environment

The NAS PHC budget was derived by a 13-man team consisting of geochemists, engineers (naval, sanitary, chemical and petroleum specialities), and geologists from four different countries (United Kingdom, Canada, Sweden, and the United States). Although most of the panel's work was done in mid-1973, the panel primarily used world petroleum production and transport statistics for 1971, supplemented by domestic oil spill data for the 1970-1972 period. The panel concluded that approximately 6.1 million metric tons of petroleum hydrocarbons entered the ocean from all sources each year (Table 1).

This estimated value of 6.1 million metric tons of PHC entering the ocean annually also approximates the amount of biogenic hydrocarbon now believed to be forming locally in the ocean each year as the result of planktonic activity. The NAS budget can also be presented in terms of gallons per year by major contributor (Table 2).

The Pollution Incident Reporting System (PIRS)

Data on actual pollution incidents occurring in United States waters are contained in the U. S. Coast Guard Pollution Incident Reporting System (PIRS). PIRS is perhaps the most comprehensive pollution incident data base in the world and includes entries on all significant discharges (as measured by volume) which occur in U.S. waters and a very extensive sample of minor discharges. The Coast Guard estimates that 90 percent of the volume of oil discharged in the United States is reported in the PIRS system. This estimate is based on the fact that

TABLE 1. Budget of Petroleum Hydrocarbons Entering the Ocean *

	Input Rate (mta**)	
	Best Estimate	Probable Range
Man-Made:		
Marine Transportation		
LOT Tankers†	0.31	0.15 - 0.4
Non-LOT Tankers	0.77	0.65 - 1.0
Dry Docking	0.25	0.2 - 0.3
Terminal Operations	0.003	0.0015 - 0.005
Bilges/Bunkering	0.5	0.4 - 0.7
Tanker Accidents	0.2	0.12 - 0.25
Non-Tanker Accidents	0.1	0.002 - 0.15
Subtotal:	2.1	
River Runoff	1.6	
Atmospheric Rainout	0.6	0.4 - 0.8
Urban Runoff	0.3	0.1 - 0.5
Coastal Municipal Wastes	0.3	
Coastal (Non-Refining) Industrial Wastes	0.3	
Coastal Refineries	0.2	0.2 - 0.3
Offshore Production	0.008	0.08 - 0.15
Natural:		
Offshore Seeps	0.6	
Total:	6.113	

* Data Source is Table 1-5 of NAS Study (1975)
** mta: millions of metric tons per annum
† LOT: Loan-On-Top Tankers Equipped with Slop Tanks

TABLE 2. Contributors to the Introduction of Petroleum Hydrocarbon Into the Ocean

Contributor	Millions of U.S. Gallons	Percent of Total (Approximate)
Marine Transportation	656	34
River Run-Off	493	26
Atmospheric Rainout	185	10
Natural Seeps	185	10
Municipalities	185	10
Industrial Wastes	154	8
Offshore Production	25	2
TOTAL	1,883	

*Assumes 308 gallons (U.S.) to 1 metric ton of crude oil.

75 percent of all reported spills are less than 100 gallons, and these spills make up only 1 percent of the total reported spill volume. In other words, dramatic increases in the reporting of minor spills would not affect volume totals to a major extent.

All PIRS records originate in reports by Coast Guard field units located in port cities and town throughout the country including our inland waterway system. The PIRS report contains information and enforcement activity relating to the discharge. The primary purpose of PIRS is to provide the Coast Guard with the capability to assess the effectiveness of actions to prevent and mitigate oil pollution.

The PIRS system as it operates today does apply to modern coastal Alaska. Evidence at hand suggests that fully 90 percent of all domestic PHC spills are reported and the cognizant data which categorically describe the nature of the discharge, the response (clean-up) of the discharge and penalty information. For each discharge reported, the following categories of information are identified:

time	– year; month, day, hour
location	– latitude/longitude for coastal areas; river mile for inland rivers
state	
waterbody	– waterbody area (Atlantic, Gulf, Pacific, Inland) and waterbody type. (bay, estuary, port territorial, contiguous, etc.)
source	– vessel, facility, pipeline, etc.
source identifier	– vessel #/industry code
cause	– immediate cause and contributing facts
operation	– type of operation source was engaged in at time of discharge
material	– 32 types of oil, and over 150 types of chemicals and other types of pollutant
quantity	– expressed in gallons for liquids, pounds for solids
affected resources	– type of resource and degree of affect
weather	– wind, sea condition, current
notifier	– indication of how Coast Guard was notified of discharge and how soon notification was made
response	– whether or not clean up action possible

Pollution Incidents in Alaskan Waters

During 1975, the PIRS system acquired data on 12,057 pollution incidents involving 15.0 million gallons of materials of which 96 percent by volume were petroleum hydrocarbons. The geographic distribution of those 10,141 incidents involving petroleum is shown in Table 3.

TABLE 3. Geographic Distribution of Oil Pollution Incidents for all U.S. Waters

	Number of incidents	% of Total	Volume in Gallons	% of Total
Atlantic Coast	2,695	26.6	1,465,689	10.2
Gulf Coast (West of Long. 83°15')	3,315	32.8	5,430,212	37.6
Pacific Coast	1,768	17.4	440,923	3.0
Great Lakes	454	4.4	307,772	2.1
Inland U.S.	1,909	18.8	6,795,001	47.1
TOTAL	10,141	100.0	14,439,597	100.0

Table 4 summarizes discharges which occurred in Alaskan waters. Alaskan spills represented less than 1 percent of the national total in 1975, indicating the relatively minor nature of the present oil pollution problem in Alaska, and the possible incompleteness of Coast Guard data in this region.

TABLE 4. Hazardous Substances Spillage, 1974-75, in Alaska Compared to Total United States

	Alaska	U.S.	Alaska % of Total
1974			
Number of Incidents	155	13,966	1.1
Volume in Gallons	181,440	16,916,308	1.1
1975			
Number of Incidents	111	12,057	0.9
Volume in Gallons	79,066	14,967,895	0.5

Additional information concerning the polluting incidents during 1975 in Alaskan waters is shown in Tables 5-8 and discussed below. These tables include spills occurring in inland waters as well as in coastal and high seas areas. As suggested in the NAS statistics, a substantial proportion on this PHC contribution finds its way to the sea as river run-off.

TABLE 5. Types of Materials Discharged in 1975 in Alaskan Waters

	Number of Incidents	% of Total	Volume in Gallons	% of Total
Crude Oil	3	2.7	222	0.3
Gasoline	7	6.3	301	0.4
Other Distillate Fuel Oil	6	5.4	2,007	2.5
Solvents	0	0	0	0
Diesel Oil	70	63.0	76,346	96.6
Asphalt or Residual Fuel Oil	0	0	0	0
Animal or Vegetable Oil	0	0	0	0
Waste Oil	13	11.7	161	0.2
Other Oil	12	10.9	29	0.0
Liquid Chemicals	0	0	0	0
Other Pollutants (sewage, dredge spoil)	0	0	0	0
Natural Substance	0	0	0	0
Other Material	0	0	0	0
Unknown Material	0	0	0	0
TOTAL	111	100.0	79,066	100.0

Several factors concerning pollution in Alaska which do not parallel national trends may be seen from these tables. Table 5 shows that all reported incidents in Alaska involved petroleum hydrocarbons. Only three incidents with a total volume of only 222 gallons or 0.3 percent of the total volume involved crude oil. Nationally, crude oil accounted for 53 percent of the total spoil discharge volume in 1975. It is expected that this presently low statewide percentage will change drastically as the production and transport of crude oil becomes a predominant aspect of the Alaskan economy.

Table 6 shows that Alaska fortunately did not have a major incident (over 100,000 gallons discharged) during calendar year 1975. The size distribution otherwise roughly parallels that of the nation; 83 percent of all discharges were under 100 gallons and one discharge accounted for most of the pollution volume. A 65,000 gallon discharge of diesel oil into a non-navigable inland waterway resulted from the rupture of a non-transportation related on-shore pipeline in February 1975. This one discharge constitutes 82.7 percent of the 1975 Alaskan reported pollution volume and dominates all the tables presented in this paper.

TABLE 6. Frequency Distribution for Pollution Incidents of Various Volumes in Alaska Waters, 1975

Volume	Number of Incidents	% of Total	Volume in Gallons	% of Total
Unknown	23	20.7		
0-9 gal.	37	33.3	99	0.1
10-49 gal.	27	24.3	574	0.7
50-99 gal.	5	4.5	265	0.3
100-499 gal.	13	11.8	2,528	3.2
500-999 gal.	0		0	
1000-2499 gal.	4	3.6	5,600	7.1
2,500-4,999 gal.	0		0	
5,000-9,999 gal.	1	0.9	5,000	6.4
10,000-49,999 gal.	0		0	
Over 50,000 gal.	1	0.9	65,000	82.2
TOTAL	111	100.0	79,066	100.0

This skewed size distribution is not unusual. In 1972, for example, almost half of the total national volume reported was the result of the flooding of the Schuykill River in Pennsylvania by Hurricane Agnes. In 1975, 68 percent of the entire reported pollution volume was due to 20 discharges over 100,000 gallons; less than 0.2 percent of the total number of reported incidents.

Tables 7 and 8 reveal two other facts that are unique to Alaska: tank vessels account for less than 1 percent of the pollution volume, and the spill volume attributed to vessel casualties if about 1/4 of the national average (7.6 percent). The fact that fishing vessels (included in Table VI under "other vessels") account for 25 percent of all reported pollution incidents is also an indicator of the uniqueness of the Alaskan economy. This economy is changing rapidly, and the authors anticipate a corresponding change in the profile of pollution statistics for Alaskan waters.

A study done for the Coast Guard by Battelle Northwest (1973) predicts how dramatic these changes might be in Alaska. Based on worldwide tanker casualties statistics, USCG PIRS data and U.S. Geological Survey information, Battelle made the following predictions:

- Approximately 3 million gallons/year spilled due to tanker casualties in Prince William Sound and the Gulf of Alaska.
- Approximately 1.8 million gallons/year discharged from the Alaskan Pipeline.
- Approximately 80,000 gallons/year from transfer operations in Valdez and lower Cook Inlet.

The study concludes that by the mid-1980's, "Practically every mile of the coastline of mainland Alaska will be subject to potential oil spills". The capability of industry and of the Federal and State government to deal with these discharges will increase significantly, but the years of insignificant levels of crude oil discharged into the Alaskan Marine Environment are coming rapidly to an end.

TABLE 7. Sources of Pollution Incidents in Alaskan Waters, 1975

	Number of Incidents	% of Total	Volume in Gallons	% of Total
Vessels				
1. Dry Cargo Ships	1	.9	5	
2. Dry Cargo Barges	2	1.8	3	
3. Tank Ships	1	0.9	400	0.5
4. Tank Barges	2	1.8	3	
5. Other Vessels	47	42.3	6,453	8.2
Total	53	47.7	6,864	8.2
Land Vehicles				
1. Highway Vehicles	3	2.7	3,105	3.9
Non-Transportation Related Facilities				
1. Onshore Bulk Storage	6	5.4	2,010	2.6
2. Offshore Production	3	2.7	170	0.2
3. Other Facilities	12	10.8	340	0.4
Total	21	18.9	2,520	3.2
Pipelines	3	2.7	65,013	82.2
Marine Facilities				
1. Onshore/Offshore Bulk Cargo Transfer	7	6.3	380	0.5
2. Onshore/Offshore Fueling	2	1.8	72	0.1
3. Onshore/Offshore Non Bulk Cargo Transfer	1	0.9	-	-
4. Other				
Total	10	9.0	452	0.6
Misc/Unknown (Including Natural Seeps)	21	19.0	1,112	1.4
Total	111	100.0	79,066	100.0

The focus of this paper has been on the oil pollution problem that exists in Alaskan waters today, rather than to predict future trends. Therefore, it furnishes a useful baseline comparison for future statistical analysis. The recent studies done for the Coast Guard by Battelle (1973) predict hydrocarbon input into the Gulf of Alaska which will result from the increased production and transport of Alaskan crude oil.

TABLE 8. Sources of Pollution Incidents* for the Entire United States, 1975

	Number of Incidents	% of Total	Volume in Gal.	% of Total
Vessels				
1. Dry cargo ships	277	2.7	21,843	0.2
2. Dry cargo barges	31	0.3	5,215	0.0
3. Tank ships	643	6.3	1,766,729	12.2
4. Tank barges	757	7.5	3,467,203	24.0
5. Combatant vessels	202	2.0	16,913	0.1
6. Other vessels	1,143	11.3	1,353,947	9.4
Total	3,053	30.1	6,631,850	45.9
Land Vehicles				
1. Rail vehicles	27	0.4	576,507	4.0
2. Highway vehicles	263	2.6	356,601	2.4
3. Other unknown vehicles	20	0.1	2,617	0.0
Total	310	3.1	935,725	6.4
Non-Transportation-Related Facilities				
1. Onshore refinery	176	1.7	145,722	1.0
2. Onshore bulk/storage	305	3.0	476,768	3.3
3. Onshore production	233	2.3	2,626,992	18.2
4. Offshore production facil.	1,243	12.3	78,217	0.5
5. Other facilities	762	7.5	567,924	4.0
Total	2,719	26.8	3,895,623	27.0
Pipelines	564	5.6	2,490,237	17.3
Marine Facilities				
1. Onshore/offshore bulk cargo transfer	250	2.5	81,203	0.6
2. Onshore/offshore fueling	74	0.7	9,388	0.0
3. Onshore/offshore nonbulk cargo transfer	19	0.2	1,326	0.0
4. Other transportation-related marine facility	80	0.8	7,239	0.0
Total	423	4.2	99,156	0.6
Land Facilities	167	1.6	200,962	1.5
Misc./Unknown	2,905	28.6	186,044	1.3
Total	10,141	100.0	14,439,597	100.0

*Data shown only for pollution incidents involving oils.

Discussion

The data and statistics used to draw inferences must be selected with care. Global PHC estimates such as the NAS figures are useful estimates of water quality, but are not geographically specific nor do they adequately describe the sources and causes of pollution incidents. USCG Pollution Incident statistics are valuable for analyzing the nature of reported incidents, but do not provide a basis for projecting trends when the extent of the production and transport of potential pollutants is rapidly changing. Nevertheless, taken together these statistics provide a useful understanding of the total problem of oil pollution.

As the PIRS data have indicated, there is still a dearth of quantified descriptions of oil spills in ice-covered Alaskan waters. Recent studies by the Massachusetts Institute of Technology (MIT) (1975) and by Lewis (1976) of Environment Canada's Frozen Sea Research Group have postulated what may happen when a major oil spill does occur in an ice-covered Arctic environment. The MIT study, for example, postulates that oil from a major supertanker rupture would probably spread under the ice and fill each underwater pocket in the ice cover up to its underwater roughness height. As a consequence, the spilled oil would be confined to a very small area as long as the spilled fluid remained in contact with the ice cover. The Canadian study also suggests that oil well blowouts in the coastal ice shear zone of the Arctic Ocean would result both in oil accumulating in pockets under the moving ice and in coming to the surface for ponding within the leads. Any oil trapped underneath at the ice-water interface would remain unweathered and come to the surface in May and June of the following year if under first-year ice and in the same months during two following years if the cover consists of multiyear ice floes. Thus development of a comprehensive statistical picture regarding the behavior of spilled oil in Arctic waters requires much more elaborate mapping and sampling practices than has been necessary to date, where major oil spills took place in "wet" water. In fact, major spills in ice-covered waters of Arctic Alaska should preferably be treated as research topics until our present concepts as to the behavior of spilled oil under such conditions are either well verified or modified as the case may warrant.

Several other studies have also been sponsored by the Coast guard to obtain a better understanding of how much crude petroleum may be released into Alaska's marine environment once there is full-scale oil production from the Prudhoe Bay and other potential oil fields. For example, the environmental impact statement (U.S. Department of Interior, 1972) for transport of petroleum out of the Valdez Harbor complex estimated that an average of 1.6 to 6.0 barrels per day could be lost during tanker transfer operations when 2 million barrels of petroleum were leaving Valdez for the "South 48." Losses from tanker casualties throughout the entire tanker route were estimated to be 140,000 barrels per year in a "worst case" situation.

References

Bates, C.C., and E. Pearson. Influx of Petroleum Hydrocarbons into the Ocean, Offshore Technology Conference Paper 239, 3, 535-544 (1975).

Boyd, B.D., C.C. Bates, and J.R. Harrald. The Statistical Picture Regarding Discharges of Petroleum Hydrocarbons In and Around United States Waters, pp. 37-53 In: Sources, Effects and Sinks of Hydrocarbons in the Aquatic Environment, Proceedings of the Symposium, American University, August 9-11, 1976, American Institute of Biological Sciences, Washington, D.C. (1976).

Battelle Memorial Institute Pacific Northwest Laboratories. Geographical Analyses of Oil Spill Potential Associated with Alaskan Oil Production and Transportation Systems, 275 pp, U.S. Coast Guard R and D Rpt. CG-D-79-74, NTIS 784-099 (1973).

Kash, D.E., et al. Energy Under the Oceans. A Technology Assessment of Outer Continental Shelf Oil and Gas Operations. 378 pp. Univ. of Oklahoma Press, Tulsa (1973).

Lewis, E. L., Oil in Sea Ice, 34 pp, Inst. Ocean Sci., Patricia Bay, Environment Canada. Unpublished Rpt.

Massachusetts Institute of Technology. Man's Impact on the Global Environment. Assessment and Recommendations for Action. Study of Critical Environmental Probe. MIT Press, Cambridge (1970).

Massachusetts Institute of Technology. Oil in the Arctic, 218 pp. U.S. Coast Guard R and D Rept. CG-D-96-75, NTIS ADA010-269 (1975).

Mathematical Sciences Northwest. Comparison of Ecological Impacts of Postulated Oil Spills at Selected Alaskan Locations, 633 pp., U.S. Coast Guard R and D Rept. CG-D-155-75, NTIS ADA-017-600 (1975).

National Academy of Sciences. Petroleum in the Marine Environment, 107 pp, Washington, D.C. (1975).

United States Coast Guard. Draft Environmental Impact Statement. For International Convention for the Prevention of Pollution from Ships, 1973, 89 pp + 6 appendices (1973).

United States Coast Guard. Polluting Incidents In and Around U.S. Waters, Calendar Year 1975. Washington, D.C. (1976).

United States Department of Interior. Final Environmental Impact Statement - Proposed Trans-Alaskan Pipeline, 6 volumes, Washington, D.C. (1972).

CHAPTER 2

HYDROCARBONS IN THE WATER COLUMN

D. G. Shaw

University of Alaska, Fairbanks, Alaska 99701

Introduction

Oil and water don't mix - the familiar saying isn't strictly true but is a good first approximation. The water solubilities of hydrocarbons are quite low (Table 1). When the local concentration of petroleum hydrocarbons exceeds the equilibrium solubility values, the stage is set for a complex sequence of physical and chemical processes which disperse the petroleum into water. Understanding the processes that bring about dispersion and knowing the physical and chemical characteristics of resulting mixtures are important prerequisites to understanding the fate and effects of petroleum hydrocarbons in marine ecosystems since the hydrocarbons' form influences their behavior in subsequent degradation and biological uptake reactions.

The conceptual relationship of hydrocarbons in the water column to the impact of petroleum on marine ecosystems is clear. The presence of organisms and petroleum hydrocarbons in the same water at the same time provides the opportunity for the petroleum's toxic potential to be expressed. But such a general concept is of little operational use to the research scientist. In order to plan and interpret experiments intelligently he needs much more detailed information about hydrocarbons in seawater. He needs to know about hydrocarbon concentrations in seawater, about rates of hydrocarbon input, dispersion and sedimentation, about chemical fractionation and reaction of petroleum in seawater, and a myriad of other factors that determine the kinds and amounts of petroleum hydrocarbons that impinge on marine organisms. Complete information of this sort is not now available and is not likely in the foreseeable future.

TABLE 1. Representative Room Temperature Solubilities of Hydrocarbons in Distilled Water.

Hydrocarbon	PPM
n-pentane	38.5^1 39.5^6
n-hexane	9.5^1 9.47^6
n-heptane	2.93^1 2.24^6
n-octane	0.66^1 0.431^6
n-decane	0.052^3
n-dodecane	0.00182^4 0.0037^7
1-hexene	50^1
1,5-hexadiene	169^1
1-hexyne	360^1
cyclohexane	55^1 66.5^6
benzene	1780^1, 1790^2, 1740^6
toluene	515^1, 627^2, 554^6
biphenyl	7.48^2, 7.45^5
naphthalene	31.3^5 34.4^2
1-methylnaphthalene	25.8^5

[1]McAuliffe (1966), [2]Bohon and Claussen (1951), [3]McAuliffe (1969), [4]Button (1976), [5]Eganhouse and Calder (1976), [6]Price (1973), and [7]Sutton and Calder (1974).

This lack of complete data provides a scientifically invigorating opportunity. In order to proceed effectively we have to construct a theoretical framework that bridges the gaps in available knowledge and identifies the key questions whose resolution is necessary to further progress. This Chapter gives such a framework in terms of the chemical and physical nature of oil and water on the molecular and larger scales.

The intent here is to review the molecular basis of the water-hydrocarbon interaction from the perspective of the degree of aggregation of hydrocarbon molecules in water. What follows is organized in terms of three degrees of aggregation:
1) true solution in the thermodynamic sense,
2) colloids, molecular aggregations less than about 1 μm, and
3) particles larger than 1 μm.

Of course, these categories are not rigidly separated; hydrocarbon aggregations exist in a continuum from individually solvated molecules to large tar balls. However, the categories are useful in that they correspond both to physio-chemical differences and to the cut off points of commonly used experimental techniques of solution, colloid and particle chemistry.

No attempt is made here to provide a encyclopedic review of hydrocarbon measurements in seawater or to discuss the analytical techniques by which such measurements are made. Both of these subjects have recently been considered elsewhere (National Academy of Sciences 1975, McAuliffe 1976). More general discussions of petroleum pollution of the marine environment are also available (*inter alia* Ruiro 1972, Blumer 1972, and Baker 1976). Hydrocarbons in the arctic are considered in a recent volume by Hood and Burrell (1976). Goldberg (1976) has recently edited a discussion of the chemistry of seawater which includes consideration of hydrocarbons.

Structure of Liquid Water

Before discussing the interaction of hydrocarbons with water, a brief review of the structure of liquid water itself is in order. The structure and energetics of the water molecule are well understood as are the properties of ice and water vapor. However, the inter-molecular interactions of water molecules in the liquid state are very difficult to treat quantitatively. In general, liquids are more difficult to study than gases or solids because the partially ordered structure of the former is harder to treat than the high structural order of crystalline solids or the randomness of gases. Yet, even by the standards of most liquids, water is difficult. The chemical and physical properties of water are so markedly unlike those of structurally similar molecules (methane, ammonia, hydrogen sulfide, hydrogen fluoride) that clearly, water molecules associate in a way, or to a degree not found in other compounds. It is well known that the mechanism for this association in the liquid state is through the formation of hydrogen bonds and that ice owes its low density structure to a tetrahedral array of water molecules held in place by hydrogen bonds (Fig. 1). Qualitatively, it is possible to picture the formation of liquid water, which is more dense than ice, as resulting from breaking or distortion of some of the tetrahedral hydrogen bonds in a way that

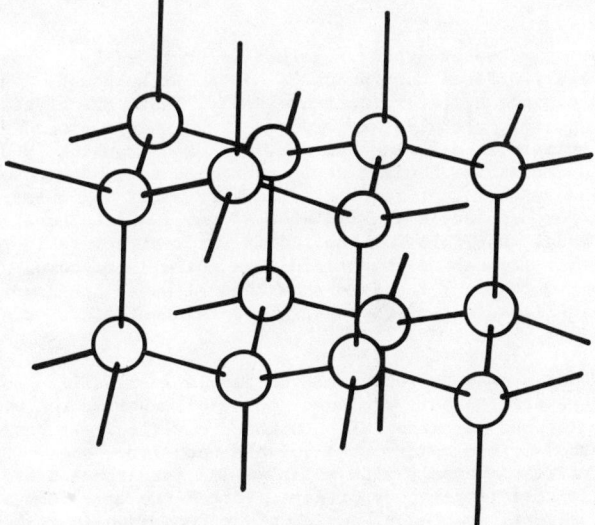

Fig. 2.1 Tetrahedral structure of ice due to hydrogen bonding. Distortions in the liquid state lead to closer packing and a denser structure.

lets the molecules pack a bit closer to one another. But no satisfactory quantitative model of the hydrogen bonding in liquid water currently exists and none seems likely to appear.

The difficulty of accounting for liquid water's structure in theoretical terms is illustrated by some results of X-ray diffraction work (Tanford, 1973). In ice, where water molecules are arranged tetrahedrally, each molecule has exactly four nearest neighbors at 2.76Å. An average water molecule in the liquid state has 4.4 nearest-neighbors at a distance which varies from 2.82 to 2.94Å as the temperature is raised from 4° to 200°C. This is good evidence that for nearest neighbors, the symmetry remains very close to tetrahedral. However, at a distance of 8Å from a reference water molecule, neighboring distances are completely randomized and evidence of any ordered structure is absent. The structure of liquid water is nearly tetrahedral, but over distances that are greater than a few molecular dimensions, the departures from perfect symmetry compound until no order is apparent at 8Å.

The difficulty of providing a quantitative model of liquid water is further complicated by the dynamic nature of its intermolecular structure. The hydrogen bonds which give rise to this structure are constantly being broken, shifted and reformed by turbulent and thermal energy. The difficulty in describing liquid water is in accounting quantitatively for this short range order and for the properties of water that derive from it. The structure of water and its solutions is the subject of a voluminous literature which will not be exhaustively reviewed here. Numerous works on this subject are available (Tanford 1973, Eisenberg and Kauzmann 1969, Kavanau 1964, Hertz 1970); the most complete single work being that of Franks (1972).

Perhaps the model of liquid water most useful as a basis for understanding hydrocarbon solubility is the bent hydrogen-bond model (Pople 1951; Bernal 1964). This model postulates that the structure of liquid water differs from the tetrahedral structure of ordinary ice by angular distortions of the hydrogen bonds. Broken hydrogen bonds are not allowed. This model has been discussed in quantitative terms and compared to other current models by Eisenberg and Kauzmann (1969, Chapters 4 and 5) and by Franks (1972, Volume 1 Chapter 14). The quantitative aspects of the theory are not necessary in the present context. What is critical is the concept of non-tetrahedral hydrogen bonds. These provide a model for the water solute interface and for the structure of water near that interface.

Hydrocarbon Solutions

Understanding hydrocarbon solubility in water requires knowledge of how the presence of a solute molecule affects the water's structure. If this effect were well enough understood, it would be possible to predict the solubility of any solute. A major step in this direction was made by Frank and Evans (1945) who observed a large decrease in entropy on the solution of non-polar solutes in water. This finding implies an ordering in the system which more than offsets the increase in entropy due to mixing. Frank and Evans' interpretation of the decrease in entropy, which is generally accepted today, is that introduction of non-polar solute molecules causes an increase in the ordering of the water in the vicinity of the solute.

A model for this structuring of water around a solute is provided by a group of substances in which water forms a highly ordered cage around a non-polar compound. These are the gas hydrates and the related organic hydrates (Jeffrey 1969). These are crystalline, stoichiometric compounds in which water molecules are ordered in polyhedra around a central molecule or ion. A well studied example is chlorine hydrate, $Cl_2 \cdot 8H_2O$ (Pauling 1960) in which Cl_2 molecules are enclosed in tetrakaidecahedral cavities formed by hydrogen bonded water molecules (Fig. 2). This phenomenon is not restricted to very small molecules. Similar hydrate formation has been shown for octadecylamine (Ralston et al. 1942). These crystalline hydrates provide the best model available for the liquid water structure in the vicinity of dissolved hydrocarbons - but it must be emphasized that there is no comparable data for the liquid state itself. On the contrary the data show that although the degree of order does increase around solutes in liquid water, there are no persistent micro-crystalline regions; the system remains dynamic.

A number of attempts have been made to relate a hydrocarbon's solubility to some of its observable properties. These efforts have produced a body of experimental evidence which emphasizes the importance of the formation of structured cavities. It has long been recognized that small hydrocarbon molecules are more soluble than large ones. Bohon and Claussen (1951) noted solubility varies inversely with molar volume for aromatic hydrocarbons. McAuliffe (1966) working with a larger body of data plotted the logarithm of solubility against molar volume for several homologous series of hydrocarbons (normal paraffins, cycloparaffins, terminal olefins, etc.). He found a linear relationship within each series but considerable scatter between series. Hermann (1972) refined this relationship by showing that the logarithm of solubility varies linearly with the calculated size of the solvent cavity just large enough to accommodate a solute molecule. This relationship holds for a

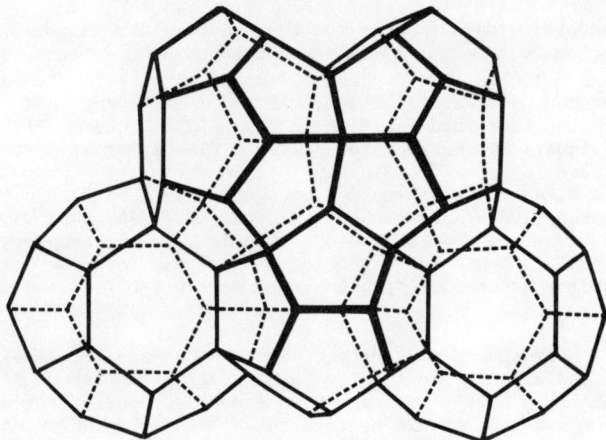

Fig. 2.2 Structure of chlorine hydrate, $Cl_2 \cdot 8H_2O$. The water structure only is shown. A chlorine molecule resides in the center of each larger cavity.

wide variety of hydrocarbons, not just members of a homologous series. Harris et al. (1973) made a yet more refined test of the theory of control of solubility by cavity surface area. These workers took into account the fact that a water molecule's size approaches that of a hydrocarbon and hence that the cavity may be somewhat larger than the closest approach distance used by Hermann. Using molecular models, Harris et al. physically packed water molecules around various alkyl and aryl groups. Their results were similar to Hermann's.

Reynolds et al. (1974) have shown that free energy of transfer (a thermodynamic quantity related to solubility) of a non-polar solute from a non-polar reference compound such as a liquid hydrocarbon to an aqueous medium is directly proportional to the surface area of the cavity created by the solute in aqueous solution, with the same proportionality constant for linear, branched and cyclic hydrocarbons. The value of the constant was estimated to be 20 to 25 cal/mole per $Å^2$ at 25°. Reynolds and co-workers point out that the relationship is empirical and that factors other than cavity size such as cavity shape may be important. This free energy of transfer, which Tanford (1973) has also called hydrophobic free energy, does not contain an entropy of mixing term but is a measure of only the internal free energy of the solute molecule and the free energy of its contacts with surrounding solvent molecules. Tanford has shown that hydrophobic free energy is equal to $RT \ln X_w$, where X_w is the mole fraction of hydrocarbon in aqueous solution at saturation. This leads to the prediction that the aqueous solubility of hydrocarbons falls exponentially as the molecular surface area increases. Using McAuliffe's (1966) aqueous solubility data for propane through octane, Reynolds et al. (1974) have shown this prediction to be correct. However, it must be noted that some data indicates that this simple relationship breaks down for larger molecules (Franks 1966).

Since petroleum is a complex mixture of hydrocarbons and other compounds, it would be useful to know how the presence of one compound affects the solubility of another. In terms of the structural model of hydrocarbon solution presented above this requires knowledge of how the ordered region of water molecules surrounding one solute molecule interacts with ordered regions around other solute molecules. It seems likely that there is some threshhold concentration below which the average spacing of hydrocarbon molecules is great enough that their ordered regions exist independently from one another. In that case there would be no effect of one solute on another. At higher concentrations interaction of the regions of ordered water molecules would take place, causing one solute to affect the solubility of another. There is no theoretical basis for predicting whether any petroleum hydrocarbons have great enough solubility to bring about this interaction or whether such interaction would enhance or depress solubility. Experimental data bearing on this point has recently been presented by Eganhouse and Calder (1976) who studied the solubilities of medium molecular weight aromatic hydrocarbons singly and in combination. A sampling of their data shown in Table 2 indicates a general trend toward small but significant mutual decreases in solubility. For instance the aqueous solubility of phenanthrene alone is 1.07 ppm; but in the presence of saturation values of naphthalene and biphenyl, phenanthrene's solubility is only 0.92 ppm. However, in some cases no change or even solubility enhancement is shown. A larger body of data for a structurally more varied group of compounds is needed before generalizations for all fractions of petroleum can be made.

Seawater, of course, contains dissolved salts. The effect of their presence on the aqueous solubility of hydrocarbons can be conceptualized in terms of a perturbation of the theory already presented. One needs to consider the effect of an electrolyte on the energy required to create a cavity for the hydrocarbon solute. If the salt lowers that energy requirement, hydrocarbon solubility increases and "salting in" is said to occur. If the energetic reverse occurs, "salting out" lowers hydrocarbon solubility. Dissolving of most salts in water is accompanied by a decrease in solution volume. This increases the internal pressure of the solution which in turn makes cavity formation energetically more difficult (Long and McDevit 1952). Gordon and Thorne (1967) and Eganhouse and Calder (1976) have shown that for natural and artificial seawater, salting out of hydrocarbons is directly proportional to salinity. The effect is fairly small: Eganhouse and Calder found that for two- and three-ring aromatic hydrocarbons the solubilities at 35.0 °/$_{\circ\circ}$ salinity were 0.6-0.7 of the values for 0.0 °/$_{\circ\circ}$ salinity.

Reports of nonvolatile hydrocarbon measurements in seawater are relatively rare compared to measurements in biota and sediment reflecting their difficulty (Brown and Huffman 1976, National Academy of Sciences 1975). However more convenient analytical methods have been developed for gaseous and volatile liquid hydrocarbons (Swinnerton and Linnenbom 1967). This has made possible extensive measurements. Swinnerton and Lamontagne (1974) reported analyses of 452 water samples from many open ocean and nearshore locations. Working in the Cariaco Trench, Reeburgh (1976) was able to obtain methane concentration profiles with depth in both water and sediment with sufficient precision to allow interpretation of its distribution in terms of advection-diffusion models.

TABLE 2. Aqueous Solubilities with 95% Confidence Intervals of Aromatic Hydrocarbons Singly and in Combination from Eganhouse and Calder (1976).

Solutes Present	Naphthalene	Solubility (ppm) Biphenyl	Phenanthrene
naphthalene	31.31 ± 0.40	-	-
biphenyl	-	7.45 ± 0.10	-
phenanthrene	-	-	1.07 ± 0.01
naphthalene biphenyl	30.54 ± 0.30 -	- 7.83 ± 0.07	- -
naphthalene phenanthrene	30.52 ± 0.58 -	- -	- 1.06 ± 0.03
biphenyl phenanthrene	- -	6.35 ± 0.06 -	- 0.91 ± 0.01
naphthalene biphenyl phenanthrene	31.45 ± 0.32 - -	- 6.36 ± 0.08 -	- - 0.92 ± 0.02

Colloidal Hydrocarbons

As pointed out in the previous discussion each molecule in hydrocarbon solution is isolated from others in its own structured cavity. However, at higher concentrations the hydrocarbon molecules are not individually solvated but aggregate to form hydrocarbon particles of colloidal size. A review of hydrophobic colloids in seawater has recently been made by Lyklema (1975).

Colloidal particles occupy the size range roughly from 10^{-9} to 10^{-6} m (1 mm to 1 μm). For petroleum hydrocarbons with molecular weights of a few hundred daltons this means aggregations of anywhere from a few to several million molecules. Particles of this size have physiochemical differences from either larger particles or individual molecules in solution. Larger oil particles rapidly float to the surface, as described by Stoke's law, when mixed into water. Colloids, however, remain in suspension for extended periods. A hydrocarbon colloid differs from a solution by the former's possession of a hydrophobic center (a hydrocarbon micro-region) into which lipids are attracted, and by the fact that colloidal particle size approaches that of microbes.

A particularly thorough treatment of the thermodynamics of colloids is that of Hall and Pethica (1967). Tanford (1973) presents a clear discussion of the thermodynamics of micelle formation. However, especially for natural systems, these treatments are so complex and our ability to make measurements is so crude, that thermodynamics has little predictive value. At present it seems more profitable to explain experimental observations in terms of a qualitative physical model.

Evidence is indirect but substantial that colloidal size hydrocarbon particles are formed in seawater under conditions of turbulent mixing, such as might be provided by wave action following an oil spill. Peake and Hodgson (1966, 1967) and Gordon et al. (1973) have mixed hydrocarbons with water. Both groups found that the quantities of hydrocarbons less than 1 μm, as determined by filtration, were greatly in excess of the values for a saturated solution. The excess above saturation levels of the hydrocarbons must be colloidal particles. Peake and Hodgson referred to these colloidal hydrocarbons as "accommodated"; Gordon called them "subparticulate". Either of these terms seems preferable to ones such as "solubilized" that hint at true solution. Peake and Hodgson (1967) report an experiment in which a mixture of 20 n-alkanes was shaken with distilled water for 16 hours. The resulting concentration of dodecane in the unfiltered water was 2 mg l^{-1}. However, the concentration dropped to 0.1 mg l^{-1} when the water was passed through a 0.45 μm filter. Button (1976) has reported that the concentration of a saturated solution of dodecane is 1.82 μg l^{-1}. Clearly the amount of dodecane in the colloidal size range can be greater than in solution but less than that which exists as particles greater than 1 μm. The amount of hydrocarbon surface area exposed to water by the various size fractions is probably just as important as the mass of hydrocarbon in determining the hydrocarbon's fate and biological effects. This enhances the importance of smaller particles since, for a fixed mass, reducing particle size by a factor of 10 increases surface area by a factor of 10.

Seawater contains organic compounds of both natural and anthropogenic origin which can form colloidal particles on their own or enhance the colloid-forming tendency of hydrocarbons. The common characteristic of these compounds, which are referred to as amphiphilic, is the possession of an ionic or polar region and a non-polar region. A classic example of such a compound is the carboxylate anion of stearic acid:

$$CH_3\ CH_2CH_2CH_2CH_2CH_2CH_2CH_2CH_2CH_2CH_2CH_2CH_2CH_2CH_2CH_2CH_2CO_2^-$$

When stearate anions aggregate to form a colloid the carboxylate groups ($-CO_2^-$) orient at the water interface and hydrocarbon-like alkyl chains form the center of the particle (Fig. 3).

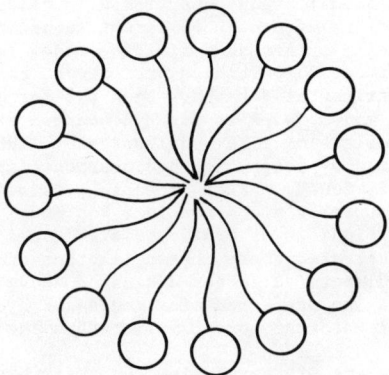

Fig. 2.3 Schematic representation of a spherical micelle composed of anions of fatty acids.

Such a colloidal particle of an amphiphilic substance is called a micelle. Numerous monographs have treated the formation and properties of micelles (*inter alia* Elworthy *et al.* 1968, Tanford 1973, Shaw 1970). If the concentration of amphiphilic molecules in an aqueous solution is raised, a concentration (called the critical micelle concentration) is reached at which the amphiphiles aggregate to form micelles.

The primary driving force for micelle formation is the lowering of cavity surface area. The area of the micelle, water interface is less than the sum of the areas of the interfaces for each of the hydrocarbon molecules solvated separately. The ionic head groups on the surface of the micelle coordinate with water molecules but repel each other electrostatically.

The maximum radius for a spherical micelle composed entirely of amphiphilic molecules is the length of the amphiphilic molecule. If this radius is exceeded, it is not possible for all of the polar head groups to be on the surface of the micelle. Larger micelles of other shapes are possible: cylinders of any length and radius equal to the amphiphiles' length, or bilayer sheets of any size (Fig. 4).

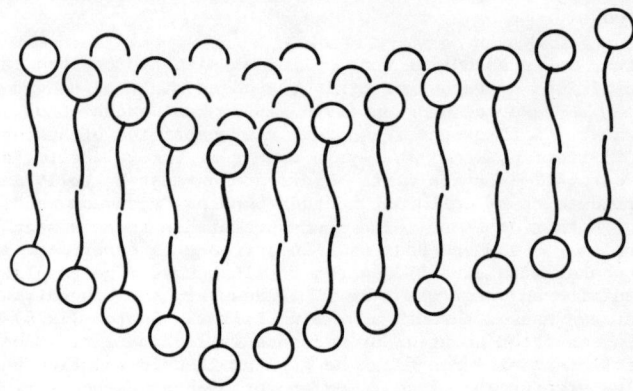

Fig. 2.4 Schematic representation of a portion of a bilayer of amphiphiles.

These size-shape constraints together with electrostatic repulsion among like charged head groups lead to preferred size modes for micelles composed only of amphiphiles (Baker 1962).

Colloidal particles that contain amphiphilic molecules and hydrocarbons can be regarded as micelles which have been swollen by the addition of hydrocarbons. (This effect is well known as the chemical basis of detergency). Thus the size-shape constraints that apply to purely amphiphilic micelles do not similarly apply when hydrocarbons or other lipids are present.

Because amphiphilic molecules are so abundant in nature (fatty acid salts, phospholipids, etc.) it is unlikely that colloidal organic particles in natural waters would be free of amphiphiles except in the immediate aftermath of abrupt substantial hydrocarbon inputs. Yet pure hydrocarbon colloids are important as a limiting model for colloids composed largely of hydrocarbons. Colloidal hydrocarbon particles are thought to be formed by wave action on marine oil spills. Some experimental data have been put forward (Peake and Hodgson 1966) that indicate preferred size modes do exist for hydrocarbon dispersions. Peake and Hodgson explain their data in terms of the micelle model of Baker (1962). However, Baker's observations deal with micelles formed by soaps, not hydrocarbon dispersions. For the soap micelles a bimodal size distribution is found. Baker interprets this finding in terms of spherical and cylindrical micelles. But Baker's observations and model for soap micelle sizes do not apply to Peake and Hodgson's findings of preferred size modes for hydrocarbons. If Peake and Hodgson's findings are accurate, there must be another explanation, perhaps involving the interplay of attractive and dispersive forces discussed above. Peake and Hodgson's data can also be interpreted as showing no preferred size modes for hydrocarbon dispersions if one assumes that a 0.22 µm filter used had pores smaller than the nominal value.

The question of whether there are preferred size modes for hydrocarbon dispersions appears to be open. There is some evidence that these modes do exist, but that evidence is by no means unassailable.

The pronounced influence that naturally-occurring amphiphiles have on enhancing the degree to which hydrocarbon-containing colloids are formed in seawater has been demonstrated by the work of Boehm and Quinn (1973). These workers measured the decrease in the concentration of non-filterable (0.5 µm) hydrocarbons when dissolved organic matter was removed from seawater by adsorption or ultraviolet photolysis. For aliphatic hydrocarbons this decrease was found to be substantial and directly related to the amount of naturally occuring dissolved organic matter removed; however no decrease was observed for aromatic hydrocarbons. Boehm and Quinn's results are strongly dependent on the length of time for which the hydrocarbons were shaken with water. No attempt was made to achieve equilibrium or steady state. Thus some of the results may have been due to kinetic effects; ascribing these results to equilibrium effects must be done with caution.

Particles

Particles greater than 1 µm constitute, in the thermodynamic sense, a separate phase which is formed spontaneously when the hydrocarbon solubility is exceeded. The presence of particles, however, does not necessarily imply that adjacent water is saturated with dissolved hydrocarbons. Particulates are a common form of hydrocarbon input: oil spills, tank washings, bilge pumpings, oil seeps, etc. Following such inputs dispersion, solution and degradation are not instantaneous. The processes of dispersion through the water as well as transfer to the atmosphere and sediments and incorporation into biota all proceed at finite rates. The interesting (and largely unanswered) questions about particulate hydrocarbons concern rates. How fast are they being added? How fast will they disperse under various environmental conditions?

Data for input rates has been assembled by SCEP (1970) and the National Academy of Sciences (1975). These two sources provide global and continental scale information, but obtaining actual or expectable input rates for particular bodies of water is often difficult or impossible. For data concerning dispersion the situation is more complex because the rate is influenced by a large number of environmental parameters which include currents, turbulence, temperature, salinity, nature of the hydrocarbon, weather conditions, the amount of suspended material in the water, the presence of grazing biota, and numerous others. In order for dispersion data to have predictive value, the effects of these parameters need to be treated individually and systematically. This work is far from complete at present.

Size distributions of particles in seawater that result from the dispersion of bulk liquid hydrocarbons have been measured under both field and laboratory conditions. Forrester (1971) found hydrocarbon particles in the water column of Chedabucto Bay, Nova Scotia, following a major oil spill in that area. Particles in the size range 5 to 85 µm were counted and measured. Numbers of particles increased with decreasing size so that hydrocarbon mass was roughly independent of particle size. Gordon et $al.$ (1973) carried out studies of size distribution of particles and subparticles formed by hydrocarbons mixed with seawater under laboratory conditions. Using a Coulter counter to determine particle sizes, they found that the greatest number of oil particles was in the size range 1 to 30 µm. In these experiments hydrocarbons (crude oil and distillate fuel oils) were added to filtered seawater and shaken briefly twice each day for two weeks. Witmer and Gollan (1973) measured the size distribution of oil droplets in discharged tanker ballast waters using optical microscopy. These workers found the most frequent droplet size to be about 10 µm. The smallest particles they were able to observe was approximately 3 µm. Lien and Phillips (1974) have also used the Coulter counter to measure droplet sizes in an oil-in-water emulsion. The emulsion used was prepared by dispersing tetralin in water in the presence of a surfactant. These workers, like Gordon et $al.$ (1973), found that the most abundant particle size observed was slightly larger than the minimum size observable.

In each of the laboratory experiments cited above, mixing of oil and water gave rise to small particles. But because of differences in mixing procedure and particle sizing techniques, different size distributions were reported in each case. No one has yet suggested a quantitative relationship between laboratory mixing and the mixing that occurs at sea.

Forrester's (1971) observation that particle abundance increased with decreasing size so that hydrocarbon mass was roughly independent of size is intriguing particularly since a similar finding for naturally occurring organic particles was made by Sheldon et $al.$ (1972). It is tempting to assume that the relationship is exact: that for logarithmically equal size intervals the aggregate mass of hydrocarbon particles is equal. If it is further assumed that there are no systematic variations in particle shape with mass, it follows that for each tenfold decrease in particle size there is a tenfold increase in hydrocarbon surface area exposed. Since molecular diffusion of hydrocarbons into the water occurs across this surface and since it is a substrate for microbes (Horn et $al.$ 1970) smaller particles may be of greater importance than their mass alone would suggest.

The fragmentation of large pelagic tar lumps into smaller particles was studied by Wade (1974) who observed that shaking tar balls with filtered seawater for 4 weeks produced microparticles with the following size distribution 1 mm - 0.25 mm, 50%; 0.25 mm - 0.3 µm, 38%; < 0.3 µm, 12%. Wade also found similar fine black flecks when samples of north Atlantic surface water were passed through glass fiber filters. Butler (1975) has developed a model of the evaporative weathering of pelagic tar which implies that physical fragmentation plays a major role in the weathering process.

The processes of chemical fractionation that accompany dispersion have also received attention. Boylan and Tripp (1971) used gas chromatography and mass spectrometry to determine the chemical identity of petroleum components that remained in seawater two hours after mixing. Their results show that the petroleum component which remained in the water phase was significantly enriched in one and two ring aromatic compounds over the bulk petroleum products which had been added. This enrichment certainly reflects the greater solubility of light aromatics than alkanes. It may also indicate a preference for aromatics in small particles and colloids. Such a preference is reasonable in view of the chemical nature of aromatics whose polarizable π electrons can interact with the polar water structure, diminishing the interfacial energy (Andrews and Keefer 1964).

Oil particles are taken for food by marine organisms. Conover (1971) has reported that following an oil spill zooplankton ingested particles of oil to such an extent that as much as 10% of the petroleum in the water column was associated with the plankton and their feces. Tar particles were found in the stomachs of 3 of 10 saurys, *Scomberesox saurus*, collected in the western Mediterranean (Horn *et al.* 1970). It is not known whether such oil ingestion is toxic or to what extent the oil is retained, metabolized or excreted.

Conclusions

In concluding this consideration of hydrocarbons in the water column, it is appropriate to consider some of the ways the ideas which have been discussed can be used in the design and interpretation of experiments dealing with the fate and effects of petroleum on marine ecosystems and organisms. The number of ways these ideas can be used is large, perhaps limited only by one's ability to perceive relationships. However, a few points are particularly worth making.

In many laboratory experiments on the effects of petroleum on marine biota, organisms are exposed to a preparation of oil in seawater. Such a preparation is typically made by mechanically stirring either crude or refined petroleum into seawater for a set length of time and then separating the oil containing water phase from the overlying oil after a specified settling time. Unless the settling time is quite long (perhaps on the order of weeks or months), this kind of procedure results in a non-equilibrium, non-steady state system containing particulate hydrocarbons, colloidal hydrocarbons and dissolved hydrocarbons. An unambiguous characterization of such a hydrocarbon preparation would have to include determination of the chemical identity and size spectrum of hydrocarbons present. If the effect of varying total hydrocarbon concentration were being studied, monitoring particle size could be critical. For instance, in studying the ingestion of oil by zooplankton, increasing hydrocarbon concentration by adding particles other than the animal's preferred food size might have little effect whereas adding oil particles of the preferred size might result in dramatically increased uptake.

Another application of the ideas presented here is the recognition of the importance of rates in determining how hydrocarbons are distributed in the water column. Following an oil spill, processes of solution and dispersion act on petroleum's hydrocarbon constituents. The chemical and size fractionation that results is controlled in part by the energetic requirements of structuring the hydrocarbon-water interface either for dissolved molecules, colloids, or particles.

Acknowledgement

The author is grateful to D. K. Button for helpful discussions of the subjects presented here and to the National Science Foundation, the National Oceanic and Atmospheric Administration and Petroleum Research Fund for support leading to this paper, which is contribution number 315 of the Institute of Marine Science, University of Alaska.

References

Andrews, L. J., and R. M. Keefer, Molecular Complexes in Organic Chemistry, Holden-Day, San Francisco, 1964, p. 196.

Baker, E. G., Distribution of Hydrocarbons in Petroleum, Bull. Am. Assoc. Pet. Geologists 46, 76-84 (1962).

Baker, J. M. (ed.), Marine Ecology and Oil Pollution, Wiley, New York, 1976.

Bernal, J. D., The Structure of Liquids, Proc. Roy. Soc. London, Series A, 280, 299-322, (1964).

Blumer, M., P. C. Blokker, E. B. Cowell, and D. F. Duckworth, Petroleum, In: E. D. Goldberg (ed.) A Guide to Marine Pollution, Gordon and Breach, New York, 1972.

Boehm, P. D., and J. G. Quinn, Solubilization of Hydrocarbons by the Dissolved Organic Matter in Seawater, Geochim. Cosmochim. Acta. 37, 2459-2477 (1973).

Bohon, R. L., and W. F. Claussen, The Solubility of Aromatic Hydrocarbons in Water, J. Am. Chem. Soc. 73, 1511-1578 (1951).

Brown, R. A., and H. L. Huffman, Jr., Hydrocarbons in Open Ocean Waters, Science, 191, 847-849 (1976).

Butler, J. N., Evaporative Weathering of Petroleum Residues: The Age of Pelagic Tar, Mar. Chem. 3, 9-21 (1975).

Button, D. K., The Influence of Clay and Bacteria on the Concentration of Dissolved Hydrocarbon in Saline Solution, Geochim. Cosmochim. Acta, 40, 435-440 (1976).

Conover, R. J., Some Relations Between Zooplankton and Bunker C Oil in Chedabucto Bay Following the Wreck of the Tanker Arrow, J. Fish. Res. Bd. Can., 28, 1327-1330 (1971).

Eganhouse, R. P., and J. A. Calder, The Solubility of Medium Molecular Weight Aromatic Hydrocarbons and the Effects of Hydrocarbon Co-Solutes and Salinity, Geochim. Cosmochim. Acta. 40, 555-561 (1976).

Eisenberg, D., and W. Kauzmann, The Structure and Properties of Water, Oxford University Press, New York, 1969, p. 296.

Elworthy, P. H., A. T. Florence, and C. G. Macfarlane, Solubilization by Surface Active Agents, Chapman and Hall, London, (1968).

Frank, H. S., and M. W. Evans, Free Volume and Entropy in Condensed Systems, J. Chem. Phys. 13, 507-532 (1945).

Franks, F., Solute-Water Interactions and the Solubility Behavior of Long Chain Paraffin Hydrocarbons, Nature, 210, 87-88 (1966).

Franks, F. (ed.), Water a Comprehensive Treatise, Plenum, New York, 4 Volumes, 1972.

Goldberg, E. D. (ed.), The Nature of Seawater, Dehlem Workshop Report, Dehlem Conference Publications, Berlin, (1975).

Gordon, D. C., P. D. Keizer, and M. J. Prouse, Laboratory Studies of the Accommodation of Some Crude and Residual Fuel Oils in Seawater, J. Fish. Res. Bd. Can., 30, 1611-1618 (1973).

Gordon, J. E., and R. L. Thorne, Salt Effects on Non-Electrolyte Activity Coefficients in Mized Electrolyte Solutions II, Artificial and Natural Seawaters, Geochim. Cosmochim. Acta., 31, 2433-2443 (1967).

Hall, D. G., and B. A. Pethica, Nonionic Surfactants, Dekker, New York, 1967.

Harris, M. J., T. Higuchi, and J. H. Rytting, Thermodynamic Group Contributions from Ion Pair Extraction Equilibria for Use in the Prediction of Partition Coefficients. Correlation of Surface Area with Group Contributions, J. Phys. Chem., 77, 2694-2703 (1973).

Hermann, R. B., Theory of Hydrophobic Bonding, II, The Correlation of Hydrocarbon Solubility in Water with Solvent Cavity Surface Area, J. Phys. Chem., 76, 2754-2759 (1972).

Hertz, H. G., Structure and Solvation Shell of Dissolved Particles, Angew. Chem. Internat. Ed., 9, 124-138 (1970).

Hood, D. W., and D. C. Burrell (eds.), Assessment of the Arctic Environment: Selected Topics, Occasional Publication No. 4, Institute of Marine Science, Univ. of Alaska (1976).

Horn, M. H., J. M. Teal, and R. H. Backus, Petroleum Lumps on the Surface of the Sea, Science, 168, 245-246 (1970).

Jeffrey, G. A., Water Structure in Organic Hydrates, Acts. Chem. Res. 2, 344-352 (1969).

Kavanau, J. L., Water and Water-Solute Interactions, Holden-Day, San Francisco, 1964, p. 101.

Lien, T. R., and C. R. Phillips, Determinations of Particle Size Distribution in Oil-in-Water Emulsions by Electronic Counting, Environ. Sci. Technol., 8, 558-561 (1974).

Lyklema, J., Interfacial Electrochemistry of Hydrophobic Colloids, In: E. D. Goldberg (ed.) The Nature of Seawater, Dehlem Workshop Report, Dehlem Conference Publications, Berlin, (1975).

Long, F. A., and W. F. McDevit, Activity Coefficients of Non-Electrolyte Solutes in Aqueous Salt Solutions, Chem. Rev., 51, 119-169 (1952).

McAuliffe, C., Solubility in Water of Paraffin, Cycloparaffin, Olefin, Acetylene, Cycloolefin, and Aromatic Hydrocarbons, J. Phys. Chem., 70, 1267-1275 (1966).

McAuliffe, C., Solubility in Water of Normal C_9 and C_{10} Alkane Hydrocarbons, Science, 163, 478-479 (1969).

McAuliffe, C., Surveillance of the Marine Environmental for Hydrocarbons, Mar. Sci. Commun. 2, 13-42 (1976).

National Academy of Sciences, Petroleum in the Marine Environment, Washington, D.C., (1975).

Pauling, L., The Nature of the Chemical Bond, 3rd Edition, Cornell University Press, Ithaca, 1960, pp. 469-471.

Peake, E., and G. W. Hodgson, Alkanes in Aqueous Systems, I, Exploratory Investigations on the Accommodation of C_{20}-C_{33} n-Alkanes in Distilled Water and Occurrance in Natural Water Systems, J. Am. Oil Chemists' Soc., 43, 215-222 (1966).

Peake, E., and G. W. Hodgson, Alkanes in Aqueous Systems, II, The Accommodation of C_{12}-C_{36} n-Alkanes in Distilled Water, J. Am. Oil Chemists' Soc., 44, 696-702 (1967).

Pople, J. A., Molecular Association of Liquids II, A Theory of the Structure of Water, Proc. Roy. Soc. London, Series A, 205, 163-178 (1951).

Price, L. C., The Solubility of Hydrocarbons and Petroleum in Water as Applied to the Primary Migration of Petroleum, University of California, Riverside Ph.D Thesis, p. 298, (1973).

Ralston, A. W., C. W. Hoerr, and E. J. Hoffman, Studies on High Molecular Weight Aliphatic Amines and Their Salts, J. Am. Chem. Soc., 64, 1516-1523 (1942).

Reeburgh, W. S., Methane Consumption in Cariaco Trench Waters and Sediments, Earth Planet. Sci. Lett., 28, 337-344 (1976).

Reynolds, J. A., D. B. Gilbert, and C. Tanford, Emperical Correlation between Hydrophobic Free Energy and Aqueous Cavity Surface Area, Proc. Nat. Acad. Sci. USA, 71, 2925-2927, (1973).

Ruivo, M. (ed.), Marine Pollution and Sea Life, Fishing New (Books) Ltd., London, 1972.

SCEP, Mans' Impact on the Global Environment: Assessment and Recommendations for Action, The MIT Press, Cambridge, 1970, p. 319.

Shaw, D. J., Introduction to Colloid and Surface Chemistry, 2nd Edition, Butterworths, London, (1970).

Sutton, C., and J. A. Calder, Solubility of Higher Molecular Weight n-Paraffins in Distilled Water and Seawater, Environ. Sci. Technol., 7, 654-657 (1974).

Swinnerton, J. W. and R. A. Lamontague, Oceanic Distributions of Low-Molecular-Weight Hydrocarbons-Baseline Measurements, Environ. Sci. Technol., 8, 657-663 (1974).

Swinnerton, J. W. and V. J. Linnenbom, Determination of C_1-C_4 Hydrocarbons in Seawater by Gas Chromatography, J. Gas Chrom., 5, 570-573 (1967).

Tanford, C., The Hydrophobic Effect, Wiley, New York, 1973, p. 200.

Wade, T. L., Measurements of Hydrocarbon, Phthalic Acid, and Phthalic Acid Ester Concentrations in Environmental Samples from the North Atlantic, M. S. Thesis, University of Rhode Island (1971) quoted in T. L. Wade and J. G. Quinn, Hydrocarbons in the Sargasso Sea Surface Microlayer, Mar. Pollut. Bull., 6, 54-57 (1975).

Witmer, F. E., and A. Gollan, Determination of Crude Oil Concentration and Size Distribution in Ship Ballast Water, Environ. Sci. Technol., 7, 945-948 (1973).

CHAPTER 3

DISPERSAL AND ALTERATION OF OIL DISCHARGED
ON A WATER SURFACE

Clayton D. McAuliffe

Chevron Oil Field Research Company
La Habra, California 90631

Introduction

Petroleum discharged on the sea surface undergoes physical, chemical, and biological alteration. Rapid physical and chemical processes include spreading and movement with wind and currents, evaporation of the volatile components, solution, emulsification, dispersion of small droplets into the water, sedimentation, spray injection into the air, and chemical oxidation. Concurrent with physical and chemical alterations are biological processes that operate most effectively after the physical processes have been initiated. Biological processes include degradation by micro-organisms, and uptake (if exposed) by larger organisms (followed by discharge, metabolism, or storage).

The fate of oil discharged to a water surface is reviewed briefly, including recent literature. No attempt is made to repeat the excellent review and summary in Chapter 3 of the National Academy of Sciences Report "Petroleum in the Marine Environment" (1975).

The only significant adverse effects from crude oil spills have been to birds, if present, and to some species of intertidal organisms where the oil stranded. Therefore methods of preventing or minimizing these effects would be beneficial.

Chemical dispersants have the potential for reducing the known adverse effects. This remedy is discussed, along with suggested research to evaluate effects from oil spills.

Fate of Spilled Oil

The dispersal and alteration processes will be influenced by a number of conditions, such as location, winds, waves, currents, water depth, temperature, salinities, organisms, nutrients, and kind of oil.

Spreading and Movement
Many oils added to calm water spread in the form of a thin, continuous layer and in a circular pattern. This tendency to spread is the result of gravity and surface tension. In open water, spreading is aided by wave motion, winds, and water currents. Other than on a calm surface, these interactions of wind and waves often elongate and distort a surface slick (Jeffrey 1973, Johanson et al, Murray 1975). Oil drift velocity is about 3 to 3.5% of the wind velocity (Johanson et al, Nelson-Smith 1973, Smith 1968). The oil slick, as it expands in size, is not uniform. Thicker patches of oil appear within the overall slick, and the largest amount of oil tends to sail with the wind at the leading edge of the slick (Jeffrey 1973, Johanson et al).

Many oils tend to spread on the sea surface at about the same rate, even though they may possess different viscosities (Johanson et al). Viscosity does have some effect on the rate of spreading. Highly viscous oils, such as Bunker C, discharged into cold water will not spread as rapidly as less viscous oil; the heavy oil does break up into separate masses. If a highly paraffinic crude oil with a high pour point, such as Minas, is discharged to sea water having a temperature lower than the pour point, the crude immediately forms semi-solid chunks, which then disperse much as solid particles disperse. Fallah and Stark (1976) have reviewed the theories and models of oil spreading and movement.

Spreading helps speed the weathering processes, by increasing the surface area of the oil exposed to air and seawater, as discussed below.

Emulsification, Dispersion, and Sedimentation
Some crude oils and petroleum products contain appreciable amounts of nitrogen, sulfur, and oxygen-containing compounds, some of which are surface-active. These compounds and their concentrations determine the formation of oil-in-water droplets or water-in-oil masses on the surface. The general tendency is to form water-in-oil emulsions. However, under agitation by wind and waves, many oils break up, and small droplets disperse into the underlying waters. Such dispersion may be assisted by bacteria-produced surface-active compounds (Boehm and Quinn 1973), or by oil attachment to particulate matter in water. Spilled oils often are not visible on the surface after 1 to 2 days.

Some crude oils have a tendency to form water-in-oil emulsions, frequently referred to as "chocolate mousse". These emulsions ultimately may become pelagic tar (tar balls) that strands on beaches. However, crude oil spills probably contribute a minor amount of oil appearing as tar balls. Most of the tar stranding on many beaches originates from washing of tanker compartments (viscous oils or waxy materials with sediment), Bunker C discharges, or bilge pumping. A portion of these larger oil pieces or globs are most likely to strand. Most pelagic tar breaks up and weathers at sea (Butler 1975, McAuliffe 1976).

As surface slicks break into oil droplets or particles and disperse in near-surface waters, most remain suspended (Brown et al 1973, Brown and Huffman 1976, Brown and Searl 1976, Butler et al 1973, Butler 1976) and move with the water. Some oil does sediment, apparently through association with inorganic mineral matter suspended in water (McAuliffe et al 1975).

Some petroleum products, such as Bunker C, have densities only slightly lower than sea water. Thus, the loss of volatile and/or soluble constituents, or attachment of mineral particles, results in an oil with a density greater than that of sea water, and it sinks. This was observed when Bunker C was spilled in Chedabucto Bay (Conover 1971) and during the San Francisco Bay tanker spill in 1971. In the case of the San Francisco Bay spill, the Bunker C oil moved from the Bay on the ebb tide into the Pacific Ocean. Some moved north to Bolinas Bay, where portions sank. Near-buoyant oil drops (pea- to grape-size) were observed by television in 30 ft of water in Bolinas Bay, moving just off the seafloor with bottom currents. Some of these balls floated to the surface, where, under higher temperatures, they melted and spread out quickly to form pancake-sized oil slicks surrounded by thin, fluorescent films (personal observation). The thin films were produced by the relatively unweathered oil released from the interiors of the semi-rigid oil balls.

Less viscous oils having higher API gravities (lower specific gravities) do not sink as large drops or globs. However, the surface slick does disappear. During the Chevron Gulf of Mexico spill, the previous day's slick could often not be found on the second day (McAuliffe et al 1975, Murray 1975). Kinney et al (1969) studied the dissipation of crude from accidental and controlled spills in the Cook Inlet with water temperatures of about 5°C. They estimated that the spill half-life was 1 day, with complete disappearance in 4 to 5 days. Johanson et al report that on the open ocean, they were unable to find small spills of a 39° API gravity crude oil on the second day, but were able to find remnants of a 24° API gravity crude oil.

Oil once dispersed and no longer observable as a surface slick appears to remain principally in near-surface waters, as particulates. Brown et al (1973), Brown and Huffman (1976), and Brown and Searl (1976) measured nonvolatile hydrocarbons along tanker routes in the Atlantic, Pacific, Indian, and Caribbean. Concentrations at 10 and 30 m were about 40% of those in surface samples. This indicates the hydrocarbons are particles (droplets?) and not in solution. These particles are altered by chemical and biological processes discussed later. Only high-density, viscous oils such as Bunker C are likely to sink and appear in bottom sediments in relatively unaltered form.

Solution and Evaporation
The first compositional changes in petroleum spilled on water are evaporation and solution of volatile components. The rates and extent of these changes depend upon the chemical and physical nature of each particular petroleum, wind velocities, sea states, and water temperatures.

   Evaporation vs solution. Evaporation and solution are simultaneous and competitive processes. Each volatile hydrocarbon will evaporate and also go into solution at rates depending upon its vapor pressure (API 1953, Doss 1943) and solubility (Button 1976, Eganhouse and Calder 1976, McAuliffe 1966 1969, Sutton and Calder 1974 1975). For each homologous series of hydrocarbons (alkane, cycloalkane, and aromatic), vapor pressures divided by solubilities are approximately constant, but with different values for the three series. As molecular weights of a class of hydrocarbons increase, vapor pressures and solubilities decrease by about the same percentages. McAuliffe (1971 1974) demonstrated this relationship in laboratory studies. Harrison et al (1975) incorporated the concept in a mathematical model and predicted that the evaporation of aromatic hydrocarbons would be

100 times faster than solution, and that the rate for alkanes would be 10,000 times higher than solution.

Although adverse biological effects have been attributed to dissolved low-molecular-weight hydrocarbons (particularly aromatics), only a few attempts have been made to measure these dissolved constituents under oil slicks. Harrison (1974) and Harrison et al (1975) attemped to verify their model by adding 4.2% isopropylbenzene (cumene) to a South Louisiana crude oil and conducting five separate 6.6-bbl spills off the south shore of Grand Bahama Island. The loss of cumene from the oil slicks was followed with time, and four water samples were collected under two of the slicks with a Van Dorn sampler. In two of the samples, cumene was detected at a concentration of about 0.3 mg/l (300 µg/l), although the method of analysis was not quantitative below 1 mg/l. The authors indicated that the samples also may have been contaminated during collection, and suggested that solution of cumene appeared to be two orders of magnitude slower than evaporation.

Johanson et al and McAuliffe (1977) report the loss of C_2 to C_{10} hydrocarbons by evaporation and solution from four 10.5-bbl spills of 24° and 39° API gravity crude oils. Water samples were analyzed by a method (McAuliffe 1971) that measured individual C_2 to C_{10} alkane, cycloalkane, and aromatic hydrocarbons with sub-µg/l (ppb) sensitivity.

Even with this more sensitive method, truly dissolved hydrocarbons apparently were not found in near-surface water (1.5 m depth) in contact with the surface slicks 15 minutes or later after the oil discharges (McAuliffe 1977). The data indicate that these low-molecular-weight hydrocarbons were present in near-surface water in concentrations of <1 µg/l. The explanation appears to be that hydrocarbons dissolved from the surface slicks quickly evaporated from water to the atmosphere.

The above discussed distribution of volatile hydrocarbons between water and air applies to all volatile hydrocarbons, regardless of molecular weight, the concentration of each in the oil phase, and degree of departure from equilibrium. However, these factors do affect the absolute and relative concentrations of the various hydrocarbons that dissolve in water. For example, if an excess of oil is mixed with water (evaporation prevented) until equilibrium is established, the water will contain the maximum dissolved concentration of each hydrocarbon. This concentration will depend upon the hydrocarbon's mole fraction in the oil and relative solubility in water.

These concentrations have been established for several crude oils and some refined products (Anderson et al 1974, McAuliffe 1977, Rice et al 1976). For six crude oils, the total C_2 to C_{10} dissolved hydrocarbons ranged from about 20 to 40 mg/l. The C_2 to C_5 alkane concentrations are often high, reflecting their moderate solubilities and relatively high concentrations in crude oils. Benzene and toluene concentrations in waters equilibrated with crude oils are approximately equal (benzene is sometimes higher), and these two hydrocarbons comprise 70 to 85% of the total dissolved aromatics. The amount of benzene and toluene each is approximately ten times higher than the individual C_8 aromatics (ethylbenzene, o-, m-, and p-xylenes) and approximately ten times higher than the sum of the trimethylbenzenes (C_9). Under nonequilibrium conditions and with evaporation prevented, the dominance of the lower-molecular-weight hydrocarbons is further accentuated over that observed for equilibrium conditions for a given class of hydrocarbons (the smaller the molecule, the more rapidly it will enter the water phase). Thus, for aromatic hydrocarbons, benzene would be in solution in higher concentration than toluene, and the relative concentrations would decrease rapidly for dimethylbenzenes and trimethylbenzenes. Similar nonequilibrium relative concentrations would be found for the alkane and cycloalkane classes of hydrocarbons.

Solution is a rate process, and concentrations of each hydrocarbon in water become progressively lower as the degree of departure from equilibrium increases. Thus, the shorter the contact time of oil and water, the lower the concentration of each hydrocarbon in water, and the higher the relative concentrations for those hydrocarbons having the lowest molecular weights.

Oil discharged on a water surface is in nonequilibrium with reference to evaporation and solution of hydrocarbons. Wind constantly renews the air over the slick, and water movement (waves and currents) renews the underlying water. If solution were important relative to evaporation, benzene at least should have been found in the water samples collected under the four controlled oil spills mentioned above. These observations indicate that hydrocarbons that dissolve in water quickly evaporate. A surface slick is nonuniform in thickness, is continually being broken, and often moves relative to the underlying water. Oil dispersed in near-surface waters is continually mixed and evaporation appears not to be inhibited.

Oil spilled on water is never in equilibrium with respect to solution of hydrocarbons. The only situations where measurable concentrations of dissolved hydrocarbons might occur are spills where oil would completely cover a small lake with a thick layer of oil, or oil is discharged in shallow nearshore waters during a storm. The rapid mixing might provide the greatest opportunity for hydrocarbons to dissolve in water. However, studies are necessary to establish actual concentrations. The above reviewed investigations show that measurable concentrations of dissolved hydrocarbons were not found in offshore waters, and probably are low in waters under most spills on the water surface.

Subsurface oil discharge would provide the greatest opportunity for hydrocarbons to dissolve into water. In the Coal Oil Point region of the Santa Barbara Channel, where an estimated 50 to 100 bbls of oil seeps each day (Allen et al 1970), Koons and Brandon (1975) found benzene and toluene each to have maximum concentrations in waters from the seep area of 0.08 µg/l (ppb). C_8 aromatics were not detected (0.01 µg/l was method sensitivity).

If oil penetrates beach sands or marshlands, then the interstitial water will have an opportunity to approach equilibrium with soluble hydrocarbons present in the oil. The content of soluble hydrocarbons (amounts and relative compositions) can vary widely, depending upon the oil and how much it had weathered prior to incorporation into the sediment.

Additional evidence that solution is a minor process, compared with evaporation, is shown by the rate of loss of benzene compared to cyclohexane. Benzene and cyclohexane have similar vapor pressures (95.5 and 97.8 mm Hg), but very different solubilities (1780 and 55 mg/l, McAuliffe 1966). If solution caused a significant fraction of hydrocarbon loss, compared to evaporation, the percentage of benzene remaining in the oil should be lower than for cyclohexane. The rate of loss during weathering was the same for these hydrocarbons from crude oil spills (McAuliffe 1977). Harrison et al (1975) found similar losses of hydrocarbons with about the same vapor pressures but different solubilities (n-C_9 and cumene). Isopropylbenzene (cumene) was completely lost in 40 to 80 min (depending on wind and wave conditions) from 5 spills of a South Louisiana crude oil discharge onto 24°C water. Normal C_9 was lost at a slightly slower rate (40-90 min). These investigators found that n-C_{12} disappeared in 3 to 8 hrs and, by association, inferred that naphthalene, which has the same boiling point as n-C_{12}, should also have been lost from the oil at approximately the same rate.

The above discussion of evaporation vs solution of volatile hydrocarbons has not addressed quantitatively the effect of wind and water movement relative to these processes. Rate of evaporation is directly proportional to wind velocity, other factors being constant. Rate of solution is likewise dependent upon removal of dissolved hydrocarbons from the water-oil interface. Thus, no wind but high water current, or the converse, would alter the ratio of evaporation to solution of volatile hydrocarbons. Normally, however, winds and currents are coupled, so that evaporation and solution will approximate a constant for each class of hydrocarbons. Obtaining data from field observations is difficult because of changing meteorological and oceanographic conditions. Fallah and Stark (1976) developed a random model to predict the residual oil volume as a function of time, the oil's initial volume and physical properties, wind velocity, and temperature.

Loss of volatile hydrocarbons from surface slicks. Evaporation and solution of volatile hydrocarbons have been measured in only two of the studies discussed above. However, additional field observations of the loss of volatile hydrocarbons from the sea surface have been made. The rate of loss of volatile hydrocarbons under various environmental conditions on the open ocean is summarized in Table 1, along with two laboratory studies. The rates of loss of volatile hydrocarbons in laboratory studies are generally less than observed in spills at sea. Regnier and Scott (1975), Table 1, found 10% of C_{12} and lower hydrocarbons remaining in a No. 2 fuel oil after 80 hrs at 20°C and air movement of 11 knots, whereas Harrison et al (1975) found loss of n-C_{12} to be complete in 3 to 8 hrs from sea surface slicks. Smith and MacIntyre (1971) found loss of volatile hydrocarbons from a No. 2 fuel oil at sea to be many times faster than in laboratory bubbler experiments. This faster loss is due to greater oil agitation by winds, waves, and currents. Ocean spills have shown marked changes in rates of loss of volatile hydrocarbons, associated with changes in winds and sea state (Harrison et al 1975, McAuliffe 1977, Smith and MacIntyre 1971). Figure 1, adapted and modified from Harrison et al (1975), shows the loss of n-C_9 to n-C_{13} hydrocarbons from a South Louisiana crude oil. The relatively rapid increase in loss of n-C_{10} to n-C_{13} is correlated to the sudden onset of whitecapping on the sea surface (arrow, Fig. 1).

Smith and MacIntyre (1971) measured the complete loss of n-C_{10} and 50% loss of n-C_{12} in 7 hrs from a 4.8 bbl No. 2 fuel oil spill 15 miles offshore. The air and water temperatures were about 5°C. The wind and waves at the time of release were calm, but the wind increased to 15 to 18 knots and seas built up to whitecaps after 4 to 5 hrs, causing an increased rate of evaporation.

McAuliffe (1977) reports the loss of C_9 and lower hydrocarbons in 4 to 8 hrs from four separate crude oil spills made 70 miles east of Boston. Weathering rates were related to differences in oil type (a 23.9°API gravity La Rosa crude oil from Venezuela, and a 39.0°API gravity Murban crude oil from Abu Dhabi), water temperatures, winds, and sea states (Table 1).

TABLE 1. Loss of Volatile Hydrocarbons from Oils on Water Surface

	Investigators						
	Harrison et al (1975)	Kinney et al (1964)	McAuliffe (1977)	Sivadier and Mikolaj (1973)	Smith and MacIntyre (1973)	Kreider (1971)	Regnier and Scott (1975)
	Open Sea Water					Laboratory	
Type of Oil	Crude	Crude	Crude	Crude	No. 2 Fuel	Crude	No. 2 Fuel
No. of Spills	5	1	4	2	1	Several	Several
Vol. Spilled, bbl	6.6	Small	10.5	0.02	4.8	450 ml[a]	
Water, °C	24	25	11-14	19-20	5	--	20
Air, °C	21-24	--	12-17	28-30	5	--	20
Wind, knots	0-18	9-12	8-24	8-12	1-18	--	11.3
Waves	Calm to Whitecaps	--	1-5 ft		Calm to Whitecaps	--	Wavelets
Time for Hydrocarbon Loss:							
C_9 and lower	40-90 min	8 hrs (10% left)	4-8 hrs	90 min?			
C_{10} and lower					7 hrs		
C_{12} and lower	3-8 hrs					24 hrs	80 hrs (10% left)

[a] Oil was added onto water in 55 gal drum outside laboratory, Richmond, California.

Extensive sampling and analysis of oil floating on the sea surface have not been extended beyond 8 hrs. Simulated "laboratory" evaporative weathering tests conducted for longer times (Krieder 1971, Smith and MacIntyre 1971) indicate that virtually all hydrocarbons smaller than C_{15} will be lost from most oils on the sea surface within a few days.

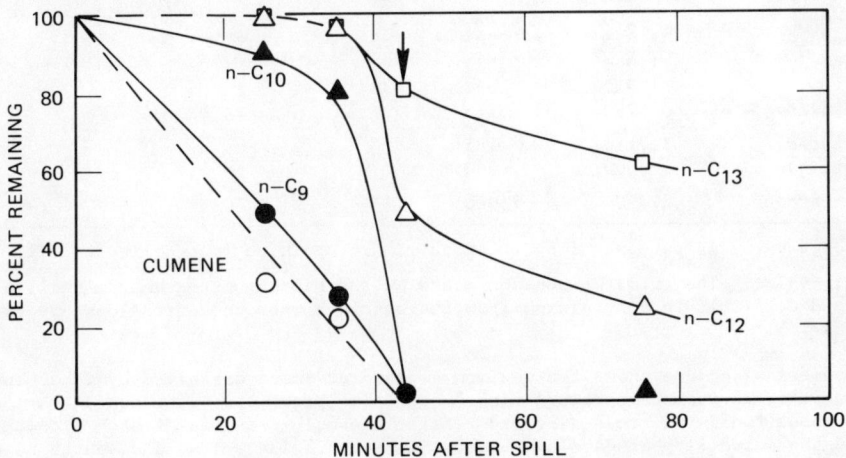

Fig. 3.1 Percent of low-boiling hydrocarbons remaining in South Louisiana crude oil slick.

Volatile hydrocarbons should be lost in proportion to their vapor pressures. However, as volatile hydrocarbons are lost, the viscosity increases, and the resulting oil may aggregate or form higher viscosity water-in-oil emulsions (McAuliffe 1973) that result in thicker bodies of oil. Diffusion then becomes slower and may be the limiting factor in the loss of volatile hydrocarbons. Regnier and Scott (1975) observed this for a No. 2 fuel oil, and indicated that the evaporation of the higher-boiling components was retarded relative to the lower-boiling fractions. Large chunks or globs of beach-stranded oil have been observed to weather slowly (Blumer and Sass 1972, McLean and Betancourt 1973).

Volatile hydrocarbons in surface waters under oil slicks. Previously it was stated that truly dissolved C_2 to C_{10} hydrocarbons were not found (<1 µg/l) in near-surface waters under oil slicks. C_2 to C_{10} hydrocarbons have been measured in the water column under oil slicks on the sea surface, and these hydrocarbons appear to be in oil droplets dispersed in near-surface waters. Of 68 water samples collected in time sequences under four separate crude oil slicks (Johanson et al), C_2-C_{10} hydrocarbons were found in only five samples collected 15 to 20 min after oil was spilled (McAuliffe 1977). The highest concentration was 60 µg/l at 5 ft under the 39° API gravity crude oil slick agitated most by wind and waves. Total C_2 to C_{10} hydrocarbon concentrations in the other four samples ranged from 2 to 16 µg/l. An example of concentrations and relative concentrations of individual hydrocarbons is shown in the last column of Table 2. Also shown in Table 2 are the equilibrium concentrations of C_2 to C_{10} hydrocarbons in sea water after mixing with an excess of oil from the spill tank, and oil collected from the sea surface slicks by a small skimmer.

TABLE 2. Hydrocarbons Dissolved in Sea Water Equilibrated With Oil Samples (39°API Gravity Crude Oil)

Source Of Oil	Tank	Water Surface		Water Column
Time After Spill	0	5-35 Min	2.5-2.75 Hrs	18 min
Hydrocarbon	Concentrations in µg/l (ppb)			
Ethane	230	0.07	0.03	
Propane	2,150	0.38	0.02(a)	
Isobutane	800	0.26	0.03	
n-Butane	2,800	1.25	0.04	
Isopentane	1,030	1.80	0.17	
n-Pentane	1,340	3.60	0.27	
Hexanes	850	7.3	0.86	
n-Hexane	500	15	3.0	
Methylcyclohexane	235	20	8.5	
Benzene	6,080	515	8.5	0.1
Cyclohexane	410	40	9.8	
n-Heptane	330	40	12.1	
Methycyclohexane	235	55	23	
Toluene	6,160	3,130	51	0.80	0.4
Ethylbenzene	825	705	85	1.3	0.5
m-, p-Xylene	1,940	1,825	94	7.3	1.1
o-Xylene	1,010	890	88	5.1	0.8
Trimethylbenzenes	750	715	95	42	1.0(b)
Total Saturates	11,110	185			
Total Aromatics	16,800	7,780		56	3
Total Hydrocarbons	27,900	7,960		56	3

(a) Underscored value is percent of hydrocarbon remaining in surface-collected oil.

(b) Uncertain value. The trimethylbenzene peaks on the gas chromatogram are small and broad, and difficult to distinguish from the baseline when concentrations are less than 3 µg/l.

The first numerical column shows the maximum concentration of dissolved hydrocarbons attainable in sea water at room temperature with evaporation prevented. The concentrations are typical of those found for crude oils and reflect the mole fraction of each hydrocarbon in the oil and the solubility of each hydrocarbon in water. The second and fourth numerical columns show the rapid weathering of C_2 to C_{10} hydrocarbons from the surface slick. The hydrocarbons were lost approximately in accordance to their vapor pressures (for oil on the sea surface for 5-35 min), as shown by the percent remaining (third numerical column).

As discussed prevously, hydrocarbons (for each class) should dissolve into water inversely to their molecular weights. This was not observed as shown by the relative concentrations in the last column of Table 2. These data (additional data are presented in McAuliffe 1977) indicate the measured hydrocarbons are those that remain in oil dispersed into water from the surface slick. In an emulsion plume of chemically dispersed 34° API gravity crude oil during the Chevron Gulf of Mexico spill, C_2 to C_{10} hydrocarbons in surface-collected waters, 900 to 1400 ft from the platform, ranged from 20 to 200 µg/l (McAuliffe et al 1975). Exposure time in water was 20 to 30 min. The relative compositions of the measured volatile hydrocarbons were also the converse of that expected for hydrocarbons truly in solution (McAuliffe, report in preparation). The chemically dispersed oil appeared to weather in a similar manner to oil naturally dispersed under slicks.

Bioassays with water soluble oil fractions. The above mentioned studies and discussion show that dissolved hydrocarbons in sea water under oil slicks is very low (<1 μg/l). Under all conditions where oil is on or in near-surface waters, concentrations of dissolved hydrocarbons are probably very low, because there is no opportunity for water to equilibrate with the oil. If measurable dissolved hydrocarbons were present in water, those with the lowest molecular weights would have the highest concentrations.

These facts therefore indicate that many bioassay tests that equilibrate fresh oil with water for a day or more are very unrealistic (Anderson et al 1974, Gilfillan 1975, Kauss and Hutchinson 1975, Rice et al 1976, Rossi et al 1976, Vanderhorst et al 1976, Winters et al 1976). The long equilibration allows naphthalenes, which some reports (Anderson et al 1974, Rice et al 1976) consider more important sources of toxicity from petroleum than the lower-molecular-weight aromatics, to attain their highest concentrations in water. The entire field of bioassay testing with petroleum products needs re-evaluation and research under conditions that simulate more realistically the exposure durations, and the compositions and concentrations of hydrocarbons likely to occur in the environment.

Microbial Modification, Organism Uptake, and Depuration
The oil that remains on the water surface or mixes down in near-surface water as the oil spreads and moves under wind and current becomes available to organisms. The degree of exposure depends upon several factors discussed in the section above on "Emulsification, Dispersion, and Sedimentation". The reader is referred to the excellent discussion of biological fates of oil in the NAS report "Petroleum in the Marine Environment" (1975). Included here will be principally supporting references.

All natural waters tested contain hydrocarbon-utilizing bacteria, yeasts, and fungi (Atlas 1973, Button 1974, Cundell and Traxler 1973, Dean-Raymond and Bartha 1975, Kinney et al 1969, Robertson et al 1973, Traxler and Cundell 1975 1976, Walker et al 1973), especially those exposed to reoccurring or continuous petroleum contamination, such as seeps and river discharges from municipal and industrial sources.

After a spill, micro-organism populations increase with the increased food source. The rate at which they modify the added oil depends upon the oil, nutrient availability, oxygen, and temperature. Oxygen is apparently not a limiting factor in most waters (Kinsey 1973, NAS 1975), but may become one in a deep, poorly mixed basin or in sediments penetrated by oil. Crude oil that sedimented in 15 m of water during a Gulf of Mexico spill underwent relatively rapid biodegradation and was gone after 11 months (McAuliffe et al 1975). Shelton and Hunter (1974) showed a slow decrease in hydrocarbon content of river sediments in laboratory investigations when the overlying water was aerobic. Anaerobic degradation of petroleum by micro-organisms appears to be very slow (Blumer and Sass 1972), or nonexistent (Davis 1967). However, Shelton and Hunter (1975) demonstrated the loss of oil from river bottom sediments with time when anaerobic conditions occurred in the overlying water. They suggested the loss may have been the result of solubility and leaching as a result of some biochemical alteration rather than actual biochemical use. Walker et al (1976) found bacteria present in marine sediment samples collected off the Atlantic coast that were capable of degrading petroleum hydrocarbons. Bacteria present in deep ocean sediments (2200 to 4400 m) accomplished significantly greater degradation in the laboratory under ambient conditions than did bacteria from coastal sediment samples.

Hydrocarbon oxidation by micro-organisms is probably the major way hydrocarbons are removed from the environment. Although very large contributions of petroleum hydrocarbons (also biogenic) have occurred through geological time, from seeps (Wilson et al 1974) and erosion of hydrocarbon-containing oil sands and source rocks, there is no evidence of hydrocarbon buildup in the ocean from these sources (McAuliffe 1976).

Biodegradation rates for hydrocarbons decrease in the order normal alkanes, branched alkanes, aromatics, and cycloalkanes (Perry and Cerniglia 1973). Polycyclic aromatic hydrocarbons, including 5 and 6 ring, have been shown to be biodegraded (Dean-Raymond and Bartha 1975, Gibson et al 1975). Major portions of the oil appear to biodegrade even in extremely cold marine environments (Atlas 1973 1974, Button 1974, Cundell and Traxler 1973, Kinney et al 1969, Robertson et al 1973, Traxler and Cundell 1975 1976). Table 3 summarizes the percent loss of crude oils at different temperatures for some of the studies. Atlas (1974) conducted additional Prudhoe Bay crude oil biodegradation studies on small ponds near Barrow, Alaska. He found the following percent losses in 3 weeks: 70 when seeded with organisms plus N and P, 40 with N and P added, and 25 when untreated. Button (1974) documented the conversion of n-C_{12} to CO_2 with in situ incubation in waters at Port Valdez (3°C); Point Barrow, through shore ice (2°C); and under an Arctic ice station (-1.7°C).

Larger organisms exposed to petroleum may ingest dispersed particulate oil or dissolved hydrocarbons if present. Exposure to dissolved hydrocarbons is usually minimal. A number of laboratory tests exposing various species of organisms to both dispersed and dissolved hydrocarbons have demonstrated relatively rapid uptake. Almost universally these studies

hydrocarbons have demonstrated relatively rapid uptake. Almost universally these studies have shown rapid depuration or metabolism, followed by depuration of metabolized products when exposure ceases (Corner et al 1973, DiSalvo and Guard 1975, Fossato and Canzonier 1976, Lee 1975, NAS 1975, Neff and Anderson 1975, Rice et al 1976). Organisms such as clams and oysters eliminate ingested hydrocarbons more slowly than do fish and other organisms with metabolic processes (livers). Zooplankton appear to ingest particulate oil, but eliminate most of it in feces (NAS 1975). Percy (1976) tested the responses of three arctic marine crustaceans to three oils, and found none attracted to crude oil. Conover (1971) noted that zooplankton ingested Bunker C fuel oil droplets during an actual spill. Thus, larger organisms may contribute to the breakup, dispersion, and destruction of petroleum hydrocarbons, complementing oxidation by micro-organisms. Although these processes may represent the major exposure pathways for macro-organisms, they are not likely to make substantial contributions to the mass balance of petroleum hydrocarbons introduced into the ocean.

TABLE 3. Biodegradation Rates of Crude Oils at Various Temperatures

Conditions	Investigators							
	Atlas (1973)		Kinney et al (1969)		ZoBell (1973)	Robertson et al (1973)		Walker and Colwell (1977)
Crude Oil	Prudhoe Bay		Cook Inlet		Prudhoe Bay	Cook Inlet		South Louisiana
	Lab	Field	Lab		Lab	Lab		Field
Temp. °C	5	25	5–12		10	−1.1	8	10	30
Nutrients Added	no	no	no	yes	no	yes	yes	no	no
Percent Lost After	21	39	60	80	Complete	61	82	90	97
Given Days	3	3	35	35	30–60	70	70	30	21

Photochemical Modification

Photochemical processes alter crude oils and partially account for the oxidation of hydrocarbons. Oxidation products have higher water solubilities than the parent hydrocarbons, which may increase biodegradation and emulsification rates, thereby promoting dispersal of spilled oil. Photo-oxidation is a relatively slow process compared to spreading and evaporation of volatile hydrocarbons. Klein and Pilpel (1974) demonstrated photo-oxidation of three crude oils irradiated with a mercury arc in a laboratory. This was found to increase spreading and the weight percent of oil solubilized in water. They demonstrated that small amounts of a photosensitizing agent (1-naphthol) increased photo-oxidation. Hansen (1975) found the principal oxidation products to be aliphatic and aromatic acids, with lesser production of alcohols and phenols.

Biodegradation rates of hydrocarbons decrease in the order: normal alkanes, branched alkanes, aromatics, and cycloalkanes. Therefore if biodegradation were the only mechanism destroying hydrocarbons in near-surface waters, the proportions of aromatic and cycloalkane hydrocarbons would increase relative to the normal and branched alkanes as these nonvolatile fractions weathered, compared to the relative concentrations in crude oils and refined products discharged from transportation sources. Brown et al (1973), Brown and Huffman (1976), and Brown and Searl (1976) measured the proportion of alkane, cycloalkane, and aromatic hydrocarbons in dispersed oil (nonvolatile fraction) from near-surface water samples collected along tanker routes. They found proportionally more cycloalkanes, but less aromatics. This suggests that aromatic hydrocarbons (known to be photosensitive) are probably being destroyed by sunlight. Ledet and Laseter (1974) found that hydrocarbons at the air-sea interface in the Gulf of Mexico consisted mostly of alkanes and cycloalkanes.

Behavior of Oil After Treatment With Chemical Dispersants

The behavior of spilled oil on a water surface can be markedly altered by the application of chemical dispersants (surfactants) to form oil-in-water emulsions. The formation of oil-in-water emulsions have been used to control oil spills, and thereby alleviate some of the known adverse effects.

Oil-in-water Emulsions

Emulsion systems are formed and are stable if sufficient emulsifying agent is present to orient at the oil-water interface to form a film. Emulsifying agents are compounds having a long hydrocarbon chain that is soluble in oil, and polar groups (such as carboxylate, sulfonate, ether, and alcohol) that are soluble in water (Becker 1965). Emulsion droplets generally exceed 0.1 μm diameter, which distinguishes them from the smaller particles of micellar solutions formed during solubilization. The emulsion droplets are not all one size, but cover a relatively large range. The amount of added surfactant determines the size distribution. Within limits, the higher the surfactant concentration, the smaller the mean droplet diameter (Becker 1965, McAuliffe 1973). Figure 2, adapted from McAuliffe (1973), shows the size distributions for oil-in-water emulsion droplets prepared from a Richfield-Kraemer crude oil (California), saline water, and percent Chevron dispersant NI-W added to the oil.

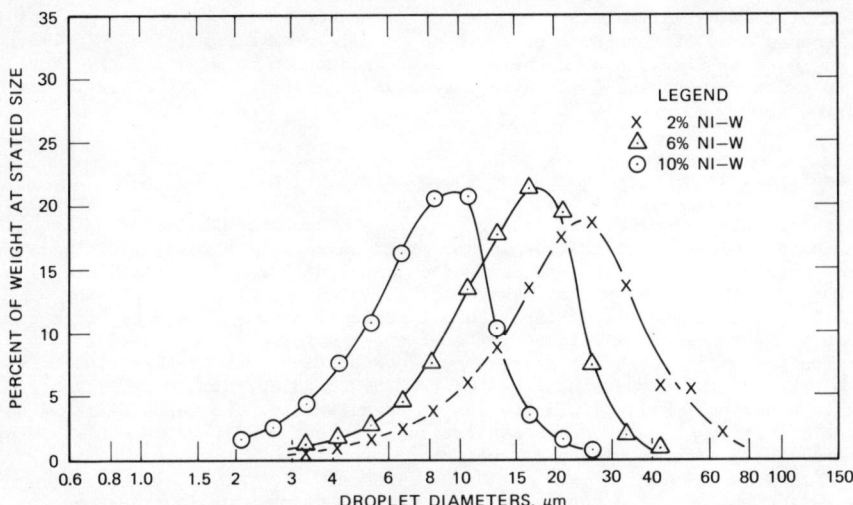

Fig. 3.2 Size distributions (Coulter counter) of o/w emulsions prepared from Richfield-Kraemer crude oil, saline water, and Chevron dispersant NI-W.

Oil-in-water emulsions can be produced from oils of all viscosities (10 to >1,000,000 cp) by the addition of chemical dispersants. Emulsions have viscosities close to that of the external phase. Thus, oil-in-water emulsions (less than 10% oil content) have viscosities close to that of water, regardless of the oil viscosity (McAuliffe 1973).

Movement and Dispersion of Emulsified Oil
An important feature of emulsification of surface oil is rapid dilution and downward mixing in near-surface waters. Thus the oil is removed from most of the wind's influence, so it does not travel as far as a surface slick. During the Chevron Gulf of Mexico spill, chemical dispersants were mixed in sea water and sprayed on the platform and surrounding water surface. The amount of dispersant was about 3.5% of the quantity of oil discharged. The dispersed oil was observed as a plume (confirmed by chemical analysis) only 1.0 to 1.5 miles (1.6 to 2.4 km) from the spill site, whereas untreated surface slicks extended 6 to 9 miles (10 to 15 km) on most days, and were found 45 to 50 miles away on two occasions. Dispersed oil in the emulsion plume was found only in near-surface waters (less than 20 ft). Water samples were collected at the surface, at 20 ft, and near bottom (40 ft). Oil content at 20 ft was less than 1 ppm, the sensitivity of the analytical method. Oil that does not travel far from the spill site lessens the need for oil spill trajectory models or refinement of existing models. Commercial or sports fishing areas temporarily restricted by surface oil slicks, would be much reduced. Although fish and shellfish may be unaffected under a slick, nets and catch can become contaminated by oil if they are brought up through the slick.

Stranded oil has adversely affected some species of intertidal organisms (Chan 1973, 1975, Cimberg et al 1973, NAS 1975, Smith 1968, Strachan 1972, Straughan 1971, 1972, Wormald 1976), coated rocky substrates, and penetrated into sandy beaches and marshland (Baker 1970 1971 1973, Burns and Teal 1971, Chan 1973 1975, Lytle 1975, Morris and Butler 1973). If the surface slick was prevented from stranding on shorelines, these adverse effects would be eliminated. Oil that penetrates into sediments biodegrades slowly, because of low oxygen concentrations (Blumer and Sass 1972), and oil in sandy beaches may subsequently be eroded and redeposited in near-shore sediments.

Stranded oil also decreases the aesthetic values of beaches. Large chunks or globs of stranded oil weather much more slowly (Blumer et al 1973), than pelagic tar (tar balls) at sea (Butler et al 1973, Butler 1975, McAuliffe 1976).

Some oils having specific gravities near that of seawater may sink, such as in the Bunker C spills in Chedabucto and San Francisco Bays. If these high-viscosity oils could be transformed into oil-in-water emulsions with small-diameter droplets, Brownian motion might keep them suspended and reduce the amount of oil that sinks.

Water-in-oil emulsions slow weathering (Harrison et al 1975, Regnier and Scott 1975) and may generate "tar balls", some of which could subsequently strand. Water-in-oil emulsions have viscosities as high as, and often much higher, than the crude oils from which they were prepared (McAuliffe 1973). For oils that form water-in-oil emulsions (Mackay et al 1973), the application of chemical dispersants to form oil-in-water emulsions should prevent the formation of "mousse", or if formed, convert it to an oil-in-water emulsion. In laboratory experiments, Zajic et al (1974) showed the bacterial production of a very active emulsifying agent from components in a Bunker C that was effective in breaking apart water-in-oil emulsion lumps.

Oil-in-Water Emulsions Accelerate Weathering and Biodegradation
Emulsification accelerates physical weathering (McAuliffe 1977). The example shown in Table 2 illustrates this. Accelerated loss of volatile hydrocarbons, including aromatics, should reduce the toxicity of oil that organisms might ingest. The possibility of synergistic toxic effects exists with the presence of chemical dispersant, but additional studies are needed. During the Chevron Gulf of Mexico spill, about 3.5 bbl of dispersant were used for each 100 bbl of oil discharged. Dispersed oil concentrations in surface water samples of the emulsion plume ranged from 2 to 60 ppm near the platform and decreased to 1 ppm at 1 mile. Thus organisms were exposed to decreasing concentrations for a short time. Static bioassays with oil plus 10% dispersant required from 300 to 3,200 ppm (mean, 1,200 ppm) dispersed oil to cause half-kill of six species of organisms in nine tests during 4-day exposures (McAuliffe et al 1975). These studies suggest that oil-in-water emulsions were unlikely to have caused adverse biological effects.

By exposing a larger surface area per unit volume of oil, dispersion increases photo-oxidation in near-surface waters (Brown et al 1973, Brown and Huffman 1976, Brown and Searl 1976, Klein and Pilpel 1974).

Bacteria operate at the oil-water interface, so increased specific surface accelerates biodegradation (Gatellier et al 1973). Not only is the oil made more readily available to bacteria, but movement of emulsion droplets through water makes oxygen and nutrients more readily available to micro-organisms.

Dispersion Lessens Oil Adhesion
The droplets in oil-in-water emulsions, if properly produced, do not stick to each other or coalesce. Glass vessels used for preparing these emulsions can be rinsed out with water, leaving the glass surface oil-free, whereas oil wets a glass surface. In like manner, oil-in-water emulsion droplets have little tendency to adhere to mineral surfaces (Canevari 1971, McAuliffe 1973). This reduces adhesion to suspended solid mineral particles in the water column, and thereby lessens the amount of oil that may sediment. This mechanism can be particularly effective when an oil slick encounters turbid water, such as at the mouth of a river. Without emulsification, the oil could sink and be concentrated in sediments at the zone where oil meets turbid water (Kolpack et al 1971).

If chemically dispersed oil from a near-shore spill stranded in the intertidal zone, it should have less tendency than untreated oil to adhere to sand, rocks, and marine plants and animals. Lowered adhesion should minimize or eliminate the smothering of intertidal marine life such as has occurred with nondispersed, partially weathered crude oil (Straughan 1971) or viscous Bunker C (Chan 1973). Most of the dispersed oil would be expected to wash out with the receding and subsequent tides. Studies are needed, however, to verify this.

Reducing the adhesion of oil to surfaces should also reduce the adverse effects on sea birds (NAS 1975). Chemically dispersed oil does not wet feathers like nondispersed oil. The small areal extent of an emulsion plume and lack of adhesion of oil droplets to feathers should minimize bird kill.

Chemical dispersants have been used extensively in Europe for over 10 years with beneficial effects (Beynon and Cowell 1974). Some of these examples of field use have been summarized by Fitzgerald (1977).

Dispersal and Alteration in Cold Waters

The processes reviewed above also apply to arctic and subarctic conditions, but the rates are usually slower. Leinonen and Mackay (1975) present mathematical models for evaporation and solution of volatile crude oil components from spills on water, on ice, and under ice.

Slick life for most oils is usually longer on colder water than on warm, due principally to increased viscosity. However, breakup by wind and waves may minimize the difference. Kinney et al (1969) found the half-life of accidental and controlled crude oil slicks on Cook Inlet (5°C) to be less than 1 day. Smith and MacIntyre (1971) could not find enough spilled No. 2 fuel oil to adequately sample after 8 hrs on 5°C water. Similar dispersal of oil slicks should occur even in near-freezing water. High density, viscous oils such as Bunker C might sink.

Evaporative loss of hydrocarbons is slower on colder water. Vapor pressures of hydrocarbons decrease approximately 5% per degree centigrade at sea water temperatures (API 1953, Doss 1943). Regnier and Scott (1975) experimentally measured the evaporation of n-C_9 to n-C_{18} alkanes from an Arctic diesel (No. 2) fuel oil at 5° to 30°C. They found the expected relationship for evaporative loss as a function of temperature, except as altered by composition changes. As the more volatile components evaporated, the residual oil became more viscous, and reduced diffusion progressively decreased evaporation rates. Evaporation of the higher boiling components was thus retarded relative to the lower-boiling fractions.

Oil discharged under ice spreads and moves much less than oil on a water surface (Glaeser 1971, Hnatiuk 1976a,b), and the volatile components do not evaporate. These limited studies have shown oil emerging from under ice during melting, as brine channels formed.

The rate of solution of hydrocarbons from oils (on the surface or under ice) into cold water is slower than in warmer waters. Diffusion decreases with decrease in temperature. Under ice the smaller areal extent of the spill (thicker layers under smooth, horizonal ice or pools in recesses under old ice) would reduce the loss of soluble constituents compared with oil on a water surface. Dissolved hydrocarbon concentrations in water immediately in contact with these localized areas might be higher, unless dissipated by currents or biodegraded.

Biodegradation rates in cold waters, while slower than in warm, are still appreciable (see section "Microbial Modification, Organism Uptake, and Depuration").

Chemical alteration by sunlight is highly variable in the Arctic and sub-Arctic. Photo-oxidation becomes significant during summer months, but very slow in winter. Oil under ice is less subject to photo-oxidation than is exposed oil.

Suggested Areas of Research

Further information on the behavior of oil spills with and without chemical dispersants should be obtained under natural conditions. These data could be collected by a series of small oil spills, with associated chemical and biological investigations. Spill sizes might vary according to location, although 10 bbl each should provide adequately affected areas. Spills could be conducted with representative crude oils (high, moderate, and low API gravity), and No. 2 and No. 6 fuel oils. Greater emphasis should be placed on crude oil spills, because they represent offshore production as well as the largest volume in worldwide transportation. To evaluate various spill conditions, controlled spills could be conducted on open ocean water, in a small tidal bay, on shorelines (rocky, sandy, and marsh), under ice, and in open-water leads in ice.

The evaporative and solution losses of hydrocarbons from oil should be measured. Analysis for total oil and dissolved hydrocarbons in water, sediment, and selected tissue samples (where appropriate) would document the fate of oils (dispersed vs nondispersed). Biodegradation and chemical alteration studies would record the longer-term fates of oil.

The above tests should determine if oil-in-water emulsification minimizes the sinking of large globs or drops of low-specific-gravity, high-viscosity oils such as Bunker C. The tests should also show whether dispersants can prevent the formation of, and if formed, the dispersal of water-in-oil emulsions ("mousse").

Many bioassay procedures with oils do not use realistic oil concentrations, dissolved hydrocarbon compositions, and exposure times for organisms. Data from the above studies would assist in designing better bioassays, but, more important, provide actual field results under controlled oil spill conditions. Such studies would substantiate the fate of spilled oil, particularly dispersed oil, for which there is a more limited scientific base for evaluation.

Summary

Oil spilled on water undergoes physical, chemical, and biological alteration.

Rapid physical and chemical processes include spreading, movement with winds and water currents, evaporation of volatile components, water-in-oil emulsification, dispersion as small droplets into the water, spray injection into the air, and sedimentation.

As the oil spreads, biological and photochemical processes start. Biological processes include degradation by micro-organisms and uptake by larger organisms. The latter is followed by discharge, metabolism, or storage. Photo-oxidation destroys hydrocarbons, especially aromatics.

Low-molecular-weight hydrocarbons (up to C_{12}) rapidly evaporate. Very small amounts of these dissolve, then quickly evaporate from near-surface waters.

Dispersal and alteration of oil in arctic and subarctic areas are the same as in warmer waters, except that the rates are usually slower. Oil discharged under ice does not spread or move as much as oil on open water.

Chemical dispersants can be used to produce oil-in-water emulsions. Dispersed oil does not travel as far as in a slick, dilutes more rapidly, and shorelines are less threatened. Evaporation, solution followed by evaporation, biodegradation, and photo-oxidation are accelerated. Dispersed oil does not adhere as readily to solid surfaces, and thereby should reduce bird kill and lessen the amount of oil that sediments.

Research should include small oil spills, with and without dispersant application. They should be conducted in different locations and under different environmental conditions to improve our estimates of evaporation, solution, photo-oxidation, biodegradation, and sedimentation. These will add to our data base for determining the fate of spilled oil, biological effects, and aid in design of bioassay testing.

References

Allen, A. A., R. S. Schlueter, and P. G. Mikolaj, Natural oil seepage at Coal Oil Point, Santa Barbara, California, Science 170, 974-977 (1970).

Anderson, J. W., J. M. Neff, B. A. Cox, H. E. Tatem, and G. M. Hightower, Characteristics of dispersion and water-soluble extracts of crude and refined oils and their toxicity to estuarine crustaceans and fish, Mar. Biol. 27, 75-88 (1974).

API. In: F. D. Rossini et al (eds.) Selected Values of Physical and Thermodynamic Properties of Hydrocarbons and Related Compounds, API Research Project 48, Carnegie Press, Pittsburgh, 1953.

Atlas, R. M., Fate and Effects of Oil Pollutants in Extremely Cold Marine Environments. U.S. Defense Documentation Center, A.D. 769895, Final Report, October 1, 1973.

Atlas, R. M., Fate and Effects of Oil Pollutants in Extremely Cold Marine Environments. Office of Naval Research, AD/A-003 554, 1974.

Baker, J. M., The effects of oils on plants, Environ. Pollut. 1, 27-44 (1970).

Baker, J. M., Refinery effluent. In: E. B. Cowell (ed.) Ecological Effects of Oil Pollution on Littoral Communities, Institute of Petroleum, London, 1971.

Baker, J. M., Biological effects of refinery effluents, Proceedings Joint Conference on Prevention and Control of Oil Spills, American Petroleum Institute, Washington, 715-724 (1973).

Becker, P., Emulsions: Theory and Practice, Reinhold Publishing Corp., New York, 1965.

Beynon, L. R. and E. B. Cowell (eds.), Ecological Aspects of Toxicity Testing of Oils and Dispersants, Applied Science Publishers, London, 1974.

Blumer, M. and J. Sass, Oil Pollution: Persistence and degradation of spilled fuel oil, Science 176, 1120-1122 (1972).

Blumer, M., M. Ehrhardt, and J. H. Jones, The environmental fate of stranded crude oil, Deep-Sea Res. 20, 239-259 (1973).

Boehm, P. D. and J. G. Quinn, Solubilization of hydrocarbons by the dissolved organic matter in sea water, Geochim. Cosmochim. Acta, 37, 2459-2475 (1973).

Brown, R. A., T. D. Searl, J. J. Elliott, B. G. Phillips, D. E. Brandon, and P. H. Monaghan, Distribution of heavy hydrocarbons in some Atlantic Ocean waters. Proceedings Joint Conference on Prevention and Control of Oil Spills, American Petroleum Institute, Washington, 505-519 (1973).

Brown, R. A. and H. L. Huffman, Jr., Hydrocarbons in open ocean waters, Science 191, 847-849 (1976).

Brown, R. A. and T. D. Searl, Nonvolative hydrocarbons along tanker routes of the Pacific Ocean, Offshore Technology Conference 1, 259-274 (1976).

Burns, K. A. and J. M. Teal, Hydrocarbon incorporation into the salt marsh ecosystem from the West Falmouth oil spill. Woods Hole Oceanographic Institution Tech. Rep. No. 71-69. Woods Hole, Mass., Unpublished manuscript, 1971.

Butler, J. N., B. F. Morris, and J. Sass, Pelagic tar from Bermuda and the Sargasso Sea, Special Publ. No. 10, Bermuda Biological Station for Research, Harvard University Printing Office, Cambridge, 1973.

Butler, J. N, Evaporative weathering of petroleum residues: The age of pelagic tar, Mar. Chem. 3, 9-21 (1975).

Button, D. K., Arctic oil biodegradation. National Technical Information Service, AD-A014 096, U.S. Department of Commerce, 1974.

Button, D. K., The influence of clay and bacteria on the concentration of dissolved hydrocarbon in saline solution, Geochim. Cosmochim. Acta 40, 435-440 (1976).

Canevari, G. P., Oil spill dispersants - Current status and future outlook, Proceedings Joint Conference on Prevention and Control of Oil Spills, American Petroleum Institute, Washington, 263-270 (1971).

Chan, G. L., A study of the effects of the San Francisco oil spill on marine organisms, Proceedings Joint Conference on Prevention and Control of Oil Spills, American Petroleum Institute, Washington, 741-783 (1973).

Chan, G. L., A study of the effects of the San Francisco oil spill on marine life. Part II: Recruitment, Proceedings 1975 Conference on Prevention and Control of Oil Pollution, American Petroleum Institute, Washington, 457-463 (1975).

Cimberg, R., S. Mann, and D. Straughan, A reinvestigation of Southern Californa rocky intertidal beach three and one-half years after the 1969 Santa Barbara Oil Spill: A preliminary report, Proceedings Joint Conference on Prevention and Control of Oil Spills, American Petroleum Institute, Washington, 697-702 (1973).

Conover, R. J., Some relations between zooplankton and Bunker C oil in Chedabucto Bay following the wreck of the tanker Arrow, J. Fish. Res. Board Can. 28, 1327-1330 (1971).

Corner, E. D. S., C. C. Kilvington, and S. C. M. O'Hara, Qualitative studies on the metabolism of naphthalene in Maia Squinado (Herbst), J. Mar. Biol. Assoc. U.K. 53, 819-832 (1973).

Cundell, A. M. and R. W. Traxler, Microbial degradation of petroleum at low temperature, Mar. Pollut. Bull. 4, 125-127 (1973).

Davis, J. B., Paraffinic hydrocarbons in the sulfate-reducing bacterium Desulfovibrio desulfuricans, Chem. Geol. 3, 155-160 (1967).

Dean-Raymond, D. and R. Bartha, Biodegradation of some polynuclear aromatic petroleum components by marine bacteria, National Technical Information Service, U.S. Department of Commerce, AD/A-006 346, 1975.

DiSalvo, L. H. and H. E. Guard, Hydrocarbons associated with suspended particulate matter in San Francisco Bay waters, Proceedings 1975 Conference on Prevention and Control of Oil Pollution, American Petroleum Institute, Washington, 169-173 (1975).

Doss, M. P. (ed.), Physical Constants of the Principal Hydrocarbons, 4th ed., The Texas Company, New York, 1943.

Eganhouse, R. P. and J. A. Calder, The solubility of medium molecular weight hydrocarbons and the effects of hydrocarbon co-solutes and salinity, Geochim. Cosmochim. Acta 40, 555-561 (1976).

Fallah, M. H. and R. M. Stark, Literature review: Movement of spilled oil at sea, J. Mar. Technol. Soc. 10, 3-11 (1976).

Fallah, M. H. and R. M. Stark, Random model of evaporation of oil at sea, Sci. Total Environ. 5, 95-109 (1976).

Fitzgerald, D. E., Rationale for utilization of dispersants, Proceedings 1977 Oil Spill Conference (Prevention, Behavior, Control, Cleanup), (In Press).

Fossato, V. U. and W. J. Canzonier, Hydrocarbon uptake and loss by the mussel Mytilus edulis, Mar. Biol. 36 243-250 (1976).

Gatellier, C. R., J. L. Oudin, P. Fusey, J. C. Lacase, and M. L. Priou, Experimental ecosystems to measure fate of oil spills dispersed by surface active products, Proceedings Joint Conference on Prevention and Control of Oil Spills, American Petroleum Institute, Washington, 497-504 (1973).

Gibson, D. T., D. M. Jerina, H. Yagi, and H. J. C. Yeh, Oxidation of the carcinogens benzo[a]pyrene and benzo[a]anthracene to dihydrodiols by a bacterium, Science 189, 295-297 (1975).

Gilfillan, E. S., Decrease of net carbon flux in two species of mussels caused by extracts of crude oil, Mar. Biol. 29, 53-57 (1975).

Glaeser, J. L., A discussion of the future oil spill problem in the Arctic, Proceedings Joint Conference on Prevention and Control of Oil Spills, American Petroleum Institute, Washington, 479-484 (1971).

Hanson, H. P., Photochemical degradation of petroleum hydrocarbon surface films on seawater, Mar. Chem. 3, 183-195 (1975).

Harrison, W., M. A. Winnik, P. T. Y. Kwong, and D. Mackay, Crude oil spills. Disappearance of aromatic and aliphatic components from small sea-surface slicks, Environ. Sci. Technol. 9, 231-234 (1975).

Harrison, W., The fate of crude oil spills and the siting of four supertanker ports, Can. Geogr. 28, 211-231 (1974).

Hnatiuk, J., An environmental research program for drilling in the Canadian Beaufort Sea, J. Can. Pet. Technol. 15, 29-36 (1976a).

Hnatiuk, J., Results of an environmental research program in the Canadian Beaufort Sea, Offshore Technology Conference 1, 221-234 (1976b).

Jeffrey, P. G., Large-scale experiments on the spreading of oil at sea and its disappearance by natural factors, Proceedings Joint Conference on Prevention and Control of Oil Spills, American Petroleum Institute, Washington, 469-474 (1973).

Johanson, E. E., J. C. Johnson, C. D. McAuliffe, and R. A. Brown, Physical and chemical weathering of crude oil slicks on the ocean. (Manuscript in preparation).

Kauss, P. B. and T. C. Hutchinson, The effects of water-soluble petroleum components on the growth of Chlorella vulgaris Beijerinck, Environ. Pollut. 9, 157-173 (1975).

Kinney, P. J., D. K. Button, and D. M. Schell, Kinetics of dissipation and biodegradation of crude oil in Alaska's Cook Inlet, Proceedings Joint Conference on Prevention and Control of Oil Spills, American Petroleum Institute, Washington, 333-340 (1969).

Kinsey, D. W., Small-scale experiments to determine the effects of crude oil films on gas exchange over the coral back-reef at Heron Island, Environ. Pollut. 4, 167-182 (1973).

Klein, A. E. and N. Pilpel, The effects of artificial sunlight upon floating oils, Water Res. 8, 79-83 (1974).

Kolpack, R. L., J. S. Mattson, H. B. Mark, Jr., and T-C Yu, Hydrocarbon content of Santa Barbara Channel sediments. pp 276-395. In: R. L. Kolpack (ed.), Biological and Oceanographical Survey of the Santa Barbara Channel Oil Spill, 1969-1970. Vol. II. Physical, Chemical, and Geological Studies, Allan Hancock Foundation, University of Southern California, Los Angeles, 1971.

Koons, C. B. and D. E. Brandon, Hydrocarbons in water and sediment samples from Coal Oil Point area, offshore California, Offshore Technology Conference 3, 513-521 (1975).

Kreider, R. E., Identification of oil leaks and spills, Proceedings Joint Conference on Prevention and Control of Oil Spills, American Petroleum Institute, Washington, 119-124 (1971).

Ledet, E. J. and J. L. Laseter, Alkanes at the air-sea interface from offshore Louisiana and Florida, Science 186, 261-263 (1974).

Lee, R. F., Fate of petroleum hydrocarbons in marine zooplankton, Proceedings 1975 Conference on Prevention and Control of Oil Pollution, American Petroleum Institute, Washington, 595-600 (1975).

Mackay, G. D. M., A. Y. McLean, O. J. Betancourt, and B. D. Johnson, The formation of water-in-oil emulsions subsequent to an oil spill, J. Inst. Pet. 59, 164-172 (1973).

McAuliffe, C. D., Solubility in water of paraffin, cycloparaffin, olefin, acetylene, cycloolefin, and aromatic hydrocarbons, J. Phys. Chem. 70, 1267-1275 (1966).

McAuliffe, C. D., Solubility in water of normal C_9 and C_{10} alkane hydrocarbons, Science 158, 478-479 (1969).

McAuliffe, C. D., GC determination of solutes by multiple phase equilibrium, Chem. Technol. 1, 46-51 (1971).

McAuliffe, C. D., Oil-in-water emulsions and their flow properties in porous media, J. Pet. Technol. 25, 727-733 (1973).

McAuliffe, C. D., Determination of C_1-C_{10} hydrocarbons in water, Marine Pollution Monitoring (Petroleum), NBS Spec. Publ. 409, U.S. Government Printing Office, Washington, 121-125, (1974).

McAuliffe, C. D., A. E. Smalley, R. D. Groover, W. M. Welsh, W. S. Pickle, and G. E. Jones, The Chevron Main Pass Block 41 oil spill: Chemical and biological investigations, Proceedings 1975 Conference on Prevention and Control of Oil Pollution, American Petroleum Institute, Washington, 555-566 (1975).

McAuliffe, C. D., Surveillance of the marine environment for hydrocarbons, Mar. Sci. Commun. 2, 13-42 (1976).

McAuliffe, C. D., Evaporation and solution of C_2 to C_{10} hydrocarbons from crude oils on the sea surface (These proceedings, 1977).

McLean, A. Y. and O. J. Betancourt, Physical and chemical changes in spilled oil weathering under natural conditions, Offshore Technology Conference 1, 249-256 (1973).

Morris, B. F. and J. N. Butler, Petroleum residues in the Sargasso Sea and on Bermuda beaches, Proceedings Joint Conference on Prevention and Control of Oil Spills, American Petroleum Institute, Washington, 521-529 (1973).

Murray, S. P., Wind and current effects on large-scale oil slicks, Offshore Technology Conference 3, 523-533 (1975).

National Academy of Sciences, Petroleum in the Marine Environment, Washington, 1975.

Neff, J. M. and J. W. Anderson, Accumulation, release, and distribution of benzo[a]pyrene-C^{14} in the clam Rangia cuneata, Proceedings 1975 Conference on Prevention and Control of Oil Pollution, American Petroleum Institute, Washington, 469-471 (1975).

Nelson-Smith, A., Oil Pollution and Marine Ecology, Plenum Press, New York, 1973.

Percy, J. A., Responses of Arctic marine crustaceans to crude oil and oil-tainted food, Environ. Pollut. 10, 155-162 (1976).

Perry, J. J. and C. E. Cerniglia, Studies on the degradation of petroleum by filamentous fungi. pp 89-94. In: D. G. Ahearn and S. P. Meyers (eds.), The Microbial Degradation of Oil Pollutants, Louisiana State University, Baton Rouge, 1973.

Regnier, Z. R. and B. F. Scott, Evaporation rates of oil components, Environ. Sci. Technol. 9, 469-472 (1975).

Rice, S. D., J. W. Short, C. C. Brodersen, T. A. Mecklenburg, D. A. Moles, C. J. Misch, D. L. Cheatham, and J. F. Karinen, Acute toxicity and uptake-depuration studies with Cook Inlet crude oil, Prudhoe Bay crude oil, No. 2, fuel oil and several subarctic marine organisms, <u>Northwest Fisheries Center Auke Bay Fisheries Laboratory</u>, Processed Report, May 1976.

Robertson, B., S. Arhelger, P. J. Kinney, and D. K. Button, Hydrocarbon biodegradation in Alaskan waters. pp 171-184. In: D. G. Ahearn and S. P. Meyers (eds.), <u>The Microbial Degradation of Oil Pollutants</u>, Louisiana State University, Baton Rouge, 1973.

Rossi, S. S., J. W. Anderson, and G. S. Ward, Toxicity of water-soluble fractions of four test oils for the polychaetous annelids, <u>Neanthes arenaceodentata</u> and <u>Capitella capitata</u>, <u>Environ. Pollut</u>. <u>10</u>, 9-19 (1976).

Shelton, T. B. and J. V. Hunter, Aerobic decomposition of oil pollutants in sediments, <u>J. Water Pollut. Control Fed</u>. <u>46</u>, 2172-2182 (1974).

Shelton, T. B. and J. V. Hunter, Anaerobic decomposition of oil in bottom sediments, <u>J. Water Pollut. Control Fed</u>. <u>47</u>, 2256-2270 (1975).

Smith, C. L. and MacIntyre, W. G., Initial aging of fuel oil films on sea water, <u>Proceedings Joint Conference on Prevention and Control of Oil Spills</u>, American Petroleum Institute, Washington, 457-461 (1971).

Smith, J. E. (ed.), <u>Torrey Canyon, Pollution and Marine Life</u>, Cambridge University Press, London, 1968.

Strachan, A., Santa Barbara oil spill: Intertidal and subtidal surveys, <u>California Marine Resources Committee, California Cooperative Oceanic Fisheries Investigations</u>, <u>Report No. 16</u>, 122-124 (1972).

Straughan, D. (ed.), <u>Biological and Oceanographical Survey of the Santa Barbara Channel Oil Spill, 1969-1970. Vol. 1. Biology and Bacteriology</u>, Allan Hancock Foundation, University of Southern California, Los Angeles, 1971.

Straughan, D., Factors causing environmental changes after an oil spill, <u>J. Pet. Technol</u>. <u>24</u>, 250-254 (1972).

Sutton, C. and J. A. Calder, Solubility of higher-molecular-weight n-paraffins in distilled water and seawater, <u>Environ. Sci. Technol</u>. <u>8</u>, 654-657 (1974).

Sutton, C. and J. A. Calder, Solubility of alkylbenzenes in distilled water and seawater at 25.0°C, <u>J. Chem. Eng. Data</u> <u>20</u>, 320-322 (1975).

Traxler, R. W. and A. M. Cundell, Petroleum degradation in low temperature marine and estuarine environments, <u>Office of Naval Research</u>, Annual Report No. 2, 1975.

Traxler, R. W. and A. M. Cundell, Petroleum degradation in low temperature marine and estuarine environments, <u>Office of Naval Research</u>, Annual Report No. 3, 1976.

Vanderhorst, J. R., C. I. Gibson, and L. J. Moore, Toxicity of No. 2 fuel oil to coon stripe shrimp, <u>Mar. Pollut. Bull</u>. <u>7</u>, 106-108 (1976).

Walker, J. D., L. Cofone, Jr., and J. J. Cooney, Microbial petroleum degradation: The role of <u>Cladosporium resinae</u>, <u>Proceedings Joint Conference on Prevention and Control of Oil Spills</u>, American Petroleum Institute, Washington, 821-825 (1973).

Walker, J. D., P. A. Seesman, T. L. Herbert, and R. R. Colwell, Petroleum hydrocarbons: Degradation and growth potential of deep-sea sediment bacteria, <u>Environ. Pollut</u>. <u>10</u>, 89-99 (1976).

Walker, J. D. and R. R. Colwell, Sampling device for monitoring biodegradation of oil and other pollutants in aquatic environments, <u>Environ. Sci. Technol</u>. <u>11</u>, 93-95 (1977).

Wilson, R. D., P. H. Monaghan, A. Osanik, L. C. Price, and M. A. Rogers, Natural marine oil seepage, <u>Science</u> <u>184</u>, 857-65 (1974).

Winters, K., R. O'Donnell, J. C. Batterton, and C. Van Ballen, Water-soluble components of four fuel oils: Chemical characterization and effects on growth of microalgae, <u>Mar. Biol</u>. <u>36</u>, 269-276 (1976).

Wormald, A. P., Effects of a spill of marine diesel oil on the meiofauna of a sandy beach at Picnic Bay, Hong Kong, <u>Environ. Pollut</u>. <u>11</u>, 117-130 (1976).

Zajic, J. E., B. Supplisson, and B. Volesky, Bacterial degradation and emulsification of No. 6 fuel oil, <u>Environ. Sci. Technol</u> <u>8</u>, 664-668 (1974).

ZoBell, C. E., Bacterial degradation of mineral oils at low temperatures. pp 153-161. In: D. G. Ahearn and S. P. Meyers (eds.), <u>The Microbial Degradation of Oil Pollutants</u>, Louisiana State University, Baton Rouge, 1973.

CHAPTER 4

BIODEGRADATION OF AROMATIC PETROLEUM HYDROCARBONS

David T. Gibson

Department of Microbiology
The University of Texas at Austin
Austin, Texas 78712

Introduction

Life on this planet is dependent on the trapping of solar energy by photosynthetic organisms. In this process the energy is utilized to convert carbon dioxide into organic molecules. Eventually living organisms release the energy as heat, and carbon dioxide is returned to the atmosphere. This simple concept may be found in any elementary biology textbook under the subject of the carbon cycle. It is now known that microorganisms are the principal agents responsible for the recycling of carbon in nature. The metabolic capabilities of this group of organisms have evolved throughout geological time and it is probably true to say that all organic compounds produced by living organisms are subject to biodegradation by microorganisms. In contrast, when a man-made compound that bears no structural resemblance to a natural product is introduced into the environment there is no valid reason for assuming that the molecule will be a candidate for microbial degradation.

Petroleum is formed from the organic material produced by living organisms. Thus, at first sight it seems probable that certain microorganisms would have evolved the metabolic capacity for the complete biodegradation of petroleum. Certainly there are many microorganisms that will grow at the expense of different crude oils. However, we must first look at the origin of hydrocarbons in the environment in order to understand the problems associated with studies on the degradation of petroleum.

Origin of Hydrocarbons in the Environment

A detailed description of Hydrocarbon formation is beyond the scope of this presentation. The following brief description is directed towards providing a framework that may be used as a rational basis for understanding how microorganisms can degrade certain aromatic constituents that are found in crude oil. Figure 1 illustrates the cycle of carbon in nature

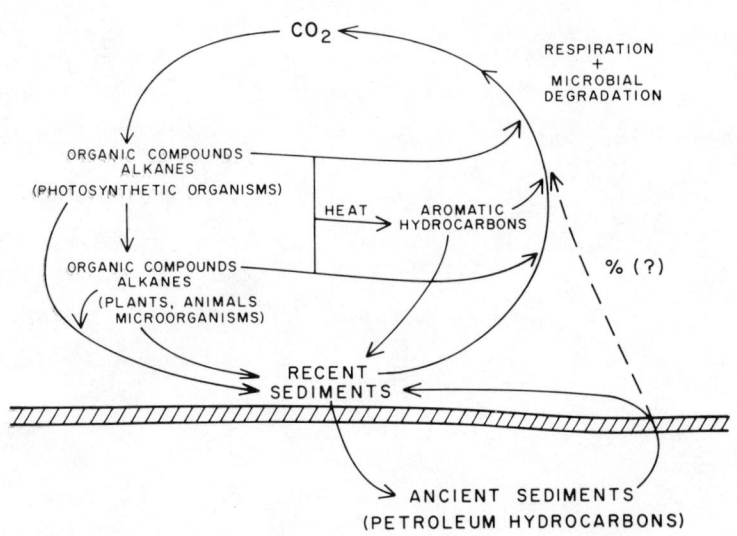

Fig. 4.1 Cycle of hydrocarbons in the environment.

with particular reference to hydrocarbon molecules. It is well known that linear and branched alkanes are produced by living organisms. These compounds are also found in petroleum. Since microorganisms have had millions of years to evolve enzyme systems that degrade this ubiquitous group of compounds it is not surprising to find that many microorganisms will degrade these compounds when they are present in different crude oils.

A polycyclic origin of certain aromatic hydrocarbons has been suggested (Borneff et al. 1968). However, most of the available evidence indicates that this class of compounds is produced by the thermal alteration of organic compounds. The types of molecules that are produced are related to the pyrolysis temperature.

The problems associated with the identification of aromatic hydrocarbons in the environment have resided at the analytical level. In recent years the development of new techniques, particularly mass spectrometry and probe scintillation, have revealed the presence of an extremely complex mixture of polycyclic aromatic hydrocarbons in different environments. The elegant studies of Blumer and his colleagues at the Woods Hole Oceanographic Institute indicate that tens of thousands of polycyclic aromatic hydrocarbons may be present in a single sediment sample (Blumer and Youngblood 1975, Giger and Blumer 1974, Blumer 1976). Even the sophisticated analytical techniques that are presently available are unable to resolve the individual components of such mixtures and the best information that has been obtained only identifies groups of compounds.

Unsubstituted and alkylsubstituted polycyclic aromatic hydrocarbons are present in most environmental samples. The former are produced by the high temperature ($2000°$ C) pyrolysis of organic material. At intermediate temperatures ($400-800°$ C), such as those found in forest fires, molecules with one to four alkyl substituents are formed (Blumer 1976). Even though the difficulties associated with identifying the individual components of these mixtures are enormous it is clear that microorganisms have been exposed to this class of compounds throughout geological time. Consequently, if the previously mentioned tenets of microbial evolution are true, microorganisms should be able to degrade these molecules.

Although the problems associated with analyzing the organic constituents of recent sediments are large they are relatively minor when compared to the problems associated with petroleum analysis. As the organic constituents of sediments sink deeper into the earth's surface they become associated with the inorganic components of the sediments, water is expelled and an organic-rich shale is generated. If the shale contains more than 0.5 percent carbon there is a good chance that, over geological periods of time the rock may generate petroleum. During this process diagenetic changes occur at temperatures of $80-150°$ C. Thus, carbon-carbon bonds are broken and reformed, reactions occur with inorganic materials and literally tens of thousands of new compounds are produced that bear little or no structural resemblance to the biosynthetically-produced starting material. This oversimplified view of petroleogenesis is important in any consideration of the biodegradation of petroleum. The changes that occur may be represented as follows:

Organic Compounds → Petroleum → Carbon

During the millions of years represented by the above sequence the solar energy trapped by living organisms is slowly released. When man discovers and releases oil from a reservoir two important consequences must be noted, (i) no two oils are the same age and therefore no two oils have the same chemical composition; (ii) a multitude of organic molecules that bear little or no structural relationship to biosynthetically-produced molecules are reintroduced into the environment.

Investigation of the microbial degradation of crude oil use must take into account the analytical limitations that are imposed by the complex nature of the starting material. Gas chromatography is frequently used as a measure of oil biodegradation. This technique is excellent for the linear alkane components of crude oils. However, in a recent article entitled "Organic Compounds in Nature: Limits of our Knowledge" (Blumer 1975) Blumer points out "The number of components in an oil sample may exceed the admittedly high resolving power of gas chromatography by another factor of 10^3 or higher! In other words: a gas chromatogram of an oil may reveal to us fewer than one tenth of one percent of all components present in the sample!" This revealing statement is particularly relevant to a consideration of aromatic constituents of petroleum. For example, the American Petroleum Institute has reported characterization studies on the heavy ends of several crude oils (Dooley et al. 1974). The types, not individual molecules, of aromatic compounds found in a Wilmington, California crude oil are given in Table 1. At this time we know almost nothing about how microorganisms interact with these groups of compounds when they are present in a sample of crude petroleum. Most of our present knowledge of the microbial degradation of aromatic hydrocarbons has been obtained with single hydrocarbon substrates and pure cultures of different microorganisms. Table 2 shows the types of molecules that have been investigated. Clearly this group of compounds represents a very small range of the types of molecules that are found in crude oil. Nevertheless, detailed studies have revealed that both eucaryotic and procaryotic microorganisms have evolved certain types of reactions for the degradation of aromatic hydrocarbons that may allow us to predict the biodegradability of some of the larger molecules that are found in crude petroleum.

TABLE 1. Hydrocarbon types in the 370° to 535° C distillate of Wilmington, California crude oil

Monoaromatic Compounds

Alkylbenzenes	Octanaphthenobenzenes
Indanes/Tetralins	Naphthalenes
Dinaphthenobenzenes	Naphthenonaphthalenes or
Trinaphthenobenzenes	Diphenylalkanes
Tetranaphthenobenzenes	Dinaphthenonaphthalenes or
Pentanaphthenobenzenes	Naphthenodiphenylalkanes
Hexanaphthenobenzenes	Teinaphthenonaphthalenes or
Heptanaphthenobenzenes	Dinaphthenodiphenylalkanes
	Tetranaphthenonaphthalenes or
	Trinaphthenodiphenylalkanes

Diaromatic Compounds

Alkylnaphthalenes	Tetranaphthenonaphthalenes or
Naphthenonaphthalenes or	Trinaphthenodiphenylalkanes
Diphenylalkanes	Pentanaphthanonaphthalenes or
Dinaphthenonaphthalenes or	Tetranaphthenodiphenylalkanes
Naphthenodiphenylalkanes	Hexanaphthenonaphthalenes or
Trinaphthenonaphthalenes or	Pentanaphthenodiphenylalkanes
Dinaphthenodiphenylalkanes	Heptanaphthenonaphthalenes or
	Hexanaphthenodiphenylalkanes

Polyaromatic Compounds

Naphthenophenanthrenens/Anthracenes	Alkyl Benzopyrenes
Dinaphthenophenanthrenes/	Alkyl Acenaphthalenes
Anthracenes	Naphthenoacenaphthalenes
Trinaphthenophenanthrenes/	Dinaphthenoacenaphthalenes
Anthracenes	Trinaphthenoacenaphthalenes
Tetranaphthenophenanthrenes/	Tetranaphthenoacenaphthalenes
Anthracenes and/or	Pentanaphthenoacenaphthalenes
Thienodibenzothiophenes	Hexanaphthenoacenaphthalenes
Pentanaphthenophenanthrenes/	Heptanaphthenoacenaphthalenes
Anthracenes and/or	Octanaphthenoacenaphthalenes
Naphthenothionodibenzothiophenes	Naphthenopyrenes and/or
Alkyl Pyrenes	Chrysenes
Alkyl Phenanthrenes/Anthracenes	Naphthenobenzopyrenes

TABLE 2. Aromatic hydrocarbons known to be oxidized by microorganisms

Monocyclic	Tricyclic
Benzene	Phenanthrene
Toluene	Anthracene
Xylenes	
Tri and Tetramethylbenzenes	
Alkylbenzenes (Linear and Branched)	Polycyclic
Cycloalkylbenzenes	
	Pyrene
Dicyclic	Benzo[α]pyrene
	Benzo[α]anthracene
Naphthalene	Dibenzo[α]anthracene
Methylnaphthalenes (Mono and Di)	Benzperylene

Degradation of aromatic hydrocarbons by eucaryotic organisms

Most aromatic hydrocarbons are toxic for experimental animals. In the case of certain polycyclic aromatic hydrocarbons carcinogenic effects have been observed. It is now generally accepted that the physiological effects of these compounds are due to the metabolites that are formed by microsomal enzyme systems. Mammals oxidize many aromatic hydrocarbons to arene oxides. The latter are often reactive electrophiles that can react with nucleophilic cell constituents (Jerina and Daly 1974). Major urinary metabolites of aromatic hydrocarbons are usually phenols that arise by isomerization of arene oxides. The mechanism of isomerization has been extensively studied and is characterized by the migration of substituents from the position finally occupied by the hydroxyl group to the adjacent carbon atom. This reaction has been termed the NIH Shift (Daly et al. 1972). Arene oxides are formed by the enzymatic incorporation of one atom of molecular oxygen into the aromatic hydrocarbon substrate. The other atom of the oxygen molecule is reduced to water. This reaction is catalyzed by cytochrome P450, six forms of which are now known to exist in rat liver (Thomas et al. 1976). Arene oxides can isomerize to phenols, undergo conjugation with glutathione or react with water to form trans-dihydrodiols. The last reaction is catalyzed by the enzyme epoxide hydratase (Oesch 1973). In fact epoxidation followed by hydration are two enzymatic reactions that are observed in the metabolism of many different aromatic hydrocarbons. The reaction sequence shown in Fig. 2, is one of several schemes that have been shown to occur during the metabolism of benzo[a]pyrene by mammalian microsomes. The final compound, 7,8-dihydro-7,8-dihydrobenzo[a]pyrene-9,10-oxide has been shown to be a powerful mutagen (Huberman et al. 1976) and also binds to DNA (Weinstein et al. 1976).

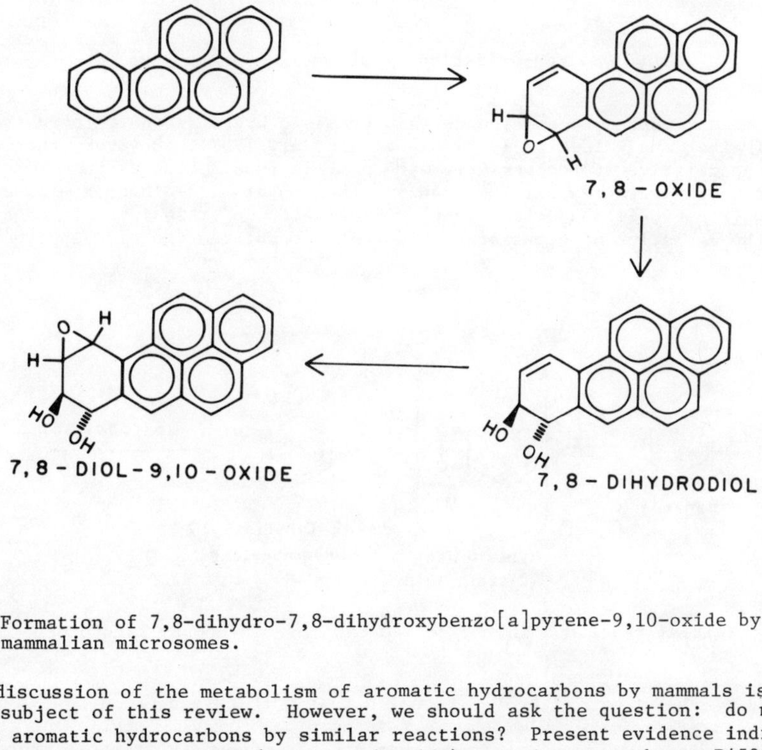

Fig. 4.2 Formation of 7,8-dihydro-7,8-dihydroxybenzo[a]pyrene-9,10-oxide by mammalian microsomes.

A detailed discussion of the metabolism of aromatic hydrocarbons by mammals is not germane to the subject of this review. However, we should ask the question: do microorganisms oxidize aromatic hydrocarbons by similar reactions? Present evidence indicates that certain eucaryotic microorganisms (yeasts and molds) contain a cytochrome P450 enzyme system (Ambike and Baxter 1970, Duppel et al. 1973, Ferris et al. 1976) and the fungus Cunninghamella bainierii oxidizes naphthalene by a similar mechanism to that reported for liver microsomes (Ferris et al. 1973).

Degradation of aromatic hydrocarbons by procaryotic microorganisms.

The ability of certain strains of bacteria to oxidize simple aromatic hydrocarbons has been observed on many occasions. Some organisms will grow at the expense of these compounds and convert them to cell mass, carbon dioxide and water. Others, although they cannot grow with a particular aromatic hydrocarbon, will oxidize such molecules provided an alternative growth substrate is available. The latter phenomenon was termed cooxidation by Leadbetter and Foster (1960) and the technique has been elegantly developed by Raymond and his colleagues (Raymond and Jamison 1971). Two examples of cooxidation are shown in Fig. 3. Growth of a Nocardia species on hexadecane in the presence of ethylbenzene led to the accumulation of phenylacetic acid in the culture medium (Davis and Raymond 1961). Under similar conditions a different strain of Nocardia oxidized p-xylene to α,α'-dimethylmuconic acid (Jamison et al. 1969). Both examples show that the degradation of aromatic hydrocarbons is accompanied by the incorporation of oxygen into the molecule.

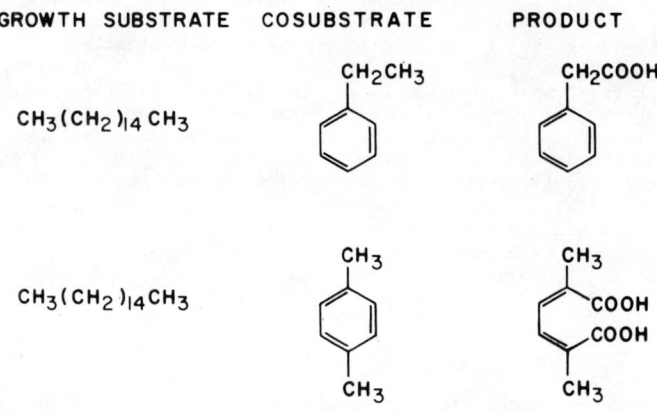

Fig. 4.3 Co-oxidation of aromatic hydrocarbons.

Bacteria have been reported to oxidize certain aromatic hydrocarbons through trans-dihydrodiols (Walker and Wiltshire 1953, Canonica et al. 1957). However, the methods used to establish the relative stereochemistry were not rigorous. Studies in our laboratory revealed that Pseudomonas putida oxidized benzene through cis-1,2-dihydroxy-1,2-dihydrobenzene as shown below (Fig. 4, Gibson et al. 1968). Elucidation of this initial reaction was facilitated by the isolation of a mutant strain that accumulated the cis-dihydrodiol in the

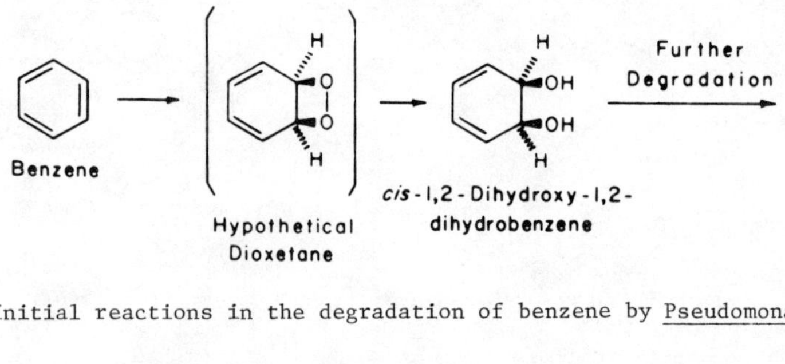

Fig. 4.4 Initial reactions in the degradation of benzene by Pseudomonas putida.

culture medium. Experiments with $^{18}O_2$ revealed that both oxygen atoms in the hydroxylated product were derived from molecular oxygen (Gibson et al. 1970). Subsequent studies revealed that the mutant organism oxidized a variety of monocyclic aromatic hydrocarbons to cis-dihydrodiols. In addition there have been several reports on the intermediacy of dihydrodiols in the bacterial degradation of aromatic hydrocarbons and related compounds (Table 3). In those cases where relative stereochemistry had been carefully investigated the hydroxyl groups have been found to have a cis-configuration.

TABLE 3. Monocyclic aromatic hydrocarbons and related compounds that are oxidized to dihydrodiols by bacteria

Substrate	Organism	Ref.
Benzene	Pseudomonas putida Acinetobacter species	Gibson et al. 1970b Högn and Jaenicke 1972
Toluene	Pseudomonas putida	Gibson et al. 1970
Chlorobenzene	Pseudomonas putida	Ziffer et al. 1976
p-Chlorotoluene	Pseudomonas putida	Ziffer et al. 1976
p-Bromotoluene	Pseudomonas putida	Ziffer et al. 1976
p-Fluorotoluene	Pseudomonas putida	Ziffer et al. 1976
Ethylbenzene	Pseudomonas putida	Gibson et al. 1973a
p-Xylene	Pseudomonas putida	Gibson et al. 1974
2-Phenylbutane	Pseudomonas putida	Baggi et al. 1972
3-Phenylbutane	Pseudomonas putida	Baggi et al. 1972
4-Phenylheptane	Pseudomonas putida	Baggi et al. 1972
tert-Butylbenzene	Achromobacter	Sorlini 1972
Biphenyl	Beijerinckia species Pseudomonas putida	Gibson et al. 1973b Catelani et al. 1971
Kynurenic acid	Pseudomonas fluorescens	Taniuchi and Hayaishi 1963
Benzoic acid	Alcaligenes eutrophus	Reiner and Hegeman 1971
3-Chlorobenzoic acid	Pseudomonas Alcaligenes eutrophus	Dorn et al. 1974 Reiner and Hegeman 1971
3,5-Dichlorobenzoic acid	Alcaligenes eutrophus	Knackmuss 1975
4-Chlorobenzoic acid	Alcaligenes eutrophus	Knackmuss 1975
3,4-Dichlorobenzoic acid	Alcaligenes eutrophus	Knackmuss 1975
3-Methylbenzoic acid	Alcaligenes eutrophus	Reiner and Hegeman 1971
3,5-Dimethylbenzoic acid	Alcaligenes eutrophus	Knackmuss 1975
4-Methylbenzoic acid	Alcaligenes eutrophus	Knackmuss 1975
3-Fluorobenzoate	Alcaligenes eutrophus	Reiner and Hegeman 1971
4-Fluorobenzoate	Alcaligenes eutrophus	Reiner and Hegeman 1971
4-Trifluoromethylbenzoate	Pseudomonas putida	Defrank and Ribbons 1976
5-Amino-4-chloro-2-phenyl-3 (2H)pyridazinone	Unidentified. Strain E	De Frenne et al. 1973
α-Methylstyrene	Unidentified. S107B1	Omori et al. 1974

The further metabolism of dihydrodiols involves dehydrogenation to form ortho-dihydroxy derivatives (catechols). It is now generally accepted that before microorganisms can cleave the aromatic nucleus two hydroxyl groups must be present (Dagley 1971). Catechols can undergo enzymatic cleavage between the hydroxyl groups (ortho-cleavage) or adjacent to one of of hydroxyl groups (meta-cleavage) as shown below (Fig. 5). The metabolic sequences initiated by these dioxygenase reactions have been the subjects of several excellent reviews (Stanier and Ornston 1973, Dagley et al. 1964) and will not be discussed in this presentation.

Fig. 4.5 Bacterial oxidation of catechol.

Aromatic hydrocarbons with fused benzene rings are also metabolized through dihydrodiol intermediates. Thus, naphthalene (Jeffrey et al. 1975) and anthracene (Jerina et al. 1976) are oxidized to cis-1(R),2(S)-dihydroxy-1,2-dihydronaphthalene and cis-1(R),2(S)-dihydroxy-1,2-dihydroanthracene respectively. cis-Hydroxylation at different sites of certain polycyclic aromatic hydrocarbons has also been observed. A mutant strain of Beijerinckia oxidized phenanthrene to cis-3,4- and cis-1,2-dihydroxy-1,2-dihydrophenanthrene (Jerine et al. 1976). The same organism catalyzed the oxidation of benzo(a)pyrene as shown in Fig. 6 (Gibson et al. 1975).

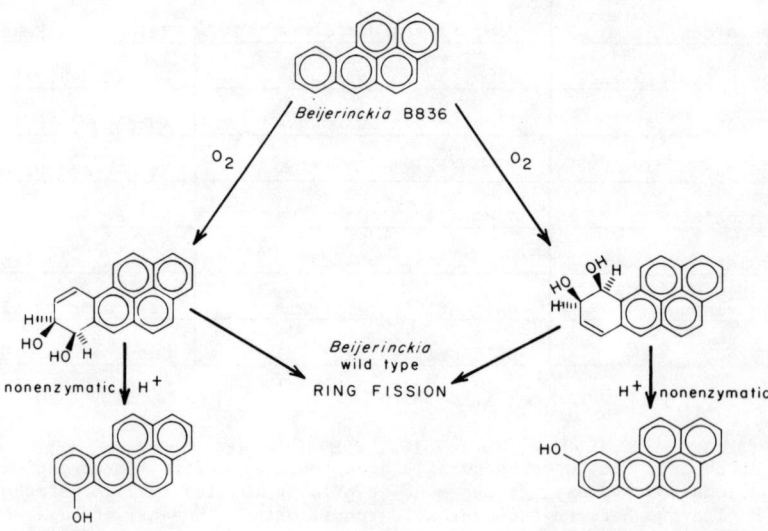

Fig. 4.6 Oxidation of benzo[a]pyrene by a Beijerinckia species.

Although trans-dihydrodiols have been reported as intermediates in the bacterial degradation of naphthalene and related compounds (Walker and Wiltshire 1953, Canonica et al. 1957) it appears that cis-hydroxylation is the principal mechanism for the initiation of degradation of polycyclic aromatic hydrocarbons. Compounds known to be oxidized through dihydrodiols are given in Table 4.

TABLE 4. Polycyclic aromatic compounds oxidized to dihydrodiols by bacteria.

Substrate	Organism	Reference
Naphthalene	*Bacillus naphthalinicum non-liquefaciens*	Walker and Wiltshire, 1953
	Pseudomonas species 53/1	Treccani et al., 1954
	Pseudomonas 53/2	Treccani et al., 1954
	Pseudomonas desmolyticum	Treccani et al., 1954
	Nocardia strain R	Treccani et al., 1954
	Nocardia species NRRL 3385	Wegner, 1973
	Pseudomonas putida	Jerina et al., 1971
		Jeffrey et al., 1975
	Pseudomonas NCIB 9816	Catterall et al., 1971
		Jeffrey et al., 1974
	Pseudomonas fluorescens	Jeffrey et al., 1975
	Pseudomonas putida, biotype B	Jeffrey et al., 1975
1-Chloronaphthalene	*Pseudomonas desmolyticum*	Walker and Wiltshire, 1955
2-Chloronaphthalene	*Pseudomonas desmolyticum*	Canonica et al., 1957
2-Methylnaphthalene	*Pseudomonas desmolyticum*	Treccani and Fiecchi, 1958
		Canonica et al., 1957
Naphthalene-2-carboxylic acid	*Pseudomonas testosteroni*	Knackmuss, 1976
Phenanthrene	*Flavobacterium* species	Colla et al., 1959
	Pseudomonas putida	Jerina et al., 1976
	Beijerinckia species	Jerina et al., 1976
Anthracene	*Flavobacterium* species	Colla et al., 1959
	Beijerinckia species	Jerina et al., 1976
Benzo[a]anthracene	*Beijerinckia* species	Gibson et al., 1975
Benzo[a]pyrene	*Beijerinckia* species	Gibson et al., 1975

Discussion

Studies with well defined systems, usually a single microorganism and a single substrate, have shown that bacteria incorporate one molecule of oxygen into aromatic hydrocarbons to form dihydrodiol intermediates. In those cases that have been rigorously investigated the hydroxyl groups have been shown to have a cis-stereochemistry. Oxidation of the dihydrodiols leads to the formation of catechols which are substrates for enzymatic cleavage of the aromatic ring. In contrast certain strains of fungi and higher organisms incorporate one atom of molecular oxygen into aromatic hydrocarbons to form arene oxides that can undergo the enzymatic addition of water to yield trans-dihydrodiols. These differences are shown in Fig. 7.

Little is known about the microbial degradation of the aromatic hydrocarbons that are present in crude oil. At this time degradation products formed during the microbial utilization of crude oil have not been identified. In addition, dibenzothiophene is the only sulfur-containing aromatic hydrocarbon that is known to be subject to microbial degradation (Yamada et al. 1968). It is possible that many of the polycyclic aromatic hydrocarbons, particuarly those that contain sulfur, that are present in crude oil are poor candidates for biodegradation.

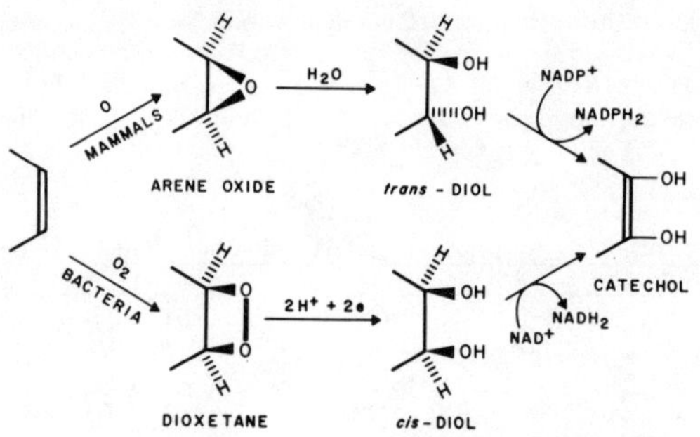

Fig. 4.7 Differences between the reactions used by eucaryotic and procaryotic organisms to initiate the oxidation of aromatic hydrocarbons.

Acknowledgements

Some of the studies described in this manuscript were carried out in colloratoration with D.M. Jerina and his associates at the National Institute of Health. Studies were supported in part by grants ES-00537, awarded by the National Institute of Environmental Health Sciences and 1 R01 CA19078 awarded by the National Cancer Institute, DHEW; F-440 from the Robert A. Welch Foundation; the Office of Naval Research, Microbiology Program, Naval Biology Project, under contract N00014-76-C-0102, NR205-008 and Contract NO1 CP 33384 awarded by the National Cancer Institute, DHEW. The author is a recipient of Career Development Award 1 K04 ES-70088 from the Institute of Environmental Health Sciences, DHEW.

References

Ambike, S. H., and R. M. Baxter, Cytochrome P450 and b_5 in Claviceps purpurea: Interconversion of P450 and P420, Phytochemistry 9, 1959-1962 (1970).

Baggi, G., D. Catelani, E. Galli, and V. Treccani, The microbial degradation of phenylalkanes, Biochem. J. 126, 1091-1097 (1972).

Blumer, M., Organic compounds in nature: limits of our knowledge, Angewandte Chemie 14, 507-514 (1975).

Blumer, M., Polycyclic aromatic hydrocarbons in nature, Scientific American 234, 34-44 (1976).

Blumer, M., and W. W. Youngblood, Polycyclic aromatic hydrocarbons in soils and recent sediments, Science 188, 53-55 (1975).

Borneff, J., F. Selenka, H. Knute, and A. Maximos, Experimental studies on the formation of polycyclic aromatic hydrocarbons in plants, Environmental Res. 2, 22-29 (1968).

Canonica, di L., A. Fiecchi, and V. Treccani, Sui prodotti di ossidazione microbica di alcune naftaline sostituite, Instituto Lombardo (Rend. Sc.) 91, 119-129 (1957).

Catelani, D., C. Sorlini, and V. Treccani, Metabolism of biphenyl by Claviceps purpurea Experientia 27, 1173-1174 (1971).

Catterall, F. A., K. Murray, and P. A. Williams, The configuration of the 1,2-dihydroxy-1,2-dihydronaphthalene formed in the bacterial metabolism of naphthalene, Biochim. Biophys. Acta. 237, 361 (1971).

Colla, C., A. Fiecchi, and V. Treccani, Ricerche sul metabolismo ossidativo microbico dell' antracene e del fenantrene, Ann. Micro. 9, 87-91 (1959).

Dagley, S., P.J. Chapman, D. T. Gibson, and J. M. Wood, Degradation of the benzene nucleus by bacteria, Nature 202, 775-778 (1964).

Dagley, S., Catabolism of aromatic compounds by micro-organisms, Adv. Microbial Physiol. 6, 1-46 (1971).

Daly, J. W., D. M. Jerina, and B. Witkop, Arene oxides and the NIH shift: The metabolism toxicity and carcinogenicity of aromatic compounds, Experientia 28, 1129-1149 (1972).

Davis, J. B. and R. L. Raymond, Oxidation of alkyl-substituted cyclic hydrocarbons by a Nocardia during growth on n-alkanes, Appl. Microbiol. 9, 383-388 (1961).

Defrank, J. J., and D. W. Ribbons, The p-cymene pathway in Pseudomonas putida PL: Isolation of a dihydrodiol accumulated by a mutant, Biochem. Biophys. Res. Commun. 70, 1129-1135 (1976).

De Frenne, E., J. Eberspacher, and F. Lingens, The bacterial degradation of 5-amino-4-chloro-2-phenyl-3(2H)pyridizinone, Eur. J. Biochem. 33, 357-363 (1973).

Dooley, J. E., D. E. Hirsch, and C. J. Thompson, Analyzing heavy ends of crude, Hydrocarbon Processing 53, 140-146 (1974).

Dorn, E., M. Hellwig, W. Reineke, and H. J. Knackmuss, Isolation and characterization of a 3-chlorobenzoate-degrading Pseudomonad, Arch. Microbiol. 99, 61-70 (1974).

Duppel, W., J. M. Lebeault, and M. J. Coon, Properties of a yeast cytochrome P450-containing enzyme system which catalyzes the hydroxylation of fatty acids, alkanes and drugs, Eur. J. Biochem. 36, 583-592 (1973).

Ferris, J. P., M. J. Fasco, F. L. Stylianopoulou, D. M. Jerina, J. W. Daly, and A. M. Jeffrey, Monooxygenase activity in Cunninghamella bainierii Evidence for a fungal system similar to liver microsomes, Arch. Biochem. Biophys. 156, 97-103 (1973).

Ferris, J. P., L. H. MacDonald, M. A. Patrie, and M. A. Martin, Aryl hydrocarbon hydroxylase activity in the fungus Cunninghamella bainierii Evidence of the presence of cytochrome P450, Arch. Biochem. Biophys. 175, 443-452 (1976).

Gibson, D. T., J. R. Koch, and R. E. Kallio, Oxidative degradation of aromatic hydrocarbons by microorganisms, I. Enzymatic formation of catechol from benzene, Biochemistry 1, 2653-2662 (1968).

Gibson, D. T., M. Hensley, H. Yoshioka, and T. J. Mabry, Formation of (+)-cis-2,3-dihydroxy-1-methyl-cyclohexa-4,6-diene from toluene by Pseudomonas putida, Biochemistry 9, 1626-1630 (1970a).

Gibson, D. T., G. E. Cardini, F. C. Maseles and R. E. Kallio, Incorporation of oxygen-18 into benzene by Pseudomonas putida Biochemistry 9, 1631-1635 (1970b).

Gibson, D. T., B. Gschwendt, W. K. Yeh, and V. M. Kobal, Initial reactions in the oxidation of ethylbenzene by Pseudomonas putida Biochemistry 12, 1520-1528 (1973a).

Gibson, D. T., R. L. Roberts, M. C. Wells, and V. M. Kobal, Oxidation of biphenyl by a Beijerinckia species, Biochem. Biophys. Res. Commun. 50, 211-219 (1973b).

Gibson, D. T., V. Mahadevan, and J. F. Davey, Bacterial metabolism of para- and meta-xylene: oxidation of the aromatic ring, J. Bacteriol. 119, 930-936 (1974).

Gibson, D. T., V. Mahadevan, D. M. Jerina, H. Yagi, and H. J. C. Yeh, Oxidation of the carcinogens benzo(a)pyrene and benzo(a)anthracene to dihydrodiols by a bacterium, Science 189, 295-297 (1975).

Giger, W., and M. Blumer, Polycyclic aromatic hydrocarbons in the environment: isolation and characterization by chromatography, visible, ultraviolet and mass spectrometry, Anal. Chem. 46, 1663-1671 (1974).

Högn, T., and L. Jaenicke, Benzene metabolism of Moraxella species, Eur. J. Biochem. 30, 369-375 (1972).

Huberman, E., L. Sachs, S. K. Yang and H. V. Gelboin, Identification of mutagenic metabolites of benzo a pyrene in mammalian cells, Proc. Natl. Acad. Sci. New York 73, 607-612 (1976).

Jamison, V. W., R. L. Raymond and J. O. Hudson. Microbial hydrocarbon co-oxidation. III Isolation and characterization of an α,α-dimethyl-cis,cis-muconic acid-producing strain of Nocardia corallina Appl. Microbiol. 17, 853-856 (1969).

Jeffrey, A. M., H. J. C. Yeh, D. M. Kerina, T. R. Patel, J. F. Davey and D. T. Gibson, Initial reactions in the oxidation of naphthalene by Pseudomonas putida, Biochemistry, 14, 575-584 (1975).

Jerina, D. M., J. W. Daly, A. M. Jeffrey, and D. T. Gibson, *cis*-1,2-Dihydroxy-1,2-dihydronaphthalene: A bacterial metabolite from naphthalene, Arch. Biochem. Biophys. 142, 394-396 (1971).

Jerina, D. M., and J. W. Daly, Arene oxides: A new aspect of drug metabolism, Science 185, 573-578 (1974).

Jerina, D. M., H. Selander, H. Yagi, M. C. Wells, J. F. Davey, V. Mahadevan, and D. T. Gibson, Dihydrodiols from anthracene and phenanthrene, J. Am. Chem. Soc. 98, 5988-5996 (1976).

Knackmuss, H. J., Über den mechanismus der biologischen persistenz von halogenierten aromatischen kohlenwasserstoffen, Chemiker Zeitung 99, 213-219 (1975).

Knackmuss, H. J., Microbiological synthesis of (+)-*cis*-1,2-dihydroxy-1,2-dihydronaphthalene-2-carboxylic acid, Angew. Chem. Int. Ed. Engl. 15, 549 (1976).

Oesch, F., Mammalian epoxide hydrases: Inducible enzymes catalyzing the inactivation of carcinogenic metabolites derived from aromatic and olefinic compounds, Xenobiotica 3, 305-340 (1973).

Omori, T., Y. Jigami, and Y. Minoda, Microbial oxidation of α-methylstyrene and β-methylstyrene, Agr. Biol. Chem. 38, 409-415 (1974).

Raymond, R. L., and V. W. Jamison, Biochemical activities of *Nocardia*, Adv. Appl. Microbiol. 14, 93-122 (1971).

Reiner, A. M., and G. D. Hegeman, Metabolism of benzoic acid by bacteria: accumulation of (-)-3,5-cyclohexadiene-1,2-diol-1-carboxylic acid by a mutant strain of Alcaligenes eutrophus Biochemistry 10, 2530-2536 (1971).

Sorlini, C., Ricerche sulla degradazione microbica del *tert*-butilbenzene, Atti XVI Congr. Soc. Ital. Microbiol. I, 405 (1972).

Stanier, R. Y., and L. N. Ornston, The β-ketoadipate pathway, Adv. Microbial Physiol. 9, 89-151 (1973).

Taniuchi, H., and O. Hayaishi, Studies on the metabolism of kynurenic acid. III. Enzymatic formation of 7,8-dihydroxykynurenic acid from kynurenic acid, J. Biol. Chem. 238, 283-293 (1963).

Thomas, P. E., A. Y. H. Lu, D. Ryan, S. B. West, J. Kawalek, and W. Levin, Immunochemical evidence for six forms of rat liver cytochrome P450 obtained using antibodies against purified rat liver cytochromes P450 and P448, Mol. Pharmacol. 12, 746-758 (1976).

Treccani, V., N. Walker, and G. H. Wiltshire, The metabolism of naphthalene by soil bacteria J. Gen. Microbiol. 11, 341-348 (1954).

Treccani, V., and A. Fiecchi, Ossidazione microbica della 2-metilnaftalina, Ann. Micr. 8, 36-44 (1958).

Walker, N., and G. H. Wiltshire, The breakdown of naphthalene by a soil bacterium, J. Gen. Microbiol. 8, 273-276 (1953).

Walker, N., G. H. Wiltshire, The decomposition of 1-chloro- and 1-bromonaphthalene by soil bacteria, J. Gen. Microbiol. 12, 478-483 (1955).

Wegner, E. H., Microbial conversion of naphthalene base hydrocarbons, U.S. Patent 3,755,080 (1973).

Weinstein, I. B., A. M. Jeffrey, K. W. Jennette, S. H. Blobstein, R. G. Harvey, C. Harris, H. Autrup and H. K. K. Nakanishi. Benzo[a]pyrene diol epoxides as intermediates in nucleic acid binding in vitro and in vivo, Science 193, 592-595 (1976).

Yamada, K., Y. Minoda, K. Kodama, S. Nakatani, and T. Akasaki, Microbial conversion of petro-sulfur compounds. Part I. Isolation and identification of dibenzothiophene-utilizing bacteria, Agr. Biol. Chem. 32, 840-845 (1968).

Ziffer, H., K. Kabuto, D. T. Gibson, V. M. Kobal and D. M. Jerina, The absolute stereochemistry of several *cis*-dihydrodiols microbially produced from substituted benzenes, Tetrahedron (submitted, 1976).

CHAPTER 5

BIOTRANSFORMATION OF PETROLEUM HYDROCARBONS IN MARINE
ORGANISMS INDIGENOUS TO THE ARCTIC AND SUBARCTIC

Donald C. Malins

Northwest and Alaska Fisheries Center
National Marine Fisheries Service
National Oceanic and Atmospheric Administration
2725 Montlake Boulevard East
Seattle, Washington 98112 U.S.A.

Introduction

A comprehensive review of accumulations and biotransformations of petroleum hydrocarbons in marine organisms was undertaken by Varanasi and Malins (1977). The literature indicates that work thus far relates primarily to the accumulation and discharge of petroleum hydrocarbons rather than to their bioconversion. In fact, our understanding of the bioconversion (metabolic fate) of hydrocarbons in petroleum-exposed marine organisms is in its infancy. Yet arguments can be made in the light of studies with mammals that certain metabolites of the polynuclear hydrocarbons may have a high innate toxicity exceeding that of the hydrocarbons themselves. Support for this argument was reviewed by Sims and Grover (1974) and discussed with respect to marine organisms by Malins (1977).

Many questions remain unanswered about possible relations between petroleum hydrocarbons and pathologic alterations (e.g., neoplastic lesions) observed in fish populations from various geographic regions (Hodgins et al. 1977, Wellings 1969). The importance of aromatic hydrocarbon metabolism is highlighted in studies of such alterations: substantial evidence exists linking the formation of certain electrophilic metabolites (notably 4 to 6 ring structures) in animal systems to irreversible alterations in genetic materials, that is DNA (Grover and Sims 1968, Arcos and Argus 1974). Biochemical interactions of this type are of particular concern as they may cause serious alterations in a host of vital life processes. Moreover, although physiological and biochemical changes are associated with the exposure of marine life to petroleum (Varanasi and Malins 1977), little is known about the role of aromatic hydrocarbon metabolites in these alterations.

Although it is possible--using sophisticated instrumental techniques--to obtain hydrocarbon profiles from marine organisms exposed to petroleum, comparable profiles of metabolic products are not obtained. Thus the degree of petroleum contamination in marine organisms is determined largely on the basis of parent hydrocarbon data, while possible complex mixtures of metabolites are essentially overlooked.

In attempting to evaluate our understanding of petroleum hydrocarbon biotransformations in marine organisms, the alkane and alicyclic fractions will not be included; rather discussion will be devoted to the aromatic hydrocarbons because of their notable potential for forming potentially toxic products *in vivo* (Sims and Grover 1974, Arcos and Argus 1974).

No effort will be made to review the relatively large volume of literature on the storage and discharge of aromatic hydrocarbons by exposed marine organisms. This matter was addressed in considerable detail previously (Varanasi and Malins 1977). An attempt will be made, however, to evaluate briefly our understanding of the biotransformation of aromatic hydrocarbons in arctic and subarctic species, and to assess deficiencies in existing knowledge, which, if not resolved, may hinder a timely understanding of the impact of petroleum operations on indigenous organisms and ecosystems.

The Ability of Animal Life to Metabolize Aromatic Hydrocarbons

Initial steps in biochemically altering the structure of aromatic hydrocarbons occur in the endoplasmic reticulum of the cell. The enzyme systems affecting the transformations are associated with the microsomes. In the laboratory, the microsomes are liberated from the endoplasmic reticulum (primarily the rough endoplasmic reticulum) via conventional cell fractionation techniques. The enzymes (oxygenases), which are NADPH (reduced pyridine nucleotide) dependent, mediate the introduction of molecular oxygen into the aromatic nucleus (Sims and Grover 1974). Often referred to as aryl hydrocarbon hydroxylases (AHH) or drug metabolizing enzymes, they operate in unison with the electron transport system (i.e., cytochrome P-450) (Fig. 1). Oxygen from the electron transport system enters the

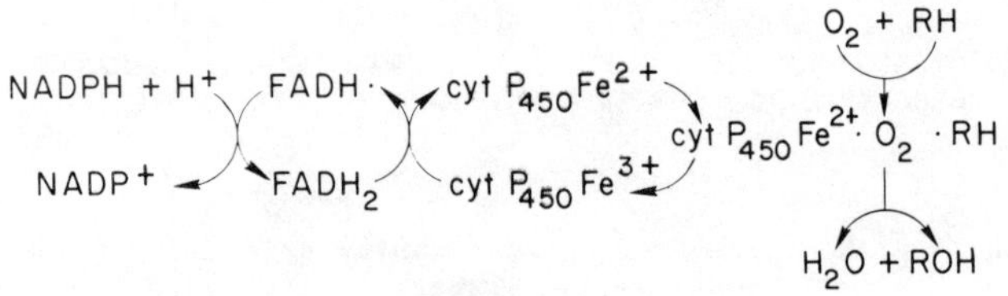

Fig. 5.1 Pathway of electron flow for hydroxylations in liver microsomes
[From White et al. 1973]

aromatic nucleus to form a phenolic structure and further combines with hydrogen to form water (Coon et al. 1973):

$$RH + O_2 + NADPH + H^+ \rightarrow ROH + H_2O + NADP^+$$

The oxygenases, which are generally responsible for the bioconversion of drugs and other xenobiotics, are believed to account for the formation of essentially all of the primary products of aromatic hydrocarbon metabolism. Although it is uncertain that the reactions mediated by the oxygenases are a prerequisite to deleterious alterations in macromolecules, such as DNA, a significant body of evidence points to such a conclusion (Sims and Grover 1974, Malaveille et al. 1975, Weinstein et al. 1976).

Presently, considerable attention is being focused on electrophilic intermediates in the formation of the phenolic derivatives. The structure of most interest is the epoxide or arene oxide. This electrophile has been variously implicated in interactions with genetic materials leading to serious impairments of life processes (Sims and Grover 1974, Malaveille et al. 1975, Weinstein et al. 1976). The enzyme epoxide hydrase has been viewed with interest because this enzyme is responsible for converting arene oxides to the corresponding phenols (Sims and Grover 1974). The reactions just described are depicted in Fig. 2 with respect to the conversion of benzo[a]anthracene to an arene oxide, phenolic products and glutathione conjugate.

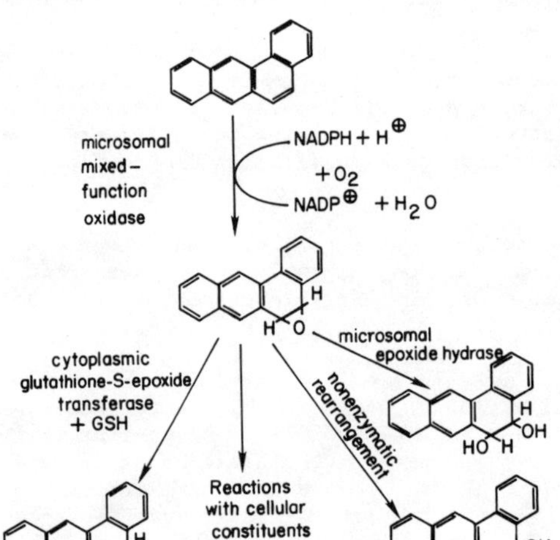

Fig. 5.2 Pathways involved in the metabolism of benzo[a]anthracene.
[From Swaisland et al. 1973]

Other reactions occur which generally increase the hydrophilic nature of the products of initial aromatic hydrocarbon oxidation. These reactions occur in vital organs (e.g., liver and kidney) and involve the conjugation of epoxides and phenols with water-soluble substances such as certain carbohydrates. The products (conjugates) include mercapturic acid derivatives, glucuronides, sulfates, and glycosides. Figure 3 depicts the conversion of 1-naphthol to the 1-naphthyl sulfate in hepatic microsomal preparations. Interest is expressed in the glucuronyl-S-transferases, which mediate the conversion of arene oxides to conjugated derivatives (Fig. 3), because the reactions involved substantially diminish the

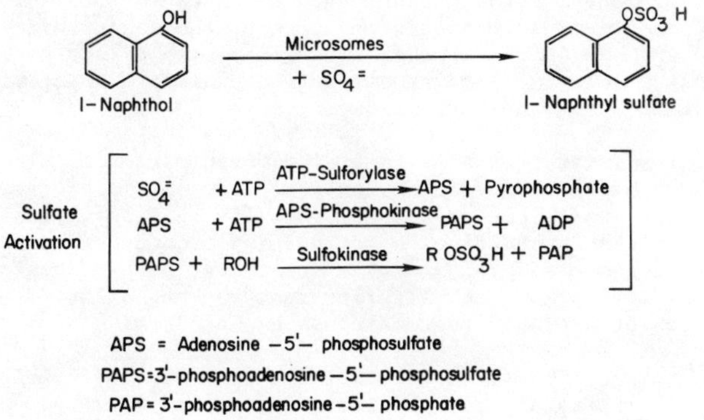

Fig. 5.3 Conjugating reactions: Sulfation of 1-naphthol.

toxic properties of the epoxide group. The enzyme systems (e.g., UDP-glucoronyl transferase and aryl sulfatase) governing the conjugating reactions have been studied extensively in mammals, but they are just beginning to be understood in marine organisms.

Oxygenase Enzyme Systems

Evidence for oxygenases in marine organisms (Aryl Hydrocarbon Hydroxylase; Table 1) has

TABLE 1. Aryl Hydrocarbon Hydroxylase (AHH) Activity in Marine Organisms

PHYTOPLANKTON		
(*Fucus* sp.)	−	Payne 1976a
ZOOPLANKTON		
copepods	+*	Lee 1975
MOLLUSCS		
mussels (*Mytilus edulis*)	− (hepatopancreas)*	Lee et al. 1972a
squid (*Illex illecebrosus*)	+ (hepatopancreas)	Payne 1976a
ECHINODERMS		
sea urchin (*Stronglyocentrotus drobachensis*)	+ (hepatopancreas)	Payne 1976a
sea star (*Asterias* sp.)	+ (hepatopancreas)	Payne 1976a
CRUSTACEANS		
lobster (*Homarus americanus*)	− (hepatopancreas)	Pohl et al. 1974, James et al. 1976
spider crab (*Maia squinado*)	+ (liver)*	Corner et al. 1973
Stage V spot shrimp larvae (*Pandalus platyceros*)	+*	Sanborn & Malins 1977
FISH		
sculpin (*Gillichthys mirabilis*)	+ (liver)*	Lee et al. 1972b
coho salmon (*Oncorhynchus kisutch*)	+ (liver)	Gruger et al. 1977a
dogfish (*Squalus acanthias*)	+ trace (liver)	Bend et al. 1977

* AHH activity is based on evidence for metabolite formation.

been reviewed in detail (Varanasi and Malins 1977, Malins 1977). The AHH's appear to be widely distributed among teleosts; however, studies of elasmobranchs and invertebrates indicate that some species possess barely detectable enzyme activity (Pohl et al. 1974, James et al. 1977, Bend et al. 1977): The little skate (*Raja erinacea*), bottlenose stingray (*Dasyatis sayi*), and dogfish shark (*Squalus acanthias*) possess only a slight ability to hydroxylate benzpyrene. Interestingly, however, certain teleosts and elasmobranchs have significant hepatic cytochrome P-450 contents which are generally close to those of rat liver microsomes (Bend et al. 1977).

Significant AHH activities were found among certain invertebrates. The spider crab (*Maia squinado*) metabolizes benzo[a]pyrene (Corner et al. 1973) and the blue crab (*Callinectes sapidus*) deposits substantial amounts of metabolites in the hepatopancreas, gill, blood, stomach, and muscle when exposed to this and other aromatic hydrocarbons (Lee et al. 1977). Yet other invertebrates, such as the mussel (*Mytilus edulis*) (Lee et al. 1972a), scallop

(*Placopecten* sp.), and snail (*Littorina littorea*) (Payne 1976a) possess virtually imperceptible AHH activities. Interestingly, planktonic crustaceans, such as copepods (Lee 1975) metabolize naphthalene and benzo[a]pyrene, whereas echinoderms, such as the sea urchin (*Strongylocentrotus drobachensis*) and the sea star (*Asterias* sp.) apparently do not possess such metabolic capabilities. Zooplankton (Lee 1975) do appear not to exhibit AHH activity. Moreover, no such activity has been found among the phytoplankton (Payne 1976a).

Overall, it appears that many differences exist in the ability of diverse marine species to degrade aromatic hydrocarbons and other xenobiotics. But a cautious view should be taken in comparing the AHH activities among species: Some organisms, normally showing virtually no enzyme activity, may exhibit detectable AHH when exposed to inducing agents. Moreover, AHH activities are commonly based on the conversion of a *single* substrate, such as benzo[a]-pyrene, to hydroxy derivatives and thus may not necessarily represent the capability of an organism to degrade the complex mixtures of aromatic hydrocarbons in petroleum.

Epoxide hydrase activity has been determined in several teleosts (Bend et al. 1977), elasmobranchs (Table 2) (James et al. 1974) and certain invertebrates (Bend et al. 1977). Styrene-7,8-oxide, benzo[a]pyrene-4,5-oxide, and octene-1,2-oxide were used as substrates with microsomal preparations from either the liver or hepatopancreas. The teleosts examined [sheepshead (*Archosargus probatocephalus*) and eel (*Anguilla rostrata*)] each exhibited significant epoxide hydrase activity with respect to styrene-7,8-oxide and benzo[a]pyrene-4,5-oxide. Moreover, the eel showed a significant epoxide hydrase activity with octene-1,2-oxide. The elasmobranchs examined [Atlantic stingray (*Dasyatis* sp.), dogfish shark (*S. acanthias*) and little skate (*R. erinacea*)] generally showed microsomal epoxide hydrase activity with each substrate; however, the little skate showed notably very low enzyme activity with styrene-7,8-oxide and benzo[a]pyrene-4,5-oxide. The invertebrates [spiny lobster (*Panulirus argus*), clam (*Mya arenaria*), blue crab (*C. sapidus*) and the rock crab (*Cancer irroratus*)] generally exhibited significant microsomal epoxide hydrase activity with respect to each substrate. The mussel (*M. edulis*) was an exception in that the epoxide hydrase activities were very low. In fact, the enzyme activity measured with respect to benzo[a]pyrene-4,5-oxide was essentially insignificant. The findings with *M. edulis* are consistent with data obtained with AHH in this species. Thus it appears certain that *Mytilus* has little capacity to effect the first stages of oxidation of aromatic hydrocarbons; however, the effects of inducing agents on the enzyme have not been explored.

Although our understanding of the distribution of epoxide hydrase in organs of marine animals is slight, Bend et al. (1977) showed with lobster (*Homarus americanus*) that significant enzyme activities are obtained from microsomal preparations of hepatopancreas, green gland, egg masses, and gills with styrene-7,8-oxide, benzo[a]pyrene-4,5-oxide, and octene-1,2-oxide. The lowest activities were obtained with gill and the highest with microsomal preparations of hepatopancreas employing styrene-7,8-oxide and octene-1,2-oxide as substrates. Despite the evidence for epoxide hydrase activity in *H. americanus*, this species shows virtually no benzo[a]pyrene activity in hepatopancreas (James et al. 1977). Generally, the capacity to degrade arene oxides is found in a number of marine species; however, the enzyme activities vary significantly among species and with the substrate employed.

Many enzyme systems are susceptible to induction or inhibition by various chemicals and other factors. Their influence on the oxidizing enzyme systems is important in assessing the ability of marine organisms to metabolize aromatic hydrocarbons and in providing some understanding of the tendency of organisms to retain potentially toxic metabolites during chronic exposures. 3-Methylcholanthrene and 1,2,3,4-dibenzanthracene induce benzpyrene hydroxylase in little skate liver (Pohl et al. 1974, Pohl et al. 1977); however, 3-methylcholanthrene administered to sheepshead (*A. probatocephalus*) neither increased cytochrome P-450 content nor induced benzpyrene hydroxylase (Bend et al. 1977). Thus, wide differences may exist in the ability of marine life to induce both cytochrome P-450 and AHH.

The question of whether epoxide hydrase is induced in marine organisms by aromatic hydrocarbons and other environmental chemicals is of considerable interest. Although very little information is available, Bend et al. (1977) were unable to induce this enzyme in sheepshead (*A. probatocephalus*) or the Atlantic stingray (*D. sabina*) with 3-methylcholanthrene; the substrate was styrene-7,8-oxide. Clearly, more work is desirable to determine whether these findings are an exception or whether they apply to other marine organisms. The induction of enzymes governing the formation of the potentially toxic arene oxides without a concomitant induction of enzymes responsible for degrading them may lead to mutagenic, neoplastic, or various cytotoxic events in marine life.

Certain algae, mollusks (snails, mussels), crustaceans (crabs, lobsters) and echinoderms (sea urchins) show little or no apparent induction of AHH on exposure to petroleum (Payne 1976a). Yet marine fish apparently do not follow this rule. The activity of AHH in *Fundulus heteroclitus* inhabiting polluted waters was significantly higher than in the same species living in relatively clean waters (Burns and Sabo 1975). Benzo[a]pyrene hydroxylase

TABLE 2. Epoxide Metabolizing Enzymes in Four Marine Species

Species	Liver		Kidney		Spiral Valve	
	Epoxide hydrase	Glutathione-S-epoxide transferase	Epoxide hydrase	Glutathione-S-epoxide transferase	Epoxide hydrase	Glutathione-S-epoxide transferase
	nmoles products formed/min⁻¹/mg protein⁻¹					
Little skate (Raja erinacea)	0.47 ± 0.11[a] n=10	2.10 ± 0.17 n=11	0.26 (0.17,0.35) n=2	2.43 (2.25,2.57) n=2	--	--
Large skate (Raja ocellata)	1.83 ± 0.24 n=5	2.55 ± 0.32 n=3	0.55 (0.48,0.62) n=2	2.43 (2.11,2.75) n=2	0.09 ± 0.01 n=3	1.79 ± 0.33 n=3
Dogfish shark (Squalus acanthias)	7.57 ± 2.39 n=4	9.3 n=1	9.61 ± 1.59 n=4	8.29 n=1	0.83 n=1	3.36 n=1
Winter flounder (Pseudopleuronecetes americanus)	1.78[b] n=1	2.06 n=1	0.56 n=1	0.3 n=1	--	--

[a] The value quoted for specific activities (with styrene oxide (8-^{14}C) as substrate) are the mean ± S.D. (N) for n > 3. Otherwise, individual values are given.
[b] Epoxide hydrase is measured in microsomes and glutathione-S-epoxide transferase in soluble fraction except in the case of winter flounder, where 10,000 x g supernatant was assayed for each enzyme.
[From James et al. 1974]

in livers and gills was induced by exposing salmonids to petroleum contaminated waters (Payne 1976b). Similar results were obtained with capelin (*Mallotus villosus*) exposed to 1 ppm of petroleum in surrounding water (Payne 1976b). Further the hepatic AHH of coho salmon (*Oncorhynchus kisutch*) were significantly induced when the animals were exposed to an initial concentration of 150 ppb of a water-soluble fraction of Prudhoe Bay crude oil for six days (Gruger et al. 1977b). These studies indicate that marine fish respond to petroleum in the environment by increasing the biosynthesis of enzymes capable of transforming or degrading aromatic hydrocarbons; however, questions remain about the influence of many other factors on the biotransformation of the benzenoid structures.

A variety of chemicals alter AHH activities in terrestrial animals: included are the polychlorinated biphenyls (PCB's) (Litterest et al. 1974), nitrosamines (Arcos and Argus 1976), trace metals (Yoshida et al. 1976), n-alkanes (Notten and Henderson 1975) and epoxides (Yang and Strickhart 1975). Because marine environments contain many chemicals, the problem of evaluating their influence either *individually* or *in concert* on the enzymes metabolizing aromatic hydrocarbons is a difficult and formidable task!

PCB's induce the drug metabolizing system in rainbow trout (*Salmo gairdneri*) (Lidman et al. 1976). Recent evidence suggests that 1 ppm of PCB's (Aroclor 1242) in the diet for 68 days results in induction of hepatic AHH in coho salmon (*O. kisutch*) (Gruger et al., 1977b); however, 1 ppm of Prudhoe Bay crude oil in the diet over the same period did not significantly alter activities of the hepatic enzymes. Thus, under some circumstances the PCB's can be more effective than petroleum in inducing hepatic AHH. Interestingly, petroleum hydrocarbons and chlorinated biphenyls, alone or together in the diet depress the AHH in chinook salmon (*O. tshawytscha*) (Gruger et al. 1977a). The influence of other environmental chemicals on enzyme systems mediating biotransformations of aromatic hydrocarbons in marine organisms is of considerable interest in future research, as well as are related effects on the retention of hydrocarbons and metabolites in tissues.

Aside from chemicals, other factors are known to influence the activities of AHH in marine organisms. Dewaide (1970) pointed out that metabolic rates of xenobiotics in hepatic microsomal enzyme preparations of rainbow trout are lowered by increasing temperature (Table 3). Thus temperature variations in marine waters may perturb the AHH of aquatic animals and thus alter their capacity to degrade aromatic hydrocarbons. Unfortunately, little information is available to permit more than a meager evaluation of the influence of arctic temperatures and other conditions on the AHH of marine organisms.

Pedersen and coworkers (1976) found substantial individual differences in AHH enzyme activities among rainbow trout obtained from several distinct geographic areas; statistically significant differences in strains were also demonstrated. In one case (Chamber Creek trout, Washington State) the range of specific activities of AHH varied by as much as a hundredfold. Large individual variations in AHH specific activities were found in laboratory (challenge) experiments with coho salmon and chinook salmon (Gruger et al. 1977a). While certain variations may be attributed to genetic differences among animals in the study by Pedersen and coworkers (1976), the influence of a host of environmental factors acting in concert must also be considered.

<u>Conjugating Enzyme Systems</u>
Although enzyme systems governing the formation of conjugated derivatives of aromatic hydrocarbons have been studied extensively in terrestrial mammals (Sims and Grover 1974), relatively little comparable information is available on marine organisms. Marine fish are capable of synthesizing mercapturic acid, glycoside, and glucuronic acid derivatives, as well as certain other conjugated structures from aromatic hydrocarbons. The supporting evidence for active enzyme systems is drawn largely from studies in which individual reaction products were identified. Aside from this indirect evidence, some data were obtained on the enzymes themselves: The hepatopancreas of *M. squinado* contains both β-glucuronidase and arylsulfatase systems (Corner et al. 1973). Moreover, glutathione S-transferase exists in certain marine fish (Bend and Fouts 1973, James et al. 1974) (Tables 2, 4, 5). These enzymes are responsible for the formation of glutathione conjugates from arene oxides; thus, they are important in the ability of organisms to convert potentially toxic epoxides to presumably less toxic conjugated derivatives. The enzymes occur in liver and kidney supernatant fractions from the rat, but our understanding of their general distribution among tissues of marine animals is slight. Bend et al. (1977) studied glutathione S-transferase activity in 176,000 x g supernatant fractions from liver and hepatopancreas of several teleosts, elasmobranchs, and invertebrates. The substrates employed were styrene-7,8-oxide, benzo[a]pyrene-4,5-oxide, and octene-1,2-oxide. All marine species examined were capable of converting these substrates to glutathione derivatives; however, wide variations in enzyme activities were often obtained between individual animals. Moreover, invertebrates [e.g., lobster (*H. americanus*); mussel (*M. edulis*); clam (*M. arenaria*); blue crab, and rock crab] showed generally lower specific

TABLE 3. Influence of Change of Environmental Temperature on Drug Metabolism[a]

Animal	Temperature (°C)	p-Hydroxylation of aniline			
		Activity per g fresh liver	Activity per mg liver protein	Activity per mg liver DNA	Activity per 100 g body-weight
Hamster (fed)	5 23	3.99 ± 0.29 (12) 2.78 ± 0.46 (13) P<0.01	0.024 ± 0.002 (12) 0.016 ± 0.003 (11) P<0.01	2.15 ± 0.34 (12) 1.38 ± 0.29 (12) P<0.01	18.87 ± 2.10 (12) 10.95 ± 2.10 (13) P<0.01
Rat (fed)	5 23	1.97 ± 0.45 (8) 1.41 ± 0.30 (8) P<0.01	0.011 ± 0.003 (8) 0.008 ± 0.002 (8) 0.05>P>0.01	0.99 ± 0.28 (8) 0.71 ± 0.15 (8) 0.05>P>0.01	8.27 ± 2.00 (8) 5.60 ± 1.40 (8) P<0.01
Roach (starved)	5 18	1.09 ± 0.46 (40) 0.62 ± 0.18 (36) P<0.01	0.0082 ± 0.0038 (37) 0.0044 ± 0.0015 (36) P<0.01	0.48 ± 0.23 (40) 0.19 ± 0.10 (36) P<0.01	2.67 ± 0.98 (32) 1.12 ± 0.60 (28) P<0.01
Trout (fed)	5 18	0.39 ± 0.08 (38) 0.31 ± 0.11 (37) P<0.01	0.0032 ± 0.0008 (38) 0.0027 ± 0.0010 (37) P<0.01	0.27 ± 0.09 (38) 0.12 ± 0.04 (36) P<0.01	0.63 ± 0.20 (26) 0.37 ± 0.15 (25) P<0.01
Trout (starved)	5 18	0.37 ± 0.07 (16) 0.17 ± 0.03 (9) P<0.01	0.0030 ± 0.0008 (16) 0.0015 ± 0.0002 (9) P<0.01	0.22 ± 0.12 (16) 0.09 ± 0.03 (9) P<0.01	0.46 ± 0.10 (16) 0.18 ± 0.05 (9) P<0.01

[a] Activity defined as μmoles p-aminophenol produced per hour; assayed with 9,000 x g supernatants derived from liver homogenates at 37°C for hamster and rat and at 25°C for roach and trout; values are means ± S.D.; the number of animals is given in parentheses. [From Dewaide 1970]

TABLE 4. Glutathoine S-Aryltransferase Activity in Liver or Hepatopancrease of Several Marine Species

Species	Specific activity [nmoles S-(2-chloro-4-nitrophenyl)glutathione formed/min/mg soluble protein][a]
Little skate *(Raja erinacea)*	6.58 ± 5.95 (4),[b]
Large skate *(Raja ocellata)*	3.14 ± 1.66 (3)
Thorny skate *(Raja radiata)*	2.69 ± 1.13 (3)
Dogfish *(Squalus acanthius)*	1.45 ± 0.55 (4)
Killifish *(Fundulus heteroclitus)*	45.1 ± 14.7 (3)
Eel *(Anguilla rostrata)*	6.70 ± 7.77 (3)
Hagfish *(Myxine glutinosa)*	5.78 ± 1.45 (3)
King of Norway *(Hemitripterus americanus)*	1.99 (1.89-2.09)[c]
Winter flounder *(Pseudopleuronectes americanus)*	1.91 ± 0.69 (3)
Mackerel *(Scomber scombrus)*	1.88 ± 0.47 (3)
Lobster *(Homarus americanus)*	6.05 ± 2.44 (6)
Crab *(Cancer borealis)*	3.55 ± 0.92 (3)

[a] Employing 1,2-dichloro-4-nitrobenzene as the substrate.
[b] Mean ± S.D. (N)
[c] Mean (range)
[From Bend & Fouts, 1973]

TABLE 5. Distribution of Glutathione S-Aryltransferase Activity in Several Organs of the Little Skate (Raja erinacea)

Organ	nmoles S-(2-chloro-4-nitrophenyl)glutathione formed:[a]	
	/min/mg soluble protein	/min/g tissue
Spiral valve mucosa	4.62[b]	209.3
Stomach lining	4.30	177.3
Liver	4.17	181.5
Pancreas	3.59	181.4
Kidney	2.13	93.1
Gill	0.67	19.3
Spleen	0.30	16.0
Heart	0	0

[a] Employing 1,2-dichloro-4-nitrobenzene as the substrate.
[b] Various organs of three skates were pooled. Values are means of two separate assays on each tissue.
[From Bend & Fouts, 1973]

enzyme activities than were obtained with either teleost fish [e.g., sheepshead (*A. probatocephalus*) and winter flounder (*Pseudopleuronectes americanus*)] or elasmobranch fish (e.g., dogfish or little skate). Of the invertebrates, the mussel, clam, blue crab, and rock crab showed relatively very low glutathione S-transferase activity with either substrate.

Glutathione S-transferase activity was studied in the liver and several other tissues of the little skate employing the epoxides used previously (Bend et al. 1977). The tissues examined were liver, kidney, testes, gill, spiral valve mucosa, spleen, heart, and pancreas. Generally the highest enzyme activities were obtained from liver and kidney with styrene-7,8-oxide and benzo[a]pyrene-4,5-oxide; the enzyme activity was notably high with respect to benzo[a]pyrene-4,5-oxide in kidney. Generally the lowest glutathione S-transferase activities were obtained with octene-1,2-oxide in each of the organs examined. The relatively low ability of the little skate to convert octene-1,2-oxide to the glutathione derivative is of interest: 1-Hexadecene is converted to the carcinogenic 1,2-epoxyhexadecane in the liver of the rat (Watable and Yamada 1975) which has a significant capacity (in comparison to the skate) to degrade epoxides via an active glutathione-S-transferase system.

The question of whether glutathione-S-transferase activity in marine fish is induced by aromatic hydrocarbons was considered (Bend et al. 1977). No evidence was found to suggest that it was induced in sheepshead and Atlantic stingray by 3-methylcholanthrene when styrene-7,8-oxide or benzo[a]pyrene-4,5-oxide were employed as substrates. Thus in relation to previous discussion of the induction of epoxide hydrase, it would appear that neither the hydrase enzyme nor the glutathione-S-transferase is appreciably affected by 3-methylcholanthrene.

Studies of the conjugating enzymes of marine organisms, although few in number, suggest that wide variations exist among individual animals as well as among different species. Further work is desirable to provide an understanding of the roles played by these enzyme systems in the bioconversion of potentially damaging structures, such as epoxides, which are formed in initial steps of aromatic hydrocarbon metabolism.

Formation and Structure of Metabolites

Concern about enzyme systems that govern the metabolism of aromatic hydrocarbons leads naturally to an interest in the formation and structure of individual metabolic products. Studies of this type have been conducted with limited numbers of marine species. A detailed discussion of the findings was included in previous reviews (Varanasi and Malins 1977, Malins 1977). *M. squinado* (Corner et al. 1973) is able to convert dietary naphthalene to 1,2-dihydro-1,2-dihydroxy naphthalene and its glycoside derivative, 1-naphthyl sulfate, 1-naphthyl glucoside, and 1-naphthyl mercapturic acid. Moreover, when *M. squinado* was dosed with 1-naphthol the urine contained 1-naphthyl glucoside and 1-naphthyl sulfate, which implies that this animal is capable of performing conjugating reactions associated with the excretion of aromatic metabolites. Although the hepatopancreas of *M. squinado* contains a considerable amount of β-glucuronidase (Corner et al. 1960), no evidence was found for glucuronide derivatives in this organ when animals were fed radioactively-labeled naphthalene.

Euphausiids, amphipods, and crab zoea were capable of converting radioactively-labeled benzpyrene, methylcholanthrene, and naphthalene to hydroxy derivatives when these hydrocarbons were presented in seawater (Lee 1975). A similar capability was not found with ctenophores or jellyfish. The amphipod (*Parathemisto pacifica*) most rapidly degraded the hydrocarbons: in 24 hours, naphthol, hydroxybenzpyrene, and hydroxy methylcholanthrene were found and identified by thin-layer chromatography (TLC). In experiments with blue crab (*C. sapidus*) radioactively-labeled benzpyrene, fluorene, naphthalene, and methynaphthalene were converted to dihydroxy and conjugated derivatives (Lee et al. 1977). Further, Sanborn and Malins (1977) showed that Stage V spot shrimp (*Pandalus platyceros*) were able to metabolize naphthalene when exposed to this hydrocarbon in flowing seawater.

Studies with coho salmon (Roubal et al. 1977) showed that radioactively-labeled benzene, naphthalene, and anthracene were readily metabolized to derivatives located in key tissues when the hydrocarbons were administered via intraperitoneal injection. Generally, the hydrocarbons themselves (identified by TLC) were retained in the order anthracene > naphthalene > benzene. The retention of aromatic hydrocarbons in salmonids may increase in relation to the number of benzenoid rings and possibly decrease in relation to increased water solubility (Roubal et al. 1977, Malins 1977). It remains to be seen whether these relations between retention of aromatic hydrocarbons in the body and ring structure generally apply to compounds having more than three fused rings.

Individual conjugated and non-conjugated metabolites formed in juvenile coho salmon receiving intraperitoneal injections of naphthalene-1-^{14}C were isolated and identified in brain, liver, heart, gall bladder and muscle (Roubal et al. 1977) (Table 6). Separations

TABLE 6. Distribution of Naphthalene Metabolites in Coho Salmon 24 Hours After Injection of 125.6 µg Naphthalene-1-^{14}C

	Concentration in dry tissue (ng/g)				
	Brain	Liver	Heart	Gall Bladder	Muscle
1-Naphthol	29	156	46	537	34
1,2-Dihydro-1,2-dihydroxy-naphthalene	<10	48	21	604	<10
1-Naphthyl sulfate/glycoside	24	12	<10	238	29
1-Naphthyl mercapturic acid	20	172	<10	1,380	<10
1-Naphthyl-β-glucuronic acid					

[From Roubal et al. 1977]

were obtained between naphthalene, 1-naphthol, 1,2-dihydro-1,2-dihydroxy naphthalene, a combined glycoside and/or sulfate fraction, 1-naphthyl mercapturic acid, and 1-naphthyl glucuronic acid by thin-layer chromatographic techniques using marker compounds. Eluted metabolites were then assayed for radioactivity. The findings of Roubal et al. (1977) demonstrated that a variety of individual metabolites of naphthalene can enter key tissues of juvenile salmonids and remain there for some time; however, it cannot be assumed from this study that similar metabolite profiles would be obtained via other routes of hydrocarbon administration, such as the diet or seawater. Nevertheless, aromatic hydrocarbons are readily absorbed across the intestinal epithelium of coho salmon: recent studies by Collier and Malins (1977) showed that orally administered naphthalene-1-^{14}C (given as a single dose in salmon oil) was absorbed into key tissues (excluding the intestinal tract) of coho salmon to the extent of about 15% of the administered dose in 16 hours. Bioconversion of naphthalene was indicated, suggesting that hydrocarbons in the diet result in metabolite accumulations in tissues. In 8 to 48 hours, gall bladder, kidney, liver, and muscle were shown to contain products of naphthalene metabolism.

No detailed information appears in the literature on the isolation and/or identification of arene oxides in marine organisms exposed to aromatic hydrocarbons. This is not surprising because the isolation and identification of such compounds in mammals has only been accomplished in recent years. Yet the question of whether arene oxides exist in key tissues of marine organisms as a result of exposure to aromatic hydrocarbons remains an intriguing and important question, particularly with regard to animals having relatively low epoxide hydrase and glutathione-S-transferase activities. Clearly, more work is needed to identify metabolite formation in marine organisms exposed to aromatic hydrocarbons.

Aromatic hydrocarbons readily bind with various biological substrates: Naphthalene will form a stable covalent complex with bovine serum albumin (Sahyun 1964) and interact with macromolecules, such as DNA (Brookes and Lawley 1964). Covalent complexes between aromatic hydrocarbons and other structures can be predicted to exist in marine waters.

The uptake of complexed and free naphthalene in seawater was studied with Stage V larvae of the spot shrimp (*P. platyceros*) (Sanborn and Malins 1977). Substantially different uptake profiles were obtained for naphthalene and naphthalene complexed with bovine serum albumin (BSA) when both chemical forms were presented at 8-12 ppb as naphthalene in flowing seawater. That is, substantially more free naphthalene was sequestered throughout the experiment (24 hours) than was naphthalene complexed with BSA. This difference is consistent with the view that decreased lipid solubility of aromatic hydrocarbons diminishes absorption and transfer of such molecules across the intestinal epithelium; however, molecular size differences and other factors may also be influential. The results with the shrimp larvae raise intriguing questions about the nature and degree of covalent binding between aromatic hydrocarbons and macromolecules, such as proteins, in marine environments. Also intriguing are questions of how such interactions alter the accumulation and bioconversion of aromatic hydrocarbons in marine forms.

Prospectus

Virtually all the information available on the bioconversion of petroleum in marine organisms is based on the hydrocarbons *per se*. A need now exists to understand more about the chemical nature of weathered petroleum and bioconversion in marine organisms of the chemically or biologically transformed products. Moreover, it should not be assumed that the low concentrations of petroleum structures in oceans occur only as free or uncomplexed molecules. Thus, it seems important to study possible physico-chemical interactions of hydrocarbons and weathered products with macromolecules (e.g., proteins) and evaluate the metabolic fate of such structures on marine life.

It is likely that a host of environmental chemicals (e.g., chlorinated hydrocarbons and metals) influence the enzyme systems responsible for the metabolism of aromatic hydrocarbons. These events would be expected to alter the retention and excretion of hydrocarbons and their metabolites in tissues of marine organisms. Such possibilities should be explored in future work.

In the quest for a better understanding of the biological effects of petroleum on marine organisms, very little has been accomplished in evaluating the extent to which hydrocarbons and their metabolites are transported through the food web or the impact of metabolic products on such fluxes. These matters seem to warrant more consideration.

Further studies in the areas described above seem necessary to resolve many of the questions relating to the bioconversion of petroleum in marine environments and the concomitant effects on life processes.

References

Arcos, J. C. and M. F. Argus, Chemical Induction of Cancer. Structural Bases and Biological Mechanisms, 360 p., Academic Press, New York, 1974.

Argus, M. F. and J. C. Arcos, Hydrocarbon-nitrosamine synergism as a possible amplifying factor in lung tumorigenesis by tobacco smoke, J. Theor. Biol. 56, 491-498 (1976).

Bend, J. R. and J. R. Fouts, Glutathione-S-aryltransferase: Distribution in several marine species and partial characterization in hepatic soluble fractions from little skate, Raja erinacea liver, Bull. Mt. Desert Island Biol. Lab. 13, 4-8 (1973).

Bend, J. R. M. O. James, and P. M. Dansett, In vitro metabolism of xenobiotics in some marine animals, Ann. N.Y. Acad. Sci., In press, 1977.

Brookes, P. and P. D. Lawley, Evidence for the binding of polynuclear aromatic hydrocarbons to the nucleic acids of mouse skin: Relation between carcinogenic power of hydrocarbons and their binding to deoxyribonucleic acid, Nature 202, 781-784 (1964).

Burns, K. A. and D. Sabo, Environmental contamination and the induction of microsomal mixed function oxidases in an estuarine fish, Fundulus heteroclitus, Fed. Proc. Abstr. 34(3), 810 (1975).

Burwood, R. and G. C. Speers, Photo-oxidation as a factor in the environmental dispersal of crude oil, Estuarine and Coastal Mar. Sci. 2, 117-135 (1974).

Collier, T. K. and D. C. Malins, Unpublished data (1977).

Coon, J. M., H. W. Strobel, and R. F. Boyer, On the mechanism of hydroxylation reactions catalyzed by cytochrome P-450, p. 92-97, IN: R. W. Estabrook, J. R. Gillette, and K. C. Leibman (eds.) Microsomes and Drug Oxidations, Williams and Wilkins, Baltimore, 1973.

Corner, E. D. S., C. C. Kilvington, and S. C. M. O'Hara, Qualitative studies on the metabolism of naphthalene in Maia squinado (Herbst), J. Mar. Biol. Assoc. U.K. 53, 819-832 (1973).

Corner, E. D. S., Y. A. Leon, and R. D. Bulbrook, Steroid sulphatase, aryl sulphatase and β-glucuronidase in marine invertebrates, J. Mar. Biol. Assoc. U.K. 39, 51-61 (1960).

Dewaide, J. H., Species differences in hepatic drug oxidation in mammals and fishes in relation to thermal acclimation, Comp. Gen. Pharmacol. 1, 375-384 (1970).

Grover, P. L. and P. Sims, Enzyme-catalyzed reactions of polycyclic hydrocarbons with deoxyribonucleic acid and protein in vitro, Biochem. J. 110, 159-160 (1968).

Gruger, E. H., Jr., M. M. Wekell, and P. A. Robisch, Effects of chlorinated biphenyl and petroleum hydrocarbons on the activity of aryl hydrocarbon hydroxylase of coho salmon (Oncorhyunchus kisutch) and chinook salmon (O. tshawytscha), In: Proceedings Symposium on Fate and Effects of Petroleum Hydrocarbons in Marine Ecosystems and Organisms, Pergamon Press, New York, 1977a.

Gruger, E. H., Jr., M. M. Wekell, P. T. Numoto, and D. R. Craddock, Induction of hepatic aryl hydrocarbon hydroxylase in salmon exposed to petroleum dissolved in seawater and to petroleum and polychlorinated biphenyls, separate and together, in food, Bull. Environ. Contam. Toxicol., In press (1977b).

Hodgins, H. H., B. C. McCain, and J. W. Hawkes, Pathology, In: D. C. Malins (ed.) Effects of Petroleum on Arctic and Subarctic Marine Environments and Organisms, Academic Press, New York, In press, 1977.

James, M. O., J. R. Fouts, and J. R. Bend, In vitro epoxide metabolism in some marine species, Bull. Mt. Desert Island Biol. Lab. 14, 41-46 (1974).

James, M. O., J. R. Fouts, and J. R. Bend, Xenobiotic metabolizing enzymes in marine fish, In: Pesticides in the Aquatic Environment: A Symposium at the XVth Internation Congress of the Entomology Society, August 1976, Washington, D. C., In press, 1977.

Karrick, N. L., Alterations in petroleum resulting from physico-chemical and microbiological factors, In: D. C. Malins (ed.) Effects of Petroleum on Arctic and Subarctic Marine Environments and Organisms, Academic Press, New York, In press, 1977.

Lee, R. F., R. Sauerheber, and A. A. Benson, Petroleum hydrocarbons: Uptake and discharge by the marine mussel Mytilus edulis, Science 177, 344-346 (1972a).

Lee, R. F., R. Sauerheber, and G. H. Dobbs, Uptake, metabolism and discharge of polycyclic aromatic hydrocarbons by marine fish, Mar. Biol. 17, 201-208 (1972b).

Lee, R. F., Fate of petroleum hydrocarbons in marine zooplankton, p. 549-553, In: Proceedings 1975 Conference on Prevention and Control of Oil Pollution, American Petroleum Institute, Washington, D.C., 1975.

Lee, R. F., C. Y. Ryan, and M. L. Neuhauser, Fate of petroleum hydrocarbons taken up from food and water by the blue crab, Callinectes sapidus, Mar. Biol., In press (1977).

Litterest, C. L. and E. J. Van Loon, Time-course of induction of microsomal enzymes following treatment with polychlorinated biphenyl, Bull. Environ. Contam. Toxicol. 11, 206-212 (1974).

Lidman, U., L. Forlin, O. Molander, and G. Axelson, Induction of the drug metabolizing system in rainbow trout (Salmo gairdneri) liver by polychlorinated biphenyls (PCBs), Acta Pharmacol. Toxicol. 39, 262-272 (1976).

Malaveille, C., H. Bartsch, P. L. Grover, and P. Sims, Mutagenicity of non-K-region diols and diol-epoxides of benz(a)anthracene and benzo(a)pyrene, Biochem. Biophys. Res. Commun. 66, 693-700 (1975).

Malins, D. C., Metabolism of aromatic hydrocarbons in marine organisms, Ann. N.Y. Acad. Sci., In press (1977).

Notten, W. R. F. and P. T. Henderson, Action of n-alkanes on drug-metabolizing enzymes from guinea-pig liver, Biochem. Pharmacol. 24, 1093-1097 (1975).

Payne, J. F., Personal communication (1976a).

Payne, J. F., Benzo(a)pyrene hydroxylase as a biochemical barometer for petroleum in the marine environment, Proc. Can. Fed. Biol. Soc. 19, Abstr. 404 (1976b).

Pederson, M. G., W. K. Hershberg, P. K. Zachariah, and M. R. Juchau, Hepatic biotransformation of environmental xenobiotics in six strains of rainbow trout (Salmo gairdneri), J. Fish. Res. Board Can. 33, 666-675 (1976).

Pohl, R. J., J. R. Bend, A. M. Guarino, and J. R. Fouts, Hepatic microsomal mixed-function oxidase activity of several marine species from coastal Maine, Drug Metab. Disposition 2(6), 545-555 (1974).

Pohl, R. J., J. R. Fouts, and J. R. Bend, Response of hepatic microsomal mixed-function oxidases in the little skate, Raja erinacea, and the winter flounder, Pseudopleuronectes americanus to pretreatment with TCDD (2,3,7,8-tetrachlorobibenzo-p-dioxin) or DBA (1,2,3,4-dibenzanthracene), Bull. Mt. Desert Island Biol. Lab., In press (1977).

Roubal, W. T., T. K. Collier, and D. C. Malins, Accumulation and metabolism of carbon-14 labeled benzene, naphthalene and anthracene by young coho salmon, Arch. Environ. Contam. Toxicol., In press (1977).

Sahyun, M. R. V., Enthalpy of binding of aromatic amines to bovine serum albumin, Nature 203, 1045 (1964).

Sanborn, H. R. and D. C. Malins, Toxicity and metabolism of naphthalene: A study with marine larval invertebrates, Proc. Soc. Exp. Biol. Med. 154, 151-155 (1977).

Sims, P. and P. L. Grover, Epoxides in polycyclic aromatic hydrocarbon metabolism and carcinogenesis, Vol. 20, p. 165-274, In: D. D. Klein, S. Weinhouse, and A. Haddow (eds.) Advances in Cancer Research, Academic Press, New York, 1974.

Swaisland, A. J., P. L. Grover, and P. Sims, Biochem. Pharmacol. 22, 1547-1556 (1973). Cited Cited in: Sims, P. and P. L. Grover, Epoxides in polycyclic aromatic hydrocarbon metabolism and carcinogenesis, Vol. 20, p. 165-274, In: D. D. Klein, S. Weinhouse, and A. Haddow (eds.) Advances in Cancer Research, Academic Press, New York, 1974.

Varanasi, U. and D. C. Malins, Metabolism of petroleum hydrocarbons: Accumulation and biotransformation in marine organisms, In: D. C. Malins (ed.) Effects of Petroleum on Arctic and Subarctic Marine Environments and Organisms, Academic Press, New York, In Press, 1977.

Watable, T. and N. Yamada, The biotransformation of 1-hexadecene to carcinogenic 1,2-epoxyhexadecane by hepatic microsomes, Biochem. Pharmacol. 24, 1051-1053 (1975).

Weinstein, I. B., A. M. Jeffrey, K. W. Jennette, S. H. Blobstein, R. G. Harvey, C. Harris, H. Autrup, H. Kazai, and K. Nakanishi, Benzo(a)pyrene diol epoxides as intermediates in nucleic acid binding in vitro, Science 197, 392-595 (1976).

Wellings, S. R., Neoplasia and primitive vertebrate phylogeny: Echinoderms, prevertebrates, and fishes--A review, Natl. Cancer Inst. Monogr. 31, 59-128 (1969).

White, A., P. Handler, and E. M. Smith, Principles of Biochemistry, 5th Ed., 1202 p., McGraw-Hill, Inc., New York, 1973.

Yang, C. S. and F. S. Strickhart, Effects of some epoxides on aryl hydrocarbon hydroxylase activity, Biochem. Pharmacol., 646-648 (1975).

Yoshida, T., Y. Ito, and Y. Suzuki, Inhibition of hepatic drug metabolizing enzyme by cadmium in mice, Bull. Environ. Contam. Toxicol. 15, 402-405 (1976).

CHAPTER 6

ACCUMULATION AND TURNOVER OF PETROLEUM HYDROCARBONS IN MARINE ORGANISMS

Richard F. Lee

Skidaway Institute of Oceanography
P. O. Box 13687
Savannah, Georgia 31406

Introduction

Marine organisms which accumulate petroleum hydrocarbons after exposure to various petroleum components are of interest because of implications to the marine environment and to human health. This review deals with the uptake, storage and discharge of petroleum hydrocarbons by marine organisms under laboratory and field conditions. Analyses of organisms collected from oil spill and chronically polluted areas are reviewed under the heading of "Bioaccumulation". The section entitled "Biological Transfer Processes" deals with the ability of animals to depurate their hydrocarbons accumulated after exposure to oil. The groups of marine organisms included are benthic algae, zooplankton, benthic invertebrates and fish. Bivalves, including mussels, clams and oysters, have received special emphasis because of the extensive amount of laboratory and field studies on the accumulation of petroleum by these mollusks.

Bioaccumulation

Marine organisms which contain petroleum hydrocarbons have generally either been exposed to a single oil spill or to an area of chronic exposure. The amounts of petroleum hydrocarbons in marine organisms collected from oceanic, chronically polluted coastal areas, polluted harbors and single oil spill have been summarized in the National Academy report (Anonymous, 1975) and in the review of Anderson et al (1974). A summary of the accumulation of petroleum hydrocarbons by marine bivalves taken from contaminated areas, either from spills or chronic pollutions, is provided in Table 1. In animals exposed to a single spill, either in the field or laboratory, attempts have been made to match the hydrocarbon composition of the animals with the composition of the petroleum product involved, e.g. fuel oil No. 2. In areas of chronic pollution, such a match is generally not possible because inputs are of a variety of different types of petroleum. Analysis of the hydrocarbon fraction from fish, oysters and crabs collected after an oil spill in Buzzards Bay, Massachusetts showed a pattern similar to the spilled fuel oil (Blumer et al., 1970a; Burns and Teal, 1971). Barnacles, crabs and benthic algae exposed to oil released during the grounding of a ship on the Washington coast also showed a hydrocarbon pattern similar to the spilled fuel oil (Clark et al., 1973).

Studies of the bioaccumulation of hydrocarbons by animals exposed to chronic oil pollution includes the following: petroleum hydrocarbons in oysters (Crassostrea virginica) from Galveston Bay, Texas, near the mouth of the Houston Ship Channel, a waterway with continual petroleum input (Ehrhardt, 1972), petroleum hydrocarbons in mussels (Mytilus edulis) from Kiel Fjord, Germany (Ehrhardt and Heinemann, 1975), petroleum hydrocarbons in clams (Mercenaria mercenaria) from Narragansett Bay, Rhode Island (Farrington and Quinn, 1973), benzo(a) pyrene in mussels (Mytilus edulis) from Vancouver Harbor, Canada (Dunn and Stich, 1975), paraffinic petroleum hydrocarbons in mussels (Mytilus galloprovincialis) from the Lagoon of Venice (Fossato and Siviero, 1974), petroleum hydrocarbons in mussels (Mytilus edulis) from San Francisco Bay (DiSalvo et al., 1975), kerosene tainted mullet (Mugil cephalus) from Brisbane, Australia (Shipton et al., 1970), petroleum tainted mullet (Mugil japanicus) from Mizushima Harbor, Japan (Ogata and Miyake, 1973).

In these studies, the content and pattern of the hydrocarbons in animals from nearby "clean" areas were compared to animals exposed to chronic oil pollution. Oysters collected at the mouth of the Houston Ship Channel, away from the mouth, and from the far end of Galveston Bay, had hydrocarbon concentrations of 237, 26 and 2 µg/g, respectively, (Ehrhardt, 1972; Anderson, R., 1975). Animals from other areas of low, but constant petroleum input do not always show a markedly higher total hydrocarbon content relative to animals from cleaner areas. In these cases, the petroleum hydrocarbons are differentiated from those of biogenic origin by methods such as gas-liquid chromatography, fluorescence spectroscopy, ultraviolet spectroscopy and mass spectrometry. Gas-liquid chromatography of biogenic hydrocarbons shows a series of well resolved peaks, generally composed of straight and branched chain alkanes. Chromatograms of the hydrocarbon fraction of organisms exposed to petroleum show a complex mixture often referred to as an unresolved envelope (Farrington and Quinn, 1973). If the oil had not been subject to degradation, then superimposed on this unresolved envelope are a series of peaks which are primarily paraffinic hydrocarbons. Ehrhardt and Heinemann (1975) used mass spectrometry in addition to gas-liquid chromatography to detect the presence of polycyclic aromatic hydrocarbons, such as fluorenes and phenanthrenes, in mussels from Kiel Fjord. They determined that mussels from these waters were concentrating aromatic hydrocarbons by a factor of 10,000. Another example of the use of polycyclic aromatic hydrocarbons as indicators of pollution, is the work of Dunn and Stich (1975) who found 42 µg/kg.

Fish collected near oil refineries can have a petroleum-like taint. Mullet collected near oil refineries in Australia had a series of hydrocarbons similar to those found in kerosene, including several alkylated naphthalenes (Shipton et al., 1970). Eel and mullet from an oil polluted harbor in Japan had an objectionable odor which was shown to be due to the presence of toluene in the fish (Ogata and Miyake, 1973). The ratio of the concentration of toluene in this fish to that in the surrounding water was 2.4:1 with levels in the mullet of up to 5 µg/g.

Biological Transfer Processes - Uptake, Storage and Discharge

The concentration of petroleum hydrocarbons in animals exposed to oil reflects the relative importance of the processes of uptake, storage and discharge. The uptake of petroleum from water, food or sediments had been shown to occur under both laboratory and field conditions. These conditions vary from large oil spills to low concentrations of a particular hydrocarbon in water, sediment or the food. Subsequent transfer of exposed animals to clean waters allows a determination of the rate and extent of the discharge processes. Depuration rates for laboratory exposed animals are generally greater than the rates for animals exposed to oil spills or chronically polluted areas. The longer period of exposure may allow entrance of hydrocarbons into stable tissue compartment, such as those with lipid stores. A summary of uptake and discharge of petroleum hydrcarbons by marine bivalves is provided in Tables 1 and 2.

Zooplankton

Copepods under oil spills have been shown to consume particles of oil which are subsequently excreted into the feces (Conover, 1971). Laboratory work with crustacean zooplankton, primarily copepods, has demonstrated that they can take up a variety of aromatic and paraffinic hydrocarbons from either food or water (Corner et al., 1976a, 1976b; Lee, 1975). The studies of Corner et al., (1976a) showed that for naphthalenes the dietary route of entry was more important than direct uptake from the water. Most of the aromatic hydrocarbons were lost during depuration experiments with half-lives of 2-3 days, but even a 28 day depuration period was not sufficient to discharge all hydrocarbon.

Sorting live zooplankton from a field sample is quite tedious because of the large number of organisms needed for analysis. A mixed zooplankton sample is not satisfactory for analysis because of the presence of suspended particles with adsorbed hydrocarbons and contamination by surface slicks during the collecting process (Mackie, et al., 1977).

Benthic Crustaceans

Various species of shrimp, crabs and lobsters rapidly take up petroleum hydrocarbons from either the water or their food (Anderson, 1975; Cox et al., 1976; Sanborn and Malins, 1976). Most of the hydrocarbons in the food were not assimilated by the tissues of the blue crab, Callinectes sapidus, but instead were immediately eliminated from the animal (Lee et al., 1976). The main site of hydrocarbon build-up in crabs was in the hepatopancreas. The metabolism of hydrocarbons by crabs facilitates their elimination through feces and urine (Corner et al., 1973; Lee et al., 1976).

After exposure to either a water extract of fuel oil or to various polycyclic aromatic hydrocarbons, shrimp and crabs were able to completely depurate the hydrocarbons after 2 to 10 days (Anderson, 1975; Lee et al., 1976; Sanborn and Malins, 1977).

Cox et al., (1975) added fuel oil No. 2 to a large pond in South Texas which contained shrimp, clams and oysters. After 38 days, the animals were transferred to clean water and 10 days of depuration were sufficient to discharge all naphthalenes from the shrimp. Clams and oysters still retained 4.0 and 2.0 µg/g of naphthalenes, respectively. Burns (1976) noted that fiddler crabs, Uca, still retained a fuel-oil like hydrocarbon pattern four years after an oil spill in the area. Possibly, the crabs took up oil from the contaminated sediments or oil released from the sediments. Blackman (1972) has observed that brown shrimp (Crangon crangon) consume sunken crude oil which remains in the foregut until the next molt.

TABLE 1. Amounts of Petroleum Hydrocarbon in Bivalves from Contaminated Areas

Species	Contamination source	Hydrocarbons analyzed	Hydrocarbon concentration in the animal ($\mu g/g$)	Half-life of hydrocarbons* (days)	Reference
oysters					
Crassostrea virginica	spill-fuel oil No.2	aliphatics & aromatics	70	-	Blumer et al., 1970b
	chronic	aliphatics & aromatics	236	-	Ehrhardt, 1972
mussels					
Mytilus edulis	spill-Bunker C	aromatics	77-103	-	Zitko, 1971
	spill-fuel oil No.2	n-paraffins	1.4	-	Clark & Finley, 1973
	chronic	n-paraffins	1.0	-	Clark & Finley, 1973
	spill	aliphatics & aromatics	400	2	DiSalvo et al., 1975
	chronic	aromatics	6-75	48-60	DiSalvo et al., 1975
	chronic	aromatics	8	-	Ehrhardt & Heineman, 1975
	chronic	aliphatics	250	4	Fossato, 1975
	chronic	benzo(a)pyrene	0.05	16	Dunn & Stich, 1976
clams					
Mercenaria mercenaria	sewage effluent	C_{16-32}	4-69	-	Farrington & Quinn, 1973
Mya arenaria	spill-fuel oil No.2	aliphatics & aromatics	26	-	Blumer et al., 1970a
scallops					
Aequipecten irradians (muscle)	spill-fuel oil No.2	aliphatics & aromatics	7-14	-	Blumer et al., 1970a

*Determined by transferring animals to clean water

TABLE 2. Uptake and Loss of Petroleum Hydrocarbons by Bivalves under Laboratory Conditions

Species	Petroleum source	Hydrocarbons Analyzed	Hydrocarbon concn in water* (μg/liter)	Exposure time (days)	Maximum hydrocarbon concn in animal (μg/g)	Hydrocarbons half-life (days)	Depuration period (days)	Depuration experiments Hydrocarbon concn in animals (μg/g)	Reference
oysters Crassostrea virginica	fuel oil No.2	aliphatics & aromatics	106	50	334	5	30	30	Stegeman & Teal, 1973
	mixture of aromatics	benz(a)anthracene	3.8	9	1.6	7	14	0.4	Table 3 of this report
		naphthalenes	77	9	52	2	14	1.2	Table 3 of this report
	fuel oil No.2	diaromatics	n.d.	4	412	5	13	2.2	R. D. Anderson, 1975
mussels Mytilus edulis	fuel oil No.2	n-paraffins	n.d.	2	110	4	35	8	Clark & Finley, 1975
	diesel fuel	aliphatics	200-400	41	400	3	35	50	Fossato & Canzonier, 1975
clams Rangia cuneata	fuel oil No.2	naphthalene	300	1	10	2	9	0.2	Anderson, 1975
		dimethylnaphthalene	200	1	1.2	3	9	0.3	Anderson, 1975
	benzo(a)pyrene	benzo(a)pyrene	30	1	7.2	2-5	20	0.1	Neff & Anderson, 1975a

*n.d., not determined

Benthic Worms

Deposit feeding worms take up organic pollutants associated with the sediments. Polychaetes, particularly <u>Capitella capitata</u>, are associated with areas of high oil input (Reish, 1971; Sanders <u>et al</u>., 1972). As a consequence, the polychaetes and quite possibly worms belonging to other phyla, have evolved enzyme systems which metabolize petroleum hydrocarbons, including aryl hydrocarbon hydroxylase (Lee, 1976; Lee <u>et al</u>., 1977). Presumably, hydrocarbon metabolism facilitates the rapid discharge of hydrocarbons observed for various species of polychaetes. Rossi and Anderson (1977) showed that the polychaete, <u>Neanthes</u>, accumulated 14 µg/g of total naphthalenes in 1 day from a fuel oil water extract containing 600 µg/liter of naphthalenes. After 12 days of depuration, naphthalenes were below detection limits (below 0.05 µg/g). Naphthalenes which entered from the sediment or food did not pass across the gut and were voided with feces. Exposure of the polychaete worms, <u>Nereis virens</u> and <u>Nereis succinea</u>, to benz(a)anthracene in sediment showed uptake into tissues and transfer to clean sediment resulted in complete depuration after 24 days (Lee and Furlong, unpublished data). Anderson <u>et al</u>., (1977) have examined uptake of Prudhoe oil by the sipunculid worm, <u>Phascolosoma agassizii</u>. Exposure to a water extract of Prudhoe crude oil for 1 day resulted in hydrocarbon uptake and a 14 day depuration resulted in a loss of all naphthalenes.

Bivalves

Bivalves, which filter large volumes of water while feeding, can take up and concentrate hydrocarbons from the water, either from solution or adsorbed to suspended particles (Anderson, 1975; Boehm and Quinn, 1976; DiSalvo, 1975; DiSalvo and Guard, 1976; Farrington and Quinn, 1973; Fossato and Canzonier, 1976; Lee <u>et al</u>., 1972a; Neff <u>et al</u>., 1976). Gill tissues of bivalves have a micellar layer on their surfaces which are responsible for the absorption of hydrophobic compounds, such as hydrocarbons (Pasteels, 1968). Laboratory experiments have demonstrated that bivalves can concentrate hydrocarbons several orders of magnitude over their concentration in the water (Table 2). Exposure of mussels to fuel oil at 200-400 µg/liter resulted in petroleum hydrocarbon concentrations of 350 µg/g (Fossato and Canzonier, 1976). Oysters exposed to fuel oil at 106 µg/liter for 50 days resulted in animals with concentrations of 335 µg/g (Stegeman and Teal, 1973). Ehrhardt (1972) found that oysters in the mouth of the Houston Ship Channel contained 237 µg/g of petroleum hydrocarbons. The maximum concentration of petroleum hydrocarbons in bivalves exposed to oil, under laboratory and field conditions, was between 300 and 400 µg/g (Tables 1 and 2). Clams, oysters and mussels differ in their rates of hydrocarbon uptake, possibly due to differences in filtering rates and amounts of lipids (Clark and Finley, 1874; Neff <u>et al</u>., 1976). Stegeman and Teal (1973) have found that oysters with high lipid content took up more fuel oil (314 µg/g) from the water than low lipid oysters (161 µg/g). Hydrocarbons because of their low water solubility appear to be retained in tissues with lipid stores.

Many of the uptake studies with marine bivalves have also involved depuration experiments. Calculations of the half-life of petroleum hydrocarbons as a result of discharge allows comparison of the results of different laboratories (Tables 1 and 2). Heavily contaminated mussels (250 µg/g) from the Lagoon of Venice released half of the hydrocarbons in 3.5 days when transferred to clean waters. However, 12% of the hydrocarbon persisted for 8 weeks (Fossato and Canzonier, 1976). Dunn and Stich (1976) allowed benzo(a)pyrene-contaminated mussels (45 µg/kg) to depurate in cleaner waters of the Vancouver Harbor and determined that this compound had a half-life of 16 days. DiSalvo <u>et al</u>., (1975) found a half-life of 48 to 60 days for the release of aromatic hydrocarbons in oil contaminated mussels from San Francisco Bay, California.

Petroleum hydrocarbons accumulated by bivalves maintained under laboratory conditions generally have a short half-life (Table 2). This may be due to the high concentrations of petroleum used and the shorter exposure period. Fossato and Canonier (1976) exposed mussels for 41 days to 200-400 µg/liter of diesel fuel on clay particles and determined a half-life of approximately 3 days. Lee <u>et al</u>., (1972) found that naphthalene accumulated by mussels for 1 day had a half-life of 1 day. Clams exposed to benzo(a)pyrene for 1 day rapidly released the compound after transfer into clean water with a half-life of 2 to 5 days (Neff and Anderson, 1975a). Oysters exposed to fuel oil for 4 days had 11.8 µg/g of mono and di- aromatic hydrocarbons and depuration experiments showed a half-life of 5 days (Anderson, R. D., 1975). Clark and Finley (1975) exposed mussels to fuel oil and after depuration, the paraffinic hydrocarbons had a half-life of 4 days. Oysters exposed to fuel oil for 50 days accumulated 325 µg/g of petroleum hydrocarbon and a 14 day depuration period decreased this to 40 g/g giving a half-life of 5 days (Stegeman and Teal, 1973).

The results of allowing oysters to take up and depurate benz(a)anthracene and naphthalenes (naphthalene, methylnaphthalene and dimethylnaphthalene) from a controlled ecosystem enclosure (ca. 2m diameter and 14m deep - 60,000 liters) are shown in Table 3(Lee, Anderson and Barswell-Clarke, unpublished data). After uptake of benz(a)anthracene and naphthalenes for 2 days followed by 14 days of depuration, the calculated half-lives of benz(a)anthracene and naphthalenes were 4 and 2 days, respectively. When uptake was extended to 4 days with 14 days of depuration, the half-lives, assuming exponential discharge of benz(a)anthracene and naphthalenes, were 7 and 2 days, respectively. Thus, a longer exposure period resulted in benz(a)anthracene being depurated at a slower rate.

Most depuration studies, whether field or laboratory, show an initial rapid discharge which results in calculated short half-lives for accumulated petroleum hydrocarbons. However, there is a small concentration of petroleum hydrocarbons which is retained for a long period after the initial rapid discharge (Clark and Finley, 1975; Lee et al., 1972a; Stegeman, 1974). Mussels from the oil-polluted Lagoon of Venice retained 30 µg/g of petroleum hydrocarbons after 56 days of depuration (Fossato and Canzonier, 1976), presumably held in tissues containing lipid stores. Paraffins are discharged at a faster rate than aromatics so that the latter are retained during long depuration experiments (Neff, et al., 1976; Stegeman, 1974). Neff et al., (1976) were not able to detect paraffins in clams after a 4 day fuel oil exposure and a 13 day depuration period. The naphthalenes were still 2.2 µg/g and their complete removal required 28 days of depuration. Mussels exposed to fuel oil still retained paraffins after 35 days of depuration so aromatic hydrocarbons are presumed to persist in these animals for very long periods (Clark and Finley, 1975).

Bivalves in oil spill areas generally depurate at slow rates which is due, in part, to continual input of oil from the sediment. Thus, shellfish from the oil spill area in Buzzards Bay, continued to show a fuel oil hydrocarbon pattern for several years after the spill (Blumer et al., 1970; Blumer and Sass, 1972; Teal and Farrington, 1976). In contrast, one year after an oil spill of the Washington coast, paraffinic petroleum hydrocarbons were not detected in shellfish of the area (Clark, et al., 1973). Our own work involved a study of mussels exposed to a small fuel oil spill (1,000 gallons) off the California coast (Lee and Benson, unpublished data). Hydrocarbons from the mussels were examined on a monthly basis until the spill, followed by analysis every 3 days for 8 weeks. Before the spill, the hydrocarbon pattern of the mussels showed only biogenic hydrocarbons which were several straight and branched chain alkanes. Two days after the spill, chromatograms of the mussel hydrocarbons showed a series of paraffin peaks superimposed on a large, unresolved envelope. In addition, the hydrocarbon fraction had absorption peaks between 210 and 360 nm suggesting the presence of polynuclear aromatics. The amount of petroleum hydrocarbon taken up was small since no increase in the total hydrocarbon content of the mussels was noted. The mussels showed evidence of petroleum hydrocarbons for 5 weeks although visible evidence of the spill was gone after 1 week. After 6 weeks, no petroleum hydrocarbons could be detected in the mussels. Thus, the persistence of petroleum hydrocarbon in bivalves after exposure to oil may depend on such factors as the size of the spill, type of oil and the amount of open ocean water circulating into the exposed area. Various species of bivalves also show differing rates of depuration. This may be due to different water filtration rates and/or to differences in rates of hydrocarbon uptake and discharge. Other workers have discussed the role of temperature, season as it relates to the physiological state of the animal (e.g. lipid content and phase of reproductive cycle) and the availability of food in hydrocarbon uptake (DiSalvo et al., 1975; Fossato and Canzonier, 1976; Stegeman, 1974; Stegeman and Teal, 1973).

Fish
Fish are assumed to be able to avoid or leave oil-polluted areas, unless behavioral patterns restrict them to a small territory. In addition, the processes of hydrocarbon metabolism and discharge result in a high turnover of accumulated petroleum hydrocarbons. Hydrocarbons can enter fish through the water or from their food. Entrance from the water is primarily through the gills, but some oil, including tar particles, can enter during drinking or feeding. The latter process would explain the presence of tar in the stomach of fish collected in the Mediterranean Sea (Horn et al. (1970). Plaice fed food-contained crude oil showed assimilation of the n-alkanes into the liver tissues, but not into the muscle (Hardy et al., 1974). The alkanes longer than C_{28} were retained while short chain alkanes were lost. Whittle et al. (1977) present evidence that retention of petroleum depends on whether uptake is through the gut or gill. When uptake is from the water, entrance is through the gills with later buildup in the liver and gall bladder (Lee et al., 1972b). Hydrocarbons entering through the food are found, after a few days, primarily in the stomach, liver and gall bladder (Corner et al., 1976b). In addition to buildup in these tissues, Anderson (1975) noted a buildup of hydrocarbons in the brain of the killifish, Fundulus similus. Urine or feces are the avenues of discharge for hydrocarbons and their metabolites. Depuration periods of 7 to 14 days generally result in discharge of hydrocarbons from all tissues.

The studies of Roubal (1977) showed that the three ringed compound, anthracene, was retained in fish longer than the naphthalene (two rings) and benzene (one ring). Uptake, distribution and discharge of benzene were studied in the anchovy and bass by Korn et al. (1976). Rapid uptake from the water occurred during a 2 day exposure with subsequent buildup in the liver and gall bladder. No benzene was detected in a fish after a 7 day depuration period.

TABLE 3. Uptake and Release of Naphthalenes and Benz(a)anthracene by Oysters in a Controlled Ecosystems Enclosure

To the top ten meters of a controlled ecosystem enclosure (ca. 2 m diameter and 15 m deep - 60,000 liters) were added 0.6 g of benz(a)anthracene and 1.5 g each of naphthalene, 1-methylnaphthalene and 2,3-dimethylnaphthalene. This would result in a nominal concentration of 110 µg/liter of total naphthalenes and 15 µg/liter of benz(a)anthracene. Oysters in wire cages were suspended at 12 meters in the enclosure. The cage had two compartments with one compartment surrounded by a 60 µm Nitex filter to filter out large particles. For depuration studies oysters were taken out after various periods of exposure and added to a second cage suspended in outside waters. Analysis of naphthalenes in water and oysters was by method of Neff and Anderson (1975b) and the results verified by gas-liquid chromatography. Benz(a)anthracene was determined by fluorescence spectroscopy of hexane extracts with excitation at 290 nm and emission at 387 nm.

Time of exposure (days)	Depuration time (days)	Cage with filter	Naphthalene oyster (µg/g)	Naphthalene water (µg/l)	Methylnaphthalene oyster (µg/g)	Methylnaphthalene water (µg/l)	Dimethylnaphthalene oyster (µg/g)	Dimethylnaphthalene water (µg/l)	Total naphthalenes oyster (µg/g)	Total naphthalenes water (µg/l)	Benz(a)anthracene oyster (µg/g)	Benz(a)anthracene water (µg/l)
2	-	yes	4.6	27	8.8	29	14.0	21	27.3	77	0.5	3.8
2	-	no	8.6	27	15.0	29	28.6	21	52.0	77	4.1	3.8
4	-	yes	0.9	17	1.6	16	2.5	11	5.1	34	1.1	1.1
4	-	no	2.7	17	5.7	16	8.8	11	17.1	34	1.7	1.1
9	-	yes	1.8	5	14.6	6	7.7	1	14.2	12	1.3	0.3
9	-	no	2.9	5	7.0	6	12.2	1	22.2	12	1.6	0.3
15	-	no	2.7	4	7.2	5	11.6	2	21.5	11	0.8	0.1
2	14	no	0.1	0	0.1	0	0.1	0	0.2	0	0.2	0
4	14	no	0.1	0	0.1	0	0.1	0	0.2	0	0.4	0
9	14	no	0.2	0	0.2	0	0.6	0	1.2	0	0.4	0

Use of Bivalves as Bioaccumulators in Petroleum Monitoring Protects

As discussed above, petroleum hydrocarbons are taken up by a variety of marine animals. Hydrocarbon metabolism and rapid discharge rates suggest that fish, worms or crustaceans would not be as useful as bivalves for monitoring petroleum input. In an oil spill area or in areas of chronic oil pollution, the hydrocarbon compositions of bivalves reflect the concentration and relative amounts of different hydrocarbons in the water. However, it should be noted that extensive depuration does occur in clean water producing a different hydrocarbon pattern as a result of removal of paraffinic and retention of aromatics. The advantages of using bivalves for bioaccumulation studies are: 1) lack of hydrocarbon metabolism in their tissues (Carlson, 1972; Lee et al., 1972a); 2) persistence of petroleum hydrocarbon after long depuration periods (Anderson, 1975, DiSalvo et al., 1972a; Rice et al., 1976); 3) ability to bioaccumulate hydrocarbons under both field and laboratory conditions (Anderson, 1975; Blumer et al., 1970; Clark and Finley, 1974; DiSalvo et al., 1975; Ehrhardt, 1972; Farrington and Quinn, 1973; Fossato and Siviero, 1974; Stegeman and Teal, 1973); 4) extensive knowledge of their biology and physiology (Harger, 1972; Jørgensen, 1966; Thompson and Bayne, 1972); 5) easy to maintain under laboratory conditions for long periods (Jørgensen, 1966; Thompson and Bayne, 1972); 6) world-wide distribution including oil polluted waters (Clark and Finley, 1973; Ehrhardt, 1972; Farrington and Quinn, 1973; Fossato and Siviero, 1974); 7) large size of animals and ease of tissue dissection.

Additional uses of monitoring studies of petroleum hydrocarbons with bivalves includes: 1) the ability to monitor carcinogenic hydrocarbon input since these hydrocarbons are difficult to analyze in the water because of their low concentration (Dunn and Stich, 1975); 2) allows discussion of biologically active hydrocarbons, i.e., hydrocarbons taken up by living organisms; 3) provides a reliable method of monitoring a petroleum clean up of an area by observing a decrease in the concentration of hydrocarbons in the animals.

Acknowledgements

Support provided by National Science Foundation, Office for International Decade for Ocean Exploration Grant OCE74-05283.

References

Anderson, J. W., Laboratory studies on the effects of oil on marine organisms: an overview, American Petroleum Institute Publication, No. 4349, American Petroleum Institute, Washington, D.C., 1975.

Anderson, J. W., R. C. Clark, and J. J. Stegeman, Petroleum hydrocarbons, pp. 36-38. G. Cox (ed.). In: Marine Bioassays Workshop Proceedings, Marine Technology Soc., Washington, D.C., 1974.

Anderson, J. W., L. J. Moore, J. W. Blaylock, D. L. Woodriff and S. L. Kiesser, Bioavailability of sediment-sorbed naphthalenes to the sipunculid worm, Phascolosoma agassizii In. D. Wolfe (ed.) Symposium on Fates and Effects of Petroleum Hydrocarbons in Marine Ecosystems and Organisms, Pergamon Press, New York, 1977. In Press.

Anderson, R. D., Petroleum hydrocarbons and oysters resources of Galveston Bay, Texas, pp. 541-548. In: Proceedings of the Conference on Prevention and Control of Oil Pollution. American Petroleum Institute, Washington, D.C., 1975.

Anonymous, Petroleum in the Marine Environment. National Academy of Science, Washington, D.C., 1975.

Blackman, R. A. A., Effects of sunken crude oil on the feeding and survival of the brown shrimp (Crangon crangon), International Council for the Exploration of the Sea, Publication C. M., 1972, K:13 (1972).

Blumer, M. and J. Sass, The West Falmouth oil spill, Reference No. 72-19, Woods Hole Oceanographic Institute, Woods Hole, Massachusetts (1972).

Blumer, M., J. Sass, G. Souza, H. Sanders, F. Grassle and G. Hampson, The West Falmouth oil spill, Reference No. 70-44, Woods Hole Oceanographic Institution, Woods Hole, Massachusetts (1970a).

Blumer, M., G. Souza, and J. Sass, Hydrocarbon pollution of edible shellfish by an oil spill, Mar. Biol., 5, 195-202, (1970b).

Boehm, P. D. and J. G. Quinn, The effect of dissolved organic matter in seawater on the uptake of mixed individual hydrocarbons and No. 2 fuel oil by a marine filter-feeding bivalve (Mercenaria mercenaria), Estuarine and Coastal Marine Sci., 4, 93-105 (1976).

Burns, K. A., Hydrocarbon metabolism in the intertidal fiddler crab, Uca pugnax, Mar. Biol. 36, 5-11, (1976).

Burns, K. A. and J. M. Teal, Hydrocarbon incorporation into the salt marsh ecosystem from the West Falmouth oil spill, Reference No. 71-69, Woods Hole Oceanographic Institution Woods Hole, Massachusetts, (1971).

Carlson, G. P., Detoxification of foreign compounds by the quahaug, Mercenaria mercenaria, Comp. Biochem. Physiol., 43B, 295-302 (1972).

Clark, R. C. and J. S. Finley, Paraffin hydrocarbon pattern in petroleum-polluted mussels, Mar. Pollut. Bull. 4, 172-176 (1973).

Clark, R. C. and J. S. Finley, Acute effects of outboard motor effluent on two marine shellfish, Environ. Sci. Technol., 8, 1009-1014 (1974).

Clark, R. C. and J. S. Finley, Uptake and loss of petroleum hydrocarbons by mussels, Mytilus edulis, in laboratory experiments, Fishery Bull., 73, 508-515 (1975).

Clark, R. C., J. S. Finley, B. G. Patten, D. F. Stefani, E. E. DeNike, Interagency investigations of a persistent oil spill on the Washington coast. Animal population studies, hydrocarbon uptake by marine ortanisms, and algae response following the grounding of the troopship General M. C. Meisa, pp. 793-808. In: Proceedings on the Joint Conference on Prevention and Control of Oil Spills. American Petroleum Institute, Washington, D. C., (1973).

Concover, R. J., Some relations between zooplankton and bunker C oil on Chedabucto Bay following the wreck of the tanker ARROW, J. Fish. Res. Bd. Can., 28, 1327-1330 (1971).

Corner, E. D. S., C. C. Kilvington and S. C. M. O'Hara, Qualitative studies on the metabolism of naphthalenes in Maia squinado (Herbst), J. Mar. Biol. Ass. U.K., 53, 819-832, (1973).

Corner, E. D. S., R. P. Harris, C. C. Kilvington and S. C. M. O'Hara, Petroleum compounds in the marine food web: short-term experiments on the fate of naphthalene in Calanus, J. Mar. Biol. Ass. U.K., 56, 121-133 (1976a).

Corner, E. D. S., R. P. Harris, K. J. Whittle, and P. R. Mackie, Hydrocarbons in marine zooplankton and fish, pp. 71-105. A. P. M. Lockwood (eds.). In: Effects of Pollutants on Aquatic Organisms, Cambridge University Press, London, 1976b.

Cox, B. A., J. W. Anderson and J. C. Parker, An experimental oil spill: the distribution of aromatic hydrocarbons in the water, sediment, and ainimal tissues within a shrimp pond, pp. 607-612. In: Proceedings of the Conference on Prevention and Control of Oil Pollution, American Petroleum Institute, Washington, D. C., 1975.

DiSalvo, L. H. and H. E. Guard, Hydrocarbons associated with suspended particulate matter in San Francisco Bay waters, pp. 169-173. In: Proceedings of the Conference on Prevention and Control of Oil Pollution, American Petroleum Institute, Washington, D. C., 1975.

DiSalvo, L. H., H. E. Guard, L. Hunter, Tissue hydrocarbon burden of mussels as potential monitor of environmental hydrocarbon insult, Environ. Sci. Tech., 9, 247-251 (1975).

Dunn, B. P. and H. F. Stich, The use of mussels in estimating benzo(a)pyrene contamination of the marine environment, Proc. Soc. Exper. Biol. Med., 150, 49-51, (1975).

Dunn, B. P. and H. F. Stich, Release of the carcinogen benzo(a)pyrene from environmentally contaminated mussels, Bull. Environ. Cont. Toxicol. 15, 398-401, (1976).

Ehrhardt, M., Petroleum hydrocarbons in oysters from Galveston Bay, Environ. Pollut. 3, 257-271, (1972).

Ehrhardt, M. and J. Heinemann, Hydrocarbons in blue mussels from the Kiel Bight, Environ. Pollut. 9, 263-282, (1975).

Fossato, V. U., Elimination of hydrocarbons by mussels, Mar. Pollut. Bull., 6, 7-10, (1975).

Fossato, V. U. and W. J. Canzonier, Hydrocarbon uptake and loss by the mussel Mytilus edulis, Mar. Biol., 36, 243-250 (1976).

Fossato, V. U. and W. J. Canzonier, Oil pollution monitoring in the Lagoon of Venice using the mussel, Mytilus galloprovincialis Mar. Biol. 25, 1-6, (1974).

Goldberg, E. D., The mussel watch - a first step in global marine monitoring, Mar. Pollut. Bull., 6, 111 (1975).

Hardy, R. P. R., Mackie, K. J. Whittle and A. D. McIntyre, Discrimination in the assimilation of n-alkanes in fish, Nature, 252, 577-578 (1974).

Harger, J. R. E., Competitive coexistence among intertidal invertebrates, Amer. Sci., 60, 600-607 (1972).

Jørgensen, C. B., Biology of Suspension Feeders, Pergamon Press, Oxford, 1976.

Korn, S., N. Hirsch and J. W. Struhsaker, Uptake, distribution and depuration of ^{14}C-benzene in northern anchovy, Engraulis mordax, and striped bass, Morone saxatis. Fish. Bull. 74, 454-551, (1976).

Lee, R. F., Fate of petroleum hydrocarbons in marine zooplankton, pp. 549-553. In: Proceedings of the Conference on Prevention and Control of Oil Pollution. American Petroleum Institute, Washington, D. C., 1975.

Lee, R. F., Metabolism of petroleum hydrocarbons in marine sediments, pp. 334-344. In: Sources, Effects and Sinks of Petroleum in the Aquatic Environment, American Institute of Biological Sciences, Washington, D. C., 1976.

Lee, R. F., G. Sauerheber and A. A. Benson, Petroleum hydrocarbons: uptake and discharge by the marine mussel, Mytilus edulis, Science, 177, 344-346 (1972a).

Lee, R. F., R. Sauerheber and G. H. Dobbs, Uptake, metabolism and discharge of polycyclic aromatic hydrocarbons by marine fish, Mar. Biol. 17, 201-208 (1972b).

Lee, R. F., C. Ryan and M. L. Neuhauser, Fate of petroleum hydrocarbons taken up from food and water by the blue crab, Callinectes sapidus, Mar. Biol. 37, 363-370, (1976).

Lee, R. F., E. Furlong and S. Singer, Metabolism of hydrocarbons in marine invertebrates. Aryl hydrocarbon hydroxylase from the tissues of the blue crab, Callinectes sapidus, and the polychaete worm, Nereis sp. In: C. S. Giam (ed.) Pollutant Effects on Marine Organisms, D. C. Heath, Lexington, Massachusetts, 1977. In Press.

Mackie, P. R., R. Hardy, K. H. Whittle, A. D. McIntyre, and R. A. A. Blackman, The alkanes of marine organisms from the U.K. and surrounding waters, Rapp. P. -V. Reun. Cons. Int. Explor. Mer. (1977). In Press.

Neff, J. M. and J. W. Anderson, Accumulation, release and distribution of benzo(a)pyrene ^{14}C in the clam Rangia cuneata, pp. 469-471. In: Proceedings of the Conference on Prevention and Control of Oil Pollution, American Petroleum Institute, Washington, D. C., 1975a.

Neff, J. M. and J. W. Anderson, An ultraviolet spectrophotometric methods for the determinations of naphthalene and alkylnaphthalenes in the tissues of oil contaminated marine animals, Bull. Environ. Cont. Toxicol. 14, 122-128 (1975b).

Neff, J. M., B. A. Cox, D. Dixit, and J. W. Anderson, Accumulation and release of petroleum-derived aromatic hydrocarbons by marine animals. Mar. Biol. (1976). In Press.

Ogata, M. and Y. Miyake, Identification of substances in petroleum causing objectionable odor in fish, Water Res. 7, 1493-1504 (1973).

Pasteels, J. J., Pinocytose et athrocytose par l'epithelium branchial de Mytilus edulis, Z. zellforsch, 92, 239-259 (1968).

Reish, D. J., Effect of pollution abatement in Los Angeles harbors, Mar. Pollut. Bull. 2, 71-74 (1971).

Rice, S. D., J. W. Short, C. C. Brodersen, T. A. Meckleburg, D. A. Mole, C. J. Misch, D. L. Cheatham, J. F. Karinen, Acute toxicity and uptake-depuration studies with cook inlet crude oil, Prudhoe Bay crude oil, No. 2 fuel oil and several subarctic marine organisms, Processed report, Northwest Fisheries Center, Seattle, Washington (1976).

Rossi, S. S. and J. W. Anderson, Bioavailability of petroleum hydrocarbons from water, sediments, and detritus to the marine annelid, Neanthes arenaceodentata. In: Proceedings of the 1977 Oil Spill Conference, American Petroleum Institute, Washington, D. C., 1977. In Press.

Roubal, W., T. Collier and D. C. Malins, Accumulation and metabolism of carbon-14 labeled benzene, naphthalene and anthracene by young coho salmon, Arch. Environ. Contam. Toxicol. 5, (1977). In Press.

Sanborn, H. and D. C. Malins, Toxicity and metabolism of naphthalene: a study with marine larval invertebrates. Proc. Soc. Exp. Biol. Med. (1977). In Press.

Sanders, H. L., J. F. Grassle and G. R. Hampson, The West Falmouth oil spill, I. Biology, Reference No. 72-70, Woods Hole Oceanographic Institution, Woods Hole, Massachusetts, (1972).

Shipton, J., J. H. Last, K. E. Murray and G. L. Vale, Studies on a kerosene-like taint in mullet (Mugil cephalus). II. Chemical nature of the volatile constitutents, J. Sci. Rd. Agric. 21, 433-436, (1970).

Stegeman, J. J., Hydrocarbons in shellfish chronically exposed to low levels of fuel oil, In: Pollution and Physiology of Marine Organsims, Academic Press, New York, 1974.

Stegeman, J. J. and J. H. Teal, Accululation, release and retention of petroleum hydrocarbon by the oyster, Crassostrea virginica, Mar. Biol. 22, 37-44, (1973).

Teal, J. M. and J. W. Farrington, A comparison of hydrocarbons in animals and their benthic habitats, Rapp. P. -V. Reun. Cons. Inst. Explor. Mer. (1977). In Press.

Thompson, R. J. and B. L. Bayne, Active metabolism associated with feeding in the mussel Mytilus edulis, J. Exp. Mar. Biol. Ecol. 9, 111-124, (1972).

Whittle, K. J., P. R. Murray, P. R. Mackie, R. Hardy and J. Farmer, Fate of hydrocarbons in fish, Rapp. P. -V. Reun. Cons. Int. Explor. Mer. (1977). In Press.

Zitko, V., Determination of residual fuel oil contamination of aquatic animals, Bull. Environ. Contam. Toxicol. 5, 559-563 (1971).

CHAPTER 7

FOOD CHAIN TRANSFER OF HYDROCARBONS

John M. Teal

Woods Hole Oceanographic Institution
Woods Hole, Massachusetts 02543

Introduction

After the publication of Rachel Carson's SILENT SPRING it was the accepted wisdom that the principal route for hydrocarbon uptake in animals was through the foodweb. Not only was foodweb uptake believed to predominate but foodweb magnification was accepted as a principle. Now there is general lack of belief in the magnification notion as far as aquatic organisms are concerned (e.g., NAS 1975) and in connection with this change in belief is the realization that uptake of hydrocarbons with food is only part of the story.

Papers related to foodweb transfer of hydrocarbons in the oceans are currently appearing in a steady stream so that many questions I ask now may well turn out to be answered in the near future.

A part of the information on foodweb transfer of petroleum hydrocarbons I will take from analogy with work that has been done with chlorinated hydrocarbons. The compounds for which there is suitable data, dieldrin, DDT and PCB's, are all similar in chemical behavior to unsubstituted hydrocarbons in having low polarity and being several orders of magnitude more soluble in lipids than in water. DDT and PCB's contain aromatic rings. Because of these properties they should behave in food chain transfers as do hydrocarbons, although they may be metabolized differently.

It will also be necessary to use data on natural hydrocarbons which must act like those from petroleum. Mention will also be made of foodweb transfer of metabolic products of hydrocarbons although there is relatively little data on this potentially important question. We can begin by asking if foodweb transfer does actually occur and what the nature of the evidence is for or against the process.

It is clear from the chemistry of hydrocarbons that they should be absorbed through the guts of animals along with the lipids in the diets, so foodweb transfer must occur. Obviously, this must be the principal mode of transfer in the case of land animals. It is likely to be the main uptake process for air-breathing marine animals such as whales and seals since direct uptake through the dead surface layers of epithelia is certainly slight unless, perhaps, the creature is actually coated with petroleum. In that case most air-breathing animals other than cetaceans will still ingest more oil in trying to clean their coats than are ever likely to absorb directly through the skin.

I will mention only one paper on marine mammals: Holden and Marsden (1967) found porpoises and seals from the British Isles and Canada contained up to two orders of magnitude more chlorinated hydrocarbon than was found in their food. They believed this was consistent with the storage in the body fat of much of these contaminants encountered by the animals over a period of two years, a case for not only foodweb uptake but for foodweb magnification.

In the case of gilled animals there is good evidence for foodweb uptake in work on natural hydrocarbons that come from the metabolic transformations of phytol: pristane and a series of related isoprenoid olefins. Production of the hydrocarbons from phytol takes place in copepods of the genus Calanus (Avigan and Blumer 1972). Blumer (1967) showed that in the gut and liver of a basking shark that had fed on Calanus there was little change in the relative concentration of these isoprenoids in spite of their differences in saturation. Presumably the shark obtains these hydrocarbons entirely from its food.

Blumer et al. (1969) and Ackman (1971) state that pristane is most abundant in fishes known to feed on Calanus, such as the basking shark mentioned above, and herring. Other fishes not feeding on Calanus, herring off New Jersey, cod and freshwater alewife (a herring), lack the high levels of pristane found in the former group. The conclusion that pristane in predators comes through their food can also come from the data of Blumer et al., (1964) on

the copepod foodweb. Predatory copepods that feed on Calanus contain on the order of a few
parts per thousand pristane (10 to 20 parts per thousand are in the Calanus) while herbivorous species contain only one-tenth as much, presumably directly from their algal food. The
latter species might synthesize small amounts of pristane but their low body levels at least
indicate the absence of a pathway of pristane uptake other than through the foodweb, e.g.,
directly from the water.

The copepod Rhincalanus nasutus, one of the herbivorous species which does not synthesize
appreciable amounts of pristane, contains large amounts of heneicosahexaene (HEH). Blumer
et al. (1970) showed the copepod accumulated HEH from several algae on which, when normalized for lipid content were very similar to the amounts present in the phytoplankton.
There was an exception in the case of one diatom from which the copepods accumulated only
one tenth the amount of HEH as in the algal food.

Ingestion of petroleum hydrocarbons has obviously occurred in these instances in which tar
particles have been found in the guts of animals or where their body contains hydrocarbons
that resemble tar balls. Horn et al. (1970) found tar balls in the stomachs of saury collected in the Mediterranean. They suggested, that since these fish are eaten by porpoises
and all larger predaceous fish, this provided for the direct introduction of tar into these
larger fishes as well. We have found evidence of tar ball ingestion in a galatheid from the
Nares Abyssal Plain in the North Atlantic (Teal 1976). Pequegnat (personal communication)
found tar particles in the stomachs of deep-living starfish. Conover (1972) observed copepods eating oil droplets. In all these cases the tar was probably eaten "by mistake" but
once eaten was definitely in the marine foodweb.

Having satisfied ourselves that foodweb transfer does occur we can move onto other questions
about the process. Do animals discriminate against uptake of some hydrocarbons in relation
to others? What is the efficiency of uptake through the gut and how important is this route
in relation to uptake via the gills from the water? Finally, are there special processes
and routes of foodweb uptake in the oceans that should have special study?

Selective Accumulation

Discrimination in uptake certainly occurs. In their studies of HEH uptake, Blumer et al.
(1970) found that of the copepods they studied, Rhincalanus was the only one that accumulated HEH. By feeding radiocarbon labelled phytoplankton (Phaeodactylum) to zooplankton,
Murray et al. (1975) showed that if one assumed that all the polyunsaturated olefins found
in their mixed crustacean plankton collections came from the algae, less than 3% of the
olefins ingested were retained in the animal tissues. This could be due to either selective
uptake or subsequent metabolism.

Selective synthesis from precursers in the food also occurs as seen by the production of
pristane from phytol in Calanus but not in Rhincalanus, Blumer et al. (1969) also showed
the probable synthesis of a variety of isoprenoid olefins from phytol. Although Rhincalanus
contains less than 0.1% pristane (Blumer 1967), the concentration of the isoprenoid olefins
in the former are an order of magnitude more abundant in relation to pristane than in the
latter.

We have shown that marsh minnows, Fundulus, from the Wild Harbor marsh oil spill area,
apparently developed the ability to discriminate against the petroleum hydrocarbons while
continuing to accumulate the n-alkanes characteristic of normal marsh plants and sediments
from their contaminated food (Burns 1975, Teal and Farrington 1976).

Selective uptake may occur between different parts of the same fish. Mackie et al. (1974)
found the n-alkanes in muscles of fish from Scotland tended to have a uniform carbon number
distribution while those in the liver show a high odd carbon predominance. Water alkanes
were similar to those in the muscles; sediment alkanes to those in the livers. Their data
are consistent with the hypothesis that liver hydrocarbons come mainly from the food,
carried by the blood directly from the gut via the hepatic portal circulation to the liver.
There they are absorbed before having the chance to enter the other tissues. The muscle
hydrocarbons instead are absorbed through the gills from the water. Blood from the gills
passes through the muscles before reaching the gut and liver. Therefore, hydrocarbons absorbed from the water tend to be localized in muscle. The only available hydrocarbon data
from pelagic animals from the open ocean (Burns and Teal 1973) have also been interpreted
on the basis of uptake from water via the gills and from food via the gut. The data of
Mackie et al. (1974) on herring to not fit this simple hypothesis since they feed on plankton which have a n-alkane distribution resembling that in the water rather than that in the
sediments, but their liver alkanes show the same odd carbon number predominance as was found
in the other fishes.

Hardy et al. (1974) fed crude oil along with food to codling and followed subsequent changes in liver and muscle n-alkanes. Control fish showed the same differences in distribution of carbon number as seen by Mackie et al. (1974). But after feeding on oil the hydrocarbon distribution in the fish's livers came to resemble that in the muscles of the control fish. Efficiency of uptake of the petroleum alkanes was very small from C-15 to 21, and at C-31 and 32, but above 20% between C-25 and C-28. They checked uptake with radiocarbon labelled hexadecane and found less than 0.4% incorporation with no indication of metabolism. (After six months of depuration, although about half of the petroleum hydrocarbons were lost, the liver alkanes continued to exhibit the same pattern.)

Mackie et al. (1976) have further analyses of n-alkanes in samples from the British Isles which, in general, show the same patterns. From all of the above data on alkane distribution I believe we can suggest that the above hypothesis about routes of uptake and distribution of alkanes need only be modified by including a process of selective uptake in the gut to account for the presently available data. If uptake in the gut is significantly more efficient for hydrocarbons from about C-20 to C-30 it would account for the data from experimental as well as environmental samples including those from Wild Harbor mentioned above. The natural alkanes abundant in marine sediments are those from plant waxes, odd carbon numbers from n-C-21 to n-C-31, with C-15 and 17 from algal sources. Animals may normally exclude other alkanes in the gut which would rise to the observed liver distributions. When exposed to petroleum pollution the liver alkanes would still be concentrated in the usual range of carbon numbers but would lose the odd carbon dominance which is lacking in petroleum. At higher levels of pollution the exclusion process might be completely overwhelmed leading to an alkane distribution resembling that of the petroleum as was observed in the fish from the Wild Harbor spill area immediately after the spill (Burns 1975). Even if the above hypothesis should turn out to be generally correct, Whittle et al. (1976) have shown on feeding radiolabelled hydrocarbons in squid to herrings that after about 2 days, 54% of labelled hexadecane and 37% of benzanthracene fed could be found in muscle lipid. In herring there is apparently rapid translocation of hydrocarbon from gut to muscle with little discrimination in uptake and little uptake by the liver. They also fed benzo[a]pyrene but found most of that still in the stomach after two days, indicating either some discrimination or lack of ability to absorb this insoluble molecule.

Hydrocarbon Uptake from Food vs. Water

If we suppose that some significant part of the hydrocarbons in an animal comes from the water via the gills we wonder how much is likely to come by this route compared with the route via the gut and food alone. A number of experiments on this question have been done with chlorinated hydrocarbons that give us a good indication of the sort of answer we can expect. Macek and Korn (1970) measured accumulation of DDT by trout from water or with food containing concentrations characteristic of Lake Michigan. They found the fish would obtain over 10 times as much DDT from food as from water.

Reinert (1972) did experiments in which guppies accumulated the same amount of dieldrin, about 4×10^{-6} g whether they were getting it from their food or through the water. When eating clean food but living in water containing 0.8 to 1.1×10^{-9} g/g dieldrin, they must have passed enough water over their gills to have oxidized the 200 Daphnia they were fed. That amount of water would have contained about 5.7×10^{-6} g dieldrin. In the other experiment in clean water but fed Daphnia containing 144×10^{-6} g/g dieldrin, they were exposed to 6×10^{-6} g dieldrin in their guts. It would seem on the basis of this experiment, that uptake by one route is about as efficient as by the other. Fish absorb over 70% of the dieldrin to which they are exposed. I made both of these calculations from data during the initial uptake phase before any sign that equilibrium was being approached. Reinert also experimented on comparing foodweb transfer directly with uptake from the water. Guppies in water containing about 2.1×10^{-9} g/g dieldrin accumulated about 22×10^{-6} g/g in their bodies. Daphnia in the same water accumulated about 32×10^{-6} g/g. Guppies in clean water fed on the contaminated Daphnia accumulated only about 2×10^{-6} g/g dieldrin or less than one-tenth through the foodweb of what they picked up directly from the contaminated water.

Uptake in perch has been modeled by Norstrom et al. (1976) for PCB using data from the Ottawa River. They assume approximately equal efficiencies (.8) for uptake of pollutants from food as for uptake of calories from food, and equal efficiencies (.75) for uptake of pollutants from water as for oxygen from water. Using values for PCB in water of 2.5×10^{-12} g/g and for fish food of 130×10^{-9} g/g, their model predicts that between 30 and 35% of the total uptake over the five year modelled life of the perch would come from the water. These predictions are for long-term equilibrium values of course, throughout the life of the animal and so include calculations of depuration.

of oxygen, which is a middle value for saturated waters on earth, then an aquatic animal will have to extract oxygen from 1.4×10^5 g water to oxidize one gram of food. If one further assumes equal efficiency of pollutant uptake from food and water as for calories and oxygen, and negligible growth, one can calculate limiting ratios for uptake from water to that from food. For Reinerts's (1972) direct comparison experiment, this ratio is 9 compared with his 10, 2.7 for the data of Norstrom et al. (1976) compared with their 0.5 and 0.14 for the data of Macek and Korn (1970). The entire range of ratios is simple to explain if the concentrations of hydrocarbons in food and water vary over the given ranges.

We can compare the pristane uptake data referred to above with pristane levels in water reported in Mackie et al. (1974). For ease of computation I will make use of the maximum concentration factor (that for guppies) of 5×10^4 (dry weight basis) Reinart (1972) calculated in his experiments. The water contained from 1 to 68×10^{-11} parts pristane, which multiplied by the concentration factor would give a maximum level in fish or copepods of $0.5 - 34 \times 10^{-6}$ g/g derived only from water uptake. The values found of over 1000×10^{-6} g/g in copepods (Blumer 1967, Blumer et al. 1969) and fishes (Ackman 1971) must have come from foodweb transport as everyone concluded they did.

I can also calculate a theoretical dieldrin uptake for Daphnia from the data of Reinert (1972) by using a respiration value from the literature (Prosser et al. 1950) since Reinert gives no feeding data for the Daphnia. Daphnia respire at a rate of about 500 mm^3 O_2/g hr, an amount of oxygen contained in 83 ml of water. After 72 hours exposure to dieldrin concentrations in water of 2.1, 4.2 and 12.8×10^{-9} g/g, the animals contained about 35, 62, and 150×10^{-6} g/g. If their efficiency of dieldrin uptake was the same as for oxygen, they should have contained 83 g water/g Daphnia hr x 72 hr x g dieldrin/g water or 12.5, 25, and 76×10^{-6} g/g. Since Daphnia actually contained 2-3 times this amount, their extraction of hydrocarbon from the water may be significantly more efficient than their extraction of oxygen. If real, this could be a part of the explanation for the different levels of pullutants found in the lower levels of food chains in different environments.

Hydrocarbon Storage and Depuration

Bacteria have been shown to accumulate hydrocarbons of various sorts with concentration factors that range from a few hundred to over fifty thousand for different compounds and species of microbes (Grimes and Morrison 1975). They apparently store hydrocarbons in their cells in unmodified form and produce more lipids when growing on hydrocarbons than on control media (Finnerty et al. 1973). The large variation in concentration factors vary from 100 to 50,000 and probably depends partly, at least, on the lipid content of the cells. Without more data, I see no reason to believe that any accumulation process is involved other than the lipid/water partitioning apparently dominant in animals. Bacteria do present yet another way in which hydrocarbons can get into the foodweb and perhaps a route associated with large amounts of lipids which could enhance their uptake and long term storage.

Depuration is the other important process influencing the levels available for transport in the foodweb. Although this is a subject for other papers in this volume, I will emphasize that if a food organism actively rids itself of pollutants it will be a poor source of them for its predators. On the other hand, long retention times have important implications for foodweb transport of hydrocarbons in marine animals. For example, Darrow and Harding (1976) showed that Calanus does not metabolize p, p'-DDT for a period of at least 8 weeks at less than 6°, and Blumer et al. (1969) demonstrated that Calanus does not metabolize pristane during starvation.

Also of significance is the long-term storage of hydrocarbons and their isolation in storage tissues. Stegeman and Teal (1973) showed that for long periods oysters retain a portion of their body burden of hydrocarbon after having been placed in clean water. Lee (1975) showed that various species of Calanus lost over 99% of what they had taken up during experimental exposure to specific petroleum compounds during depuration, but they did not lose the last amount even during 28 days of depuration. He suggested that this remainder might not be lost during the life of the animal. This type of long term storage has often been reported for even warm-blooded vertebrates, especially in the case of birds and pesticides (TIE 1972), in which the animals' metabolism would be expected to be most active in removing the pollutants.

Another potentially important aspect of storage in the bodies of prey organisms is the form of compound stored, whether as the parent compound or as the metabolite. Lee (1975) found that in Calanus plumchrus 33% of the benzo[a]pyrene in the animals was present as the metabolite after 1 day of depuration and 60% after 10 days, while 88% of naphthalene found in Euchaeta japonica after 1 day of depuration was present as the metabolite. All these experiments involved 4 days of exposure before depuration began. In experiments with only 1 day of uptake Corner et al. (1976) found 94% of the naphthalene remaining in Calanus helgolandicus after 24 hours of depuration was present as the parent compound, although most of that lost to the water was as the metabolite. Since the metabolites of these aromatics are both more toxic and more water soluble than their parents, in exactly what form the compounds move through the foodweb could be very important on the effect they will have.

The question of foodweb transport to fishes would seem to be largely a matter of uptake and accumulation of hydrocarbons from the environment by food organisms, and of the transport and availability of food in different parts of the marine environment. Many aspects of uptake from food are still not known. How does uptake of different types of hydrocarbons differ from that of alkanes on which most of the available data is based at present? Does uptake from food depend on the abundance or composition of food available? A fat predator, one with high levels of blood lipids, would have both more uptake and more storage capacity. But in contrast, an excess of food might mean that it would move more rapidly through the predator's gut with resulting poor assimilation efficiency and lower uptake rate. There is the very large matter of extending the arguments I have made from data on fish to the lower trophic levels of the foodweb. Detailed data on uptake efficiencies by various pathways are needed for members of many more and differing types of marine animals. An approximation of the necessary data could be made from available physiological data on feeding and respiration but I have not undertaken this task here as I expect that data from experiments with hydrocarbons will soon be available.

Oceanic Foodweb Transfer

I would finally like to consider some special routes of foodweb transfer that may be of special importance in the ocean. We often think of the open ocean as having little input from land other than by trash jettisoned from passing ships or by fallout of particles or volatile substances from the atmosphere. Craddock (1969), however, reports an instance in which a species of midwater fish was feeding mostly on insects. Apparently the insects were blown from land during the autumn, landed on the sea surface and were consumed by a Gonichthys that feeds at the surface at night. This fish migrates to midwater depths during the day and so could provide a mechanism of rapid transport for land-derived materials into the midwater depths. I have some evidence from radar studies on ships at sea that this influx of insects from land onto the sea is a regular occurrence, at least in autumn when the studies were done. Of course the daily migration of scattering layer organisms provides a mechanism whereby pollutants from the surface layers will be transported deeper into the oceans.

Sinking of other materials, including such pollutants as tar particles, must also provide a mechanism of transport of hydrocarbons into the deep sea. And even though we cannot as yet see these directly in the sediments at their average concentration, as Farrington points out in another paper in this volume, we have already seen evidence of them in bottom animals. There is other evidence that trash from the surface may be normally treated as "food" by animals living in the very impoverished deep benthic environment. Haedrich and Henderson (1974) found besides a variety of vegetables, insect parts and bird feathers, strips of rubber and plastic jar cap liners in the stomach of Coryphaenoides armatus, one of the more common deep sea rattail fish. Clarke and Merrett (1972) found sheep and cow bones and thin pieces of rubber in the stomach of a shark at over 1200 meters. Another fish from similar depths contained large pieces of whale blubber signifying that hydrocarbons concentrated from the foodweb by this air breather do reach the deep ocean.

Finally, I would like to mention the transport of materials to the deep ocean in the form of fecal pellets. Honjo (1976) has shown that the flux of fecal pellets is the major pathway for coccoliths reaching the deep ocean bottom. Wiebe et al. (1976) have evidence for the role of fecal pellets in the transport of other materials including chlorinated hydrocarbons to the bottom. Since this may be a major means of transport of organic matter into the deep sea it is important to know to what extent the feces of zooplankton contain hydrocarbons from either their food or from surface waters if the role of this transport to the deep sea is to be understood. The feeding of bottom animals also produces feces which may be an important source of food for other benthos, and which, if hydrocarbons are concentrated in them as organic matter often is, may be an indirect mechanism by which foodweb transport of hydrocarbons occurs.

Research Needs

It should be obvious from the above that I believe we have more gaps than knowledge about the foodweb transfer of hydrocarbons in the oceans. Still, although we can construct reasonable models for hydrocarbon uptake by fish, some crustacea and bivalves, we as yet have relatively little data on food uptake by other marine organisms. Hydrocarbon cycling by such animals as salps, which contact large amounts of water in their feeding and produce large amounts of feces, could be especially important. The role of transport on fecal matter in the oceans in general, deserves more study.

Fatty acids have been used to study marine foodwebs (e.g., Jeffries (1975). Studies like those of Blumer and colleagues indicate that similar information might be gained by investigating the cycling of natural hydrocarbons but this remains a promise. The question of selective uptake is especially interesting in this connection and also has a great deal of importance for the distribution of pollutant hydrocarbons in the oceans.

Lastly, most of the data I have had to review is based on studies of n-alkanes. More effort is certainly needs on the aromatics which are the compounds with greatest biological effects under most conditions. I believe we shall also gain some valuable information on hydrocarbon cycling by looking at the cyclic alkanes which are the least biologically active of the more mobile fractions of petroleum.

This research was supported by the Victoria and the Laurel Foundations.

References

Ackman, R. G., Pristane and other hydrocarbons in some freshwater and marine fish oils. Lipids 6, 520-522 (1972).

Avigan, J. and M. Blumer, On the origin of pristane in marine organisms. J. Lipid Res. 9, 350-352 (1968).

Blumer, M., Hydrocarbons in digestive tract and liver of a basking shark. Science 156, 390-391 (1967).

Blumer, M., M. M. Mullin and R. R. L. Guillard, A polyunsaturated hydrocarbon (3, 6, 9, 12, 15, 18-heneicosahexaene) in the marine foodweb. Mar. Biol. 6, 226-235 (1970).

Blumer, M., M. M. Mullin and D. W. Thomas. Pristane in the marine environment. Helgol. Wiss. Meeresunters. 10, 187-201 (1964).

Blumer, M., J. C. Robertson, J. E. Gordon and J. Sass, Phytol-derived C_{19} di- and tri-olefinic hydrocarbons in marine zooplankton and fishes. Biochemistry 8, 4067-4074 (1969).

Burns, K. A., Distribution of hydrocarbons in a salt marsh ecosystem after an oil spill and physiological changes in marsh animals from the polluted environment. MIT-WHOI Joint Program Ph.D. Thesis (1975).

Burns, K. A. and J. M. Teal. Hydrocarbons in the pelagic Sargassum community. Deep-Sea Res. 20, 207-211 (1973).

Clarke, M. R. and N. Merrett., The significance of squid, whale and other remains from stomachs of bottom-living deep-sea fish. J. Mar. Biol. Assoc. U.K., 52, 599-603 (1972)

Conover, R. J., Some relations between zooplankton and bunker C oil in Chedabucto Bay following the wreck of the tanker ARROW. J. Fish. Res. Bd. Canada 28, 1327-1330 (1972).

Corner, E. D. S., R. P. Harris, C. C. Kilvington and S. C. M. O'Hara. Petroleum compounds in the marine food web: short-term experiments on the fate of naphthalene in Calanus. J. Mar. Biol. Assoc. U.K., 56, 121-133 (1976).

Craddock, J. E., Neuston fishing. Oceanus, 15(1), 11-12 (1969).

Darrow, D. C. and G. C. Harding. Accumulation and apparent absence of DDT metabolism by marine copepods, Calanus spp., in culture. J. Fish. Res. Bd. Canada, 32, 1845-1849 (1976).

Finnerty, K. R., R. S. Kennedy, P. Lockwood, B. O. Spurlock and R. A. Young., Microbes and petroleum: perspectives and implications. The Microbial Degradation of Oil Pollutants. Center for Wetland Resources. LSU-SG-73-01, 105-125 (1973).

Grimes, D. J. and S. M. Morrison., Bacterial bioconcentration of chlorinated hydrocarbon insecticides from aqueous systems. Microb. Ecol. 2, 43-59 (1975).

Haedrich, R. L. and N. R. Henderson. Pelagic food of Coryphaenoides armatus, a deep benthic rattail. Deep-Sea Res. 21, 739-744 (1974).

Hardy, R., P. R. Mackie, K. J. Whittle and A. McIntyre., Discrimination in the assimilation of n-alkanes in fish. Nature 252, 577-578 (1974).

Holden, A. V. and K. Marsden., Organochlorine pesticides in seals and porpoises. Nature, 216, 1274-1276 (1967).

Honjo, S., Coccoliths: production, transportation and sedimentation. Mar. Micropalaentology 1, 65-79 (1976).

Horn, M., J. M. Teal and R. H. Backus. Petroleum lumps on the surface of the sea. Science 168, 245-246 (1970).

Jeffries, H. P., Diets of juvenile Atlantic menhaden (Brevoortia tyrannus) in three estuarine habitats as determined from fatty acid composition of gut contents. J. Fish. Res. Bd. Canada 32, 587-592 (1975).

Lee, R. F., Fate of petroleum hydrocarbons in marine plankton. Proc. Joint Conf. Prevent. Control of Oil Spills, 1975, 549-553 (1975).

Mackie, P. R., J. Hardy, K. J. Whittle, A. D. McIntyre and R. A. A. Blackman. The alkanes of marine organisms from the UK and surrounding waters. In A. D. McIntyre and K. Whittle (eds.), Petroleum Hydrocarbons in the Marine Environment. Rapp. P.-v. Reun. Cons. int. Explor. Mer. 171, in press (1976).

Mackie, P. R., K. J. Whittle and R. Hardy, Hydrocarbons in the marine environment. I. n-alkanes in the Firth of Clyde. Est. Coast. Mar. Sci. 2, 359-374 (1974).

Macek, K. J. and S. Korn, Significance of the food chain in DDT accumulation by fish. J. Fish. Res. Bd. Canada 27, 1496-1498 (1970).

McIntyre and K. Whittle (eds.), Petroleum Hydrocarbons in the Marine Environment. Rapp. P.-v. Reun. Cons. int. Explor. Mer. 171, in press (1976).

Murray, J., A. Stagg, A. B. Thomson, K. J. Whittle, and P. R. Mackie, On the origin of hydrocarbons in marine organisms. In A. D. McIntyre and K. Whittle (eds.) Petroleum Hydrocarbons in the Marine Environment. Rapp. P.-v Reun. Cons. int. Explor. Mer. 171, in press (1976).

NAS, Petroleum in the marine environment. Report of the U. S. National Academy of Sciences, Washington, D. C. (1975).

Norstrom, F. J., A. E. McKinnon and A. S. W. deFreitas, A bioenergetics-based model for pollutant accumulation by fish. Simulation of PCB and methylmercury residue levels in Ottawa River yellow perch (Perca flavescens). J. Fish. Res. Bd. Canada 33, 248-267 (1976).

Prosser, C. L. (ed.), Comparative Animal Physiology. Sanders. Philadelphia (1950).

Reinert, R. E., Accumulation of dieldrin in an alga (Scenedesmus obliquus), Daphnia magna and the guppy (Poecilia reticulata). J. Fish. Res. Bd. Canada 29, 1413-1418 (1972).

Stegeman, J. J. and J. M. Teal, Accumulation, release and retention of petroleum hydrocarbons by the oyster Crassostrea virginica. Mar. Biol. 22, 37-44 (1973).

Teal, J. M., Hydrocarbon uptake by deep sea benthos. pp. 358-371. In Sources, Effects and Sinks of Hydrocarbons in the Aquatic Environment. AIBS, Washington, D. C. (1976).

Teal, J. M. and J. W. Farrington, A comparison of hydrocarbons in animals and their benthic habitats. In A. D. McIntyre and K. Whittle (eds.) Petroleum Hydrocarbons in the Marine Environment. Rapp. P.-v. Reun. Cons. int. Explor. Mer. 171, in press (1976).

TIE, Man in the Living Environment. Institute of Ecology Report of Workshop on Global Ecological Problems (1972).

Wiebe, P. H., S. H. Boyd and C. Winget, Sedimentation trap for use of above deep-sea floor with preliminary results of its use in the Tongue of the Ocean, Bahamas. J. Mar. Res. 34, 341-354 (1976).

CHAPTER 8

COMPARATIVE OIL TOXICITY AND COMPARATIVE ANIMAL SENSITIVITY

Stanley D. Rice, Jeffrey W. Short, and John F. Karinen

Northwest and Alaska Fisheries Center Auke Bay Laboratory
National Marine Fisheries Service, NOAA
P.O. Box 155, Auke Bay, AK 99821

Introduction

Laboratory bioassays have been used to describe or quantitate the ability of crude and refined oils to kill marine life. The principal objective of earlier studies was to estimate oil concentrations that were definitely lethal to marine animals, and in some cases there have been attempts to predict environmental effects on the basis of these studies. More recently, two objectives of laboratory toxicity testing have emerged. First, acute toxicity studies have been used to determine and compare the toxicities of different crude and refined oils to marine life, and the toxicities of individual components of oils. In these studies, the quantity of oil or oil component necessary to kill 50% of the standard test species exposed under a constant set of conditions (median lethal concentration, LC50) is equated to the toxicity of the oil or oil component. Secondly, acute toxicity data have been used to determine and compare the sensitivities of different species or the different life stages of a given species. The goal here has been to identify species or life stages that are especially sensitive or vulnerable to oil. Here the LC50 has been used as an index of species sensitivity.

There has been a wide variety of results from these studies, and some apparent contradictions, mainly because of the absence of standard exposure and analytical methods but also because of difficulties involving quantitation of the oil used and its instability in the test waters. The amount of oil used to prepare a test solution is invariably much more than the amount of oil that enters into the test water and the composition of oil in the test water is invariably different from the composition of the parent oil. The composition and amount of oil that enters the water column is a function of many factors, such as mixing energy and duration, temperature, and salinity. Once in the water, oil components are subject to evaporation and microbial oxidation, which can appreciably alter the composition and concentration of oil constituents with time, even at low temperatures. Analytical methods for measuring the amount of oil in the test water have only recently been widely available, and all of these methods are less than ideal.

The scope of this review is limited to studies dealing with the ability of crude and refined oils to kill marine animals. Emphasis is given to the more recent quantitative studies that were not available to earlier reviewers (Evans and Rice 1973; Moore and Dwyer 1974; National Academy of Science 1975). This review covers (1) the behavior of oil in water; (2) the methodology problems associated with bioassays; (3) the comparative toxicity of oil-water mixtures, oils, and components of oils; and (4) the comparative sensitivity of different life stages and species.

Behavior of Oil in Water

Bioassays involving crude or refined oils are fundamentally different from bioassays involving other substances (such as heavy metals) in one major respect. Oil bioassays attempt to evaluate the effects of a mixture of many different toxic compounds together rather than a single pure compound. There is a wide spectrum of physical and biological properties associated with these various compounds which affects the rate and amount of these compounds transported into water. Once in the water, the persistence of some of the toxic compounds in oil is influenced by a variety of processes such as biodegradation, evaporation, etc.

A basic knowledge of physical processes that affect both the transport of oil into an aqueous phase and the persistence of oil transported is essential to evaluate results of bioassay tests. In this section, we discuss some of the factors that transport oil into and out of the water column. All of these processes occur simultaneously, and the relative importance of these processes varies from one compound or class of compound to another. Because of the complexities involved, simple bioassay methods do not work well for investigating toxic effects of oil.

Factors that Affect the Quantity of Oil Transported into Water

Oils can become associated with an aqueous phase in a variety of different ways, such as emulsion, dispersion, or accommodation (Peake and Hodgson 1966); or some of the constituent compounds of an oil may dissolve, forming a true solution. The solubility of oil compounds in water varies considerably with the class of compound and is an important factor that determines the toxicity of oil-water solutions. Some hetero compounds such as pyridine are completely miscible with water. Benzene is the most soluble aromatic hydrocarbon at about 1,800 ppm in water. The solubility of other aromatic hydrocarbons decreases with increasing degree of alkyl substitution and number of aromatic rings. Aliphatic hydrocarbons are among the least soluble hydrocarbons, with solubility decreasing sharply with increasing carbon number (McAuliffe 1966, 1969).

The amount of the soluble fraction of oil that enters the water phase is mainly determined by mixing energy, mixing duration, and the viscosity of the oil. Turbulence (or mixing energy) was found to have a pronounced effect on the amount of both particulate and sub-particulate oil going into the water phase (Gordon et al 1973). In similar studies at our laboratory we have been unable to detect concentrations of 10 ppb in water 1 cm beneath a slick that is gently layered on the water surface (Taylor and Karinen, this symposium). Gentle mixing of oil in seawater for 20 hr will generate water-soluble fractions with oil concentrations in seawater from about 1 to 10 ppm (Anderson et al. 1974, Rice et al. 1976a). Violent mixing can produce oil concentrations in seawater in the hundreds of parts per million, with much of the oil present as dispersed droplets.

The amount of time that oil and water are mixed is as important as mixing energy in determining the quantity of oil that enters the water phase. Using gentle mixing, the amount of oil that enters the water phase steadily increases for over 30 hr (Gordon et al. 1973, Anderson et al. 1974, Percy and Mullin 1975, Rice et al. 1976a).

The viscosity of the oil also affects the amount of oil that enters the water phase, because more mixing energy is required to mix thick, viscous crude oil. We have observed that the relatively viscous Prudhoe Bay crude oil yields WSF's that are about half the concentration of those from Cook Inlet crude oil when mixed under identical conditions (Rice et al. 1976a).

There are several other less well-studied factors that affect the amount of oil entering the water phase. There is evidence that polar hydrocarbon derivatives are generated from oil by photo-oxidation (Lysyj and Russell 1974). These polar hydrocarbons tend to dissolve into solution from an oil slick, which raises the total concentration of oil-derived hydrocarbons with time. In addition, pH (Kauss et al. 1973) and salinity (Rice et al. 1975) affect the amount of oil entering the water-phase.

Changes in temperature influence the transport of oil into water because changes in temperature change the viscosity of the oil, the solubility of oil components, and the stability of emulsified or suspended oil. The viscosity of oil increases as temperature decreases, thus at low temperatures more mixing energy or time is required to transport oil into the aqueous phase. As temperature decreases, the solubility of the non-volatile oil components decreases, but the solubility of the volatile components increases. Finally, emulsions and suspensions are more stable at lower temperatures. These conflicting effects make it difficult to predict the overall effect of temperature on the amount of oil transported into the aqueous phase.

Factors Affecting the Composition of Oil in Water

The composition of oil transported into the aqueous phase is also strongly dependent on compound solubilities, mixing energy, mixing duration, and oil viscosity. The composition of oil in water may or may not be similar to the composition of the parent oil, depending on how the oil is associated with the water. When oil is mixed violently with water, many dispersed droplets having a composition similar to that of the parent oil are formed. When oil is mixed slowly, the bulk of the hydrocarbons transported into water is composed of the more soluble hydrocarbons, unlike the composition of the parent oil. For example, Bean et al. (1974) found water-soluble fractions to have compositions quite unlike the parent oil. They found increases in IR absorption at 3,000 to 3,100 cm^{-1} for WSF's, indicating significant increases in the relative concentration of aromatic hydrocarbons in the WSF.

An "aromatic enrichment factor" (AEF) has been used by Anderson et al. (1974) to evaluate the degree to which the composition of an oil-water solution differs from the parent oil. The AEF is the ratio of the concentration of aromatic compounds to n-paraffins in the oil-water mixture, divided by the ratio of the concentration of aromatic compounds to the n-paraffins in the parent oil. A dispersion from a turbulent mix will result in an AEF of 1-3, indicating that the composition of oil in water is about the same as the parent oil. In contrast, the AEF will be higher in oil-water solutions prepared with less turbulence. Aromatic enrichment factors of 10-125 and similar magnitude have been reported for WSF's prepared with slow, gentle mixing of Kuwait, Prudhoe Bay, and Cook Inlet crude oils (Anderson et al 1974, Short et al 1975, Rice et al 1976a).

In addition to solubility, mixing energy, mixing duration, and oil viscosity, there are undoubtedly other factors that affect the compositon of oil in water. For example, the solubility of many compounds is influenced by pH (Kauss et al 1973), salinity, and temperature. Removal of selected hydrocarbons from solution by biodegradation, evaporation, photochemical oxidation, etc. will change the compositon of oil in water.

The fact that the amount and composition of oil that is transported into distilled water or seawater is strongly dependent on the method used to prepare the oil-water mixture, emphasizes the need for analytically determining the amount of oil actually transported into the aqueous phase. There have been many studies of static bioassays that report only the volume of oil used to prepare the oil-water test mixture. The concentrations of oil that the test species were actually exposed to in these studies are almost completely unrelated to the amount of oil used to prepare the test solutions, so that these studies are of limited value.

Factors Affecting the Persistence of Oil in Water
After the oil has been transported into the water, several factors cause hydrocarbons to be lost, resulting in changes in both concentration and composition of the oil solution or dispersion. Cheatham et al. (In prep.) demonstrated that the losses of total aromatics from WSF's of crude oil were significant, and that the rate of loss was less at low temperatures (Fig. 1).

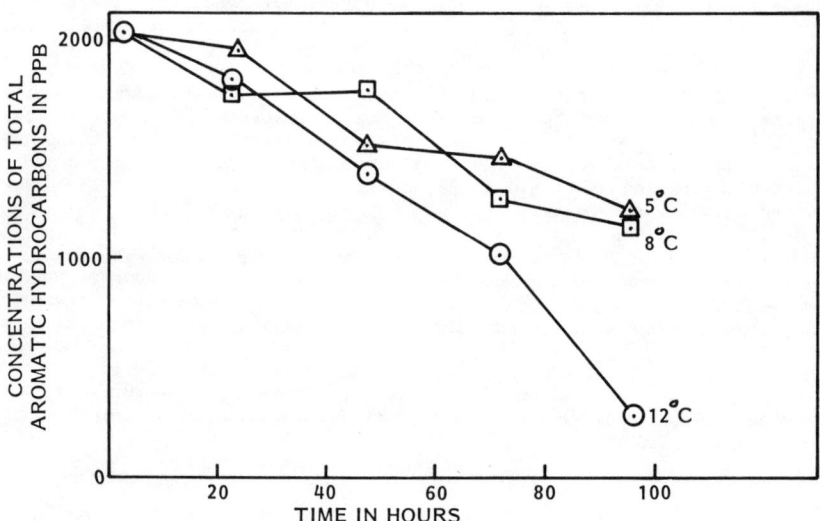

Fig. 8.1. Concentration of total aromatic hydrocarbons from water-soluble fractions of Cook Inlet crude oil measured by gas chromatography at 24-hr intervals. Solutions were nonaerated, and held at either 5°, 8°, or 12°C (from Cheatham et al. In prep.).

Evaporation causes significant losses of low molecular weight aromatics. The low molecular weight aromatics all have significant vapor pressures at laboratory temperatures, although vapor pressures decrease with decreasing temperature. The evaporation of these light aromatics in static bioassay systems is a well known phenomenon.

Both paraffinic and aromatic hydrocarbons are susceptible to microbial oxidation, although several studies (Kator et al. 1971) have indicated that paraffins are oxidized by microbes more easily than aromatic hydrocarbons. Further, dinuclear aromatic hydrocarbons are lost from solutions primarily by biodegradation, while mononuclear aromatics are lost primarily by evaporation; both processes occur at faster rates at higher temperatures (Cheatham et al. In prep.).

Chemical and photo-oxidation can also change the concentration and composition of oil in water. Aromatic hydrocarbons in particular are susceptible to photo-oxidation. Finally, significant quantities of oil can be separated from the water when dispersed droplets coalesce into larger droplets and form a layer of oil at the surface.

Methodology Problems Associated with Bioassays

Since oil-water solutions are so complex and difficult to work with, researchers have had to develop various modifications of standard bioassay methods. The following discusses the general problems encountered in oil toxicity testing, and specific problems associated with comparisons of oil toxicity.

Preparation of Oil-Water Mixtures

Different investigators have used a wide range of methods to prepare oil-water mixtures. For example, mixing methods range from virtually no mixing (layering oil on top of water) to very turbulent mixing for prolonged periods. As noted previously, the mixing turbulence and duration strongly affects the amount and composition of oil entering the water phase. Direct comparisions between results of different investigators cannot be made when different mixing methods have been used.

Individual investigators have tended to justify their choice of mixing methods on the basis of analogy to probable environmental situations. This strikes us as a fundamentally faulty approach. Several different mixing regimes may occur in the environment, so that it is impractical to try to simulate all of them. Acute bioassays cannot provide more than a hazy idea of what is likely to occur in the environment, nor were they intended to do so.

Methods of Exposing Animals to Oil

There are numerous practical problems in oil toxicity tests that have led to a variety of specific exposure techniques. For example, sometimes the test animals are passed directly through a surface film of oil into the oil-water mixture, and sometimes the animals are added to oil-water dispersions or WSF's. Some investigators have aerated the oil-water mixtures, which will affect results since aeration accelerates the evaporation of the toxic light aromatics. Many different designs of exposure contains have been used because the requirements of various animals differ. For example, most intertidal animals have to be trapped in the exposure container to prevent them from crawling out (controls included).

Death versus Moribundity

Since animals have a variety of structures and activity levels, dead and moribund animals may be difficult to distinguish from live animals. Usually, the lack of visible motion is used to define death, but with some slow-moving animals, death may not be distinguishable until some necrosis occurs, and this requires extending the time of the observation period. Some animals depress their metabolism and respiratory activities when stressed, and will appear dead, but can recover when placed in clean water. On the other hand, delayed mortality (deaths occurring after the exposure period has terminated), may occur in some species, such as molluscs (Swedmark 1973).

Delayed mortality leads to the problem of identifying moribund animals in contrast to dead animals. Moribund animals are severely affected and destined to die, usually within a few days. Moribund animals will die a physiological death and are not to be confused with affected animals that can live but would be "ecologically dead" because of increased vulnerability to predation or environmental stresses. If moribund animals are counted as alive at the end of the exposure (e.g. 96 hr), the calculated median lethal concentrations will be high, implying a level of tolerance that is erroneous. Crustacean larvae have been observed to be particularly slow in dying, and will exist in a moribund state for up to several days before the larvae become necrotic (Brodersen et al. In press).

Each species that is tested must be observed carefully and with attention to its biology. Although there can be no universal guidelines for observing all animals in toxicity tests, post-exposure observations should be made to determine if affected animals are moribund or if they will recover.

Effects of Temperature on Oil Toxicity Tests

Temperature can affect the results of oil toxicity tests in a complex manner. Percy and Mullin (1975) found that the amphipod Onisimus affinis was consistently more sensitive to oil when exposed at 8°C than at 0°C, but that oil-water solutions prepared at 0°C were consistently more toxic than solutions prepared at 22°C. Korn et al (In prep.) attempted bioassays at different temperatures with pink salmon fry and shrimp exposed to crude oil WSF's, naphthalene, and toluene and found opposing results. Survival of pink salmon fry was decreased when exposed at low temperatures, whereas shrimp survival was decreased when exposed at higher temperatures. Temperature can affect toxicity measurements in two basic ways: (1) by affecting oil toxicity, and (2) by affecting animal sensitivity. Oil toxicity is increased at lower temperatures, since toxic aromatics persist longer at lower temperatures (Percy and Mullin 1975, Cheatham et al In prep.). Animal sensitivities will be affected by changes in temperature since rates of hydrocarbon uptake, metabolism, and excretion will be altered. Because the effects of temperature on oil toxicity and animal sensitivity may oppose each other, it is conceivable that the order of potency of different oils could vary with temperature. Indeed, Ottway (1971) found that the relative toxicities of 10 different crude oils changed in a complicated way between 4°C and 26°C.

Chemical Analysis

Until the 1970's, comparison of results between experimenters has been especially difficult because most experimenters have not measured the concentrations of oil present in their test solutions. In these cases, results have usually been reported as the volume of oil used to prepare the test solution. Reporting oil concentrations as the "volume added" will underestimate the toxicity of oil, because most of the oil exists as a separate phase above the water phase. As previously discussed, the amount of oil that actually enters the water phase depends strongly on the method of mixing, and the volume of oil used to prepare the test solution is an inadequate measure of the amount of oil actually in the test solution.

Several chemical methods have recently been used to determine the concentration of oil in the water phase. Infrared spectrophotometry (Anderson et al 1974, Bean et al 1974, Rice et al 1975, 1976a, 1976b), ultraviolet spectrophotometry (Brenniman et al 1976, Rice et al 1976a, Caldwell et al This symposium), fluorescence spectroscopy (Gordon et al 1973, Percy and Mullin 1975, Wells and Sprague 1976), and gas chromatography (Bean et al 1974, Benville and Korn In press, Cheatham et al In prep., Korn et al In prep.) have been used as analytical methods. These analytical methods all have the advantage of providing a measure of the amount of oil actually in the test water, as opposed to the amount of oil used to prepare the test water. Each of these analytical methods also has different practical advantages.

The infrared (IR) method of Gruenfeld (1973) is relatively simple, but is much more sensitive to paraffinic hydrocarbons than to aromatics. In addition, standards for this method are necessarily arbitrary because the composition of oil in the water phase is not the same as the parent oil. Although the IR method quantitates paraffins that presumably are not toxic, their concentrations are usually proportional to concentrations of toxic compounds.

Ultraviolet (UV) spectrophotometry and fluorescence spectroscopy are especially sensitive methods for detecting aromatic hydrocarbons, which are presumed to be the most toxic fraction of crude and fuel oils. However, the UV and fluorescence methods are completely insensitive to paraffinic hydrocarbons, and measure different aromatics with different sensitivities, making standards somewhat arbitrary.

Analysis by gas chromatography (GC) or high pressure liquid chromatography (HPLC) is more complicated than the UV and fluorescence methods, but it also provides more detailed results. Both GC and HPLC separate individual compounds from mixtures, can measure paraffins and aromatics with nearly equal sensitivity, and have standards that are not arbitrary. However, because most individual compounds are separated by these methods, the amount of data produced by a single analysis can be formidable. Analyses by GC and HPLC are relatively expensive since fewer samples can be analyzed per day.

Gas liquid chromatography (GLC) and high pressure liquid chromatography (HPLC) are probably the most appropriate analytical methods for oil toxicity investigations. Both GLC and HLPC are able to separate and measure the concentrations of individual aromatic hydrocarbons present in oil-water mixtures. As discussed later, there is general agreement that aromatic hydrocarbons are responsible for most (if not all) of the toxicity of oil-water mixtures. Thus, using GLC or HPLC, one can attempt to correlate changes in the toxicity of oil-water mixtures with changes in chemical (i.e. aromatic) composition in order to assess the relative toxicity of each of the aromatics present (Bean et al. 1974).

All of the previously described methods involve extraction of oil components into a low polarity solvent, such as hexane or methylene chloride. Since a low polarity solvent may not efficiently extract polar components of oil in the WSF, polar compounds may not be measured effectively. These polar compounds may form a significant proportion of oil-derived material in the water phase, especially if the water has been in contact with oil in the presence of light for a prolonged period (Lysyj and Russell 1974). Detailed measurement of these polar compounds is a formidable analytical problem, and the contribution made by these polar derivatives to toxicity has not been evaluated.

Statistical Analysis
Statistical analysis of bioassay data is required for meaningful comparisons between tests. Unfortunately, only a few investigators have determined 95% fiducial intervals so that statistically significant differences could be determined.

Probit analysis (Finney 1976) is the most appropriate statistical method for analyzing bio-assay data because (1) it provides an accurate determination of the concentration of oil necessary to kill 50% of the test animals (LC50, LD50, or TLm), (2) it provides a measure of how much the toxicity of oil increases with increasing oil concentration, and (3) 95% fiducial intervals for LC50's can be calculated.

Other methods for calculating LC50's have been used which do not require as much data as the probit method, but the certainty of the calculated LC50 is reduced. The most commonly used of these methods is that of Doudoroff et al. (1951) where the percent mortality is plotted against the log of the oil concentration, and a line is fitted to the plot. The antilog of the oil concentration corresponding to 50% mortality is the estimate of the LC50. This method will work even when exposure to oil concentrations results in only zero and complete mortality. However, the accuracy of the method in such cases can be very poor, and there is no way to estimate confidence in the LC50. In addition, an underlying assumption of Duodoroff's method is that percent mortality is a linear function of the log of the oil concentration. A detailed plot of percent mortality versus the log of the oil concentration will usually yield a sigmoid curve, if enough observations are made at oil concentrations resulting in less than 100% mortality but more than zero mortality in the test animals.

A comparatively simple method that does not have the drawbacks of Doudoroff et al's (1951) method is that of Spearman and Kärber (Finney 1976). This method requires only that oil concentrations resulting in zero and 100% mortality in the test animals be known, although it also makes efficient use of data where some concentrations have some survivors. Confidence intervals can be calculated for LC50's determined by this method, and the applicability of the method is not affected by assumptions regarding the tolerance distribution of test animals.

Neither probit analysis nor the method of Spearman and Kärber determine variability in toxicity test results from any source other than the test organisms. Specifically, we have found that variation in the toxicity of water-soluble fractions or dispersions of a given crude or refined oil mixed under identical conditions significantly exceeds the variation expected from the test organisms. Therefore, in order to obtain statistically reliable data when other sources of variation are involved, such as different WSF preparations, a bioassay should be conducted on each WSF preparation and the LC50's analyzed using Student's t-test. This procedure would then include all sources of variation.

Specific Methodology Requirements Associated with Comparing Oil Toxicities or Animal Sensitivities
At least three specific methodology requirements must be met when comparing the toxicities of different oils or comparing the sensitivities of different species.

First, a reference toxicant is needed when comparing oil toxicities. Different investigators have compared oil toxicities, but the validity of comparing results without a reference toxicant is questionable when the tests are not conducted simultaneously with the same population of test animals under identical conditions. Use of a reference toxicant will permit valid comparisons by verifying that different sub-populations of animals are equivalent in sensitivity, or that minor modifications of procedures between different researchers are insignificant.

The toxicant DSS (dodecyl sodium sulfate), suggested as a standard reference toxicant by LaRoche et al (1970), has been used in some tests; however, it is evidently not a satisfactory standard for use over a wide range of temperatures. Rice et al (1976a) were unable to generate reliable data on DSS toxicity to some shrimp because a precipitate involving DSS forms readily above 30 ppm at 30 °/oo salinity and temperatures less than $15^{O}C$. They concluded that DSS is a poor reference toxicant.

Although the bioassay method is appropriate for comparisons of toxicity, few investigators have used the same test species, exposure conditions, or appropriate reference toxicant. It is therefore impossible to directly compare results between investigators. However, conclusions made by each investigator are usually valid, and one can compare conclusions formed by different investigators.

The second requirement when comparing oil toxicities is a standard test species. Comparisons of toxicity differences between several oils require that the sub-populations of animals used in the tests be of equal sensitivity. This is presumably assured when the same test species is used, although "reference toxicant" tests will verify that the sensitivities of the sub-populations have not been affected by other factors, such as sex, health, etc. Several investigators have used more than one species, and found the relative toxicities of different oils (or different methods of mixing a given oil) are generally not significantly affected by the choice of test species. Anderson et al (1974) tested four different oils each mixed with water in two different ways (WSF's and dispersions), and determined the toxicity of each oil-water mixture to six different marine animal species. Within each mixing method, the order of toxicity was not significantly different for the six animals. Rice et al (1976a) conducted a similar study and found that the LC50's of the eight species studied were generally within the same order of magnitude regardless of the kind of oil or mixing method used. Percy and Mullin (1975) found that the order of potency of four different crude oils was different for each of the three marine invertebrates they tested. However, they did not determine the concentrations of oil in the doses they used to determine the toxicity of these oils, so that these results may be due to variation in potency between preparations of their test solutions. When different investigators use different species to compare oil toxicities, the relative toxicities of different oils can be compared, but absolute values from one study cannot be compared to values from another study unless the same species is used. For the purpose of comparing toxicities of different oils, the use of a standard test species is most desirable.

Third, accurate comparisons of sensitivities of different species require that oil concentrations remain stable during the exposure. However, in static tests the oil concentrations usually decline with time, since aeration, biodegradation, temperature, and other factors affect the persistence of petroleum hydrocarbons in seawater. Nevertheless, most toxicity studies to date have used static exposures rather than flowthrough exposures because static exposures are relatively easy and inexpensive, and the technology for flowthrough bioassays with oil has been complicated, costly, and slow to develop.

There are two reasons why declining oil concentrations during static exposure cause difficulties in comparing the sensitivities of different species. First, the effective exposure period is short, possibly only a few hours, depending on the rate of oil concentration declines. Consequently, estimates of toxic oil concentrations are likely to be high. Since the rate of decline is likely to vary between tests, the effective period of exposure is also likely to vary between static tests. Second, some species accumulate the toxicants at faster rates than others, and are likely to be the species that will die faster than others. Animals that slowly approach equilibrium with toxic solutions may not be as tolerant as they seem because lethal concentrations may decline to sublethal concentrations before the animals have achieved equilibrium. Sensitivities of fast- and slow-dying species can be compared if they are tested long enough in flowthrough exposures where the toxicant concentrations are held constant during exposure. The value of flowthrough tests depends on constant exposure to stable concentrations, which must be verified by chemical analyses.

Comparative Toxicity of Oil-Water Mixtures, Oils, and Components of Oils

Historically, bioassays were developed to determine the biological potency of substances such as drugs and insecticides. For these purposes, the bioassay method requires that a standard set of exposure conditions be maintained and that standard test species be used whose sensitivity does not change between tests. If the standard species sensitivity does change between tests, then reference bioassays may be conducted with a standard reference toxicant to identify and correct for this change in sensitivity. If the above requirements are met, then the bioassay method will yield valid comparisons of toxicity. The bioassay method is thus appropriate for determining relative toxicity differences.

Toxicity Differences Between Water-Soluble Fractions of Oil and Oil-Water Dispersions

The toxicity of oil-water mixtures depends in part upon the way the oil is associated with the water. Experiments comparing the toxicity of water-soluble fractions and dispersions of oil suggest that the toxicity of an oil is due to the soluble compounds contained in that oil, and not due to compounds in dispersed droplets. The chemical composition of the droplets is probably very similar to that of the parent oil, except that the droplets probably contain slightly lower concentrations of soluble compounds because of losses of these compounds due to equilibration with the water. If oil dispersions are less than or equally as toxic as WSF's (which contain far fewer dispersed droplets of oil), the toxicity must be due to the soluble fractions of oil.

Oil dispersions are slightly less toxic than WSF's of a given oil when oil concentrations are analyzed by IR. Rice et al. (1976a) tested five species with dispersions and WSF's of two crude oils and found that toxicities of dispersions were always less than, but within an order of magnitude of, the toxicities for corresponding WSF's. Anderson et al. (1974) did not analytically determine oil concentrations in the doses they used to determine the toxicity of dispersions. However, they did provide data relating the amount of oil used to prepare a dispersion to the amount of oil found in a typical dispersion (as determined by IR). After calculating the amount of oil found in toxic dispersions, we estimate that the toxicity differences between the dispersions and WSF's studied by Anderson et al. ranged from nil to about an order of magnitude, with the dispersions being consistently less toxic.

Dispersions and WSF's seem to be equally toxic when oil concentrations are measured by UV. Rice et al. (1976a) compared toxicities of dispersions and WSF's of two crude oils to each of 6 test species. They found significant differences in only 4 of 12 cases when oil concentrations were determined by UV. In 3 of these 4 cases, the dispersions were slightly more toxic than the corresponding WSF. The lack of significant differences in toxicity suggests that the toxicity of oil is due to the water-soluble compounds.

In conclusion, all these results suggest that oil toxicity is due to chemical toxicity of soluble aromatics, rather than physical toxicity of dispersed droplets.

Toxicity Differences Between Different Oils
Toxicities vary between oils, which is to be expected because the concentration and composition of individual hydrocarbons within the oil vary. The refined oils are generally considered to be more toxic than crude oils, since smaller volumes of refined oils are needed to kill 50% of the test animals in a given length of time. The increased toxicity of refined oils, as measured on a volume added basis, is primarily caused by two factors. Refined oils often have concentrations of aromatic hydrocarbons, and refined oils are usually less viscous than crude oils thus requiring less mixing energy for toxic concentrations to be mixed into the water.

We believe that direct toxicity comparisons between different oils is a simplistic approach, because so many factors other than composition differences between oils can affect the observed results. We now know that the toxicity of a given oil is influenced by mixing method, temperature, salinity, etc.

We believe that any oil, whether crude or refined, is best considered as a source of toxic compounds. Based on a knowledge of the toxicity of all the compounds contained in an oil and the concentration of these compounds, one should be able to predict the toxicity of any oil-water mixture. (This may not be so formidable a task, since the solubility of most toxic compounds in oil is so low that they probably do not make significant contributions to toxicity.) With this approach, the toxicity of an oil would be evaluated in terms of (1) the concentrations of toxic compounds it contained, and (2) the physical characteristics of the oil (such as viscosity) that would promote the transfer of these toxic compounds into solution when a standard oil-water preparation method is used.

Comparative Toxicity of Different Oil Components
Although there is general agreement that aromatic hydrocarbons are responsible for the toxic effects of crude oils and refined products, the relative importance of various aromatic hydrocarbons to toxicity is not clearly defined. Therefore, there has been an increasing interest in determining the degree of contribution of individual aromatics to the toxicity of oil-water solutions.

There have been several studies on the toxicity of individual hydrocarbons (summarized in Table 1). Studies by Neff et al. (1976), Benville and Korn (In press), and Caldwell et al. (This symposium) compare the toxicities of mono- and di-aromatic compounds to four marine species. Even though the species tested are quite different, and include tests with larvae, the following three results are quite consistent: mono-aromatics are the least toxic, acute toxicity increases with increasing molecular size until the 4- and 5-ring compounds are reached, and alkylation of the aromatic nucleus seems to increase the toxicity of the parent compound. Thus, in both the benzene and naphthalene series, toxicity increases with increasing degrees of alkylation. The toxicities reported for m-, o-, and p-xylene for shrimp, bass, and crab larvae suggest that the position of alkyl substitution on the aromatic ring may influence toxicity (Benville and Korn In press, Caldwell et al This symposium). The most toxic hydrocarbon evaluated by Neff et al (1976) was 1-methylphenanthrene. Considering the toxicity of the compounds they studied and the concentrations of these compounds in toxic WSF's, Neff et al (1976) concluded that much of the toxicity of most crude and refined oils was due to the mono- and dinuclear aromatics.

TABLE 1

Comparative toxicity of different aromatic hydrocarbons, expressed in 96-hr LC50's with concentrations in ppm. Asterisk (*) indicates that toxic concentrations were above solubility limits.

AROMATIC H.C.	96-hr LC50's in ppm					
	POLYCHAETE[1]	SHRIMP[2]	CRAB LARVAE[3]	SHRIMP[4]	BASS[5]	GOLDFISH[6]
Benzene	----	27	108	20	5.8-10.9	----
Toluene	----	9.5	28	4.3	7.3	22.8
Ethyl benzene	----	----	13	0.5	4.3	----
Tri-methyl benzene	----	5.4	5.1	----	----	12.5
Xylene	----	7.4	----	----	----	16.9
m-	----	----	12	3.7	9.2	----
o-	----	----	6	1.3	11.0	----
p-	----	----	----	2.0	2.0	----
Naphthalene	3.8	2.4	> 2			
Methyl naphthalene	----	1.1	1.6			
Di-methyl naphthalene	2.6	0.7	0.60			
Tri-methyl naphthalene	2	----	----			
Phenanthrene	0.6	----	----			
Methyl phenanthrene	0.3	----	----			
Fluorene	1	----	----			
Fluoranthrene	0.5	----	----			
Chrysene	*	----	----			
Benzo(a)pryene	*	----	----			
1,2,5,6-Dibenzanthracene	*	----	----			

[1] Neff et al (1976). *Neanthes arenaceodentata*
[2] Neff et al (1976). *Palaemonetes pugio*
[3] Caldwell et al (This symposium). *Cancer magister*, Stage I larvae
[4] Benville and Korn (In press). *Crago franciscorum*
[5] Benville and Korn (In press). *Morone saxatilis*
[6] Brenniman et al (1975). *Carassius auratus*

The solubility of individual aromatic hydrocarbons decreases with increasing methyl substitution and number of rings, and this is reflected in most WSF's (Caldwell et al This symposium). The concentrations of mono-aromatics, such as benzene, are usually greater in WSF's than naphthalene and other larger aromatics. Since the toxicity of aromatic hydrocarbons increases as the concentrations in the WSF decrease, it is difficult to identify which specific compounds are most responsible for the toxicity of WSF's. The toxicity of oil-water solutions is probably due to contributions from both the more soluble, less toxic mono-nuclear aromatics and the less soluble, more toxic dinuclear aromatics. The relative importance of the two classes will likely depend on several factors, such as their relative concentrations in the parent oil, temperature, mixing characteristics, etc.

Since the solubilities of the larger polynuclear aromatic hydrocarbons are so low, these compounds probably contribute little to the acute toxicity of oil-water solutions. Neff et al. (1976) found several polynuclear aromatics that were not acutely toxic in four days at the maximum possible concentrations (solutions were 100% saturated). However, these compounds may contribute to long-term damage if they accumulate to significant concentrations in the tissues after lengthy exposure.

Several investigators have given evidence suggesting that mono- and dinuclear aromatic hydrocarbons account for much of the toxicity of oil-water solutions, and there have been correlations of adverse effects with tissue concentrations of aromatic hydrocarbons. However, it is premature to conclude that the aromatic hydrocarbons are solely responsible for the acute toxicity of oil-water solutions. The presence and toxicity of polar hydrocarbon derivatives or polar oxidation products of oil hydrocarbons have generally been ignored, because they are more difficult to identify and measure.

Comparative Sensitivity of Different Life Stages and Species

When assessing the potential impact of spill oil, it is crucial to know if some species or life stages are more sensitive than others to oil toxicity. As previously discussed, comparisons of animal sensitivities to oil are most valid when tested with flowthrough exposures during which the oil concentrations remain constant. Unfortunately, most studies have used static exposures, in which the oil concentrations decline with time, and slow-dying animals may appear more tolerant than they actually are to four-day exposures. In spite of these limitations with static exposures, we attempt to compare sensitivities of different life stages and species to oil.

Comparing Sensitivities of Different Life Stages

Survival of a species is dependent on the survival of each developmental stage. In general, eggs are assumed to be more tolerant than other life stages to oil exposures because of the protection given by the surrounding chorion. In contrast, larvae have been assumed to be more sensitive to oil pollution, probably because of their high mortality in response to natural environmental stresses. In an earlier review, Moore and Dwyer (1974) concluded that larvae are approximately an order of magnitude more sensitive than adults to oil toxicity. Although their conclusions are consistent with general assumptions of larval sensitivity, support for their conclusions is based on literature where oil concentrations had to be "normalized" for comparative purposes.

Comparisons of results from early studies on the sensitivity of eggs to oil are severely limited since exposure concentrations are reported in quantities of oil added. The sensitivities of fish eggs and larvae of Black Sea turbot (Mironov 1967), herring eggs incubated under an oil film (Kühnhold 1970), and cod eggs incubated in the water-soluble fraction of crude oil (Kühnhold 1974) have been reported. In addition to mortality, effects on developmental rates and developmental abnormalities which would presumably cause death at a later stage have been observed. In more recent quantitative studies, Struhsaker et al (1974) found herring eggs to be more tolerant than larvae. Similarly, Rice et al (1975) found salmon eggs to be quite resistant to the crude oil WSF, and that sensitivity increased after hatching until yolk absorption was complete.

Some investigators have also attempted to document the sensitivity of larvae of various species to oil but the concentrations are given as "volume added" and comparisons are arbitrary. Crude oil concentrations as low as 1,000 ppm are reported to retard development of Crassostrea larvae (Renzoni 1975). Concentrations of 100 ppm of oil are fatal to larvae of lobster (Wells 1972), shrimp, and crab (Mironov 1968). In addition, 10 ppm or less produced a delayed mortality and resulted in marked changes in color and behavior in lobster larvae (Wells 1972). Chia (1973) exposed 14 species (5 phyla) of pelagic larvae to 0.5% diesel oil and found that larger larvae generally lived longer than smaller larvae. The large range in reported lethal concentrations (10 to 1,000 ppm) may reflect differences in the toxicities of crude oils or the sensitivities of species, but is more likely due to differences in preparation of the oil-water mixtures. As stated previously, the amount of oil in solution is particularly influenced by mixing energy and duration (Gordon et al 1973, Anderson et al 1974, Rice et al 1976a). All of the above studies on larvae were quantitated in terms of "volume of oil added" rather than by quantitative analytical measurements that allow comparisons.

Recent studies employing analytical measurement of oil concentrations in the test exposures have found the sensitivities of larvae and juveniles to vary considerably, with sensitivity depending on life stage and species (summarized in Table 2). Neff et al, (1976) found larvae of grass shrimp to be more sensitive than adults, while post-larvae of brown shrimp were more resistant than early or late juveniles. They concluded that generalizations cannot be made about larvae versus adult sensitivity, and this is borne out by other studies. The early stages of a polychaete worm were found to be more resistant than later juvenile stages and adults to crude oil and No. 2 fuel oil (Rossi and Anderson 1976). Stage I lobster larvae are more sensitive (4-day LC50 of 0.86 mg oil/liter) than stages III and IV (LC50 of 4.8 mg/liter) (Wells and Sprague 1976). Brodersen et al (In press) determined sensitivities of stage I larvae of four species of shrimp, two species of crab, and of six larval stages of coonstripe shrimp. Sensitivities differed considerably, but stage III and VI larvae of coonstripe shrimp were the most sensitive (4-day LC50's of 0.35 and 0.24 ppm). Brodersen et al concluded that the stage I larvae tested were all slightly more sensitive than adults (larvae were 1.2-4.9 times more sensitive). They concluded that larvae are slow to die, and that comparisons of sensitivities should be based on oil concentrations that cause moribundity rather than mortality during a four-day test.

The animals most sensitive to oil appear to be crustacean larvae but this may only be apparent because of test methodology differences, or because these larvae have been studied more than any other phylum. Brodersen et al (In press) report a 96-hr LC50 for moribundity of 0.24 ppm of crude oil for stage VI larvae of coonstripe shrimp; Wells and Sprague (1976) report an LC50 of 0.14 ppm of oil after exposing lobster larvae for 30 days; and Sanborn and

Malins (1977) report an LC50 of less than 8-12 ppb of naphthalene after 36-hr exposures for larvae of spot shrimp and Dungeness crab. The test by Wells and Sprague was quite long (30 days) compared to most larval bioassays. The observations on shrimp larvae by Brodersen et al were of moribundity rather than death during the exposure. Although the observations of Wells and Sprague (1976) during the 30-day test with lobster were of death, they may be analogous to the previous observations on moribundity in shrimp larvae (Brodersen et al In press) since there was ample time during the 30-day tests for moribund lobster larvae to die. The tests by Sanborn and Malins (1976) were short in duration (36 hr) but were conducted with a pure and highly toxic compound under continuous-flow conditions.

TABLE 2

Comparison of larval sensitivities to crude oil and No. 2 fuel oil. The reported LC50's are from 96-hr static tests, except the 30-day test by Wells and Sprague, and the flowthrough, 36-hr tests by Sanborn and Malins.

Study and Species	Stage	Range of LC50 (ppm)	
		Crude oils	No. 2 fuel oil
Rossi and Anderson (1976)			
Polychaete	Juveniles	15-19.8	4-8.4
	Adults	12.5-17.6	2-4.2
Neff et al (1976)			
Grass shrimp	Larvae	--	1.2
3 sp. shrimp	Post larvae	--	1.4-6.6
3 sp. shrimp	Juveniles, adults	--	1.0-3.7
Wells and Sprague (1976)			
American lobster	Stage I-IV	0.8-4.9	--
	Stage I	0.14 (30-day LC50)	
Brodersen et al (In press)			
6 sp. shrimp and crab	Stage I	0.9-1.8	
Coonstripe shrimp	Stage I-VI	0.2-1.8	
Mecklenburg et al (This symposium)			
King crab	Stage I	2.0	
	Stage I molting	1.3	
Coonstripe shrimp	Stage I and II	4-7.9	
	Stage I molting	0.9	
Sanborn and Malins (1977)			
Spot shrimp	Stage I and V		Naphthalene ≤ 12 ppb

Previous studies with crustacean larvae exposed to oil (Wells 1972, Katz 1973) have suggested that molting larvae were more sensitive, since mortalities often increased during molting. Mecklenburg et al (This symposium) confirmed that molting larvae of coonstripe shrimp and king crab were more sensitive than non-molting larvae. Molting larvae of coonstripe shrimp were about five times more sensitive to oil than non-molting larvae.
Similarly, adult tanner crabs exposed to oil just prior to molting died during the molting process (Karinen and Rice 1974). Emery (1970) also found molting larvae of two crustaceans to be more sensitive to creosol than non-molting larvae. Emery suggested that tolerance is reduced during molting because of the toxicant-laden fluid that is taken up rapidly prior to molting to create enough hydrostatic pressure to split the exoskeleton. Since all molting crustaceans are probably more sensitive to oil exposure than non-molting crustaceans, comparison of sensitivities will be more valid if sensitivities are determined for crustaceans when they are all in the same period of the molt cycle, such as intermolt. However, isolating intermolt larvae for toxicity tests can be a problem, especially in warmer climates where larvae of some species molt more frequently and the intermolt periods are shorter.

Juveniles of non-crustaceans, such as polychaetes, are apparently more tolerant than adults to oil (Rossi and Anderson 1976). Polychaete juveniles grow by adding additional segments, and do not molt like crustaceans. Since there is no molting process, polychaete juveniles do not have a rapid uptake of fluid associated with growth. These differences in growth and sensitivity suggest that molting animals are more vulnerable to oil, probably because of increased permeability during the molting process.

At this time, generalizations about greater sensitivity to oil during early life stages cannot be made, since a general pattern has not emerged from the literature. More life stages from several groups need to be tested, with the bulk of the early literature being of little value. It seems likely that sensitive stages will be those that are already under a certain amount of natural stress. Larvae in general would seem to qualify, since they typically have a high natural mortality. If larvae are more sensitive, then species from a colder climate may be particularly vulnerable because of the relatively long period spent as developing larvae.

Although some questions remain about the relative sensitivity of larvae and adults exposed to oil, larvae are probably more vulnerable than adults. Larvae that are weakened by oil, though not killed outright, may become easy prey while adults so affected are afforded some protection by their greater size and their better-developed exoskeletons. Furthermore, some adults and juveniles will probably avoid contaminated water, since they can detect the oil and have the motor ability to avoid the area (Rice 1973). Larvae may lack both of these abilities. Lack of avoidance behavior has been observed in larvae of three species of marine fish (Kühnhold 1970) and in herring and crab larvae (Rice et al 1976a). The interactions in the environment are complex, and while sensitivities to oil can be extrapolated from the laboratory to the ecosystem, vulnerability and survival of larvae exposed to oil in the natural environment are another matter.

Comparing Sensitivities of Different Species
The sensitivities of several marine species have been tested in quantitative static or flow-through tests (summarized in Table 3). There have been differences between the studies, such as the oil used, mixing techniques, analytical methods, temperatures, and exposure methods (static or flowthrough). In spite of these differences, the ranges of LC50's overlap considerably between most studies, suggesting that sensitivities of most animals are fairly similar.

The study by Rice et al. (1976b) tested 27 species of marine fish and invertebrates, and permits the best comparisons of species sensitivities since methods, temperature, etc., were all similar. Fish and shrimp were usually among the more sensitive species tested, while intertidal animals were generally more tolerant. Intertidal animals are probably more tolerant to static exposures because they can temporarily insulate themselves from the exposures, at least until the concentrations in the static exposures have declined to sublethal levels. The intertidal limpets and chitons were more sensitive than the other intertidal animals, but this may have been due to damage occurring when they were collected (pried off the substrate).

The sensitivities of cold-water fish and shrimp (Rice et al. 1976b, Korn et al. In prep) appear greater than sensitivities of similar species from warmer climates (Anderson et al 1974, Neff et al 1976). The differences in sensitivity are consistent, but are not large. Rice et al (1976b) speculate that the cold-water species may appear more sensitive because lower temperatures increase the persistence of toxic aromatic hydrocarbons, even though there are differences in oils and species between the studies. This speculation is compatible with other studies. Cheatham et al (In prep.) have shown that water-soluble fractions at lower temperatures have increased persistence of aromatic hydrocarbons because of decreased losses from evaporation and biodegradation. Korn et al (In prep.) tested the effect of temperature on the toxicity of toluene, naphthalene, and the water-soluble fraction of crude oil to shrimp and fish. They found that with increasing temperature, toxicity increased for shrimp but decreased for fish. They concluded that temperature affects complex interactions in three ways: (1) the persistence of aromatic hydrocarbons is increased at lower temperatures, (2) temperature affects an animal's sensitivity by changing the rates of hydrocarbon uptake, metabolism, and excretion, and (3) temperature may act as a synergistic stress at cold or warm extremes.

TABLE 3

Comparison of adult sensitivities to crude oils and No. 2 fuel oil. All tests were static, except the flowthrough tests (FT) by Battelle. Most of the LC50's reported for the Battelle studies are estimates that have been calculated from Battelle's raw data.

Study and species tested	Temperature range (°C)	96-hr LC50's (ppm)	
		Crude oils	No. 2 fuel oil
Battelle/Sequim 1973-76			
2 fish sp. (FT)[1]	8	15-65	--
Coonstripe shrimp (FT)[2,3]	10-11	6.6-24.9	0.8
Coonstripe shrimp[1]	8	1.3-4.9	--
Texas A&M 1974-76			
4 crustacean sp.[4,5]	18-22	6->19.8	1.3-4.9
3 fish sp.[4]	18-22	5.5-19.8	3.9-6.3
3 polychaete sp.[6]	20	9.5-12.5	2.3-2.7
Auke Bay Lab 1976[7]			
4 fish sp.	3.6-10	1.2-2.9	0.8-2.1
6 crustacean sp.	3.5-5.4	0.6-4.2	0.5-1.7
4 limpet and chiton sp	3.9-7	3.6-9.6	0.4-5.0
12 invertebrate sp.	3.6-10	>3.1-14.7	>0.9-5.6

[1] Bean et al 1974
[2] Vaughan 1973
[3] Vanderhorst et al 1976
[4] Anderson et al 1974
[5] Neff et al 1976
[6] Rossi et al 1976
[7] Rice et al 1976

Conclusions

Methods
1. Toxicity tests with oil require some form of chemical analysis to determine what concentrations of oil are in the water. Gas chromatography or HPLC is preferable since these methods can measure concentrations of individual aromatic hydrocarbons.

2. Statistical analysis generating confidence intervals is more informative than graphic estimates of the LC50.

3. If oil toxicities are compared, reference toxicants and the same test species should be used.

4. If species sensitivities are compared, flowthrough tests with stable oil concentrations during exposure should be used.

Comparative Toxicity
1. Crude and refined oils are best considered as sources of toxic compounds with toxicity depending on the concentration of toxic compounds in the oil and on physical factors, such as temperature and viscosity of the oil, which affect the transport of petroleum hydrocarbons into the water.

2. Refined oils are generally considered more toxic than crude oils on a volume added basis because refined oils often have higher concentrations of aromatic hydrocarbons and are usually less viscous than crude oils.

3. The toxicity of oils is apparently due to the soluble compounds in the water rather than dispersed droplets.

4. The toxicity of aromatic hydrocarbons increases with the number of rings and with the degree of alkyl substitution. The solubility of these compounds decreases with these factors, so that the relative importance of individual aromatic hydrocarbons to toxicity of water-soluble fractions is unknown. Mono- and dinuclear aromatics are probably the most important classes of compounds, accounting for most of the toxicity in water-soluble fractions.

Comparative Sensitivity
1. No conclusions can be made concerning egg sensitivities because not enough quantitative data exist.

2. Crustacean larvae appear more sensitive than most adults. Molting animals are more sensitive than non-molting animals.

3. Sensitivity data for larvae or juveniles from other groups are generally lacking, except for polychaete juveniles, which are more tolerant than adults.

4. Adult sensitivities are fairly similar when data from static tests are compared. Intertidal animals appear more tolerant in short-term, static exposures, probably because of their ability to temporarily insulate themselves from the environment.

5. Cold-water species probably have sensitivities that are equivalent to the sensitivities of species from warmer water, but cold-water species may be more vulnerable to oil toxicity because toxic aromatic hydrocarbons persist longer at colder temperatures.

6. The preceeding conclusions are based on static exposures. Because of the deficiencies of the static exposure method (declining dose), conclusions about sensitivity differences may be altered considerably when tests with flowthrough exposures have been conducted.

Recommendations for Future Research
1. A high standard of quantitative methodology (flowthrough tests, chemical analyses, and statistical analyses) should be used in future toxicity tests.

2. There is little need to test the toxicity differences between different oils. Future research should: (a) determine if aromatics account for most, if not all, of the toxicity of WSF, and (b) determinate the relative importance of individual mono- and dinuclear aromatics to the acute toxicity under various conditions.

3. Additional sensitivity data are needed for eggs, larvae, and juveniles of several groups of animals such as molluscs, echinoderms, etc. since virtually no information on these animals is available.

4. Several species need to be retested with flowthrough exposures, so that (a) better estimates of sensitivity can be measured, and (b) sensitivities can be compared between species.

5. A few selected species should be tested for longer periods of time, so that the relationship between oil concentrations that are toxic for short and long exposures can be determined.

References

Anderson, J. W., J. M. Neff, B. A. Cox, H. E. Tatem, and G. M. Hightower, Characteristics of dispersions and water-soluble extracts of crude and refined oils and their toxicity to estuarine crustaceans and fish, *Mar. Biol. 27*, 75-88 (1974).

Bean, R. M., J. R. Vanderhorst, and P. Wilkinson, Interdisciplinary study of the toxicity of petroleum to marine organisms, Battelle Pacific Northwest Laboratories, Richland, Washington, 31 pp. (1974).

Benville, P. E., and S. Korn, The acute toxicity of six monocyclic aromatic crude oil components to striped bass (*Morone saxatilis*) and bay shrimp (*Crago franciscorum*), Calif. Fish Game (In press).

Brenniman, G., R. Hartung, and W. J. Weber, Jr., A continuous flow bioassay method to evaluate the effects of outboard motor exhausts and selected aromatic toxicants on fish, *Water Res. 10*, 165-169 (1976).

Brodersen, C. C., S. D. Rice, J. W. Short, T. A. Mecklenburg, and J. F. Karinen, Sensitivity of larval and adult Alaskan shrimp and crabs to acute exposures of the water-soluble fraction of Cook Inlet crude oil. In: *Proceedings, 1977 Oil Spill Conference (Prevention, Behavior, Control, Cleanup)*, American Petroleum Institute, Washington, D.C., In press.

Caldwell, R. S., E. M. Caldarone, and M. H. Mallon, Effects of a seawater-soluble fraction of Cook Inlet crude oil and its major aromatic components on larval stages of the Dungeness crab, *Cancer magister* Dana. In: *Proceedings, NOAA-EPA Symposium on Fate and Effects of Petroleum Hydrocarbons*, Pergamon Press, Oxford, 1977.

Chia, F., Killing of marine larvae by diesel oil, *Mar. Pollut. Bull. 4*, 29-30 (1973).

Cheatham, D. L., R. S. McMahon, S. J. Way, J. W. Short, and S. D. Rice, Effects of temperature, volatility, and biodegradation on the persistence of aromatic hydrocarbons in seawater, Manuscr. in prep.

Doudoroff, P., B. G. Andersen, G. E. Burdick, P. S. Galtsoff, W. B. Hart, R. Patrick, E. R. Strong, E. W. Surber, and W. M. Van Horn, Bio-assay methods for the evaluation of acute toxicity of industrial wastes to fish, *Sewage Ind. Wastes 23*, 1380-1397 (1951).

Emery, R. M., The comparative acute toxicity of creosol to two benthic crustaceans, *Water Res. 4*, 485-491 (1970).

Evans, D. R., and S. D. Rice, Effects of oil on marine ecosystems: A review for administrators and policy makers, *Fish. Bull., U.S. 72*, 625-638 (1974).

Finney, D. J., *Probit Analysis*, 3d edition, 333 pp., Cambridge University Press, London, 1971.

Gordon, D. C., Jr., P. D. Keizer, and N. J. Prouse, Laboratory studies on the accommodation of some crude and residual fuel oils in sea water, *J. Fish. Res. Board Can. 30*, 1611-1618 (1973).

Gruenfeld, M., Extraction of dispersed oils from water for quantitative analysis by infrared spectrophotometry, *Environ. Sci. Technol. 7*, 636-639 (1973).

Karinen, J. F., and S. D. Rice, Effects of Prudhoe Bay crude oil on molting tanner crabs, *Chionoecetes bairdi*, *Mar. Fish. Rev. 36*, 31-37 (1974).

Kator, H., C. H. Oppenheimer, and R. J. Miget, Microbial degradation of a Louisiana crude oil in closed flasks and under simulated field conditions. In: *Joint Conference on the Prevention and Control of Oil Spills,* pp. 287-296, American Petroleum Institute, Washington, D. C. 1973.

Katz, L. M., The effects of water soluble fraction of crude oil on larvae of the decapod crustacean *Neopanope texana* (Sayi), *Environ. Pollut. 5*, 199-204 (1973).

Kauss, P., T. C. Hutchinson, C. Soto, J. Hellebust, and M. Griffiths, The toxicity of crude oil and its components to freshwater algae. In: *Proceedings of Joint Conference on Prevention and Control of Oil Spills,* pp. 703-714, American Petroleum Institute, Washington, D.C., 1973.

Korn, S., D. A. Moles, and S. D. Rice, Effects of temperature on the acute toxicity of toluene, naphthalene, and the water-soluble fraction of Cook Inlet crude oil to pink salmon and shrimp, Manuscr. in prep.

Kühnhold, W. W., The influence of crude oils on fish fry, FAO Technical Conference on Marine Pollution and Its Effects on Living Resources and Fishing, Rome, Italy, Dec. 1970, FIR: MP/70/E-64 (1970). (See also, M. Ruivo (ed.), *Marine Pollution and Sea Life*, pp. 315-318, Fishing News (Books) Ltd., Surrey, England, 1972.)

Kühnhold, W. W., Investigations on the toxicity of seawater-extracts of three crude oils on eggs of cod (*Gadus morhua* L.), *Ber. Komm. Meeresforsch. 23*, 165-180 (1974).

LaRoche, G., R. Eisler, and C. M. Tarzwell, Bioassay procedures for oil and oil dispersant toxicity evaluation, *J. Water Pollut. Control Fed. 42*, 1982-1989 (1970).

Lysyj, I., and E. C. Russell, Dissolution of petroleum-derived products in water, *Water Res. 8*, 863-868 (1974).

McAuliffe, C., Solubility in water of paraffin, cycloparaffin, olefin, acetylene cycloolefin, and aromatic hydrocarbons, *J. Phys. Chem. 70*, 1267-1275 (1966).

McAuliffe, C., Determination of dissolved hydrocarbons in subsurface brines, *Chem. Geol. 4*, 225-233 (1969).

Mecklenburg, T. A., S. D. Rice, and J. F. Karinen, Molting and survival of king crab (*Paralithodes camtschatica*) and coonstripe shrimp (*Pandalus hypsinotus*) larvae exposed to Cook Inlet crude oil water-soluble fraction. In: *Proceedings, NOAA-EPA Symposium on Fate and Effects of Petroleum Hydrocarbons*, Pergamon Press, Oxford, 1977.

Mironov, O. G., The effect of small concentrations of oil and oil products on developing eggs of the Black Sea turbot, *Probl. Ichthyol. 7(3)(44)*, 1-5 (1967).

Mironov, O. G., The variability of larvae of some crustaceans in marine water polluted with oil products, *Zoological Journal 48*, 1734-1737 (1969).

Moore, S. F., and R. L. Dwyer, Effects of oil on marine organisms. Critical assessment of published data, *Water Res. 8*, 819-827 (1974).

National Academy of Sciences, *Petroleum in the Marine Envrionment. Workshop on Inputs, Fates, and the Effects of Petroleum in the Marine Environment*, 170 pp., Natl. Acad. Sci., Washington, D. C., 1975.

Neff. J. M., J. W. Anderson, B. A. Cox, R. B. Laughlin, Jr., S. S. Rossi, and H. E. Tatem, Effects of petroleum on survival, respiration and growth of marine animals. In: *Sources, Effects, and Sinks of Hydrocarbons in the Aquatic Environment*, pp. 515-539, American Institute of Biological Sciences, Washington, D. C., 1976.

Ottway, S., The comparative toxicities of crude oil. In: E. B. Cowell (ed.), *The Ecological Effects of Oil Pollution on Littoral Communities*, pp. 172-180, Elsevier Publishing Co., New York, 1971.

Peake, E., and G. W. Hodgson, Alkanes in aqueous systems. 1. Exploratory investigation on the accommodation of C_{20}-C_{33} n-alkanes in distilled water and occurrence in natural water systems, *J. Am. Oil Chem. Soc. 43(4)*, 215-222 (1966).

Percy, J. A., and T. C. Mullin, Effects of crude oils on Arctic marine invertebrates, Canadian Fisheries and Marine Service, Dept. of the Environment, Beaufort Sea Tech. Rep. 11, 167 pp. (1975).

Renzoni, A., Influence of crude oil derivatives and dispersants on larvae, *Mar. Pollut. Bull. 4*, 9-12 (1973).

Rice, S. D., Toxicity and avoidance tests with Prudhoe Bay oil and pink salmon fry. In: *Proceedings of the Joint Conference on Prevention and Control of Oil Spills*, pp. 667-670, American Petroleum Institute, Washington, D. C., 1973.

Rice, S. D., D. A. Moles, and J. W. Short, The effect of Prudhoe Bay crude oil on survival and growth of eggs, alevins, and fry of pink salmon, *Oncorhynchus gorbuscha*. In: *Proceedings, 1975 Conference on Prevention and Control of Oil Pollution*, pp. 503-507, American Petroleum Institute, Washington, D. C., 1975.

Rice, S. D., J. W. Short, C. C. Broderson, T. A. Mecklenburg, D. A. Moles, C. J. Misch, D. L. Cheatham, and J. F. Karinen, Acute toxicity and uptake-depuration studies with Cook Inlet crude oil, Prudhoe Bay crude oil, No. 2 fuel oil and several subarctic marine organisms, NWAFC Processed Report, 90 pp., Northwest and Alaska Fisheries Center Auke Bay Laboratory, National Marine Fisheries Service, NOAA, P. O. Box 155, Auke Bay, AK 99821, 1976a.

Rice S. D., J. W. Short, and J. F. Karinen, Toxicity of Cook Inlet crude oil and No. 2 fuel oil to several Alaskan marine fishes and invertebrates. In: *Sources, Effects, and Sinks of Hydrocarbons in the Aquatic Environment*, pp. 394-406, American Institute of Biological Sciences, Washington, D.C., 1976b.

Rossi, S. S., and J. W. Anderson, Toxicity of water-soluble fractions of No. 2 fuel oil and South Louisiana crude oil to selected stages in the life history of the polychaete, *Neanthes arenaceodentata*, *Bull. Environ. Contam. Toxicol. 12*, 18-24 (1976).

Rossi, S. S., J. W. Anderson, and G. S. Ward, Toxicity of water-soluble fraction of four test oils for the polychaetous annelids *Neanthes arenaceodentata* and *Capitella capitata*, *Environ. Pollut. 10*, 9-18 (1976).

Sanborn, H. R., and D. C. Malins, Toxicity and metabolism of naphthalene: A study with marine larval invertebrates. In: *Proc. Soc. Exp. Biol. Med.* 154, 151-155 (1977).

Short, J. W., S. D. Rice, and D. L. Cheatham, Comparison of the standard method for oil and grease determination with an infrared spectrophotometric method on known toxic water-soluble fractions of oils. In: *Assessment of the Arctic Marine Environment: Selected Topics*, pp. 451-462, Maple Press, York, Pennsylvania, 1976.

Struhsaker, J. W., M. B. Eldridge, and T. Echeverria, Effects of benzene (a water-soluble component of crude oil) on eggs and larvae of Pacific herring and northern anchovy. In: F. M. Vernberg and W. B. Vernberg (eds.), *Pollution and Physiology of Marine Organisms*, pp. 253-284, Academic Press, New York, 1974.

Vanderhorst, J. R., E. I. Gibson, and L. J. Moore, Toxicity of No. 2 fuel oil to coonstripe shrimp, *Mar. Pollut. Bull.* 7, 106-107 (1976).

Vaughan, B. E. (ed.), Effects of oil and chemically dispersed oil on selected marine biota-- a laboratory study. Battelle Pacific Northwest Laboratories, Richland, Washington, API Pub. No. 4191, 1973.

Wells, P. G., Influence of Venezuelan crude oil on lobster larvae, *Mar. Pollut. Bull.* 3, 105-106 (1972).

Wells, P. G., and J. B. Sprague, Effects of crude oil on American lobster (*Homarus americanus*) larvae in the laboratory, *J. Fish. Res. Board Can.* 33, 1604-1614 (1976).

CHAPTER 9

RESPONSES TO SUBLETHAL LEVELS OF PETROLEUM HYDROCARBONS:
ARE THEY SENSITIVE INDICATORS AND DO THEY CORRELATE WITH TISSUE CONTAMINATION?

J. W. Anderson

Battelle, Pacific Northwest Division
Marine Research Laboratory
Sequim, Washington 98382

Introduction

The term "sublethal effect" has come to mean the abnormal response of an organism produced by concentrations of various pollutants that are below the levels required to kill 50% of the organisms in 96 hr (96 hr LC_{50}, 96 hr TLM). Since lethal concentration is the best term to use in tests on aquatic organisms (as contrasted with an injected dose), toxicity results will be expressed as LC_{50} values. Many acute toxicity tests utilize a 96 hr period and generally the exposure water is not changed (static conditions). Toxicity tests are sometimes run for extended periods when water is changed frequently and food is supplied, and these tests yield more meaningful values. There is considerable disagreement regarding the use of toxicity data, since it is thought that these pollutant concentrations are well above the levels which in the real world are likely to be harmful to various marine organisms. As the findings of recent investigations are reviewed in this chapter, we will again turn to this question and perhaps be able to suggest an answer.

In an earlier paper (Anderson et al., 1974a), a list of sublethal studies was proposed and various aspects of the list were discussed. We can now reflect on advances regarding sublethal studies since 1974. One of the prime reasons for advances in the study of effects of petroleum hydrocarbons has been the development of new and better techniques for analysis of these compounds in water, tissues, and sediments. Most studies concerning acute or sublethal exposures conducted prior to 1972 generally lacked the analytical information necessary to relate findings of one study to any other. However, the general use of advances in gas chromatography (GC), ultraviolet (u.v.) (Neff and Anderson, 1975), florescence spectrophotometry (Gordon et al., 1974), and the GC-mass spectrometric system (Warner, 1976) have greatly enhanced the quality of studies on biological effects. Of prime consideration in measuring an organism's response to pollutant exposure should be the careful analysis of the hydrocarbon concentration contacting the organism and the concentration accumulated in its tissue during the exposure. The topics of accumulation and release of petroleum hydrocarbons are discussed in greater detail elsewhere in these Proceedings; however, a discussion of the effects of these compounds must consider the concentration in the water and in the tissues of exposed organisms.

Before concentrating on the types of sublethal studies that have been examined, I would first like to set the stage by discussing what we have to gain by careful observation of acute toxicity data. One must collect these data before conducting more sophisticated assays of the condition of organisms. The more toxicity information gathered the better will be the design of future experimentation. Indeed, it is impossible to determine the significance or the sensitivity of a sublethal response without comparison to a value such as the 96 hr LC_{50}, which also provides investigators with comparative information on compounds, species or life stages. Comparisons are only valid when the same chemical and physical conditions are present in each test. If these conditions vary, inter-investigation and inter-laboratory comparisons are extremely difficult, particularly when the toxicant is petroleum. Not only do the numerous procedures for preparation of the test hydrocarbon solution lead to strikingly different mixtures, but the original petroleum product may vary in endless ways. Even the oil produced from a single well-head will change significantly in composition with time. Faced with these problems, the best one can do is to utilize the same batch of oil (from one source at one time) to test the sensitivity of several species or life stages of the same species, under the same experimental conditions. The researcher then has a sound basis on which to compare his own data from different experiments, and if the same conditions and petroleum products are used by other investigators, data can be compared between laboratories. Before about 1973, no such inter-laboratory comparisons could be made, but Reference Oils (Anderson et al., 1974b) have been used since that time and greater attention has been paid to the characteristics and water-born quantities of hydrocarbons used in biological investigations.

Measuring the concentration acutely toxic to the most sensitive life stage of a given species is one approach to determining the level of a pollutant which will reduce the population. A concentration that is sublethal to an adult animal is generally assumed to produce mortality in its eggs, larvae or young. Estimates have been made which indicate that early life stages of a species may suffer significant mortality at concentrations that are 1 to 3 orders of magnitude less than those toxic for adults (Moore and Dwyer, 1974). We have examined the sensitivity of selected life stages of several species using the same hydrocarbon solutions as those in physiological studies (Cox, 1974; Tatem, 1975; Anderson, 1975; Rossi and Anderson, 1976). Therefore, evaluation is possible of differences between life stages, species, and the various means of determining abnormal physiological, behavioral or growth responses.

Investigations to elucidate the effects on organisms of sublethal doses of petroleum hydrocarbons and other pollutants can be categorized as physiological studies, behavioral studies, and investigations of growth and reproduction, all of which this paper discusses. The histological effects of exposure are discussed in detail by Dr. Hawkes in these Proceedings.

Physiological studies include a wide variety of different types of investigations which might range from the level of enzymes through cells to whole organisms. The metabolism of an organism can be measured in numerous ways, and one of the most popular and most readily available techniques is the measurement of oxygen consumption. Two other physiological functions of marine organisms that could be discussed under the category of physiological investigations are osmoregulation, including ionic regulation, and feeding and nutrition. The relatively few investigations regarding the effects of hydrocarbons on the behavioral response of marine organisms will be discussed. The final category, which will receive emphasis in this paper, is that of reproduction and growth, since I believe that these are perhaps the most valuable parameters in determining the effects of petroleum hydrocarbons on marine organisms.

Results and Discussion

The results shown on the following pages were derived from organisms exposed at room temperature 20±1°C in synthetic seawater (Instant Ocean) at either 15 or 20 parts per thousand ($^0/_{00}$) salinity, containing petroleum compounds in either the form of a water extract (WSF) or a dispersion (OWD). Details regarding the specific reference oils used, the preparation of the test solutions, and the characteristics may be found in Anderson *et al.* (1974b). This paper also demonstrated the significantly greater toxicity produced by No. 2 Fuel oil, as compared to 2 crude oils and Bunker C residual oil. Several other recent studies have verified this general pattern of toxicity.

Comparative Toxicity
Larvae, young (postlarvae), and adult grass shrimp (*Palaemonetes pugio*) were exposed to water-soluble fractions (WSF) of No. 2 Fuel oil to determine the 96 hr LC_{50} values for each stage. Adults were also tested with naphthalenes (naphthalene + methylnaphthalenes + dimethylnaphthalenes), which represent approximately 30% of the total hydrocarbons in the WSF (Fig. 1).

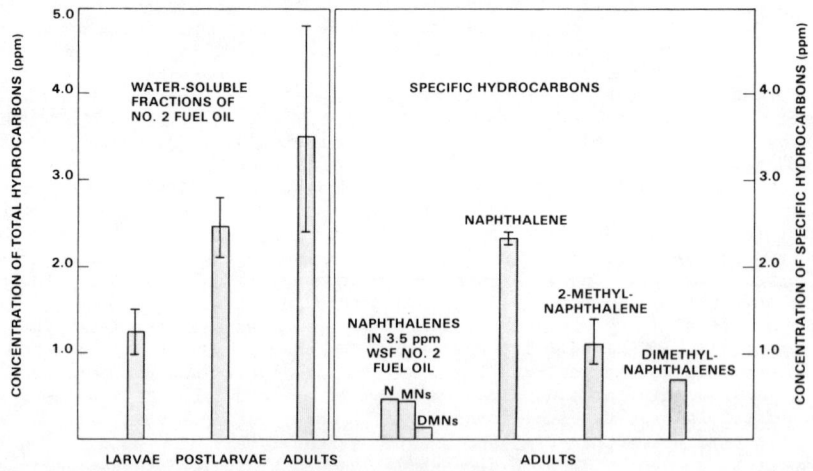

Fig. 9.1 Concentrations of WSFs or specific hydrocarbons producing 50% mortality of *Palaemonetes pugio* life stages in 96 hr. Vertical bars represent standard deviations. (Redrawn from Tatem, 1975).

The term total naphthalenes (TN) also refers to the sum of the concentrations of these compounds. Note that the 96 hr LC_{50} values for postlarvae and adults are not statistically different from one another (overlap of standard deviations). However, the level that is toxic to larvae is significantly smaller than the other 2 values, representing approximately 1/3 the concentration necessary to produce the effects on the juveniles and adults. The concentration of naphthalenes present in a 3.5 ppm WSF of No. 2 Fuel oil is also shown in Fig. 1. Though we assume that these compounds are responsible for at least a majority of the toxicity exhibited by this No. 2 Fuel oil WSF, I will not attempt to prove this point by mathematical means. It can be seen that there is an increase in the toxicity of the compounds as alkylation of the naphthalene structure increases. That is, the order of most toxic to least is dimethylnaphthalenes (DMN) > methylnaphthalenes (MN) > naphthalenes (N).

Similar tests were conducted using the commercially important brown shrimp (*Penaeus aztecus*) and white shrimp (*P. setiferus*) Fig. 2). The sensitivity of brown shrimp to the WSF of No. 2 Fuel oil increased by approximately 2 times while the size of the individuals increased by approximately 2 orders of magnitude. The brown shrimp appeared to become more sensitive as they grew, whereas the white shrimp increased slightly in their resistance to these same hydrocarbons over a size increase of approximately 200-fold. In neither of these species was there a dramatic change in the tolerance to these petroleum hydrocarbons, even though the organisms increased considerably in size. *Penaeus aztecus* shows sensitivity to specific naphthalene compounds in the same order as *Palaemonetes* (Fig. 1): DMN > MN > N. *Palaemon Palaemonetes* and *Penaeus aztecus* exhibit 96 hr LC_{50} values at between 3 and 4 ppm of the WSF of No. 2 Fuel oil and at about 1 ppm or less for methyl- and dimethylnaphthalenes.

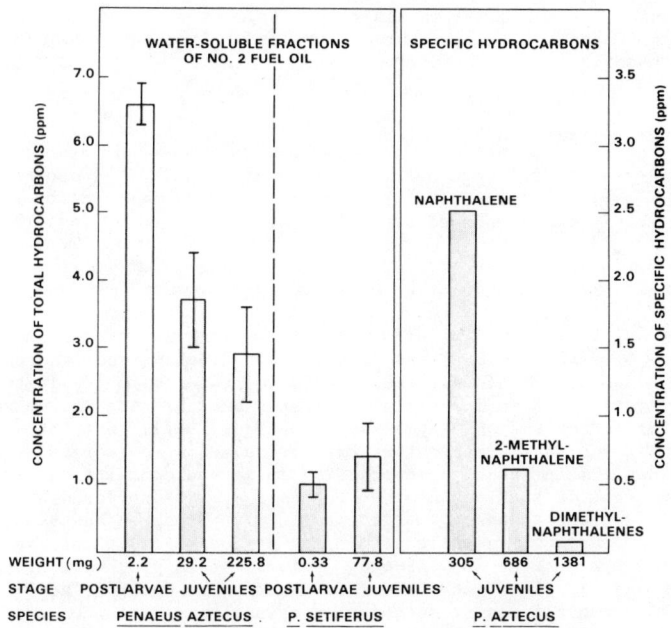

Fig. 9.2 Concentrations of WSFs or specific hydrocarbons producing 50% mortality of *Penaeus aztecus* or *P. setiferus* life stages in 96 hr. Vertical bars represent standard deviations. (Redrawn from Cox, 1974)

Figure 3 condenses information regarding the tolerance of various life stages of the polychaete *Neanthes arenaceodentata* (Rossi and Anderson, 1976). At numerous periods during the growth of these organisms, groups of animals were taken for 96 hr tests to determine the relative sensitivity of each life stage. The results shown in this figure again illustrate the greater toxicity of extracts from No. 2 Fuel oil as compared to a crude oil (South Louisiana). Several other interesting conclusions can be drawn from these data, including the general finding that these organisms decreased in tolerance as they increased in age. An exception to this pattern was the tolerance of gravid females as compared to adult males. On exposure to both fuel oil and crude oil WSFs, organisms increased in sensitivity by approximately a factor of 2 over a 50-day growth period. Worms that were 40 days or older

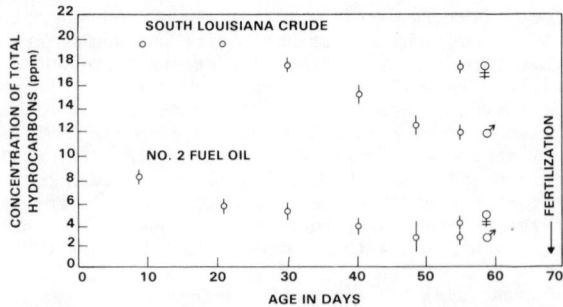

Fig. 9.3 Concentrations of No. 2 Fuel oil and So. Louisiana Crude oil WSFs producing 50% mortality of *Neanthes arenaceodentata* life stages in 96 hr. Vertical lines indicate standard deviations and ♀ represents gravid females. (From Rossi and Anderson, 1976)

exhibited 96 hr LC_{50} values of between 3 and 4 ppm for No. 2 Fuel oil WSFs which compare closely to those values exhibited by *Palaemonetes* and the 2 species of penaeid. The greater resistance of gravid females appears to stem from the large lipid pool provided by the eggs which serves to sequester the naphthalenes and remove them from the active tissue of the adult animal (Rossi and Anderson, 1976). Larvae developing from these contaminated eggs appear to lose the contamination after several days.

Respiration

The respiratory rate of aerobic organisms clearly represents the metabolic activity of that animal. The rate of oxygen consumption by an organism has been used for some time to determine the extent of stress by various natural (temperature/salinity) and man-made perturbations. A problem in any study regarding the effect of a pollutant on the rate of oxygen consumption of an animal is the abundance of endogenous and exogenous factors producing alterations in the metabolic rate of an organism (Newell, 1970; 1973; Prosser, 1973). As will be illustrated later, the extent of modification in rate produced from changes in nutritional state may be as great as pollutant-produced alternations.

Earlier studies examining the effects of effluents on the respiratory rate of fish and crustaceans were those of Steed and Copeland (1967) and Wohlschlag and Cameron (1967). *Cyprinodon variegatus* (sheepshead minnow) exhibited a suppression in respiratory rate when exposed to low concentrations of an industrial effluent, but the rate increased above the control level at higher concentrations (Steed and Copeland, 1967). When this species was exposed to WSFs of No. 2 Fuel oil, respiratory rate increased with increasing concentration, whereas Bunker C residual oil produced a suppression in oxygen consumption (Anderson *et al.*, 1974c). The metabolic rate of pinfish (*Lagodon rhomboides*) was depressed when exposed to a petrochemical waste at extreme temperatures (10 and 30°C). Brocksen and Bailey (1973) measured the respiratory rates of juvenile salmon and striped bass at intervals during a 96 hr exposure to 5 and 10 ppm benzene in a flowing system. At 10 ppm, salmon respiration increased, but after 10 days in clean water the rate returned to control level. *Littorina littorea* (marine snails) exhibited a stimulation in respiratory rate when exposed to a dispersion of Bunker C oil (Hargrave and Newcombe, 1973). Two other molluscs, *Mytilus* and *Modiolus* (bivalves), also responded to exposure to crude oil (WSF) by increasing their metabolic rate (Gilfillan, 1975). Avolizi and Nuwahyid (1974), however, reported depression in respiratory rates of the bivalves *Brachiodontes* and *Donax* exposed to a light Arabian Crude oil. The amphipod, *Onisimus affinis*, generally responded to low levels of oil-water dispersions (OWDs) by a suppression in oxygen consumption, but at higher concentrations respiration increased (Percy, in these Proceedings).

In two recent publications, the investigators examined parameters which indirectly relate to metabolic rate, even though oxygen consumption was not measured. Anderson *et al.* (1976) showed a correlation between the heart beat rates of embryonic estuarine fish (*Fundulus similus*, *F. heteroclitus*, and *Cyprinodon variegatus*) and the concentrations of petroleum hydrocarbons in the media. Extracts of No. 2 Fuel oil and South Louisiana Crude oil reduced the rate of heart beat in proportion to the concentration of hydrocarbons. Heart beat rate was not a sensitive measure of sublethal effects because approximately the same WSF concentrations producing reduced rates (0.2 to 0.5 ppm TN) resulted in 50% or greater embryonic mortality. The determination of hydrocarbon effects on breathing (opercular movements) and coughing rates of pink salmon fry during short exposures (3-22 hr) showed that abnormal responses occurred at concentrations approximating 30% of the 96 hr LC_{50} (Rice *et al.*, 1976).

There was a close correlation between increased rate of breathing and the content of naphthalenes in the WSFs of 3 oils, when expressed as per cent of the 96 hr LC_{50}. The greatest increase in breathing rate was produced by a 3 hr exposure to the WSF of No. 2 Fuel oil (about 0.6 ppm TN), but rates dropped during later phases of the exposure (up to 22 hr), regardless of the oil or concentration. Of the aromatic compounds measured, methyl- and dimethylnaphthalenes were the only hydrocarbons remaining in tissues after 96 hr of exposure.

It is obvious that one cannot come to a clear understanding of the effects of petroleum hydrocarbons on the respiratory rates of marine organisms from the findings discussed above, but perhaps the results of the following studies, utilizing precisely the same hydrocarbon preparations, will shed some light on the matter. In extensive bioassay studies (Anderson et al., 1974b), the mysid crustacean, *Mysidopsis almyra*, consistently demonstrated the greatest sensitivity to the various petroleum solutions. The respiratory rate of this mysid was measured during or shortly after exposure to different sublethal concentrations of the OWDs and WSFs. In general, animals exposed to the WSFs for 2 hr before respiratory measurements were made in clean seawater did not exhibit respiratory rates significantly different from controls.

Since there was the possibility that the organisms were respiring at abnormal rates during exposure to the WSFs but returned to normal rates after transfer to clean water, oxygen consumption rates were measured during exposure of the mysids to various concentrations of the WSF of No. 2 Fuel oil. All concentrations tested (5 to 30% WSF) induced respiratory responses, which were significantly higher than the controls (Fig. 4A). There was a sharp and significant rise in respiratory rate between the 15 and 20% exposure levels, followed by a sharp drop in oxygen consumption by mysids exposed to the 30% WSF.

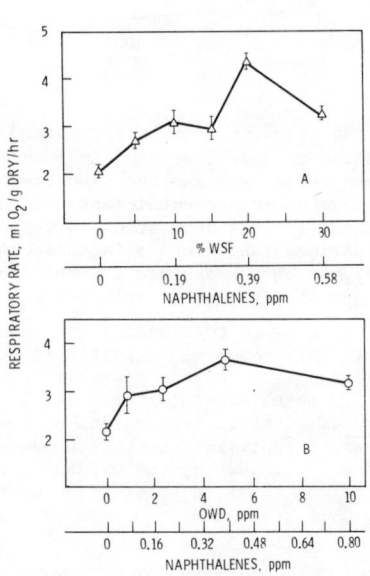

Fig. 9.4 Relationship between respiratory rate of mysids, *Mysidopsis almyra*, and the hydrocarbon content of exposure water. Vertical bars represent standard errors. (From Anderson, 1975)

A. Measured during exposure to WSFs of No. 2 Fuel oil (5 animals/container and 5 containers/concentration).

B. Measured in clean seawater after a 2 hr exposure to OWDs of No. 2 Fuel oil (10 animals/container and 3 containers/concentration).

To test the effect of an OWD on the respiratory rate of the mysids, groups of animals were exposed for 2 hr to OWD concentrations between 1 and 10 ppm of the No. 2 Fuel oil. After exposure, 30 mysids from each exposure concentration, including controls, were placed in respiration chambers (10/chamber) with clean water and their oxygen consumption rates measured. As before, the respiratory rates of the mysids increased gradually to a maximum of between 1.5 and 2 times control levels as the exposure concentration increased, and then dropped at the highest exposure concentration (Fig. 4B). It should be noted that, in an attempt to elicit an effect, the highest concentrations of OWDs and WSFs used approximated the 24 hr LC_{50} values for this species (Anderson et al., 1974b). The drop in respiratory rate at the highest exposure levels may therefore be indicative of the near lethal condition.

When the levels of total naphthalenes in both solutions are examined, one finds that the concentration producing the highest respiratory rates (both cases) is approximately 0.4 ppm. These data lend some support to the thesis that naphthalenes are the most significant compounds in fuel oil extracts or dispersions. Unfortunately, tissues were not analyzed for content of naphthalenes, but subsequent experiments with other species were designed to correlate response with body burden of naphthalenes.

Postlarval brown shrimp, *Penaeus aztecus*, were exposed to a 30% WSF of No. 2 Fuel oil. There was a close correlation between increased rate of breathing and the content of naphthalenes in WSFs of 3 oils, when expressed as per cent of the 96 hr LC_{50}. The greatest increase in breathing rate was produced by a 3 hr exposure to the WSF of No. 2 Fuel oil (about 0.6 ppm TN), but rates dropped during later phases of the exposure (up to 22 hr), regardless of the oil or concentration. Of the aromatic compounds measured, methyl- and dimethynaphthalenes were the only hydrocarbons remaining in tissues after 96 hr of exposure.

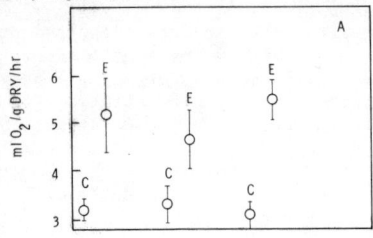

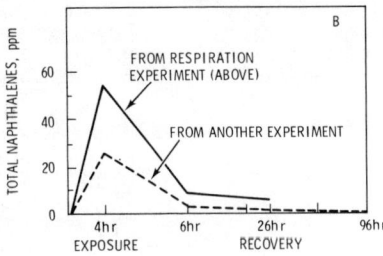

Fig. 9.5 Relationship between the respiratory rate of postlarval *Penaeus aztecus* and the rates of accumulation and release of naphthalenes. Determinations were made during a 4 hr exposure to a 30% WSF of No. 2 Fuel oil (0.85 ppm TN), and after 6 and 26 hr in clean seawater. (Redrawn from Cox, 1974)

A. Vertical bars represent standard deviations, when n=12 for both exposed (E) and control (C) animals.

B. Each data point represents the analysis of a pooled sample of 15 to 20 postlarvae. After 4 hr of exposure (as in respiratory phase, A), a large group of postlarvae were transferred to clean seawater.

Uptake and release rates were recorded for an additional group of postlarvae, which were treated exactly as those used in respiratory determinations (Fig. 5B). These data (solid line) and those from another experiment (dashed line) show that a 26 hr recovery period in clean seawater is not sufficient time for complete release of naphthalenes. Apparently, a minimum of 96 hr of depuration is required to obtain animals with only background levels of contamination. While other factors, including concentrations in the tissues of other hydrocarbons, may be responsible for the increased rate of respiration, the contamination level of 2-6 ppm TN might well explain the stimulated respiratory rate.

Further studies on respiration were conducted with the grass shrimp, *Palaemonetes pugio*. Using several groups of organisms in both respiratory and uptake and release phases of the experiments, it was again possible to correlate the rate of oxygen consumption with tissue content of total naphthalenes. Initially, the response to a 48 hr period of starvation was measured (Fig. 6A). Suppression of respiratory rate was clearly exhibited by these shrimp to an extent equivalent to that produced by exposure for 5 hr to 0.1 ppm TN from an OWD of No. 2 Fuel oil. During the exposure, shrimp accumulated a total of 2.2 ppm of naphthalenes, but on return to clean water released these compounds to background levels in 72 hr (Fig. 6B).

Fig. 9.6 Relationship between the respiratory rate of *Palaemonetes pugio* and the accumulation and release of naphthalenes. Shrimp were exposed for 5 hr to an OWD of No. 2 Fuel oil containing 3 ppm TH and 0.1 ppm TN, during which respiration was measured. (Redrawn from Anderson, 1975)

A. The same 8 groups of 12-14 shrimp were used in all respiratory rate determinations over a period of 10 days. Vertical bars represent standard deviations.

B. A large group of shrimp were exposed to the same hydrocarbon solution for 5 hr before transfer to clean seawater for 72 hr. Analyses of tissues at both 5 and 72 hr were from 2 sets of 10 animals each.

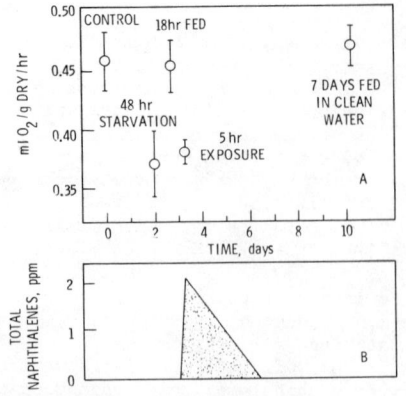

Since it was not possible to predict the point at which depuration would be complete, respiratory rate was not again measured until 7 days after exposure. At this time there was no tissue contamination by naphthalenes, and oxygen consumption rates had returned to the control level.

Summarizing these experiments, one can say that for mysids there appears to be a correlation between extent of increased respiratory rate and the concentration of naphthalenes in the water. Brown shrimp exhibiting tissue contamination by naphthalenes also demonstrated enhanced levels of oxygen consumption. When grass shrimp were held in clean water for a period longer than that required to reduce accumulated naphthalenes to background levels, respiratory rate decreased to that of control animals. Although these data do not definitely demonstrate a cause-and-effect relationship, they strongly implicate the involvement of naphthalenes content in the expression of an abnormal physiological response.

Ionic and Osmotic Regulation

While many truly marine species of invertebrates tolerate a rather narrow range of salinity (stenohaline), intertidal and estuarine species have developed adaptations to cope with a changing ionic environment (euryhaline). For reviews of the mechanisms and extent of adaptation to different ionic and osmotic environments, the reader is referred to Kinne (1967, 1971, 1975), Potts and Parry (1964), and Prosser (1973). It is possible that environmental pollutants might inhibit or in some way interfere with the adaptive mechanisms of regulation in marine organisms and thus reduce their ability to tolerate stressful salinities.

It has been shown that DDT inhibits the Na^+, K^+ and Mg^+ ATPases of excised intestine of saltwater-adapted eels, *Anguilla rostrata*, producing a disruption in normal osmoregulation (Janicki and Kinter, 1971). In a later report (Kinter *et al.*, 1972), DDT and the PCB, Aroclor 1221, were found to reduce the ability of *Fundulus heteroclitus* to osmoregulate, as determined by elevated total osmotic concentration and Na^+ and K^+ levels in the blood. Roesijadi *et al.* (1974) found that the porcelain crab, *Petrolisthes armatus*, was capable of regulating blood chloride levels at salinities between 7 and 35 $^0/_{00}$, even when exposed to 50 ppb of mercury (as $HgCl_2$). In additional studies, Roesijadi *et al.* (1976) demonstrated a reduction in blood chloride regulation by juvenile grass shrimp (*Palaemonetes pugio*) exposed to 10 ppb Aroclor 1254. This chlorinated hydrocarbon reduced the ability of the shrimp to acclimate to a range of chloride concentrations equivalent to a salinity range of 1 to 31 $^0/_{00}$. There were more dramatic differences between control and exposed animals (10 ppb), which included mortality, when they were rapidly transferred to salinity extremes (1 and 31 $^0/_{00}$).

Penaeid shrimp have been previously examined for their capability of osmoregulate (Williams, 1960; McFarland and Lee, 1963). To my knowledge, the research conducted by Cox (1974), which is reviewed in Anderson *et al.* (1974a), is the only attempt to determine the effect of petroleum hydrocarbons on the ionic regulation of a crustacean. Acclimated brown shrimp (*Penaeus aztecus*) were shown to regulate blood chloride levels above the environment at salinities below 22 $^0/_{00}$S and demonstrate hypochloride regulation at higher salinities (Anderson *et al.*, 1974a). Exposure to a 20% WSF of No. 2 Fuel oil (about 0.4 ppm TN) for 10 hr at the acclimation salinity of 20 $^0/_{00}$S did not affect the steady-state chloride regulation of these shrimp. However, when animals were rapidly transferred from 20 $^0/_{00}$S water to a higher (30 $^0/_{00}$S) and a lower (10 $^0/_{00}$S) salinity, the fluctuations in blood chloride levels of exposed shrimp were greater than those of control animals (Fig. 7A-B).

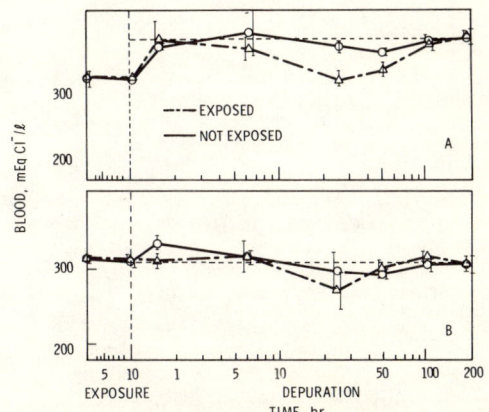

Fig. 9.7 Relationship between the rate of blood chloride regulation of *Penaeus aztecus* and the uptake and release of naphthalenes. Shrimp were exposed to a 20% WSF of No. 2 Fuel oil (1.3 ppm TH and 0.4 ppm TN) for 10 hr in 20 $^0/_{00}$S seawater before transfer to clean water at either 10 $^0/_{00}$S or 30 $^0/_{00}$S. Each blood chloride value represents the mean of 3 replicates, each from the pooled blood of 3 shrimp. Vertical bars represent standard deviations and the dashed line indicates the final acclimation level. (Redrawn from Anderson *et al.*, 1974a, and Cox, 1974)

A. Blood chloride concentrations of control (solid line) and exposed (broken line) transferred to a salinity of 30 $^0/_{00}$.

B. Blood chloride concentrations of shrimp transferred to 10 $^0/_{00}$.

At most time intervals, the blood chloride levels of control and exposed animals were not significantly different, but at 24 hr the concentration of chloride in the blood of exposed shrimp transferred to 30 $^0/_{00}$S was considerably less than that of control animals. The uptake and release rates of exposed shrimp (Fig. 7C) demonstrate that tissue levels dropped from 7 ppm TN to about 0.5 ppm (30 $^0/_{00}$S) after 24 hr in clean water, but only after 96 hr had the content decreased to below the level of detection (<0.1 ppm). At the 96 hr time interval, control and exposed shrimp exhibited nearly identical blood chloride levels.

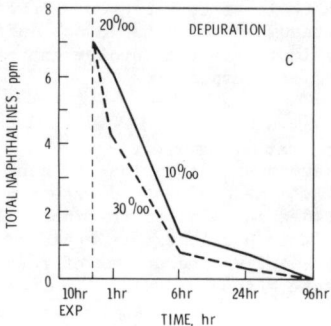

Fig. 7C. After blood samples were taken, the tissues of the exposed shrimp (9 per time interval) were analyzed for content of total naphthalenes. Analyses were conducted after the 10 hr exposure and at 1, 6, 24 and 96 hr in clean water.

While marine bivalves generally conform to the osmotic concentration of the bathing medium, the rates at which adjustments occur may vary between species. For a recent discussion of the mechanisms involved in the passive equilibrium of bivalves with environmental salinity, refer to Pierce and Greenberg (1973). Anderson and Anderson (1976) studied the effects of exposure to No. 2 Fuel oil and South Louisiana Crude oil on the rate of blood chloride adjustment by the oyster, *Crassostrea virginica*. Oysters exposed to rather high concentrations (1% v/v) of these oils for 4 days were transferred from 20 $^0/_{00}$S to 10 and 30 $^0/_{00}$S, and the pericardial fluid was analyzed for chloride concentration at 3 and 10 days (Fig. 8A). Few significant differences between control and exposed animals were noted at 3 days after transfer, and none were exhibited at the 10 day interval. Only the oysters exposed to fuel oil failed to conform to the chloride concentration of the medium after 3 days. This partial impairment of normal adjustments in water and chloride ions is small compared to the substantial accumulation of petroleum hydrocarbon during the 4-day exposure (Fig. 8B). While a sharp decrease in the level of hydrocarbon contamination (12 to 2.5 ppm TN) occurred over a period of 7 days in clean water, there was still a relatively high

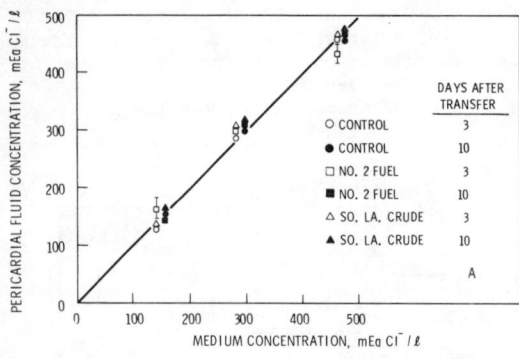
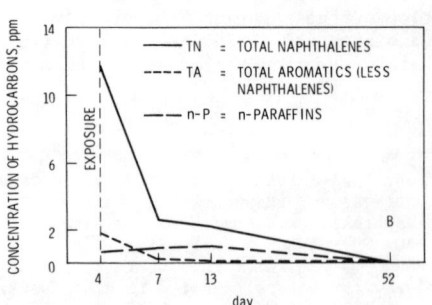

Fig. 9.8. Relationship between the adjustment of pericardial fluid chloride concentration by *Crassostrea virginica*, and the accumulation and release of petroleum hydrocarbons. Oysters were exposed to 1% (v/v) OWDs of No. 2 Fuel oil or South Louisiana Crude oil for 96 hr before transfer from 20 $^0/_{00}$S to either 10 or 30 $^0/_{00}$ salinities. (Redrawn from Anderson, R. D., 1975, and Anderson and Anderson, 1975)

A. Each point represents the mean pericardial chloride concentration of 3 oysters. The solid line indicates the isochloride line, and vertical bars are shown for standard deviations greater than 10%.

B. Content of petroleum hydrocarbons in oyster tissues exposed to a 1% OWD of No. 2 Fuel oil for 96 hr and then held in clean water for 52 days.

concentration of these compounds associated with the tissue at the 13 day interval. If one extrapolates from the depuration curve, the difference in tissue content on days 3 and 10 was approximately 4.5 ppm TN. It is at least possible that this decrease in the level of contamination may correspond to the final adjustment of blood chloride levels to the normal acclimation concentration.

Behavior

For an excellent review of the factors that must be taken into account to utilize behavior as a measure of effect from a sublethal concentration of a pollutant, the reader is referred to Olla (1974). The contributors to the chapter on "Behavioral Measures of Environmental Stress" discussed the selection of test organisms, the prerequisite of field observations, experimental design, and the integration of behavioral observations with standard bioassays. Examination of the extensive literature review reveals that only a few investigators have studied the effects of oils or petroleum hydrocarbons on the behavior of marine animals.

Jacobson and Boylan (1973) showed that a rather low concentration of water-soluble fraction from kerosene inhibited the chemotaxis of the snail, *Nassarius obsoletus*. The feeding behavior of the lobster, *Homarus americanus*, was affected by relatively small amounts of crude oil in seawater (Atema and Stein, 1974). In a more recent paper, Reimer (1975) discussed the specificity of the feeding response of the zoanthid (Cnidaria) *Palythoa variabilis* tested with proline and some analogs. "Marine Diesel" and "Bunker C oils" also elicited feeding reactions but inhibited discrimination between organic and inert particles for a period of 3 to 5 days. Percy (1976) reported that the Arctic marine amphipod, *Onisimus affinis*, avoided oil-tainted food as well as oil masses, but this response was decreased by pre-exposure to oil and the use of weathered oil. The isopod, *Mesidotea entomon*, which is more resistant to oil exposure, exhibited no preference for clean food and did not avoid oil masses.

Animals exposed to acute levels of pollutants often exhibit gross behavioral abnormalities which may predict the onset of death (Olla, 1974). In earlier bioassays (Anderson *et al.*, 1974b), it has been noted that crustaceans swim in spiral patterns and fish often demonstrate loss of equilibrium by swimming vertically or in an inverted position. While these observations are obviously not comparable to the sophisticated type required to determine sublethal effects, they were clearly correlated with the concentration of hydrocarbons in the water. Since the techniques were available to determine tissue content of naphthalenes in fish, Dixit and Anderson (1977) studied the correlation between organ content of total naphthalenes and the stage of behavioral abnormality leading to death.

A large group of *Fundulus similus* was exposed to a lethal concentration (6 ppm TH and 1.9 ppm TN) of the 100% WSF from No. 2 Fuel oil in two 20-gal aquaria. When 5 or more fish exhibited the same abnormality, they were removed, rinsed, blotted, and the organs were pooled for analysis by the method of Neff and Anderson (1975). The stages, progressively from "normal" to death, are shown in Fig. 9, in addition to the concentration of total naphthalenes present in 4 organs. Other organs and tissues analyzed generally possessed lower

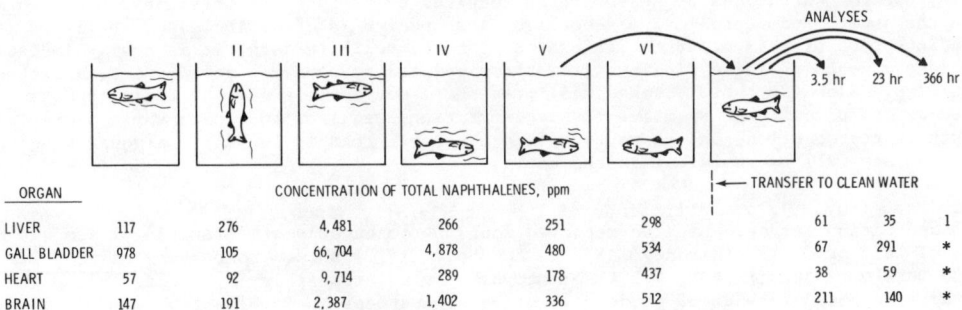

ORGAN	CONCENTRATION OF TOTAL NAPHTHALENES, ppm						TRANSFER TO CLEAN WATER		
	I	II	III	IV	V	VI	3.5 hr	23 hr	366 hr
LIVER	117	276	4,481	266	251	298	61	35	1
GALL BLADDER	978	105	66,704	4,878	480	534	67	291	*
HEART	57	92	9,714	289	178	437	38	59	*
BRAIN	147	191	2,387	1,402	336	512	211	140	*

* LESS THAN THE LIMIT OF DETECTION (0.5 ppm)

Fig. 9.9. Relationship between the behavioral stage of *Fundulus similus* and the content of naphthalenes in 4 organs. Five fish at each stage were selected from a large group exposed to an acute concentration of No. 2 Fuel oil WSF (6.3 ppm TH and 1.9 ppm TN). Organs from all 5 were pooled for analyses at each behavioral stage during exposure (about 4 hr) and at 3 intervals of depuration in clean water. The schematic of the stages of stress proceeding from "normal" (I) to death (VI) and the corresponding content of naphthalenes are shown. Stage V illustrates slight opercular movements but not swimming, and fish transferred at this stage to clean water were held for 3.5, 23 and 366 hr before analyses. (Redrawn from Dixit and Anderson, 1977)

levels of contamination than the liver, gall bladder, heart and brain shown in Fig. 9. This schematic diagram illustrates the decrease in mobility and the loss of equilibrium exhibited by the fish over a period of a few hours. Individual variability was observed as one group of fish was swimming upside-down at the surface (Stage III), while several others in the same tank were still swimming "normally" (Dixit and Anderson, 1977). From earlier investigations (Lee et al., 1972; Roubal et al., 1976), the high levels of naphthalenes found in liver and gall bladder were not surprising. The former organ is the major site of hydrocarbon metabolism, and the latter functions in the storage and elimination of the parent compounds, in addition to their metabolites. As these compounds are certainly being transferred about the body by the circulatory system, the relatively high concentration found in the heart was to be expected. Perhaps the most interesting aspect of these data is the fact that the brain possessed concentrations higher than heart and liver at all stages except II and III, including the depuration periods of 3.5 and 23 hr. Since loss of equilibrium was first exhibited by Stage II fish, containing 191 ppm TN in the brain, it might be assumed that levels in excess of about 200 ppm interfere with the locomotor and regulatory capabilities of the fish. After 3.5 hr in clean water, fish which had returned to "normal" swimming activity still possessed 211 ppm in their brain tissue, so this level may be very near the threshold concentration. It is difficult to explain the general decrease in contamination levels from Stage IV to VI (death), but partitioning of the hydrocarbons back to the water must be responsible for the majority of the release.

The overall results of this study illustrate the capability of fish to transport hydrocarbons to organs such as the liver and gall bladder for detoxification and excretion. Even though fish at Stage V were under extreme stress and near death, transfer to clean water for periods of 23 hr to 366 hr (15 days) resulted in recovery to "normal" swimming and feeding activities and complete depuration of naphthalenes (within 15 days). Of course, in the field, predation and/or disease may well have eliminated fish exhibiting such abnormalities before they had the opportunity to recover. It is hoped that this type of investigation will serve as a model for future studies utilizing more natural exposure conditions, lower levels of hydrocarbons, longer exposure periods, and more detailed analyses of hydrocarbon contamination. A combination of sophisticated approaches for measuring behavioral responses and the best analytical techniques for determining levels of hydrocarbons in the water (or food) and the animals will surely produce much needed information on threshold levels of contamination in the marine environment and mechanisms of attack on marine organisms.

Growth and Reproduction

It is likely that any environmental stress, which may or may not be detected by short-term physiological measurements, would be eventually expressed in the growth and reproduction of an organism. In discussing the sublethal effects of pollutants, Sprague (1971) suggested that growth should be determined in all chronic expsures. Until recently, the effects of oil and detergents on the reproductive capabilities of marine invertebrates had not been examined to any extent (Davis, 1972), and as noted above, sensitive analytical methods had not been developed.

Many of the studies concerning larval or juvenile stages of marine organisms describe primarily the concentrations of hydrocarbons required to produce 50% mortality, but information on the growth and reproductive success of the species is often included. While it is not the intention of this paper to discuss, at length, toxicity data, it is nearly impossible to separate aspects of these findings from those related to abnormal development, growth or reproduction. Rice (in these Proceedings) presents a comparative view of toxicity data. It is necessary to place in perspective the concentrations required to reduce some parameter of growth or reproduction, and as noted earlier, a level that is lethal to a young stage of an organism may well be sublethal to the adult.

At calculated (nominal) concentrations of 1 ppm crude oil, Mironov (1969) observed deaths of crab and shrimp larvae. He also reported zooplankton capable of tolerating 1 ppm but not 100 ppm of "oil products" (Mironov, 1972). Kuhnhold (1972) showed that young herring larvae (Clupea) were more sensitive to oil than embryos. In a later study (Kuhnhold, 1974), cod eggs (Gadus) exposed to WSFs of crude oils at total hydrocarbon levels of 0.015 to 3.5 ppm exhibited abnormalities in embryonic development. He also reported lethal effects at concentrations between 1 and 12 ppm TH. This range of acute toxicity compares closely with the results of Rice et al. (1975), as they found 6 ppm TH from Prudhoe Crude oil was lethal to pink salmon fry in 96 hr. The growth of salmon fry was decreased after 10 days exposure to 0.73 ppm TH. Constant exposure to 0.3 - 0.4 ppm TH (from crude oil) for 60 days reduced the growth rate of larval amphipods, Gammarus oceanicus (Linden, 1976). Mecklenburg et al. (in these Proceedings) studied the tolerance of intermolt and molting king crab (Paralithodes camtschatica) and coonstripe shrimp (Pandalus hypsinotus) larvae to WSFs of Cook Inlet Crude oil. Mortality increased with the length of exposure up to 24 hr, but not beyond. Stage VI larvae of the shrimp were more sensitive than the earlier stages, and larvae molting from Stage I to II were more sensitive than either intermolt stage. Molting was also a more sensitive stage for the king crab larvae, but the differential was not as large. Concentrations of total hydrocarbons from the WSFs which resulted in significant mortality ranged from 0.24 to about 1.65 ppm (exposures 24 hr or longer). Caldwell et al. (in these Proceedings) reported an extensive amount of toxicity data for larvae of the Dungeness crab (*Cancer magister*), and included some information on the effects of Alaskan Crude WSFs, benzene and naphthalene on the rate of larval growth. First, second and third stage zoea grew at a slower rate when exposed to 0.13 ppm naphthalene and 5 ppb WSF (measured as naphthalene). The value of 5 ppb naphthalene in the WSF converts to 4% of the full strength extract, 0.22 ppm total aromatics (TA) measured and 6 ppb TN. The only effect on the final size reached (first to fifth stage zoeae) was a slightly larger length attained by those exposed to 8.3 ppb naphthalene from WSF (= 6.8% WSF = 0.38 ppm TA measured = 10 ppb TN).

Wells and Sprague (1976) exposed lobster (*Homarus americanus*) larvae to dispersions of Venezuelan Tia Juana Crude oil. The 96 hr LC_{50}s for first-stage larvae was 0.86 ppm TH, while the third and fourth-stage larvae were more tolerant (4.9 ppm). Larvae exposed for 30 days exhibited an LC_{50} of 0.14 ppm, which was also the concentration that retarded development and about the same level producing decreased feeding. The ratio of "safe" to lethal concentrations was about 0.03. Sediments containing as much as 1740 ppm did not significantly reduce growth or survival of post-larval lobsters, but caused the animals to dig more burrows. Gravid female grass shrimp, *Palaemonetes pugio*, exposed to 1.4 ppm TH (0.6 ppm TN) from No. 2 Fuel oil for 72 hr produced significantly fewer larvae (Tatem, in these Proceedings). Larvae, which hatched in clean water after the females were exposed, averaged about 40 per female at concentrations between 0 and 0.3 ppm TN, but only 9 per female at 0.6 ppm TN. Carr and Reish (in these Proceedings) determined 4-day and 28-day LC_{50} values for 5 species of polychaetes using WSFs of No. 2 Fuel oil and South Louisiana Crude oil. The order from most to least sensitive to No. 2 Fuel oil WSF was *Ophryotrocha puerilis* > *O. sp.* > *Ctenodrilus serratus* > *Cirriformia luxuriosa* > *Capitella capitata*. The range of LC_{50} values on 28-day tests for these species was 1.4 to 7.3 ppm TH from No. 2 Fuel oil, while those tested with So. Louisiana Crude WSF produced a range of 7.9 to 17.8 ppm TH. Fuel oil extracts reduced the reproductive capability of *Ophryotrocha sp.* and *Ctenodrilus serratus* at between 1 and 2 ppm TH, while about 10 ppm TH from So. Louisiana Crude were required to produce a significant reduction. At 1.9 ppm TH, the WSF of the crude oil produced a larger total population of *Ophryotrocha sp.*, which is another example of a type of enrichment observed in some studies.

Studies conducted in the laboratory of Dr. J. M. Neff and mine over the past 4 years have concerned several different species, but always the same oils and exposure preparations. It is therefore relatively easy to compare the results of these investigations, and an attempt will be made to relate these to findings of others. It should first be noted that in some earlier studies the growth of small juvenile brown shrimp (*Penaeus aztecus*) and oysters (*Crassostrea virginica*) was not affected by exposures to petroleum hydrocarbons (Anderson *et al.*, 1974a; Anderson, R. D., 1975). Both of these exposure conditions involved relatively short periods in contact with the oil or WSFs, and it was possible that sufficient time was available for the organisms to recover to normal activity levels.

Tatem (in these Proceedings) reported reduced hatching success of *Palaemonetes pugio* when gravid females were exposed for 72 hr to 0.6 ppm TN, but in addition he examined the effect on larval growth when they were hatched in clean water and exposed to 0.7 ppm TH (= 0.3 ppm TN) from No. 2 Fuel oil WSF. Growth rate of exposed larvae was suppressed during the 12 days of exposure (12 to 24 days after hatching). Since some mortality occurred on day 24, the larvae were transferred to clean water for 5 days before re-exposure for 2 days. After this final short exposure, the larvae were held for 10 days in clean water and final size measurements were taken. After the initial exposure (12 days), there was clear evidence of reduced growth, but the growth rate of exposed larvae during the last 17 days was greater than that of control animals.

Neff *et al.* (1976) determined the effect of WSFs of No. 2 Fuel oil on the larval development of the mud crab, *Rithropanopeus harrisii*. Fig. 10 shows the survival curves for these zoeae, continuously exposed to concentrations of total hydrocarbons between 0 and 1.26 ppm (0 - 0.4 ppm TN). While there was some reduction in survival at 0.64 ppm TH, a sharp increase in mortality occurred at 0.95 ppm (0.3 ppm TN). The time at which mortality for these larvae increased rapidly (4-6 days) is approximately the period for the first zoeal molt. As can be seen in this figure, the length of time required to reach the megalopa stage was in general shorter for those exposed to low concentrations of WSF, but longer for those surviving exposure to higher concentrations. It was also noted by the investigators that the synchrony of molting exhibited by control and low concentration groups was partially abolished in those exposed to higher levels. In this study, as in an earlier report on the effects of hydrocarbons on crab larvae (Katz, 1973), the first zoeal stage was the most sensitive period of development. As suggested by Epifanio (1971), changes in the larval molting rate are apparently a sensitive measure of pollutant stress in decapod crustaceans.

The effects of WSFs of No. 2 Fuel oil on the growth of larvae and juvenile polychaetes, *Neanthes arenaceodentata*, were reported by Rossi (1976) and Anderson (1976). As shown in Fig. 11, polychaete larvae exposed from an age of 9 days to 31 days to 0, 5 and 8% WSFs grew at the same rate. Larvae exposed to a 16% WSF (1.0 ppm TH and 0.3 ppm TN) exhibited a reduction in size, which became evident after day 21. Those worms transferred to clean water from

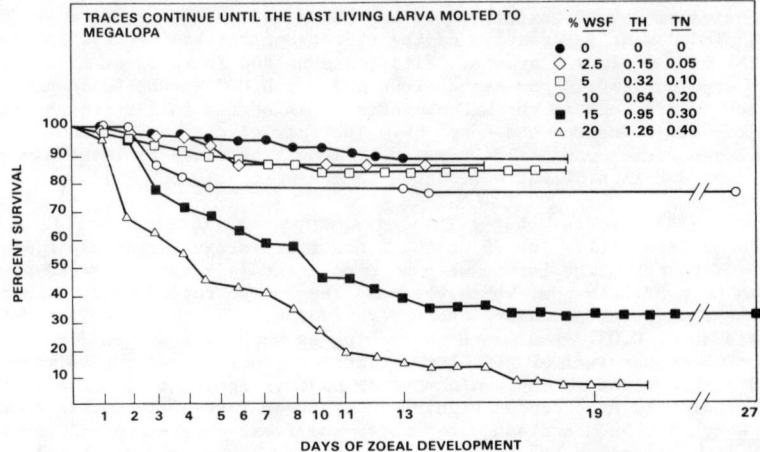

Fig. 9.10. Viability of *Rithropanopeus harrisii* larvae exposed to WSFs of No. 2 Fuel oil. (From Neff *et al.*, 1976)

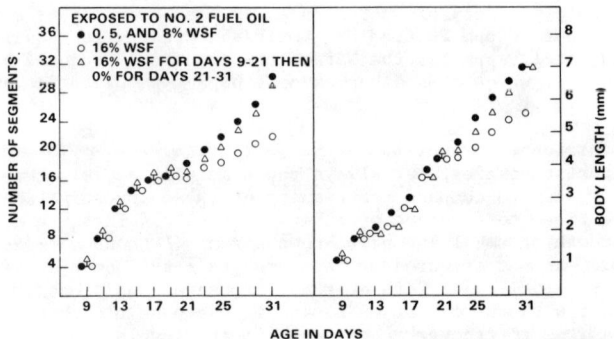

Fig. 9.11. Growth of *Neanthes arenaceodentata* larvae under control and exposed conditions. (From Rossi, 1976)

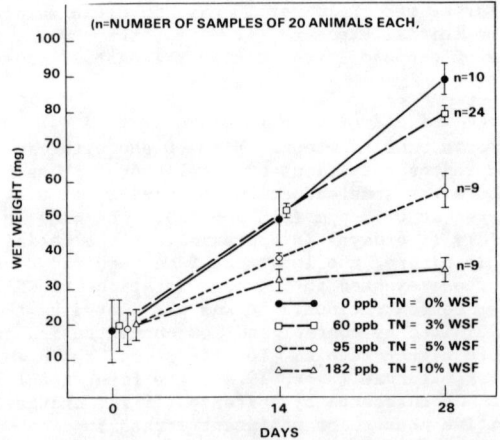

Fig. 9.12. Growth of *Neanthes arenaceodentata* juveniles under control and exposed conditions. (From Anderson, 1976)

16% WSF on day 21 recovered to the extent that their final size was not significantly different from control animals.

Results of exposures to juvenile worms are illustrated in Fig. 12, where growth was measured over a 28-day period (Anderson, 1976). Suppression in growth rate increased as hydrocarbon concentrations increased from 60 to 182 ppb TN. The reduction in growth produced by the lowest exposure level was not apparent at 14 days, but was significant after 28 days.

Summary and Conclusions

Summarizing the various studies regarding the effects of petroleum hydrocarbons on physiological parameters, including growth and reproduction, of marine organisms is indeed a difficult task. One must first omit from consideration any study which does not report measured levels of hydrocarbons in the water, since all available data indicate gross differences between the solubility and dispersing characteristics of different oils and the extent of mixing caused by different types of agitation.

Hydrocarbon Mixtures

The infrared analysis of total hydrocarbons at least provides some indication of the amount of petroleum in the water phase, and these determinations become quite meaningful when data on the compositional characteristics of the specific extract are available. To make accurate comparisons between the effects on different species and/or studies, a knowledge of the constituents in the exposure media is necessary. Measurement of total aromatics generally requires use of a gas chromatograph, while total naphthalenes may be determined by ultraviolet spectrophotometry (Neff and Anderson, 1975).

In attempting to relate the findings of one study to that of another, one must evaluate the limitations of the methods used to quantitate hydrocarbons in the water and consider the differences between oils. Water extracts of crude oils generally produce higher total hydrocarbon values than fuel oils, since they contain a higher proportion of low molecular weight soluble compounds. Monocyclic aromatic compounds (C_{12} or less) are relatively abundant in water mixtures from crude oils, while refined or residual oils and their water extracts contain a higher proportion of di- and tri-aromatics. Naphthalenes are di-aromatic compounds which have been used as tags in exposures to WSFs and OWDs and also in specific compound exposures over the past few years (Anderson, 1975). The 100% WSF of No. 2 Fuel oil contains about 6 ppm TH, of which 90% are aromatics and 30% are naphthalenes. However, these compounds only represent 2.2% of the total aromatics and 1.3% of the total hydrocarbons in the WSF of South Louisiana Crude oil. While it would appear that naphthalenes would contribute less to the toxicity of crude oils than fuel oils, the relative volatility of hydrocarbons in the different mixtures should be considered. Toluene, benzenes and xylenes, which are prominent in crude oil extracts, are more volatile than naphthalene and particularly the alkylnaphthalenes.

Growth and Reproduction

Table 1 summarizes several very recent investigations on the effects of petroleum hydrocarbons on egggs, larvae or juveniles of selected marine species. Several different oils and preparations were used, but many of the studies involve the well characterized WSF of No. 2 Fuel oil (No. 2 FO). The exposure periods are relatively long, providing more meaningful evaluations of effects. The concentrations of both total hydrocarbons (TH) and total aromatics (TA) shown to reduce growth and/or survival range from about 0.2 to 10 ppm. This variability is not surprising, considering the differences between exposure media which were noted above. Much less variability is observed when the levels of total naphthalenes (TN) producing adverse effects are considered. Where data are available, the values fall between the range of 0.1 to 0.6 ppm TN. It is not difficult to assume that this range represents the critical or threshold level for animals exposed to No. 2 Fuel oil; however, for reasons already discussed, it is probably less valid to describe the effects of crude oils on a basis of naphthalenes. The values for these compounds do fall within the range of concentrations derived from fuel oil tests, but further research is needed to define the most toxic components of crude oil mixtures.

Adults versus Larvae

Comparisons of tests with larvae and adults of the same species generally show the greater sensitivity of larval stages. The earliest larval stages are not always the most sensitive, but the period of crustacean molting consistently seems to be a critical stage at which mortality is generally greatest, regardless of pollutant. The differential between lethal concentration (96 hr LC_{50} values) for larvae and adults may, in some cases, be quite small, and sensitivity may even increase with age (*Penaeus aztecus* and *Neanthes arenaceodentata*). Available data indicate caution should be used in attempting to extrapolate larval sensitivity to petroleum hydrocarbons from the results of tests with adults. A major problem in interpreting present findings is the lack of consistency regarding exposures of adults and larvae. The latter are generally held in contaminated water until metamorphosis to postlarval or juvenile stages, while adults are usually tested for only a few days (96 hr). In order to

TABLE 1. Summary of Effects of Petroleum Hydrocarbons on the Growth and Reproduction of Marine Animals

Species	Oil	Exposure (days)	Concentration (ppm)[1] TH	TN	TA	Growth or Reproduction Parameter	Reference
FISH							
Fundulus similus	So.La.C[2]	20	16.6	0.2	9.7	4% hatch of eggs.	Anderson et al., 1976.
Cyprinodon variegatus	No. 2 FO[3]	7	2.0	0.6	1.7	0% hatch of eggs.	Anderson et al., 1976.
Oncorhynchus gorbuscha	Prudhoe Crude	10	0.7			Reduced growth rate of pink salmon fry.	Rice et al., 1975.
DECAPODS							
Cancer magister	Alaskan Crude	60			0.2	Reduced survival of zoeae on longterm exposure.	Caldwell et al., 1977.
Rithropanopeus harrasii	No. 2 FO	27	1.0	0.3	0.9	Reduced survival and extended development to megalopa.	Neff et al., 1976.
Palaemonetes pugio	No. 2 FO	12	0.9	0.3	0.8	Reduced growth rate of larvae.	Tatem, 1977.
		3	1.4	0.6		Reduced viability of eggs from exposed gravid females.	
Paralithodes camtschatica	Cook Inlet Crude	1-4	1.6			Inhibition of molting in larvae.	Mecklenburg et al., 1977.
Pandalus hypsinotus	Cook Inlet Crude	1-4	1.2			Inhibition of molting in larvae.	
Homarus americanus	Venezuelan Crude	30	0.14			30 day LD$_{50}$ and retarded development of larvae.	Wells & Sprague, 1976.
AMPHIPODS							
Gammarus oceanicus	Crude	60	0.3-0.4			Reduced growth rate of larvae.	Linden, 1976.
POLYCHAETES							
Neanthes arenaceodentata	No. 2 FO	22	1.0	0.3	0.9	Reduced growth rate of larvae.	Rossi, 1976.
	No. 2 FO	28	0.3	0.1		Reduced growth of juveniles by 30%.	Anderson, 1976.
Ctenodrilus serratus	No. 2 FO	28	2.2	0.5	1.4	Reduced survival and reproduction.	Carr & Reish, 1977.
	So.La.C	28	9.9	0.2	7.0		
Ophryotrocha sp	No. 2 FO	28	1.3	0.3	0.9		
	So.La.C	28	9.9	0.2	7.0		

[1]TH, total hydrocarbons; TN, total naphthalenes; TA, total aromatics.

[2]South Louisiana Crude Oil, API reference oil.

[3]High aromatic (38%) No. 2 Fuel Oil, API reference oil.

derive comparable toxicity data, both adults and larvae should be exposed in constant-flowing systems, where the levels and composition of hydrocarbons are well characterized and controlled.

Physiological versus Growth Studies

Measurements of respiratory rates and chloride ion regulation during or after exposure to No. 2 Fuel oil indicated that a range of 0.1 to 0.85 ppm TN produced responses that differed significantly from those of control animals. Tissue analyses showed that during the periods when abnormal responses were exhibited, the animals contained measurable amounts of naphthalenes. Since it is relatively difficult to hold animals under conditions of constant hydrocarbon exposure during determination of physiological effects, most studies have been limited to a period of a few hours to a few days. This is perhaps the most significant difference between physiological studies and analyses of growth and development, where exposures have been continued for weeks or months. Recognizing the need to expose larval stages for the total period of their development, investigators frequently renewed the hydrocarbon solutions, thus enhancing the chance of bioaccumulation and decreasing the probability of hydrocarbon evaporation, degradation or release. It is therefore not surprising that long-term studies on growth of larvae indicate adverse effects at levels of about 0.1 to 0.3 ppm TN, while short-term investigations on physiological responses appear to be less sensitive (0.3 to 1.0 ppm TN) indicators of sublethal effects.

Lethal versus Sublethal Studies

The usual method of comparing results from acute toxicity tests to those derived from physiological or growth studies is to examine the ratio of the concentration producing an abnormal response to the 96 hr LC_{50} value for the species or life stage (if possible). By converting the data presented above on No. 2 Fuel oil WSF to ppm total naphthalenes, it is possible not only to derive these ratios, but also ratios based on the duration of exposure. Table 2 demonstrates that without considering the duration of exposure, the "sensitivity factors" for 3 crustaceans and a polychaete range between 0.06 and 0.8. There is no clear separation between these factors on a basis of short and long-term tests. However, when the duration of exposure in both acute and sublethal tests is multiplied by the concentration, strikingly different values are produced. There is a sharp separation between factors derived from long-term growth and reproduction studies and physiological investigations of 10 hr or less duration. On the basis of exposure concentration alone, the values are considerably higher than the 0.01 factor which is often used to estimate a safe environmental level from 96-hr toxicity data. By including time as a factor in the ratio, values from short-term studies are near, or less than the 0.01 level, whereas long-term exposures produce much higher values. Perhaps the best explanation for these findings is the transitory nature of physiological responses elicited during or soon after exposure to hydrocarbons. In nearly all examples discussed above, organisms demonstrated their capability to recover when transferred to clean water. There are numerous other instances in which organisms have reversed abnormal behavior or released a portion of the accumulated hydrocarbons as the concentration of hydrocarbons in the medium gradually decreased from volatilization (Cox, 1974; Tatem, 1975); personal observations). The responses exhibited during respiratory measurements were in fact significant, but it is quite possible that these crustaceans would be capable of adapting after a certain period, as they do on transfer to a new temperature or salinity. It is clear that such a reversal is possible when the source of contamination is removed, but data on physiological responses over extended exposure periods are not available.

TABLE 2. Relationship Between Results of Lethal and Sublethal Studies[1]

Animal	Response	Sublethal(S) Time x Conc. (hr x ppm)	Lethal(L) Time x Conc. (hr x ppm)	Ratio Conc.(S)/Conc.(L)	T x C(S)/T x C(L)
PHYSIOLOGY					
Penaeus aztecus	Respiration	4 x 0.85 = 3.4	96 x 1 = 96	0.8	0.035
	Cl⁻	10 x 0.4 = 4.0		0.4	0.042
Mysidopsis almyra	Respiration	2 x 0.1 = 0.2	48 x 0.9 = 43.2	0.1	0.005
Palaemonetes pugio	Respiration	5 x 0.1 = 0.5	96 x 1.2 = 115.2	0.1	0.004
GROWTH AND REPRODUCTION					
Palaemonetes pugio	Viability of offspring	72 x 0.6 = 43.2		0.5	0.375
	Larval growth	288 x 0.3 = 86.4	96 x 0.4 = 38.4	0.8	2.25
Neanthes arenaceodentata	Juvenile growth	672 x 0.09 = 63.8	96 x 1.5 = 144	0.06	0.443
	Larval growth	528 x 0.3 = 158.4	96 x 2.0 = 192	0.2	0.825

[1]All concentrations represent measured amounts of napthalenes from the WSF of No. 2 Fuel oil.

The ratios produced from long-term growth and reproduction studies are relatively large (Table 2). These data represent the capabilities of larvae and juveniles to tolerate constant exposure to compositionally consistent hydrocarbon contamination, and in the final analysis these concentration-time factors produced a significant reduction in growth.

Tissue Contamination and Turnover Rates

Natural seepage of petroleum hydrocarbons has always represented a significant influx to the marine environment (Wilson *et al.*, 1974), and the total input from all sources has been estimated at 6.2 million metric tons per year (National Academy of Sciences, 1974). It is not surprising, therefore, to find that many species of marine organism possess the capability of metabolizing (detoxifying) natural products such as petroleum hydrocarbons. There is evidence of detoxification pathways in polychaetes, crustaceans and fishes (Lee, in these Proceedings). One must therefore consider the flux of hydrocarbons within these organisms, which is a dynamic process probably correlated with the transitory nature of observed abnormal physiological responses. It is reasonable to suppose that variations between the findings of sublethal studies can be accounted for by close examination of the flux rates of hydrocarbons between the media and the organisms. Physiological effects on polychaetes, crustaceans and fish are relatively short-lived if the animals are transferred to clean water. Tissue contamination, and perhaps sublethal effects, are somewhat longer lasting in bivalves perhaps because enzymatic pathways (detoxification) are either lacking or present at significantly lower activity levels. The eventual depuration of hydrocarbons from bivalves in periods of a few weeks to about 2 months may either be the result of gradual partitioning back to the water or very low rates of metabolic turnover. I believe that careful examination of the results of sublethal studies, which have included detailed chemical analyses of water and tissues, will illustrate close correlation between extent of tissue contamination of di- and triaromatic hydrocarbons and the magnitude of the given abnormal response. Constant exposure of larval crustaceans has produced, thus far, the lowest threshold response levels of petroleum hydrocarbons, and it is quite likely that tissue contamination levels of these organisms are comparable to those of adults briefly exposed to acute concentrations. Only when these possible relationships between exposure time, exposure concentration, accumulated concentrations in tissues, hydrocarbon composition, and phylogenetic position of the species are more thoroughly examined, can we begin to understand and predict the effects of petroleum hydrocarbons on marine organisms.

Recommendations

The strongest recommendation I would make is for the absolute necessity of combining the most valid experimental approaches available to biologists with the expertise of an analytical chemist. A great deal of innovative and significant biological research will be wasted if we do not take advantage of the recent advances in analytical methodology and equipment to produce correlations between biological responses and hydrocarbon concentrations and fluxes.

It is quite evident that marine organisms accumulate a range of petroleum hydrocarbons from solution or dispersed droplets. To my knowledge, only two investigations have examined the availability of hydrocarbons that are sorbed to sediments (Rossi, 1976; Anderson *et al.*, in these Proceedings). Both of these involved detritus-feeding worms and the results indicated little or no accumulation, in addition to rapid release correlated with egestion. There is a need to examine a range of benthic deposit-feeding organisms for possible bioavailability of various petroleum hydrocarbons. In areas receiving chronic or occasional high inputs of hydrocarbons, sediments contain much higher levels than the overlying water mass. We must determine the biological significance of these present and future hydrocarbon sinks.

I do not wish to eliminate any biological parameter from consideration as a useful tool to measure the effects of sublethal concentrations of hydrocarbons; however, it appears that behavior, growth, and reproduction may be the most sensitive and meaningful responses. Inputs from chemoreceptor organs may be altered at quite low concentrations, and any interference with the system would likely result in adverse effects on feeding and reproduction. Growth reflects the total metabolic activity of an organism and thus significant alterations in respiratory rate or feeding activity will be expressed in the growth and reproductive cycle of an organism. Reports of growth studies involving exposure to hydrocarbons have often included observations on decreased feeding, which in turn may well be linked to chemoreception of contaminated food.

I believe it is clear that because of the time-dependent variability of hydrocarbon mixtures and the detoxification capabilities of many marine species, fruitful investigations of threshold levels must incorporate long-term exposure systems. Exposures should provide a constant supply of well characterized hydrocarbon mixtures and extend for at least a period of weeks, if not months. Emphasis should be placed on subjecting large cultures of organisms to contamination for generations, such that data on growth, reproduction, behavior and physiology can all be considered in relation to hydrocarbon fluxes. This type of research would necessitate maintenance of organisms under relatively natural conditions, and would provide valuable information on effects of hydrocarbons on more than one generation.

Finally, as to information required to evaluate the impact of petroleum hydrocarbons in the Arctic environment, I can see no major differences, from the standpoint of biological investigations, from the general knowledge needed for temperate zones. Many species found in Alaskan waters have ranges extending to Oregon, and subtidal temperature regimes are not dramatically different over this geographical range. Present results tend to indicate similar tolerances for temperate and Alaskan species. The most significant differences to be encountered are likely to be related to the chemical and physical behavior of oil and thus the characteristics of the hydrocarbons coming in conteact with the organisms.

References

Anderson, J. W., J. M. Neff, and S. R. Petrocelli (1974a) Sublethal effects of oil, heavy metals and PCBs on marine organisms, in: Survival in Toxic Environments, M.A. Q. Khan and J. P. Bederka, Jr., ed., Academic Press, New York.

Anderson, J. W., J. M. Neff, B. A. Cox, H. E. Tatem and G. M. Hightower, Characteristics of dispersions and water-soluble extracts of crude and refined oils and their toxicity on estuarine crustaceans and fish, Marine Biol. 27, 75 (1974b).

Anderson, J. W., J. M. Neff, B. A. Cox, H. E. Tatem and G. M. Hightower (1974c), The effects of oil on estuarine animals: Toxicity, uptake and depuration, respiration, in: Pollution and Physiology of Marine Organisms, F. J. and W. B. Vernberg, ed., Academic Press, New York.

Anderson, J. W. (1975), Laboratory Studies on the Effects of Oil on Marine Organisms: An Overview, A.P.I. Publication No. 4249.

Anderson, J. W. (1976), Effects of petroleum hydrocarbons on the growth of marine organisms, in: Petroleum Hydrocarbons in the Marine Environment, A. D. McIntyre and K. Whittle, ed. Rapp. P.-v Reun., Cons. Int. Explor. Mer, 171 (in press).

Anderson, J. W., D. B. Dixit, G. S. Ward and R. S. Foster (1976), Effects of petroleum hydrocarbons on the rate of heart beat and hatching success of estuarine fish embryos, in: Pollution and Physiology of Marine Organsisms, II, F. J. and W. B. Vernberg, ed., Academic Press, New York (in press).

Anderson, J. W., L. J. Moore, J. W. Blaylock, D. L. Woodruff, and S. L. Kiesser (1977) Bioavailability of sediment-sorbed naphthalenes to the sipunculid worm, Phascolosoma agassizii, in: These Symposium Proceedings.

Anderson, R. D. (1975), Petroleum hydrocarbons and oyster resources of Galveston Bay, Texas, in: Proceedings of 1975 Conference on Prevention and Control of Oil Pollution, A.P.I., E.P.A., and U.S.C.G.

Anderson, R. D., and J. W. Anderson, Effects of salinity and selected petroleum hydrocarbons on the osmotic and chloride regulation of the American oyster, Crassostrea virginica, Physiological Zoology 48, 420 (1976).

Atema, J., and L. S. Stein, Effects of crude oil on the feeding behavior of the lobster Homarus americanus, Environmental Pollution 6, 77 (1974).

Avolizi, R. J., and M. Nuwayhid, Effects of crude oil and dispersants on bivalves, Marine Pollution Bulletin 5, 149 (1974).

Brockson, R. W., and H. T. Bailey (1973, Respiratory response of juvenile chinook salmon and striped bass exposed to benzene, a water-soluble component of crude oil, in: Proceedings of Joint Conference on Prevention and Control of Oil Spills, A.P.I., E.P.A., and U.S.C.G.

Caldwell, R. S., E. M. Caldarone and M. H. Mallon (1977), Effects of a seawater-soluble fraction of Alaskan Crude oil and its major aromatic components on larval stages of the Dungeness crab, Cancer magister Dana, in: These Symposium Proceedings.

Carr, R. S., and D. J. Reish (1977), The effect of petrochemicals on the survival and life history of polychaetous annelids, in: These Symposium Proceedings.

Cox, B. A. (1974), Responses of the marine crustaceans Mysidopsis almyra Bowman, Penaeus aztecus Ives, and Penaeus setiferus (Linn.) to petroleum hydrocarbons, Texas A&M University, Ph.D. Dissertation.

Davis, C. C. (1972), The effects of pollutants on the reproduction of marine organisms, in: Marine Pollution and Sea Life M. Ruivo, ed., Fishing News (Books) Ltd., Rome.

Dixit, D., and J. W. Anderson (1977), Distribution of naphthalenes within exposed Fundulus similus and correlations with stress behavior, in: Proceedings of 1977 Oil Spill Conference, A.P.I., E.P.A., and U.S.C.G., New Orleans, March 1977 (In Press).

Epifanio, C. E., Effects of Dieldrin in seawater on the development of two species of crab larvae, Leptodius floridanus and Panopeus herbstii, Marine Biology 11, 356 (1971).

Gilfillan, E. S., Decrease of net carbon flux in two species of mussels caused by extracts of crude oil, Marine Biology 29, 53 (1975).

Gordon, D. C., Jr., P. D. Keizer and J. Dale, Estimates using fluorescence spectroscopy of the present state of petroleum hydrocarbon contamination in the water column of the northwest Atlantic Ocean, Mar. Chem. 2, 251 (1974).

Hargrave, B. T., and C. P. Newcombe, Crawling and respiration as indices of sublethal effec effects of oil and a dispersant on an intertidal snail Littorina littorea, Journal of the Fisheries Research Board of Canada 30, 1789 (1973).

Hawkes, J. (1977), Morphological abnormalities produced by hydrocarbon exposure, in: These Symposium Proceedings.

Jacobson, S. M., and D. B. Boylan, Effect of seawater soluble fraction of kerosene on chemotaxis in a marine snail, Nassarius obsoletus, Nature 241, 213 (1973).

Janicki, R. H., and W. B. Kinter, DDT inhibits Na^+, K^+, Mg^{++} - ATPase in intestinal mucosa and gills of marine teleosts, Nature 233, 148 (1971).

Katz, L. M., The effects of water soluble fraction of crude oil on larvae of the decapod crustacean Neopanope texana (Sayi), Environ. Poll. 5, 199 (1973).

Kinne, O. (1967), Physiology of estuarine organisms with special reference to salinity and temperature: General aspects, in: Estuaries, G. H. Lauff, ed., AAAS Publication No. 83, Wash., D.C.

Kinne, O., ed. (1971) Marine Ecology. A comprehensive Integrated Treatise on Life in Oceans and Coastal Waters. Volume 1, Part 2: Salinity, Wiley Interscience, New York.

Kinne, O., ed. (1975) Marine Ecology. A Comprehensive Integrated Treatise on Life in Oceans and Coastal Waters. Volume 2, Part 1: Physiological Mechanisms, Wiley Interscience, New York.

Kinter, W. B., L. S. Merkens, R. H. Janicki and A. M. Guarino, Studies on the mechanisms of DDT and polychlorinated biphenyls: Disruption of osmoregulation in marine fish, Environ. Hlth Perspec. exp. Iss. 1, 169 (1972).

Kuhnhold, W. W. (1972), The influence of crude oils on fish fry, in: Marine Pollution and Sea Life, M. Ruivo, eds., Fishing News (Books) Ltd., Rome.

Kuhnhold. W. W., Investigations on the toxicity of sea water extracts of three crude oils on eggs of cod (Gadus morhua L.), Sonderdruck aus Bd. 23, 165 (1974).

Lee, R. F., R. Sauerheber, and G. H. Dobbs, Uptake, metabolism and discharge of polycyclic aromatic hydrocarbons by marine fish, Marine Biology 17, 201 (1972).

Lee, R. F., (1977) Accumulation and turnover of hydrocarbons in marine organsism, in: These Symposium Proceedings.

Linden, O., Effects of oil on the amphipod Gammarus oceanicus, Environmental Pollution 10. 239 (1976).

McFarland, W. N., and B. D. Lee, Osmotic and ionic concentrations of penaeidean shrimps of the Texas Coast, Bull. Mar. Sci. Gulf. Carribb. 13, 391 (1963).

Mecklenburg, T. A., S. D. Rice and J. F. Karinen (1977), Effects of Cook Inlet Crude oil water-soluble fraction on survival and molting of king crab (Paralithodes camtschatica) and coonstripe shrimp (Pandalus hypsinotus) larvae, in: These Symposium Proceedings.

Mironov, O. G., Visability of larvae of some crustacea in the sea water polluted with oil products, Zoo. Zh. 48, 1734 (1969) (English abstract).

Mironov, O. G. (1972), Effect of oil pollution of flora and fauna of the Black Sea, in: Marine Pollution and Sea Life, M. Ruivo, eds., Fishing News (Books) Ltd., Rome.

Moore, S. F., and R. L. Dwyer, Effects of oil on marine organisms: A critical assessment of published data, Water Research 8, 819 (1974).

National Academy of Sciences (1975), Petroleum in the Marine Environment; Workshop on Inputs, Fates, and the Effects of Petroleum in the Mairne Environment, Airlie, Virginia May 21-25, 1973.

Neff, J. M., and J. W. Anderson, Ultraviolet spectrophotometric methods for the determination of naphthalene and alkylnaphthalenes in the tissues of oil-contaminated marine animals, Bull. of Environ. Contam. & Toxic. 14, 122 (1975).

Neff, J. M., J. W. Anderson, B. A. Cox, R. B. Laughlin, Jr., S. S. Rossi, and H. E. Tatem (1976), Effects of petroleum on survival, respiration and growth of marine animals, in: Sources, Effects & Sinks of Hydrocarbons in the Aquatic Environment, AIBS, Washington, D.C.

Newell, R. C. (1970), Biology of Intertidal Animals, American Elsevier, New York.

Newell, R. C., Factors affecting the respiration of intertidal invertebrates, Amer. Zool. 13, 513 (1973).

Olla, B., ed. (1974), Behavioral bioassays, in: Marine Bioassays Workshop Proceedings G. Cox, convener, Marine Technology Society, Washington D.C.

Percy, J. A., Responses of Arctic marine crustaceans to crude oil and oil-tainted food, Envir. Poll. 10, 155 (1976).

Percy, J. A. (1977), Effects of dispersed crude oil upon the respiratory metabolism of an Arctic marine amphipod, Onisimus (Boekisimus) affinis, in: These Symposium Proceedings.

Pierce, S. K., and M. J. Greenberg, The Initiation and control of free amino acid regulation of cell volume in salinity-stressed marine bivalves, J. of Exper. Biol. 59, 435 (1973).

Potts, W. T. W., and G. Parry (1964), Osmotic and Ionic Regulation in Animals, Pergamon Press, Oxford.

Prosser, C. L., ed. (1973), Comparative Animal Physiology, W. B. Saunders, Philadelphia.

Reimer, A. A., Effects of crude oil on the feeding behavior of the zoanthid Palythoa variabilis, Environmental and Physiological Biochemistry 5, 258 (1975).

Rice, S. D., D. A. Moles, and J. W. Short (1975), The effect of Prudhoe Bay Crude oil on survival and growth of eggs, alevins and fry of pink salmon Oncorhynchus gorbuscha, in: Proceedings of 1975 Conference on Prevention and Control of Oil Pollution, A.P.I., E.P.A. and U.S.C.G.

Rice, S. D., R. E. Thomas, and J. W. Short (1976), Effect of petroleum hydrocarbons on breathing and coughing rates and hydrocarbon uptake-depuration inpink salmon fry, in: Pollution and Physiology of Marine Organisms, II, F. J. Vernberg and W. B. Vernberg, eds. Academic Press, New York (in press).

Rice, S. D. (1977), Comparative toxicities of petroleum hydrocarbons in marine organisms, in: These Symposium Proceedings.

Roesijadi, G., S. R. Petrocelli, J. W. Anderson, B. J. Presley and R. Sims, Survival and chloride ion regulation of the porcelain crab Petrolisthes armatus exposed to mercury, Marine Biology 27, 213 (1974).

Roesijadi, G., J. W. Anderson, S. R. Petrocelli and C. S. Giam, Osmoregulation of the grass shrimp Palaemonetes pugio exposed to polychlorinated biphenyls (PCBs). I. Effect of PCBs on chloride and osmotic concentrations and chloride and water exchange kinetics, Marine Biology (In press).

Rossi, S. S. (1976), Interactions between Petroleum Hydrocarbons and the Polychaetous Annelid, Neanthes arenaceodentata: Effects on Growth and Reproduction; Fate of Diaromatic Hydrocarbons Accumulated from Solution or Sediments, Texas A&M University, Ph.D. Dissertation.

Rossi, S. S., and J. W. Anderson, Toxicity of water-soluble fractions of No. 2 Fuel oil and South Louisiana Crude oil to selected stages in the life history of the polychaete, Neanthes arenaceodentata, Bull. of Environ. Contam. & Toxic. 16, 18 (1976).

Roubal, W. T., T. K. Collier and D. C. Malins, Accumulation and metabolism of Carbon-14 labeled benzene, naphthalene, and anthracene by young Coho salmon (Oncorhynchus kisutch), Archives of Environ. Contam. & Toxic. (in press).

Sprague, J. G., Meaurement of pollutant toxicity to fish. III. Sublethal effects and "safe" concentration, Water Research 5, 245 (1971).

Steed, D. L., and B. J. Copeland, Metabolic responses of some estuarine organisms to an industrial effluent, Contrib. in Mar. Sciences 12, 143 (1967).

Tatem, H. E. (1975), The Toxicity and Physiological Effects of Oil and Petroleum Hydrocarbons on Estuarine Grass Shrimp Palaemonetes pugio Holthuis, Texas A&M University, Ph.D. Dissertation.

Tatem, H. E. (1977), Accumulation of naphthalenes by grass shrimp: Effects on respiration, hatching, and larval growth, in: These Symposium Proceedings.

Warner, J. S., Determination of aliphatic and aromatic hydrocarbons in marine organisms, Anal. Chem. 48, 578 (1976).

Wells, P. G., and J. B. Sprague, Effects of crude oil on American lobster (Homarus americanus) larvae in the laboratory, J. of the Fisheries Research Board of Can. 33, 1604 (1976).

Williams, A. B., The influence of temperature on osmotic regulation in two species of estuarine shrimps (Penaeus), Biol. Bull. 119, 560 (1960).

Wilson, R. D., P. H. Monoghan, A. Osanik, L. C. Price and M. A. Rogers, Natural marine oil seepage, Science 184, 857 (1974).

Wohlschlag, D. E., and J. N. Cameron, Assessment of a low level stress on the respiratory metabolism of the pinfish (Lagodon rhomboides), Contributions in Marine Science 12, 160, (1967).

CHAPTER 10

THE EFFECTS OF PETROLEUM HYDROCARBON EXPOSURE
ON THE STRUCTURE OF FISH TISSUES

Joyce W. Hawkes

Northwest and Alaska Fisheries Center
National Marine Fisheries Service
National Oceanic and Atmospheric Administration
2725 Montlake Boulevard East
Seattle, Washington 98112 U.S.A.

Introduction

Within the sphere of sublethal effects of contaminants on fish there are numerous structural changes which range from gross anomalies to subtle subcellular defects. At the organismal level, fish may exhibit severe anatomical alterations, such as a curvature of the spine which has been noted in the larvae of Pacific herring (*Clupea pallasi*) when exposed to benzene (Struhsaker et al 1974). At the tissue level, lesions may develop on the skin or gills when toxic materials are present in water or in the intestine when the contaminant is introduced via the food. In addition, a tissue or organ may exhibit no grossly observable changes but the cells may be altered and this may, in turn, affect the overall function of the organism.

Relatively few papers in the literature address the subject of morphological effects of contaminants, and in those there are differences in experimental designs, including a variety of exposure levels and types of contaminants. In spite of the differences, some subcellular changes appear to have patterns of similarity that are common to several species and contaminants.

The intent of this paper is to selectively review the information on cellular and subcellular changes in fish after exposure to petroleum as well as to present previously unpublished data, and to suggest approaches for integrating information from several divergent disciplines involved. In our laboratory, both freshwater and marine fish were exposed to petroleum: rainbow trout (*Salmo gairdneri*) were fed Prudhoe Bay crude oil, and coho salmon (*Oncorhynchus kisutch*), starry flounder (*Platichthys stellatus*), and English sole (*Parophrys vetulus*), were exposed to the water-soluble fraction of Prudhoe Bay crude oil in seawater. The experimental conditions for previously unpublished work are included in each section on tissue changes for the specific study.

Identical microscopic techniques were followed for all of our experiments: tissue samples were excised from freshly sacrificed fish and fixed in 0.75% glutaraldehyde, 3% formalin, 0.5% acrolein in 0.1 M sodium cacodylate buffer with 0.02% $CaCl_2 \cdot H_2O$, 0.02 M s-collidine, and 5.5% sucrose (Hawkes 1974). The tissues designated for examination with the light microscope (LM), or for transmission electron microscopy (TEM), were post-fixed in osmium tetroxide in the same buffer, dehydrated in an ethanol series, and embedded in plastic (Spurr 1969). Sections were cut at 0.5 µ, stained with toluidine blue or a trichrome (MacKay and Mead 1970) for LM. For TEM, sections were cut with a diamond knife and stained with lead citrate, uranyl acetate, again with lead citrate, and examined with a Philips 301 microscope*. For scanning electron microscopy (SEM), the samples were dehydrated after the initial fixation, critically point-dried, coated with gold-palladium, and examined with an AMR-1000 microscope.

Routes of Entry of Contaminants

Waterborne petroleum may enter a fish through the skin or gills by passing directly across the interfacing cellular membrane. In marine fish, which compensate for the dehydrating effect of their high salinity environment by gulping water and selectively secreting salts, contaminants may ingress through the intestine. Comparative studies of cadmium-exposed

---

*Trade names referred to in this publication do not imply endorsement of commercial products by the National Marine Fisheries Service.

mummichogs (*Fundulus heteroclitus*), adapted to freshwater and to seawater, established that the gills were damaged in the freshwater mummichogs (Voyer et al 1975, Gardner and Yevich 1970) and the intestinal epithelium became swollen and eventually edematous in seawater-adapted fish (Voyer 1975). There was some necrosis and hypertrophy in the gills of these fish but the areas of damage were infrequent and less severe than in the freshwater-exposed individuals (Gardner and Yevich 1970). If entry routes and sites of damage of petroleum parallel those of cadmium in mummichogs, the skin, gills, and intestines should be the tissues most likely to indicate initial waterborne petroleum effects. From these tissues, contaminants may enter the circulatory system and thereby affect other organs as indicated by uptake of ^{14}C-labeled petroleum components in liver, gallbladder, kidney, muscle, and brain (Anderson 1977).

In some instances, petroleum or its metabolic products enter marine organisms via the food-chain (Teal 1977), or by direct ingestion of tar balls (Conover 1971), and the alimentary canal is a primary site of contact and potential damage. If the contaminants cross the stomach or intestinal cell membranes and gain access to the circulatory system, again they can affect other organs.

Laboratory studies have been designed to test the effects of petroleum from both water- and food-borne sources on skin, gills, intestine, liver, and kidney in marine pelagic and benthic species and in freshwater fish. A number of parameters have been considered in these studies and the succeeding sections will be organized according to the exposure regime and affected tissue and will include the experimental design as well as a review of the pertinent literature. I will refer to trace metal and chlorobiphenyl contamination where these compounds appear to elicit a common response to that of petroleum hydrocarbons.

Tissue Changes

Gill
When toxic contaminants are waterborne, the gills are a site of damage that is easily assayed. Certain trace metals, such as cadmium, nickel, and lead, which may be associated with petroleum, damage gills by causing sloughing of epithelial cells in both marine and freshwater fish (Voyer et al 1975, Schwiger 1957, Haider 1964). Excessive mucus production is also characteristic of exposure of fish to trace metals (Varanasi et al 1975) and is thought to cause death by suffocation (Gardner 1975, Haider 1964). Similarly, petroleum hydrocarbons elicited epithelial sloughing and discharge of mucous glands in marine fish captured near an oil slick off the Texas and Louisiana coast (Blanton and Robinson 1973). The secretory cells of the pseudobranch of the Atlantic silverside (*Menidia menidia*) degenerated after exposure to 0.14 mg/l Texas-Louisiana crude oil (Gardner 1975) and to 100 ppm waste motor oil (Gardner 1975); light micrographs revealed severe vacuolization of the entire pseudobranch but no data on the adjacent gill was presented.

The effects of phenol, a relatively minor component of crude oil (Keith 1976), have been documented for several freshwater species. In goldfish (*Carassius auratus*) 10 ppm phenol did not damage the gills but did affect liver, kidney, and spleen (Vishnevetskii 1961). Rainbow trout sloughed epithelial cells and developed a general inflammation of the gills after exposure to 6.5 to 9.6 mg/l phenol (Mitrovic et al 1968). Another freshwater fish, the bream (*Abramis brama*), was exposed to lower concentrations of phenol for a longer duration with similar destruction of the gill epithelial cells and, in addition, suffered damage to the blood vessels and extravasation of blood in the gill lamellae (Waluga 1966b). Fourteen species of fish were sampled from phenol-polluted portions of the Rhine and Elbe Rivers; their gills revealed discharged mucous glands and generalized inflammation (Reichenbach-Klinke 1965).

In our laboratory, coho salmon and starry flounder were exposed to 100 $\pm$ 90 ppb of the water-soluble fraction of crude oil (WSF) for five days in a flow-through marine system (Roubal et al 1977). The gills of exposed fish developed lesions which reflected the loss of the surface cells or the first two to three layers of cells (Figs. 1, 2, & 3). Immature mucous glands below the surface were exposed when the surface sloughed and their contents in some instances were exuded, which could account for some of the excess mucous secretion observed in other studies (Blanton and Robinson 1973). The area of sloughing varied from filament to filament: 10 to 30 cells were lost in the smaller lesions and, in a few cases, the surface of an entire gill filament lost its outermost layer of cells. In both experimental and control coho salmon a gill ecto-parasite was observed, a monogenetic trematode (*Gyrodactylus* sp.) which needed to be considered in determining if the cellular damage could be the sole result of the parasite. The opishaptor of *Gyrodactylus* has hooks for attachment to the gill or skin surface and the attachment does not appear to particularly damage the fish cells (Fig. 4 & 5). Heavy infestations, however, can cause a diseased state in fish (Mellon 1928) and should be accounted for in reporting lesions, especially with routine histological techniques where serial sections would be needed to even ascertain the presence of the parasite. In the present study, the unexposed fish had

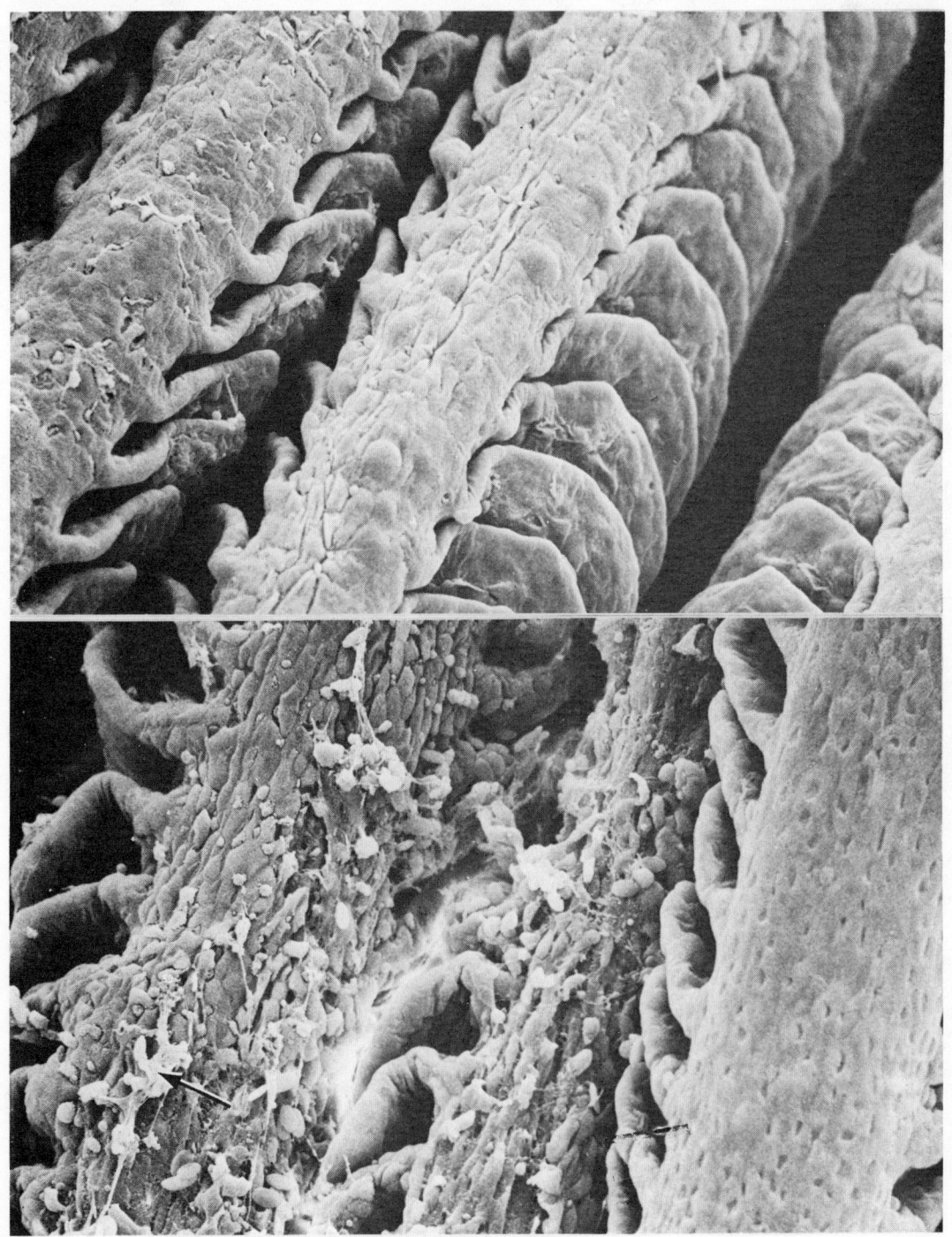

Plate I

Fig. 10.1. (Upper) SEM of gill from an untreated coho salmon. X 370.

Fig. 10.2. (Lower) SEM of gill from an WSF (10 ppb) petroleum-exposed coho salmon. One filament appears undamaged and the two on the left show severe cellular disruption. Many surface cells have sloughed and there is an abundance of exuded mucus (arrow). X 370.

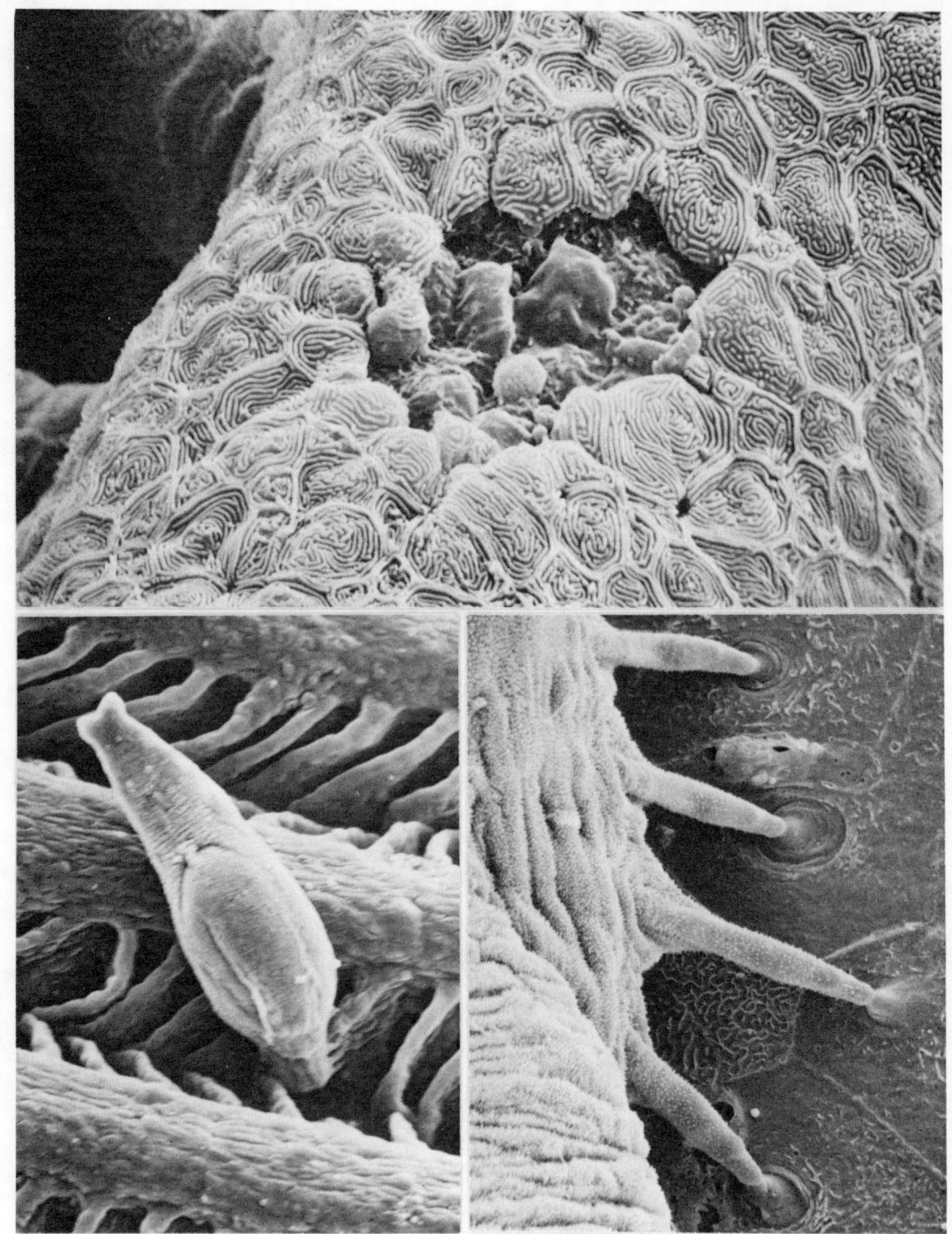

Plate II

Fig. 10.3. (Upper) Surface cells of the gill of starry flounder after exposure to 10 ppb WSF with SEM. The cells with microridges appear normal but a group of cells have sloughed. The surface of the underlying cell layer is evident. X 2,000.

Fig. 10.4. (Lower left) The trematode, Gyrodactylus sp. on the gill of coho salmon with SEM. X 320.

Fig. 10.5. (Lower right) Higher magnification of the opishaptor of Gyrodactylus sp. The attachment to the skin does not appear to harm the surface cells of the salmon gill. X 1,400.

Gyrodactylus and did not have lesions. In this case, the infection was probably low enough to not adversely affect the host. An area for future research could include a study of the relationship between degree of infection with specific parasites or pathogens and possible synergistic or antagonistic effects of particular contaminants, which may, in combination, severely affect survival of fish.

Skin
Changes in the mucous glands and epithelium of skin have been observed in fish exposed to phenols and petroleum hydrocarbons. Not only were there more mucous glands but they were distended in phenol-treated bream (Waluga 1966a). In several species of fish sampled from the phenol-contaminated Rhine and Elbe Ribers, the epidermis was swollen and inflamed (Reichenbach-Klinke 1965).

In an early experiment in our laboratory, English sole were exposed to the water-soluble fraction (WSF) of Prudhoe Bay crude oil: 10 ml crude oil in 1 liter of seawater was stirred for 20 hours, allowed to stand for three hours, and the bottom, clear fraction was removed. The WSF was then mixed with seawater to make a 13% WSF solution. The glass tanks were aerated and maintained between $10°$ and $13°C$.

In the skin samples of English sole taken five days after the WSF had been added (Figs. 6 & 7), many of the mucous glands were completely empty. In a repeat experiment, skin samples were taken from three different body locations during a two-hour to five-day time study. Results were inconclusive because there was a great deal of variability in both the numbers of glands and in the number that had discharged their contents. More extensive studies are projected to understand the normal rate of mucous exudation and to precisely define alterations of that process with increasing concentrations of petroleum.

Liver
The liver, richly bathed in blood through its portal system, is a primary target for toxic materials carried by the circulatory system and is a major site of detoxification. Exposure to phenol and No. 2 fuel oil from both external and internal sources can cause liver changes which range from gross color differences (Cardwell 1973, Waluga 1966a) to subcellular alterations such as proliferation of the endoplasmic reticulum (Sabo and Stegeman 1977). Storage and secretion products may also be altered: glycogen and lipid depletion has been reported in *Fundulus heteroclitus* collected near an oil spill (Sabo et al 1974, Sabo and Stegeman 1977). Glucose and acetate metabolism indicated that the total amount of lipid synthesized was the same in exposed and unexposed fish; however, the classes of lipids were different. The amount of triglyceride was lower in the livers of oil-contaminated fish and the phospholipid levels were elevated. Consistent with the biochemical results, there were fewer lipid droplets in the liver after petroleum exposure and the endoplasmic reticulum, a membranous system rich in phospholipids, was exceptionally abundant (Sabo and Stegeman 1977). It is apparent that both morphological and subtle biochemical changes occurred in the exposed fish, but the relationship of these findings to survival is still a topic for future research.

Halogenated hydrocarbons, including certain pesticides and polychlorinated biphenyls, also alter liver glycogen and lipid storage. In 10 freshwater and 3 marine species of fish, glycogen and lipid reserves were depleted, except following chlorobiphenyl exposure where there was pronounced fatty infiltration in the liver (Couch 1974, Couch et al 1974). This effect also occurs in other animal forms; for example, when the halogenated hydrocarbon pesticides, p,p'-DDT, α-chlordane, heptachlor, and endrin, were administered to rats orally, there was a reduction in liver glycogen and hepatic enzymes, including glucose-6-phosphatase (Singhal et al 1976).

In our laboratory, depletion of energy-storage products and infiltration of hepatic blood vessel by connective tissue were found in the livers of rainbow trout that received Prudhoe Bay crude oil in their diets. In one study, 2 ml of crude oil was dissolved in trichlorotrifluoroethane and adsorbed to 1 kg of Oregon moist pellets (OMP). The solvent was allowed to evaporate and the OMP was fed to the trout at 2% of body weight for five days per week. Assuming no evaporative loss of petroleum components, each fish received 11 mg of crude oil per day and the average weight of the trout was about 90 g. The controls were treated exactly the same as the experimental trout except that crude oil was omitted from the solvent-treated food. Two fish were sampled from each of four replicates of both experimental and control groups.

After two weeks of feeding, with no mortality, there were dramatic differences in the levels of energy reserves in the liver: the hepatocytes of control fish were full of glycogen whereas those of the experimental fish had virtually none. These changes were evident in 0.5 μ sections stained with toluidine blue. The polychrome method (MacKay and Mead 1970), which stains mucopolysaccharide moieties bright red when the cytoplasm is blue, was used on 1.0 μ sections to differentiate glycogen deposits in the cells (Figs. 8 & 9). The sections for TEM analysis showed the same disparity (Figs. 10 & 11). Proliferation

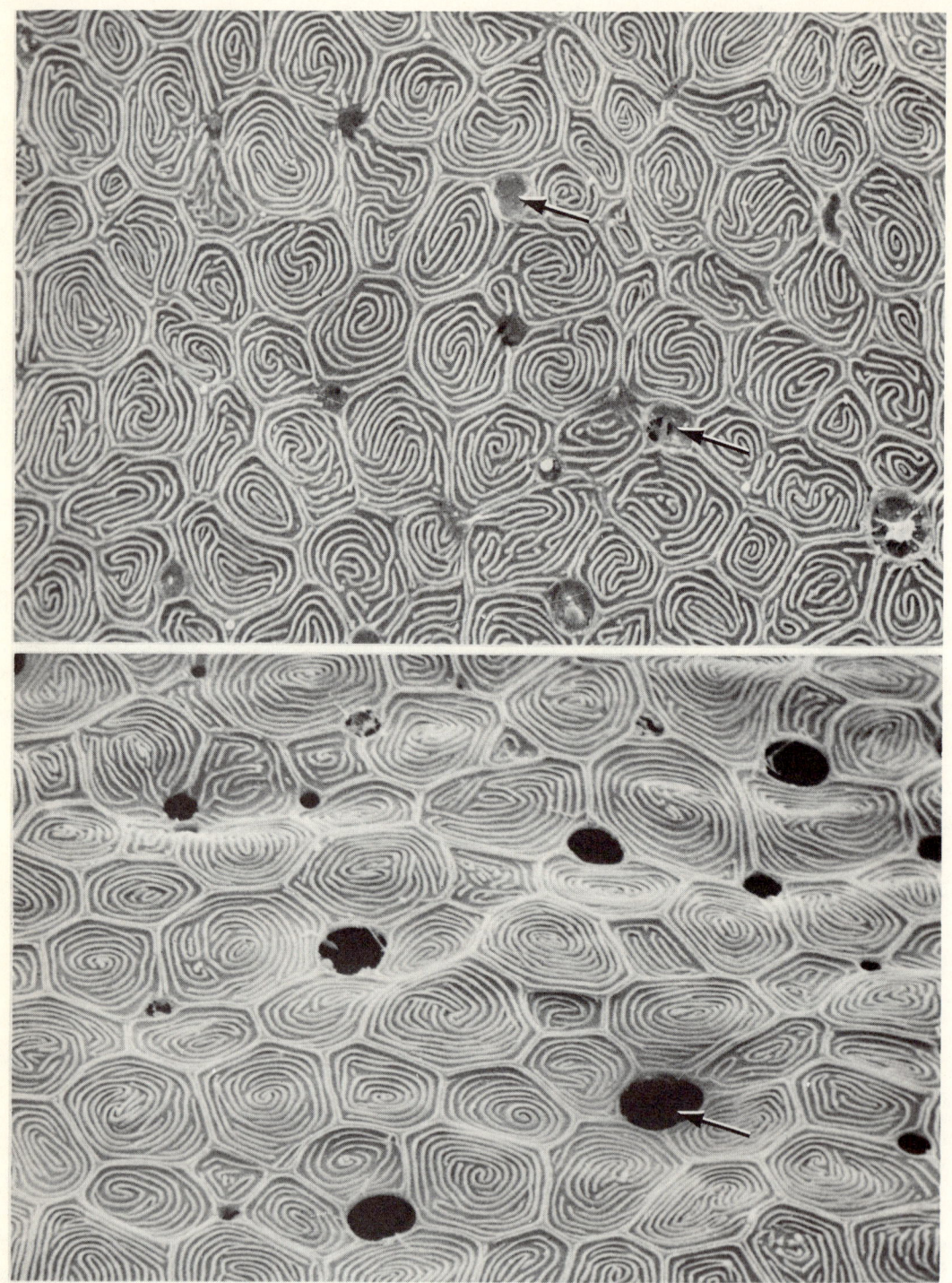

Plate III

Fig. 10.6. (Upper) SEM of the skin surface of an untreated English sole. The mucous glands (arrows) are liberally scattered throughout the normal skin between the filament-containing cells which are sculptured on their outer surface by micro-ridges. X 3,500.

Fig. 10.7. (Lower) English sole, five days after exposure to a 13% solution of WSF. The mucous glands (arrow) are numerous and conspicuously open. These fish felt more slimy than the untreated. X 3,500.

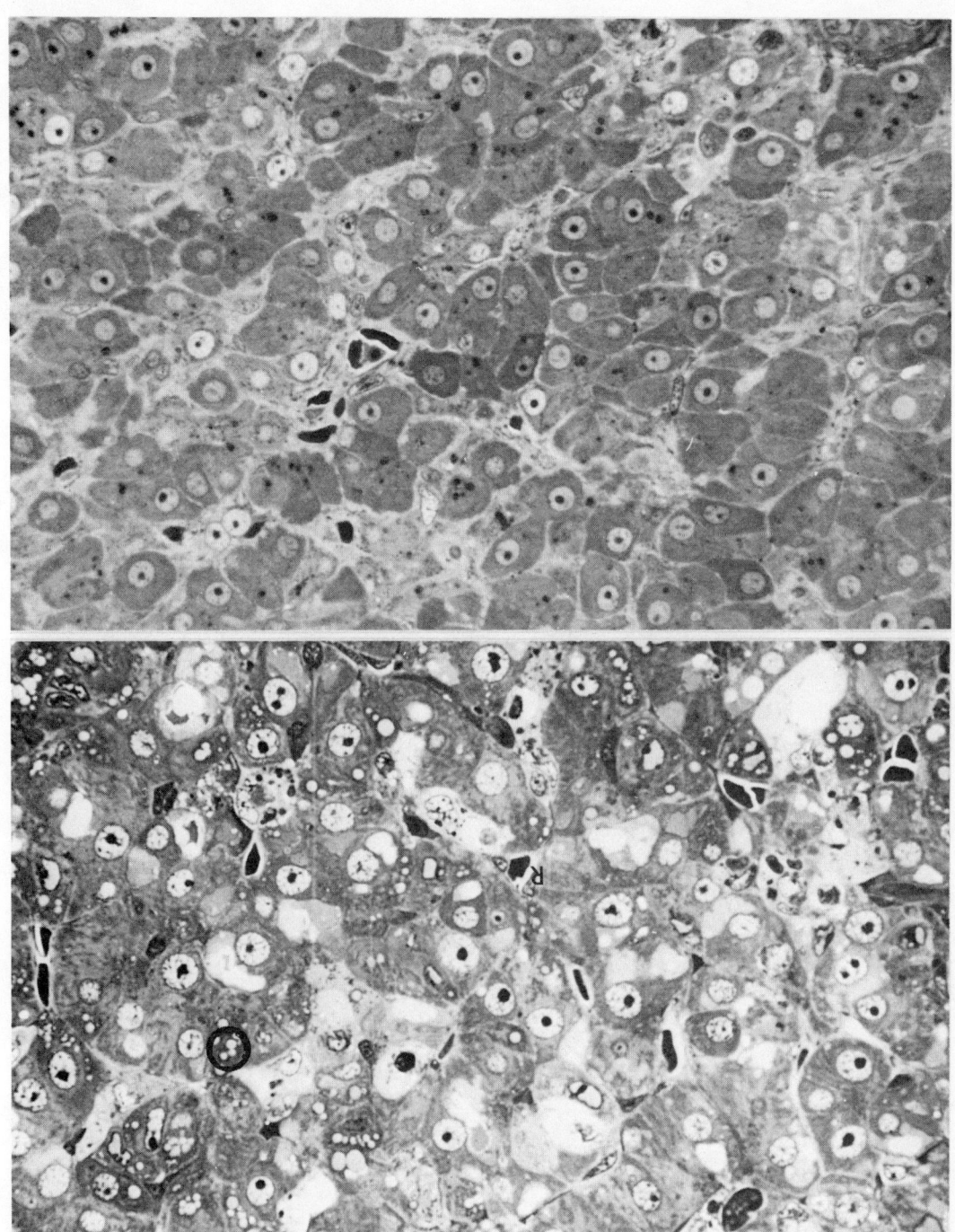

Plate IV

Fig. 10.8. (Upper) Light micrograph of sections of the liver of untreated rainbow trout. This and the accompanying micrograph were photographed from sections prepared with a polychrome stain adapted for plastic sections. The glycogen deposits are bright red against the blue background of the cytoplasm. Parts of the larger glycogen reserves were lost during preparation. Lipid droplets (circled) are abundant and red blood cells (R) are often apparent in the small hepatic vessels or sinusoids. X 620.

Fig. 10.9. (Lower) Companion micrograph to Fig. 10.8 of petroleum-fed rainbow trout (see text). No glycogen or lipid deposits are seen in this section and the cytoplasm is an overall blue tone. X 620.

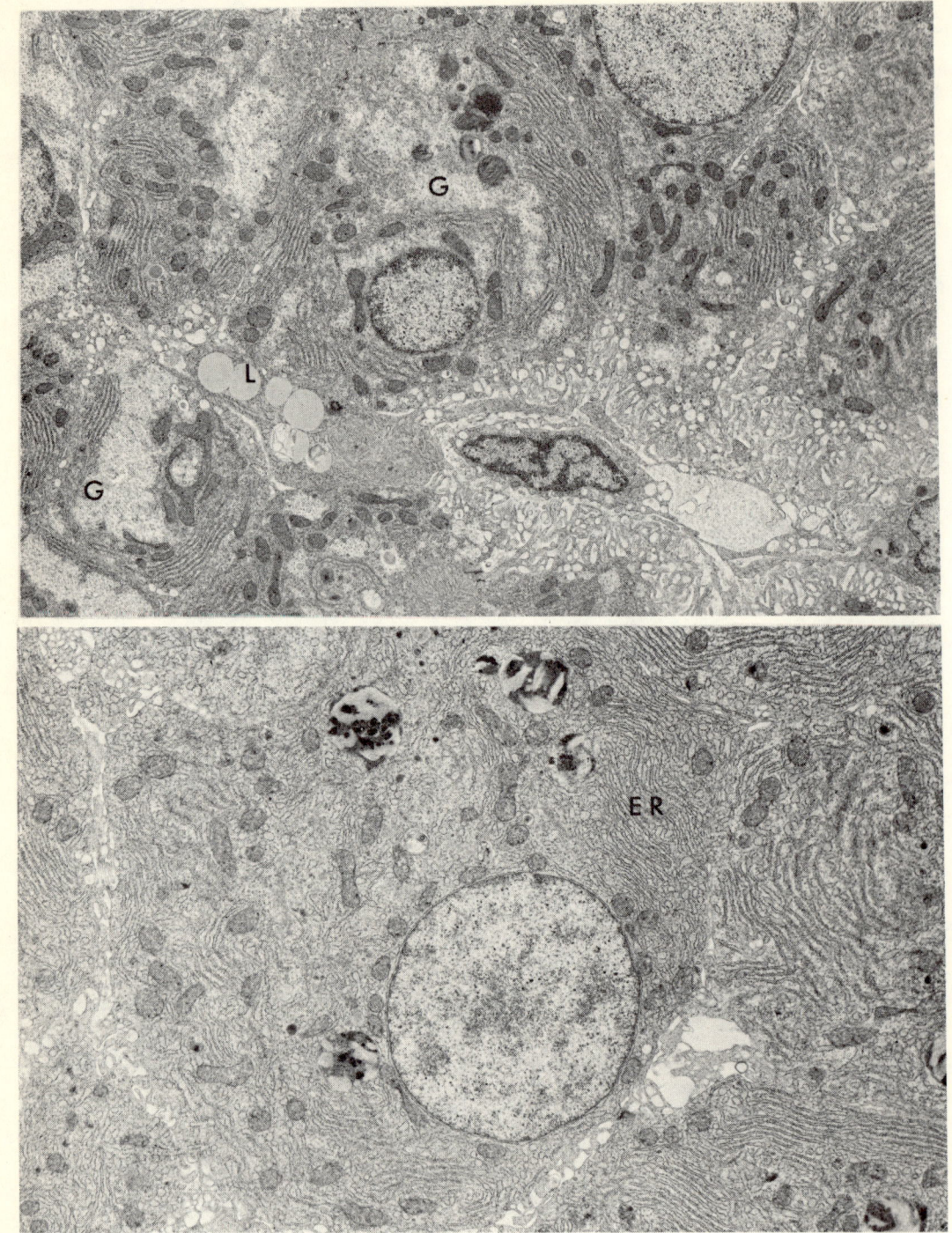

Plate V

Fig. 10.10. (Upper) TEM of thin sections of untreated rainbow trout liver. The hepatocytes are rich in glycogen (G) and lipid (L). X 6,400.

Fig. 10.11. (Lower) Thin section of liver from a rainbow trout fed petroleum for 75 days (see text). In addition to a lack of glycogen in the hepatocytes, the endoplasmic reticulum (ER) has proliferated to an extreme degree. X 7,500.

of the endoplasmic reticulum was evident (Fig. 11) and cochlear ribosomes, a common feature of rapidly synthesizing cells (i.e., in embryos), were apparent (Fig. 12).

In a longer feeding experiment with the same parameters, all the fish gained weight and no mortalities were observed for 75 days, at which time tissue samples were taken. At the termination of the experiment, the control fish had gained an average of 95.5% in body weight; the oil-fed fish 70.5%. The glycogen in the liver of test fish showed the same striking differences as in the above experiment. Small amounts of glycogen were present and could be seen with the electron microscope (Fig. 13). However, the glycogen stores were so minute that only a rare cell showed differential staining with the polychrome method for light microscopy. In addition, lipid reserves were reduced in the oil-fed fish.

A common response to severe cellular damage is replacement of the necrotic regions with connective tissue. Fibrotic infiltration occurred in the livers and kidneys of rabbits treated with 160 ppm cadmium chloride in their drinking water for 200 days (Stowe et al 1972). The gallbladders of these rabbits were hyperplastic and the hyperplasia was considered by Stowe and colleagues (1972) to be "a response to collagen formation which, in turn, appears to be a reaction to chronic cell injury." Connective tissue replacement has been reported after cadmium exposure in the testicular tissue of fish; the germinal epithelium of goldfish was mostly destroyed after intraperitoneal injections of 10 mg cadmium chloride per kg of body weight and few fish produced sperm (Tafanelli and Summerfelt 1975).

In our laboratory, a group of rainbow trout were fed Prudhoe Bay crude oil prepared in the same manner as described above but with less oil so that the dose was 17 mg crude oil/kg body weight/day, again assuming no evaporative loss. The fish were exposed from July 1975 through August 1976 and sampled at the time of spawning in August 1976. These fish were used for a study concerning maturation and reproduction (Hodgins et al 1977) and sampled for microscopy at the time of spawning. An abnormal increase of collagen around the liver sinusoids was noted using both light microscopy with a connective tissue stain (MacKay and Mead 1970) and conventional electron microscopy. Work is in progress to better define the extent of fibrosis and possible adverse effects. Such a response, however, is indicative of cell injury and may prove to be a useful gauge of liver damage.

Eye Lens
The same trout that developed liver abnormalities after exposure to crude oil in the diet also had enlarged eye lenses (Table 1) which were abnormally soft. Relatively mild pressure permanently compressed the lenses of exposed fish into an amorphous mass, whereas the control lenses returned to their normal geometry after the same pressure.

TABLE 1. Volume of Eye Lenses from Trout Fed Crude Oil for One Year

Group	$\bar{x}$ (mm^3)	Sx	S$\bar{x}$	N
Control	100.9	19.2	9.6	4
Oil-treated	226.10	81.2	33.1	6

The lens is composed of ribbon-like filaments which interdigitate and form a sphere. The filaments have simple projections on their broad surfaces which plug into pits on the adjacent fiber in addition to a complex interlocking series of protuberances on their thin side (Fig. 14). After treatment with petroleum, the fiber structure changed: the broad surface was wrinkled and the interdigitating projections were not smooth and regular as in untreated fish (Fig. 15). The fibers looked shriveled, as if the fixative was hyperosmotic, suggesting that the increase in size might be due to hydration of the lens rather than to increased mass resulting from cell proliferation or cell secretory activity. To test the hydration hypothesis, lenses were removed from normal rainbow trout, measured, and placed in a dilution series of "Dulbecco's" saline and distilled water. Hydration occurred at slightly different rates but in approximately five hours there was an 80% increase in volume which stabilized until the termination of the experiment at 45 hours. A series of experiments are underway to define factors contributing to lens enlargement in fish and the implications for vision and behavior.

Other Organs
Gross observation and limited histology have been done on spleen (Cardwell 1973, Waluga 1966a), kidney (Waluga 1966a), and portions of the circulatory system (Waluga 1966a, 1966b) of fish exposed to either No. 2 fuel oil or phenol. In the spleen of coho salmon, exposure to these compounds produced color changes which are related to destruction of red blood cells; normally black-red splenic tissue becomes light tan (Cardwell 1973). In addition,

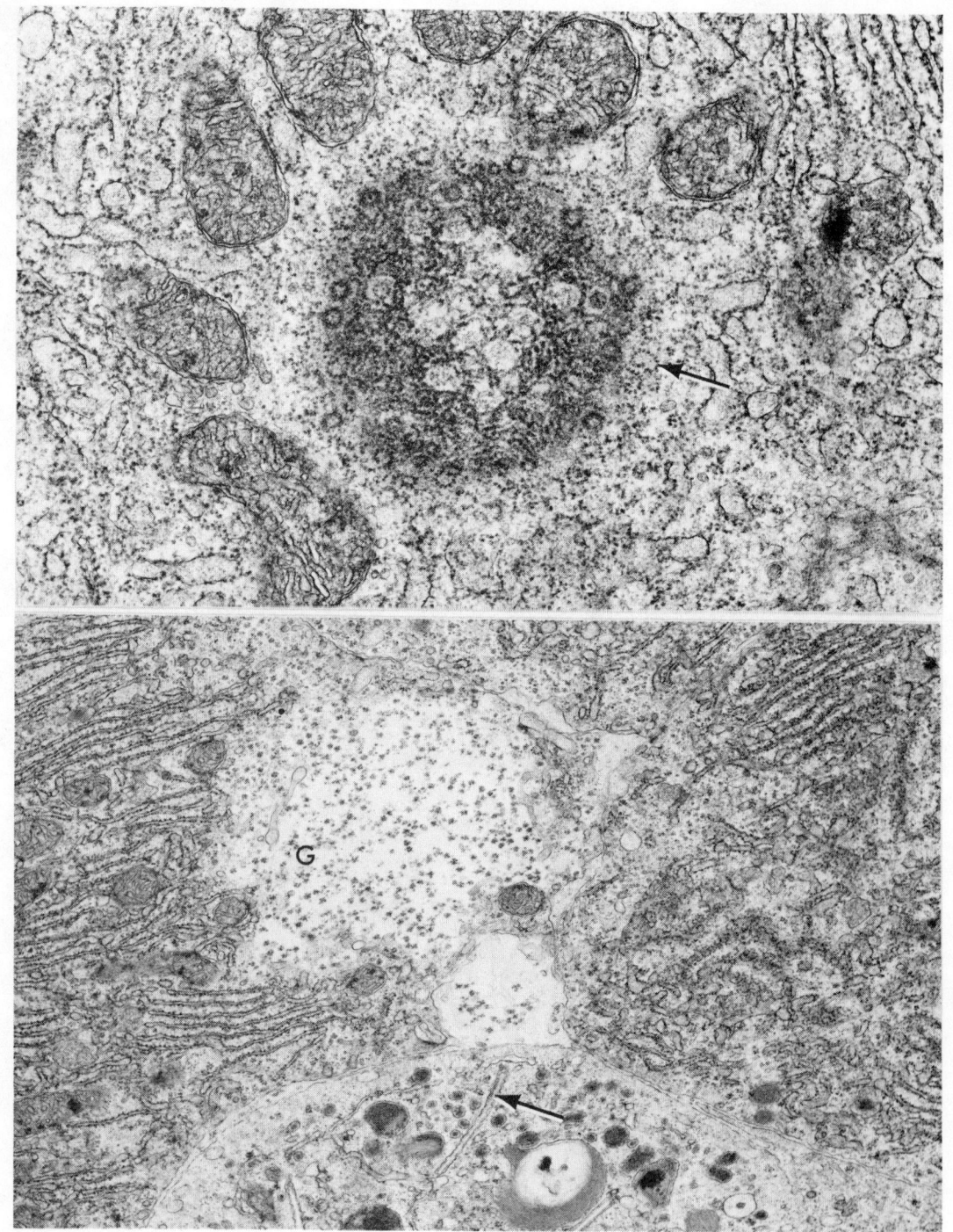

Plate VI

Fig. 10.12. (Upper) Thin section of liver from rainbow trout fed petroleum as above. There was an abundance of cochlear ribosomes (arrow) in the petroleum-fed fish. The mitochondria appear unchanged by petroleum exposure. X 26,000.

Fig. 10.13. (Lower) TEM of liver from rainbow trout fed petroleum as above. Small pools of glycogen (G) were found after considerable searching with the electron microscope. The deposits were small enough that they did not show in the thick sections used for light microscopy. In addition, there are unusual crystalline inclusions (arrow) of unknown significance in some of the hepatocytes. X 16,000.

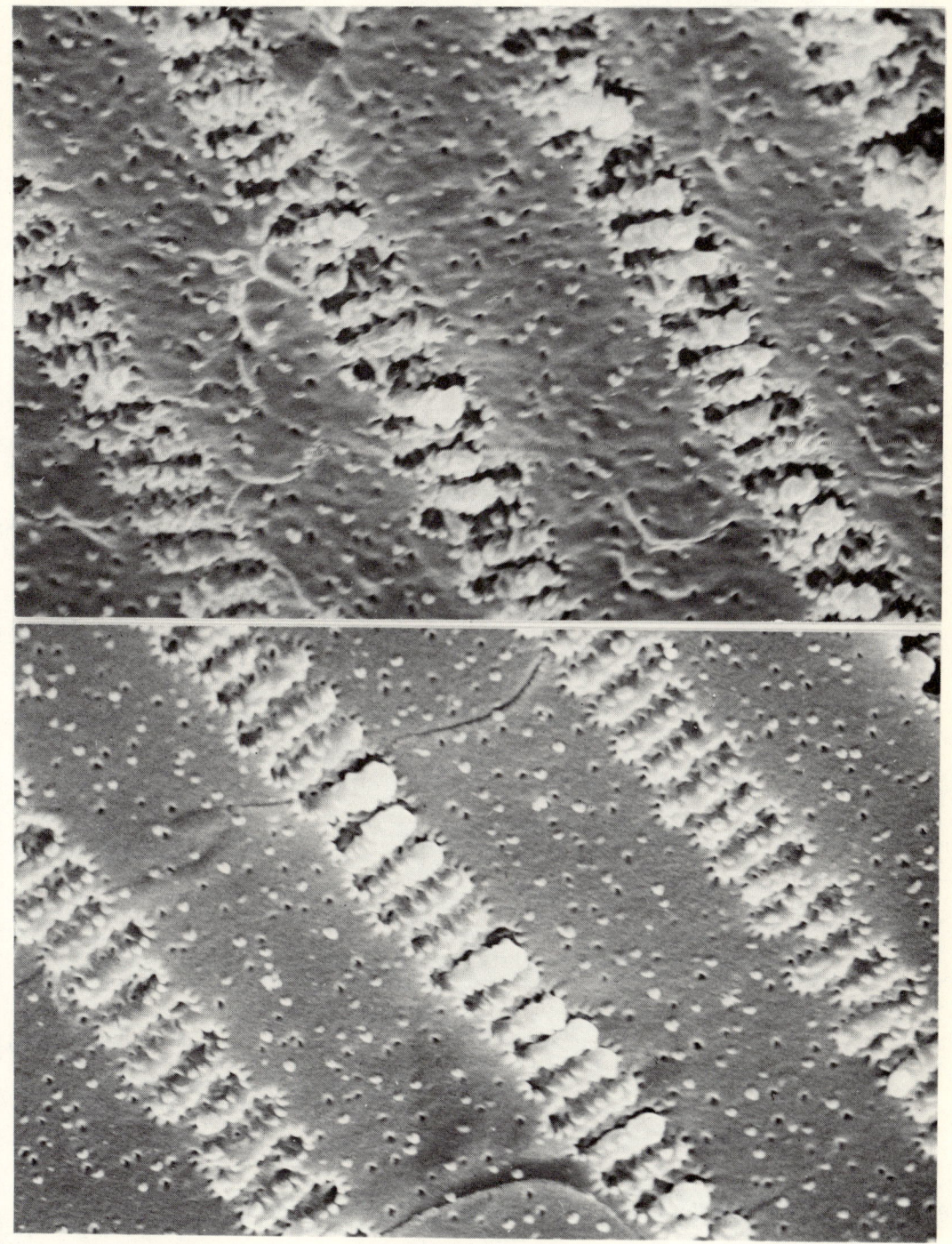

Plate VII

Fig. 10.14. (Upper) SEM of fibers from the eye lens of an untreated rainbow trout. The fiber surface, exclusive of connecting projections and pits, is smooth. The lateral interdigitations are quite regular. X 3,100.

Fig. 10.15. (Lower) Lens fibers from a rainbow trout fed 17 mg crude oil/kg body weight per weekday for about one year (see text). These fibers are from an enlarged lens and appear undulated and irregular. X 3,100.

when bream (*A. brama*) were treated with 2.6 mg/l phenol, there was an overall kidney discoloration and the kidney blood vessels were distended and filled with masses of blood (Waluga 1966a).

Certain trace metals, such as cadmium and lead, damage fish and mammalian tissues. Not only is cadmium deleterious to the testes (Parizek 1957, Singhal et al 1976, Sangalang and O'Halloran 1972), but to the nervous system (Gabbiani 1966, Gabbiani et al 1967), liver, and kidney. Granulomas have been found in the livers of goldfish after injections of cadmium (Tafanelli and Summerfelt 1975) and necrotic regions were found in the kidneys of cunners (*Tautogolabrus adspersus*) exposed to 48 ppm cadmium chloride in seawater (Newman and MacLean 1974). However, lower levels of cadmium in seawater (5 mg/l) did not produce lesions in *Fundulus* after a year's exposure (Gardner 1975). (For an additional review of the literature on petroleum hydrocarbon and trace metal effects at the cellular level see: Hodgins et al 1977.)

Summary and Recommendations

Although there are few reports on the cellular effects of exposure to petroleum, subcellular changes do occur, can be monitored, and are useful as one approach for determining the existence or severity of damage to marine organisms. The relative newness of ultrastructural studies as applied to petroleum and aquatic species means that many of these results are preliminary. There are, of course, several histological studies at the light microscope level that have provided knowledge about gross structural alterations. By continuing both light and electron microscope studies, enough information has been obtained to warrant summarizing both the effects of petroleum on the structure of certain tissues of aquatic species and to suggest probable implications of these changes.

Fish gills have shown epithelial sloughing within a few days after exposure to the water-soluble fraction of crude oil. With lesions present on the gill surface, a fish may have an increased susceptibility to bacterial or fungal infection and reduced respiratory function. In addition, excessive mucus exudation may lead to respiratory stress and eventual suffocation.

Several changes in liver structure have been reported after petroleum exposure. The striking increase in rough endoplasmic reticulum which has been observed probably reflects increased enzyme synthesis consonant with the detoxification function of the liver.

Fibrosis is an alteration that indicates a degree of damage severe enough that cells have died and been replaced, not with new cells, but with connective tissues. Such changes in fish livers have been clearly observed after petroleum exposure. In addition, depletion of liver glycogen indicates potentially reduced liver function and lowered energy reserves. There is a point at which such alterations would affect the survival of fish. Long-term survival and stress-response studies correlated with a degree of morphological liver alterations appear appropriate.

The enlargement of the lens found in trout after long-term petroleum feeding should produce extreme myopia; resulting vision-related behavior difficulties would be expected. Certainly there is much work yet to be done to check the repeatability of these results with additional species and at lower levels of petroleum. Behavioral tests for visual acuity will be an important correlative experiment.

In conclusion, the problem of determining effects of long-term exposure to low levels of petroleum in the aquatic environment is difficult. There is a problem in defining both the qualitative and quantitative aspects of petroleum, which is not a single entity but a large number of hydrocarbons and associated trace metals. There is also the difficulty of the wide variation in potentially affected aquatic species. These organisms are diverse; at the present time only meager data is available on normal characteristics of many species. It is evident that an intensive multidisciplinary approach is needed to define normal structures and functions, abnormalities resulting from sublethal exposure to petroleum, and consequent implications. Clearly, morphological studies at the tissue, cellular, and subcellular levels should be integral investigations of contaminant effects on marine biota.

Acknowledgements

Mr. D. Weber, Drs. H. Hodgins and T. Roubal, generously shared experimental fish. SuzyAnn Gazarek and Carla M. Stehr contributed excellent technical assistance. In addition, Drs. H. Hodgins, D. C. Malins, A. K. Sparks, and Mr. D. Weber critically reviewed the manuscript and their input is greatly appreciated.

References

Anderson, J. W., Sublethal effects of petroleum hydrocarbons. In: Proceedings of Fate and Effects of Petroleum Hydrocarbons in Marine Ecosystems and Organisms, Pergamon Press, New York, 1977, In press.

Anderson, J. W., R. C. Clark, and J. Stegeman, Contaminant bioassays: Petroleum hydrocarbons. In: G. V. Cox (Workshop Chairman), Marine Bioassays Workshop Proceedings 1974, p. 36-75, Mar. Technol. Soc., Washington, D.C., 1974.

Blanton, W. G. and M. C. Robinson, Some acute effects of low-boiling petroleum fractions in the cellular structures of fish gills under field conditions. In: D. G. Ahearn and L. P. Meyers (eds.) The Microbial Degradation of Oil Pollutants, p. 265-273, Publ. No. LSU-SG-73-01, Louisiana State University, Center for Wetland Resources, Baton Rouge, Louisiana, 1973.

Cardwell, R. C. Acute toxicity of No. 2 diesel oil to selected species of marine invertebrates, marine sculpins and juvenile salmon, Ph.D. thesis, University of Washington, Seattle, 1973.

Conover, R. J., Some relations between zooplankton and Bunker C oil in Chedabucto Bay following the wreck of the tanker Arrow, J. Fish. Res. Board Can. $\underline{28}$, 1327-1330 (1971).

Couch, J., Histopathological effects of pesticides on the livers of fishes. In: The Pathology of Fishes, p. 559-584, University of Wisconsin Press, Madison, 1974.

Couch, J., G. Gardner, J. C. Harshbarger, M. R. Tripp, and P. P. Yevich, Evaluation of responses in marine organisms. Histological and physiological evaluations in some marine fauna. In: Marine Bioassays. Workshop Proceedings, pp. 156-173, Am. Petrol. Inst., Environ. Protection Agency, and Mar. Technol. Soc., Washington, D.C., 1974.

Gabbiani, G., Action of cadmium chloride on sensory ganglia, Experientia $\underline{22}$, 261-262, (1966).

Gabbiani, G., D. Baic, and D. Deziel, Toxicity of cadmium for the central nervous system, Exp. Neurol. $\underline{18}$, 154-150 (1967).

Gardner, G. R., Chemically induced lesions in estuarine or marine teleosts. In: W. E. Ribelin and G. Migaki (eds.) The Pathology of Fishes, p. 657-693, University of Wisconsin Press, Madison, 1975.

Gardner, G. R. and P. P. Yevich, Histological ahd hematological responses of an estuarine teleost to cadmium. J. Fish. Res. Board Can. $\underline{27}$, 2185-2196 (1970).

Haider, G., Heavy metal toxicity to fish. I. Lead poisoning of rainbow trout (Salmo gairdeneri) and its symptoms, Z. Angew Zool. $\underline{51}$, 347-368 (1964).

Hawkes, J. W., The structure of fish skin. I. General organization, Cell and Tissue Res. $\underline{149}$, 147-158 (1974a).

Hodgins, H. O., W. D. Gronlund, J. L. Mighell, J. W. Hawkes, and P. A. Robisch, Effect of crude oil on trout reproduction. In: Proceedings of Fate and Effects of Petroleum Hydrocarbons in Marine Ecosystems and Organisms, Pergamon Press, New York, 1977, In press.

Keith, L. N., Identification and analysis of organic pollutants in water, p. 563, Ann Arbor Science Publishers, Inc., Ann Arbor, Michigan, 1976.

Newman, M. W. and S. A. MacLean, Physiological response of the cunner, Tautogolabrus adspersus, to cadmium. VI. Histopathology, NOAA Tech. Rept. NMFS SSRE-681, p. 27-33 (1974).

MacKay, G. R. and M. L. Mead, A simple dichromatic stain for plastic embedded tissues, Proc. 28th EMSA Meeting, p. 296-297 (1970).

Mellen, I., The treatment of fish disease, Zoopathologica $\underline{2}$, 1-31 (1928).

Parizek, J., The destructive effect of cadmium ion on testicular tissue and its prevention by zinc, J. Endocrinol. $\underline{15}$, 56-63 (1957).

Reichenbach-Klinke, H., Der Phenolgehalt des Wassers in seiner Auswirkung auf den Fishorganismus, Arch. Fischereiwiss. $\underline{16}$, 176 (1965).

Roubal, W. T., D. H. Bovee, T. K. Collier, and S. I. Stranahan, Flow-through system for chronic exposure of aquatic organisms to seawater-soluble hydrocarbons from crude oil: construction and applications. In: 1977 Oil Spill Conference (Prevention, Behavior, Control, Clean-Up), Am. Petrol. Inst., Washington, D.C., 1977.

Sabo, D. J. and J. J. Stegeman, Some metabolic effects of petroleum hydrocarbons in marine fish. In: A Calabrese and J. F. Vernberg (eds.) Pollution and Physiology of Marine Organisms II, Academic Press, New York, 1977, In press.

Sabo, D. J., J. J. Stegeman, and L. S. Gottleib, Petroleum hydrocarbon pollution and hepatic lipogenesis in the marine fish Fundulus heteroclitus, Fed. Proc. 34(3), 810, Abst. (1975).

Sangalang, G. B. and M. J. O'Halloran, Cadmium-induced testicular injury and alterations of androgen synthesis in brook trout, Nature 240, 470-471 (1972).

Schwiger, G., The toxic action of heavy metals salts on fish and organisms on which fish feed, Arch. Fischereiwiss. 8, 54-78 (1957).

Singhal, R. L., Z. Merali, and P. D. Hrdina, Aspects of the biochemical toxicology of cadmium, Fed. Proc. 35(1), 75-80 (1976).

Spurr, A. R., A low viscosity epoxy resin embedding medium for electron microscopy, J. Ultrastruct. Res. 26, 31-43 (1969).

Stowe, H. D., M. Wilson, and R. A. Goyer, Clinical and morphologic effects of oral cadmium toxicity in rabbits, Arch. Pathol. 94, 389-405 (1972).

Struhsaker, J. W., M. B. Eldridge, and T. Echeverria, Effects of benzene (a water-soluble component of crude oil) on eggs and larvae of Pacific herring and Northern anchovy. In: F. J. Vernberg and W. B. Vernberg (eds.) Pollution and Physiology of Marine Organisms, p. 253-284, Academic Press, New York, 1974.

Tafanelli, R. and R. C. Summerfelt, Cadmium-induced histopathological changes in goldfish. In: W. E. Ribelin and G. Migaki (eds.) The Pathology of Fishes, University of Wisconsin Press, Madison, 1975.

Teal, J., Food web transport of petroleum hydrocarbons. In: Proceedings of Fate and Effects of Petroleum Hydrocarbons in Marine Ecosystems and Organisms, Pergamon Press, New York, 1977, In press.

Varanasi, U., P. A. Robisch, and D. C. Malins, Structural alterations in fish epidermal mucus produced by water-borne lead and mercury, Nature 258, 431-432 (1975).

Vishnevetskii, F. E., Pathomorphology of fishes poisoned with phenol and water-soluble components of crude oil, coal tar and mazut (an experimental study), Tr. Astrakh. Gos. Zapov. 5, 350-352 (1961).

Voyer, R. A., Effect of dissolved oxygen concentration on the acute toxicity of cadmium to the mummichog, Fundulus heteroclitus (L), at various salinities. Trans. Am. Fish. Soc. 104, 129-134 (1975).

Waluga, D., Phenol effects on the anatomico-histopathological changes in bream (Abramis brama L.), Acta Hydrobiol. 8(1), 55-78 (1966a).

Waluga, D., Phenol induced changes in the peripheral blood of the breams, Abramis brama (L) Acta Hydrobiol. 8(2), 87-95 (1966b).

CHAPTER 11

THE EFFECTS OF PETROLEUM HYDROCARBONS ON MARINE POPULATIONS AND COMMUNITIES

A. D. Michael

University of Massachusetts Marine Station
Box 128 Lanesville Station
Gloucester, Massachusetts 01930

Introduction

We will limit our discussion to the effects of oil as they concern living organisms in their natural surroundings, i.e. field studies of communities and populations. A great deal of literature has been published on oil spills and general ecological effects of oils. Much of the information is fragmentary, speculative and based on inconclusive data. Perhaps only three or four spills have had substantial attention from the scientific community. Some of these might be the Torrey Canyon, Santa Barbara, West Falmouth and Chedabucto incidents. In all of these and other cases, funding has limited the scope and nature of the work. The inadequacies of the literature have been pointed out by others (e.g. Boesch et al. 1974, National Academy of Sciences 1975). Part of the problem is the state of the art of the science involved and part is due to difficulty in obtaining suitable funding to follow studies to their logical conclusion. There has been a tendency to consider an area "recovered" when effects are no longer visible.

There are perhaps two aspects to the problem of oil spills. One is the science of the problem and the second is the social or economic significance. I do not mean to suggest that these two attributes are mutually exclusive. In fact I feel that they are closely related but for the purposes of discussion I will address them separately.

A large section of the human community is primarily concerned about the effects of oil spills and chronic discharges on commercial fisheries and esthetics i.e. the immediate practical aspects. Whereas adverse affects of oil on the sex life of an alga may be of great interest to a scientist, it is of little concern to the administrator unless a clear relationship between the problem for the alga and the harvest of some commercial species can be demonstrated. The practical person (not that all scientists are impractical!) wants to see demonstrable effects in the short term and will probably not recognize the value of maintaining funding over the long term. Even when an area is effectively wiped out by a spill some would argue that this is insignificant if it represents only a small portion of a total ecological system, e.g. part of a bay or a section of a marsh.

Some serious limitations are associated with this latter strategy. One is the state of the art of the science. Many areas of the science of marine ecology are still in fairly elementary stages. Consider for example secondary production by the benthos which is important in supporting many fisheries. There are no definitive studies in this topic. Our knowledge of the process is limited and certainly not firmly based on an established theoretical and statistical framework. This stacks the odds against anyone trying to measure the effect of oil or any possible contaminant on such an important community function. Unfortunately the same is true of most areas in marine ecology. While scientists may feel confident of the relationship between cause and effect and the significance of such, the possibility of demonstrating this in a manner to convince the practical school is markedly reduced. Agencies responsible for oil research have not considered it their responsibility to provide support for the basic research which must eventually fill the gap.

The second problem with the practical short term approach is that short-term studies will identify only major changes. Minor alterations may appear to be insignificant but such changes can be cumulative and very significant over the long term. Populations and communities in marine environments undergo large natural variations - often much larger than the changes one expects to measure in monitoring communities under stress. There is an urgent need to gain a better understanding of the subtle sequence of events that eventually leads to community collapse. This can only be achieved by careful long term documentation of a few selected situations and the knowledge of the processes involved. We must try to escape our present limitations in being able to simply document the decline of an area. The need is for predictability. There are many environmentally degraded bays and harbors around the country where one hears local people speak of how they were able to harvest shellfish and swim 10 or 15 years earlier. Yet recent studies in those areas produce little understanding of how or why these systems have declined so that they are no longer suitable for such activities.

There have been several thorough reviews of the literature on ecological effects of oil within the last few years (e.g. Boesch et al 1974; National Academy of Sciences 1975). The additions to the literature since then do not justify a complete restatement. After a few brief comments on some general aspects I propose to discuss some recent work and the status of our current research knowledge.

Plankton

There has been no strong evidence for major damage to plankton communities as a result of an oil spill. Some effects on phytoplankton were observed after the Torrey Canyon incident but none on the zooplankton. Much data has been accumulated in the laboratory and it has been reported that water extracts of various crude and refined oils can inhibit growth (Kauss et al, 1973; Nuzzi, 1973). Concentrations of hydrocarbons in the test tanks for such experiments were often unrealistically high and so the results may not have much ecological application. Gordon and Prouse (1973), in an attempt to simulate concentrations that might occur in polluted coastal areas found stimulation of photosynthesis at low levels (10-30 ppb). Concentrations of 60-200 ppb suppressed photosynthesis. It seems possible that photosynthesis could be either stimulated or suppressed in areas of chronic pollution or the immediate vicinity of a major spill.

The two factors operating to minimize effects of oil on plankton are 1) the fact that fractions of oil which enter the water column disperse rapidly so that concentrations are usually very low and 2) plankton populations typically have rapid regeneration rates and usually cover large geographic areas. Even if one assumed 100% mortality at the site of a spill it would be difficult to demonstrate the significance of effect on the overall population. Jeffries and Johnson (1975) documented the rapid recovery of plankton communities within weeks after a major collapse in Raritan Bay.

Field evidence on chronic pollution affecting plankton is lacking. Decreased diversity of phytoplankton and zooplankton communities in the Houston ship channel may be related to petroleum hydrocarbon contamination but this cannot be proven since many other toxic wastes are also present. Potential ecological effects on plankton communities might occur in the case of a major oil spill or chronic contamination in a fairly enclosed bay or estuary which had a unique community. Persistent exposure within a restricted area could lead to a change in community structure and subsequent impact on the food chain. The probability of this happening appears to be low. Larval and juvenile stages of benthic invertebrates and demersal fish species which spend only a short time in the plankton are probably more vulnerable than phytoplankton or holoplankton. If juvenile stages are more susceptible to the effects of oil and if reproduction occurs only once or twice a year, an oil spill at the critical time could take on much more significance. There is no documentation of such an event in the literature.

Benthos

Intertidal
Much of the emphasis in early oil spill studies was on the intertidal biota. Concern for the esthetic aspects of beaches and the fact that mortality due to oil could be readily seen along the shore line provided the impetus. Oil in the intertidal zone can affect the biota by smothering, fouling or directly poisoning organisms.

Physical effects of oil, particularly obvious with heavier fuel and crude oils, include smothering, abrasion, removal and alteration of the substrate. Different species will survive for varying reasons. At Santa Barbara, a tall species of barnacle projected above the oil layer and fared much better than a short sessile one. Algae with a natural coating of mucus were not as heavily damaged as an intertidal surf grass (Nicholson et al, 1971; Foster et al, 1971; Neushel, 1970). The large intertidal algae are generally resistant to the toxic effects of oil but if oil does adhere to the blades the sheer weight will often cause them to tear off.

Toxic effects of oil on the intertidal biota were quite marked after the Tampico Maru and West Falmouth incidents. In both cases the oil spilled was a light fuel oil and almost all the intertidal organisms were killed (North et al, 1965; Blumer et al, 1971). Several factors serve to minimize the effects of oil on intertidal biota. Oil stranded on the shore is subject to the physical elements (wind, waves, light, etc.) which facilitate its breakdown. The intertidal zone is a rigorous environment and the plants and animals there are more hardy than their subtidal relatives. The speed with which oil breaks down and is removed from the system is highly dependent on the energy of the system and the type of oil. With exposed sandy beaches and rocky shores, physical extremes favor breakdown processes. Crude oils can however weather to an asphaltic composition which may persist for years. In sheltered environments any type of oil will last much longer as indicated at West Falmouth (Michael et al. 1975) and Chedabucto (Scarrat and Ziko. 1972).

Wetlands are unique intertidal environments. They are generally considered to be of high ecological value because of their productivity or their use as nursery grounds or wildlife refuges. They are protected areas of low physical energy where oil will persist. Spills have contaminated both mangrove swamps and salt marshes. There is qualitative evidence that mangrove trees have died as a result of oil spills (Diaz-Piffer 1962; Rutzler and Sterrer 1970). A great deal more information exists for salt marshes and it appears that susceptibility depends on the amount and type of oil, the plant species and the time of year (Boesch et al. 1974). At best marsh plants can survive single oilings in light to moderate doses whereas chronic exposure or heavy doses can kill the plants and induce erosion of the substratum. The fauna of marshes also appears to be very sensitive.

Subtidal
It is now known that oil has reached subtidal sediments in most of the major oil spills (Kolpack et al. 1971; Blumer et al. 1971; J. E. Smith 1968; Vandermeulen and Gordon 1976). Qualitative observations after the Tampico Maru and Torrey Canyon spills revealed dead clams, snails, crustaceans and echinoderms. The West Falmouth spill was the first in which a detailed study of the subtidal benthos was made. Components of fuel oil were found in the subtidal sediments at 11 meters depth for four years after the West Falmouth spill. The process by which oil reaches the sea floor is not fully understood but the primary mechanism is probably through adsorption to small sediment particles in the water column and subsequent sinking. Evidence is already accumulating that subtidal habitats are likely to show long term effects of oil exposure. The persistence of oil will again be dependent on the energy of the system. In coarse sediments where strong currents predominate, oil will flush out more readily than in depositional environments with soft sediments. An interesting facet is that total hydrocarbon content of coarse and fine sediments differ greatly. In some recent samples I collected from the New York Bight, two samples taken side by side in coarse and fine sediments, varied by 1,000 ppm. The sand sample had 300 µg/g total hydrocarbons whereas the mud sample had 1,300. This could either be due to the larger potential surface area for attachment in the case of the fine material or the fact that the greater interstitial space in the sand allows the oil to leach out.

Fisheries

There are at least three ways oils spills or chronic exposure can affect fisheries:

1. Loss of fishing time or gear.
2. Tainting of the catch.
3. Direct destruction of the fishery.

In the aftermath of a spill, the risk of fouling gear or of catching tainted fish reduces fishing effort which is a very real although probably short term loss. The most probable effect and longest lasting is the tainting problem. Any fishery where the animals are in direct contact with the sea floor is vulnerable if oil reaches the sediments. There have been a variety of instances where oil has contaminated commercial species, e.g. Mead and Sorensen (1970), Nelson-Smith (1972), Vale et al. (1970), Blumer et al. (1970). Even though oysters and other molluscan species survive and grow in affected areas the extra effort and expense in moving these to clean areas for depuration reduces the income from the fishery. The difficulty in identifying the causal factors in the decline of a fishery are well known. The problems range from sampling statistics, overfishing and, often, a variety of possible contaminants. This should not, however, be used as a convenient excuse for acquitting any contaminant from probable cause. Direct toxic effects on an entire fishery of finfish which live in the water column and whose populations cover large areas are not probable. The most vulnerable aspects of such fisheries are during the reproductive and early stages. Many species concentrate in small areas at these times and contamination of the critical place at the particular time could have serious consequences. Although the probability of this is low, the potential for long term disruption exists since many fisheries are age-class dominated and once the pattern is distrubed, it could take several years to return to normal.

Birds

Bird mortality following oil spills has been quite considerable in several cases. Between 40,000 and 100,000 birds were reported killed by the Torrey Canyon spill (Smith 1968); about 3,000 by the Santa Barbara spill (Straughan 1971), and 7,000 by the San Francisco Bay incident (Chan 1973). These numbers include significant proportions of the populations of some species within the affected areas. Birds that spend most of their lives on the surface

are the most vulnerable, e.g. auks, penguins and diving sea ducks. They are weak fliers and
dive rather than fly when disturbed. Since these birds are also highly gregarious, it is
possible for a small slick to cause large casualties (National Academy of Sciences, 1975).
Fouling of the feathers is the primary problem. Birds lose insulation and bouyancy and may
therefore die through heat loss or simply sink and drown. The ingestion of oil as birds
attempt to preen disturbs digestive processes, and a small amount taken internally can cause
death.

The practice of discharging oily wastes from tankers at sea produces casualties continually
(Boesch et al. 1973). There is a variety of circumstantial evidence that oil has contributed to the decline of populations of certain species. Westfall (1969) suggests that the
jackass penguin which has a restricted distribution around the coast of South Africa has
suffered significant losses. The breeding populations of puffins, guillemots and razorbills
on the Island of Rouzie on the Brittany coast were reduced by 80-88 percent in the year
following the Torrey Canyon spill (Goethe 1968) although additional factors besides oil may
be involved. Auks and penguins have low replacement rates and high early mortality from
natural causes. They lack the potential to replace losses caused by increased adult mortality (National Academy of Sciences, 1975).

The potential for ecological damage from spills or even chronic contamination exists for
many bird species. The probability of this happening for any species depends on whether its
range is restricted or its population numbers are low. Most aquatic species may be affected by a spill but are not likely to suffer losses beyond the ability of the population to
recover. There are a variety of special cases however and these should be watched carefully
in the future.

Mammals

Although it is possible for mammals to be affected by oil, this is most likely to happen on
an individual basis rather than at the population level. It would be difficult to determine
clear evidence of significant losses unless a spill affected the area in which a small restricted population lived. We should however be aware of the latter possibilities and watch
for long term consequences of chronic contamination over large areas.

Spills Versus Chronic Contamination

Large oil spills draw much public attention and probably have had the greater share of
research funds. A single inoculation of oil into an area will eventually disappear since
there are biological, chemical, and physical breakdown processes. In some cases virtually
all the oil will disappear quickly whereas in others, components of oil and biological
effects may be traced for years after a spill. The ecological consequences of a spill in
any one particular area can vary dramatically. Some of the factors which determine impact
are; oil dosage; oil type; oceanographic conditions; meteorological conditions; turbidity;
season and biota type (National Academy of Sciences, 1975). A single factor such as wind
direction can be of overwhelming importance. Oil blown ashore and driven into low energy
areas may persist for several years with severe effects on the biota whereas the same oil,
driven out over the open sea by offshore winds may produce little measurable damage if the
oil does not reach the sea floor.

There is perhaps a tendency to assume that we know a good deal about oil spills now since
there is a considerable volume of literature. This attitude emanates from the supposition
that if one cannot see effects, they do not exist. I hold the view that we still have much
to learn about the consequences of spills and would prefer to see fewer, more comprehensive
studies, rather than the perpetuating cursory overviews which do not enhance our understanding.

I do not wish to understate the possible ecological dangers of larger spills but at present
I feel that the chronic contamination issue deserves more attention and may be the more
serious ecological problem. One of the critical problems is that most areas of chronic
contamination from petroleum hydrocarbons are subject to a wide variety of other pollutants.
This provides a strong argument for following oil spills to their logical conclusion, i.e.
continuing studies when oil levels are low and effects slight. It is in this way that we
can learn what contribution oil may make to changes in areas of chronic input. We can
apparently do little about some areas where oil contamination will continue. Mueller and
Jeris (1975) estimate that some 870 metric tons of oil and grease enter the New York Bight
each day. A high percentage of this is through wastewater and runoff. That is a quantity
of about 300,000 metric tons per year and it is obvious that much of this is cycled through
the system or we would have an oilfield in the bight by now. Oil is accumulating in the
sediments however. Farrington (1975) reported quantities of 1,810 µg/g at the dump site
(total hydrocarbons) and some of my own samples from depths of 30 meters have been in the
1,000 to 1,300 µg/g range. We should be learning the dynamics of the cycling of petroleum
hydrocarbons, the rate at which they are accumulating and their biological effects in

order to make reasonable predictions about the consequences of increased dosages in such systems.

Recent Work

Grassle and Grassle (1975) documented short term selection for a single genotype in a species of polychaete worm following an oil spill. They ranked species for their degree of "opportunism" according to their response to disturbed conditions. On the basis of these results and theoretical considerations they suggested that initial response to disturbed conditions, ability to reproduce rapidly, large population size, early maturation, and high mortality were all features of opportunistic species. These species are generalists which can take advantage of niches left open after disturbance. They are present in all communities but typically in low numbers in climax communities where they are at a competitive disadvantage compared to the specialists. When disturbances occur, either natural or manmade, they quickly become dominant.

Electrophoretic studies of malate dehydrogenase loci of the most opportunistic species, Capitella capitata, a small polychaete, indicated selection for a single genotype in a large population following the spill. This illustrates the genetic flexibility of such species and suggests that dominance of such species in contaminated areas may be in part a matter of adaption and not simply higher tolerance of individuals. Numbers of Capitella capitata rose to about $200,000/m^2$ within three months after the spill at West Falmouth and then gradually declined over the subsequent year as other species began to colonize. Michael et al. (1975) correlated the decrease in the numbers of opportunistic species and individuals with the disappearance of oil from the subtidal sediments in the 3rd and 4th years after the spill. The subtidal stations where oil was detected (stations 10, 9, 31) had higher percentages of opportunistic species and individuals compared to the controls (20, 35). Both marshes, station IV and Sippewisset control station, had many more opportunists since these are more extreme habitats. But the percentages were much higher in the contaminated marsh (IV) in 1973 and dropped to levels closer to the control site within a year as hydrocarbon levels decreased. Oil had disappeared from the subtidal sediments at stations 9 and 10 by 1974 and there was a corresponding decrease in opportunists. Station 31 was heavily contaminated and showed no real change in either hydrocarbon content or opportunists. The presence of opportunists in areas known to be contaminated with a variety of compounds including petroleum hydrocarbons, e.g. New Haven harbor and New York Bight, suggests that these types of responses are typical for communities under a variety of stresses.

Gilfillan et al. (1976) developed energy budgets for populations of the softshell clam Mya arenaria in Casco Bay Maine. One of the populations was found in an area contaminated by No. 6 fuel oil which leaked from the "Tamano" in 1969. Animals were taken from this and a control site into the laboratory where respiration, assimilation and filtration were measured. The tests were repeated each month for a year. Monthly estimates for carbon flow were calculated. The same trends were observed in both populations; a large positive carbon flow in summer was followed by a small net loss of carbon in the winter and a large negative flow during the spring. The oiled population however, gained carbon at only 50% of the rate seen in the unoiled group.

This work provides a very important link between laboratory and field data. Although the animals are tested in the laboratory the effects are based on long term exposure in the field. This in effect correlates Gilfillan's earlier work (1975) in which water soluble extracts of crude oils produced lowered assimilation and filtering rates in laboratory tests on mussels.

Field data on mortalities of Mya arenaria were collected following a fuel oil spill at Searsport, Maine (Dow and Hurst, 1976). An original crop of 157 metric tons of clams existed in the Long Cove area. Mortality was observed immediately after the spill but the decline in numbers continued for three years. Twenty five percent died within four months; 55% were dead some 17 months later and 86% mortality was reported some 3½ years later. Only 22 metric tons survived as of August 1974. Since it takes 5 to 6 years to produce a marketable crop the long term loss as a fishery is significant.

Wormald (1976) sampled the meiofauna of a sandy beach in Hong Kong for 14 months after a spill of diesel oil. This is the first time the meiofauna has been studied quantitatively and over an extended period of time following a spill. The meiofauna was initially almost totally destroyed. Nematodes reappeared within one month but the harpacticoid copepods did not return for 8 months when the amount of oil had decreased by 50%. Normal populations had returned to the area within 10 to 15 months following the spill. The lower shore was recolonized earlier than the upper level probably because of more rapid dispersion of oil.

TABLE 1. Opportunistic Species as a Component of the Fauna after a Spill at West Falmouth, (from Michael et al. 1975)

Station	1973	1974	1973 and 1974 Combined
Opportunistic species expressed as % of total species.			
35*	4.3%	4.2%	4.3%
20*	3.8%	4.5%	4.1%
10	10.2%	4.6%	7.3%
9	9.5%	4.9%	7.4%
31	27.6%	28.1%	27.8%
IV	31.6%	25.5%	27.7%
Sipp.*	17.1%	20.2%	19.0%
Opportunistic individuals expressed as % of total individuals.			
35*	62.4%	20.7%	53.2%
20*	21.1%	8.3%	15.4%
10	74.3%	45.1%	65.6%
9	80.6%	39.9%	72.9%
31	54.8%	79.6%	70.1%
IV	40.4%	29.6%	32.2%
Sipp.*	9.2%	19.4%	15.3%

*Control stations, Sipp. - Sippewisset Control

Ecological Effects

The literature emphasizes the differing tolerances shown by various species. The factors controlling the survival of species in nature can be grouped into two major areas. One is the physical factors of the environment and the second includes biological factors such as competition and predation. In addition to direct toxicity resulting in death, sublethal effects of oil may lower the species tolerance to the physical extremes of the environment or reduce its competitive ability within the community.

From a community standpoint we can expect to see structural changes in a community as each species drops out or is replaced by another more opportunistic one. As contamination levels increase there will be a loss of further species with an eventual reduction in species diversity in the absolute sense (total number of species). Increasing contamination will produce a series of changes which is the reverse of the sequence of events documented at West Falmouth. Immediately after the spill there was a very high level of contamination and almost total mortality. At first there was an influx of a single opportunistic species. As the levels of oil decreased more species settled and became established and the proportion of opportunists decreased. In a variety of field studies in the New England area I have seen the same pattern indicated by various levels of contamination. In the extreme case such as the upper reaches of New Haven harbor or at station 31 at West Falmouth (Michael et al. 1975) there are virtually no resident populations. Benthic grab samples contain a few individuals of very few species. Most of the specimens are juveniles which do not survive. Samples from the same sites taken a month later do not often produce the same species. At the opposite end of the scale, with light levels of contamination, species numbers and the total number of individuals are high but the community is dominated by opportunists. From a biological point of view it is easy to distinguish these communities from the normal climax community found in an undisturbed area.

In some communities the biological dominance of one particular species may have an overwhelming effect on community structure, as for example, with macroalgae in the role of primary producers or starfish as predators. The sequence of events precipitated by the loss of one of these species may differ. I would emphasize the need for the community approach in future studies on the ecological effects of oil. It is the only way to learn something about community processes such as production and nutrient regeneration. We need to know that structural changes mean in terms of basic community processes. Is a community that is dominated by opportunists more productive, or less productive? What is the stability of communities showing structural changes due to stress? The answers to questions such as these will bring us closer to the point where we can make a realistic evaluation of the significance of a spill or chronic contamination.

Summary and Conclusions

I have summarily dismissed many aspects of the problems of oil contamination in the earlier comments. This is on the basis of a lack of evidence and certainly not through a lack of concern. The issue of ecological effects is fundamental to the whole aspect of environmental monitoring. If we do not collect the sort of information which will give definitive answers in areas of uncertainty we will continue to lack clear evidence on questions of community variability and causal relationships under spill conditions. I feel strongly about improving the state of the art and would urge that we make serious efforts to detect and identify early signs of environmental decay. This means working with long term, low level effects. If we fail to develop predictability at this level, our alternative is to document the decline in each case as an after-the-fact exercise. I would suggest priority be given to research in the following areas:

1. Long term studies on a few spills.
2. Studies on the dynamics of systems where chronic contamination exists.
3. Combined field and laboratory studies addressing the significance of sublethal effects.

From the basic science aspect I would hope for more effort to understand community processes with the intention of attacking the questions of significance. We should expand on the major findings of some recent papers. Grassle and Grassle (1975) have made a major contribution to our understanding of the biology of changes we see in benthic communities under stress. Gilfillan et al. (1975) produced evidence of a change in a basic community process (production) due to exposure to oil. The field data by Dow and Hurst (1976) indicate the potential for significant losses in a clam fishery although in that particular case the area was not an active fishery because of other contaminants. Wormald (1976) has reported the first long term quantitative data on the response of a meiofauna community to a spill.

Commenting on potential problems in the cold water environments Boesch et al. (1974) noted that (1) cold temperatures do not permit rapid evaporation of aromatics, thus allowing more of these toxic hydrocarbons to enter in solution in sea water even though the solubility of these compounds is lower at low temperatures; (2) the rate of bacterial degradation and other processes of weathering are comparatively slower at very cold temperatures; and (3) the marine biota of polar regions are generally long lived, have low reproductive potentials and do not have wide ranging dispersal stages (Dunbar, 1968). The persistence of oil and effects in three cold-temperate situations at West Falmouth, Chedabucto Bay and the Straits of Magellan (Metula spill; Straughan - personal communication) seem to support the contention that spills in cold waters will have more severe ecological effects. I would also note that weather conditions in these areas are less favorable and often unsuitable for the technology used in containment and clean up. I would hope that oil production and development in the northwestern United States proceeds with due caution.

References

Blumer, M., Souza, G., and Sass, J., Hydrocarbon Pollution of an Edible Shellfish by an Oil Spill. Marine Biology 5, 195-202 (1970).

Blumer, M., Sanders, H. L., Grassle, J. F., and Hampson, G. R., A Small Oil Spill. Environment 13(2), 1-12 (1971).

Boesch, D. F., Hershner, C. H., and J. H. Milgram, Oil spills and the marine environment. A report to the energy policy project of the Ford Foundation. 114 pp. Ballinger Publishing Co. (1974).

Chan, G. L., A Study of the Effects of the San Francisco Oil Spill on Marine Organisms. In: Proceedings of a Joint Conference on Prevention and Control of Oil Spills, Washington, D.C., March 13-15, 1973, pp. 739-781. Washington, D.C., American Petroleum Institute, 1973.

Diaz-Piferrer, M., The Effects of an Oil Spill on the Shore of Guanica, Puerto Rico. (Abstract). Association of Island Marine Laboratories, 4th Meeting, Curacao, 12-13, (1962).

Dow, R. L. and J. W. Hurst, Jr., The Ecological Chemical and Histopathologic Evaluation of an Oil Spill Site. Part I. Ecological Studies. Mar. Pollut. Bull. 6, 164-166 (1976).

Dunbar, M. J., Ecological Development in Polar Regions. A study in Evolution. 119 pp. Prentice-Hall, Englewood Cliffs, N.J. (1968).

Farrington, J. W., Some Problems Associated with the Collection of Marine Samples and Analysis of Hydrocarbons. Unpublished Manuscript, Woods Hole Oceanographic Institute, 74-24 (1974).

Foster, M., Neushul, M., and Zingmark, R., The Santa Barbara Oil Spill: Part 2--Initial Effects on Intertidal and Kelp Bed Organisms. Environmental Pollution 2, 115-134 (1971).

Gilfillan, E. S., D. Mayo, S. Hanson, D. Donovan and L. C. Jiang, Reduction in Carbon Flux in Mya arenaria caused by a spill of No. 6 Fuel Oil. Marine Biology 37, 115-123 (1976).

Goethe, F., The Effects of Oil Pollution on Populations of Marine and Coastal Birds. Helgolander wiss. Meeresunters 17, 370-74 (1968).

Gordon, D. C., and Prouse, N. J., The Effects of Three Oils on Marine Phytoplankton Photosynthesis. Marine Biology 22, 329-333 (1973).

Grassle, J. F., and J. P. Grassle, Opportunistic Life Histories and Genetic Systems in Marine Benthic Polychaetes. J. Mar. Res. 32, 253-84 (1974).

Jeffries, H. P. and W. C. Johnson II, Petroleum, Temperature and Toxicants: Examples of Suspected Responses by Plankton and Benthos on the Continental Shelf. In: B. Manowitz editor, Effects of Energy Related Activities on the Atlantic Continental Shelf. Brookhaven National Laboratory (1975).

Kauss, P., Hutchinson, T. C., Soto, C., Hellebus, J., and Griffiths, M., The Toxicity of Crude Oil and its Components to Freshwater Algae. pp. 703-714. In: Proceedings of a Joint Conference on Prevention and Control of Oil Spills, Washington, D.C., March 13-15, 1973. Washington, American Petroleum Institute (1973).

Kolpack, R. L., Mattson, J. S., Mark, H. G., Jr., and Yu, T., -C. Hydrocarbon Content of Santa Barbara Channel Sediments. pp. 276-295. In: R. L. Kolpack, editor, Biological and Oceanographical Survey of the Santa Barbara Channel Oil Spill 1969-1970. Vol. 2. Allan Hancock Foundation, University of Southern California (1971).

Mead, J. W. and Sorensen, P. E., The Economic Cost of the Santa Barbara Oil Spill. In: Holmes, R. W. and DeWitt, F. A. editors. Santa Barbara Oil Spill Symposium, Santa Barbara, December 16-18, 1970. University of California (1970).

Michael, A. D., C. Van Raalte and L. S. Brown, Long Term Effects of an Oil Spill at West Falmouth. In: Proc. API-EPA Conf. Prevention and Control of Oil Pollution, San Francisco, 1975, American Petroleum Institute, Washington D.C. (1975).

Mueller, J. A. and J. S. Jeris, Contaminant Inputs to the New York Bight. Report for the Marine Ecosystems Analysis Program. New York Bight Project, National Oceanographic and Atmospheric Administration (1975).

National Academy of Sciences, Petroleum in the Marine Environment; Workshop on Inputs, Fates and Effects of Petroleum in the Marine Environment, Airlie, Virginia, May 21-25, (1973), (1975).

Nelson-Smith, A., Oil Pollution and Marine Ecology. 260 pp. Elek Science, London, (1972).

Neushul, M., Jr., Effects of Pollution on Populations of Intertidal and Subtidal Organisms. Paper presented at Santa Barbara oil symposium, Offshore Petroleum Production--An Environmental Inquiry, Santa Barbara, California, December 16-18, 1970.

Nicholson, N. L., and Climberg, R. L., The Santa Barbara Oil Spills of 1969: A Post-Spill Survey of the Rocky Intertidal. pp. 325-399. In: D. Straughan, editor, Biological and Oceanographical Survey of the Santa Barbara Channel Oil Spill 1969-1970. Vol. 1. Allan Hancock Foundation, University of Southern California (1971).

North, W. J., Neushul, M., Jr., and Clendenning, K. A., Successive Biological Changes Observed in a Marine Cove Exposed to a Large Spillage of Oil. Symposium Commision International Exploration Scientifique Mer Mediterranee, Monaco, 1964, 335-354 (1965).

Nuzzi, R., Effects of Water Soluble Extracts of Oil on Phytoplankton. pp. 809-813. In: Proceedings of a Joint Conference on Pollution and Control of Oil Spills, Washington D.C., March 13-15, 1973, Washington, American Petroleum Institute (1973).

Rutzler, K., and Sterrer, W., Oil Pollution: Damage Observed in Tropical Communities along the Atlantic Seaboard of Panama. BioScience 20, 222-224 (1970).

Scarratt, D. J., and Zitko, V., Bunker C Oil in Sediments and, Benthic Animals from Shallow Depths in Chedabucto Bay, N. S. <u>Fisheries Res. B. Canada</u> <u>29</u>, 1347-1350 (1972).

Smith, J. E., editor. <u>"Torrey Canyon" Pollution and Marine Life</u>: A Report by the Plymouth Laboratory. xiv 196 p. Cambridge University Press (1968).

Straughan, D., Oil Pollution and Sea Birds, pp. 307-312. In: D. Straughan, editor, <u>Biological and Oceanographical Survey of the Santa Barbara Channel Oil Spill 1969-1970</u>. Vol. 1. Allan Hancock Foundation, University of Southern California (1971).

Vale, G. H., Sidhu, G. S., Montgomery, W. A., and Johnson, A. R., Studies of a Kerosene-like Taint in Mullet (<u>Mugil</u> <u>cephalus</u>). <u>Sci. Food and Agric.</u> <u>21</u>, 429-432 (1970).

Vandermeulen, J. H., and D. C. Gordon, Jr., Reentry of 5 year old stranded Bunker C Fuel Oil from a low-energy beach into the water, sediments and biota of Chedabucto Bay, Nova Scotia. <u>J. Fish. Res. Bd. Can.</u> <u>33</u>(9), 2002-2010 (1976).

Westfall, A., Jackass Penguins. <u>Mar. Pollut. Bull.</u> <u>14</u>, 2-7 (1969).

Wormald, A. P., Effects of a Spill of Marine Diesel Oil on the Meiofauna of a Sandy Beach at Picnic Bay, <u>Hong Kong. Environ. Pollut.</u> <u>11</u>, 117-130 (1976).

CHAPTER 12

EFFECTS OF CERTAIN PETROLEUM PRODUCTS ON REPRODUCTION AND
GROWTH OF ZYGOTES AND JUVENILE STAGES OF THE
ALGA FUCUS EDENTATUS DE LA PYL
(PHAEOPHYCEAE: FUCALES)

Richard Lewis Steele, Ph.D.

Environmental Research Laboratory
Environmental Protection Agency
South Ferry Road
Narragansett, Rhode Island 02882

Abstract

A method has been devised to utilize easily obtainable eggs of Fucus as a bioassay. The effects of various petroleum products on growth and early development of the zygote has been studied. A crude oil had much less effect on Fucus growth than did No. 2 fuel oil or two jet fuels. When exposure occurred immediately prior to and during release of gametes, no germination or growth occurred with any of the oil types, even at the lowest concentration used.

Key words: Algae, Crude oil, Fucus, No. 2 fuel oil, Reproduction

Introduction

An organism to be used as a bioassay tool becomes more valuable if it is the type of organism that can either be maintained in the laboratory or can be procured from nature at any time. When maintained in the laboratory it should be hardy enough to grow in culture, but sensitive enough to react to levels of stress likely to be encountered in nature. Preferably, the organism should also be of ecological or economic importance.

The biology of Fucus is well known in the literature (Fritch, 1945), and more recently studies have been published dealing with egg polarity (Moss, 1965), wound healing and regeneration (Moss, 1968), subcellular cytology (McCully, 1868), and hormonal effects (Russel, 1973). However, studies dealing with the toxicology of petroleum products to Fucus and other marine algae have been more or less limited to the antecedent effects of oil spills (Straughan, 1971; Cowell and Baker, 1969).

Fucus has many of the qualities already mentioned that would make it valuable as a bioassay organism. The purpose of this paper is to determine the sensitivity of Fucus to various petroleum products and to develop a bioassay using this species as a bioassay tool. In the event of shoreline contamination by petroleum products, Fucus, being a high intertidal organism would be subject to repeated immersion in the water mass containing the contaminant. Fucus is generally the dominant organism in the high intertidal zone and the effect of the possible obliteration of this zone on the lower zones is not known.

Materials and Methods

Fucus plants were collected from various locations in and around Narragansett Bay (i.e., Camp Varnum (a National Guard installation), the dock of the Environmental Research Laboratory, Narragansett, R.I., and Monahans Cove, Narragansett, R.I.). Several species were used, including Fucus vesiculosus, Fucus edentatus, and Fucus distichus. These species represent both monoecious and dioecious types. In deciding which species to use, little difference was noted in response among the various species, so Fucus edentatus was used and was indicative of the response of the other species.

Methods of procurement of eggs and sperm were evaluated (Pollock, 1970; McLachlan et al. (1971) and a technique was devised that is applicable to all species tested. The method consists of the following: Fucus plants were collected from the environment and receptacles (fertile plant tips) that appeared most erumpent and mature, even to the point of being partially eroded, produced the highest numbers of viable eggs and sperm. Receptacles were rinsed in sterile, charcoal filtered* seawater at 30 ppt salinity, and were placed in a moist chamber over night. The moist chamber consisted of large, 6 inch (15.24 cm) diameter, plastic petri dishes (Falcon Plastic)** containing filter paper of the same diameter moistened with sterile seawater. These plates were placed in a 12°C culture chamber over night. Receptacles were then placed in sterile charcoal filtered seawater the next morning. Eggs and sperm were immediately released, and fertilization occurred within 15-20 minutes. Zygotes were immediately pipetted into culture dishes (60 x 15 mm) plastic petri dishes with 2 mm square grids-Falcon Plastics) while keeping the eggs suspended in seawater by stirring. Densities of zygotes in each dish were adjusted so that by counting the zygotes in two rows of grids, ca. 30 were measured. Each dish represents ca. 30 counts and each datum given in the tables represents the average of 3-4 dishes. The toxicant was introduced by allowing the zygotes to settle on the bottom of the dish, the seawater removed, and replaced with seawater containing the toxicant at the proper levels 24 hours after dispensing into culture dishes. In a few cases, the tips were pretreated with the toxicant and in these cases the above sequence was modified to suit the situation.

Observation and counting of the zygotes was done with a Unitron inverted microscope (Model BMIC). The parameter measured for these experiments was increased in length of the juvenile plants after 12 days growth.

The growth medium for the assays was, in all cases, sterile charcoal filtered seawater. Light intensity for all assays was held at 400 ft-c. with continuous light in all tests. Except for the first series of experiments to determine the best temperature and salinity for the rest of the assays, the temperature and salinity adhered to were 18°C and 30 o/oo.

For the various tests, the petroleum product at concentrations ranging from 0-2000 ppm was equilibrated with seawater, proper dilutions made, and added to the cultures. The oil-seawater mixture was analyzed by infrared spectrophotometry (Perkin Elmer Model 621 Infrared Spectrophotometer) to determine the amount of dissolved product causing toxicity to Fucus.

Results

The first experiments were designed to test the simultaneously effects of temperature and salinity on growth. To do this, various temperature regimes and salinity gradients were arranged so that a matrix was developed that would show the effects of temperature and salinity.

These data are presented to indicate growth obtained at various combinations of salinities and temperatures. Experiments to determine toxicity of oil and oil fractions were run at 18°C and 30 o/oo salinity, since these were near optimum conditions for growth as indicated in Table 1, but since most of the seawater collected in Narragansett Bay is ca. 30 o/oo no advantage could be seen in going to a higher salinity.

Data presented in Table 2 show the effects of No. 2 fuel oil; jet fuels JP-4 and JP-5; and Willamar crude (a domestic crude) on growth of Fucus edentatus juvenile plants from the zygote stage. Small amounts of oil(s) (200 ppb or less) caused an apparent stimulation of growth of the juvenile plants; however, above this level there was an increasingly deleterious effect on growth. Zygotes germinated at 20 ppm of No. 2 fuel oil, JP-4 and JP-5, but did not grow. Concentrations between 2 and 20 ppm were not tried so the actual amount allowing germination but not growth is not known. Growth at 2.0 ppm was only slightly reduced from the control in all cases and it is likely that a curve of response can be developed between 2.0 and 20 ppm. As indicated in Table 2, the crude oil tested is not as toxic as either the JP-4, JP-5 or No. 2 fuel oil. At 20 ppm the growth was somewhat reduced from that of the control and at 200 ppm the growth was only about half that of the control.

All of the above experiments were done following methods as described. In a variation of this procedure the above sequence was followed until time to place the receptacles in seawater for egg and sperm release. At this point, the receptacles were placed directly into the various levels of oil and eggs and sperm were allowed to release. Results of this experiment were dramatic in the fact that no germination or growth occurred, even at the lowest levels of oil (Table 3).

---

*Cartridge filtration through Commercial Filter Corporation honeycomb would filters, 15 μ porosity, and .22 μ porosity pleated Gelman filters. All apparatus is plastic.

**Mention of specific products herein does not constitute endorsement by the U.S. Environmental Protection Agency.

TABLE 1. Effects of various temperature and salinity combinations on growth of juvenile plants of Fucus edentatus. Numbers shown are μ in length of juvenile plants after 12 days.

Salinity o/oo	Temperature °C				
	6	12	18	24	30
6	102	190	0	0	0
12	153	273	341	140	0
18	165	419	586	331	0
24	151	383	645	400	0
30	153	380	704	406	0
36	147	395	701	400	0
42	122	285	712	365	0
48	102	260	601	342	0
54	0	254	458	295	0
60	0	0	0	0	0

TABLE 2. Effects of No. 2 fuel oil, JP-4, JP-5 and Willamar crude oil on growth of Fucus edentatus zygotes. Numbers shown are μ in length of juvenile plants after 12 days.

added	(analyzed)	No. 2 Oil	JP-4	JP-5	Willamar Crude
control 0.0	(no data)	694	704	700	776
2ppb	(no data)	702	712	701	764
20ppb	(no data)	748	732	726	748
200ppb	(no data)	696	687	693	750
2ppm	(0-30 ppb)	671	635	655	756
20ppm	(1-3 ppm)	0	0	0	734
200ppm	(18-28ppm)	0	0	0	399
2ppt	(45-50ppm)	0	0	0	0

TABLE 3. Results of treatment of receptacles in oil solutions during gamete release. Numbers shown are μ in length of juvenile plants after 12 days.

added	(analyzed)	No. 2 Oil	JP-4	JP-5	Willamar Crude
control 0.0	(no data)	699	701	704	710
.2ppb	(no data)	0	0	0	0
2.0ppb	(no data)	0	0	0	0
20.0ppb	(no data)	0	0	0	0
200.0ppb	(no data)	0	0	0	0
2.0ppm	(0-30 ppb)	0	0	0	0
20.0ppm	(1-3 ppm)	0	0	0	0

Because no fertilization occurred, the eggs lost viability and completely disintegrated by the sixth day of the experiment. In another series of experiments, Fucus receptacles were allowed to stand in various concentrations of oil 5 hours before being placed in the moist chamber over night. Completing the experiment in the normal manner by replacing the water the next day with sterile seawater, the deleterious effects on the sperm were not found. However, growth of the juvenile plants was affected. Subsequent observations showed that g growth of the zygotes approximated that shown in Table 2, even though the plants were not in the oil solutions during or after fertilization.

Discussion

The purpose of this study was to determine the sensitivity of Fucus to oil products and to develop a bioassay using Fucus. It appears that zygotes of Fucus, at least in the methodology developed, were not as sensitive to oil as some microalgae that have been studied (Pulich, et al, 1974) but have about the same or slightly more sensitivity than some animals (Eisler, 1975). Although the toxicity values obtained in these previous studies were not directly applicable to Fucus (doubling times in the former and LC_{50} values in the latter), similar values can be derived. However, when Fucus receptacles were placed in the oil solutions during gamete release, all germination ceased even at very minute quantities of oil. This was apparently a reaction of the sperm, which are apparently much more sensitive than eggs to oil. The mechanism of this toxicity, whether surfactant or biochemical, is not understood. Without viable sperm, fertilization did not occur, thereby eliminating any subsequent events in the life cycle.

The deleterious effect of low level oil pollution on the reproductive cycle of Fucus can easily be visualized, especially chronic repeated pollution such as that occurring around marinas and similar installations.

Fucus occurring in the upper intertidal zone, has evolved as a plant able to withstand extremes of temperature, salinity, dessication, etc. Possibly this toughness of the zygote is an indication of the niche that Fucus occupies. It is also highly likely that the egg, being 50 μ and larger, contains enough stored nutrients that it is not initially dependent on the exterior environment. The sperm, being much smaller, 4-6 μ, does not have these stored nutrients. In the first two weeks of growth of the zygote, no additional nutrients are required in the seawater for growth. Immersion of the receptacles for 5 hour previous to being put in a moist atmosphere did not interfere with maturation of the sperm that were released the next morning, but did have a toxic effect on subsequent development of the zygotes. This is paradoxical when considering that immersion during gamete release completely destroys sperm viability.

Although they have not been tested yet, juvenile plants may be more sensitive after 2-3 weeks growth than the initial zygote soon after fertilization. Presumably, at this later stage, the juvenile plant is less dependent on nutrients stored in the egg.

There appears to be little difference in toxicity between No. 2 fuel oil, and JP-4 and JP-5 to Fucus zygotes. A concentration of 1-2 ppm results in some toxicity with a complete kill at 20 ppm. However, the crude oil (Willamar Crude) was not nearly as toxic. Little toxicity was detected below 10 ppm. Results such as this are not surprising, and are to be expected, as toxicity of different crude oils can vary as much as 2 orders of magnitude (Eisler and Kissel, 1975).

References

Cowell, E.B. and J.M. Baker, Recovery of a salt marsh in Pembrokeshire, southwest Wales, from pollution by crude oil, Biological Conservation 1, 291-96 (1969).

Eisler, R., Acute toxicities of crude oils and oil-dispersant mixtures to Red Sea fishes and invertebrates, Israel Journ. of Zool. 24, 16-27 (1975).

Eisler, R. and G.W. Kissel, Toxicities of crude oils and oil-dispersant mixtures to juvenile rabbit fish, Siganas rivulatus, Trans. Am. fish. Soc. 104, 571-78 (1975).

Fritch, F.E., The Structure and Reproduction of the Algae, Vol. 1, Macmillan Co., New York, 1945.

McCully, M.E., Histological studies on the genus Fucus II. Histology of the reproductive tissue, Protoplasma 66, 205-30 (1968).

McLachlan, J., Effects of temperature and light on growth and development of embryos of Fucus edentatus and F. distichus spp. distichus, Can. Journ. Bot. 52, 943-51 (1974).

Moss, B.L., Apical dominance in Fucus vesiculosis, New Phytol. 64, 387-92 (1965).

Moss, B.L., The transition from vegetative to fertile tissue in Fucus vesiculosis, Br. Phycol. Bull. 3(3), 567-73 (1968).

Pollock, E.G., Fertilization in Fucus, Planta 92, 85-90 (1970).

Pulich, W.M., K. Winters and C. Van Baalen, The effects of a No. 2 fuel oil and two crude oils on growth and photosynthesis of microalgae, Marine Biology 28, 87-94 (1974).

Russel, G., the Phaeophyta: A synopsis of some recent developments, Oceanogr. Mar. Biol. Ann. Rev. 11, 45-88 (1973).

Straughan, D., What has been the effect of the spill on the ecology in the Santa Barbara Channel, Biol. and Oceanogr. Survey of Santa Barbara Channel Oil Spill 1, 401-26 (1970).

CHAPTER 13

EFFECT OF CRUDE OIL ON TROUT REPRODUCTION

H. O. Hodgins, W. D. Gronlund, J. L. Mighell, J. W. Hawkes, and P. A. Robisch

Northwest and Alaska Fisheries Center
National Marine Fisheries Service
National Oceanic and Atmospheric Administration
2725 Montlake Boulevard East
Seattle, Washington 98112   U.S.A.

Abstract

Prudhoe Bay crude oil was incorporated into the diet (1 g oil/kg food) of adult rainbow trout during sexual maturation to assess the effects of long-term petroleum exposure on salmonid fish reproductive success. Parallel control fish received identical rations, except without added petroleum. When the fish reached maturity, six to seven months after initiation of petroleum exposure, a total of 31 test and 10 control crosses were made. Mean survival through hatching was 86% and 90% for test and control eggs, respectively; the difference was non-significant ($P = 0.10$). Test and control males were virtually identical in fertility. Mean survival from hatching to swim-up fry stage of development was 76% for test and 91% for control fish; again the difference was not significant ($P = 0.10$). In addition, no gross morphological or histological abnormalities were observed in offspring of petroleum-fed fish. The results of these studies were, therefore, that there was no significant impairment of reproductive success detected from this type of dietary exposure to petroleum.

Key words: Fertility, hatching success, histology, Prudhoe Bay Crude, rainbow trout

Introduction

Although in several studies there have been reports of the effects of crude oil and refined petroleum products on viability and of structural and functional abnormalities of eggs, sperm, and juveniles of aquatic species (Mironov 1969a, 1969b, Morrow 1974, Renzoni 1975, Rice et al 1975, Struhsaker 1974), we are aware of no information on the consequences of petroleum exposure on sexual maturation of fish. Interference by petroleum with sexual maturation processes could result in infertile gametes and teratogenic effects on progeny. These kinds of effects were demonstrated for trout exposed to DDT. In lake trout (*Salvelinus namaycush*), adults were exposed to DDT as a result of its use in insect control and, although they remained apparently unaffected, toxic quantities of DDT or its metabolites were deposited in eggs, resulting in high mortality of offspring (Burdick, *et al* 1964). In studies with brook trout (*Salvelinus fontinalis*), DDT fed to maturing adults also resulted in lower survival of offspring (Macek 1968).

Previous studies in our laboratory have elucidated endocrinological and biochemical mechanisms of maturation of salmonid fishes (Gronlund 1969, Gronlund and Hodgins 1970, Gronlund et al 1973). We have shown that initiation of sexual maturation takes place well in advance of outward signs of maturity and that the synthesis and storage of gonadal material is occurring at least six months prior to spawning. During this period of maturation, anadromous fishes are actively feeding at sea; as they approach estuarine waters and natal streams, the risk of exposure to pollutants increases.

The present study was initiated to test the effects of chronic dietary exposure to crude oil on salmonid fish reproduction. Rainbow trout (*Salmo gairdneri*) were used that were maturing for the first time and which were representative of steelhead trout, the anadromous form of the species. Trout were exposed to large quantities of crude petroleum incorporated into the diets, beginning in early summer of 1975 and continuing past spawning in the winter of 1976. These animals were compared with control animals maintained on a diet without added petroleum. The primary objective of the study was to evaluate effects of long-term dietary petroleum exposure on reproductive success, represented by successful hatching and by survival of normal-appearing alevins. If reproduction was impaired by the large amounts of dietary petroleum in these exploratory studies, a more in-depth examination of threshold doses and mechanisms involved was to be undertaken.

Materials and Methods

Experimental Animals

Fish used were 3-year-old rainbow trout of Cape Cod strain, obtained in June of 1975 from the Washington State Department of Game Hatchery in Spokane, Washington. At the beginning of the study the fish measured 41 to 53 cm in fork length and weighed 1.0 to 1.8 kg.

Upon arrival at the Northwest and Alaska Fisheries Center of the National Marine Fisheries Service (NMFS), Seattle, Washington, the fish were randomly placed in approximately equal numbers in one or the other of two adjacent circular fiberglass tanks (1.8 m diameter) continuously supplied with dechlorinated city water at 30 l/min and maintained at a depth of 0.8 m. Water temperature was controlled at $11 \pm 1^\circ C$ and artificial light was maintained to correspond to a natural light cycle.

Petroleum Exposure

Petroleum-containing food was routinely prepared in the following manner: Two kg of 1/4 inch diameter Oregon moist pellets were placed in a 4 l. glass beaker. Two g (2.6 ml) of Prudhoe Bay crude oil were mixed with 148 ml of FREON$^{(R)}$ TF[1] solvent (trichlorotrifluoroethane) and poured over the food. The food and oil were thoroughly mixed and the food was spread over porcelain-covered metal trays for 90 min of air drying in a fume hood. The food was then weighed into daily aliquots, sealed in plastic bags, and frozen until used. Food for control experiments was prepared identically except the crude oil was omitted.

Fish were fed the above diets at a rate of 1.5 to 2.5% of body weight (wet weight of food) each week day starting in July 1975 and continuing through August 1976.

Maturity Assessment and Spawning Procedures

In late November 1975, all fish were examined for sex and degree of maturity. Subsequent biweekly and then weekly examinations were made. As females ripened they were spawned within one day of examination and eggs were fertilized using standard trout-culture methods (Leitritz 1959). Ripe males were consistently available for the duration of the spawning period from January through February, 1976. All eggs from control fish were divided into equal aliquots and one aliquot was fertilized with sperm from one test male and the other with sperm from one control male. Ten of the test females were similarly treated; eggs from the remaining test females were fertilized with test sperm only. A total of 31 test and 10 control crosses were made, that is, crosses utilizing different fish. Eggs were incubated in Heath trays (Heath Tecna Plastics, Inc.) at 7 to $9^\circ C$. Mortality data were collected through the yolk-sac resorption developmental stage and statistically analyzed using the Mann-Whitney modification of Wilcoxon's sum of ranks test (Langley 1971).

Chemical Analyses

Samples of adult tissue, eggs, and alevins were collected and frozen for later analysis for petroleum. Some samples were collected at the time of spawning, others were obtained four to five months after spawning. The analytical procedures followed methods of Warner (1976) utilizing alkaline digestion, solvent extraction, and silica gel column chromatography. To reduce losses of volatile compounds of Prudhoe Bay crude oil, the alkaline digestion procedure was modified as follows: 6 ml of 4N NaOH were added to 10 g of sample, which was digested at $30^\circ C$ for a minimum of 16 hours. Column chromatographic fractions were analyzed by spectrofluorometry.

Fraction III from the modified Warner method, containing the fluorescent aromatic compounds, was concentrated to 2.0 ml, which were analyzed using an Aminco-Bowman spectrofluorometer, with a Model 4-8912 radio mode accessory (American Instrument Company, Silver Springs, Maryland).

Dilute solutions of Prudhoe Bay crude oil (0.1 µg/ml to 10.0 µg/ml) in methylene chloride: petroleum ether (20:80 v/v) were used as standards for the spectrofluorometric quantitation of the samples. The maximum excitation wavelength and maximum emission wavelength for Prudhoe Bay crude oil were found to be 262 nm and 364 nm, respectively.

All solvents used were either Burdick and Jackson "Distilled-in-glass" grade or Mallinckrodt "Nanograde." All glassware used in the preparation of samples for spectrofluorometric analyses was given special cleaning by immersion in boiling concentrated HNO_3 overnight, then rinsing in distilled water and drying at $120^\circ C$.

---

[1]/Trade names referred to in this publication do not imply endorsement of commercial products by the National Marine Fisheries Service.

Histology
Samples were collected for histology and prepared in the following manner: Fry were fixed in 0.75% glutaraldehyde, 3% formaldehyde, 1% acrolein in 0.1 M sodium cacodylate buffer with 0.02% $CaCl_2 \cdot H_2O$, 0.02 M S-collindine and 5.5% sucrose (Hawkes 1974). After a buffer wash the samples were post-fixed in 2% osmium tetroxide dehydrated in an ethanol series and embedded in Spurr plastic (Spurr 1969).

Sections were cut with glass knives at 0.75 to 1.0 µ thickness and stained with Richardson's stain.

Results

Mortality
Totals of 12 control (not fed petroleum) and 48 test (fed petroleum) fish were available for reproduction studies. Due to holding facility failure an additional 33 control fish died two months prior to spawning which resulted in fewer control crosses than anticipated. There was a substantial post-spawning mortality in the petroleum-fed group in which 15 fish died one to three months after spawning; all of these animals were heavily infected with fungus. None of the controls were similarly affected.

Maturation
The first males were in spawning condition by mid-December 1975, and the first females were ripe two to three weeks later (Fig. 1   ). Although the first ripe fish were from the test group, there appeared to be no pronounced acceleration or retardation of maturity related to petroleum exposure. Eggs were collected from ripe females starting on January 6, 1976, and collections continued weekly through February 17, 1976.

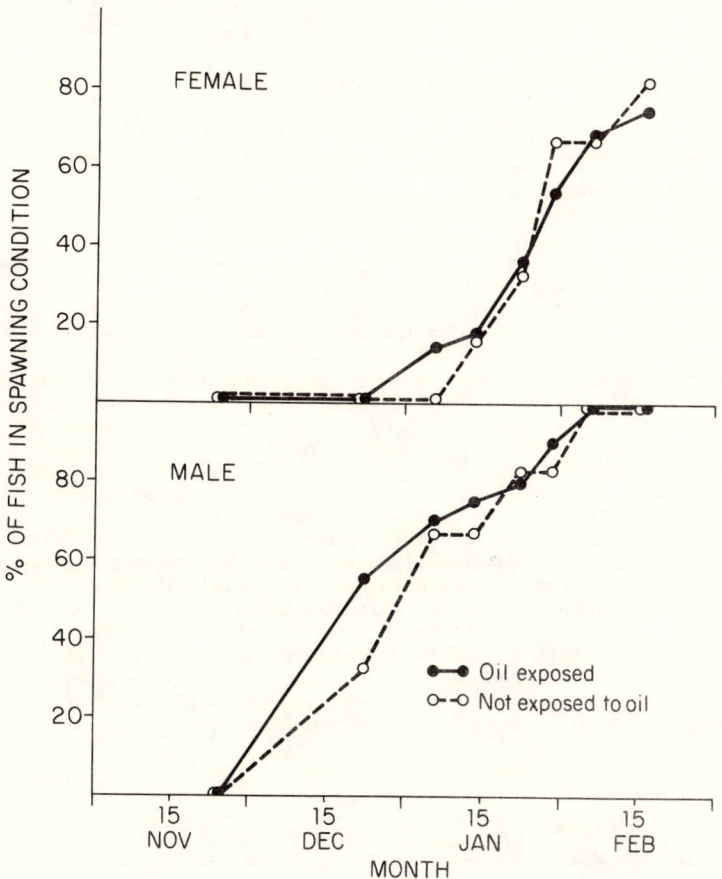

Fig. 13.1 Timing of maturation for petroleum-exposed and non-petroleum exposed rainbow trout.

Reproductive Success

No significant difference (n_A=14, R=197, P=0.10) was observed in hatching success (percent survival) among crosses in which sperm was used from petroleum-fed and non-petroleum-fed males (Table 1). One particular cross did, however, result in very low survival (5.1%) and slightly lowered the average percent hatching success of eggs fertilized with sperm from a non-petroleum-fed male.

TABLE 1. Survival of Eggs Fertilized with Sperm from Petroleum-Exposed and Non-Petroleum-Exposed Male Rainbow Trout through Hatching

Female	% Survival through hatching	
	Crossed with petroleum-exposed male	Crossed with non-petroleum-exposed male
Non-petroleum exposed	96.8	96.7
	81.3	77.3
	95.7	96.3
	89.9	87.7
	88.5	94.4
Petroleum exposed	99.4	99.6
	98.2	98.1
	98.6	95.2
	95.9	95.2
	36.9	37.3
	58.9	5.1
	95.8	98.3
	94.7	94.5
	78.5[a]	72.5[b]
	$\bar{x}$ = 86.4	$\bar{x}$ = 82.0
	$s_{\bar{x}}$ = 18.0	$s_{\bar{x}}$ = 27.7

[a] Pool of eggs from two females.
[b] Pool of eggs from three females.

Hatching success ranged from 32.4% to 99.5% for eggs from petroleum-exposed females and from 79.2% to 96.8% for non-petroleum-exposed eggs (Table 2), but the respective means of 86.4% and 90.3% were not significantly different (n_A=5, n_B=15, R=41.5, P=0.10). Eggs from two test females with 32% and 37% survival lowered the average survival of the test group.

Average survival of alevins was higher, although not significantly (n_A=4, n_B=10, R=19, P=0.10), for control than for test fish (Table 3). Again, low survival occurred in one petroleum-exposed group.

Chemical Analyses

Interference from fluorescing compounds prevented precise quantitation of Prudhoe Bay crude oil from adult trout muscle and eggs; only qualitative and semiquantitative results were possible. An emission maximum (364 nm) superimposed on the background of fluorescing compounds was observed for all samples from fish fed Prudhoe Bay crude oil; this maximum was not observed for any of the samples from fish fed the control diet (Fig. 2). The ratios of the average relative intensities at an excitation wavelength of 262 nm and an emission wavelength of 364 nm of petroleum-fed fish to control fish for muscle tissue and eggs were 2.8:1 and 3.8:1, respectively. A total of 14 analyses of muscle and 7 analyses of eggs from petroleum-fed fish and four analyses of muscle and two of eggs from control fish were performed. The background of fluorescing compounds was sufficiently high for the control and petroleum-impregnated food so that no definitive results could be obtained via spectrofluorometry.

Histology

No alevins with gross deformities were observed as a result of oil exposure. There was also no indication of abnormality in low magnification (X70) cross sections of anterior and posterior body areas in three fry from a petroleum-exposed female/petroleum-exposed male cross. At higher magnifications (X690), blood cells, muscle, mucous glands, and kidney tubules appeared normal. There were, however, instances of potentially deleterious histological abnormalities of eye lenses and livers of adult trout, exposed to petroleum in these studies. These anomalies will be discussed in another paper (Hawkes 1976).

TABLE 2. Survival of Eggs from Petroleum-Exposed and Non-Petroleum-Exposed Female Rainbow Trout through Hatching

% Survival for eggs from non-petroleum-exposed trout	% Survival of eggs from petroleum-exposed trout
96.8	96.8
79.2	98.4
96.0	98.9
88.4	98.5
91.3	99.5
$\bar{x}$ = 90.3	98.1
	97.0
s_x = 7.1	95.6
	37.2
	32.4
	97.0
	94.6
	75.6[a]
	89.8[b]
	86.5[c]
	$\bar{x}$ = 86.4
	s_x = 21.9

[a] Pool of eggs from 2 females.
[b] Pool of eggs from 3 females.
[c] Pool of eggs from 4 females.

TABLE 3. Survival of Alevins from Petroleum-Exposed and Non-Petroleum-Exposed Female Rainbow Trout

% of offspring surviving from hatching to swim-up	
Non-petroleum exposed	Petroleum exposed
99.0	97.4[a]
99.4[a]	96.2[a]
89.1	90.3[a]
76.3	87.3[c]
$\bar{x}$ = 91.0	81.1[a]
	79.8[b]
s_x = 10.9	74.1
	68.0
	61.7[a]
	25.0[a]
	$\bar{x}$ = 76.1
	s_x = 21.4

[a] Pool of offspring from 2 females.
[b] Pool of offspring from 3 females.
[c] Pool of offspring from 4 females.

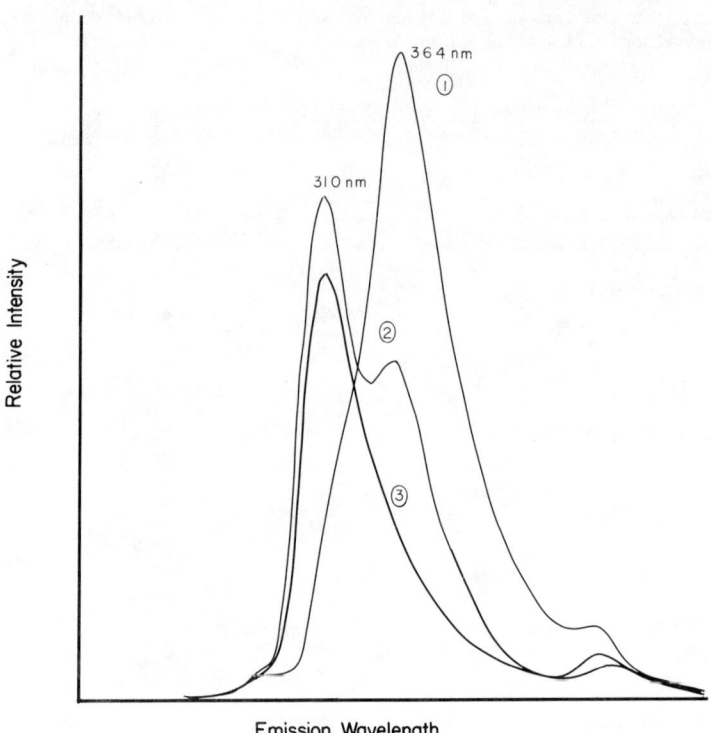

Fig. 13.2 Typical spectrophotofluorometric curves from the analyses of standard Prudhoe Bay crude oil and extracts of fish eggs.
Excitation at 262 nm
Solvent: methylene chloride in petroleum ether, 20:80 (v/v)
Curve No. 1 - Prudhoe Bay crude oil (10 μg/ml) standard
Curve No. 2 - Oil-fed fish egg extract
Curve No. 3 - Control fish egg extract

Discussion

The quantities of petroleum components consumed by test fish almost surely exceeded that which would be encountered in natural food supplies; however, it was our intention to examine an extreme case of exposure. The fish readily consumed the petroleum-impregnated food and continued to grow and develop. Although there were no mortalities of petroleum-fed fish prior to spawning, the post-spawning mortality of petroleum-exposed trout with fungus infections suggests some possible interaction between petroleum exposure and recovery from spawning. It is also possible, however, that the differential mortality between the test and control groups may have been related to a greater density of fish in the test tank compared to that in the control tank.

There was no significant impairment of hatching success related to the petroleum exposure. Survival percentages of 86 to 90% compare well with survivals of 90 to 95% for the hatchery program from which the fish were obtained (M. Albert, personal communication, 1976), as well as with published values for other studies using rainbow trout (Anon. 1973). However, eggs from two of the test females had low survivals, and it may be that certain individual fish were adversely affected by the petroleum exposure.

There is no indication that the dietary petroleum exposure had any effect on male fertility. In only one case, the hatching survival of eggs was greatly different for eggs fertilized with sperm from both best and control fish; in fact, the lowest survival was associated with a control male (Table 1). Of course, many other behavioral and physiological aspects of natural reproduction were not examined in these studies. Clearly, activities such as homing, mate selection, redd-building behavior, and territoriality could be disrupted by petroleum consumption and contribute to poor reproductive success in the natural environment.

The fluorescence spectra associated with the trout muscle indicated that certain fluorescing compounds were mobilized from the food through the circulatory system in the fish, and localized in the tissues. Similarly, as shown in Fig. 2, evidence suggests that trout are capable of transporting certain fluorescent hydrocarbons into eggs when the fish are exposed to petroleum in food.

There is no evidence from these studies to suggest that a chronic dietary exposure to concentrations of the less volatile components of Prudhoe Bay crude oil that are likely to occur in the environment would result in reproductive failure of rainbow or steelhead trout. The histological abnormalities of eye lenses and livers observed in adult fish exposed to petroleum are potentially deleterious, however. New studies are in progress to determine if the eye and liver changes develop in young fish of the same stock fed either the same large quantity of Prudhoe Bay crude oil or 1% of that amount.

Acknowledgement

We are indebted to Linda Rhodes and Teresa Scherman for expert technical assistance.

References

Anon., Effects of oil and chemically dispersed oil on selected marine biota -- A laboratory study, Battelle, Pacific Northwest Laboratories Report for the American Petroleum Institute on Environmental Affairs, Am. Petrol. Inst. Publ. No. 4191, p 2.4-2.5 (1973).

Burdick, G. E., E. J. Harris, H. J. Dean, T. M. Walker, J. Skea, and D. Colby, The accumulation of DDT in lake trout and the effect on reproduction, Trans. Am. Fish. Soc. 93(2) 127-136 (1964).

Gronlund, W. D., Biological assay and partial characterization of the gonadotropic factors of the pituitary gland of Pacific salmon (Oncorhynchus), M.S. thesis, University of Washington, 57 p (1969).

Gronlund, W. D. and H. O. Hodgins, Use of pituitary glands to determine maturity in salmon, Int. North Pac. Fish. Comm. Annu. Rep., 1968, 101-103 (1970).

Gronlund, W. D., H. O. Hodgins, and E. A. Blood, Pituitary gonadotropic activity and ovarian antigens for predicting age at maturity of high seas sockeye salmon, Int. North Pac. Fish Comm. Annu. Rep., 1971, 101-107 (1973).

Hawkes, J. W., Morphological effects of hydrocarbon exposure. In: Proceedings of the Symposium on Fate and Effects of Petroleum Hydrocarbons in Marine Ecosystems and Organisms, November 10-12, 1976, Seattle, Washington. Pergamon Press, New York, In Press.

Hawkes, J. W., The structure of fish skin. I. General Organization. Cell Tissue Res. 149, 147-158, (1974).

Langley, R., Practical Statistics, p 166-178, Dover Publications, New York, 1971.

Leitritz, E., Trout and salmon culture (hatchery methods), State of California Department of Fish and Game, Fish Bull. 107, 169 p (1959).

Macek, K. J., Reproduction in brook trout (Salvelinus fontinalis) fed sublethal concentrations of DDT, J. Fish. Res. Board Can. 25(9), 1787-1796 (1968).

Mironov, O. G., The development of some Black Sea fishes in sea water polluted by petroleum products, Vopr. Ikhtiol. 9(6), 1136-1139 (1969a).

Mironov, O. G., Viability of larvae of some crustacea in sea water polluted with oil products, Zool Zh 48(11), 1734-1737 (1969b).

Morrow, J. E., Effects of crude oil and some of its components on young coho and sockeye salmon. U.S. Environmental Protection Agency Report No. EPA-660/3-73-018, 37 p (1974).

Renzoni, A., Toxicity of three oils to bivalve gametes and larvae, Mar. Pollut. Bull. 6(8), 125-128 (1975).

Rice, S. D., D. A. Moles, and J. W. Short, The effect of Prudhoe Bay crude oil on survival and growth of eggs, alevins, and fry of pink salmont, *Oncorhynchus gorbuscha*. pp. 503-507. In: Proceedings of 1975 Conference on Prevention and Control of Oil Pollution, American Petroleum Institute, Washington, D.C.

Spurr, A. R., A low viscosity epoxy resin embedding medium for electron microscopy. J. Ultrastruct. Res. 26, 31-43, (1969).

Struhsaker, J. W., M. B. Eldridge, and T. Echeverria, Effects of benzene (a water-soluble component of crude oil) on eggs and larvae of Pacific herring and northern anchovy, p. 253-284. In: F. J. Vernberg and W. B. Vernberg (eds.) Pollution and Physiology of Marine Organisms, Academic Press, New York, 1974.

Warner, J. S., Determination of aliphatic and aromatic hydrocarbons in marine organisms, Anal. Chem. 48(3), 578-583 (1976).

CHAPTER 14

THERMAL CONDUCTANCE OF IMMERSED PRINNIPED AND SEA OTTER PELTS
BEFORE AND AFTER OILING WITH PRUDHOE BAY CRUDE

G. L. Kooyman, R. W. Davis and M. A. Castellini

Physiological Research Laboratory
Scripps Institution of Oceanography
University of California, San Diego
La Jolla, California 92093

Abstract

Thermal conductance (C) of the sea otter and several species of pinniped pelts was determined during immersion, after oiling, and after cleaning. A (C) of 7 Watts $\cdot$ Meter^{-2} $\cdot$ $°C^{-1}$ for the sea otter pup was the lowest measured in all controls. The highest was 58 W $\cdot$ M^{-2} $\cdot$ $°C^{-1}$ for the California sea lion. Most affected by oiling was the sea otter pup in which (C) doubled. Least affected was the sea lion in which no change in (C) occurred. Washing slightly reduced (C) of the adult otter and fur seal. The results indicate that even a light oiling would have marked detrimental effects on the thermoregulatory abilities of otters and fur seals at sea. The thermal effects of oiling on other adult pinnipeds while at sea would be slight.

Key words: Fur seal, sea lion, walrus, Phocidae, lanugo, groom, sea otter thermal conductance, thermoregulation

Introduction

The insulative quality of the fur of arctic and temperate mammals has been assessed (Scholander et al., 1950a; Hammel, 1955). Its remarkable effectiveness is exemplified by the arctic fox, *Alopex lagopus*, whose thermal neutral zone extends below -40°C (Scholander et al., 1950b). Aquatic mammals and birds, however, must face a more unconventional challenge than that of keeping warm just in air. They must also reduce heat loss in water where the heat conduction is tens of times greater.

Most marine mammals have a subcutaneous blubber layer which, combined with a remarkable peripheral vascular structure and an exquisite control of blood flow through the blubber and to the skin, results in an ideal substance for insulation. Not only is it a poor thermal conductor, but under circumstances of heat loading warm blood can pass through this insulator and dissipate heat at the skin's surface. Furthermore, the blubber functions simultaneously as a store for high energy fats. An important disadvantage to some species is the increased bulk it adds. The amphibious groups such as the pinnipeds are awkward when ashore. Perhaps because a thick layer of blubber would be a serious impediment to mobility, freshwater species such as muskrat, beaver and river otter rely primarily if not wholly on fur for insulation. This is the case for some marine mammals as well, i.e., fur seals and the sea otter. Why these marine mammals rely on fur as the primary barrier to heat loss in water is an interesting evolutionary question. Fur has some important disadvantages which include: 1) A considerable ammount of energy is expended in grooming. 2) In times of heat loading the heat cannot be dissipated over the entire body surface because of the fur barrier and this loss must be accomplished by way of the bare flippers. Consequently, these animals have a narrow thermal tolerance in aerial environments.

The characteristics of heat flux in aquatic animals have been the subject of several studies. The thermal conductance (C) of seal blubber was studied by Scholander et al., (1950a), Hart and Irving (1959) and Bryden (1964). Flensed blubber was found to conduct about the same as asbestos, and the blubber of the living animal with its flow of blood was about 50% higher (Hart and Irving, 1959). Changes in (C) of dry and immersed pelts of beaver and polar bear also have been determined by Scholander et al., (1950a). The influence of flow rate, guard hairs and undercoat on (C) has been assessed by Frisch et al., (1974).

With the widespread use of supertankers to transport oil, and the proliferation of platforms for offshore drilling and pumping of oil, interest in the effects of oil on the insulative properties of marine birds and mammals has developed. Several recent studies have been addressed to this problem. McEwan and Koelink (1973) have studied mallard ducks and scaup. The metabolic response of muskrats before and after oiling has been tested (McEwan et al., 1974). Most recently the effects of crude oil on ringed seals has been investigated (Smith and Geraci, 1975).

The purpose of this paper is to describe the effects of immersion, oiling and immersion, and cleaning with detergent and immersion on (C) of the pelts of several species of pinnipeds and the sea otter.

Materials and Methods

Fresh, blubber-free pelts from several species of pinniped were collected and kept frozen until the time for the heat flux measurements. The pelts studied are listed in Table 1. The pelts were mounted on wooden frames with enough stretching so that there was little slack in the skin. Thickness, measured by dial calipers, was difficult to estimate due to uneven fur loft.

TABLE 1. Thermal conductance of immersed pinniped pelts

P=pup; SA=subadult; & A=adult. Bracketed samples are from the same animal.

Family	Genus & Species[1]	Age	Pelt Thickness (Cm)	Fur Thickness (Cm)	Conductance ($W \cdot M^{-2} \cdot °C^{-1}$)
Mustelidae	Enhydra lutris	P	2.2	1.9	7
	" "	A	1.1	0.7	26
	" "[2]	A	2.1	1.7	23
	" "	A	1.3	0.9	22
	" "	A	2.6	2.3	22
Phocidae	Erignathus barbatus	A	1.2	-	27
	Phoca groenlandica	A	0.4	-	52
	Lobodon carcinophagus	A	0.8	-	37
	Hydrurga leptonyx	SA	0.6	-	34
	Leptonychotes weddelli	P[3]	1.5	0.4	28
	" "	P[3]	1.5	0.4	28
Otariidae	Zalophus californianus	A	0.4	-	58
	Callorhinus ursinus	SA	1.0	0.5	26
	" "	SA	0.8	0.4	26
	" "	P	1.2	0.3	40
Odobenidae	Odobenus rosmarus	SA	2.5	-	15

(1) In accordance with the Marine Mammal Commission nomenclature list.
(2) Same fur sample as previous measurements but combed and fluffed.
(3) Pup long haired fur (lanugo).

The heat flux measurements were made with a Beckman-Whiteley heat flow transducer, Model T200-3. The transducer consists of a silver-constantan thermopile sandwiched between thin bakelite plates. Heat flow through the thermopile generates an electromotive force due to the difference in temperature between the thermocouple junctions of the thermopile. The output was measured with a Leeds and Northrup Model 8686 potentiometer.

Mounted on the bottom side of the transducer was a brass chamber with the edges and bottom surface insulated with 5 cm of styrofoam. Water at 37°C (< 0.1°C variation) was circulated from a Thermomix 1420 circulator. Over the top surface of the transducer was placed a 30 by 30 cm water bath with a thin (0.1 mm thick) sheet of plastic as the bottom of the container. The mounted pelts were bonded to the plastic sheet by means of a thin film of grease (Crisco). The pelt was then held tightly in place with four mounting brackets which pressed against the wooden frame. A Lauda/Brinkmann circulator K-2/RD pumped water into and out of the bath through two manifolds which distributed the water evenly across the pelt. Flow rate through the bath was 25 cm · min^{-1}. The water depth was about 3 cm. The water temperature selected was usually about 12°C, and the variation in temperature through the course of a run was 0.1°C. Plate (skin temperature) and water temperatures were read with a Leeds and Northrup 8686 potentiometer whose reference temperature source was a distilled water and ice slush. Conductance was computed as the heat flow per unit area divided by the difference in plate and water temperatures, and it was expressed in W · M^{-2} · °C^{-1}. In order to achieve thermal stability a run lasted from 8 to 12 hours.

After the initial run the pelts were "squeezed" dry and 10 to 20 ml (.02 ml · cm^{-2}) of Prudhoe Bay Crude oil was painted into the fur, except for the sea otter pup which was thoroughly drenched. The paint strokes ran with the grain on the fur. The pelt was left sitting for about 5 min and then rinsed with fresh water for 15 sec. It was then placed in the water bath.

Results

The fur of the three long-haired animals that we studied: the Weddell seal pup, *Leptonychotes weddelli*, fur seal, *Callorhinus ursinus*, and sea otter, *Enhydra lutris*, had quite different appearances from each other. The Weddell seal was a woolly, rather disheveled looking fur that wetted rapidly when immersed. The sea otter pelt had a woolly, loose appearance superficially, but a very dense underfur or wool. Kenyon (1969) cites an examination of the fur by Scheffer in which hair densities of 101,000 fibers · cm^{-2} were estimated. The fur did not lay as flat as that of the fur seal. It did seem more water repellent.

The fur seal pelt was a dense, smooth, orderly looking fur that was a water resistant barrier. Water penetrated slowly into the underfur. The wetting seemed to be hastened if the guard hairs were parted. We found that if the fur were pressed hard and the pressure advanced with the grain of the fur water was forced out and the underfur appeared dry. After watching fur seals groom we suspect that process may achieve similar results. The texture and density of hair fibers of the fur seal pelt were noted by Scheffer (1962). He found that the hair density of the mature pelt was 57,000 hairs · cm^{-2}, and that of the pup fur was 9000 hairs or fibers · cm^{-2}. He notes that 75 to 80% of the pup fur was underhair which is coarser than underfur or wool of older animals. Scheffer also noted that when it rains the pups soak to the skin. Considering these differences in physical characteristics the greater (C) of the pup pelt compared to that of the subadults is expected (Table 1).

TABLE 2. Thermal conductance of oiled and immersed pelts.

Symbols are the same as Table 1.

Subject	Age	Pelt Thickness (Cm)	Conductance (W · M^{-2} · °C^{-1})	Multiple of Control
E. lutris	P	1.1	15	2.1
" " (1)	A	-	26	-
" " (2)	A	-	29	1.3
" " (2)	A	-	26	1.1
E. barbatus	A	1.2	27	1.0
L. weddelli	P	0.9	42	1.5
Z. californianus	A	0.4	56	1.0
C. ursinus	SA	1.0	53	2.0
" "	SA	0.8	45	1.7
" "	P	0.7	54	1.4

(1) Naturally oiled with heavy crude.
(2) Compared to combed and fluffed fur.

Of all the sea mammal pelts tested the best insulator, or the one in which (C) was the least was that of the sea otter pup (Table 1). Its conductance value was 7 $W \cdot M^{-2} \cdot °C^{-1}$. The next lowest (C) was the skin of the walrus, *Odobenus rosmarus*. Several species had about the same (CO) values of between 20-30 $W \cdot M^{-2} \cdot °C^{-1}$. These were the adult sea otters, subadult fur seals, Weddell seal pups, and the bearded seal, *Erignathus barbatus*. The highest (C) was recorded from the California sea lion, *Zalophus californianus*, pelt of 58 $W \cdot M^{-2} \cdot °C^{-1}$.

The most profound effects of oiling were on the sea otter pup and the subadult fur seal in which (C) increased 2.1, and 1.7 to 2.0 times, respectively (Table 2).

The (C) of the naturally oiled sea otter, an animal that apparently swam through an oil slick and whose partially oiled carcass was later found, was nearly the same as the pelt we oiled with Prudhoe Crude. However, the density of the two oils was different. The Prudhoe Crude was much lighter, was not tarry and did not clump the fur. Oil caused no change in (C) of the bearded seal or California sea lion.

The most improvement due to cleaning was in the fur seal, but the (C) was still 1.5 times greater than the control (Table 3). However, the pup fur seal pelt had a lower (C) after cleaning than the control. This was probably because the loft due to fluffing of the fur after cleaning and drying was better in the cleaned than in the control pelt. The (C) of the adult sea otter pelt was about the same as the control. The pup sea otter pelt deteriorated and no measurements were possible. There was no change in the sea lion.

TABLE 3. Thermal conductance of cleaned and immersed pelts. Cleaning agent was Basic-H detergent.† Symbols the same as Table 1.

Subject	Age	Pelt Thickness (Cm)	Conductance ($W \cdot M^{-2} \cdot °C^{-1}$)	Multiple of Control
E. lutris	A	1.9	21	0.9*
" "	A	2.6	20	0.9*
Z. californianus	A	0.4	56	1.0
C. ursinus	SA	1.0	38	1.5
" "	SA	0.9	40	1.5
" "	P	1.4	34	0.9

*Compared to combed and fluffed fur.
†Shaklee Corporation

Discussion

The pelt of the nearly hairless walrus was a poor conductor because the skin was so thick, nearly 5 cm. In the live animal any blood flowing through the skin would increase its (C). No doubt the most important thermal barrier would be the thick subcutaneous blubber layer.

The (C) through the skin and pelage of the earless seals (phocids) and the sea lion were high compared with the sea otter and fur seals. This is not surprising considering that seals and sea lions possess a thick, subcutaneous layer which serves as the main barrier to heat loss. It was found previously that (C) through skin, fur and 4 to 5 cm of blubber was about 3.5 $W \cdot M^{-2} \cdot °C^{-1}$ in the ringed seal, *P. hispida* (Scholander et al., 1950a). This is a tenth of the (C) we measured in the skin and fur only of seals. However, as we mentioned earlier Hart & Irving (1959) determined that blubber in the live animal is not so effective an insulator. (C) is about 50% higher than the value previously determined by Scholander et al., (1950a).

For a short time after birth all species of polar seal pups possess a long haired pelt called lanugo. This coat functions as the main barrier to heat loss until a thick blubber layer develops as the pup nurses. Until the blubber layer develops the pups usually do not enter the water. If they were to do so the pelt quite likely would wet through and there would be a large heat loss such as that which we measured of 28 $W \cdot M^{-2} \cdot °C^{-1}$ for Weddell seals. Considering the thermal gradient across the skin and fur would be about 37°C; that is the difference between the body temperature and the sea water temperature, this represents a total output of over 1000 W · M^{-2}. Estimating that a 27 Kg pup has a surface area of about 0.93 M^2 (Meeh's surface area equation, SA = 10 W$^{0.67}$, Drent & Stonehouse, 1971), the total heat loss would be 930 W. This is about twelve times the predicted heat production based on body weight (Hart & Irving, 1959). The newborn pups probably can not sustain such a high metabolic rate and thus they could not tolerate immersion long.

Thermal conductivity results seem unrealistically high for control pelts of the sea otter and fur seal. Based on calculations similar to those of the Weddell seal the (C) of the pelt is more than five times greater than the expected heat production of the animal. Apparently some important property of the pelt is lost after removal from the animal. The important missing element may be grooming. Without this activity water may leak into the fur, and is not removed. Thus, any agent that increases the wettability of the fur will increase (C) of the pelt.

The various pelts studied might be grouped into three major categories: 1) The sparsely furred, wettable pelts of sea lions and seals, which depend wholly upon a thick blubber layer for insulation rather than upon fur. 2) The fur seal pelt is a dense fur that may gradually wet unless groomed. This fur is probably the sole insulation even though some subcutaneous fat is present. 3) The sea otter has no subcutaneous fat and the pelage is the barrier to heat loss. With this grouping in mind, Fig. 1 summarizes the effects of oiling and cleaning of adult sea otter and sea lion and subadult fur seal.

Thermal conductance is very high in the sea lion pelt. Oiling and washing do not alter the insulative properties much. These results are consistent with the nature of their fur. Such an effect would be similar in all those species in which blubber was the primary insulator.

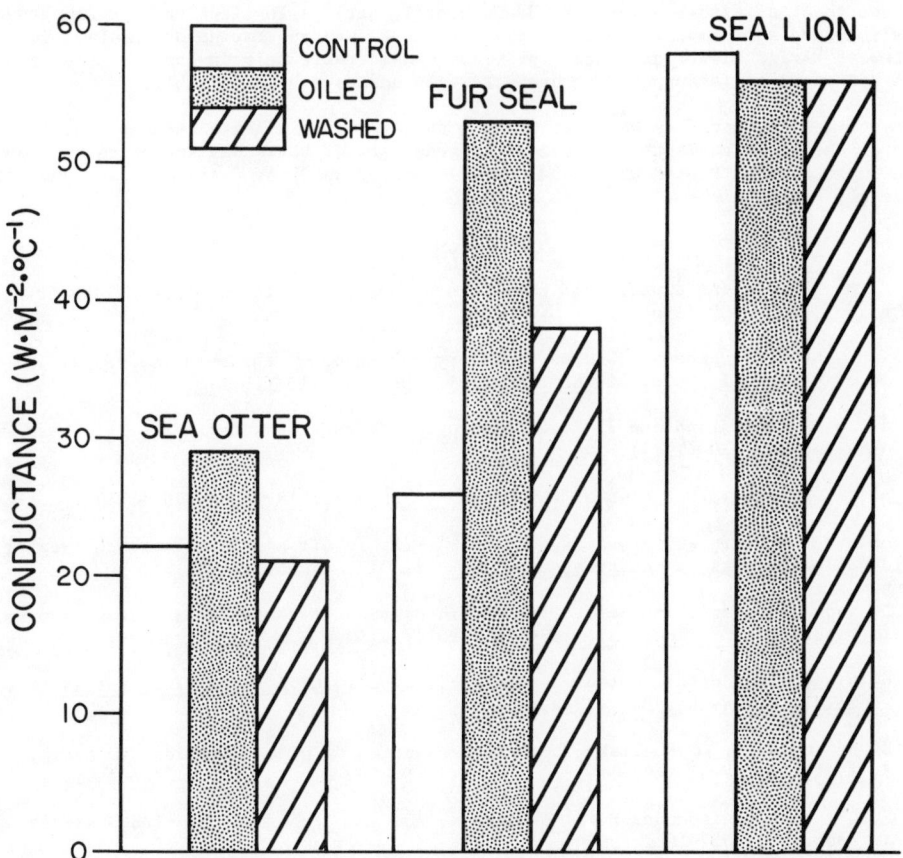

Fig. 1. Thermal conductance of sea otter, fur seal and sea lion pelts during normal immersion, after oiling, and after cleaning.

The increase in (C) in the sea otter and especially the fur seal after oiling (Fig. 1, Table 2) is a serious degradation of the fur. If the animals are unable to reverse this effect by some means, such as grooming, they probably could not endure cold water immersion long. Washing the fur, particularly that of the fur seal does not decrease (C) much.

Conclusions

The method of determining heat flow through pelts that was described in this study is an uncomplicated way of obtaining relative information about conductance. A shortcoming of this technique is that it tells nothing about what effects the behavior of the animal may have on conductance, especially grooming. Therefore, the relative effects of oiling and cleansing may be substantially different in the living animal.

A mild oiling of the pelt with a light crude oil does not increase heat loss during immersion significantly in seals and sea lions. The (C) of the pelts of the otter and fur seal were unrealistically high even in the controls. This is suspected to be due to loss of water repellency for some unknown reason. These results emphasize the importance of pelt integrity and the calamitous effects of its loss. Oiling increases the (C) of the pelt presumably by reducing water repellency. The consequences would be gravely serious for these species if they were not able to overcome quickly the oiling affects. Removing the oil from the fur with a detergent does not improve the quality much.

Acknowledgements

We wish to thank those people who contributed pelts to this study. They are: Dr. L. H. Cornell, Sea World of San Diego; Dr. D. D. Hammond, Oceanic Park, Ltd., Hong Kong; Dr. D. H. Kerem, Aba Khousky School of Medicine, Haifa, Israel; and Dr. Karl Schneider, Alaska Department of Fish & Game. Dr. H. T. Hammel, Scripps Institution of Oceanography, generously loaned equipment for this study. Dr. R. L. Gentry, Marine Mammal Division of the Northwest Marine Fisheries Center, provided considerable help in many aspects of the research and without his support the project would not have been possible.

This project was supported by National Oceanic and Atmospheric Administration Outer Continental Shelf Environmental Assessment Program. R. W. Davis was supported by Biomed research support grant PHS RR 05665-09 to Dr. J.H. Moxley and M. A. Castellini was supported by a Regent's Fellowship.

References

Bryden, M. M., Insulating capacity of the subcutaneous fat of the southern elephant seal, Nature 203, 1299-1300 (1964).

Drent, R. H. and B. Stonehouse, Thermoregulatory responses of the Peruvian penguin, Spheniscus humboldti, Comp. Biochem. Physiol. 40A, 689-710 (1971).

Frisch, J., N. A. Oritsland and J. Krog, Insulation of furs in water, Comp. Biochem. Physiol. 47A, 403-410 (1974).

Hammel, H. T., Thermal properties of fur, Amer. J. Physiol. 182, 369-376 (1955).

Hart, J. S. and L. Irving, Energetics of harbor seals in air and in water with special consideration of seasonal changes, Can. J. Zool. 317, 447-457 (1959).

Irving, L. and J. S. Hart, The metabolism and insulation of seals as bare-skinned mammals in cold water, Can. J. Zool. 35, 497-511 (1957).

Kenyon, K. S., The sea otter in the eastern pacific ocean, U.S. Dept. Int. Bur. Sport. Fish. & Wildlife Res. 68, 352 (1969).

McEwan, E. H. and A. F. C. Koelink, The heat production of oiled mallards and scaup, Can. J. Zool. 51, 27-31 (1973).

McEwan, E. H., N. Aitchison and P. E. Whitehead, Energy metabolism of oiled muskrats, Can. J. Zool. 52, 1057-1062 (1974).

Scheffer, V. B., Pelage and surface topography of the northern fur seal, Bur. Coml. Fish. 64, 206 pp (1962).

Scholander, P. F., V. Walters, R. Hock and L. Irving, Body insulation of some arctic and tropical mammals and birds, Biol. Bull. 99, 225-236 (1950a).

Scholander, P. F., R. Hock, V. Walters, F. Johnson and L. Irving, Heat regulation in some arctic and tropical mammals and birds, Biol. Bull. 99, 237-258 (1950b).

Smith, T. G. and J. R. Geraci, The effect of contact and ingestion of crude oil on ringed seals of the Beaufort sea, Beaufort Sea Tech. Rep. No. 5 (1975).

CHAPTER 15

EFFECTS OF EXTERNAL APPLICATIONS OF FUEL OIL
ON HATCHABILITY OF MALLARD EGGS

Peter H. Albers

U. S. Fish and Wildlife Service
Patuxent Wildlife Research Center
Laurel, Maryland   20811

Abstract

An experiment was performed to determine the toxicity of oil to incubating eggs. Number 2 fuel oil, a mixture of 9 paraffin compounds, and propylene glycol were applied to the surface of artificially incubated mallard (*Anas platyrhynchos*) eggs. Seven groups of 50 eggs each were treated with 1, 5, 10, 20, and 50 µl of fuel oil, 50 µl of the paraffin mixture, and 50 µl of propylene glycol. Fifty untreated eggs served as a control. Microliter syringes were used to apply the liquid around the air cell end of the egg on the 8th day of incubation. Embryonic mortality was significantly greater ($P \leq 0.01$) in all oil treated groups and the paraffin mixture group than in the control group. Most of the embryonic mortality for the oiled eggs occurred within 72 hours of treatment. Hatching and post-hatching (4 weeks) weights of the ducklings in all treatment groups were not significantly different ($P > 0.01$) from the control. Thus, the transfer of even small quantities of oil to the egg surface is sufficient to reduce hatchability.

Key Words: oil; egg; hatchability; embryo; age; weight; duckling.

Introduction

Crude oil and refined oil spillages and discharges kill large numbers of marine and wetland birds (National Academy of Sciences 1975). Presumably many birds are also contaminated with sublethal amounts of petroleum hydrocarbons. Oil adhering to the plumage of these birds may be transferred to eggs during incubation. The effects of external applications of oil on avian eggs are not well known; however, results of previous studies indicate that eggs contaminated by crude or refined oil seldom hatch (Gross 1950, Rittinghaus 1956, Hartung 1965, Kopischke 1972, Birkhead et al. 1973). Embryonic mortality may be caused by: (1) penetration of oil through the shell and shell membrane, (2) reduction of gaseous transfer through the shell, and (3) abnormal incubation behavior of the oiled adult. Abnormal gaseous transfer and reduced incubation temperatures are known to increase hatching mortality and malformation of embryos (Romanoff 1972).

This study was performed to determine (1) the effects of external applications of oil on the hatching success and post-hatching survival of birds and (2) whether the toxicity of oil or the blockage of egg shell pores is primarily responsible for these effects.

Methods

Domestic mallard eggs were candled to remove infertile eggs after 8 days of incubation in a commercial incubator. The fertile eggs were randomly assigned to 7 experimental groups of 50 each and returned to the incubator. Six groups of eggs were treated 2-3 hours later with 5, 10, 20, and 50 µl of No. 2 fuel oil[1], 50 µl of a mixture of 9 paraffin compounds[2], and 50 µl of propylene glycol, respectively. All treatment substances were at room temperature (21°C). The 7th group was an untreated control. The paraffin mixture served as a coating substance representing the aliphatic portion of No. 2 fuel oil. Propylene

---

[1]Obtained from the American Petroleum Institute.

[2]Mixed in equal proportions (volume): pentadecane, hexadecane, heptadecane, octadecane, nonadecane, 2,2,4,6,6-pentamethylheptane, 2,2,4,4,6,8,8-heptamethylnonane, 2,6,10,14-tetramethylpentadecane, decahydronapthalene.

glycol is non-toxic when taken internally (The Merck Index 1968) and was used as a highly viscous substance for coating the egg shell.

Test substances were applied by microliter syringe in 5 or 10 µl quantities around the air cell end of eggs placed upright in the incubator trays. The substances were allowed to spread freely over the egg. Applications were made in the following manner: 5 µl = 1 location, 10 µl = 2 locations of 5 µl, 20 µl = 4 locations of 5 µl, and 50 µl = 5 locations of 10 µl. This was done to permit an increase in surface coverage proportional to the increase in volume of liquid.

The amount of egg surface area affected by each treatment was determined by applying test substances stained with Oil Blue N[3] biological stain. The substances penetrated the egg shell and spread outward through the shell and shell membrane. Five extra eggs in each of the 6 treatment groups were used at the same time that the experimental application was made. After 24 hours in the incubator the stain pattern was copied on tracing paper. The area of the pattern was measured with a plastic acreage grid (356 dots per sq in). The total surface area of each egg was estimated by

$$S = kV^{2/3}$$ where $k = 4.831$ and $V =$ volume of egg
(Romanoff and Romanoff 1963)

A reference collection of normal mallard embryos was made from extra eggs. Embryos were preserved in 5 percent formalin. All eggs were candled at 3-4 day intervals to identify dead embryos. Eggs with dead embryos were opened and the embryos were aged by comparison with the reference collection and the descriptions of Hanson (1954) and Caldwell and Snart (1974). Eggs which failed to hatch by the 30th day of incubation were opened and the embryos were aged.

A second experiment was performed because of the high embryonic mortality caused by the smallest application of fuel oil (5 µl). A group of 50 eggs from a new batch of mallard eggs was treated with 1 µl of fuel oil and compared to an untreated control group of 50 eggs. Experimental procedures were the same.

Ducklings were weighed 24 hours after hatching and examined for gross external malformations. The birds were separated by treatment group and weighed at weekly intervals for 4 weeks (days 8, 15, 22, and 29). Hatching occurred over a 48 hour period; therefore, some of the ducklings were a day younger or older than the weighing date.

Extra treated eggs, extra control eggs, and a sampling of the ducklings were frozen at various stages of the experiment for future hydrocarbon residue analysis.

Results

The hatching success of all groups of eggs treated with fuel oil was significantly lower than the control group in both experiments ($P \leq 0.01$; Table 1). Five microliters of fuel oil killed approximately half of the eggs treated. Fuel oil in amounts of 20 µl or more per egg produced almost 100 percent mortality. The paraffin mixture caused a much smaller, although still significant, decrease in hatching success than the equivalent amount of fuel oil (50 µl). Propylene glycol did not significantly alter hatchability.

The percentage of egg surface (shell and shell membrane) affected by the fuel oil treatments increased proportionally with the volume applied (Table 2). The surface areas affected by 50 µl of fuel oil and 50 µl of the paraffin mixture were similar, but the surface area affected by 50 µl of propylene glycol was only equivalent to the surface area affected by 10 µl of fuel oil. Yet, embryonic mortality was much higher for the fuel oil treatments than for the propylene glycol or paraffin mixture.

A one-way analysis of variance on the 8 groups of ages of dead embryos was significant ($P \leq 0.01$; Table 3). Most of the mortality in the groups treated with fuel oil occurred within 72 hours of treatment. Eggs treated with the paraffin mixture or propylene glycol and the controls lived significantly ($P \leq 0.05$) longer. No external malformations were noted during the aging of the dead embryos.

Two eggs with live embryos, 1 each from the 5 µl and 20 µl fuel oil groups, were mistakenly opened on the 20th day of incubation. The embryos were at the same stage of development as 2 extra untreated embryos which were used for the reference collection but they were considerably smaller (13.7 g and 14.7 g versus 24.8 g and 25.5 g).

---

[3]National Biological Stains & Reagents Dept., Allied Chemical Corp.

TABLE 1. Hatching Success of Fertile Mallard Eggs Treated on the 8th Day of Incubation

Treatment	Amount (µl)	Number of eggs	Hatched	Failed to hatch	Percent hatching success
Control*	--	50	44	6	88
Propylene glycol	50	50	40	10	80
Paraffin mixture	50	50	36	14	72**
No. 2 fuel oil	5	49	22	27	45**
No. 2 fuel oil	10	50	6	44	12**
No. 2 fuel oil	20	49	1	48	2**
No. 2 fuel oil	50	50	0	50	0**
Control*	--	50	49	1	98
Oil	1	50	34	16	68***

* Two separate experiments

** $P \leq 0.01$, Chi-square test, df = 1

*** $P \leq 0.01$, Binomial test

TABLE 2. Dispersal of No. 2 Fuel Oil Applied to Incubating Mallard Eggs. N = 5

Treatment	Amount (µl)	Percent egg surface affected	
		Mean	S.D.
Propylene glycol	50	12.6	3.0
Paraffin mixture	50	34.1	6.6
No. 2 fuel oil	1	1.3	0.7
No. 2 fuel oil	5	5.0	2.7
No. 2 fuel oil	10	11.5	2.0
No. 2 fuel oil	20	20.2	4.8
No. 2 fuel oil	50	32.1	6.8

The hatching weights of the ducklings in the treatment groups were not significantly different from each other ($P > 0.01$) or from the control group ($P > 0.05$) in either of the experiments (Table 4). The 1 duckling hatched in the 20 µl fuel oil group was omitted from the analysis.

TABLE 3. Time of Embryonic Mortality in Mallard Eggs Treated
on the 8th Day of Incubation

Treatment	Amount (µl)	Number of embryos	Mean age at death (days)
Controls	--	7	23.7*
Propylene glycol	50	10	24.2*
Paraffin mixture	50	14	21.6*
No. 2 fuel oil	1	16	10.8
No. 2 fuel oil	5	27	11.1
No. 2 fuel oil	10	44	10.1
No. 2 fuel oil	20	48	9.4
No. 2 fuel oil	50	50	8.5

* Significantly older than any of the fuel oil treated groups; pair-wise contrasts ($P \leq 0.05$), Scheffe procedure, significant ($P \leq 0.01$) one-way analysis of variance.

TABLE 4. Body Weight (g) of Mallard Ducklings

Treatment	Amount (µl)	Days after hatch				
		1	8**	15	22	29
Controls	--	43.3±3.7(44)*	107.5±19.8(31)	289.7±46.0(31)	566.3±47.7(31)	738.0±62.3(30)
Propylene glycol	50	41.3±4.1(40)	119.0±19.3(30)	299.2±37.4(30)	561.8±51.6(30)	733.0±55.2(30)
Paraffin mixture	50	41.1±3.1(36)	113.4±17.4(28)	286.3±39.0(28)	571.4±65.9(28)	756.8±69.0(28)
No. 2 fuel oil	5	43.0±3.7(22)	100.0±15.6(22)	309.6±40.6(22)	605.0±67.8(21)	756.2±83.5(21)
No. 2 fuel oil	10	41.5±3.9(6)	82.0±21.1(6)	259.7±70.1(6)	531.7±95.2(6)	670.0±96.7(6)
Controls	--	41.8±3.0(49)	111.7±17.9(48)	297.1±33.8(48)	540.1±46.4(48)	716.7±62.1(48)
No. 2 fuel oil	1	42.3±4.1(34)	113.2±17.6(34)	306.2±45.1(34)	546.8±63.9(34)	743.5±78.9(34)

* Mean ± standard deviation (sample size).

** Significant ($P \leq 0.01$), one-way analysis of variance for the first experiment; significant pair-wise contrasts ($P \leq 0.05$), Scheffe procedure, between propylene glycol and 10 µl fuel oil, paraffin mixture and 10 µl fuel oil, and propylene glycol and 5 µl fuel oil.

The post-hatching weights of ducklings in the first experiment were significantly different ($P \leq 0.01$) from each other only on the 8th day after hatching (Table 4). Significant pair-wise contrasts ($P \leq 0.05$) occurred between the propylene glycol group and the 10 µl fuel oil group, the paraffin mixture group and the 10 µl fuel oil group, and the propylene glycol group and the 5 µl fuel oil group. None of the treatment groups were significantly different from the control group. The post-hatching weights of the 1 µl fuel oil ducklings were never significantly different from the control ducklings in the second experiment ($P > 0.01$). The 10 µl fuel oil ducklings were the lightest ducklings throughout the 4 week post-hatching period.

Nearly all of the post hatching mortality indicated in Table 4 (31 of 33 deaths) was due to 1 night of accidental drownings. No external malformations, diseases, or abnormal behavior were noted in any of the ducklings during the 29 day period.

Discussion

My results demonstrate that even in very small amounts, external contamination by No. 2 fuel oil will severely reduce hatchability of mallards eggs. The small reduction in hatching success resulting from application of a paraffin mixture indicates that the embryonic toxicity of the fuel oil is primarily due to substances other than the paraffins. Saturated compounds such as the paraffins have been reported to be less toxic to marine invertebrates and fish than the aromatic compounds (Moore and Dwyer 1974). The failure of propylene glycol to cause an increase in mortality and the large difference in mortality between 50 µl of paraffin mixture and 50 µl of fuel oil indicate that gaseous transfer was not seriously affected (Tables 1 and 2). Interruption of normal gaseous transfer is probably not a cause of embryonic mortality when only portions of the egg are coated with viscous materials.

Fuel oil killed the mallard embryos sooner after the 20 µl and 50 µl applications (12 to 36 hrs) than after the 1 µl, 5 µl, and 10 µl applications (48 to 72 hrs). Embryos that died after treatment with propylene glycol and the paraffin mixture, and the embryos of the controls that died were considerably older than the dead embryos of the fuel oil treated groups. Disruption of the temporal sequence of embryonic development may have affected the accuracy of aging dead embryos from treated eggs, particularly as the difference between the aging date and the treatment date increased. The size of the 2 live fuel oil treated embryos on the 20th day of incubation shows that fuel oil may retard growth when mortality is not immediate. Seventy-nine percent of the embryos that died were aged on the 12th day of incubation; 4 days after treatment. The remaining 21 percent of the age estimates were made later and the accuracy may have been affected by a disruption of embryonic development. At best, the age estimates indicate the date of death, and, at worst, they indicate the developmental stage at the time of death.

Ducklings from fuel oil treated eggs exhibited no gross external malformations or behavioral abnormalities. Hatching weights and weekly weight changes were relatively poor indicators of embryonic exposure to fuel oil. However, the slightly lighter post-hatching weights of the 6 ducklings from the 10 µl fuel oil treatment and the low weight gain, between day 1 and day 8, of ducklings from the 5 and 10 µl fuel oil treatment suggest that fuel oil contamination during incubation may have some effect on the ducks after hatching.

Growth inhibition in ducklings hatched from oiled eggs may indicate the presence of internal malformations or sublethal physiological damage. Additional chronic experiments will help to define long term survival and reproductive success of survivors from oiled eggs.

Acknowledgements

I thank Michael P. Dieter and Stana Federighi for reviewing a preliminary draft of this paper.

References

Birkhead, T. R., C. Lloyd, and P. Corkhill, Oiled seabirds successfully cleaning their plumage, Brit. Birds 66, 535-537 (1973).

Caldwell, P. J. and A. E. Snart, A photographic index for aging mallard embryos, J. Wildl. Manage. 38, 298-301 (1974).

Gross, A. O., The herring gull-cormorant control project, Proc. Int. Ornith. Congr. 10, 533-536 (1950).

Hanson, H. C., Criteria of age of incubated mallard, wood duck, and bob-white quail eggs, Auk 71, 267-272 (1954).

Hartung, R., Some effects of oiling on reproduction of ducks, J. Wildl. Manage. 29, 872-874 (1965).

Kopischke, E. D., The effect of 2,4-D and diesel fuel on egg hatchability, J. Wildl. Manage. 36, 1353-1356 (1972).

Moore, S. F. and R. L. Dwyer, Effects of oil on marine organisms: a critical assessment of published data, Water Res. 8, 819-827 (1974).

National Academy of Sciences, Petroleum in the Marine Environment. Workshop on inputs, fates and the effects of petroleum in the marine environment. Ocean Affairs Board, Washington, D. C., 1975.

Rittinghaus, H., Etwas uber die indirekte verbreitung der olpest in einem seevogelschutzgebiete, Ornithol. Mitteil. 8, 43-46 (1956).

Romanoff, A. L., Pathogenesis of the Avian Embryo, Wiley, Interscience, New York, 1972.

Romanoff, A. L. and A. J. Romanoff, The Avian Egg, Wiley, New York, 1963.

The Merck Index, P. G. Stecher (ed.), Merck & Co., Rahway, New Jersey, 1968.

CHAPTER 16

EFFECTS OF EXTERNAL APPLICATIONS OF NO. 2 FUEL OIL ON COMMON EIDER EGGS

Robert C. Szaro and Peter H. Albers

U. S. Fish and Wildlife Service
Patuxent Wildlife Research Center
Laurel, Maryland 20811

Abstract

Because eggs of marine birds may be exposed to oil adhering to the feathers of adult birds, a study was undertaken to determine the effects of oil contamination. Two hundred common eider eggs (*Somateria mollissima*) were divided into four experimental sets of 50 each. Two sets were treated with No. 2 fuel oil in amounts of 5 µl and 20 µl; a third with 20 µl of propylene glycol, a neutral blocking agent. The fourth set served as a control. Hatching success was 96 percent for the eggs treated with 20 µl propylene glycol, 96 percent for the controls, and 92 percent for the eggs treated with 5 µl oil hatched. Only 69 percent of the eggs treated with 20 µl of oil survived: a significant reduction in hatchability ($P \leq 0.05$). Mean hatching weights for all sets were statistically equal. Thus, oil pollution may significantly increase embryonic mortality in marine birds.

Key Words: No. 2 fuel oil; common eider; egg; hatchability; embryonic mortality; duckling weight.

Introduction

The Common Eider (*Somateria mollissima*) is a circumpolar marine species in the northern hemisphere (Bellrose 1976) that may be subjected to varying levels of oil pollution. Eiders commonly breed on near-shore islands in areas being increasingly utilized by oil tankers. A number of researchers have shown the high vulnerability of eiders to oil pollution (Greenwood & Keddie 1968, Greenwood et al. 1971, Joensen 1972, Milne & Campbell 1973). Reed (1975) suggests that eiders must be subject to periodic die-offs due to oil pollution. For example, at the Sands of Forvie, Scotland in 1968 and 1970 when relatively large numbers of young eiders were fledged, projected increases in the population failed to occur, apparently because of oil pollution at the wintering grounds (Milne 1974).

Oil may have further deleterious effects by adhering to the plumage of the birds and transfering to the eggs during laying and incubation (Hartung 1965). The effects of external applications of oil on the eggs of marine birds are not well known but previous studies indicate that hatchability is greatly reduced (Gross 1950, Birkhead et al. 1973). Knowledge of the effects of external contamination of eider eggs by oil is needed to evaluate the full impact of oil pollution on marine bird populations.

Methods and Materials

Two hundred common eider eggs were collected in May 1976 on an island in East Casco Bay, Maine, transported to the Patuxent Wildlife Research Center, Laurel, Maryland and placed in a commercial incubator. After allowing 12 hours for the eggs to reach normal incubation temperature, they were randomly divided into four groups of 50 eggs each. Number 2 fuel oil (API Reference Oil III) was applied to two groups of eggs in 5 µl and 20 µl amounts. Twenty microliters of propylene glycol was applied to a third group as a neutral coating substance (Albers 1977). The fourth group was an untreated control. The oil and propylene glycol were applied with a microliter syringe to the middle portion of eggs placed on their side in the incubator trays and allowed to spread freely over the egg. The 5 µl applications were made at one location and the 20 µl applications were made at four locations of 5 µl each (Albers 1977).

The eggs were candled periodically after treatment to identify dead embryos. Eggs with dead embryos were opened and the embryos were aged according to the descriptions of Gorman (1974) and by comparison with known-age mallard (*Anas platyrhynchos*) embryos. The age at treatment for ducklings that hatched was estimated assuming a normal 26 day incubation period of eider eggs (Gorman 1974). Dead embryos were also examined for gross external malformations. Infertile eggs were removed from the experiment. Eider ducklings were weighed and examined for gross external malformations within 24 hours of hatching.

Results

The 20 µl oil application significantly ($P \leq 0.05$) decreased eider egg hatchability (Table 1), but 20 µl of propylene glycol and 5 µl of oil had no significant effects.

TABLE 1. Hatching Success of Common Eider Eggs

Treatment	N	Hatched	Percent
Control	49	47	96
Propylene glycol	49	47	96
No. 2 fuel oil (5 µl)	48	44	92
No. 2 fuel oil (20 µl)	48	33	69*

* $P \leq 0.05$, Chi-square, df = 1

The age of the eggs at treatment varied from a mean of 12.2 days in the control set to a mean of 14.3 days in the 5 µl oil set (Table 2); this difference was not significant.

TABLE 2. Age of Common Eider Eggs at Treatment

Treatment	N	Mean	S.D.	Range
Control	49	12.2	5.3	0 - 24
Propylene glycol	49	13.8	5.1	3 - 23
No. 2 fuel oil (5 µl)	47*	14.3	5.7	3 - 24
No. 2 fuel oil (20 µl)	48	12.4	7.1	0 - 24

* One of the 48 treated eggs was too decomposed to determine the date of treatment.

Nevertheless, the survival of oil treated embryos was strongly dependent on age. Embryos that died averaged 3.8 days of age at treatment, whereas surviving embryos were treated at an average age of 14.3 days (Table 3). For example, the mean age at treatment of the dead embryos treated with 20 µl of oil (4.3 days) indicated that the younger embryos were much more sensitive to oil than older ones.

There was no significant difference in mean hatching weight of the ducklings in the four groups (Table 4), nor were morphological deformations visible in any of the ducklings.

TABLE 3. Comparison of Hatching Success with Age at Treatment of Common Eider Eggs

All Embryos

	N	Age (Days)		
		Mean	S.D.	Range
Died	22	3.8*	4.6	0 - 11
Hatched	171	14.3	4.9	3 - 24

20 μl Oil Embryos

Died	15	4.3*	5.4	0 - 11
Hatched	33	16.1	4.1	7 - 24

* Ages of dead and hatched embryos after treatment were significantly different ($P \leq 0.05$); one-way analysis of variance.

TABLE 4. Hatching Weights of Common Eider Ducklings

Treatment	N	Weight (g)		
		Mean	S.D.	Range
Control	47	82.1	6.4	65 - 98
Propylene glycol	47	82.5	8.4	64 - 100
No. 2 fuel oil (5 μl)	44	81.5	6.7	66 - 96
No. 2 fuel oil (20 μl)	33	80.8	6.7	67 - 95

Discussion

The hatchability of common eider eggs was significantly reduced by external applications of 20 μl of oil. However, eider embryos appear to be less sensitive to oil than similarly treated mallard embryos. The hatchability of mallard eggs treated after 8 days of incubation with 5 μl of oil was 44 percent less than the control (Albers 1977) compared with a 4 percent reduction in the hatchability of mixed-age eider eggs. Similarly, the hatchability of mallard eggs treated with 20 μl of oil was 86 percent less than the control compared with a 27 percent reduction in the hatchability of the eider eggs.

The differences in hatchability of eider and mallard eggs may be partially due to the differences in stage of incubation at the time of treatment. Albers (unpublished data) found that mortality from oil applied to incubating mallard eggs was inversely related to the age of the embryos. Common eider embryos were also particularly sensitive to small amounts of oil applied during the early stages of development. Thus, higher mean treatment dates for the eider eggs (12.2 - 14.3 days) than the mallard eggs (8 days) may have resulted in less mortality. A further consideration is the size of the two types of eggs; eider eggs are approximately twice the volume of mallard eggs. Therefore, the dose per unit volume for eider eggs was about half that of the mallard eggs.

The level of oil used to demonstrate lethality in incubating eider eggs (20 µl) may be well below levels that might be encountered in the environment. Hartung (1963) estimated that 3.5 g represented an average amount of oil found on dead lesser scaup (*Aythya affinis*) under natural conditions. Transfer of microliter amounts of oil during critical periods of incubation could seriously affect seabird populations by decreasing the hatchability of fertile eggs.

Acknowledgements

We thank Carl Korschgen, Bill Snow, and Howard Mendell for their help in obtaining the eider eggs used in this study. Michael Dieter and Stana Federighi kindly reviewed the preliminary manuscript.

References

Albers, P. H., Effects of external application of fuel oil on hatchability of mallard eggs, (in press), (1977).

Birkhead, T. R., C. Lloyd, and P. Corkhill, Oiled seabirds successfully cleaning their plumage, Brit. Birds 66, 535-537 (1973).

Bellrose, F. G., Ducks, Geese, and Swans of North America, Stackpole Books, Harrisburg, Pa., 1976.

Gorman, M. L., Criteria for ageing embryos of the eider, Wildfowl 25, 29-32 (1974).

Greenwood, J. J. D., R. J. Donally, C. J. Feare, N. J. Gordon, and G. Waterston, A massive wreck of oiled birds: northeast Britain, winter 1970, Scot. Birds 6, 235-250 (1971).

Greenwood, J. J. D. and J. P. F. Keddie, Birds killed by oil in the Tay estuary, March and April 1968, Scot. Birds 5, 189-196 (1968).

Gross, A. O., The herring gull-cormorant control project, Proc. Int. Ornith. Congr. 10, 533-536 (1950).

Hartung, R., Ingestion of oil by waterfowl, Papers Mich. Acad. Sci. 48, 49-54 (1963).

Hartung, R., Some effects of oiling on reproduction of ducks, J. Wildl. Manage. 29, 872-874 (1965).

Joensen, A. H., Studies on oil pollution and seabirds in Denmark 1968071, Dan. Rev. Game Biol. 6, 1-32 (1972).

Milne, H., Breeding numbers and reproductive rate of eiders at the Sands of Forvie National Nature Reserve, Scotland, Ibis 116, 135-154 (1974).

Milne, H. and L. H. Campbell, Wintering seaducks off the east coast of Scotland, Bird Study 20, 153-172 (1973).

Reed, A., Migration, homing, and mortality of female breeding eiders Somateria mollissimi dresseri of the St. Lawrence estuary, Quebec, Ornis. Scand. 6, 41-47 (1975).

CHAPTER 17

THE EFFECT OF PETROLEUM HYDROCARBONS ON THE
SURVIVAL AND LIFE HISTORY OF POLYCHAETOUS ANNELIDS

Robert Scott Carr* and Donald J. Reish

Department of Biology, California State University
Long Beach, Long Beach, California 90840

Abstract

The toxicity of seawater-soluble fractions of No. 2 fuel oil and South Louisiana crude oil to five species of polychaetous annelids (Capitella capitata, Cirriformia spirabrancha, Ctenodrilus serratus, Ophryotrocha puerilis, and Ophryotrocha sp.) was determined. The results of these 28-day bioassays indicate a wide variability in sensitivity among the species tested with the two oils. The 28-day LC_{50} values ranged from 1.4 ppm of initial total hydrocarbons for O. puerilis exposed to No. 2 fuel oil to 17.8 ppm for C. capitata exposed to South Louisiana crude oil. The water soluble fractions of No. 2 fuel oil were more toxic to all species tested than South Louisiana crude oil.

The effects of petrochemicals on reproduction were measured for Ctenodrilus serratus and Ophryotrocha sp. Using the number of offspring produced as a measure of effect, a significant suppression in reproduction was noted at concentrations of 2.2 ppm and 1.3 ppm fuel oil to Ctenodrilus and Ophryotrocha, respectively, and at 9.9 ppm of South Louisiana crude oil for both species. A significant reproductive stimulation, however, was observed for Ophryotrocha sp. exposed to a concentration of 1.9 ppm of South Louisiana crude oil.

Key words: bioassay, crude oil, No. 2 fuel oil, naphthalenes, polychaete, toxicity, reproduction

Introduction

Modern day oil pollution disasters stimulated an awareness concerning the deleterious effects produced when marine organisms are exposed to petroleum hydrocarbons. The physical presence of oil stranding in littoral zones and on associated organisms is the most obvious harmful effect of oil spilled near shore. A more subtle effect, influencing not only littoral but sublittoral organisms as well, is produced by the leaching of water-soluble petroleum hydrocarbons from crude or refined petroleum products. Moore and Dwyer (1974) have reviewed and tabulated the studies which have been conducted with marine organisms and petrochemicals. Literature concerning the toxicity of oil to polychaetes is limited to the observations of George (1971) and Straughan (1971) on the effects of oil spills on survival and reproductive capacity. The studies by Rossi et al., (1976) and Rossi and Anderson (1976) are the only recent investigations which have dealt with the toxicity of oils to laboratory-reared polychaetes. They used the water soluble fractions (WSF) of four different oils, including the two used in the present study to measure 96-hr LC_{50} values. For Neanthes arenaceodentata these values were 2.7 and 12.5 mg/l with No. 2 Fuel oil and South Louisiana crude oil, respectively, while Capitella capitata were slightly more sensitive (2.3 and 12.0 mg/l, respectively). They found the 4-segmented larva, a non-feeding stage, was more tolerant to oils than the other stages and suggested oil sequestered in the yolk of the developing young and rendered them more tolerant. There appeared to be little variation in susceptibility between these polychaetes to the oils tested.

The purposes of the present study were to determine the long-term toxic effects of the WSF of two of these oils, No. 2 fuel oil and South Louisiana crude, to additional species of polychaetes and to determine the effect of these oils on the reproduction of two species possessing a short life history.

---

* Present address: Department of Biology, Texas A&M University, College Station, Texas 77843

Materials and Methods

The five species of polychaetes utilized were selected to represent various sizes and reproductive strategies. Ophryotrocha sp. and O. puerilis are minute protandric hermaphrodites. Ctenodrilus serratus is a minute species which reproduces asexually by transverse budding. Capitella capitata is a small species in which the female incubates the developing eggs in her tube. Cirriformia spirabrancha is a large species which spawns countless gametes into the water. All species used in this study were reared in the laboratory and all but C. spirabrancha had passed through at least one complete generation under laboratory conditions.

The oils used in this study were South Louisiana crude oil and No. 2 fuel oil which are originally supplied by the American Petroleum Institute and redistributed by Dr. Jack W. Anderson of Texas A&M University. The procedure for preparation of the water-soluble fraction of the oils was similar to the method employed by Anderson et al, (1974). Nine parts millipore filtered seawater were stirred with one part oil on a magnetic stirrer for 20 hours at a slow speed. After stirring, the aqueous phase was siphoned off and utilized immediately in experiments. A detailed liquid-gas chromatography analysis of these oils and their WSF had been carried out earlier by Dr. J. Scott Warner, Battelle Laboratories (Anderson et al., (1974). It should be noted that the concentrations listed in all tables are initial total hydrocarbon values derived from the same preparation of solutions from the same oils. The only different was that the extracts utilized seawater, whereas Anderson et al., (1974) obtained water-soluble fractions with artificial seawater (Instant Ocean).

All experiments were conducted at $19.5 \pm 0.5°C$ with millipore filtered seawater placed in 20 ml Stender dishes. These vessels were sealed with ground-glass lids, which strongly decreased evaporative loss as evidenced by the fact that salinity only increased by about 8 parts per thousand over the 28 days (initial salinity was 34 ppt). While some evaporation of volatile hydrocarbons no doubt occurred during the 28 day tests, the decrease is likely to be significantly less than those reported from bioassays with open containers (Anderson et al., (1974; Rossi et al., (1976). Each dish contained 15 ml of test solution. Four specimens were placed in each dish in the case of both species of Ophryotrocha and Ctenodrilus, two per dish with Capitella and one per dish with Cirriformia. All worms were fed powdered (<0.061 mm) green alga, Enteromorpha crinita. Approximately 0.3 ml of a 1 g/100 ml seawater solution of this powdered alga was fed only at the commencement of each experiment. Specimens were examined at 4, 7, 14, 21 and 28 days, at which time each experiment was terminated. Because of their rapid life histories, reproduction occurred in Ophryotrocha sp. and Ctenodrilus serratus during the experimental period. The results were analyzed statistically by either the Litchfield and Wilcoxon (1949) or by the Mann-Whitney (Wilcoxon, 1955) methods.

Results

The 96-hr LC_{50} results for the five species of polychaetes tested with the two oils are in Table 1. It is apparent that there is a wide range in response to the two oils tested. Capitella capitata and Cirriformia spirabrancha appear to be relatively resistant to exposure to the WSF or these two oils. The two species of Ophryotrocha were the most sensitive. The WSF of No. 2 fuel oil was more toxic than the water-soluble components of South Louisiana crude oil on a percent and total hydrocarbon concentration basis for all the species tested after a 96-hr period. The LC_{50} results at 28 days are shown in Table 1 for the five species tested. The long-term toxicity data present a more complex situation. Again, the two species of Ophryotrocha were the most sensitive to the No. 2 fuel oil at 28 days of exposure. Cirriformia spirabrancha was the most sensitive species tested to the long-term effects of South Louisiana crude oil. Ctenodrilus serratus, which is relatively sensitive to the effects of WSF of No. 2 fuel oil, was one of the more resistant species to the long-term effects of South Louisiana crude oil. Overall, Capitella capitata was the most resistant polychaete to the toxic effects of oil.

A state of narcosis, characterized by lethargic movement and loss of body pigmentation, was often evident in all the species preceeding death. An unusual condition was frequently observed in C. spirabrancha exposed to 40 percent or more of the WSF of South Louisiana crude oil developed large swellings or growths in mid or posterior body regions. Whether these protuberances were caused by a blockage of the gut, swellings of coelomic fluids, or were manifestations of neoplastic growth was not ascertained. Their occurrence, however, invariably resulted in death within several days.

The data showing the effects of these oils on reproduction in Ctenodrilus serratus and Ophryotorcha sp. are given in Tables 2-5. The general results are similar regardless of the species or type of oil. There was a low concentration at which the number of offspring produced may be reduced in number but was not statistically significant from the controls. Next, there was a higher concentration at which there was a statistically significant

TABLE 1. The 96-hour and 28-day TLm(ppm) of the WSF of two oils to five species of polychaetes.

SPECIES OF POLYCHAETES

Type of Oil	Ctenodrilus serratus		Ophryotrocha sp.		Ophryotrocha puerilis		Capitella capitata		Cirriformia spirabrancha	
	96 hr	28 day	96 hr	28 day	96 hr	28 day	96 hr	28 day	96 hr	28 day
No. 2 Fuel Oil										
TLm	4.1	2.6	2.9	2.4	2.2	1.4	>8.7	7.3	>8.7	5.7
95%C.I.[a]	3.6-4.6	---	---	2.1-2.8	2.1-2.3	1.2-1.6	---	6.6-8.0	---	4.6-6.9
S.F.[b]	1.51	---	---	1.66	1.26	1.47	---	1.36	---	1.56
So.La. Crude Oil										
TLm	>19.8	15.8	12.9	10.9	17.2	---	>19.8	17.8	>19.8	7.9
95%C.I.[a]	---	---	12.1-13.7	---	16.5-17.8	---	---	16.5-19.2	---	---
S.F.[b]	---	---	1.23	---	1.20	---	---	1.27	---	---

[a] Confidence interval
[b] Slope function

TABLE 2. Effect of WSF of No. 2 fuel oil on reproduction of Ctenodrilus serratus.

Percent stock WSF	Conc.[a]	96-hour percent survival	14-day		21-day		28-day		t[b]
			No. Worms	No. Budding	No. Worms	No. Budding	No. Worms	No. Budding	
Control	0	100	91	16	200	2	210	0	-
1	0.09	97.5	55	17	187	1	192	1	0.74
5	0.44	100	50	16	146	4	174	0	1.88
10	0.87	100	51	15	157	3	173	0	1.56
25	2.2	100	27	2	105	7	137	0	2.85[c]
50	4.4	2.5	0	0	0	0	0	0	-
100	8.7	0	0	0	0	0	0	0	-

[a] Initial hydrocarbon concentration ppm
[b] T-test comparing control means at 28 days with the means of the experimental populations (10 dishes per concentration with four worms per dish initially).
[c] Significant at the .05 level (two-tailed test).

TABLE 3. Effect of WSF of South Louisiana crude oil on reproduction of Ctenodrilus serratus.

Percent stock WSF	Conc.[a]	96-hour percent survival	14-day		21-day		28-day		U Stat.[b]
			No. Worms	No. Budding	No. Worms	No. Budding	No. Worms	No. Budding	
Control	0	100	37	1	180	3	284	6	-
5	0.99	100	33	1	135	3	251	14	59.5
10	1.98	100	37	3	155	3	250	17	69
25	4.95	100	39	2	151	3	243	6	54
50	9.9	100	39	0	139	2	175	3	93
75	14.9	100	37	1	118	5	140	5	95
100	19.8	95	29	0	16	4	42	0	98.5

[a] Initial hydrocarbon concentration ppm
[b] Mann-Whitney U-test comparing the control population at 28 days with the experimental populations (10 dishes per concentration with four worms per dish initially). U statistic at the .05 level of significance is 73.

reduction in reproduction. And, finally, there was a still higher concentration at which the adult may survive but not reproduce.

TABLE 4. Effect of WSF of No. 2 fuel oil on reproduction and survival of Ophryotrocha sp.

Percent stock WSF	Conc.[a]	n	96-hour percent survival	14-day percent survival	28-day Adults surviving	Juveniles & encapsulated juveniles	Total	U Stat.[b]
Control	0	40	100	100	40	54	94	–
5	0.44	40	100	100	37	15	52	58.5
10	0.87	40	100	100	38	12	50	66.5
15	1.3	40	100	95	21	0	21	93
20	1.7	40	100	100	27	0	27	86
25	2.2	40	100	95	26	0	26	82.5
30	2.6	40	100	67.5	17	0	17	89.5

[a] Initial hydrocarbon concentration ppm
[b] Mann-Whitney U-test utilized in analysis of data comparing total adults, juveniles and encapsulated juveniles of experimental populations with the control population. U statistic at the .05 level of significance is 73.

Statistically significant suppression of reproduction in Ophryotrocha sp. occurred at 1.3 mg/l concentration of No. 2 fuel oil as compared to 2.2 for Ctenodrilus. Suppression occurred at 9.9 mg/l South Louisiana crude oil for both species. While not statistically significant, the trend towards reduction in number of individuals produced occurred at the lowest concentration of oils tested in three of the four experiments. A significant increase in the number of offspring produced was noted for Ophryotrocha sp. exposed to 1.98 mg/l South Louisiana crude.

TABLE 5. Effect of WSF of South Louisiana crude oil on reproduction and survival of Ophryotrocha sp.

Percent stock WSF	Conc.[a]	n	96-hour percent survival	14-day percent survival	28-day Adults surviving	Juveniles & encapsulated juveniles	Total	U Stat.[b]
Control	0	40	100	100	40	71	111	–
5	0.99	40	100	100	40	89	129	52
10	1.98	40	100	100	39	129	168	76.5[c]
25	4.95	40	100	100	40	64	104	56
50	9.9	40	72.5	55	22	26	48	76
75	14.9	40	52.5	47.5	17	26	43	79.5
100	19.8	40	7.5	5	2	0	2	100

[a] Initial hydrocarbon concentration ppm
[b] Mann-Whitney U-test utilized in analysis of data comparing total adults, juveniles and encapsulated juveniles of experimental populations with the control population. U statistic at the .05 level of significance is 73.
[c] Significant increase in reproduction.

Discussion

Polychaetes are affected by exposure to the WSF of petrochemicals. The degree of susceptibility appears to be species dependent as evidenced by the wide range of LC_{50} values (Table 1). The toxicity of these petrochemicals on organisms is dependent upon its water-soluble hydrocarbon composition. No. 2 fuel oil contains a higher proportion of di and triaromatic compounds than South Louisiana crude oil, which is high in low molecular weight aromatics and produces a more concentrated total hydrocarbon extract (19.8 mg/l, Anderson et al., 1974).

Naphthalene compounds have been shown to decrease an order of magnitude in concentration over a 96-hr period from the WSF of No. 2 fuel oil (Rossi, et al., 1976). In this study the rate of volatilization should be significantly less, as the bioassay containers were covered by ground-glass stoppers. It is, however, possible that after 28 days the hydrocarbon content was in some cases negligible. The results of the 28 day tests more specifically relate to the long-term effects from chronic exposure, regardless of the rate of decrease in water hydrocarbon content. Tables 2-5 demonstrate the substantial effects on reproduction occurring after the 96 hr acute tests. In many cases no significant mortality had been exhibited at 96 hr, but a significant reduction in the total numbers of individuals was observed at the 28 day termination. A significantly decreased reproductive output was noted at 1.3 and 2.2 mg/l No. 2 fuel oil to Ctenodrilus and Ophryotrocha sp., respectively, and at 9.9 mg/l South Louisiana crude to both species.

The 96 hr LC_{50} values reported by Rossi et al., (1976) fall within the range of values determined in this study. Albeit, Rossi and associates used C. capitata from the same stock colony as was utilized in the present study and employed similar experimental procedures, the 96 hr LC_{50} values of 2.3 and 12.0 mg/l are considerably less than the values of >8.7 and >19.8 for No. 2 fuel oil and South Louisiana crude, respectively, reported herein. The difference may be accounted for by the use of Instant Ocean seawater by Rossi et al., (1976) as opposed to natural seawater in these experiments. Evidence is beginning to accumulate regarding the binding of natural organics with pollutants, thus reducing the toxicity of the solution.

The difference in the concentration of oils between the 96 hr LC_{50} and statistical significant suppression of reproduction in these polychaetes was small (0.5 order of magnitude or less) in comparison to the over two orders of magnitude found in N. arenaceodentata exposed to chromium (Oshida et al., 1976). This difference between oils and chromium may be attributed to the fact that the oils are naturally occurring organic materials formed of many elements whereas chromium is not. Rossi, et al., (1976) suggested that oils could be sequestered in polychaete worms rendering the compounds less toxic. Orton (1925) observed Ophryotrocha puerilis feeding on hardened oil, but it was not known whether or not is was able to utilize the oil as a source of energy. Lee (1977) discusses the recent evidence regarding the capability of polychaetes to metabolize petroleum hydrocarbons.

This is the first time that the effect of oils has been studied over an entire life cycle in a polychaete. It would be of interest to determine if such a slight difference between the 96 hr LC_{50} and a significant reduction in reproductive capacity exists, not only in other species of polychaetes, but other organisms as well. The value of such studies would be increased if hydrocarbon accumulation were also examined.

Acknowledgements

This research was supported by Research Grant R-800962 from the U.S. Environmental Protection Agency.

References

Anderson, J. W., J. M. Neff, B. A. Cox, H. E. Tatem, and G. M. Hightower, Characteristics of dispersions and water-soluble extracts of crude and refined oils and their toxicity to estuarine crustaceans and fish, Marine Biol. 27, 75-88 (1974).

George, J. D., The effects of pollution of oil and oil-dispersants on the common intertidal polychaetes, Cirriformia tentaculata and Cirratulus cirratus, J. Appl. Ecol. 8, 411-420 (1971).

Lee, R. F., Accumulation and turnover of petroleum hydrocarbons in marine ogranisms. This volume (1977).

Litchfield, J. T. and F. Wilcoxon, A simplified method for evaluating dose effect experiments, J. Parmac. exp. the. 96, 99-113 (1949).

Moore, S. F. and R. L. Dwyer, Effects of oil on marine organisms; a critical assessment of published data, Water Res. 8, 819-827 (1974).

Orton, J. H., Possible effects on marine organisms of oil discharged at sea, Nature, Lond. 115, 910-911 (1925).

Oshida, P. S., A. J. Mearns, D. J. Reish, and C. S. Word, The effects of hexavalent and trivalent chromium on Neanthes arenaceodentata (Polychaeta: Annelida), So. Calif. Coastal Water Res. Project. TM 225.

Rossi, S. S. and J. W. Anderson, Toxicity of water-soluble fractions of No. 2 fuel oil and South Louisiana crude oil to selected stages in the life history of the polychaete, Neanthes arenaceodentata, Bull. Environ. Contam. & Toxic. 16, 18 (1976).

Rossi, S. S., J. W. Anderson and G. S. Ward, Toxicity of water-soluble fractions of four test oils for the polychaetous annelids, Neanthes arenaceondentata, Bull. Environ. Contam. & Toxic. 16, 18-24 (1976).

Straughan, D., Breeding and larval settlement of certain intertidal invertebrates in the Santa Barbara chennal following pollution by oil. pp. 223-224 In: D. Straughan (ed.) Biological and Oceanographical Survey of the Santa Barbara Channel Oil Spill. Allan Hancock Foundation, University of Southern California, Los Angeles, (1971).

Wilcoxon, R., Handleiding voor de Toets van Wilcoxon. Mathematical Center, Amsterdam, Rept. S176 (M65) (1955).

CHAPTER 18

CYTOLOGICAL DAMAGE IN MERCENARIA MERCENARIA EXPOSED TO PHENOL

C. R. Fries and M. R. Tripp

University of Delaware
Newark, Delaware 19711

Abstract

Adult clams were exposed to various concentrations (1, 10, 100, 1000, 10000, 25000, 50000 parts per billion) of phenol in artificial seawater (25 °/oo) for 24 hours. Control animals were placed in artificial seawater only. Gills, gut, digestive gland and blood cells (hemocytes) were damaged. Basophilic tissue staining was evident at the lower concentrations of phenol; moderate necrosis and sloughing of ciliated epithelial layers were seen at higher concentrations. Blood sinuses were distended and contained precipitated hemolymph. In the gill, only the chitinous supporting rods remained at 50000 PPB phenol.

Electron microscopy also shows damage to epithelial cells. Cell and nuclear membranes remain intact at low concentrations; intracellular organelles (lysosomes, mitochondria, etc.) are in various stages of disintegration. At 10000 PPB cell membranes rupture. Hemocytes show extensive intracellular damage; at 1000 PPB and above only the hyaline cells remain intact.

Key words: Phenol, cytology, Mercenaria mercenaria, epithelial cells, histology, electron microscopy.

Introduction

The equilibrium between a host and its parasites is the result of the interaction of the parasite's invasive properties and the host's defenses. If the host is vertebrate it can respond with nonspecific cellular and chemical factors as well as specifically induced humoral immunoglobulins. Invertebrates, on the other hand, do not have the capacity to produce immunoglobulins. Molluscan defense mechanisms involve epithelial barriers, non-immunoglobulin humoral factors and wandering cells which can encapsulate or phagocytose the invading microbe and digest it intracellularly (Tripp, 1970, 1974).

Physiological mechanisms (homeostatic) exist to keep the internal environment of an animal in a state of functional equilibrium. Defense mechanisms are examples of homeostatic mechanisms. Stress, be it physically, biologically, psychologically, or chemically induced, may disturb that equilibrium. Disease may occur when the invading microbe directly (i.e. toxin) immobilizes the defense system or the defense system itself fails. At times stress may alter the defense system temporarily so that an opportunistic microorganism may cause disease whereas normally it could not. For example, Jeffries (1972) reported the susceptibility of Mercenaria mercenaria to Polydora sp., a polychaete rarely associated with this clam, following environmental hydrocarbon exposure.

In the environment polluting hydrocarbons appear to act as chemical stressors. Surveys show changing sizes of animal populations (Jeffries, 1972, and Mackin, 1961) in the presence of hydrocarbons. Toxic effects are apparent in individual animals. Blumer et al. (1970) reported aromatic fractions incorporated into the general lipid pools of Crassostrea virginica and Aequipecten irradians. Lee et al. (1972) found the gill of Mytilus edulis to be the most important organ in the uptake of petroleum hydrocarbons; long-term effects were seen only in the "gut" (defined as all but the mantle, gill, and adductor muscle). Jeffries (1972) has shown "tarlike irritants" from hydrocarbon pollution collecting in the epithelial tissue with long-term effects in the renal sac and amoebocytes of chronically exposed clams accumulating tarry materials; no correlation with defense was shown.

The toxicity of oil to molluscs varies with the type of oil, concentration, length of exposure, variable environmental parameters and the physiological condition of the mollusc. Blumer et al. (1970) has indicated there is no hydrocarbon detoxification by Crassostrea or

Aequipecten. Stegeman and Teal (1973) found that C. virginica could undergo partial depuration when placed in fresh sea water; aromatic fractions were retained. Aryl-hydroxylase has been found essentially lacking in oysters (C. virginica) by Lee (1976) and Anderson (1976, personal communication). Lee et al. (1972) concluded that Mytilus edulis could not metabolize hydrocarbons and that most were purged. Carlson (1972) showed that Mercenaria was not able to detoxify several insecticides but could reduce p-nitrobenzoic acid to p-aminobenzoic acid.

The defense system of M. mercenaria consists primarily of structural (epithelial) barriers and wandering phagocytic cells. This report describes cytological damage in clams exposed to phenol (an aromatic hydrocarbon). Detailed studies concerning changes in phagocytic ability and hemolymph components will be published later.

Materials and Methods

Adult clams (M. mercenaria) were obtained from Rehoboth Bay, Delaware in October 1974 and April 1975. Experiments were conducted within 24 hours of collection of the animals. Groups of six animals with notched shells were placed in glass aquaria containing 2 liters of phenol solutions in aerated artificial sea water ($25^{o}/oo$) at $20^{o}C$. Initial phenol concentrations were 0, 1, 10, 100, 1000, 10000, 25000, and 50000 parts per billion (ppb). After 24 hours those clams which were to be examined by light microscopy were placed in Bouin's fixative. Four hours later the animals were shucked and transverse sections (0.5 cm. thick) were cut from each clam to include digestive gland, gonad, and heart. These pieces were fixed for an additional 24 hours in fresh Bouin's then embedded in Paraplast by routine histological procedures. Sections (10µm) were stained with Harris alum hematoxylin and eosin.

Tissue for electron microscopy was minced (1 mm pieces) and fixed in 3% glutaraldehyde in 0.2 M sodium cacodylate buffer containing 3mM $CaCl_2$ for one hour at $4^{o}C$. Post-fixation involved 1% osmium tetroxide in 0.2M sodium cacodylate buffer containing 3mM $CaCl_2$ (one hour) and 0.5% uranyl acetate (12 hours). The tissue was dehydrated and embedded in Epon. Silver sections were stained with lead citrate and uranyl acetate. Sections were viewed and photographed with a Zeiss EM-9.

Hemolymph was drawn from the heart with a 26G needle attached to a tuberculin syringe. The hemolymph was placed in a conical glass centrifuge tube containing an equal volume of 3% glutaraldehyde in 0.2M sodium cacodylate buffer containing 3mM $CaCl_2$. After centrifugation, the resulting pellet was processed as above.

Blood cells were also processed for light microscopy. The hemolymph was collected from the heart as described above. Large drops of hemolymph were placed on clean glass coverslips and allowed to adhere to the glass for 1/2 hour. Excess hemolymph was rinsed off with sea water. The cells were fixed in methanol, stained in Giemsa, and then mounted in Permount for examination.

Results

Although all tissues in the higher concentrations showed some damage, gill, gut, digestive gland and blood cells were found to exhibit the greatest amount.

Normal gill filaments with ciliated epithelium surround a water tube and contain blood sinuses (Fig. 1). Clams treated with phenol (1000 ppb) have filaments with sloughing epithelial cells (Fig. 2) and precipitated hemolymph (Fig. 3). At the highest concentrations tested (25,000 and 50,000 ppb) general disintegration of the demibranch occurs with only the chitinous rods retaining structural integrity (Figs. 4 and 5).

Electron micrographs of normal gill epithelial cells reveal a nucleus with peripheral chromatin, nucleolus, cytoplasm with typical mitochondria, electron dense bodies, Golgi apparatus, vesicles, basal bodies, rootlets, microvilli and other parts of cilia. After exposure to intermediate phenol concentrations loss of cell components is evident (Figs. 7 and 8); at higher concentrations membranes are lysed (Fig. 9).

The typical organization of normal digestive gland tissue (Fig. 10) is disrupted by phenol treatment (Fig. 11). Electron microscopy of normal cells shows considerable endoplasmic reticulum (mostly vesicular) and Golgi apparatus (Fig. 12). Digestive gland cells of phenol-treated (100 ppb) clams contain altered endoplasmic reticulum as well as Golgi; mitochondria appear normal (Fig. 13).

Normal hindgut (Fig. 14) consists of ciliated epithelial cells. The hindgut undergoes sloughing of these epithelial cells following phenol (10000 ppb) treatment (Fig. 15).

Normal hemocytes capable of phagocytosis possess a typical nucleus and contain phagocytosed yeast with vacuoles (Fig. 16). Following phenol treatment (50000 ppb) massive lysis of hemocytes was noted (Fig. 17). Ultrastructure of hemocytes from phenol-treated clams shows progressive loss of organelles (Figs. 19 and 20) when compared with normal (Fig. 18).

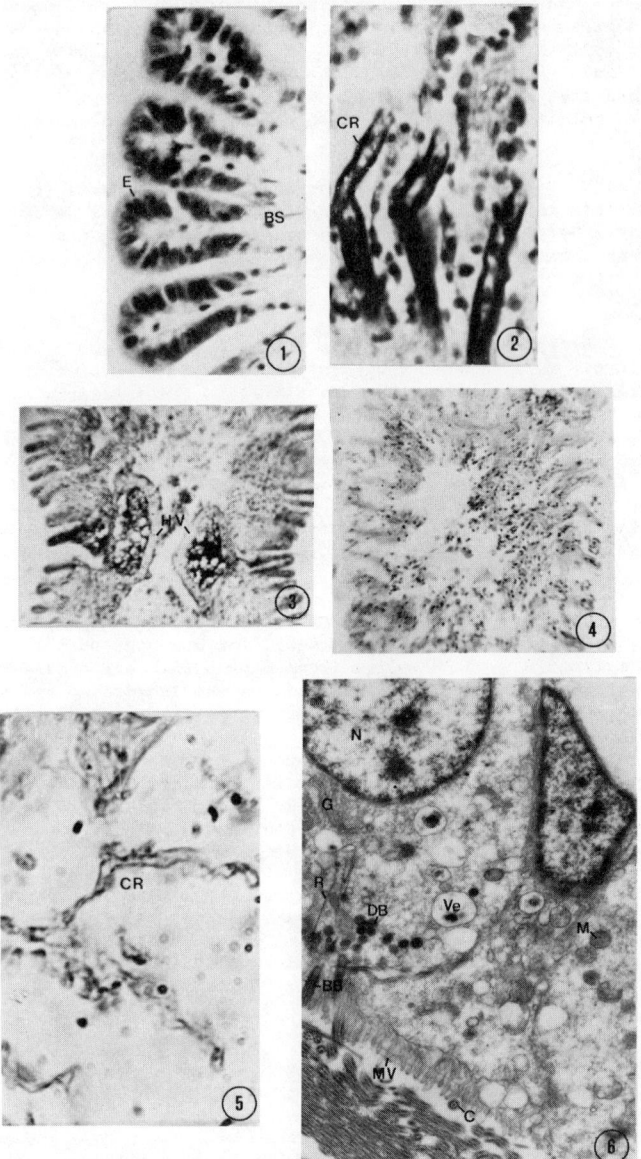

Plate I

Fig. 18.1 Normal Gill. Light Micrograph (LM; 750x). Filament with ciliated epithelial (E) cells surrounding a blood sinus (BS).

Fig. 18.2 Treated Gill: 1000 ppb. (LM; 750x). Epithelial cells at the tips of the filaments are sloughed off. Chitinous rods (CR) are apparent.

Fig. 18.3 Treated Gill: 10000 ppb. (LM; 140x). Damaged demibranch. Hemolymph vessels (HV) contain precipitate.

Fig. 18.4 Treated Gill: 25000 ppb. (LM; 140x). General disintegration of the demibranch.

Fig. 18.5 Treated Gill: 50000 ppb. (LM; 750x). Chitinous rods (CR) and little connective tissue remain.

Fig. 18.6 Normal Gill: Electron Micrograph (EM; 9000x). Epithelial cells from the base of a filament. The nucleus of the epithelial cell contains mostly peripheral chromatin. A nucleolus is evident in most cells. The cytoplasm contains mitochondria (M), electron dense bodies (DB), Golgi (G) apparatus, numerous vesicles (Ve) as well as basal bodies (BB), rootlets (R), microvilli (MV) and other parts of the cilia (C).

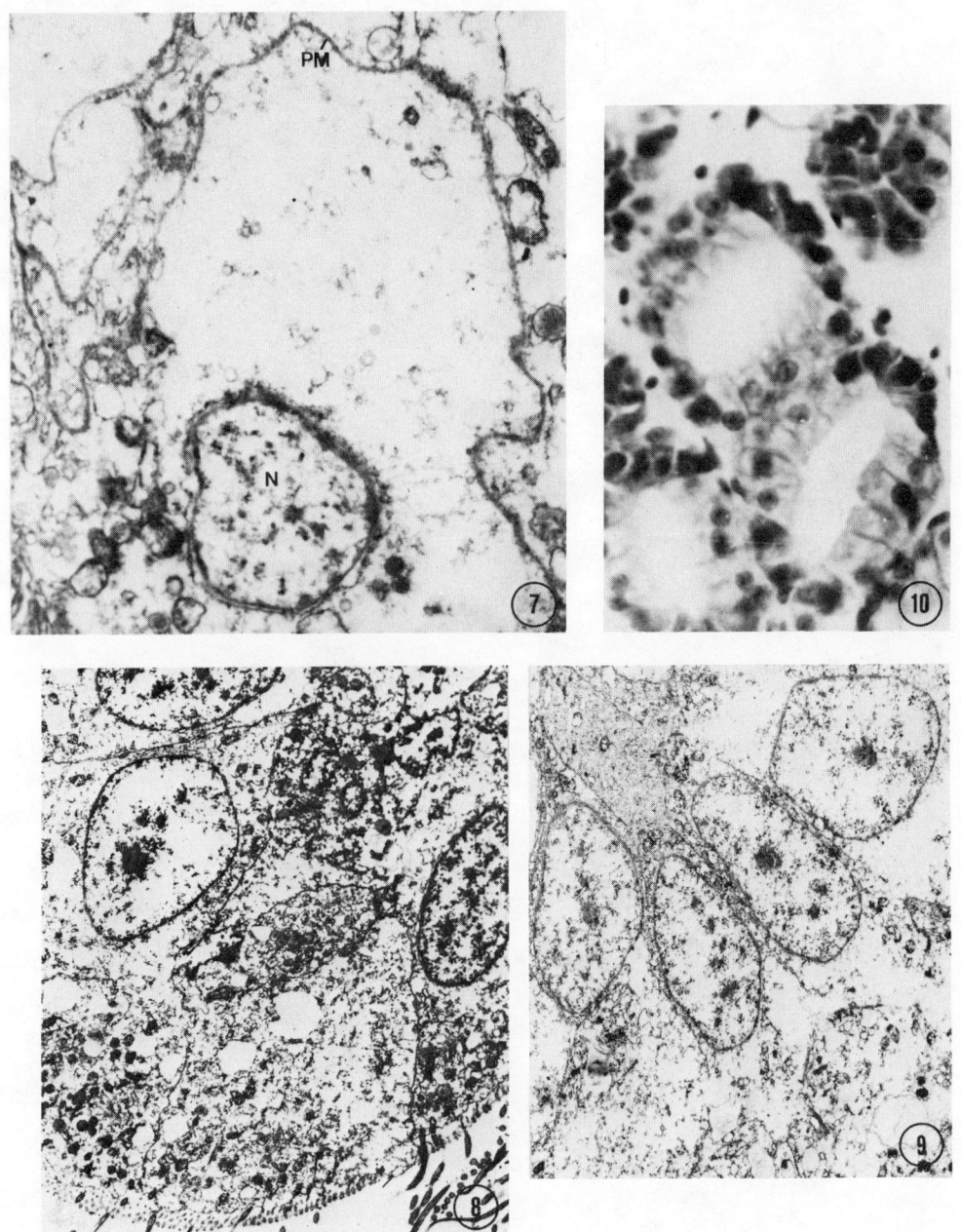

Plate II

Fig.18.7 Treated Gill: 1000 ppb. (EM; 17500x). Small epithelial cell at the base of a filament. Nucleus (N) and plasma membranes (PM) remain intact. Cytoplasm essentially devoid of components; remnants of what were probably mitochondria can be seen.

Fig.18.8 Treated Gill: 100 ppb. (EM; 4500x). Cytoplasm of the epithelial cells at the tip of the filament shows a general loss of cytoplasmic components; mitochondria and cell membranes remain intact.

Fig.18.9 Treated Gill: 10000 ppb. (EM; 4500x). Disintegration of epithelial cells at the tip of the filament. Cell membranes have lysed; few cytoplasmic components are seen.

Fig.18.10 Normal Digestive Gland. (LM; 750x). Typical tubules consisting of epithelial cells.

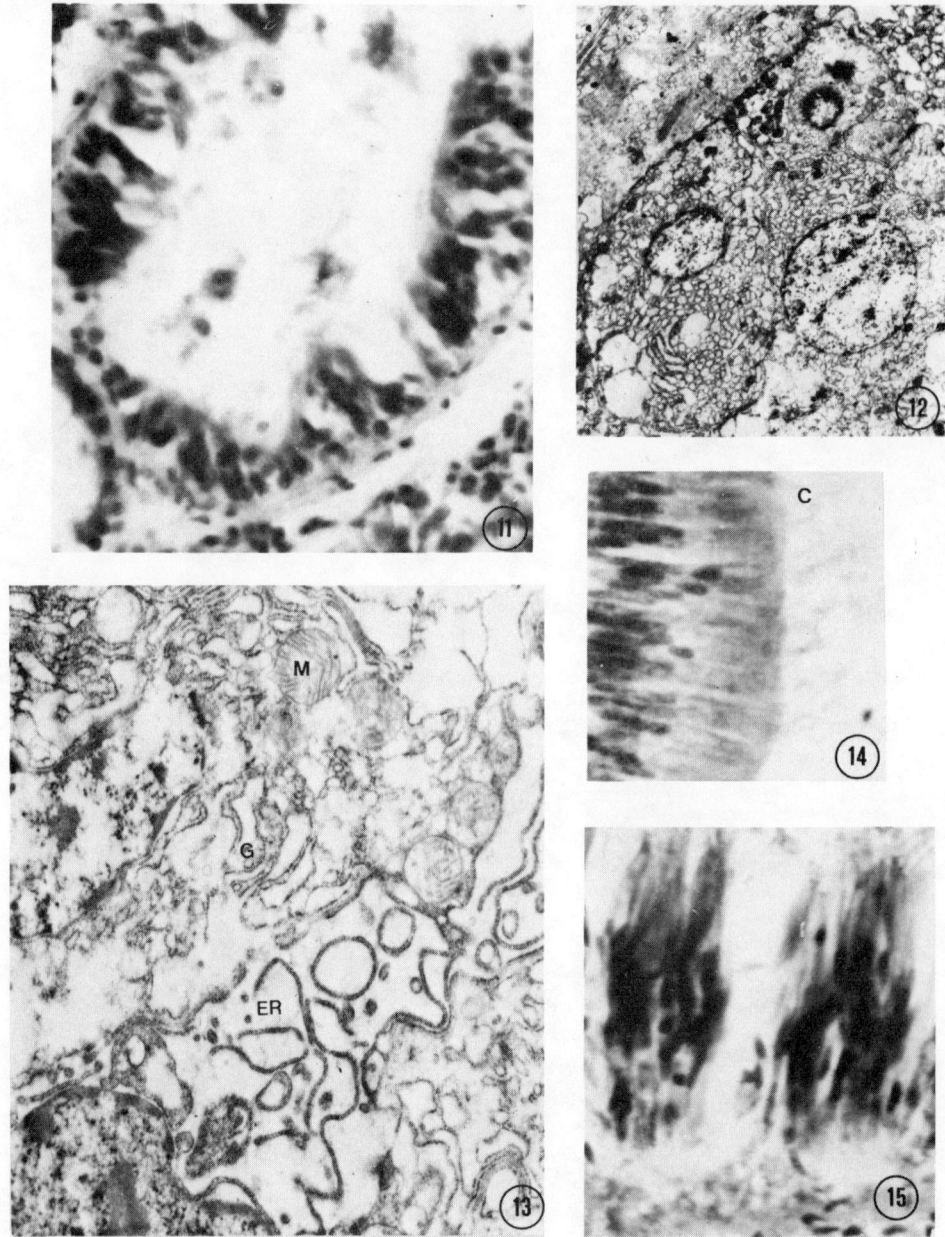

Plate III

Fig. 18.11 Treated Digestive Gland: 10000 ppb. (LM; 750x). Disorganization of the epithelium of a digestive tubule.

Fig. 18.12 Normal Digestive Gland: (EM; 4500x). The edge of normal digestive tubule. Epithelial cells contain great quantities of endoplasmic reticulum (vesicular form predominently and Golgi.

Fig. 18.13 Treated Digestive Gland: 100 ppb. (EM; 17500x). Two tubule (epithelial) cells. The endoplasmic reticulum (ER) is altered as is the Golgi (G) in the adjacent cell. Only remnants of the cell membrane can be seen. The mitochondria (M) appear normal.

Fig. 18.14 Normal Hindgut: (LM; 750x). Typical ciliated (C) columnar epithelium lining the lumen of the gut.

Fig. 18.15 Treated Hindgut: 10000 ppb. (LM; 750x). Separation (sloughing) of epithelial cells from the adjacent connective tissue.

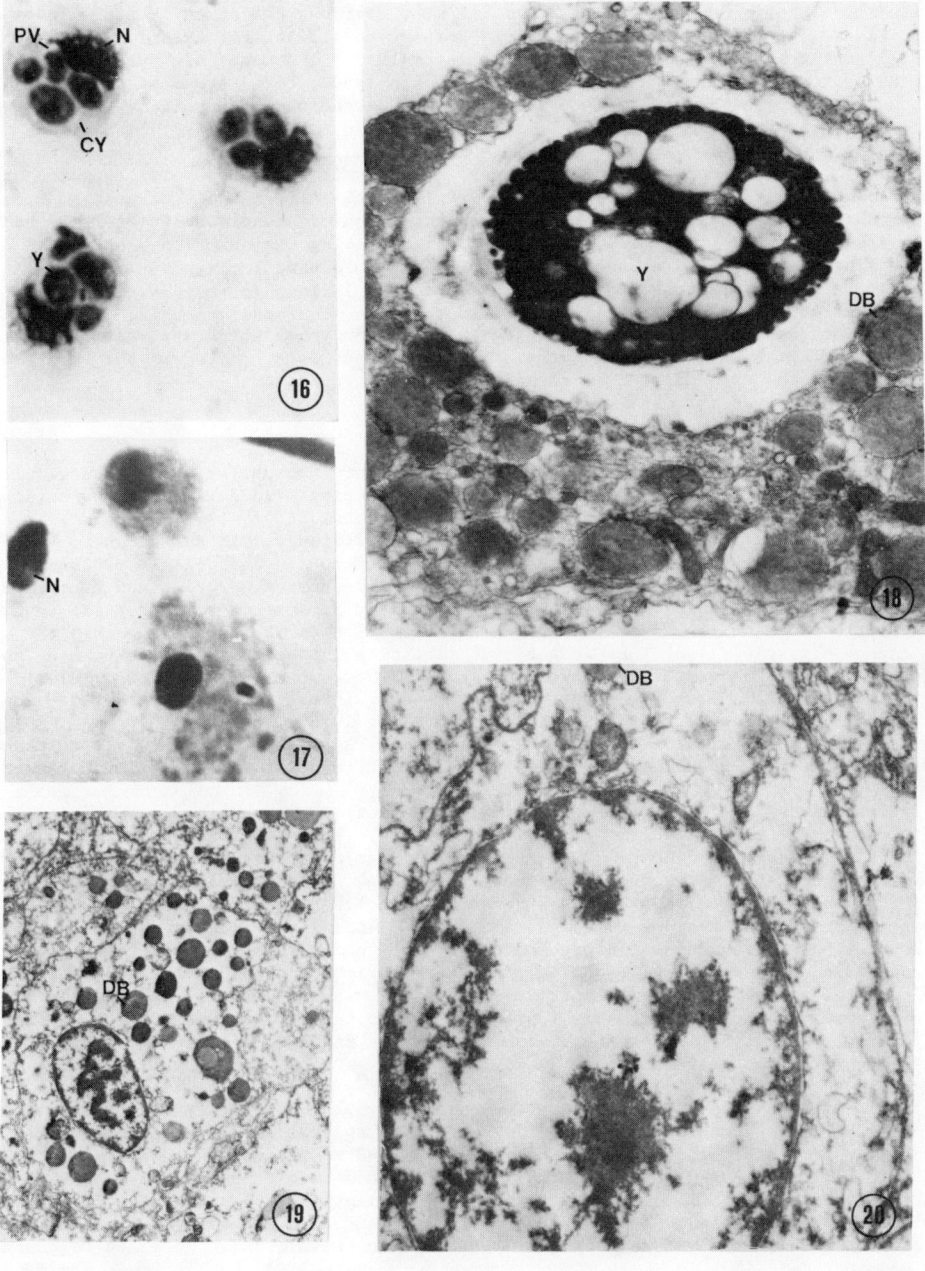

PLATE IV

Fig. 18.16 Normal Hemocytes: (LM; 1850x). Cells have phagocytosed yeast (Y). Nucleus (N), cytoplasm (CY) and phagocytic vacuoles (PV) surrounding the yeast are apparent.

Figure 18.17 Treated Hemocytes: 50000 ppb. (LM; 1850x). Disintegrating hemocytes. Many have lysed; only the nuclei (N) remain.

Fig.18.18 Normal Hemocytes: (EM; 17500x). Hemocyte has phagocytosed a yeast (Y) particle. Note numerous electron dense bodies (DB). See Cheng and Foley (1975).

Fig. 18.19 Treated Hemocytes: 100 ppb. (EM; 4500x). Granulocytes from heart hemolymph. Most cytoplasmic organelles have disintegrated; electron-dense bodies (DB) remain.

Fig.18.20 Treated Hemocytes: 10000 ppb. (EM; 17500x). Granulocyte in gill tissue. Few cytoplasmic organelles remain. Electron-dense bodies (DB) remain.

Phenol has been shown to induce cytological damage in Mercenaria mercenaria particularly in tissues with epithelial cells. The gill is the most obviously affected. At the lower concentrations (1 → 100ppb) nuclei and cell membranes remain intact; cytoplasmic components begin to disintegrate. At moderate concentrations (1000 → 10000ppb) although the demibranch still retains its basic structure, cell membranes are lysed, cilia are lost and general sloughing of the epithelial cells is evident. At the highest concentrations (25000 → 50000ppb) the demibranch loses its integrity; in general, only the chitinous rods remain. In the digestive gland and hindgut the response is basically the same as in the gill. At the lower concentrations (1 → 100ppb) cytoplasmic components disintegrate; at the higher concentrations (>10000ppb) sloughing of epithelial cells and disintegration of the tissue occurs. In general, the greater the concentration of phenol the greater the damage to all tissues observed. Care must be taken in interpreting the observed "damage" in tissues from experimental animals; one cannot ignore possible phenol-fixative induced artifacts. Experiments are needed to determine if phenol alone is responsible for lysis of cell membranes or if phenol initially disrupts organelle membranes which in turn are responsible for the release of enzymes (i.e. lysosomal enzymes) which synergistically cause the disintegration of the cell membrane. The latter is suggested by the early loss of cytoplasmic components as seen in the electron micrographs.

Discussion

Physiological functions are largely dependent on structure; morphological (cytological) changes usually imply altered (frequently abnormal) function. In filter feeding bivalves the gill is involved in food sorting and respiration (Galtsoff, 1964). Loss of cilia and epithelial cells from filaments may reduce filtering efficiency. If damage is such that the hemolymph sinuses are ruptured loss of hemolymph may occur. In the digestive gland the epithelial cells of the tubules and ducts are involved in both absorption and secretion. Loss of cell membranes, Golgi, and endoplasmic reticulum may affect both of those functions. The phagocytic blood cells wander throughout the open circulatory system and into the tissues. In the digestive gland they phagocytose particles in the lumen of the tubules then migrate into the tissue itself. The phagocytic cells are the sites of intracellular digestion; their damage is apparent whether in the tissue or in the circulatory system. The hindgut functions in removal of metabolic wastes; sloughing of the epithelium may impede this process. In general, loss of feeding, metabolism, and waste removal would be expected following the observed phenol-induced damage to the gastrointestinal system of Mercenaria.

With the loss of epithelial barriers an animal is potentially susceptible to any invading microorganism. The microorganism need not be pathogenic; many microorganisms can become opportunistic in a stressed host. A phenol-damaged clam may represent such an animal. Water drawn in the inhalant siphon flows over the gills. Microorganisms may have the opportunity to penetrate the open wounds induced by phenol. Microorganisms also may be able to penetrate at any point along the digestive tract. Once inside the animal the phagocytic cells would treat an invader as foreign material to be phagocytosed and disposed of. Foley and Cheng (1974) have shown all blood cells in Mercenaria to be phagocytic. Damage to these cells indicates probable reduced clearance of invading microorganisms. The precipitate in the blood vessel of the gill probably represents precipitated hemolymph (Fries and Tripp, unpublished results). If soluble hemolymph factors are involved in defense, they may not be available after phenol treatment.

Cytological damage as described here may indicate abnormal metabolic function (possibly accompanied by lowered resistance) as well as abnormal epithelial barriers, phagocytic cells and hemolymph factors needed for defense. The end result may be an animal more susceptible to infectious disease. Animals severely damaged with phenol have been observed to live more than five weeks after phenol treatment with no mortality (Fries and Tripp, unpublished results). The questions "Are these treated animals more susceptible to infection in their natural environment?" and "How do these animals continue to function?" remain unanswered.

Acknowledgements

This research was funded by grants from NOAA, NSF and DIMER.

References

Blumer, M., G., Souza, and J. Sass, Hydrocarbon pollution of edible shellfish by an oil spill, Mar. Biol. 5, 195 (1970).

Carlson, C., Detoxification of foreign organic compounds by the quahog, Mercenaria mercenaria, Comp. Biochem. Physiol. 43B, 295 (1972).

Cheng, T. C., and D. A. Foley, Hemolymph cells of the bivalve mollusc Mercenaria mercenaria: an electron microscopical study, J. Invert. Pathol. 26, 341 (1975).

Foley, D. A., and T. C. Cheng, A quantitative study of phagocytosis by hemolymph cells of the pelecypods Crassostrea virginica and Mercenaria mercenaria, J. Invert. Pathol. 25, 189 (1975).

Galtsoff, P., The American Oyster, Crassostrea virginica (Gmelin), Fishery Bull. Fish Wild. Serv. U.S. Gov't. (1964).

Jeffries, H. P., A stress syndrome in the hard clam Mercenaria mercenaria, J. Invert. Pathol. 20, 242 (1972).

Lee, R. F., E. Furlong, and S. Singer, Detoxification systems in marine invertebrates. Aryl hydrocarbon hydroxylase from tissues of the blue crab Callinectes sapidus and the polychaete Nereis sp., Proceedings of Biological Effects Program Workshop, College Station, Texas, May 16-19, 1976 (1976).

Lee, R., R. Sauerheber, and A. A. Benson, Petroleum hydrocarbons: uptake and discharge by the marine mussel Mytilus edulis, Science 177, 344 (1972).

Mackin, J. G., Studies on oyster mortality in relation to natural environments and to oil fields in Louisiana. Inst. Mar. Sci., Univ. of Texas 7, 3 (1961).

Stegeman, J. J., and J. M. Teal, Accumulation, release and retention of petroleum hydrocarbons by Crassostrea virginica, Mar. Biol. 22, 37 (1973).

Tripp, M. R., Defense mechanism of mollusks, J. Reticulo. Soc. 7, 173 (1970).

Tripp, M. R., Molluscan immunity, Ann. N.Y. Acad. Sci. 234, 23 (1974).

CHAPTER 19

INTERACTIVE EFFECTS OF TEMPERATURE, SALINITY SHOCK AND CHRONIC
EXPOSURE TO NO. 2 FUEL OIL ON SURVIVAL, DEVELOPMENT RATE AND
RESPIRATION OF THE HORSESHOE CRAB, LIMULUS POLYPHEMUS

R. B. Laughlin, Jr. and J. M. Neff

Department of Biology, Texas A&M University
College Station, Texas   77843

Abstract

Eggs and the first three instars of the horseshoe crab, Limulus polyphemus were exposed to 0, 5, 10, 25 and 50% water-soluble fractions (WSF) of No. 2 fuel oil. Rearing salinity and temperatures were 32 ppt and 20°, 25°, and 30°C, respectively. Hatching success was generally good at all temperatures and WSF concentrations, except 50% WSF, 25°C. Survival of larvae decreased with decreasing temperature and increasing WSF concentration. The significance of differences in growth and development rates were difficult to determine because of high variability in these traits. These appear to be the least sensitive indicators of petroleum hydrocarbon stress in L. polyphemus. Respiration rates for control and WSF-exposed first-tailed staged animals from each temperature were measured at salinities of 32, 20 and 10 ppt. The respiration rates of controls decreased with decreasing salinity and temperature. The respiration rate of oil-exposed animals was significantly higher than that of controls at nearly all salinity/temperature combinations. The interaction of salinity and WSF exposure on respiratory rate was highly significant. The data suggest that the ability to adapt to new environmental conditions rather than maintenance of any given equilibrium metabolic state is an important parameter to consider when evaluating the impact of petroleum on estuarine organisms.

Key words: Limulus polyphemus, No. 2 fuel oil, larval bioassay, respiration rate, salinity stress

Introduction

There has been a growing concern about the possible effects of petroleum and petrochemicals on marine organisms and ecosystems because of the large quantities of these materials that are introduced into estuaries and near shore oceanic waters from natural and anthropogenic sources (Wilson et al., 1974/National Academy of Science, 1974). Recent investigations in our laboratory have dealt with toxicity to estuarine animals of various crude and refined oils (Anderson et al., 1974a), uptake and release of petroleum hydrocarbons (Anderson et al, 1974b; Neff et al., 1976b) and the elucidation of physiological responses which might be sensitive indicators of sublethal pollutant stress (Neff et al., 1976a). Respiration and larval growth and development have been cited as sensitive indicators of petroleum hydrocarbon-induced stress.

This paper deals with the effects of temperature, salinity shock and exposure to water-soluble fractions (WSF) of No. 2 fuel oil on survival, development rate, growth and respiration of the horseshoe crab, Limulus polyphemus. Respiratory rates of first-tailed stage larvae were determined at the rearing salinity and under conditions of osmotic shock. This strategy gives a measure of the steady-state respiratory response to WSF exposure as well as an indication of the metabolic cost of salinity acclimation by animals exposed to petroleum hydrocarbons.

Materials and Methods

Limulus polyphemus eggs were collected in the Indian River south of Rockledge, Florida. The eggs were allowed to remain in the sand 24 hours after the pair of adults were observed depositing them. They were then removed and shipped via air freight to Texas.

Embryos and juveniles were exposed continuously throughout development to water-soluble fractions of No. 2 fuel oil (38% aromatics, API reference oil III). The water-soluble fractions were prepared by the method of Anderson et al., (1974a). One volume of No. 2 fuel oil was gently stirred over 9 volumes of 32 ppt Instant Ocean (Aquarium Systems, Inc.) for 20 hrs. The resulting 100% WSF was diluted to exposure concentrations of 5, 10, 25, and 50%. The control consisted of 32 ppt Instant Ocean.

An experiment was run concurrently with the oil exposure to determine the rate of disappearance of naphthalenes from the WSF's at the different exposure temperature. Rearing bowls of the same size and containing the same volume of WSF as in the rearing experiments, but without animals, were used at the three temperatures and three WSF concentrations. Periodic samples were analyzed spectrophotometrically for naphthalenes (Neff and Anderson, 1975).

Rearing Methods
Exposure to petroleum hydrocarbons began on the tenth day after the eggs were deposited. Eggs were exposed to WSF in groups of 15 eggs per bowl with 5 bowls per condition. Thus 75 animals were exposed to each condition of temperature and WSF concentration. The eggs and larvae were censused daily for hatches, mortalities and molts. Six days each week, the exposure solutions were replaced. After the eggs hatched, all larvae of a given stage at each condition were kept in the same rearing bowl. Freshly-hatched Artemia nauplii were given to feeding stages. Animals were maintained in incubators at 20°, 25° and 30^oC throughout the experiment.

Neff and Giam (1975) described the development of Limulus as follows: the eggs hatch to the non-feeding trilobite stage, notable for the complete absence of a tail. The trilobite molts to the first stage in which the tail is much shorter allometrically than in the adult form. In the second-tailed stage, the tail length is slightly longer proportionally, and the third-tailed stage has a tail length which allometrically resembles that of miniature adults.

To assess long term effects of WSF exposure, survival by stage, development rate and growth growth were measured. The number of animals which molted to the following stage determined at each condition of WSF exposure and temperature. The day each organism molted to a certain stage was recorded. Data are expressed with day 0 as the day the eggs were deposited. Size is defined as the body length from the forward margin of the prosoma to the tip of the median spine on the posterior edge of the opisthosoma. This length was determined using a Wild M-7 stereo microscope with an ocular micrometer.

Respiration Measurements
When at least 36 animals at each condition had reached the first-tailed state, respiration measurements were performed. For 24 hrs. before respirometry, animals were not fed. Those that had molted within 24 hrs. were not used, and after respirometry, animals were isolated to be sure they did not molt. Data for those which did molt were not used in the analysis.

During the respiration study, separate groups of animals were exposed to a salinity of 32 ppt, 20 ppt or 10 ppt. The animals were reared at 32 ppt so respiration rates for this group relfect a steady-state response at each temperature and WSF combination. Other groups were exposed either to 20 ppt or 10 ppt to assess a salinity stress response. For all temperatures, enough animals survived to determine respiration rates after exposure to 0, 5% and 10% WSF. Only at 30^oC did enough animals survive to determine respiration rates for 25% WSF exposure.

Respiration rates were measured by a potentiometric method employing a Radiometer blood-gas analyzer. First-tailed stage animals were blotted on tissue paper to remove algae or other adhering material from the exosketeton. Then they were put individually into 5 ml all-glass syringes. These were full of air-saturated Instant Ocean of the proper salinity. Care was taken to exclude all air bubbles. An initial pO_2 was determined by injection of an aliquot of water into the blood gas analyzer. The syringe was sealed and put into a water bath adjusted to the appropriate rearing temperature. At the end of one hour, pO_2 was determined again. The process was repeated for the second hour, after refilling the syringe with new air-saturated water. The amount of oxygen used by the animals was computed from the difference in pO_2, the volume of the respirometer (syringe), the barometric pressure and Bunson coefficient for the salinity and temperature of the water. The animals were blotted dry on tissue, and weighted to the nearest 0.1 mg on a Mettler balance. Respiration rate is expressed as ml O_2/g wet weight/hr.

Results

The concentration of total naphthalenes in the WSF of No. 2 fuel oil dropped substantially during the 24 hours between exposure solution changes (Fig. 1). The largest drop in naphthalenes concentration occurred in the first five hours after introduction of the WSF into the bowls. The rate of loss was highest in the 25% WSF and proportionately lower for the 10 and 5% WSF's. Mean percent losses of naphthalenes from the 25, 10, and 5% WSF's in 24 hours were 79, 74 and 73% respectively. At each WSF concentration, initial (five hour) losses of naphthalenes were greatest at 30°C, intermediate at 25°C and lowest at 20°C. The data for 25% WSF measured at 5 hrs showed the largest difference among mean concentrations, but the differences were not significant (F=2.89, df=2, 5). After 24 hours, naphthalenes concentrations were very similar, and not necessarily correlated with temperature. At all sampling times, the naphthalenes concentrations differed significantly in the three dilutions of the WSF.

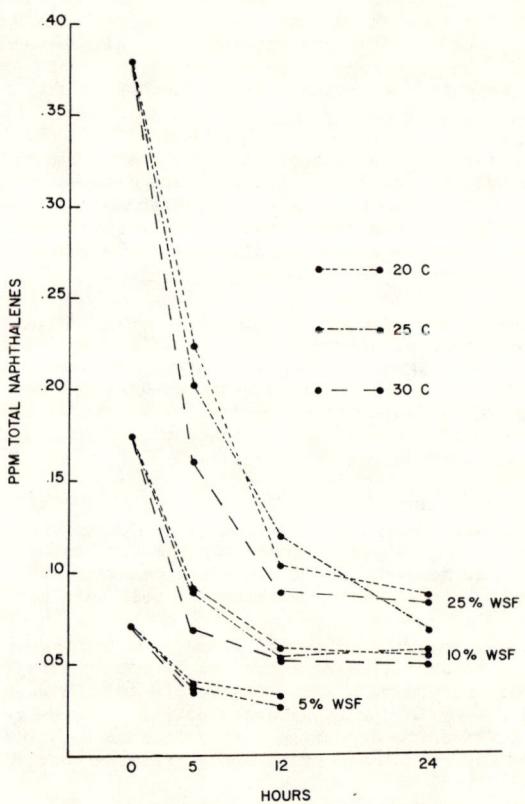

Fig.19.1 Rate of loss of total naphthalenes from three concentration of the WSF of No. 2 fuel oil at 20, 25 and 30°C. Conditions were the same as those in the exposure regime, except that animals were not present in the WSF. Each point is the mean of three determinations. For values at 5 hrs, 25°C, F = 2.89 df = 2,5 ns.

Survival

Survival to the trilobite stage generally decreased with decreasing temperature and increasing WSF concentration at 20 and 25°C (Fig. 2). The nonfeeding stages (egg and trilobite) were particularly susceptible to these parameters. No eggs hatched 25°C, 50% WSF, and hatching success was better than 65% in both 25 and 50% WSF but all the trilobites died before molting to the first-tailed stage. Those exposed to the 50% WSF survived about two months, but were moribund throughout this period, responding weakly and spasmodically to prodding with forceps. Individuals exposed to 25% WSF at 20°C behaved in a similar manner but survived for about four months. At 20 and 25°C, survival of the control, 5 and 10% WSF exposure groups was better than 50% for development to the second-tailed stage (20°C) or third-tailed state (25°C). At 20°C, more than 65% of the eggs hatched at all WSF exposure concentrations. Survival to the third-tailed stage was similar for the controls, 5 and 10% WSF exposure groups (better than 55%). Groups exposed to 25 and 50% WSF showed survival to the third-tailed stage of 35 and 20%, respectively.

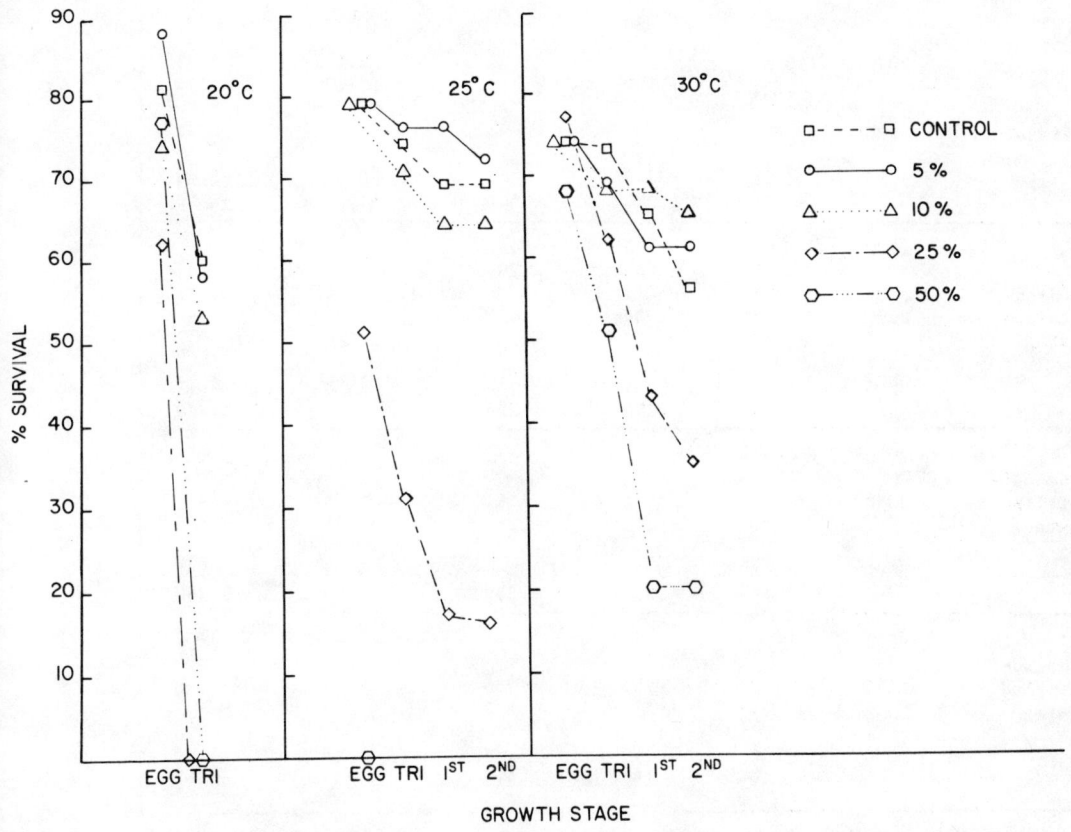

Fig.19.2 Survival of <u>Limulus polyphemus</u> during chronic exposure to WSFs of No. 2 fuel oil at 20, 25 and 30°C and 32 ppt salinity. Points represent the percentage of animals which survived the indicated stage and molted to the succeeding one.

Size
Temperature alone had no effect on mean size of individuals in any instar at any WSF concentration. With respect to interactions, temperature and WSF concentration had relatively small effect on the mean size of different larval instars of L. polyphemus (Table 1). At any temperature, there was a trend toward decreasing size with increasing WSF exposure concentration. However, animals exposed to 25°C, 25% WSF and 30°C, 50% WSF showed the greatest mean growth within their respective temperature groups by the third-tailed stage.

Development Rate
Temperature had a greater effect than WSF concentration on the development rate of L. polyphemus (Fig. 3). The mean time to hatching and mean duration of larval instars decreased with increasing temperature. At 20°C less than 50% of the crabs had molted to the second-tailed stage in 150 days, whereas surviving crabs at 25 and 30°C were all at least in the third-tailed stage by this time. At all exposure temperatures, mean time of hatching increased slightly with increasing WSF concentrations. This effect was greatest at 25°C, 25% WSF. At 20 and 25°C the mean time to the molt from the trilobite to the first-tailed stage also increased with increasing WSF concentration. This trend was not seen in the 30°C animals. For later molting stages, the trend was reversed and animals exposed to the highest WSF concentrations at 25 and 30°C had mean intermolt periods equal to or less than

TABLE 1. The Size of Limulus Polyphemus by Stage and WSF Concentrations are Shown. Animals at 20°C had not Molted, as a Group, Past the Second-Tailed Stage. No Determinations were made for Trilobites at 25°C. All Lengths are Given in mm.

		Water Soluble Fractions				
		0%	5%	10%	25%	50%
Trilobite	20°C	3.50 ±.225	3.54 ±.180	3.52 ±.158	3.47 ±.217	3.53 ±.144
	25°C	--	--	--	--	--
	30°C	3.48 ±.229	3.46 ±.200	3.44 ±.203	3.34 ±.312	3.43 ±.203
First-tailed stage	20°C	4.13 ±.177	4.05 ±.205	3.96 ±.209	--	--
	25°C	4.34 ±.244	4.34 ±.158	4.18 ±.222	3.96 ±.206	--
	30°C	4.23 ±.258	4.19 ±.290	4.14 ±.254	4.22 ±.210	4.19 ±.275
Second-tailed stage	20°C	--	--	--	--	--
	25°C	5.78 ±.129	5.74 ±.193	5.61 ±.282	5.59 ±.344	--
	30°C	5.59 ±.389	5.46 ±.339	5.34 ±.354	5.38 ±.260	5.64 ±.360
Third-tailed stage	20°C	--	--	--	--	--
	25°C	7.52 ±.351	7.31 ±.323	7.36 ±.293	7.75 ±.504	--
	30°C	7.30 ±.646	7.19 ±.139	7.05 ±.410	7.18 ±.343	7.48 ±.382

those of the controls. Differences in development rate were statistically significant at the three temperatures, but were not statistically significant at different WSF concentrations at any temperature.

Respiratory Rate and Salinity Stress

Respiratory rates for the first-tailed stage crabs at the rearing salinity (32 ppt) rose with increasing acclimation temperature (Figure 4a). The Q_{10} for the temperature interval 20-30°C was 2.1 for the control animals. With the exception of animals at 25°C, 5% WSF and 30°C, 25% WSF, all WSF exposure groups had higher mean respiratory rates than the controls at the same temperature.

When animals were transferred from 32 ppt and 20 or 10 ppt S immediately before respiratory measurements were made, the stimulatory effect of WSF exposure was increased especially at 20 and 30°C (Figure 4b, c). The temperature sensitivity of respiratory rates of control animals was reduced by salinity shock. Values for Q_{10} in the temperature range 20-30°C were 1.3 and 1.8 at 20 and 10 ppt S, respectively. At both 20 and 10 ppt S, animals exposed to 5% WSF had greatly elevated mean respiratory rates at 20° and 30°C, with the former rates higher than the latter. At 25°C, mean respiratory rates were elevated only slightly over those of the controls. At an exposure concentration of 10% WSF, mean respiratory rates were higher than those of controls and rose in a nearly linear fashion with increasing temperature in both the 20 and 10 ppt salinity shock groups. The 20-30°C Q_{10} values for the 20 and 10 ppt S groups were 2.1 and 2.8 respectively. Thus, animals exposed to 10% WSF at 20 and 10 ppt S had a temperature sensitivity of respiratory rate similar to that of control animals at the rearing salinity of 32 ppt S, but much higher than those of control animals at 20 and 10 ppt S. Animals exposed to 25% WSF were available in sufficient numbers for respiration studies in only the 30°C group. In these, respiratory rate was highest at 10 ppt S, intermediate at 20 ppt S and lowest at 32 ppt S.

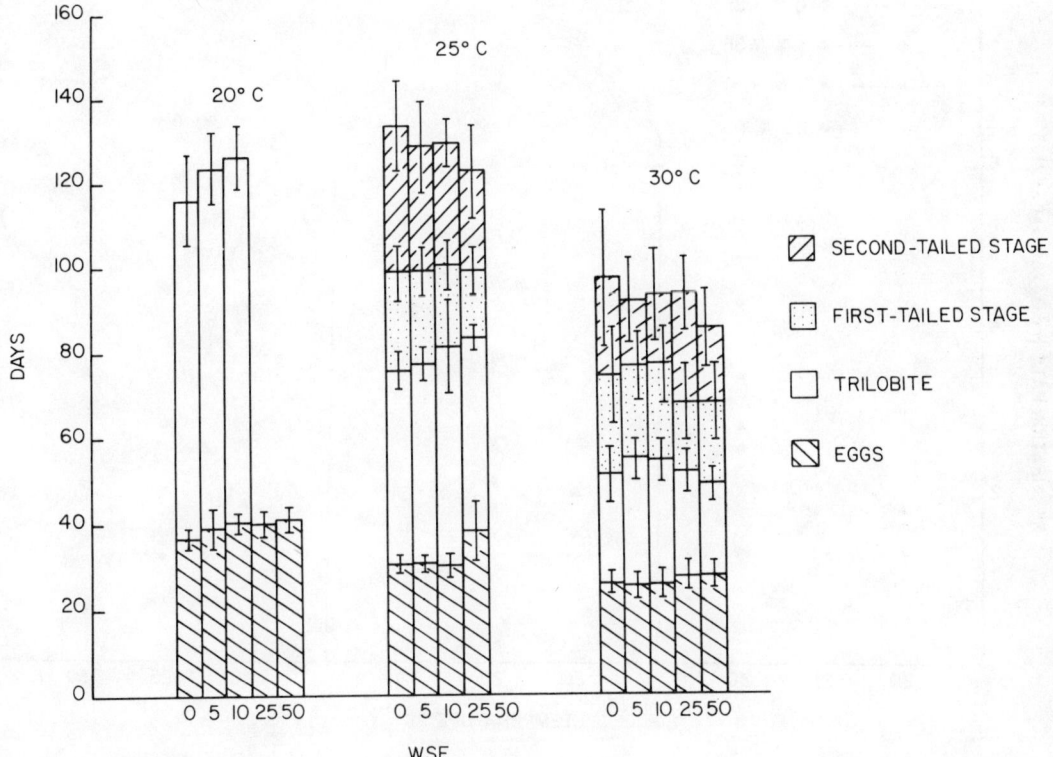

Fig. 19.3 Mean intermolt periods (development rates) of <u>Limulus polyphemus</u> exposed to 5 concentrations of the WSF of No. 2 fuel oil at 3 temperatures at 32 ppt salinity. No data are avilable for 25°C, 50° WSF. Vertical lines represent standard deviations.

An analysis of variance for the respiration data is shown in Table 2. The interaction of salinity and WSF exposure on respiratory rate is highly significant. However, the main effect of temperature and the interaction of temperature and WSF exposure on respiratory rate is not significant at the $P = 0.05$ level.

Discussion

In this discussion we will concentrate on a comparision of steady-state effects of WSF exposure on <u>Limulus</u>, and the ability of exposed animals to make adjustments to new salinity conditions. In the environment, conditions are not usually stable for very long. It may be more important for animals to maintain an ability to cope with changing conditions rather than with any given steady-state.

Survival is the crudest method of assessing the impact of pollutant on the organism. <u>Limulus</u> larvae are more tolerant to petroleum hydrocarbon exposure than larvae of the mud crab, <u>Rhithropanopeus harrisii</u> (Neff <u>et al</u>., 1976a). They are also more tolerant than many adults of marine organisms that have been tested (Anderson <u>et al</u>., 1974a). Temperature had

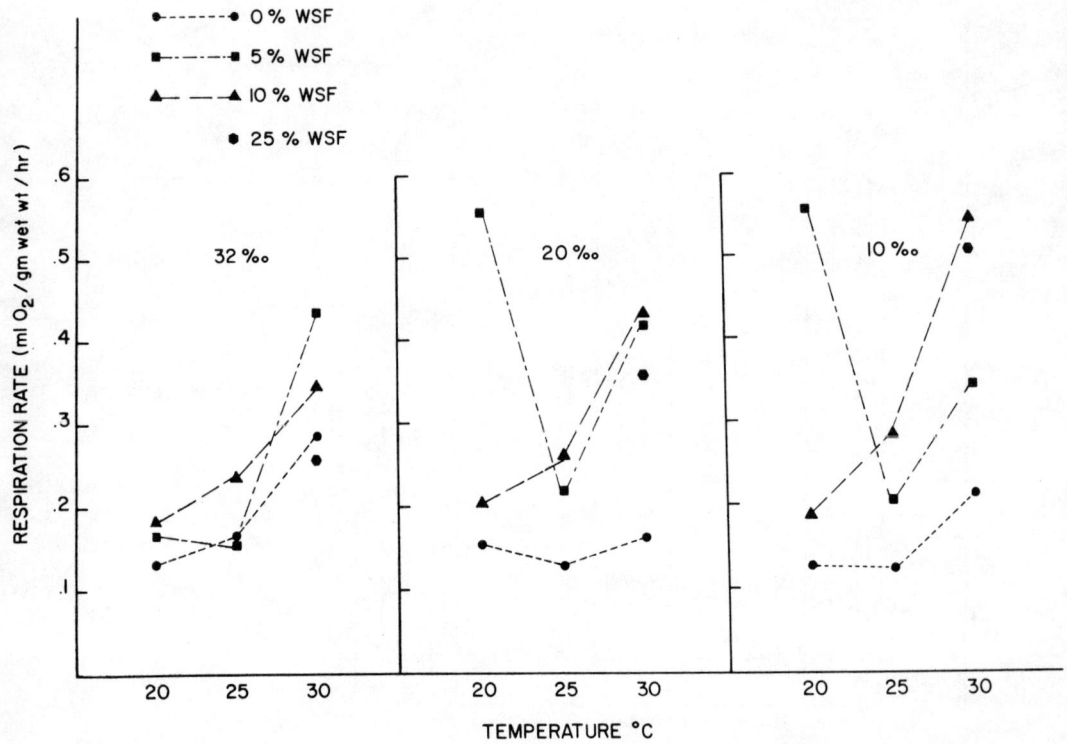

Fig. 19.4 Respiratory rates of first-tailed stage larvae of Limulus polyphemus exposed to 0, 5, 10 and 25% WSF of No. 2 fuel oil at 20, 25 and 30°C at 32 ppt salinity (A) or immediately after transfer to 20 (B) or 10 ppt (C) salinity sea water. Animals exposed to 25% WSF were available in sufficient numbers for respiration measurements at only 20°C.

an effect on the toxicity of petroleum hydrocarbons to Limulus. Sensitivity to petroleum increased with decreasing temperature. Although naphthalenes and presumably other hydrocarbons appeared to be lost from the WSF's more rapidly at high than at low temperatures, the mean naphthalenes concentrations at different temperatures were not significantly different from each other at each sampling time. Therefore, the influence of temperature on the toxicity of oil to Limulus was not due to differences in actual exposure concentrations. Fucik and Neff (1977) showed that the clams Rangia cuneata and Protothaca staminea accumulated naphthalenes from a WSF more rapidly at low than at high temperatures. Hydrocarbon accumulation data are not available for the eggs, larvae and juveniles of Limulus. However, if an inverse relationship between temperature and hydrocarbon uptake rate also occurred in this species, it could explain the toxicity data obtained.

TABLE 2. Analysis of Variance for Factors Examined Affecting
Mean Respiration Rates of Limulus Polyphemus

P Source	df	F	P>F
Regression	6	20.47	0.001
Error	353		
Corrected total	359		
Interaction Analysis Source	df	F	P>F
T	1	0.23	.6315
S	1	7.76	.0056
WSF	1	6.43	.0177
TxS	1	8.12	.0046
TxWSF	1	0.43	.5111
SxWSF	1	23.56	.0001

T = Temperature; S = Salinity; WSF - WSF of No. 2 Fuel Oil

Both development rate (Epifanio, 1973) and larval size (Laughlin, et al., 1977) have been shown to be sensitive indices of sublethal pollutant stress in crustaceans. However, the high variability in these traits displayed by Limulus limited their usefulness as indices of stress. Mean intermolt periods and mean size of different instars were not significantly affected by any WSF concentration. However, the former was significantly affected by temperature. Carapace dimensions may not be a reliable index of biomass accretion in this species, since, at the molt, the size of the new carapace is influenced by the magnitude of the osmotic influx of water (Baumberger and Olmstead, 1928). Factors affecting water flux rates and osmoregulatory capacity could influence carapace size increments in such a way that they are not a true measure of biomass accretion. The respiration data showed that oil exposed animals had a greater respiratory response than the controls to salinity stress. At the stress salinities used in the present investigation (20 and 10 ppt S) Limulus is a hyperosmotic regulator (McManus, 1969; Robertson, 1970). The data suggest that oil-exposed animals experienced difficulty in osmotic and ionic regulation which might result in an alteration in dimensional increase at the molt. Support for this hypothesis comes from the studies of Morrow et al. (1975), who showed that exposure to crude oil components resulted in an increase in the membrane permeability of the gills in coho salmon Onchcorhynchus kisutch.

Exposure to oil WSF's, even at 32 ppt salinity where Limulus is an osmotic conformer, always caused an increase in mean respiratory rates. A similar respiratory stimulation ocurred in juvenile Limulus during chronic exposure to sublethal concentrations of polychlorinated biphenyls and polychlorinated naphthalenes (Neff and Giam, 1976). Thus respiratory stimulation may represent a generalized stress response in this species. A pollutant-mediated increase in respiratory rate might reduce the proportion of assimilated carbon available for biomass production (net growth efficiency) as has been shown for mussels exposed to extracts of crude oil (Gilfillan, 1975). This would partially explain some of the size differences seen in Limulus chronically exposed to low concentrations of oil.

Our data suggest a nested effects phenomenon of pollutant action. At low WSF concentrations under equilibrium conditions, the response is a generalized stress reaction. In this case, net growth efficiency may have been decreased. At increasingly toxic concentrations, specific systems or functions appeared to be targeted. Here the osmoregulatory capacity during the molt might have been affected.

The usefulness of multifactorial experiment lies in its predictive capacity for animals in the environment. Alderdice (1973) called attention to a "response ridge" often seen in these experiments. That is, a metabolic response rises to a maximum level with increasing exposure to a stress, but continued increase in stress causes a decline in the level of the response due to a finite response capacity. There was a ridge at 10% WSF at 30°C, but it was decreased to 5% WSF at 20°C. No ridge phenomenon was apparent under experimental conditions at 25°C. This temperature was the prevailing environmental temperature for the parent population in the Indian River, Florida. These results indicate that Limulus was more tolerant to WSF exposure and salinity shock at temperatures near the environmental norm than at lower and higher temperatures.

This experiment has shown the Limulus polyphemus is relatively tolerant to petroleum hydrocarbons, but that tolerance decreases with temperature. Growth and development rate were not as good indicators of stress for this specieis as they were for the crustacean larvae which have been examined. While petroleum hydrocarbons did increase the mean respiration rates for animals under constant salinity conditions, acute exposure to dilute media caused a greater increase in respiration rate which was proportional to the degree of osmotic stress. Interference with the ability to respond to a change rather than to maintain an equilibrium state may be a more realistic function to examine when assessing a pollutant's impact on an organism.

Acknowledgements

This research was supported by grant No. ID075-04890 from NST, International Decade of Oceanic Exploration, Biological Effects Program to J. M. Neff. We would like to acknowledge Gene Young for help in taking care of the animals and Dr. R. J. Freund for freely giving his time and advice for the statistical analysis.

References

Alderice, D. F., Response of marine poikilotherms to environmental factors acting in concert. In O. Kinne, ed. Marine Ecology I (3) 1959, New York, Wiley Interscience (1972).

Anderson, J. W., J. M. Neff, B. A. Cox, H. E. Tatem and G. M. Hightower, Characteristics of dispersions and water-soluble extracts of crude and refined oils and their toxicity to estuarine crustaceans and fish, Mar. Biol. 27, 75 (1974a).

Anderson, J. W., J. M. Neff, B. A. Cox, H. E. Tatem and G. M. Hightower, The effects of oil on estuarine animals: Toxicity, uptake and depuration, respiration. In F. J. Vernberg and W. B. Vernberg (eds.), Pollution and Physiology of Marine Organisms, Academic Press, New York (1974b).

Baumberger, J. P. and J. M. D. Olmsted, Changes in the osmotic pressure and water content of crabs during the molt cycle, Physiol. Zool. 1 531 (1928).

Epifanio, C. E., Effects of dieldrin in seawater on the development of two species of crab larvae, Leptodius floridanus and Panopeus herbstii, Mar. Biol. 11, 356 (1971).

Fucik, K. W. and J. M. Neff, Effects of temperature and salinity on naphthalene uptake in the temperature clam Rangia cuneata and the boreal clam Prothaca staminea. In this volume.

Gilfillan, E. S., Decrease of net carbon flux in two species of mussels by extracts of crude oil, Mar. Biol. 29, 53 (1975).

Laughlin, R. B., Jr., J. M. Neff and C. S. Giam, Effects of polychlorinated biphenyls, polychlorinated naphthalenes and phthalate esters on larval development of the mud crab Rhithropanopeus harrisii. In Workshop on Biological Effects of Pollutants on Marine Organisms, College Station, Texas, 16-19 May, 1976. National Science Foundation, International Decade of Oceanic Exploration (1976, in press).

McManus, J. J., Osmotic relations in the horseshoe crab Limulus polyphemus, Amer. Midland Natur. 81, 569 (1969).

Mangum, C. P., C. E. Booth, P. L. DeFur, N. A. Heckel, R. P. Henry, L. C. Oglesby and G. Polites, The ionic environment of hemocyanin in Limulus polyphemus, Biol. Bull 150, 453 (1976).

Morrow, J. E., R. L. Gritz and M. P. Kirton, Effects of some components of crude oil on young Coho Salmon, Copeia, 2, 326 (1975).

Neff, J. M. and J. W. Anderson, An ultraviolet method for the determination of naphthalene and alkylnaphthalenes in the tissues of oil-contaminated marine animals, Bull. Environ. Contam. Toxicol. 14, 122 (1975).

Neff, J. M., J. W. Anderson, B. A. Cox, R. B. Laughlin, Jr., S. S. Rossi and H. E. Tatem, Effects of petroleum on survival, respiration and growth of marine animals. In symposium on "Sources, Effects and Sinks of Hydrocarbons in Aquatic Environment". 9-11 August, (1976a). To be published by AIBS.

Neff, J. M., B. A. Cox, D. Dixit and J. W. Anderson, Accumulation and release of petroleum-derived aromatic hydrocarbons by four species of marine animals, Mar. Biol. (1976b, in press).

Neff, J. W. and C. S. Giam, Effects of Aroclor® 1016 and Halowax® 1099 on juvenile horseshoe crabs Limulus polyphemus, In Calabrese, A., F. P. Thurberg and F. J. Vernberg (eds.) Pollution and Physiology of Marine Organisms, New York, Academic Press (1976, in press).

Robertson, J. D., Osmotic and ionic regulation in the horseshoe crab Limulus polyphemus (Linnaeus), Biol. Bull. 138, 157 (1970).

Wilson, R. D., P. H. Monoghan, A. Osanik, L. C. Price and M. A. Rogers, Natural marine oil seepage, Sci. 184, 857 (1974).

CHAPTER 20

EFFECTS OF DISPERSED CRUDE OIL UPON THE
RESPIRATORY METABOLISM OF AN ARCTIC MARINE
AMPHIPOD, *ONISIMUS (BOEKISIMUS) AFFINIS*

J. A. Percy

Arctic Biological Station, Fisheries and Marine Service
Fisheries and Environment Canada, Ste. Anne de Bellevue, Quebec, Canada

Abstract

Short-term lethality is an unsuitable criterion for assessing the ecological effects of pollutants. A variety of sublethal physiological effects may impair an organism's ability to function normally and lead to a reduction or elimination of sensitive populations in a polluted area. The effects of exposure to sublethal concentrations of dispersed crude oils upon the respiratory metabolism of a marine amphipod have been examined. At low oil concentrations metabolism is significantly depressed but with increasing concentration a reversal of the response occurs. A possible explanation for this complex response is presented. The effects of other factors, such as oil type, presence of dispersants, nutritional state of the animals and weathering of the oil, upon the metabolic response are also considered.

Key words: Respiration, Crude oil dispersions, Amphipod, Dispersant

Introduction

Although lethal bioassays are widely used in pollution studies they provide only a crude first approximation to the impact of pollutants upon animal populations. Rapid death is too gross a criterion for assessing deleterious ecological effects. Increasingly, attention is turning to a variety of subtle, sublethal effects that may severely impair an organism's ability to function normally, without killing it outright. Only recently has there been a general appreciation of the wide variety of physiological, biochemical and behavioral processes that may be adversely influenced by pollutants. The diversity and complexity of the processes involved make it difficult to differentiate between ecologically significant and merely incidental effects. It is likely that significant ecological consequences will result from a number of the sublethal effects acting in concert. Such an interacting complex of diverse biological effects has been termed a "toxic response syndrome" (Warner, 1965).

Although it is important to consider this syndrome approach, it is also necessary to examine individual potentially significant sublethal effects. Metabolic rate has been widely used as a sensitive indicator of changes in physiological state during exposure to environmental stresses. However, results of the few studies dealing with effects of petroleum upon metabolism appear contradictory and suggest an underlying complexity in the response that defies generalization at present. Petroleum products have been shown to both stimulate (Gilfillan, 1973; Hargrave and Newcombe, 1973; Brocksen and Bailey, 1973) and depress metabolism (Avolizi and Nuwayhid, 1974; Dunning and Major, 1974). Although much of the variability is undoubtedly attributable to differences in both petroleum and species, it is nevertheless clear from the individual studies that other biological and physical factors play a significant role in determining both the direction and magnitude of the response. The present study deals with the effects of exposure to dispersions of a number of crude oils upon the respiratory metabolism of the arctic marine amphipod *Onisimus (Boekisimus) affinis* H. J. Hansen.

Methods

Animals were collected in the brackish water (17-18°/$_{oo}$) Eskimo Lakes adjacent to the Mackenzie Delta in the Northwest Territories at a site previously described (Percy, 1975). Animals were held in a circulating seawater system at a temperature (7°±1°C) and salinity (17 ± 1°/$_{oo}$) similar to that in the natural habitat.

Northern oils used in the study include Pembina, Atkinson Point and Norman Wells crudes. Physical characteristics of these oils have been described by Keevil and Ramseier (1975):

	Pour Pt (°C)	Specific Gravity	Viscosity Centipoise (16°C)
Atkinson Pt.	-45	0.919	61
Pembina	-10	0.857	13
Norman Wells	-50	0.833	6.5

Venezuela crude (Tijuana light) was employed in some instances for comparative purposes. The oils were kept in tightly capped 2 ml vials stored below 5°C and a fresh vial was used for each set of dispersions. Four concentrations of dispersed oil were prepared for experimental exposures:

1. Control: seawater only, no oil.
2. Light: 25 µl of oil per 500 ml of seawater (50 ppm)
3. Medium: 250 µl of oil per 500 ml of seawater (500 ppm)
4. Heavy: 1000 µl of oil per 500 ml of seawater (2000 ppm)

These values represent nominal concentrations of oil added to the seawater. Actual concentrations of dispersed oil, concentration variability and changes in concentration during the exposure period are considered in detail in Percy and Mullin (1975). This fluorimetric data is summarized in Table 1.

TABLE 1

Fluorimetric estimates of concentrations of crude oils dispersed in seawater following standardized preparation and exposure procedures. $\bar{X}_i$ = mean concentration (ppm) of oil in dispersions immediately after preparation and settling (i.e. initial exposure level); $\bar{X}_f$ = mean concentration of oil in dispersions after 24 hours at standard exposure conditions (i.e. final exposure level); % loss = percent of the oil lost from the dispersions during the 24 hour exposure period. (N = 7-8 replicates for each dispersion concentration and sample period.)

	Dispersion	$\bar{X}_i$ ± S.D.	$\bar{X}_f$ ± S.D.	% loss
Atkinson Pt.	Light	18.2 ± 1.9	10.1 ± 1.8	44.5
	Medium	169.5 ± 28.1	13.4 ± 2.0	92.0
	Heavy	>800 ±	180.5 ± 33.1	--
Norman Wells	Light	13.3 ± 3.2	6.0 ± 2.2	54.9
	Medium	26.1 ± 6.0	9.8 ± 3.9	62.6
	Heavy	476.3 ± 85.2	34.4 ± 8.9	92.8
Pembina	Light	20.9 ± 2.3	19.4 ± 4.5	7.2
	Medium	51.2 ± 19.8	19.4 ± 10.4	62.1
	Heavy	267.7 ± 22.9	72.0 ± 17.6	73.1
Venezuela	Light	16.1 ± 2.1	11.3 ± 0.8	29.8
	Medium	21.9 ± 2.8	19.4 ± 6.0	11.4
	Heavy	643.0 ± 103.0	93.7 ± 19.1	85.7

Dispersions were prepared at room temperature. For light and medium dispersions the oil was added to 500 ml of filtered seawater (salinity 17-18°/оо) in 0.95 l glass jars and shaken on a reciprocating shaker (280 excursions/min) for 1 hour. For heavy dispersions the oil was added to 500 ml of filtered seawater and dispersed in a Waring blender for 5 minutes. The oil/Corexit mixtures were prepared using a 1:1 ratio by volume. The resulting dispersions were transferred to separatory funnels and left undisturbed for varying periods depending upon the particular oil (Atkinson Point, 180 min; Venezuela and Pembina, 120 min; Norman Wells, 90 min). These times were selected on the basis of the time required for the rapidly coalescing portion of the dispersed oil to separate out (Percy and Mullin, 1975). The lower 450 ml of seawater was drawn off into 0.71 l glass jars which served as exposure vessels. The dispersions were equilibrated in an 8°C water bath before adding the animals. Gentle aeration was provided via pasteur pipets. Animals were not fed during the exposure period, which lasted for 24 hours unless otherwise indicated.

Oil was weathered in 3 ml quantities in shallow open dishes held at 10°C and stirred continuously. With Norman Wells crude, the weight loss after 96 hrs was 35%. With Pembina crude the weight loss was 26% after 96 hours. Dispersions were prepared as outlined above, using the same volumes of weathered oil.

Seawater extracts of the oils were prepared by placing 100 ml of the oil in a 2 l flask with 1 l of filtered seawater of salinity 17-18°/oo. The water was stirred vigorously with a magnetic stirrer for 6 hours at a standard speed. The mixture was then allowed to stand for two hours in a separatory funnel. The seawater extract was drawn off and filtered (Whatman no. 1). This stock solution was used to prepare dilutions containing nominal concentrations of oil of 10, 100, 1000, 5000 and 10,000 ppm.

Respiration was measured at 8°C with a Gilson submarine respirometer using standard 15 ml flasks. Five animals were placed in each flask with 5 ml of filtered seawater of salinity 17°/oo. The centre well was charged with 0.2 ml of 20% KOH. The shaking rate was 72 oscillations per minute. Flasks were equilibrated for 45 minutes and readings taken at 1 hour intervals for 5 to 6 hours. Animals were rinsed in distilled water, dried at 70°C for 24 hours and weighed. Metabolic rates were expressed as $\mu l O_2$/mg dry weight/hr. The statistical significance of differences in mean respiration rates was determined by students' t test.

Cell free homogenates were prepared from adult animals exposed to the oil dispersions for 24 hours. Each sample consisted of 12 animals homogenized in a motor driven Potter-Elvehjem tissue homogenizer held in an ice bath. The homogenization medium consisted of 3 ml of the sucrose medium of Peterson and Anderson (1969) supplemented with yeast extract (10 gm/l). The chilled homogenate was transferred to the respirometer flasks along with an additional 2 ml rinse. Metabolic rates were determined at 14°C.

Results

Following 24 hours exposure to the oil dispersions the metabolic rate of the amphipods exhibited considerable variability in response (Fig. 1). The treatment appeared to

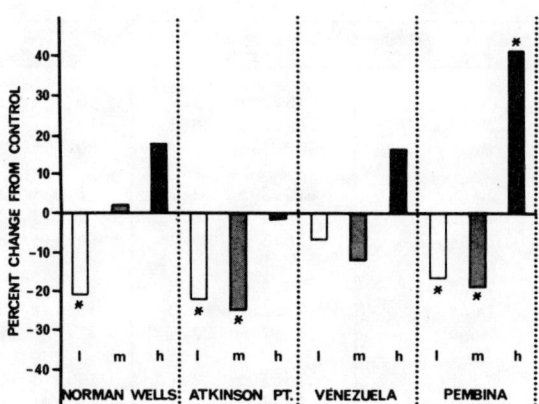

Fig. 20.1 Effect of 24 hr exposure to light, medium and heavy dispersions of crude oils upon the respiratory metabolism of *Onisimus*. Asterisk indicates difference significant at $p < 0.05$ level. (N = 8 in each group.)

increase, decrease or have little significant effect upon respiration, depending upon oil type and concentration. However, closer examination revealed a general trend occurring consistently in each experimental series. In each there was an initial depression of metabolism, ranging from 7 to 22%, in light dispersions, followed by a reversal of the depression as the oil concentration increased. Heavy dispersions of three of the oils actually stimulated respiration by 16 to 41%. The concentration at which the reversal occurred, and the magnitude of the initial depression and subsequent increase of metabolism differed among the different oil types. This variability may be attributable to the marked differences in physical and chemical characteristics of the oils.

Amphipods exposed to the oil in combination with the dispersant Corexit 8660 exhibited a slightly different metabolic response-concentration relationship (Fig. 2). Corexit alone at concentrations comparable to the oil depressed respiration by about 30%. There was no evidence of a reversal of the inhibition. However, the reversal of the metabolic depression with increasing concentration was apparent in animals exposed to the Corexit and crude oil

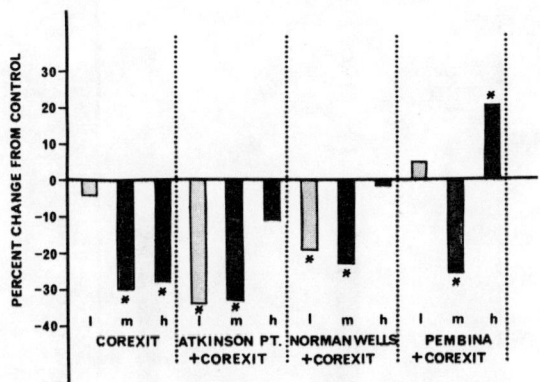

Fig. 20.2 Effect of 24 hr exposure to light, medium and heavy concentrations of Corexit, alone and in combination with crude oils, upon the respiratory metabolism of *Onisimus*. Asterisk indicates difference significant at $p < 0.05$ level. (N = 8 or 9 in each group.)

simultaneously. The metabolic rate was consistently depressed following exposure to light and medium dispersions of all three oils with Corexit, while at higher concentrations the metabolic depression was less pronounced. However, the relative magnitude of the reversal was considerably less than that occurring with the oil alone. With oil/dispersant mixtures significant stimulation of metabolism occurred only with Pembina crude and even then not to the degree observed with this oil alone. With both the oils alone and with the oil/dispersant mixtures the relative magnitude of the reversal of respiratory depression decreased in the sequence:

Pembina > Norman Wells > Atkinson Point

In general, the maximum respiratory inhibition induced by the oil/Corexit mixture was slightly greater than that induced by the oil alone.

Weathering of the oil for periods up to 96 hours prior to preparing dispersions did not reduce its metabolic impact significantly (Table 2). In the case of Norman Wells crude, in fact, weathering may slightly increase the metabolic impact.

TABLE 2

Influence of exposure to light dispersions (25 µl/500 ml) of fresh and weathered crude oils on the respiration of *Onisimus affinis*. Results expressed as percentage difference from controls (Δ%).

Oil type	Weathering time	Δ%	P
Norman Wells	Fresh	-12.7	< 0.025
	24 h	-19.2	< 0.025
	72 h	-17.7	< 0.025
	96 h	-17.5	< 0.025
Pembina	Fresh	-21.4	< 0.005
	96 h	-16.2	< 0.01

Metabolic rates of cell free homogenates prepared from amphipods exposed to dispersions of Norman Wells oil were consistently higher than those of unexposed animals (Fig. 3). There was no evidence of a depression of metabolism comparable to that occurring in intact animals. The increase ranged from 10% to 45%, with no consistent trend with concentration apparent. Comparable studies with the other oils have not yet been conducted.

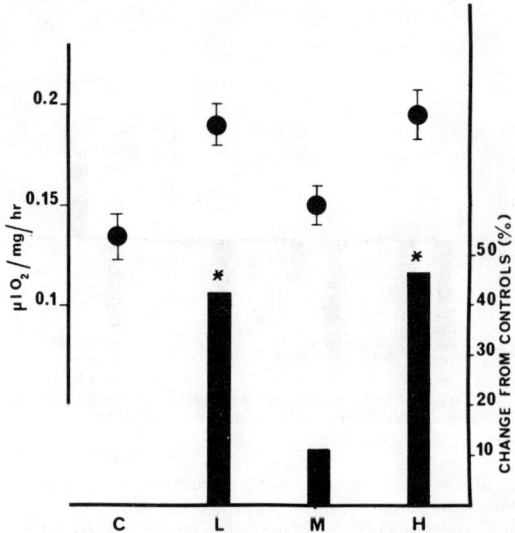

Fig. 20.3 Effect of exposure to light, medium and heavy dispersions of Norman Wells crude upon the respiration of cell free homogenates of *Onisimus*. Asterisk indicates difference significant at $p < 0.05$ level. (N = 7-9 in each group.)

Exposure of the amphipods to seawater extracts of Norman Wells and Atkinson Point crudes (filtered to remove particulate oil) results in a marked stimulation of respiration (Fig. 4). The maximum stimulation was 38% for Norman Wells crude and 30% for Atkinson Point crude. There was no evidence of a depression of metabolism at any of the concentrations tested. The slight decrease in stimulation at the highest oil concentration may represent an incipient narcotization effect.

Exposure to light dispersions of Norman Wells crude for longer than 24 hours (with the dispersion changed each 24 hours) did not result in a further significant depression of respiration relative to controls, at least for periods up to 96 hours (Fig. 5). During the first 24 hours of experimental exposure, respiration of both control and oil exposed animals declined markedly. This was clearly a starvation effect. Prior to use the animals were fed ad libitum, but they were not fed during the exposure periods. In many marine animals metabolic rate increases considerably immediately following feeding (Newell, 1970). During the 24 hour exposure period without food the respiration rate dropped sharply to a non-feeding level. Between 24 and 96 hours respiration declined only slightly in both control and exposed animals. The general metabolic trend is similar in the two groups, although the curve for oil exposed animals is displaced downward as a consequence of the pollutant stress. After 96 hours the animals were placed in fresh dispersions or clean seawater, fed *ad libitum* and the metabolic rate determined 24 hours later. The difference in respiration rate between control and exposed animals was greatly increased following feeding. Respiration of control animals returned almost to the pre-test level, while that of exposed animals rose only slightly.

Discussion

The results indicate that the metabolic response to crude oil is more complex than the simple unidirectional inhibition or stimulation suggested by some studies. All four oils tested evoked the same basic pattern of response, although differences in detail were evident. Low concentrations of dispersed oil in seawater depressed metabolism significantly. As the oil content of the dispersion increased inhibition was reversed and the metabolic rate approached the control level and even exceeded it in some cases. A comparable reversal

of inhibition occurred in animals exposed to oil-Corexit mixtures, but not in those exposed
to Corexit alone. The data of Avolizi and Nuwayhid (1974) for the bivalves *Brachidontes* and
Donax reveals a similar reversal of metabolic inhibition with increasing oil concentration.
In this instance, however, the metabolic rate approaches the control level after maximum
inhibition but never actually exceeds it.

How is such a reversal in metabolic response to be interpreted? The following is a
tentative scheme that is consistent with the presently available data and is amenable to
further testing. Oxygen utilization by an organism involves two distinct metabolic
components. Basal metabolism reflects routine maintenance processes of the organism at
rest. A further significant fraction of the oxygen demand is attributable to locomotion and
other forms of activity. A discussion of the concept of basal and active metabolism as it
applies to marine invertebrates is presented in Newell (1970). It is useful to consider
these metabolic components separately in relation to the observed effects of exposure to
crude oils.

It has been demonstrated that the locomotory activity of *Onisimus* is severely impaired
following 24 hours exposure to concentrations of dispersed crude oil similar to that used
in the present study (Percy and Mullin, 1977). It is not clear whether the effect is
attributable to physical or chemical processes. Such a decline in activity with increasing
oil concentration would be paralleled by a decline in respiratory metabolism. With complete
cessation of locomotory activity respiration would stabilize at the basal level. The
observed reduction in respiration following exposure to low concentrations of dispersed oil
may reflect this decline in activity. The depression of respiration following exposure to
light dispersions occurs even if the oil is weathered (with a loss of 1/4 to 1/3 of its
original weight). This implies either that the metabolic response is a result of some
physical action of the particulate oil, or that it has a chemical basis and is attributable
to certain of the heavier oil components that remain after weathering.

The effect of petroleum upon the basal component of metabolism is more difficult to
ascertain. It is possible to examine relative changes in cellular metabolism without the
complications of a superimposed activity component by measuring the oxygen uptake of cell
free homogenates. One assumption implicit in all such homogenate studies is that although
the absolute respiration rate of homogenized tissue may be far different from the in situ
rate, it nevertheless, closely reflects relative changes occurring in situ. Metabolic rates
of homogenates of animals exposed to crude oil dispersions were significantly higher than
control values. There was no evidence of a depression of metabolism similar to that
observed in oil exposed intact animals. Furthermore, it was noted that in animals exposed
to seawater extracts of the oil, from which particulate oil had been removed, the metabolic
rate was significantly stimulated. The degree of stimulation increased with increasing oil
concentration, although there was some indication of a decrease in the stimulation at very
high concentrations. The homogenate results suggest that cellular basal metabolic processes
are stimulated by the oil, while the oil extract results suggest that some of the more
soluble oil components may be responsible for the stimulation. It has been shown that
benzene, a soluble hydrocarbon that penetrates readily into tissues, stimulates the
respiration of fish (Brocksen and Bailey, 1973; Struhsaker et al., 1974). It was suggested
that the increased respiration reflects a requirement for more oxygen to metabolize the
benzene. Higher concentrations or extended exposure times resulted in a decline in
metabolic rate, a possible narcotic effect arising from accumulation of benzene in the
tissues. Similarly Gilfillan (1973) observed that low concentrations of seawater extract of
crude oil (containing chiefly soluble components) stimulated the metabolism of mussels. At
higher concentration the degree of stimulation decreased. If the soluble components of
crude oil that penetrate into tissues stimulate basal metabolism, then the basal rate would
rise with exposure to increasing oil concentrations up to the point at which narcosis
occurs. Anderson (1975), in studies on the uptake of napthalenes by shrimp tissues, noted
a possible relationship between the presence of the hydrocarbons in the tissues and the
increased respiration rate.

If the above hypothetical effects on activity metabolism and on basal metabolism are
combined, as they would be in the animal, the observed respiratory response would be the
resultant of the inhibitory and stimulatory effects. A reversal of metabolic inhibition
would occur if activity is inhibited at low dispersion concentrations while basal metabolism
is maximally stimulated by the higher concentrations of soluble components present in the
heavier dispersions. Corexit depresses metabolism. A similar effect was observed in marine
snails (Hargrave and Newcombe, 1973). In the oil-Corexit mixtures, the depressant effect of
the Corexit may counter the stimulatory effect of the soluble oil components on basal
metabolic processes and thus account for the observed reduction in magnitude of the reversal
of respiratory depression. This scheme appears to account for some of the observed
metabolic responses subsequent to oil exposure and may also explain some of the apparently
contradictory results reported in the literature. However, it is still highly speculative
and must be considered a tentative model designed primarily to stimulate additional testing.

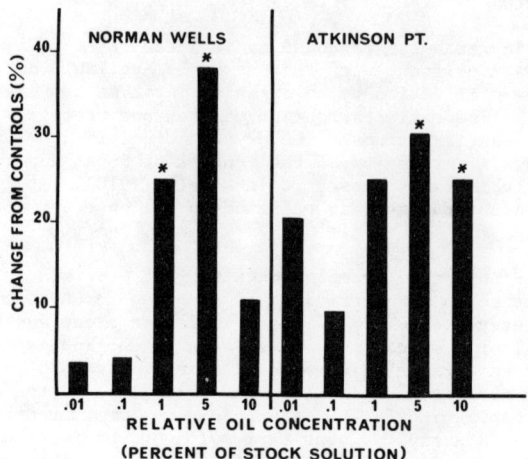

Fig. 20.4 Effect of exposure to various concentrations of seawater extracts of Norman Wells and Atkinson Point crude oils upon the respiration of *Onisimus*. Asterisk indicates difference significant at $p < 0.05$ level. (N = 13-22 for each group.)

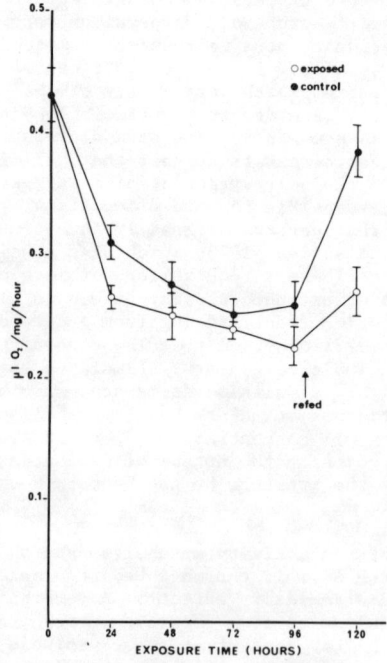

Fig. 20.5 Effect of exposure to light dispersions of Norman Wells crude for various periods upon the respiration of *Onisimus*. Animals not fed between 0 hr and 96 hr, but refed after 96 hr. Vertical lines indicate standard errors of means. (N = 8-14 in each group.)

In applying this scheme to other species it is important to keep in mind that the locomotory response to oil exposure may vary from species to species. In the present study, the locomotory activity of *Onisimus* is depressed following exposure to oil dispersions. In contrast, crawling of the snail *Littorina* increased following exposure to petroleum and respiration was stimulated (Hargrave and Newcombe, 1973). Anderson (Pers. comm.) has observed that when mysids are exposed to petroleum they tend to swim continuously while the shrimp *Palaemonetes* tends to settle to the bottom of the tank. The respiration rate of the mysids increases following exposure with no evidence of an initial depression of metabolism at low concentrations of petroleum (Anderson, 1975). In contrast, respiration of *Palaemonetes* is depressed following exposure to petroleum and "the greatest respiratory depression appeared to occur following exposure to the lowest concentration" (Anderson et al., 1974). In terms of the above hypothesis these results suggest that in those species that respond to the presence of oil by an increase in activity the stimulatory effects on activity metabolism and basal metabolism will be additive; i.e. both will tend to elevate the respiration rate and there will be no initial depression of respiration. In contrast, in those species that respond to the presence of oil with a decrease in locomotory activity there will be an initial depression of metabolism at low concentrations followed by a reversal of the depression at higher concentrations as cellular metabolic processes become increasingly stimulated. Clearly, in studying the respiratory response of animals to petroleum it is important to consider the behavioral response. I would recommend that in future studies of petroleum effects on respiration both oxygen consumption and activity levels be measured simultaneously, or under comparable experimental conditions. In addition, a much clearer understanding is required of the behavioral responses of organisms to oil exposure, particularly in relation to concentration and duration of exposure.

The routine starving of the animals during oil exposure reduced the metabolic rate of both control and oil exposed animals significantly, presumably by depressing cellular metabolic processes. Upon refeeding, the respiration of control animals returned to normal while that of oil exposed animals rose only slightly. This nutritional effect implies that in the natural habitat, where food is continuously available, the actual difference in metabolic rate between oil exposed and unexposed animals would be even greater than that indicated by the laboratory exposure tests. It is clear that petroleum may modify respiration rates both directly by influencing activity and cellular metabolism and indirectly by influencing nutritional state. Differences in nutritional state may account for some of the variability in metabolic responses to oil reported in the literature.

References

Anderson, J. W., Laboratory studies on the effects of oil on marine organisms: an overview, American Petroleum Institute Publication No. 4249 (1975).

Anderson, J. W., J. M. Neff, B. A. Cox, H. E. Tatem and G. M. Hightower, The effects of oil on estuarine animals: Toxicity, uptake and depuration, respiration. pp. 285-310. In: F. J. Vernberg and W. B. Vernberg (eds.) Pollution and Physiology of Marine Organsisms, Academic Press, New York, 1974.

Avolizi, R. J. and M. Nuwayhid, Effects of crude oil and dispersants on bivalves. Mar. Poll. Bull. 5, 149-152 (1974).

Brocksen, R. W. and H. T. Bailey, Respiratory response of juvenile chinook salmon and striped bass exposed to benzene, a water-soluble component of crude oil. pp. 783-791. In: Proceedings, Joint Conference on the Prevention and Control of Oil Spills, American Petroleum Institute, Washington, D.C., 1973.

Dunning, A. and C. W. Major, The effects of cold seawater extracts of oil fractions upon the blue mussel Mytilus edulis. pp. 349-366. In: F. J. Vernberg and W. B. Vernberg (eds.) Pollution and Plysiology of Marine Organisms, Academic Press, New York, 1974.

Gilfillan, E. S., Effects of water-soluble fractions of crude oil on carbon budgets in two species of mussels. pp. 691-695. In: Proceedings, Joint Conference on the Prevention and Control of Oil Spills, American Petroleum Institute, Washington, D.C., 1973.

Hargrave, B. T. and P. Newcombe, Crawling and respiration as indices of sublethal effects of oil and a dispersant on an intertidal snail, Littorina littorea, J. Fish. Res. Board Can. 31, 1789-1792 (1973).

Keevil, B. E. and R. O. Ramseier, Behavior of oil spilled under floating ice. pp. 497-501. In: Proceedings, Joint Conference on the Prevention and Control of Oil Spills, American Petroleum Institute, Washington, D.C., 1975.

Newell, R. C., Biology of Intertidal Animals, Elsevier, New York, 1970.

Percy, J. A., Ecological physiology of arctic marine invertebrates. Temperature and salinity relationships of the amphipod, Onisimus affinis, H. J. Hansen, J. Mar. Biol. Ecol. 20. 99-117 (1975).

Percy, J. A., and T. C. Mullin, Effects of crude oils on arctic marine invertebrates, Beaufort Sea Project Technical Report No. 11, Environment Canada, Victoria, B.C., 1975.

Percy, J. A., and T. C. Mullin, Effects of crude oil on the locomotory activity of arctic marine invertebrates, Mar. Poll. Bull. (1977, in press).

Struhsaker, J. W., M. B. Eldridge and T. Echeverria, Effects of benzene (a water-soluble component of crude oil) on eggs and larvae of pacific herring and northern anchovy. pp. 253-284. In: F. J. Vernberg and W. B. Vernberg (eds.) Pollution and Physiology of Marine Organisms, Academic Press, New York, 1974.

Warner, R. E., Formal discussion of paper by M. Fujiya. pp. 325-329. In: E. A. Pearson (ed.) Advances in Water Pollution Research, Vol. 3, Pergamon Press, New York, 1965.

CHAPTER 21

ACCUMULATION OF NAPHTHALENES BY GRASS SHRIMP:
EFFECTS ON RESPIRATION, HATCHING, AND LARVAL GROWTH

H. E. Tatem

U. S. Army Engineer Waterways Experiment Station
Vicksburg, Mississippi 39180

Abstract

Estuarine grass shrimp, Palaemonetes pugio, were exposed in artificial seawater to petroleum hydrocarbons (PH) from a No. 2 fuel oil. Exposure water and organisms were sampled periodically for naphthalene concentrations; water initially contained 2.6 ppm PH and 0.55 ppm total naphthalenes (TN). Naphthalene levels in the exposure solution decreased rapidly while concentrations in the shrimp tissue increased dramatically. After 6 hr, tissue levels of methylnaphthalene were 150 times greater than water levels. Depuration of hydrocarbons was rapid and began during the exposure period; however, complete depuration did not occur. Dimethylnaphthalenes remained in exposure water and organism tissues to the greatest extent.

Respiratory rates of eight groups of shrimp were measured frequently over a two-week period. An average control rate of 0.46 ml O_2/g/hr was established. Exposure to over 3.0 ppm PH for 5 hr resulted in a lower average rate. A 48-hr starvation period also lowered this rate. Seven days after exposure, respiratory rates had returned to control levels.

A single 72-hour exposure of gravid female shrimp to PH (1.44 ppm) had a detrimental effect on larval hatching. Control females released an average of 45 larvae compared to 9 larvae for exposed females. A concentration of 0.72 ppm PH did not have a significant effect on larval hatching.

Grass shrimp larvae continuously exposed to PH (0.85-0.52 ppm) weighed significantly less than control animals after 12 days. Removal of larvae to clean seawater resulted in accelerated growth.

Exposure of adult grass shrimp to sublethal levels of PH or naphthalenes intially resulted in increased activity and disoriented swimming patterns. Animals which survived exposures seemed to recover completely when placed in uncontaminated seawater.

Key Words: Palaemonetes, petroleum hydrocarbons, naphthalenes, sublethal effects

Introduction

Production and transportation of oil products result in the continuous contamination of the coastal environment by soluble petroleum hydrocarbons (PH). PH's have been found in coastal waters at concentrations of 10-20 ppb (Brown et al. 1973), but refined oil spills in confined estuarine areas may result in much higher concentrations. For example, Cox et al. (1975) found over 300 ppb aromatic hydrocarbons in a small estuarine pond after an experimental oil spill.

Recently techniques have been developed to identify and quantify the biologically important aromatic hydrocarbons in water, tissue and sediment samples (Warner 1974, Neff and Anderson 1975). Techniques are also available to determine total PH in water samples (API 1958). These developments have enabled researchers to determine the toxicity of PH to various marine organisms (Anderson et al. 1974a, Rossi et al. 1976, Tatem et al., Wells and Sprague 1976, Vanderhorst et al. 1976) and to study sublethal effects (Anderson et al. 1974b, Linden 1976). As a result of these and other studies, it has been determined that the aromatic naphthalene hydrocarbons are the PH which are accumulated and retained to the greatest extent by organisms exposed to oil-seawater mixtures. Naphthalenes including the monomethyl and dimethyl analogues have been found relatively soluble in seawater and highly toxic to estuarine animals (Anderson et al. 1974b).

Possible sublethal effects of PH on aquatic organisms include tainting of edible species, changes in behavior and metabolic rates, and inhibition of growth and reproduction. Two reviews (Evans and Rice 1974, Moore and Dwyer 1974) have noted the meager data available on the sublethal effects of PH.

Methods

Grass shrimp were collected at different times by seine or hand net from Spartina marsh areas of the Galveston Bay estuarine system. Animals were returned to the laboratory and held for two weeks in charcoal-filtered artificial seawater (Instant Ocean). All experiments were conducted in artificial seawater adjusted to 15 ppt salinity at $21°±1°C$ temperature.

Accumulation and Depuration

Two oil-seawater mixtures were prepared, using refined No. 2 fuel oil (38% aromatics), according to established procedures (Anderson et al. 1974a). Either oil-water soluble fractions (WSF) or oil-water dispersions (OWD) were used for various experiments. Fresh mixtures were prepared for each experiment and added to additional seawater in glass aquaria. Approximately 30 min after addition of the PH, water samples were collected for infrared (IR) analysis of total PH (API 1958). Water samples were also taken for analysis of naphthalenes by the ultraviolet (UV) spectrophotometric technique of Neff and Anderson (1975). The experimental UV spectra were compared to naphthalene standards in hexene and absorbance converted to ppm naphthalene compounds by solution of simultaneous equations (Neff and Anderson 1975). Approximately 140 shrimp were removed from the holding tanks; 20 were rinsed and frozen as control tissue and the remainder placed in the exposure aquarium. Exposure water was aerated gently but not filtered. Organisms were observed closely during the exposure period and water and tissue samples were collected at various time intervals. Tissue samples consisted of the pooled tissues of 10-15 organisms, sufficient for two or three replicate samples. After the exposure period was over, additional seawater was added to the aquarium and charcoal filtration of the water was begun. Water and tissue samples continued to be taken for analysis.

Respiration

Eight groups of 12-14 shrimp each were held in static chambers with water changed daily. All shrimp were medium-size (2.0-3.5 cm) adults. Oxygen consumption of each of the groups was determined six times over a 14-day period. The oxygen runs were conducted in light at the same time each day; a YSI oxygen meter was used to measure O_2 concentration in ppm in the test chamber initially and 5 hr later. Test chambers were flasks containing 730 ml of seawater fitted with rubber stoppers through which were inserted two glass tubes. The oxygen concentration of a large batch of seawater was determined. Flasks were then filled by siphoning, organisms were introduced, and stoppers set in place. Wet weight of the shrimp was measured after each oxygen run. Oxygen consumption of each group was determined and these values were averaged. Shrimp were not forced to maintain a specific level of activity during the tests. Oxygen consumption was measured twice to obtain control values. Consumption for each group was again determined after a 48-hr starvation period. Shrimp were then fed and their respiratory rates were again measured. The following day, oxygen consumption of the groups was determined during exposure to No. 2 fuel OWD for 5 hr and then again after a seven-day recovery period in clean seawater with food. The results were converted from the change in ppm of oxygen over time to ml O_2 used per g (wet weight) per hr (Tatem 1975). Another group of shrimp was exposed to a similar OWD mixture in a static system for 5 hr and then transferred to PH-free seawater. Tissue samples were collected at various times to assess accumulation and depuration of naphthalenes.

Larval Hatching and Growth

Eighteen gravid female grass shrimp collected from the field were separated into three groups for exposure to No. 2 fuel oil WSF. Six additional gravid females were used as controls. Shrimp were held individually and exposed to WSF containing 1.44, 0.72, or 0.24 ppm PH. The total naphthalenes (TN) present were 0.62, 0.31, or 0.10 ppm, respectively. Exposure solutions were not aerated and animals were exposed once for 72 hr. No hatches occurred during the exposure period. After this period shrimp were transferred to clean artificial seawater, which was renewed every two days until hatching. As the larvae hatched they were observed, counted, and placed in dishes for rearing. A representative sample of larvae from each group, including controls, was weighed approximately 45 days after hatching.

In a separate experiment, larvae from two Palaemonetes females were reared for 12-14 days Sixty larvae (0.4-0.5 cm) were divided among four petri dishes. Ten animals were chosen from the 60 and individual wet weights determined. Two dishes were established as controls while larvae in the other two dishes were exposed daily to No. 2 fuel oil WSF. Both groups were fed Artemia nauplii daily. Analyses of water samples showed that the initial exposure concentration was 0.85 ppm PH and 0.34 ppm TN. The WSF was stored at 5°C and used for three days in the preparation of exposure solutions, after which a new WSF was prepared. Samples of exposure solutions after three days revealed concentrations of hydrocarbons as low as 0.52 ppm PH and 0.27 ppm TN. Twelve days after the exposures began, the weights of the controls and experimentals were determined. Some experimental larvae were moribund. All organisms were placed in clean seawater for five additional days and again weighed. Then the experimental animals were exposed one final time to the WSF. The larvae were left in the exposure solution for two days and then transferred to clean seawater. After ten days the weights of all remaining postlarvae were determined.

Results

Accumulation and Depuration

The data showing the accumulation and depuration of naphthalenes by the grass shrimp are presented in Fig. 1. Shrimp were exposed for 24 hr to No. 2 fuel oil WSF which initially contained 2.6 ppm PH and 0.55 ppm TN. There were no mortalities due to exposure to the PH.

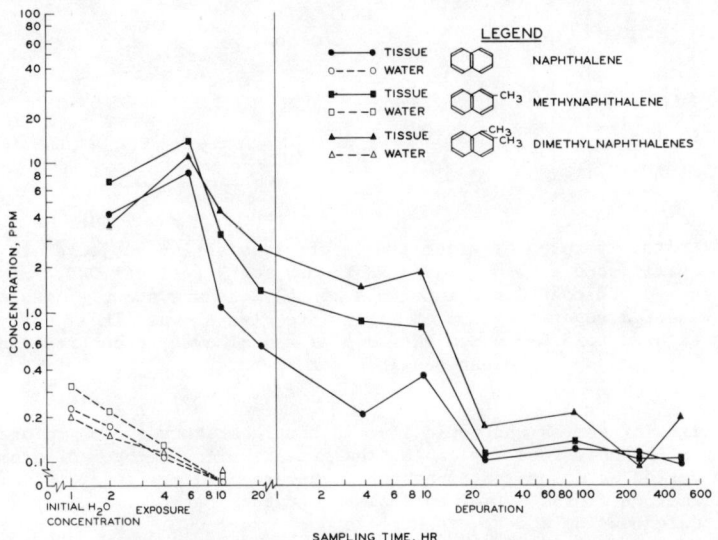

Fig. 21.1 Accumulation and release of naphthalenes by grass shrimp exposed for 24 hr to PH from a No. 2 fuel oil WSF.

Shrimp were unusually active during the first 6-8 hr of the experiment. Hydrocarbon concentrations in the exposure water decreased rapidly in a linear fashion such that no measurable concentration remained after 12-14 hr. Hydrocarbon accumulation by the grass shrimp was very rapid during the first few hours of exposure, yet depuration began as the water concentrations decreased to approximately 0.1 ppm of the individual naphthalenes. After 2 hr of exposure, water levels of the naphthalene compounds were approximately 0.2 ppm, while tissue concentrations were in the range of 4.0-7.0 ppm. At 6 hr, methylnaphthalenes (MN) were present in the water at a concentration of 0.1 ppm, while shrimp tissue contained over 15.0 ppm MN. After 6 hr of exposure, both naphthalene (N) and MN decreased in the tissues rapidly, whereas the dimethylnaphthalenes (DMN) were depurated to a lesser extent. At the end of the exposure period, shrimp contained approximately 3.0 ppm DMN, 1.5 ppm MN, and 0.5 ppm N. After 24 hr in clean charcoal-filtered seawater, tissue burdens had decreased to 0.1-0.2 ppm of the naphthalene compounds. The grass shrimp remained contaminated at these low levels for 18 days. Shrimp available for tissue samples were exhausted at this point. Background concentrations of hydrocarbons found in control tissues have been subtracted from the values shown in Fig. 1. Additional experiments of a similar design have revealed essentially complete depuration of naphthalene compounds by grass shrimp in one to four days (Tatem 1975, Anderson 1975).

Effects on Respiration

An experiment was formulated to test the effects of accumulated PH on the overall metabolic rate of grass shrimp as determined by oxygen consumption. Since accumulation of PH produced an effect on the activity level of grass shrimp, it seemed likely that respiratory rates would also be affected. Earlier respiration experiments had suggested that the level of nutrition of the test organisms had an important effect on respiratory rates. Therefore, the effect of 48 hr without food was also examined as part of the experiment. The data are presented in Fig. 2. They show similar average respiratory rates for two control tests conducted during the first two days of the experiment. The two-day starvation period produced a rather dramatic lowering of the average respiratory rate considering the fact that the

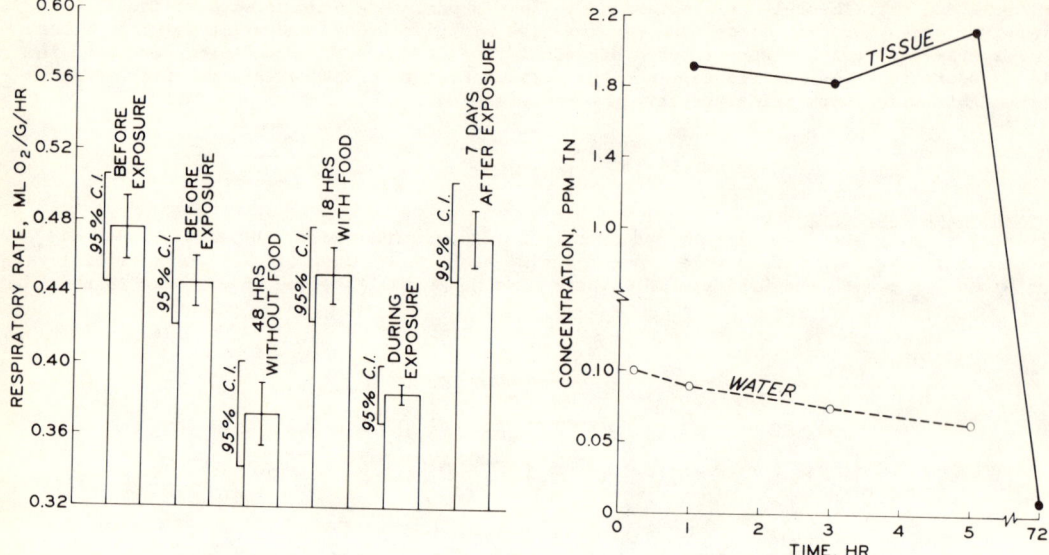

Fig. 21.2 Respiratory rates of eight groups of grass shrimp subjected to a 48-hr starvation period and to 3.0-3.6 ppm PH from a No. 2 fuel oil OWD. Standard errors and 95% confidence intervals of the data are shown. Tissue analyses of a separate group of shrimp are also shown. These shrimp were exposed to 3.0-3.6 ppm PH for 5 hr and allowed to depurate in clean seawater for 72 hr.

shrimp showed no signs of stress and could live in the laboratory for up to one week without food. After 18 hr with ample food available, the grass shrimp "recovered" from the effects of the starvation period as demonstrated by the higher respiratory rate. Also, these data were further confirmation of the normal control respiratory rate. The eight groups of grass shrimp were then tested during a 5-hr exposure to No. 2 fuel OWD (3.0-3.6 ppm PH). At the same time a separate group of 75 grass shrimp were exposed to a similar concentration of PH for 5 hr. Water and tissue samples were collected during the 5-hr exposure period and again after a 72-hr depuration period. The results (Fig. 2) show that at the end of the exposure period shrimp contained over 2.0 ppm naphthalenes after background levels had been subtracted. The naphthalenes were eliminated from the tissues after 72 hr in clean seawater. Respiratory measurements demonstrated that the accumulated PH lowered the average respiratory rate; however, a final run seven days after the exposure showed that this rate had returned to control levels.

Effects on Hatching and Larval Growth
The single 72-hour exposure of gravid female grass shrimp had an obvious detrimental effect on larval hatching at the highest exposure concentration. These data are provided in Fig. 3. No larvae hatched during the exposure period. Females were placed in uncontaminated seawater after exposure and larvae began hatching one week later. Control females produced an average of 45 larvae, while females exposed to 1.44 ppm PH for 72 hr produced an average of only 9 healthy larvae. Females exposed to 0.24 ppm PH hatched an average of 48 larvae, while those exposed to 0.72 ppm PH produced an average of 35 larvae. Larvae which hatched from the exposed females seemed to develop and grow normally. After 45 days, a representative sample of control larvae weighed 11 mg, while a sample of larvae from exposed females weighed 8-9 mg.

The deleterious effects of a continuous exposure of Palaemonetes larvae to PH are shown in Fig. 4. There was a definite effect as revealed by the average weights taken 12 days after the exposures began. Statistical analysis of the data found that the weights of experimental larvae were significantly different (P<.001) from those of the controls.

After the 12-day exposure period, some experimental larvae were immobile; therefore, both experimentals and controls were held in clean seawater for 5 days. Growth rates were similar during this period. The experimental shrimp were exposed again for two days to similar PH concentrations. After this final exposure, they were held and fed for ten days in uncontaminated seawater before final weight determinations. These data showed that the experimental larvae were recovering. They grew at an accelerated rate compared to the rate shown during the 12 days of exposure. Statistical analysis of the weights of the controls and experimentals revealed that they were not significantly different (P<.05) after the ten days in clean seawater.

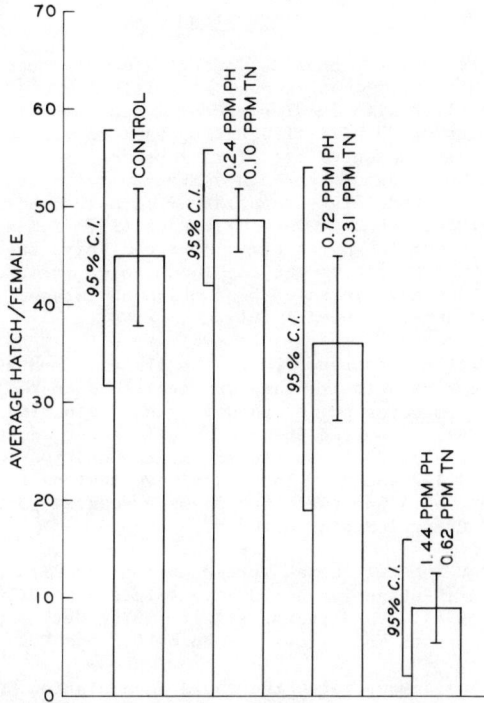

Fig. 21.3 Hatching success of <u>Palaemonetes</u> larvae after gravid female grass shrimp were exposed to PH from a No. 2 fuel oil WSF for 72 hr. Standard errors and 95% confidence intervals are shown. Larvae hatched in uncontaminated seawater.

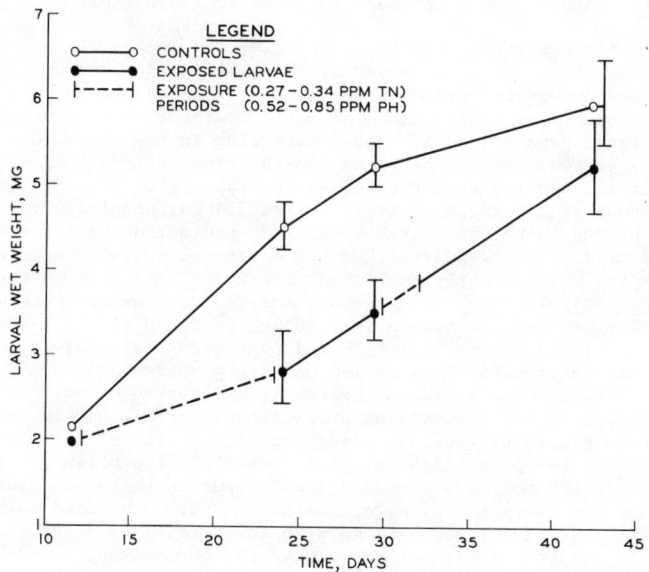

Fig. 21.4 Growth of grass shrimp larvae exposed for a total of 14 days to PH from a No. 2 fuel oil WSF. Each data point represents the mean weight of 10-12 larvae. Standard deviations of the data are shown.

Discussion

The potential sublethal effects of PH on aquatic organisms are complex and highly variable. Edible organisms may become tainted when exposed at fairly low concentrations. Complex chemosensory mechanisms involved with feeding and/or mating behavior may be impaired. More importantly, the accumulation of PH will affect important physiological processes such as growth and reproduction unless exposures are of an acute nature. Experiments with lobster larvae, Homarus americanus, have demonstrated a decrease in food consumption at a PH concentration of 0.19 ppm and have found first-stage larvae more susceptible to PH than older larvae (Wells and Sprague 1976). Recent work by Linden (1976) with the amphipod, Gammarus oceanicus has shown larvae several hundred times more sensitive than adult animals. The sublethal effects due to exposure of the amphipods to PH included impaired swimming performance and light reaction and decreased production of larvae. Also, larval growth was affected during chronic exposures (Linden 1976).

Although there have been little data reported on the toxicity and sublethal effects of chronic PH exposure on estuarine animals, data are available on the acute toxicity of No. 2 fuel oil OWD and WSF to Palaemonetes pugio and some other estuarine organisms (Anderson et al. 1974a, Rossi et al. 1976). Reported 96-hr LC50 values for adult grass shrimp were 3.0 ppm PH for OWD and 3.5 ppm PH for WSF. Larvae are more sensitive to the No. 2 fuel oil WSF giving a 96-hr LC50 value of 1.2 ppm PH. Individual naphthalene compounds are lethal to adult grass shrimp in the 0.7-2.4 ppm range for 96-hr exposures (Tatem et al. manuscript submitted to Estuarine and Coastal Marine Science).

The four-day LC50 values for lobster larvae were 0.86 ppm for first-stage larvae and 4.90 ppm for third- and fourth-stage larvae; 30-day values were 0.14 ppm for first-stage larvae (Wells and Sprague 1976). Vanderhorst et al. (1976) determined the 96-hr LC50 value for coon stripe shrimp exposed to No. 2 fuel oil to be 0.8 ppm PH.

The present investigation has demonstrated the rapid accumulation and depuration of naphthalene compounds from an oil-seawater mixture by grass shrimp. Palaemonetes are similar in this respect to the penaeid, Penaeus aztecus (Cox, 1974). It has been noted that, in general, estuarine fish and crustaceans accumulate naphthalenes very rapidly and begin depuration immediately, whereas molluscs accumulate the hydrocarbons more slowly with depuration taking many days (Neff et al. 1976). Accumulation with the grass shrimp was most rapid during the first 2 hr of exposure, and DMN was the compound which usually reached the highest tissue concentration and was the last to be released. Depuration began during the exposure period as the naphthalenes in the water decreased to levels near 0.1 ppm.

The most likely site for the transfer of PH from water to organism tissues and vice versa is the respiratory membrane. Data discussed by Anderson et al. (1974c) have shown that accumulated naphthalenes are associated with the digestive gland and the gill regions of shrimp until their complete release. Since aromatic hydrocarbons are volatile and are released from the surface, organisms which spend more time in the water column might be expected to accumulate greater amounts of PH in shorter time periods compared to more sessile benthic animals. It is possible that the release of naphthalenes is due in part to metabolism of these compounds by the grass shrimp. Lee (1975) has concluded that many crustacean groups including copepods, amphipods, crab zoea, and euphausiids have the ability to metabolize naphthalene as well as other polynuclear hydrocarbons. Further evidence of PH metabolism by crustaceans has been recently published (Lee 1976). Also, Corner et al. (1976) have reported that Calanus helgolandicue released accumulated naphthalene in a water-soluble form other than naphthalene, indicating metabolism. However, Burns (1976) has examined the fiddler crab for its ability to metabolize PH and concluded that these crustaceans cannot significantly increase their rate of oxidation of foreign hydrocarbons. In the majority of these various accumulation studies, whether metabolism occurred or not, exposed organisms retained some percentage of the accumulated hydrocarbons for an indefinite time period. Hydrocarbons were retained in tissues or organs containing significant amounts of lipids. Molluscs may contain greater percentages of lipid materials than fish or crustaceans. This could partially explain why molluscs accumulate and depurate petroleum hydrocarbons at a much slower rate than crustaceans. Also, molluscs have not been found to be able to metabolize these compounds. Lee (1976) presents data on the metabolism of PH by blue crabs and discusses the general subject of metabolism of foreign chemicals by invertebrate and vertebrate marine animals. It is suggested that blue crabs do not retain PH since these lipid-soluble xenobiotics (foreign chemicals) are quickly metabolized into water-soluble compounds (Lee 1976).

Various studies have shown altered respiratory rates by estuarine animals due to exposure to PH. Cox (1974) found that exposure of brown shrimp and mysids to PH always resulted in changed rates; however, in most cases the changes were not significant. The present work has revealed a significant depression of respiratory rate due to exposure to a rather high concentration of PH (3.0-3.6 ppm) as well as a 48-hr starvation period. A fairly rapid recovery was demonstrated seven days after exposure when tissue analyses showed release of the naphthalenes. These data plus other unreported experiments (Tatem 1975) have not found respiration to be a sensitive indicator of the sublethal effects of PH on estuarine organisms.

A single exposure of gravid female grass shrimp to PH had a detrimental effect on larval hatching and daily exposure of larvae delayed normal growth significantly. The gravid females were exposed to three concentrations of PH. Only the highest concentration, 1.4 ppm PH (41% of the 96-hr LC50 value), produced a significant effect, while 0.7 ppm PH had only a slight effect. Growth of the grass shrimp larvae was inhibited at an average PH concentration of 0.67 ppm PH or 19% of the 96-hr LC50 value.

There is evidence that PH's are stored in tissues which contain large amounts of lipids (Stegeman and Teal 1973, Rossi and Anderson 1976). Exposure of the gravid females probably resulted in some PH being sequestered in the egg masses. Since PH incorporated in lipid-rich tissues tends to remain in place, this could explain why larval hatches were affected in clean water up to two weeks after the exposure.

The reason why larval growth was adversely affected by the continuous exposure to PH is related to behavior. Larval grass shrimp attempt to ingest any particulate matter contacted in the water column. Here larvae were fed live _Artemia salina_ nauplii. In order for the larvae to obtain food, they had to remain active in the water column, contact a nauplius, grasp, and ingest it. The more time a single larva spent swimming about or swimming toward a light source, the more likely food was to be encountered. Larvae chronically exposed to PH were not nearly as active as control animals. Also, the food for the experimental group was contaminated to some extent by the PH.

The experiments discussed in the present paper have presented evidence that estuarine grass shrimp exposed to sublethal concentrations of PH are adversely affected. These sublethal effects are ecologically significant and could result in reduction of populations in the field. The recent papers of Linden (1976) and Wells and Sprague (1976) have demonstrated similar sublethal effects and have found larval development and growth of larvae to be sensitive, measurable parameters which were affected by low levels of PH. Very little work has been published on the chronic toxicity of PH to aquatic organisms. However, Wells and Sprague (1976) have stated that the ratio of "safe" to acutely lethal concentrations of PH was about 0.03. Since many estuarine organisms have given acute LC50 values in the range of 1.0-2.0 ppm, this means that estuarine populations, especially invertebrates, could be harmed if water concentrations of PH remained above 30-50 ppb for any length of time. It should be noted that historically most toxicity studies have been conducted with organisms capable of adapting to laboratory conditions, not the most sensitive species. Also, Wells and Sprague (1976) were conservative in their determinations of harmful concentrations of PH for lobster larvae since they used initial concentrations for their calculations. From the trends indicated by the present work plus the recent data of others, it seems evident that PH's have the capacity to adversely affect aquatic life at concentrations of less than 0.5-1.0 ppm. Chronic exposure to PH may produce adverse effects at much lower concentrations.

References

American Petroleum Institute, Determination of volatile and non-volatile oil material. Infrared spectrometric method No. 733-58 (1958).

Anderson, J. W., J. M. Neff, B. A. Cox, J. E. Tatem, and G. M. Hightower, Characteristics of dispersions and water-soluble extracts of crude and refined oils and their toxicity to estuarine crustaceans and fish, Mar. Biol. 27, 75-88 (1974a).

Anderson, J. W., J. M. Neff, B. A. Cox, J. E. Tatem, and G. M. Hightower, The effects of oil on estuarine animals: toxicity, uptake and depuration of respiration. In: Pollution and physiology of marine organisms, pp. 285-310. Ed. by F. J. Vernberg and W. B. Vernberg, Academic Press, New York, 1974b.

Anderson, J. W., R. C. Clark, and J. J. Stegeman, Petroleum hydrocarbons. IN: Marine bioassays workshop proceedings, pp. 36-75. Ed. by F. A. Cross and T. W. Duke, G. V. Cox, Chairman, Marine Technology Society, Washington, D. C., 1974c.

Anderson, J. W., Laboratory studies on the effects of oil on marine organisms: an overview, American Petroleum Institute Pub. No. 4249, 70 pp., 1975.

Brown, R. A., T. D., Searl, and J. J. Elliott, Distribution of heavy hydrocarbons in some Atlantic Ocean waters. In: Proceedings of Joint Conference on Prevention and Control of Oil Spills, pp. 505-519. American Petroleum Institute, Washingtion, D. C., 1973.

Burns, K. A., Hydrocarbon metabolism in the intertidal fiddler crab Uca pugnax, Mar. Biol. 36, 5-11 (1976).

Corner, E. D. S., R. P. Harris, C. C. Kilvington, and S. C. M. O'Hara, Petroleum compounds in the marine food web: short-term experiments on the fate of naphthalene in Calanus, J. Mar. Biol. Ass. U.K. 56, 121-133 (1976).

Cox, B. A., Responses of the marine crustaceans Mysidopsis almyra Bowman, Penaeus aztecus Ives and P. setiferus (Linn.) to petroleum hydrocarbons. Ph.D. Dissertation, Texas A&M University, College Station, Texas, 167 pp. 1974.

Cox, B. A., J. W. Anderson, and J. C. Parker, An experimental oil spill: the distribution of aromatic hydrocarbons in the water, sediment and animal tissues within a shrimp pond. In: Proceedings of Joint Conference on Prevention and Control of Oil Spills, pp. 607-612. American Petroleum Institute, San Francisco, 1975.

Evans, D. R., and S. D. Rice, Effects of oil on marine ecosystems: a review for adminstrators and policy makers, Fish. Bull. 72, 625-638 (1974).

Lee, R. F., Fate of petroleum hydrocarbons in marine zooplankton. In: Proceedings of Joint Conference on Prevention and Control of Oil Spills, pp. 549-533. American Petroleum Institute, San Francisco, 1975.

Lee, R. F., Fate of petroleum hydrocarbons taken up from food and water by the blue crab, Callinectes sapidus, Mar. Biol. 37, 363-370 (1976).

Linden, O., Effects of oil on the amphipod Gammarus oceanicus, Environ. Pollut. 10, 239-250 (1976).

Moore, S. F., and R. L. Dwyer, Effects of oil on marine or ganisms: a critical assessment of published data, Water Res. 8, 819-827 (1974).

Neff, J. M., and J. W. Anderson, An ultraviolet spectrophotometric methods for the determination of naphthalene and alkylnaphthalenes in the tissues of oil-contaminated marine animals, Bull. Environ. Contam. and Tox. 14, 122-128 (1975).

Neff, J. M., B. A. Cox, D. Dixit, and J. W. Anderson, Accumulation and release of petroleum derived aromatic hydrocarbons by four species of marine animals, Mar. Biol. (In Press).

Rossi, S. S., J. W. Anderson, and G. S. Ward, Toxicity of water-soluble fractions of four test oils for the polychaetous annelids, Neanthes arenaceodentata and Capitella capitata, Environ. Poll. 10, 9-18 (1976).

Rossi, S. S., and J. W. Anderson, Toxicity of water-soluble fractions of No. 2 fuel oil and South Louisiana crude oil to selected stages in the life history of the polychaete, Neanthes arenaceodentata, Bull. Environ. Contam. and Tox. 16 (In Press).

Stegeman, J. J., and J. M. Teal, Accumulation, release ans retention of petroleum hydrocarbons by the oyster, Crassostrea virginica, Mar. Biol. 22, 37-44 (1973).

Tatem, H. E., The toxicity and physiological effects of oil and petroleum hydrocarbons on estuarine grass shrimp, Palaemonetes pugio Holthuis. Ph.D. Dissertation, Texas A&M University, College Station, Texas, 133 pp. 1975.

Tatem, H. E., B. A. Cox, and J. W. Anderson, The toxicity of oils and petroleum hydrocarbons to estuarine crustaceans. (Submitted to Estuar. and Coastal Mar. Sci.)

Vanderhorst, J. R., C. I. Gibson, and L. J. Moore, Toxicity of No. 2 fuel oil to coon stripe shrimp, Mar. Poll. Bull. 7, 105-106 (1976).

Warner, J. S., Quantitative determination of hydrocarbons in marine organisms, National Bureau of Standards Special Publication 409, 195-196 (1974).

Wells, P. G., and J. B. Sprague, Effects of crude oil on American lobster Homarus americanus larvae in the laboratory, J. Fish. Res. Board Can. 33, 1604-1614 (1976).

CHAPTER 22

EFFECTS OF A SEAWATER-SOLUBLE FRACTION OF COOK INLET
CRUDE OIL AND ITS MAJOR AROMATIC COMPONENTS
ON LARVAL STAGES OF THE DUNGENESS CRAB,
CANCER MAGISTER DANA

R. S. Caldwell, E. M. Caldarone and M. H. Mallon

Department of Fisheries and Wildlife
Marine Science Center
Oregon State University
Newport, Oregon 97365

Abstract

Larval stages of the Dungeness crab, *Cancer magister* Dana, were exposed continuously to dilutions of Cook Inlet crude oil water-soluble fraction (WSF) of seawater solutions of naphthalene or benzene for periods lasting up to 60 days. Effects on survival, duration of larval development and size were employed as indicators of toxic effects. The lowest concentration of the WSF at which toxic effects were seen was 4.0% of the full strength WSF (0.0049 mg/l as naphthalene or 0.22 mg/l as total dissolved aromatics). The lowest concentration at which toxic effects were observed with naphthalene was 0.13 mg/l and with benzene was 1.1 mg/l.

The concentration of aromatic hydrocarbons in the WSF were inverseley related to the degree of alkylation in each of the benzene and naphthalene families, but the acute toxicity of the 12 compounds was directly related to the degree of alkyl substitution. In addition, naphthalene and its derivatives were more toxic than benzene and its derivatives, but less concentrated in the WSF. Because of these relationships, the individual aromatic compounds, contributed approximately equally to the acute toxicity of the WSF. The collective toxicity of these compounds tested individually accounted for only 8.45% of the WSF acute toxicity. Since benzene contributed a greater fraction of the WSF toxicity in the chronic experiments (approximately 30%) it is suggested that the toxicity of this compound may involve a different mechanism in long term exposures than in acute tests.

Key words: Crustacea, zoeae, *Cancer magister*, petroleum, aromatic hydrocarbons, crude oil, benzene, naphthalene, toxicity.

Introduction

Studies of the biological effects of oil pollution in marine waters have intensified in recent years. Recent reviews have summarized much of this work (Nelson-Smith 1970, 1973). Although the developmental stages of organisms are often more sensitive to toxicants than adults, relatively little work with oil or components of oil has yet been done on these forms. Struhsaker *et al.* (1974) have studied the effects of benzene on developmental stages of two marine fish. Wells and Sprague (1976) have found that emulsions of Venezuelan crude oil were toxic to lobster larvae at concentrations as low as 0.14 mg/l. Katz (1973) reported that cultured larvae of the crab, *Neopanope texana* showed reduced survival when exposed throughout the zoeal stages to a seawater-soluble fraction of Venezuelan crude oil. In this paper we report on the results of a study that was carried out to test the toxic effects of a seawater-soluble fraction of Cook Inlet crude oil and its major aromatic components on larval stages of the Dungeness crab, *Cancer magister* Dana.

Materials and Methods

Long term exposures of Dungeness crab larvae to dilutions of a 1% (1 part crude oil to 100 parts seawater) Cook Inlet crude oil water-soluble fraction (WSF) and to benzene and naphthalene dissolved in seawater were conducted in flowing water laboratory culture systems (Buchanan *et al.* 1975). In the first of two tests series, zoeae were the progeny of a female crab obtained off the coast of Oregon and in the second, they were the progeny of a female obtained from Auke Bay, Alaska. Toxicant exposures were begun within a few hours of hatching of the larvae. Zoeae were cultured at 13°C (10.5-14.2°C) in filtered sterile seawater 29-34 °/oo) saturated with air. The photoperiod was 11 hr of darkness and 13 hr of light. In each test the initial numbers of larvae, 25 to a culture container, were approximately 300 in the control treatment and 100 in each of the toxicant exposure treatments. During the culture period the larvae were transferred to clean culture containers three times per week and fed newly hatched nauplii of San Francisco brine shrimp, *Artemia salina*. Crab larvae reaching the fifth zoeal stage in the first test were fed one week old brine shrimp larvae. Mortality and molt data were recorded daily and the lengths of the cephalothorax, measured across the greatest lateral dimension beginning at the base of the rostral spine, of the surviving larvae were taken at the termination of each experiment to evaluate effects on growth. The criterion of death was development of an opaque appearance to the larvae.

Stock solutions of WSF, benzene and naphthalene used in the long term tests were prepared daily using a procedure similar to that of Anderson et $al.$ (1974). The materials to be tested, 180 ml of crude oil, 3.0 ml of benzene and 360 mg of naphthalene, were added to the surface of 18 l. of seawater in each of three 19 l. pyrex bottles, and the contents were stirred nonturbulently with magnetic stirrers for 20 hours at $13°C$. After allowing the contents to remain undisturbed for an additional 3 hr, these stock solutions were transferred without contamination by surface materials to the Mariotte bottles of the diluter system through glass delivery tubing. Test solutions (Table 1) were obtained by a continuous flow serial dilution of the stock toxicant solutions.

TABLE 1. Toxicant Concentrations Employed in the Chronic Toxicity Bioassays

Toxicant	Test I		Test II	
	Concentration (mg/l.)	N	Concentration (mg/l.)	N
Crude Oil WSF*				
low	0.0022 ± 0.0010**	5	0.0013 ± 0.0003	10
high	0.0083 ± 0.0019	11	0.0049 ± 0.0005	10
Benzene				
low	0.18***		0.17 ± 0.04	11
medium	1.1 ± 0.1	4	1.2 ± 0.2	11
high	7.0 ± 1.6	8	6.5 ± 0.4	11
Naphthalene				
low	0.019 ± 0.001	3	0.021 ± 0.004	10
high	0.17 ± 0.03	13	0.13 ± 0.02	10

*As naphthalene.
**Mean ± one standard deviation.
***Estimated by dilution.

The acute toxicity of the WSF and twelve of its major aromatic hydrocarbon components to newly hatched first instar zoeae was determined in 96-hr static bioassays at $13°C$ in 30 °/oo seawater. Larvae were obtained from an Alaskan female crab and were not fed during the experiment. Stock solutions of individual aromatic hydrocarbons were prepared daily by non-turbulent mixing at $13°C$ of each hydrocarbon with 900 ml of sterile 30 °/oo seawater. Stock solutions of the WSF were prepared as in the long term tests. The concentrations of the stock solutions, siphoned from the bottom of the vessels, were determined, and the solutions were diluted to provide logarithmic series of toxicant concentrations. At least 50 larvae were used per test concentration. Deaths were recorded at 24-hr intervals prior to renewal of the test solutions. The criterion of death was the same as in the chronic bioassays. The acute toxicity of each compound is expressed as 48-hr and 96-hr LC_{50}s, determined according to the graphical interpolation method of the American Public Health Assocation et $al.$ (1971). The LC_{50}s are based on the initial concentrations of toxicants in the test solutions. The concentrations of all of the toxicants tested were found to decline by about 25 to 50% within the 24-hr exposure period.

Routine monitoring of seawater toxicant concentrations in both chronic and acute experiments was performed by UV absorption methods employing a Bausch and Lomb Spectronic 505 spectrophotometer. Extraction of toxicants from seawater was made into UV-quality n-hexane with efficiencies greater than 95% and absorbances at the measuring wavelengths greater than 0.3. The concentrations of the hexane solutions of pure aromatic hydrocarbons were determined by comparison of their absorbances at appropriate wavelength maxima with those of standard solutions. The wavelength maximum employed for benzene was 255 nm; that for naphthalene was 221 nm. The concentration of the WSF was routinely monitored by measurement of the 221 nm peak in hexane extracts of seawater and expressed as naphthalene.

Detailed analyses of the aromatic hydrocarbon composition of the WSFs were performed by gas-liquid chromatography using a Hewlett-Packard 5700 series gas chromatograph equipped with a flame ionization detector and temperature programming. Hexane extracts of the WSF employing 1:10 and 1:100 ratios of hexane to seawater were chromatographed without additional treatment. Extraction efficiencies in all determinations were greater than 95%. Extracts were chromatographed on 2 sets of columns: 6'x1/8" stainless steel packed with 10% SE-30 on 100/120 mesh chromosorb W HP or 10% TCEP [1,2,3-tris(2-cyanoethoxy)propane] on 80/100 mesh chromosorb P AW. Nitrogen was used as the carrier gas at 20 ml/min with each column. The injection port and detector temperatures were 250°C and 300°C, respectively. Separations on 10% TCEP were performed isothermally at 80°C, whereas temperature programming was employed with the 10% SE-30 columns. The programming rate was 8°C/min with an initial temperature of 80°C held for 8 min, and a final temperature of 260°C. Qualitative and quantitative analyses were accomplished by reference to pure standards of each hydrocarbon.

Cook Inlet crude oil was purchased from Shell Oil Company. Reagent grade naphthalene was obtained from J.T. Baker Chemical Company, chromatoquality m- and o-xylene and reagent grade toluene from Matheson, Coleman and Bell and UV-grade n-hexane and benzene from Burdick and Jackson Laboratories, Inc. Other aromatic hydrocarbons were purchased from Chem Service, Inc., West Chester, Pa. Column packing materials were purchased from Hewlett-Packard, Inc. and Supelco, Inc.

Statistical differences in mean duration of larval development to each stage and in larval sizes were determined with Student's t-test.

Results

Dungeness crab zoeae reared in dilutions of the WSF and in naphthalene seawater solutions in the initial long term tests did not exhibit a higher mortality rate than control zoeae even at the highest concentrations tested, 0.0083 mg/l. as naphthalene for the WSF and 0.17 mg/l. naphthalene (Fig. 1). Similarly, the highest naphthalene concentration tested in the second experiment, 0.13 mg/l., had no effect on survival, but larvae exposed to the highest concentration of the WSF, 0.0049 mg/l. as naphthalene, exhibited a slightly poorer survival than control animals (Fig. 2). At the end of the test on day 40, only 20% of the original number of larvae in this treatment survived compared with 36% of the control zoeae.

Exposures of larvae to benzene at concentrations of 1.1 and 7.0 mg/l. in the first test and 1.2 and 6.5 mg/l. in the second test had substantial effects on survival (Figs. 1 and 2). In each of these treatments an accelerated mortality rate first became apparent after day 10 and was coincident with the timing of the first molt in control larvae. At day 10, the condition of these larvae appeared to be poor. At the lower two concentrations, 1.1 and 1.2 mg/l., the mortality rates were less than at 7.0 and 6.5 mg/l. benzene, but even at these lower concentrations the majority of deaths occurred before day 20. Larvae exposed to 0.18 and 0.17 mg/l. benzene in the two tests survived as well as the controls.

In contrast to its adverse effect on survival, benzene, even at lethal concentrations (1.1 and 1.2 mg/l.), had little effect on the duration of larval stages (Table 2). The only exception was a slight delay (P=0.05) in the mean time to the second molt of larvae exposed to 1.1 mg/l. benzene. In addition, benzene did not affect the size of larvae surviving the tests (Table 3).

Neither the WSF nor naphthalene appeared to influence the duration of larval development in the first test which employed Oregon larvae, but exposure of Alaskan larvae to the highest concentrations of each of these toxicants in the second series of tests resulted in significant delays to each of the first, second and third molts (Table 2). As an example of the distribution of molting through time, percentage molted per day in the 0.13 mg/l. naphthalene treatment and in the controls, respectively, for days 11 to 16 was: 0, 28, 52, 14, 5, 2; and 18, 48, 30, 3, 1, 0. By the time of the third molt, the mean delay of larvae exposed to the highest concentration of the WSF was 2.9 days and that of zoeae exposed to 0.13 mg/l. naphthalene was 1.5 days. No significant delays were observed at the lower concentrations of these two toxicants.

Naphthalene did not affect the size of larvae surviving the exposures compared with controls in either test even at the highest concentrations (Table 3). In the tests with the WSF, the only significant difference between treatments and controls was with the highest concentration of the WSF in the first test with Oregon larvae. In this group the small number of surviving larvae were larger than the controls by 0.20 mm, averaging 2.74 mm across the cephalothorax (Table 3).

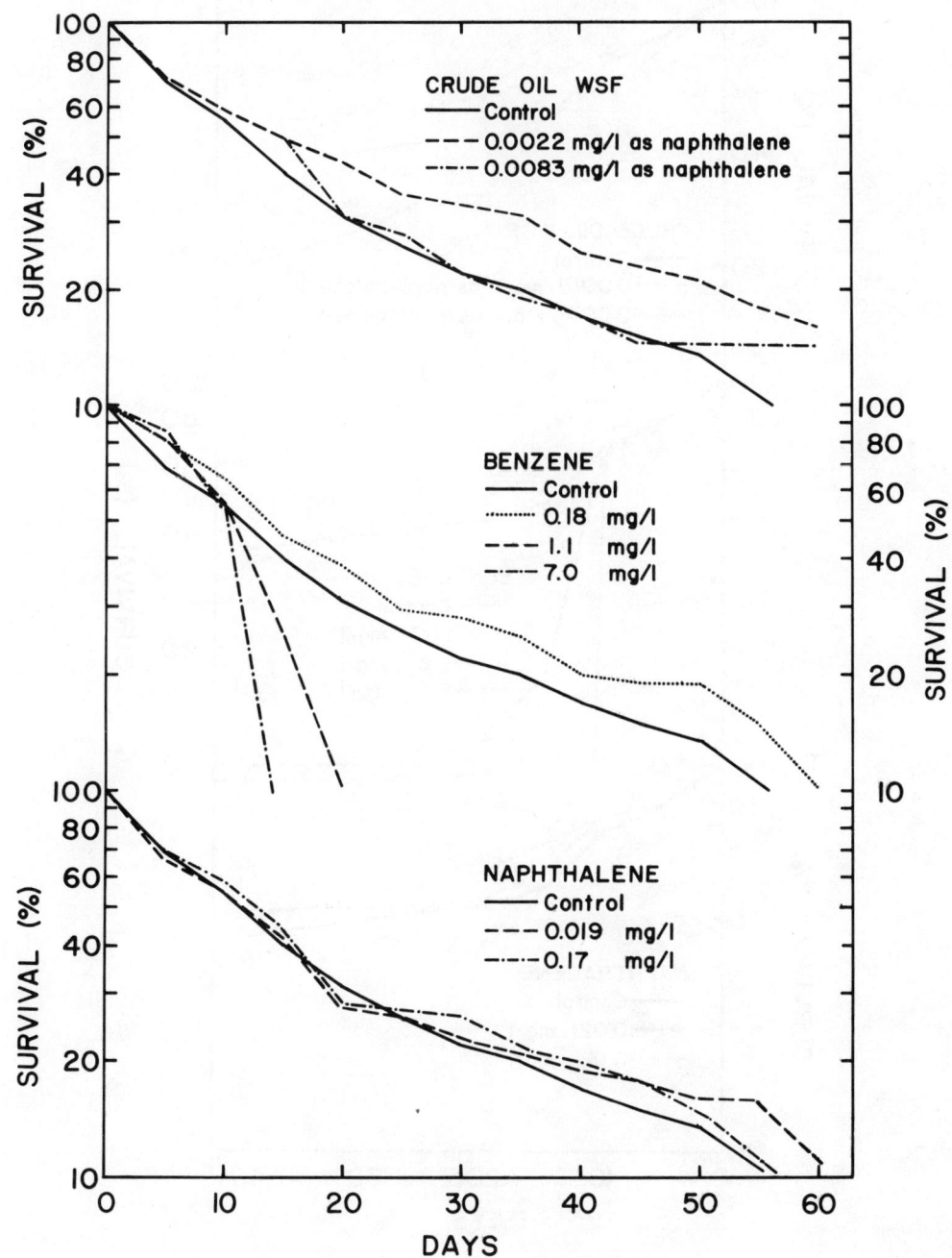

Fig. 22.1 Survival of *C. magister* zoeae exposed continuously to different concentrations of crude oil WSF, benzene and naphthalene. Results are from the first long-term experiment which employed the progeny of an Oregon female crab.

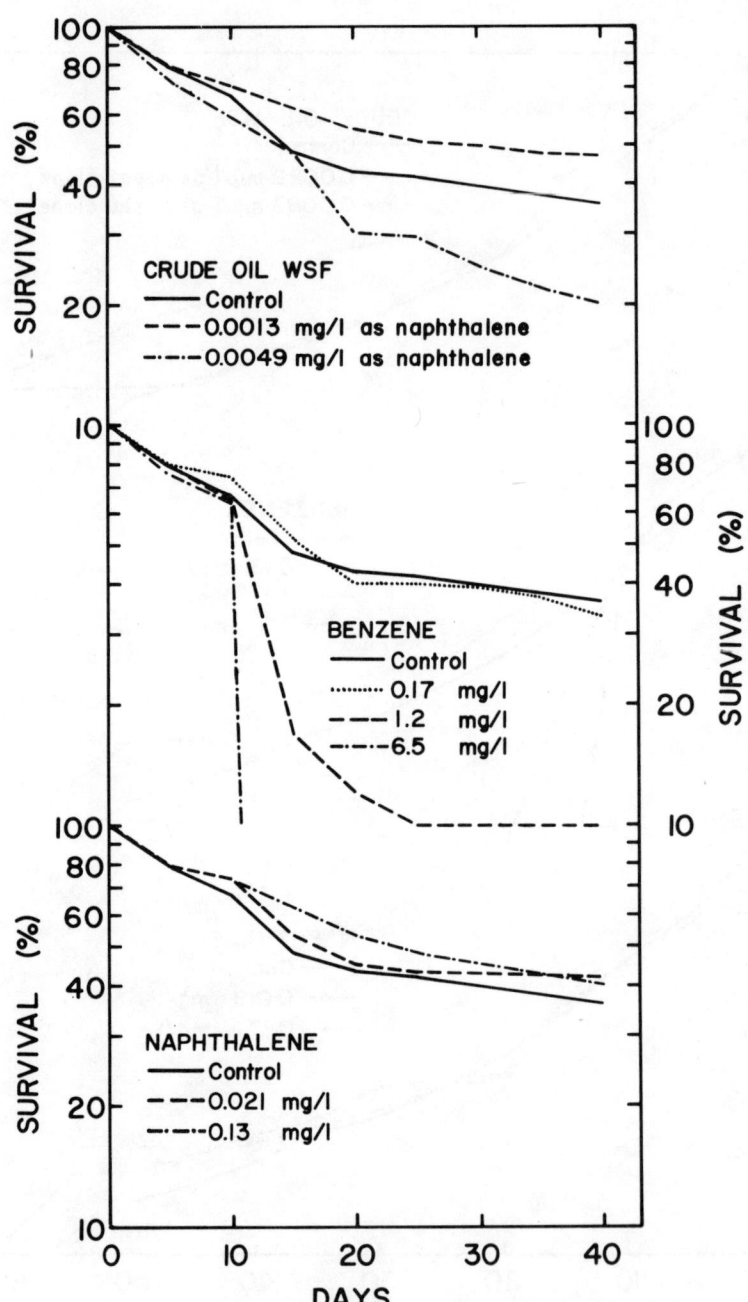

Fig. 22.2 Survival of *C. magister* zoeae exposed continuously to different concentrations of crude oil WSF, benzene and naphthalene. Results are from the second long-term experiment which employed the progeny of an Alaskan female crab.

TABLE 2. Effect of the crude oil WSF, benzene and naphthalene on the duration of larval development to the end of the first, second, third and fourth zoeal stages

Toxicant (mg/l.)	First stage			Second stage			Third stage			Fourth stage		
	N	Duration (days)	t	N	Duration (days)	t	N	Duration (days)	t	N	Duration (days)	t
					TEST I							
SW control	107	12.4 ± 1.1*		68	22.1 ± 2.2		38	33.8 ± 2.2		27	49.1 ± 3.6	
Crude oil WSF												
0.0022	45	12.4 ± 1.8	-0.12	31	21.5 ± 1.3	1.19	23	34.6 ± 2.8	-1.31	16	49.6 ± 3.1	-0.45
0.0083	35	12.1 ± 0.7	1.45	23	22.0 ± 1.2	0.21	15	33.9 ± 1.4	-0.18	8	47.0 ± 1.2	1.65
Benzene												
0.18	37	12.4 ± 1.0	-0.09	25	21.9 ± 1.5	0.37	20	34.8 ± 2.2	-1.68	9	49.7 ± 1.7	-0.42
1.1	12	12.5 ± 0.9	-0.32	6	24.5 ± 1.8	-2.61**	0	---	---	0	---	---
7.0	0	---	---	0	---	---	0	---	---	0	---	---
Naphthalene												
0.019	33	12.9 ± 1.2	-2.37**	24	22.7 ± 2.1	-1.24	16	34.8 ± 2.4	-1.51	10	50.3 ± 3.6	-0.86
0.17	32	12.6 ± 1.4	-0.68	22	22.0 ± 1.2	-0.02	15	34.3 ± 2.6	-0.78	4	48.2 ± 2.6	0.48
					TEST II							
SW control	186	12.2 ± 0.8		120	21.6 ± 1.8		117	32.3 ± 2.3				
Crude oil WSF												
0.0013	67	12.4 ± 0.9	-1.58	51	21.7 ± 1.8	-0.27	48	32.6 ± 2.4	-0.80			
0.0049	44	13.6 ± 1.3	-8.83***	30	23.1 ± 2.6	-3.60***	24	35.2 ± 3.4	-4.88***			
Benzene												
0.17	54	12.1 ± 1.0	1.05	39	21.3 ± 1.7	1.04	39	31.7 ± 2.1	1.45			
1.2	31	12.0 ± 0.8	0.26	11	21.8 ± 1.3	-0.30	9	32.6 ± 1.7	-0.29			
6.5	0	---	---	0	---	---	0	---	---			
Naphthalene												
0.021	63	12.2 ± 0.9	0.16	54	21.5 ± 1.5	0.60	52	31.4 ± 1.6	2.38			
0.13	64	13.0 ± 0.9	-6.60***	43	23.1 ± 1.8	-4.65***	42	33.8 ± 2.8	-3.15***			

*Mean ± one standard deviation.
**Value of Student's t significant at the 0.05 level.
***Value of Student's t significant at the 0.01 level.

TABLE 3. Effect of the Crude Oil WSF, Benzene and Naphthalene on the Size of the Cephalothorax of Surviving Fifth Stage Zoeae at the Termination of the Second Chronic Exposure Test

Toxicant (mg/l.)	N	Size (mm)	t
		TEST I	
SW control	5	2.54 ± 0.12*	
Crude oil WSF			
0.0022	8	2.41 ± 0.05	1.85
0.0083	6	2.74 ± 0.14	-2.59**
Benzene			
0.18	6	2.50 ± 0.12	0.48
1.1	0	---	----
7.0	0	---	----
Naphthalene			
0.019	3	2.52 ± 0.22	0.15
0.17	2	2.50 ± 0.10	0.32
		TEST II	
SW control	105	1.86 ± 0.08	
Crude oil WSF			
0.0013	45	1.84 ± 0.10	0.80
0.0049	17	1.85 ± 0.08	0.31
Benzene			
0.17	31	1.87 ± 0.09	0.38
1.2	8	1.79 ± 0.08	1.60
6.5	0	---	----
Naphthalene			
0.021	24	1.88 ± 0.10	-0.80
0.13	38	1.84 ± 0.09	1.09

*Mean ± one standard deviation.
**Value of Student's t significant at the 0.05 level.

Since the aromatic hydrocarbons are often considered to be the most toxic components of crude oil (Boylan and Tripp 1971, Moore and Dwyer 1974, Struhsaker et al. 1974), we decided that an analysis of these compounds in our WSF together with a determination of their acute toxicities to first stage crab larvae might be valuable in identifying the major toxicants in the oil WSF. Table 4 lists the concentrations of 12 of the principal aromatics in our WSF. Benzene, at 3.15 mg/l., was the most concentrated of the aromatic hydrocarbons analyzed. There were lesser amounts of the alkylated benzenes; their concentrations bearing an inverse relationship with the degree of alkylation. The concentrations of naphthalene and its alkylated derivatives in the WSF were also inversely related to the degree of side chain substitution, but, in contrast to benzene, the concentration of naphthalene was only 0.06 mg/l.

Table 5 lists the acute toxicities of the same 12 aromatic hydrocarbons and the WSF to first stage zoeae. The 96-hr LC_{50} for the WSF was 0.038 mg/l. as naphthalene, a concentration 31% of the full strength WSF (Table 4). Of the individual aromatic components, benzene was the least toxic with a 96-hr LC_{50} of 108 mg/l., while dimethylnaphthalene (a mixture of isomers) was the most toxic with a 96-hr LC_{50} of 0.60 mg/l. Within both the benzene and naphthalene series, the acute toxicities of these compounds exhibited a clear pattern of increasing toxicity with increasing side chain substitution. Furthermore, naphthalene and its alkylated derivatives were more toxic than benzene and its derivatives. Although only one ethyl-substituted aromatic was tested, the results suggest that ethyl-substitution results in greater toxicity than methyl-substitution; the 96-hr LC_{50} for toluene (methyl-benzene) was 28 mg/l. compared with a 96-hr LC_{50} of 13 mg/l. for ethyl-benzene.

TABLE 4. Concentration of the Principal Aromatic Hydrocarbons
in Cook Inlet Crude Oil WSFs*

Compound	Concentration (mg/l.)
Benzene	3.15 ± 0.43**
Toluene	1.52 ± 0.18
m-, p-xylenes	0.30 ± 0.05
o-xylene	0.19 ± 0.03
Ethylbenzene	0.13 ± 0.02
Naphthalene	0.06 ± 0.02
1,2,4-trimethylbenzene	0.06 ± 0.01
2-methylnaphthalene	0.03 ± 0.01
1-methylnaphthalene	0.03 ± 0.01
Dimethylnaphthalenes	0.02 ± 0.01
1,3,5-trimethylbenzene	0.01 ± 0.01
1,2,4,5-tetramethylbenzene	0.01 ± 0.00
Total measured aromatics	5.52 ± 0.69
WSF concentration as naphthalene (by UV absorption method)	0.122 ± 0.013

*Full strength WSFs prepared as described in the Materials and Methods.
**Mean ± one standard deviation. N = 3.

Discussion

In our studies, the 96-hr LC_{50} for Cook Inlet crude oil WSF was found to be 31% of the stock WSF. Larval cultures were not affected by continuous exposures to WSFs representing 1.8 and 1.1% of the stock fractions (0.0022 and 0.0013 mg/l. as naphthalene, respectively). The initiation of toxic effects during chronic exposures appeared to occur near 4.0% of the full strength WSF (0.0049 mg/l. as naphthalene) since survival was slightly reduced (Fig. 2) and a retardation of developmental rate was noted (Table 2) in the second experiment at this concentration, but the size of surviving larvae was not affected (Table 3). No effects were seen, however, in the first test at 6.8% of the WSF (0.0083 mg/l. as naphthalene). The difference in response of larvae in the two tests could represent differences in the sensitivity of Oregon and Alaskan populations of the crab to the WSF. Such variability of response may, however, be expected even within a single population as has been seen with the larvae of *Callinectes sapidus* for the pesticide mirex (Bookhout and Costlow 1975).

Since the sum of concentrations of the measured aromatics in our WSF is 5.52 mg/l. (Table 4), the initially toxic concentration of 4.0% of the stock WSF would be equivalent to 0.22 mg/l. as aromatics. Anderson *et al.* (1974), in studies with WSFs of two crude oils, found that the concentration of aromatics was approximately 50% of the total dissolved hydrocarbons. If a similar relationship exists for our WSF, we would expect the total dissolved hydrocarbons in our WSF at the initially toxic concentration to be approximately 0.44 mg/l. The use of this value allows us to compare our results with those of Katz (1973) and Wells and Sprague (1976) who studied the toxicity of Venezuelan crude oil to larval crabs and lobsters, respectively.

Katz (1973) found that the larvae of *N. texana* exposed to full strength 1% WSF from the day of hatching, experienced about 60% mortality between days 2 and 5 of culture, but thereafter experienced only a low rate of dying until the termination of the experiment on day 14. Larvae exposed to the WSF beginning with the second zoeal stage on day 4 until the end of the experiment, however, did not die at a significantly greater rate than the control larvae. Although Katz did not employ dilutions of his WSF, his results with delayed exposures suggest that the full strength fraction was close to a lethal threshold. The concentration of the WSF was reported to be 4 ppm as total hydrocarbons, about 10 times our estimate of the initially toxic level of total dissolved hydrocarbons in our WSF.

Our results are not as easily compared with those of Wells and Sprague (1976) since they employed dispersions of oil in water rather than WSFs. They found, however, that the 30-day LC_{50} for lobster larvae and the threshold of retardation of larval development were both about 0.14 mg/l. for larvae starting the test as first instars; a concentration remarkably similar to our estimate (0.44 mg/l. as total dissolved hydrocarbons) of the concentration of Alaskan crude oil WSF initially toxic to Dungeness crab larvae.

TABLE 5. Acute Toxicity of the Crude Oil WSF and the Principal Aromatic Hydrocarbon Compounds of the WSF to the First Instar Zoeae

Toxicant	LC50 (mg/l.)	
	48-hr	96-hr
Crude oil WSF (as naphthalene)	0.082	0.038
Benzene	>347.	108.
Toluene (methylbenzene)	170.	28.
Ethylbenzene	40.	13.
m-xylene (1,3-dimethylbenzene)	33.	12.
o-xylene (1,2-dimethylbenzene)	38.	6.0
1,2,4-trimethylbenzene	17.	5.1
Mesitylene (1,3,5-trimethylbenzene)	13.	4.3
Durene (1,2,4,5-tetramethylbenzene)	>3.1	2.1
Naphthalene	--	>2.0
1-methylnaphthalene	8.2	1.9
2-methylnaphthalene	5.0	1.3
Dimethylnaphthalene (mixture of isomers)	3.1	0.60

Little information is available in the published literature on the toxicity of individual chemical species found in crude oil WSFs that might allow a reconstruction of the toxicity of the whole fraction. Our acute toxicity studies established that the toxicity of aromatic hydrocarbons to zoeae increased with the degree of alkylation within each of the benzene and naphthalene families of compounds. Additionally, the naphthalene derivatives were substantially more toxic than the benzene derivatives. McAuliffe (1966) has shown that the alkylation of benzene results in decreasing water solubility; at 20°C the solubilities of benzene, toluene, o-xylene, ethylbenzene and 1,2,4-trimethylbenzene are 1780, 515,175,152 and 57 ppm, respectively. These data suggest that the relative toxicity of the various aromatic compounds is largely related to their ability to partition into the lipophilic components of the larvae. A similar conclusion was reached by Currier (1951) who found that the herbicidal effectiveness of benzene and several of its methylated derivatives to plants was inversely correlated with the water to paraffin oil partition coefficients. Presumably the more lipophilic compounds could more readily penetrate cell membrane barriers.

The concentration of an individual aromatic hydrocarbon in the WSF is a function both of the water solubility of the compound and its concentration in the parent oil. Probably due largely to the water solubility characteristics of the compounds, the concentrations of the specific aromatics in our WSF tended to decrease with increasing alkyl-substitution of the aromatic rings. As a result the most toxic compounds were found in the lowest concentrations and consequently the fractional toxicities of the compounds were all similar based on acute toxicity experiments (Table 6). Since the aromatics are generally considered to be the most toxic components in crude oils (Boylan and Tripp 1971), Moore and Dwyer 1974, Struhsaker et al. 1974), it is curious that we have been able to account for only 8.45% of the toxicity of the whole WSF by these compounds. These results suggest either that the alkanes and cycloalkanes are more toxic relative to the aromatics than previously supposed, that other very highly toxic compounds which are present only in low concentration occur in the WSF, or that the toxicity of the aromatics, and possibly other compounds, is more than additive in the whole fraction. At present we do not have sufficient information to evaluate these several alternatives.

Although benzene accounted for only 0.91% of the acute toxicity of the WSF, this compound contributed a higher fraction of the toxicity of the WSF in long-term tests. We found in chronic exposures that 1.1 mg/l. benzene markedly affected larval survival, but 0.18 mg/l., did not (Fig. 1). Therefore, the threshold of benzene toxicity lies between these two concentrations, say 0.44 mg/l., the logarithmic average of 1.1 and 0.18 mg/l. However, at the initial toxic concentration of 4.0% of the stock WSF in chronic tests the concentration of benzene would be only 0.13 mg/l. Benzene could, therefore, account for approximately 30% of the toxicity of the WSF in the chronic toxicity experiments. Employing the same reasoning for naphthalene where 0.13 mg/l. may be taken as the initial toxic concentration for this compound (Table 2), this diaromatic compound would account for only 1.8% of the threshold toxicity of the WSF. While the latter percentage is in fair agreement with the

TABLE 6. Fractional Toxicity of Specific Aromatic Hydrocarbons in the Cook Inlet Crude Oil WSF

Compound	Percent of the 96-hr toxic concentration of each compound*	
	In full strength WSF (0.122 mg/l. as naphthalene)	In the 96-hr toxic concentration of the WSF (0.038 mg/l. as naphthalene)
Benzene	2.92	0.91
Toluene	5.43	1.69
Ethylbenzene	1.00	0.31
m-xylene	2.50	0.78
o-xylene	3.17	0.99
1,2,4-trimethylbenzene	1.18	0.37
1,3,5-trimethylbenzene	0.23	0.07
1,2,4,5-tetramethylbenzene	0.48	0.15
naphthalene	<3.00	<0.93
1-methylnaphthalene	1.58	0.49
2-methylnaphthalene	2.31	0.72
Dimethylnaphthalene	3.33	1.04
Total	27.13%	8.45%

*Percent = $\dfrac{\text{concentration of hydrocarbon in WSF}}{\text{96-hr LC50 of hydrocarbon}} \times 100$

fractional toxicity of naphthalene as determined in the acute toxicity experiments (Table 6), that for benzene is not, suggesting that the mechanism of benzene toxicity may be different in acute and chronic exposures. Such a suggestion is consistent with our observation that the death of benzene exposed larvae in the long term tests was conspicuously associated with the time of molting in control larvae.

Acknowledgements

This research was supported by NOAA Outer Continental Shelf Environmental Assessment Program under Task Order No. 3, Contract No. 03-5-022-68. Technical Paper No. 4377, Oregon Agricultural Experiment Station. The authors are grateful to Dr. J. Karinen, NMFS, Auke Bay, Alaska for collection and shipment of several Alaskan female crabs.

References

American Public Health Association, American Water Works Association and Water Pollution Control Federation, Standard Methods for the Examination of Water and Wastewater, 13th ed, Amer. Public Health Assoc., Washington, D.C., 1971.

Anderson, J. W., J. M. Neff, B. A. Cox, H. E. Tatem and G. M. Hightower, Characteristics of dispersions and water-soluble extracts of crude and refined oils and their toxicity to to estuarine crustaceans and fish, Marine Biology 27, 75-88 (1974).

Bookhout, C. G. and J. D. Costlow, Jr., Effects of mirex on the larval development of blue crab, Water, Air, Soil Pollut. 4, 113-126 (1975).

Boylan, D. B. and B. W. Tripp, Determination of hydrocarbons in seawater extracts of crude oil and crude oil fractions, Nature 230, 44-47 (1971).

Buchanan, D. V., M. J. Myers and R. S. Caldwell, Improved flowing water apparatus for the culture of brachyuran crab larvae, J. Fish. Res. Bd. Can. 32, 1880-1883 (1975).

Currier, H. B., Herbicidal properties of benzene and certain methyl derivatives, Hilgardia, 20, 383-406 (1951).

Katz, L. M., The effects of water soluble fraction of crude oil on larvae of the decapod crustacean Neopanope texana (Sayi), Environ. Pollut. 5, 199-204 (1973).

McAuliffe, C., Solubility in water of paraffin, cycloparaffin, olefin, acetylene, cycloolefin, and aromatic hydrocarbons, J. Phys. Chem. 70, 1267-1275 (1966).

Moore, S. F. and R. L. Dwyer, Effects of oil on marine organisms: a critical assessment of published data, Water Res. 8, 819-827 (1974).

Nelson-Smith, A., The problem of oil pollution of the sea, Adv. Marine Biol. 8, 215-306 (1970).

Nelson-Smith, A., Oil Pollution and Marine Ecology, Plenum Press, N.Y., 1973.

Struhsaker, J. W., M. B. Eldridge and T. Echeverria, Effects of benzene (a water-soluble component of crude oil) on eggs and larvae of Pacific herring and northern anchovy, pp. 253-284, In: F. J. Vernberg and W. B. Vernberg (eds.) Pollution and Physiology of Marine Organisms, Academic Press, New York, 1974.

Wells, P. G. and J. B. Sprague, Effects of crude oil on American lobster (Homarus americanus) larvae in the laboratory, J. Fish. Res. Bd. Can. 33, 1604-1614 (1976).

CHAPTER 23

MOLTING AND SURVIVAL OF KING CRAB (PARALITHODES CAMTSCHATICA) AND COONSTRIPE SHRIMP (PANDALUS HYPSINOTUS) LARVAE EXPOSED TO COOK INLET CRUDE OIL WATER-SOLUBLE FRACTION

T. Anthony Mecklenburg, Stanley D. Rice, and John F. Karinen

Northwest and Alaska Fisheries Center Auke Bay Fisheries Laboratory
National Marine Fisheries Service, NOAA
P.O. Box 155, Auke Bay, AK 99821

Abstract

Larvae of coonstripe shrimp and king crab were exposed to solutions of the water-soluble fraction (WSF) of Cook Inlet crude oil in a series of bioassays on intermolt stages I and II and the molt period from stage I to stage II. Molting larvae were more sensitive than intermolt larvae to the WSF, and molting coonstripe shrimp larvae were more sensitive than molting king crab larvae. When molting larvae were exposed to high concentrations of the WSF (1.15-1.87 ppm total hydrocarbons) for as little as 6 hr, molting success was reduced by 10-30% and some deaths occurred. When larvae were exposed to these high concentrations for 24 hr or longer, molting declined 90-100% and the larvae usually died. The lowest concentrations tested (0.15-0.55 ppm total hydrocarbons) did not inhibit molting at any length of exposure, but many larvae died after molting. Median lethal concentrations (LC50's) based on 144 hr of observation for molting coonstripe shrimp and 120 hr for molting king crab were much lower than the 96-hr LC50's, showing that the standard 96-hr LC50 is not always sufficient for determining acute oil toxicity. Although our LC50's for intermolt larvae are higher than levels of petroleum hydrocarbons reported for chronic and spill situations, some of our LC50's for molting larvae exposed 24 hr and longer are similar to or below these environmental levels. Comparisons of sensitivity to oil between different crustacean species or life stages should be based on animals tested in the same stage of the molt cycle, such as intermolt.

Key Words: Molting, crustaceans, larvae, *Paralithodes camtschatica*, *Pandalus hypsinotus*, crude oil, Alaska.

Introduction

The breeding and larval stages of marine invertebrates are considered to be the most sensitive to natural environmental stresses (Thorson 1950). In crustacean larvae this sensitivity is compounded during molting. Factors contributing to a high natural mortality in molting crustaceans include increased permeability and altered ionic regulation, the mechanical process of casting off the old exoskeleton, and increased predation while the cuticle is still soft and locomotion is slowed down (Lockwood 1967, Hagerman 1973). Manmade stresses such as pollution from offshore oil production and transoceanic transport of crude and refined oils could impose an additional burden. Pollution by petroleum hydrocarbons in Alaskan surface waters would be particularly damaging to crustacean larvae in the early spring when the larvae are released from the females and undergo several rather closely spaced molts. Cold temperature is another factor to consider in arctic and subarctic waters because aromatic hydrocarbons persist longer (Atlas and Bartha 1972, Cheatham et al. 1976) and some animals retain accumulated hydrocarbons for longer periods of time (Short and Rice In prep.) than would be the case in warmer waters.

The sensitivity of molting crustacean larvae to oil pollution has been reported for only a few species. Oil was more toxic to molting larvae than intermolt larvae in tests with a lobster, *Homarus americanus* (Wells 1972, Wells and Sprague 1976), and the crabs *Neopanope texana* (Katz 1973) and *Rhithropanopeus harrisii* (Neff et al. 1976). Sublethal effects on these crustaceans included both delayed and stimulated molting.

For this study we chose two crustaceans that are important Alaskan fishery resources, the king crab, *Paralithodes camtschatica*, and the coonstripe shrimp, *Pandalus hypsinotus*, and exposed their larvae to various concentrations of the water-soluble fraction of Cook Inlet crude oil. Separate bioassays were conducted on larvae during intermolt stages I and II and during the molting period from stage I to stage II. We wished to determine sensitivity differences between the intermolt and molt stages and the effects of various concentrations and lengths of exposure to the WSF on molting success and survival.

Methods

Mixing and Analysis of Oil Solutions

Solutions of the water-soluble fraction of Cook Inlet crude oil were prepared at ambient seawater temperatures (3°-5°C) and salinity (29 °/oo) by the methods of Anderson et al (1974) as modified by Rice et al (In press). Samples of the test solutions were taken at the beginning of each exposure and whenever a solution was replaced with a fresh solution. The samples were measured by infrared spectrophotometry (IR) at a wavelength of 2930 cm^{-1} (Gruenfeld 1973). At this wavelength IR measures paraffinic hydrocarbons, but not aromatics. Optical densities were converted to ppm total hydrocarbons by comparing the optical densities to prepared standards. Gas chromatographic characterization of a typical WSF of Cook Inlet crude oil has been given by Rice et al (1976).

Collection and Rearing of Larvae

Ovigerous coonstripe shrimp and king crab were caught in pots or collected by divers and held in flowing seawater tanks at the Kasitsna Bay and Auke Bay laboratories. Individual females were isolated in rearing containers just before hatching and release of the larvae. When the larvae were released the females were removed from the containers. The water in the rearing containers was aerated continuously and changed daily. Larvae were fed daily on detritus and laboratory-reared *Artemia*.

Bioassays

The bioassays were conducted in sealed 200-ml and 500-ml glass jars submerged in a waterbath chilled to ambient seawater temperature. There were replicate jars for each of 7 to 12 concentrations of the WSF. There were 10-20 larvae in each jar and the tissue weight/volume ratio in the test jars never exceeded 1 g/liter. Oxygen concentrations measured at the end of the exposure periods were above 70% saturation.

Observations of death and molting success were made without removing the larvae from the test jars. This eliminated handling of the larvae, which could be detrimental. We observed the larvae in each test jar by holding the jar horizontally above a 24-in convex magnifying mirror. Larvae were recorded as dead when there was no visible motion.

Intermolt stage I king crab and coonstripe shrimp larvae were bioassayed within 24 hours of release from the females. The first intermolt stage lasted about 10 days, so we could be sure that molting would not begin during these bioassays. Bioassays on molting larvae were begun 1 or 2 days before the larvae started to molt to stage II. Intermolt stage II coonstripe shrimp larvae were bioassayed 3 days after molting was completed. The second intermolt stage also lasted 10 days. No bioassays were conducted on intermolt stage II king crab larvae.

For intermolt larvae we conducted static 96-hour bioassays (single-dose with declining concentration). Three separate bioassays were conducted on coonstripe shrimp intermolt stage I larvae, and one bioassay each for coonstripe shrimp intermolt stage II and king crab intermolt stage I.

We conducted several bioassays on molting larvae, and each bioassay had a different exposure period: 6, 24, 48, and 96 hr for coonstripe shrimp, and 6, 12, 24, 48, and 72 hr for king crab. In exposures longer than 24 hr the old WSF solution was replaced with fresh solution every 24 hr. After the exposure periods the larvae were returned to uncontaminated seawater for further observations. The total length of time for each test, including the exposure period and the period in clean seawater, was 120 hr for king crab larvae and 144 hr for coonstripe shrimp larvae.

Statistical Analysis

The acute toxicity of the Cook Inlet crude oil WSF to intermolt and molting larvae was expressed as the median lethal concentration (LC50), which is the concentration of WSF causing death in 50% of exposed larvae in a given amount of time. For intermolt larvae we give the standard 96-hr LC50. For molting larvae we give 96-hr LC50's which, although they are each based on 96 hr of observation, involve 24 hr of exposure and 72 hr in clean seawater.

The LC50's and 95% fiducial limits were calculated by probit analysis (Finney 1971) where possible. Where the data were insufficient for probit analysis we determined LC50's by the methods of Spearman-Karber (Finney 1971) and Doudoroff et al (1951). Abbott's formula (Finney 1971) was used to compensate for control deaths.

Results

Molting larvae of both coonstripe shrimp and king crab were significantly more sensitive to the Cook Inlet crude oil WSF than were intermolt larvae of either species (LC50's with nonoverlapping fiducial limits, Table 1). Molting coonstripe shrimp larvae were four to eight times more sensitive to the WSF than were intermolt stages I and II. The tolerance of king crab larvae during molting was not decreased to the extreme observed in coonstripe shrimp larvae.

TABLE 1. The 96-hr LC50's for Molting and Intermolt Larvae of Coonstripe Shrimp, Pandalus hypsinotus, and king crab, Paralithodes camtschatica, Exposed to the WSF of Cook Inlet Crude Oil. The 95% Fiducial Limits are Given in Parentheses.

Larval stage	96-hr LC50's in ppm of total hydrocarbons	
	Coonstripe shrimp	King crab
Intermolt larvae:		
Stage I	7.94* (6.05-9.50)	2.00 (1.60-2.60)
Stage II	4.06 (3.22-5.11)	-- --
Molting larvae:		
Stage I-Stage II	0.95 (0.87-1.03)	1.33 (1.23-1.45)

*Mean of three separate bioassays.

As the WSF concentrations and exposure periods increased, the number of coonstripe shrimp larvae that molted to stage II declined (Fig. 1). Control coonstripe shrimp larvae for each bioassay completed molting to stage II, with 95-100% molting success, by the sixth day or 144 hr after the beginning of each bioassay. At the lowest concentration tested, 0.25 ppm total hydrocarbons, there was little or no effect on molting success even when the larvae were exposed to replenished WSF solutions for as long as 96 hr. The next highest concentration, 0.63 ppm total hydrocarbons, did not affect molting success when the larvae were exposed only 6 hr, but was progressively more effective in inhibiting molting with each increased duration of exposure. The highest concentrations, 1.15 and 1.37 ppm, severely inhibited molting in exposures of 24 hr or longer. Molting success was only about 10% in these longer exposures to high concentrations and did not drop below 10% even when the WSF solutions were renewed daily.

Molting success in king crab larvae exposed to the WSF followed essentially the same pattern of decline (Fig. 2). Control groups for king crab larvae molting to stage II reached a maximum molting success of 37% by the end of 120 hr of observation. Exposure of king crab larvae to the two lowest concentrations, 0.15 and 0.55 ppm total hydrocarbons, for 6 and 12 hr resulted in some stimulation of molting compared to controls. Longer exposures to these low concentrations had little effect, even when the larvae were exposed for 72 hr to periodically-renewed solutions. The 1.20 ppm concentration had no effect on molting success at exposures less than 48 hr, but 48- and 72-hr exposures completely inhibited molting. The highest concentrations, 1.65 and 1.87 ppm, progressively inhibited molting with increased exposure time until the 48- and 72-hr exposures which again resulted in zero molt.

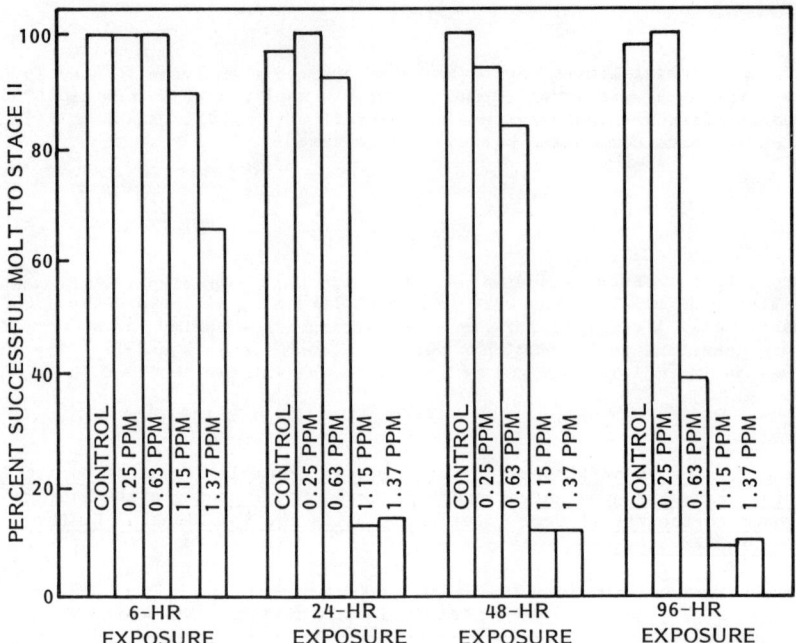

Fig. 23.1. Molting success of larvae of coonstripe shrimp, *Pandalus hypsinotus*, exposed for increased lengths of time to increased concentrations of the WSF of Cook Inlet crude oil. After the exposure time listed, the exposure water was replaced with clean seawater to make a total of 144 hr of observation for each test. Molting success at 144 hr = stage II larvae/initial total of stage I larvae.

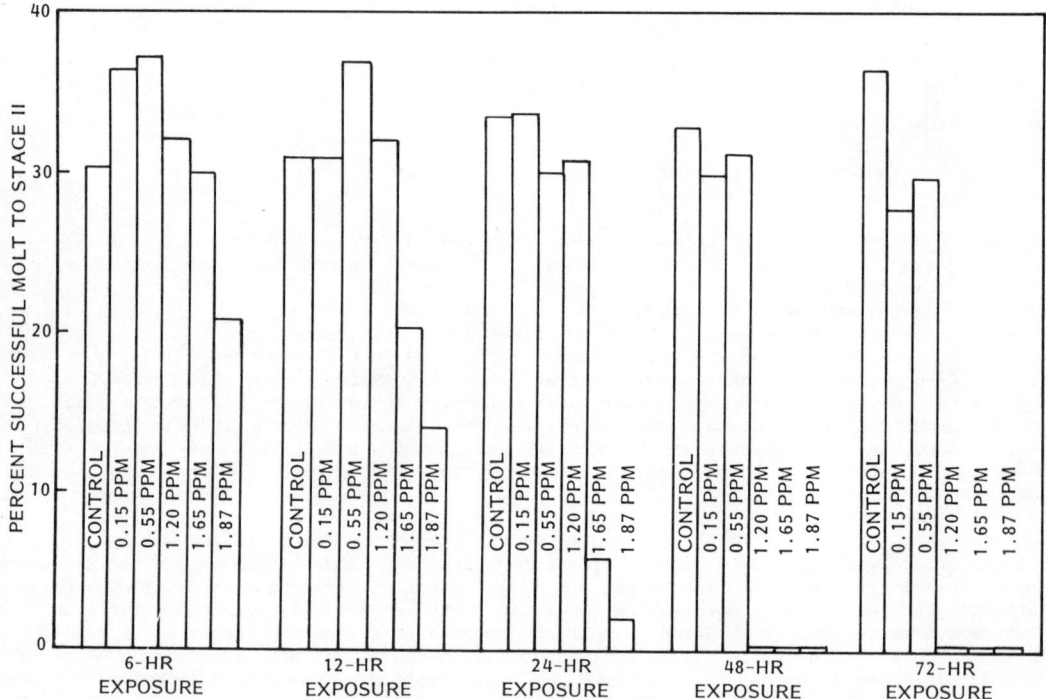

Fig. 23.2. Molting success of larvae of king crab, *Paralithodes camtschatica*, exposed for increased lengths of time to increased concentrations of the WSF of Cook Inlet crude oil. After the exposure time listed, the exposure water was replaced with clean seawater to make a total of 120 hr of observation for each test. Molting success at 120 hr = stage II larvae/initial total of stage I larvae.

Deaths at stage I or stage II in the bioassays on molting larvae often did not occur until late in the tests. This is shown by comparing the LC50's for deaths occurring by 96 hr with the LC50's for deaths occurring by the end of the observation period (Table 2). There were no significant numbers of deaths by 96 hr when the larvae were exposed for only 6 hr, but deaths occurring within the next 48 hr yielded an LC50 of 2.24 ppm total hydrocarbons for coonstripe shrimp. (Observations were not continued as long for king crab.) For coonstripe shrimp larvae exposed 24 hr and longer the 96-hr LC50's were all similar, averaging 0.96 ppm, and also very similar in king crab which averaged about 1.40 ppm total hydrocarbons. The progressively more toxic effects of these longer exposures to periodically-renewed WSF solutions did not show up until sometime after 96 hr, when the larvae were in clean seawater. For example, for coonstripe shrimp larvae exposed for 96 hr, the LC50 dropped from 0.96 ppm at 96 hr to 0.24 ppm at 144 hr.

TABLE 2. The LC50's for Larvae of Coonstripe Shrimp and King Crab Exposed to the WSF of Cook Inlet Crude Oil during the Molting Period from Stage I to Stage II. After the Duration of Exposure listed, the Exposure Water was Replaced with Clean Seawater to make a Total of 144 Hr for each Test on Coonstripe Shrimp and 120 Hr for King Crab. The 95% Fiducial Limits are given in Parentheses; NS Indicates Insignificant Number of Deaths.

Duration of oil exposure	LC50's in ppm of total hydrocarbons			
	Coonstripe shrimp		King crab	
	96-hr LC50	144-hr LC50	96-hr LC50	120-hr LC50
6 hr	NS	2.24	NS	NS
12 hr	No bioassay		4.75 --	1.75 (1.72-1.78)
24 hr	0.95 (0.87-1.03)	0.62 (0.35-0.75)	1.33 (1.23-1.45)	1.38 (1.08-1.61)
48 hr	0.98 (0.83-1.15)	0.50 (0.42-0.62)	1.45 (1.35-1.50)	0.76 (0.66-0.88)
72 hr	No bioassay		1.37 (1.20-1.55)	0.93 --
96 hr	0.96 (0.80-1.15)	0.24 (0.15-0.37)	No bioassay	

Discussion

Larvae of coonstripe shrimp and king crab were more sensitive to the WSF of Cook Inlet crude oil during the molting period from stage I to stage II than during intermolt stages I and II (Table 1), which agrees with previous studies on molting crustacean larvae exposed to different oils. In two of the earlier studies (Wells 1972, Katz 1973) the oil concentrations were not chemically quantified so the reported lethal concentrations cannot be compared to ours. Lethal concentrations for molting stage I-II larvae reported in the other two studies (Neff et al. 1976, Wells and Sprague 1976) are based on different oils and different types of bioassays than ours, but are useful to mention for the purpose of a broad comparison. The 96-hr LC50 for the first larval molt of the crab Rhithropanopeus harrisii was between 0.63 and 0.94 ppm (WSF of No. 2 fuel oil; Neff et al 1976) and for the lobster Homarus americanus it was 0.86 mg/liter (oil-water dispersion of crude oil; Wells and Sprague 1976). These concentrations are similar to our 96-hr LC50's of 0.95 ppm for molting coonstripe shrimp larvae and 1.33 ppm for molting king crab larvae in that all of the concentrations are lower than the 96-hr LC50's reported for nonmolting larvae, juveniles, and adults exposed to various oils (reviewed by Wells and Sprague 1976).

Molting coonstripe shrimp larvae were about four times more sensitive to the WSF than molting king crab larvae (Table 1), suggesting that as much difference in sensitivity to oil exists between crustacean species during molting as has been reported for species in intermolt stages. These differences in sensitivity to oil also occur intraspecifically at different life stages (Neff et al 1976, Brodersen et al In press) and, judging from the available literature, at different molts. The six intermolt stages of coonstripe shrimp larvae differ in their sensitivities to oil with, for example, stage III being more sensitive than stages I and II (Brodersen et al In press); this suggests that coonstripe shrimp larvae molting at the later stages could also be more sensitive than we observed for the first molt. However, larvae of *Neopanope texana* were more sensitive during the first molt than the later molts (Katz 1973). In *Homarus americanus* (Wells and Sprague 1976) and *Rhithropanopeus harrisii* (Neff et al 1976) most of the deaths also occurred during the first molt, but the tests on these species were long-term (30 or more days) and did not include separate tests for the different molt periods, so the sensitivities of later molts were not determined. The available data are insufficient to determine if the most sensitive molt period differs with species or is usually the first larval molt, and if any one group of crustaceans (such as shrimp) is more sensitive than another (such as crab) during molting.

The effects on molting success (Figs. 1 and 2) and mortality (Table 2) that occurred when larvae of coonstripe shrimp and king crab were exposed for increasing periods of time to increasing concentrations of the WSF of Cook Inlet crude oil were basically the same for each species and are summarized as follows. The lowest concentrations of the WSF (0.15-0.55 ppm total hydrocarbons) did not cause molting success to decline at any length of exposure, but did cause some stimulation of molting in king crab larvae after short exposures. Although molting success did not decline, at long exposures to the low concentrations many of the larvae died after molting, as reflected in the low 144-hr and 120-hr LC50's. Exposure of larvae to the highest concentrations of the WSF (1.15-1.87 ppm) for only 6 and 12 hr reduced molting success by 10-30% and resulted in some deaths of both molted and nonmolted larvae. When exposed to these high concentrations for 24 hr or longer, 90-100% of the larvae were inhibited from molting and usually died, yielding LC50's that were equal to or well below the high concentrations tested. The stimulation of molting we observed in king crab larvae has also been reported for larvae of the crab *Rhithropanopeus harrisii* when exposed to low concentrations of No. 2 fuel oil (Neff et al 1976). On the other hand, we did not observe any delays in molting relative to controls; long-term studies like those reported on crabs and lobster (Katz 1973, Neff et al 1976, Wells and Sprague 1976) would be needed to determine if the nonmolted larvae which survived our tests would have eventually molted to stage II or later stages.

In our bioassays on molting larvae, many larvae died after 96 hours, so the progressively more toxic effects of increased exposure times and increased concentrations of the Cook Inlet crude oil WSF were not as clear from the 96-hr LC50's as they were from the 144-hr and 120-hr LC50's (Table 2). Delayed mortality (i.e., deaths occurring after exposure) has also been noted for some other crustacean larvae exposed to toxic pollutants (Buchanan et al 1970, Brodersen et al In press). These data indicate that a bioassay lasting only 96 hr is not always sufficient to yield an accurate determination of acute toxicity.

Greater sensitivity to environmental stresses during molt than intermolt periods is a general phenomenon among crustaceans (Lockwood 1967) which should also be reflected in tests on crustaceans molting at any life stage, and in tests with pollutants other than oil. Exposures of tanner crab, *Chionoecetes bairdi*, to crude oil near the time of molting resulted in reduction in molting success and autotomizing of limbs (Karinen and Rice 1974). The greater sensitivity of molting crustacean larvae during exposures to insecticides (Buchanan et al 1970, Epifanio 1971) and creosol (Emery 1970) has also been reported.

Increased sensitivity to oil during molting is probably related to the physiological changes associated with the molting process. Rises in blood osmotic pressure and changes in permeability to seawater (Lockwood 1967, Hagerman 1973) could result in heightened concentrations of toxic hydrocarbons in the tissues. Depressed metabolism has been observed in molting adult blue crab, *Callinectes sapidus* (Lewis and Haefner 1976), and in oil-exposed king crab, *Paralithodes camtschatica* (Mecklenburg and Rice In prep.), and may impair the ability of molting larvae to metabolize and excrete the oil-derived hydrocarbons accumulated in the tissues.

Because molting has a significant influence on the sensitivity of crustaceans to oil, comparisons of sensitivities between life stages and species will be most valid if based on animals tested in the same stage of the molt cycle, such as intermolt. Testing the sensitivity of intermolt larvae to oil may be difficult for many crustacean species because the intermolt periods are often as short as 2 or 3 days. Thus, larvae would begin to molt during the standard 96-hr bioassay.

Measurements of hydrocarbons from chronically polluted areas or oil spills in Alaskan waters are not yet available. Levels of petroleum hydrocarbons reported for other areas, harbors and tanker routes as well as the open ocean (reviewed by Brown et al 1973, Wells and Sprague 1976), are very low compared to our LC50's for intermolt and molting larvae of king crab and coonstripe shrimp. However, the concentration of 0.8 mg/liter reported for a small oil spill in Nova Scotia (Gordon et al 1973) is higher than some of the LC50's we found for molting larvae exposed 24 hr and longer.

Conclusions

(1) For both coonstripe shrimp and king crab, larvae molting from stage I to stage II were more sensitive to the WSF of Cook Inlet crude oil than were larvae in intermolt stages I or II. From this we further conclude that comparisons of the sensitivities of crustaceans to oils should be based on tests on animals in the same stage of the molt cycle, such as intermolt.

(2) In tests on molting coonstripe shrimp and king crab larvae, as the concentrations and lengths of exposure to the WSF of Cook Inlet crude oil increased, molting success decreased and deaths increased. This pattern was not as clear from the 96-hr LC50's as it was from the 120-hr and 144-hr LC50's. From this we also conclude that the standard 96-hr bioassay is not always long enough for accurate determinations of the sensitivities of crustacean larvae to oils.

References

Anderson, J. W., J. M. Neff, B. A. Cox, H. E. Tatem, and G. M. Hightower, Characteristics of dispersions and water-soluble extracts of crude and refined oils and their toxicity to estuarine crustaceans and fish, *Mar. Biol. 27*, 75-88 (1974).

Atlas, R. M., and R. Bartha, Biodegradation of petroleum in seawater at low temperatures, *Can. J. Microbiol 18*, 1851-1855 (1972).

Brodersen, C. C., S. D. Rice, J. W. Short, T. A. Mecklenburg, and J. F. Karinen, Sensitivity of larval and adult Alaskan shrimp and crabs to acute exposures of the water-soluble fraction of Cook Inlet crude oil. In: *Proceedings, 1977 Oil Spill Conference (Prevention, Behavior, Control, Cleanup)*, American Petroleum Institute, Washington, D.C., In press.

Brown, R. A., T. D. Searl, and J. J. Elliott, Distribution of heavy hydrocarbons in some Atlantic Ocean waters. In: *Proceedings of Joint Conference on Prevention and Control of Oil Spills*, pp. 505-519, American Petroleum Institute, Washington, D.C., 1973.

Buchanan, D. V., R. E. Milleman, and N. E. Stewart, Effects of the insecticide Sevin on various stages of the Dungeness crab, *Cancer magister, J. Fish. Res. Board Can. 27*, 93-104 (1970).

Cheatham, D. L., R. S. McMahon, S. J. Way, J. W. Short, and S. D. Rice, Effects of temperature, volatility, and biodegradation on the persistence of aromatic hydrocarbons in seawater. Paper presented at NOAA-EPA Symposium on Fate and Effects of Petroleum Hydrocarbons, Seattle, Wash., November, 1976, Manuscr. in prep.

Doudoroff, P., Andersen, B. G., Burdick, G. E., Galsoff, P. S., Hart, W. B., Patrick, R., Strong, E. R., Surber, E. W., and W. M. Van Horn, Bio-assay methods for the evaluation of acute toxicity of industrial wastes to fish, *Sewage Ind. Wastes 23*, 1380-1387 (1951).

Emery, R. M., The comparative acute toxicity of creosol to two benthic crustaceans, *Water Res. 4*, 485-491 (1970).

Epifanio, C. E., Effects of dieldrin in seawater on the development of two species of crab larvae, *Leptodius floridanus* and *Panopeus herbstii, Mar. Biol. 11*, 356-362 (1971).

Finney, D. J., *Probit Analysis*, 3d edition, 333 pp., Cambridge University Press, London, 1971.

Gordon, D. C., Jr., P. D. Keizer, and N. J. Prouse, Laboratory studies on the accomodation of crude and residual fuel oils in sea water, *J. Fish. Res. Board Can. 30,* 1611-1618 (1973).

Gruenfeld, M., Extraction of dispersed oils from water for quantitative analysis by infrared spectrophotometry, *Environ. Sci. Technol. 7*, 636-639 (1973).

Hagerman, L., Ionic regulation in relation to the moult cycle of *Crangon vulgaris* (Fabr.) (Crustacea, Natantia) from brackish water, *Ophelia 12,* 141-149 (1973).

Karinen, J. F., and S. D. Rice, Effects of Prudhoe Bay crude oil on molting tanner crabs, *Chionoecetes bairdi, Mar. Fish. Rev. 36*, 31-37 (1974).

Katz, L. M., The effects of water soluble fraction of crude oil on larvae of the decapod crustacean *Neopanope texana* (Sayi), *Environ. Pollut. 5,* 199-204 (1973).

Lewis, E. G., and P. A. Haefner, Jr., Oxygen consumption of the blue crab, *Callinectes sapidus* Rathbun, from proecdysis to postecdysis, *Comp. Biochem. Physiol. 54A*, 55-60 (1976).

Lockwood, A. P. M., *Aspects of the Physiology of Crustacea*, 325 pp., W. H. Freeman, San Francisco, 1967.

Mecklenburg, T. A., and S. D. Rice, Effects of Cook Inlet crude oil, benzene, and naphthalene on heart rates of the Alaskan king crab, *Paralithodes camtschatica* (Tilesius), (In prep.).

Neff, J. M., J. W. Anderson, B. A. Cox, R. B. Laughlin, Jr., S. S. Rossi, and H. E. Tatem, Effects of petroleum on survival, respiration and growth of marine animals. Paper presented at AIBS Symposium on Sources, Effects, and Sinks of Hydrocarbons in the Aquatic Environment, Washington, D.C., August, 1976.

Rice, S. D., J. W. Short, C. C. Brodersen, T. A. Mecklenburg, D. A. Moles, C. J. Misch, D. L. Cheatham, and J. F. Karinen, Acute toxicity and uptake-depuration studies with Cook Inlet crude oil, Prudhoe Bay crude oil, No. 2 fuel oil and several subarctic marine organisms, NWAFC Processed Report, 90 pp. Northwest and Alaska Fisheries Center Auke Bay Fisheries Laboratory, National Marine Fisheries Service, NOAA, P. O. Box 155, Auke Bay, AK 99821, 1976.

Rice, S. D., R. E. Thomas, and J. W. Short, Effects of petroleum hydrocarbons on breathing and coughing rates, and hydrocarbon uptake-depuration in pink salmon fry. In: F. J. Vernberg and W. J. Vernberg (eds.) *Proceedings, Symposium on Pollution and Physiology of Marine Organisms*, Academic Press, New York, In press.

Short, J. W., and S. D. Rice, Accumulation, retention, and depuration of petroleum-derived hydrocarbons by four species of Alaskan marine animals (scallops, shrimp, king crab, and salmon, (In prep.).

Thorson, G., Reproductive and larval ecology of marine bottom invertebrates, *Biol. Rev. (Camb.) 25*, 1-45 (1950).

Wells, P. G., Influence of Venezuelan crude oil on lobster larvae, *Mar. Pollut. Bull. 3*, 105-106 (1972).

Wells, P. G., and J. B. Sprague, Effects of crude oil on American lobster (*Homarus americanus*) larvae in the laboratory, *J. Fish. Res. Board Can. 33*, 1604-1614 (1976).

CHAPTER 24

RESPONSE OF THE CLAM, MACOMA BALTHICA (LINNAEUS), EXPOSED TO PRUDHOE BAY CRUDE OIL
AS UNMIXED OIL, WATER-SOLUBLE FRACTION, AND OIL-CONTAMINATED SEDIMENT IN THE LABORATORY

Tamra L. Taylor and John F. Karinen

Northwest and Alaska Fisheries Center Auke Bay Fisheries Laboratory
National Marine Fisheries Service, NOAA
P. O. Box 155, Auke Bay, AK 99821

Abstract

The small clam, Macoma balthica (Linnaeus 1758), will likely be subjected to oil slicks layered on the mud and to water-soluble fractions of crude oil or oil-contaminated sediment. Groups of adult clams in or on their natural sediment were exposed in flow-through aquaria at 7°–12°C to various concentrations of Prudhoe Bay crude oil layered on the mud surface, the water-soluble fraction (WSF) of the crude oil, and oil-treated sediment (OTS).

Gentle settling of crude oil over clam beds had negligible effects on clams observed for 2 months. Water-soluble and oil-treated sediment fractions of Prudhoe Bay crude oil inhibited burrowing and caused clams to move to the sediment surface. Responses were directly proportional to concentrations of the WSF or amount of OTS. The 1-hr and 72-hr effective median concentrations of the WSF for the responses of burrowing by unburied clams and surfacing by buried clams were 0.234 and 0.367 ppm naphthalene equivalents respectively. The interpolated amount of OTS needed for a 50% surfacing response within 24 hr was 0.67 g OTS cm^{-2}.

Although short-term exposures of clams to the WSF of crude oil and OTS caused few deaths, behavioral responses of clams to oil may be of great importance to their survival in the natural environment. In these laboratory tests, many of the clams recovered, but in nature clams that come to the sediment surface may be eaten by predators or die from exposure.

Key words: Macoma balthica, crude oil, Alaska.

Introduction

The small clam, Macoma balthica, is a dominant species in silty sediments from the upper intertidal in bays to about 37 m in quiet offshore areas. It is found in the North Atlantic, Arctic, and North Pacific oceans as far south as San Francisco Bay (Coan 1971).

Shaw et al., (1976) proposed M. balthica as an indicator of oil pollution in the sediment environment based on (1) wide distribution, (2) detrital deposit feeding in addition to filter feeding, and (3) indications from field tests that it is sensitive to oil. Reduced survival of M. balthica experimentally exposed to stranded Prudhoe Bay crude oil on the Valdez, Alaska intertidal mudflat was found in Shaw et al's study, but questions of how toxic oil is to M. balthica and the behavioral responses to oil remained to be answered. Were they affected by the water-soluble fractions or was their deposit feeding behavior responsible for sensitivity?

We conducted laboratory exposures with M. balthica to answer these questions and to determine the effects of Prudhoe Bay crude oil on survival and behavior.

Experimental Procedure

Macoma balthica were exposed to oil three ways. The first was a simulated natural environment in which crude oil was poured over clam beds to simulate stranding of oil on a tideflat. The second and third were bioassays with the water-soluble fraction (WSF) and oil-treated sediment (OTS). Water temperature ranged between 7° and 12°C. Response parameters examined were death and burrowing behavior.

Experimental Animals and Sediment
Adult M. balthica averaging 1 cm in length and their natural sediment were collected from a mudflat 200 yeards south of the public launching ramp at Amalga Harbor near Eagle River northwest of Juneau, Alaska. The area has had only limited use and is regarded as relatively free of oil contamination.

Exposure to Stranded Crude Oil

Beds of M. balthica in simulated tideflat aquaria were acclimated one month, exposed to stranded crude oil, and observed 60 days for behavioral response or mortality. There were four flow-through aquaria (200 x 50 cm) each containing approximately 2,000 clams in a 6-cm-deep layer of mud at a density comparable to natural populations. Low tides were simulated daily with an average of 3 hr of exposure.

Three aquaria were treated with Prudhoe Bay crude oil and one left untreated as a control. Oil was poured on the water during the falling tide on five successive days at three concentrations: 1.2, 2.4, and 5.0 µl oil cm^{-2} day^{-1}. The oil settled on the mud as the water receded but on the incoming tide was floated and flushed from the aquaria.

Exposure to Water-Soluble Fraction of Crude Oil

The water-soluble fraction was prepared by slowly mixing 1% Prudhoe Bay crude oil in seawater (1 liter oil:100 liters seawater) for 20 hr at ambient seawater temperatures (10°-12°C). Rheostats controlled the stirring speed of motor-driven propellers so that oil particles stayed in the upper third of the container. After 20 hr of mixing, the mixture stood for 3 hr to allow the oil slick to separate from the WSF. The WSF was then siphoned from below the slick. Based on the analysis of this stock solution by ultraviolet (UV) spectrophotometry, the WSF was diluted to the concentrations desired (Rice et al., In press).

Samples of the WSF were analyzed for aromatic hydrocarbons by UV spectrophotometry with a hexane extraction technique (Neff and Anderson 1975). Concentrations of naphthalene and substituted naphthalenes were estimated by reading the UV optical density (UVOD) at 221 nm. This method detects concentrations of naphthalene greater than 0.015 ppm but is not specific for naphthalene. Substituted naphthalenes and some monoaromatics also contribute to the absorbance at 221 nm; however, no attempt was made to measure the contribution of each group separately (as did Neff and Anderson 1975). Concentration due to absorbance at 221 nm was expressed as naphthalene equivalents by calibrating measured absorbances with absorbances of known dilutions of naphthalene in hexane, using the equation:

$$\text{ppb as naphthalene (naphthalene equivalents)} = \frac{\text{Sample UVOD 221 nm}}{0.000899591}$$

The constant (8.99591×10^{-4}) includes the slope value from the standard curve for naphthalene in hexane and a dilution factor. Efficiency of naphthalene extraction in calibration tests ranged from 91% to 95%. Concentration values for total naphthalenes by this method are somewhat higher than total naphthalenes as determined by gas chromatography but are usually within 30% when concentrations in reshly prepared solutions (3 hr) are compared (Rice et al 1976).

Macoma balthica's response to the WSF was tested two ways: (1) clams already buried in a 3-cm-deep layer of sediment were exposed to the WSF and observed for burrowing activity, and (2) marked clams (to distinguish from ones buried prior to WSF introduction) were placed in WSF on top of the sediment and observed as they burrowed. Clams were exposed to 100 liters of recirculating WSF for 6 days (beginning Nov. 10), then held in recirculating seawater for 4 days. Doses of the WSF were replaced at 48-hr intervals within the exposure period to compensate for the natural loss of aromatics (Cheatham et al 1976). There were five WSF doses and a seawater control. Sample sizes for each dose were 200 initially buried clams and 40 initially unburied clams. The number used to identify the strength of WSF dose is the average of the ppm naphthalene equivalents measured on days 0, 2, and 4 when the WSF's were replenished. The average doses were 0.019, 0.036, 0.081, 0.160, and 0.302 ppm naphthalene equivalents.

The data recorded were numbers of clams on the surface. Initially unburied clams were counted at 10-min intervals from the time of introduction through 170 min, then daily for the remainder of the experimental period. Initially buried clams were counted daily.

Data on the response of the clams were analyzed by computerized probit analysis (Finney 1971) and expressed as effective median concentration (ECm) with 95% confidence intervals. The ECm is defined as the dose causing a response in 50% of the test animals for a given length of exposure. The slope function is used to predict the dose (ppm naphthalene equivalents) with a 95% confidence interval at which the burrowing rates of 10% of exposed clams would be significantly reduced from the normal rate of control clams at 60 min. The data were adjusted with Abott's formula (Finney 1971) to correct for partial response by control clams.

Exposure to Oil-Treated Sediment

Oil-treated sediment was prepared by mixing 1 volume Prudhoe Bay crude oil with 2 volumes dry sediment and 10 volumes seawater in an oscillating shaker for 1 hr. The mixture was allowed to settle for 1 hr and the water decanted. Clean seawater was mixed an additional 30 min with the sediment, allowed to settle 1 hr and the liquid again decanted. Sediment was then put in suspension with seawater and added to test aquaria. Uncontaminated sediment for control was made in the same manner, except no oil was added. Amount of sediment added is expressed as grams dry weight.

Successive washing of the sediment limited exposure to only oil firmly adsorbed by the sediment. Similar sediment was analyzed for oil content by the gravimetric and gas chromatographic method of MacLeod et al., (1976).

Macoma balthica's response to OTS was tested by exposing clams buried in a 3-cm-deep layer of untreated mud to a new layer of treated sediment falling over them. The amount of OTS in the new layer was varied experimentally to form three doses: 0.10, 0.25, and 0.50 g OTS cm^{-2}. Each dose of OTS had a control dose of the same amount of non oil-treated sediment. The sample size in each dose was 190 clams. Fresh seawater flowed through the exposures except during the first 24 hr when treated sediment was added and allowed to settle.

Exposure to treated sediment began Sept. 3, 1975, and lasted 30 days. The response data were recorded daily as counts of dead clams and clams on the surface.

Data on response by surfacing were analyzed by probit analysis (Finney 1971) and expressed as the calculated amount of sediment (g cm^{-2}) at which 50% of the clams will move to the surface within a specified period of time (ECm) with a 95% confidence interval. These data were adjusted with Abbott's formula (Finney 1971) to correct for partial response from the control clams. The relationship between surfacing response and square of the amount of OTS was determined by linear regression (Snedecor 1956). Data on response by dying were analyzed by comparison of two observed proportions (Natrella 1966).

Results

Exposure to Stranded Crude Oil

Mortality of control or exposed clams was not significant during the 60-day observation period after oil exposure. The only indication that the exposed clams experienced any stress was reduced siphon activity during the time that oil slicks were on the mud. Clams continued to feed in the test aquaria following incoming tidal flooding. The quantity of oil in the interstitial water following the five applications was below the detection limits of our analytical method.

Exposure to Water-Soluble Fraction

Exposure to the WSF caused some buried clams to come out of the sediment but no clams died within 10 days. The data show response proportional to dose for the higher doses (Fig. 1). There was little effect from concentrations below 0.081 ppm naphthalene equivalents. All control clams remained buried throughout the test period. The calculated dose at which 50% of the clams surfaced within 3 days (ECm) is 0.367 (0.317-0.411) ppm naphthalene equivalents. Under the same conditions the ECm for response within 5 days is 0.323 (0.288-0.363) ppm naphthalene equivalents. Almost all of the clams that had surfaced reburied themselves within the test period, following addition of clean water. The concentrations causing 50% response in 3 and 5 days, 0.367 and 0.323 ppm, represent 97% and 85% respectively of a 100% WSF containing 0.379 ppm naphthalene equivalents. This 100% WSF is near the saturation level obtained from a 1% v/v oil/water mixture.

Exposure to the WSF inhibited the rate of burrowing of some unburied clams. Burrowing rate decreased in proportion to WSF concentration for the two higher doses (Fig. 2). By day 7 in the experiment, at least 97% of the clams were buried. The calculated dose at which the rate of burrowing of 10% of exposed clams would be significantly reduced from the rate of the control clams at 60 min in 0.044 (0.010-0.088) ppm naphthalene equivalents. The calculated doses at which 50% of the initially unburied clams will fail to burrow (ECm) within 60 and 170 min are 0.234 (0.175-0.310) ppm and 0.222 (0.181-0.272) ppm naphthalene equivalents.

Exposure to Oil-Treated Sediment

Clams moved to the surface in proportion to the amount of oil-treated sediment added (Table 1). In our heaviest dose (OTS 0.5 g cm^{-2}) 29% of the clams moved to the surface within 24 hr. In the intermediate dose (OTS 0.25 g cm^{-2} 10% moved to the surface within 24 hr, while 6% of the clams at the lowest dose (OTS 0.10 g cm^{-2}) surfaced. Less than 2% of the control clams came to the surface within this period of time. Responses in the three control doses of treated sediment were equal. The amount of OTS, calculated by probit, that it would take under conditions of the experiment to stimulate 50% of the clams to move to the surface within 1 day is 0.67 (0.58-0.76) g cm^{-2}.

Response to oil-treated sediment within 24 hr appeared linear with respect to oil-treated sediment weight squared (Fig. 3). The computed line slope is 112.95 with standard deviation of 3.07, $P < 0.005$.

Total extractable oil in the OTS as determined by gravimetric procedure was 312 µg/g dry weight of sediment. Total saturated and unsaturated hydrocarbons were 154 and 178 µg/g dry weight respectively. Data on hydrocarbon content as determined by gas chromatography and lithological-chemical characteristics of the OTS were also obtained and are available from the authors.

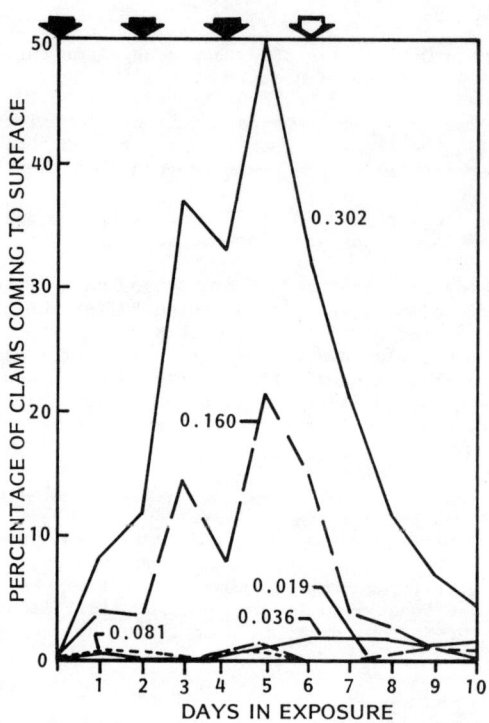

Fig. 24.1. Response of buried <u>Macoma balthica</u> exposed to concentrations of the WSF of Prudhoe Bay crude oil (given in ppm naphthalene equivalents) in a recirculating water-sediment system. The percentage of clams that responded by coming to the sediment surface is graphed. Control clams made no response throughout. Solid arrows = days WSF added or changed; open arrow = the day clean seawater was added.

TABLE 1. Response of Clams to Oil-Treated Sediment

Dose of oil-treated sediment		Percentage of clams at the surface--		
		After 24 hr	After 72 hr	After 120 hr
0.50 g cm^{-2}	Exposure	28.7	22.9	20.7
	Control	0.5	2.1	3.0
0.25 g cm^{-2}	Exposure	9.8	12.4	13.5
	Control	0	0.5	0
0.10 g cm^{-2}	Exposure	5.8	3.1	3.7
	Control	1.5	2.6	2.0

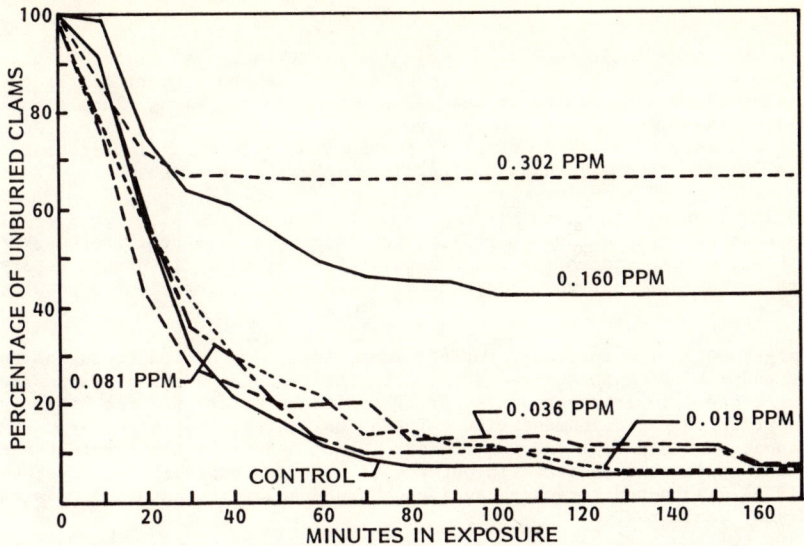

Fig. 24.2. Response of unburied clams to concentrations of the Prudhoe Bay crude oil WSF (given in ppm naphthalene equivalents) in a recirculating water-sediment system. The percentage of clams that remained on the surface is graphed.

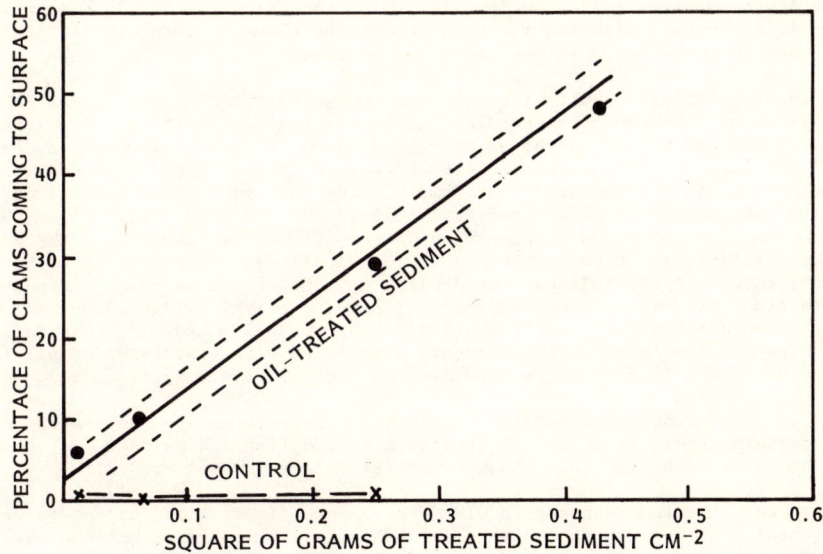

Fig. 24.3. Percentage of clams that responded to 24-hr exposure to oil-treated sediment by coming to the surface vs. grams of sediment added cm^{-2} squared. Broken lines for oil-treated sediment = ±1 standard deviation.

Discussion

The results of our study indicate that M. balthica is quite resistant to short-term oil exposure. Although we have shown a definite behavioral response of M. balthica to water-soluble and oil-treated sediment fractions, no deaths occurred among clams exposed to either for short periods (4-6 days). The apparent resistance during short-term exposure is consistent with the observations (Rice et al In press) that intertidal invertebrates are generally more resistant to oil than subtidal invertebrates and fish.

Delayed mortality occurred in clams after 30 days of exposure to 0.5 g cm^{-2} of OTS. Other investigators have observed delayed mortality following oil exposure. Swedmark et al (1973) report it in bivalves exposed to oil dispersants. Rice et al., (in press) report from unpublished data that scallops continue to die up to 4 weeks following 96-hr exposures to WSF's of crude oil.

In our WSF experiments with initially buried clams, the clams showed trends of recovering from exposure and reburying themselves (Fig. 1). The recovery shown by depressions in the curves on days 2 and 4 is attributed to loss of toxicants by the WSF and relief from stress for the clams prior to replenishment with a fresh dose later those same days. Replacement with fresh seawater on day 6 of the experiment did not appear to accelerate the recovery rate, indicating again that decrease in potency of WSF was responsible for recovery. Percy and Mullin (1975) also noted a reduced avoidance response by the amphipod Onisimus affinis as oiled sediment weathered.

Field and laboratory data provide evidence that concentrations of oil under spills are similar to concentrations used in our WSF experiments. Gordon et al (1973) reported that immediately under (0.25 m) an oil slick resulting from a small spill (1,000 barrels), total oil concentrations (measured by flourescence spectroscopy) approaching 0.80 ppm were observed 2 days after the spill. Laboratory data collected on water-soluble preparations of Prudhoe Bay crude oil indicate that total naphthalenes represent about 28% of the total oil present in the water-soluble fraction (Rice et al 1976). Based on these data, total naphthalene concentrations occurring at depths of less than 1 m under an oil spill are probably within the 0.200-0.300 ppm range. Concentrations would presumably be even higher within 1-2 cm of the slick's lower surface. Tidal action will bring the slick and highly concentrated WSF layer under it in contact with the mud surface and the clams residing in it, therefore responses equal to those noted in our study can be expected.

The factor causing inhibited burrowing of clams and surfacing of buried clams is not determined by this study. We can only speculate as to the reasons, which may be different for the two cases (WSF and OTS). Inhibited burrowing rates of initially unburied clams in the presence of WSF could be (1) a narcotic effect exerted on the clams' nervous or muscular systems, or (2) exposure stimulated the clams to close up, protecting themselves until oil concentration was reduced (Stegeman and Teal 1973). Surfacing by buried clams could be because (1) a compound or class of compounds in the oil may be penetrating the sediment and acting as an irritant, (2) chemical compounds in the oil may change the pH within the sediment, (3) oxygen in the sediment may decline to concentrations below minimum requirements of the clams because of increased biological or chemical oxidation, or (4) bacterial populations associated with oil contamination may be producing hydrogen sulfide or similar irritants. The decline in surfacing response associated with weathering of oil (Fig. 1) seems to indicate an effect directly dependent on the oil (i.e., see 1 above).

Hiatt et al., (1963) reported that specific chemical structures are responsible for the ability of certain chemicals to act as irritants toward fish and oxidation of these structures changes their potency. Such changes may take place in weathered oil.

It is of interest that the response to OTS (Fig. 3) was linear with respect to weight of sediment squared. The reason for this relationship is not apparent, but the observation may prove useful in evaluating the response of clams and other benthic organisms to oil-contaminated detrital material settling on them.

There was a difference between the results of our exposure to stranded crude oil and the field work of Shaw et al., (1976). In the field study M. balthica were exposed in situ to stranded Prudhoe Bay crude oil daily for 5 days while exposed at low tide. Two rates of oiling were used, 1.2 and 5.0 µl oil cm^{-2} day^{-1}. Shaw et al observed significant mortalities at the higher oiling rate 2 days after the last oiling. Deaths remained significant throughout the 60-day experimental period. Differences in behavior of oil under field conditions at Port Valdez versus our aquarium situation explain the contrasting results. Water in the Valdez study area carries a heavy sediment load (Myren and Pella 1977) which would have mixed with the oil and settled over the clams. In our aquaria virtually no mixing energy was applied to the oil, water, or sediment; therefore, no sediment was raised and no detectable amount of oil dissolved into the water phase. Secondly, we did not achieve close contact of the oil with the sediment because instead of percolating through the sediment, as the oil was said to do in Shaw et al's (1976) experiment, it drained across the surface and settled on elevations or rested on pools of water in the depressions.

An important observation relative to comparing our work to Shaw et al's (1976) is that only a small percentage (6%) of the clams that eventually died in our OTS exposures were not visible at the surface. If this trend is the same under natural conditions, it is possible that there was a greater effect of the oil on the clams in the Valdez field experiment of Shaw et al than is indicated by their data, since the aluminum containment frames used to define samples in their study were not covered. Clams may have died, floated away, or been taken by predators and would not be recorded in Shaw et al's data.

Other investigators have reported the surfacing response of clams following oil spills and the avoidance of oiled sediments during toxicity studies. Dow and Hurst (1975) report such a response from Mya arenaria to an oil spill in Long Cove, Searsport, Maine, where many clams surfaced from their burrows and died. Percy and Mullin (1975) reported that when the amphipod Onismus affinis was presented a choice between clean sediment and oil-contaminated sediments it overwhelmingly rejected the oiled sediment.

Surfacing response has also been reported with other chemicals. Karinen et al., (1967) reported that by the second day after treatment of aquaria with carbaryl, juvenile clams, nemertean worms, and polychaetes appeared on the surface of the mud. Similar observations were made when carbaryl was applied to a mud flat in Yaquina Bay at Newport, Oregon (Karinen, unpublished data).

Our results indicate that the impact an oil spill will have on M. balthica depends upon how the oil is introduced and mixed into the clam's environment. The amount of mixing applied to the sediments and/or seawater will determine to how much, if any, oil the clams are exposed. If there is essentially no mixing associated with a spill, such as we had with our stranded crude oil exposure, effects will probably be negligible. However, under the usual environmental conditions encountered when oil spills move onshore, mixing energy is likely sufficient to bring oil or its WSF in contact with buried clams, therefore the movement of clams to the surface following such spills would be expected.

Although several questions regarding the responses of M. balthica to hydrocarbons remain to be answered, the results of this study lead us to agree with Shaw et al (1976) that the relative absence of M. balthica may be useful as an indicator of oil pollution. The responses of clams proportional to oil dose as observed throughout our study suggest they are good bioassay organisms and well suited for use in environmental impact studies. Few of the clams die upon short-term exposure to Prudhoe Bay crude oil treated sediments or water-soluble fractions, but their presence on the surface would enhance predation.

Summary

1. The impact of oil on M. balthica depends on how the oil is introduced and mixed into sediments and/or seawater.

2. Macoma balthica respond to stress from the Prudhoe Bay crude oil water-soluble fraction and oil-treated sediment by burrowing to the surface, thereby exposing themselves to adverse environmental conditions or predators.

3. At least 94% of the clams that are going to die burrow to the surface before dying.

4. The calculated dose in naphthalene equivalents at which 50% of initially buried clams would respond to the oil water-soluble fraction by burrowing to the surface with 3 days (ECm) is 0.367 (0.317-0.411) ppm.

5. The calculated dose in naphthalene equivalents at which 50% of initially unburied clams will fail to burrow within 60 min (ECm) is 0.234 (0.175-0.310) ppm.

6. The calculated amount of oil-treated sediment that it would take to stimulate 50% of initially buried clams to move to the surface with 24 hr is 0.67 (0.58-0.76) g cm^{-2}.

Acknowledgements

These experiments were jointly funded by an Environmental Protection Agency grant through Dr. Howard Feder of the University of Alaska and by the Bureau of Land Management through the Outer Continental Shelf Environmental Assessment Program of the National Oceanic and Atmospheric Administration.

Dr. Feder was responsible for the design of the stranded crude oil test and provided much encouragement during the remainder of the work. We thank Dr. Richard Myren for assistance in in designing the aquaria for the intertidal exposures and providing suggestions relative to the biology of *M. balthica*, Jeffrey Short and D. Loren Chestham for assistance in analytical procedures, Dr. Stanley Rice for providing and coordinating assistance, and others at the Auke Bay Fisheries Laboratory for assisting in the construction and mud collection phase of these experiments. We also thank Dr. William D. MacLeod and Donald W. Brown of the NOAA National Analytical Facility, Northwest and Alaska Fisheries Center, Seattle for analysis of oil-treated sediment.

References

Anderson, J. W., J. M. Neff, B. A. Cox, H. E. Tatem, and G. M. Hightower, Characteristics of dispersions and water-soluble extracts of crude and refined oils and their toxicity to estuarine crustaceans and fish, *Mar. Biol. 27*, 75-88 (1974).

Cheatham, D. L., R. S. McMahon, S. J. Way, J. W. Short, and S. D. Rice, Effects of temperature, volatility, and biodegradation on the persistence of aromatic hydrocarbons in seawater. Paper presented at NOAA-EPA Symposium on Fate and Effects of Petroleum Hydrocarbons, Seattle, Wash., November, 1976, Manuscr. in prep.

Coan, E. V., The northwest American Tenninidae, *Veliger 14*, Suppl. (1971).

Dow, R. L., and J. W. Hurst, Jr., The ecological, chemical, and histopathological evaluation of an oil spill site, *Mar. Pollut. Bull. 6*, 164-166 (1975).

Finney, D. J., *Probit Analysis*, 3d edition, 333 pp., Cambridge University Press, London, 1971.

Gordon, D. C., Jr., P. D. Keizer, and N. J. Prouse, Laboratory studies of the accommodation of some crude and residual fuel oils in seawater, *J. Fish. Res. Board Can. 30*, 1611-1618 (1973).

Hiatt, R. W., J. J. Naughton, and D. C. Matthews, Relation of chemical structure to irritant responses in marine fish, *Nature 172*, 904-905 (1953).

Karinen, J. F., J. G. Lamberton, N. E. Stewart, and L. C. Terriere, Persistence of carbaryl in the marine estuarine environment. Chemical and biological stability in aquarium systems, *J. Agric. Food Chem. 15*, 148-156 (1967).

MacLeod, W. D., D. W. Brown, R. G. Jenkins, L. S. Ramos, and V. D. Henry, A pilot study on the design of a petroleum hydrocarbons baseline investigation for northern Puget Sound and Strait of Juan de Fuca. Completion report submitted to Puget Sound Energy-related Research Project, Marine Ecosystem Analysis Program, Environmental Research Laboratory, Appendix A. Northwest and Alaska Fisheries Center, National Marine Fisheries Service, NOAA, 2725 Montlake Boulevard East, Seattle, Wash. 98112, November 1976.

Myren, R. T., and J. J. Pella, Natural variability in distribution of an intertidal population of *Macoma balthica* subject to potential oil pollution at Port Valdez, Alaska, *Mar. Biol.* (In press, 1977).

Natrella, M. G., *Experimental Statistics*, U. S. Nat. Bur. Stand. Handb. 91, 1966.

Neff, J. M., and J. W. Anderson, An ultraviolet spectrophotometric method for the determination of naphthalene and alkylnaphthalenes in the tissues of oil-contaminated marine animals, *Bull. Environ. Contam. Toxicol. 14*, 122-128 (1975).

Percy, J. A., and T. C. Mullin, Effects of crude oils on arctic marine invertebrates. Beaufort Sea Technical Report 2, 167 pp. Beaufort Sea Project, Dept. of the Environment, Victoria, B. C., 1975.

Rice, S. D., J. W. Short, C. C. Brodersen, T. A. Mecklenburg, D. A. Moles, C. J. Misch, D. L. Cheatham, and J. F. Karinen, Acute toxicity and uptake-depuration studies with Cook Inlet crude oil, Prudhoe Bay crude oil, No. 2 fuel oil and several subarctic marine organisms. NWAFC Processed Report, 90 pp. Northwest and Alaska Fisheries Center Auke Bay Fisheries Laboratory, National Marine Fisheries Service, NOAA, P. O. Box 155, Auke Bay, AK 99821, 1976.

Rice, S. D., J. W. Short, and J. F. Karinen, Toxicity of Cook Inlet crude oil and No. 2 fuel oil to several Alaskan marine fishes and invertebrates. In: *Proceedings of AIBS Symposium on Sources, Effects, and Sinks of Hydrocarbons in the Aquatic Environment*, Washington, D.C., August 1976, In press.

Shaw, D. G., A. J. Paul, L. M. Cheek, and H. M. Feder, *Macoma balthica*: an indicator of oil pollution, *Mar. Pollut. Bull.* 7, 2931 (1976).

Snedecor, G. W., *Statistical Methods Applied to Experiments in Agriculture and Biology, with Chapter 17 on Sampling by W. G. Cochran*, 5th edition, 534 pp., Iowa State University Press, Ames, 1956.

Stegeman, J. J., and J. M. Teal, Accumulation, release and retention of petroleum hydrocarbons by the oyster *Crassostrea virginica*, *Mar. Biol.* 22, 37-44 (1973).

Swedmark, M., A. Granmo, and S. Kollberg, Effects of oil dispersants and oil emulsions on marine animals, *Water Res.* 7, 1649-1672 (1973).

CHAPTER 25

LONG TERM BIOLOGICAL EFFECTS OF BUNKER C OIL IN THE INTERTIDAL ZONE

Martin L. H. Thomas

Department of Biology
University of New Brunswick
Saint John, New Brunswick, Canada

Abstract

In February, 1970 a large spill of Bunker C oil occurred in Chedabucto Bay, Nova Scotia, Canada when the tanker "*Arrow*" grounded. Oil from the tanker has persisted for over six years on rocks and in intertidal sediments on the shores of the bay. During this period mortalities of common species in all major communities on both exposed and sheltered shores have occurred. On rocky shores, the dominant fucoid algae suffered heavy initial mortalities which were more severe at high tidal levels. Recolonization has proceeded from lower to higher levels but has not yet occurred in the high tide zone. Delayed recolonization appears to be related to long term toxicity. In salt-marsh and sheltered lagoonal communities, the dominant grass, salt marsh cord grass, suffered heavy mortality delayed one year from the initial spill, recovery commenced two years later and is proceeding steadily. Soft-shell clams in lagoonal sediments have shown persistent mortalities proportional to oil content of sediments. This pattern appears to be a result of direct toxicity, environmental change caused by oil and sub-lethal metabolic effects.

Key words: Chedabucto Bay, *Fucus*, Bunker C Oil, oilspill, *Spartina*, *Mya arenaria*

Introduction

The tanker "*Arrow*" carrying 2,800,000 gal of Bunker C oil struck Cerberus Rock in Chedabucto Bay, Nova Scotia, Canada on Feb. 4, 1970. Details of the disaster and the cargo are given in the Report of the Task Force-Operation oil (Anon, 1970). More than half the cargo escaped from ruptured tanks and much of this was driven by prevailing winds on to the mainly rocky, deeply indented coastline along the northern side of the bay. This shore also contains a number of extensive shallow lagoons many of which trapped large quantities of oil (Owens, 1971).

Biological observations of intertidal areas began on Feb. 7, 1970 and initially covered most of the bay. It was found that shores in the Isle Madame area (Fig. 1), were most seriously contaminated and seven study locations in varied habitats were selected in this area.

The most serious problem encountered in this study was that the marine biota were previously undescribed. Initial observations at non-oiled areas showed that intertidal communities were similar to those described for this part of the region by Stevenson and Stevenson (1954 and Bousfield (1960). Lack of detailed information has however, prevented showing changes in the populations of any but the commonest species. This problem was confounded by the subsequent oiling of sites originally chosen as unoiled controls and the inability to examine most shores closely due to ice cover. Earliest results were summarized by Thomas (1973).

Study Locations and Sampling Schedule
Fig. 1 shows the shorelines of Chedabucto Bay and the locations of study locations. At each study location a belt transect 2 m wide and running at right angles to the shoreline, from extreme high to extreme low tide level, was established. The upper end of this transect was permanently marked on bedrock or by driving a stake. Each location was surveyed to determine its cross-sectional profile with reference to mean low tide level, using the method detailed in Thomas (1974). The sites were located as follows: Station 1, Janvrin Is. causeway (S. side); Station 2, S. shore of Creighton Is.; Station 3, Arichat Habor at Arichat Church; Station 4, Arichat Harbor at Petit Barachois; Station 5, near Petit de Grat; Station 6, N. side of Janvrin Lagoon; Station 7, S. side of Janvrin Lagoon. Observations

were carried out 6 times in 1970, 3 times in 1971, 2 times in 1972 to 1974 and yearly thereafter. One sampling has always been in late July or August and prior to 1975 one was always in April or May.

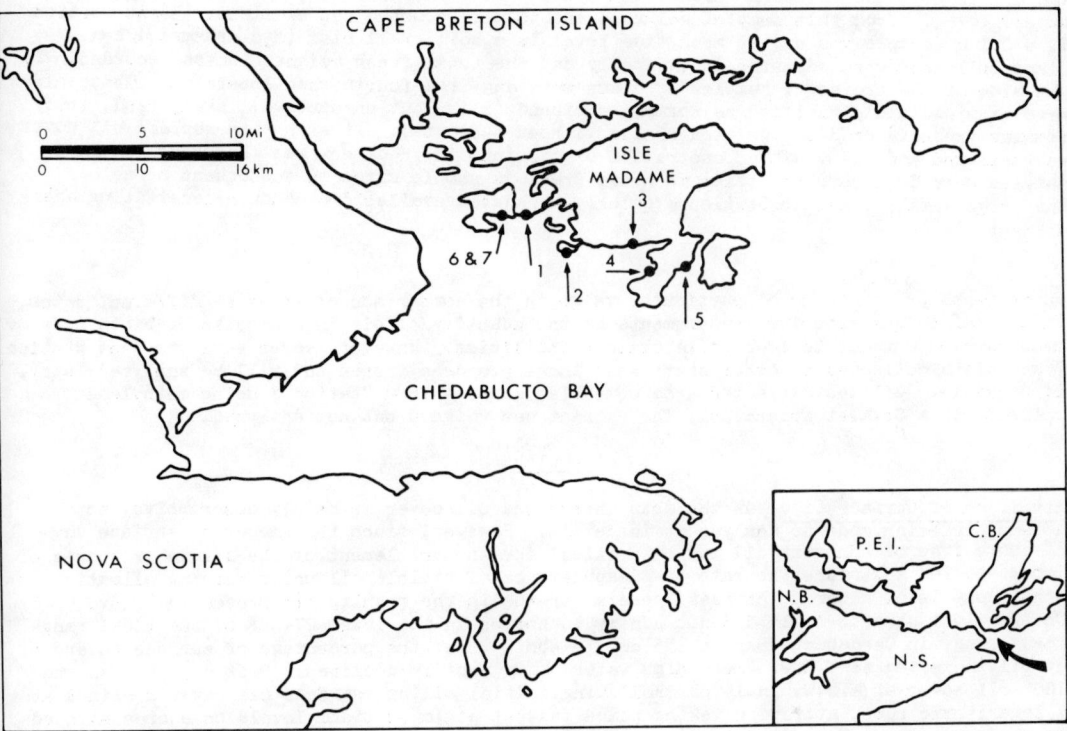

Fig. 25.1. The Chedabucto Bay area of Nova Scotia, Canada, showing the general study area and the location of stations. The inset shows the location of Chedabucto Bay in the Maritime Provinces of Canada.

Characteristics of study stations. Table 1, shows width of the shore and general substrate type for each station.

TABLE 1. Characteristics of the Shore at Study Stations

Station	Shore width m.	Substrate Type
1	40	Broken rock and boulder
2	21	Mainly bedrock, sand at HW
3	19	Broken rock and gravel
4	14	Bedrock at HW to broken rock at LW[1]
5	50	Mainly broken rock, sand at HW
6	25	Muddy sand at HW to sandy mud at LW
7	70	Muddy sand at HW to sandy mud at LW

[1] Abbreviations are: HW, high water; LW, low water

Observations. The following observations were made on each visit to each station: (1) The horizontal zones occupied by all common species were determined by measurement from the marked upper limit of the transect. Vertical zones were determined from the cross-sectional profile. (2) The abundance of the four commonest fucoid algae, *Fucus spiralis*, *Fucus vesiculosus*, *Fucus serratus* and *Ascophyllum nodosum*, where present, was measured by counting plants within $1m^2$ quadrats placed in specific locations. (3) Each entire transect and several specific locations on each were photographed in color. (4) General observations were made on the condition, abundance and distribution of biota. (5) The surface area covered by oil throughout the transect was estimated using quadrats and grids of various sizes in the field and by reference to photographs later. Station 5, initially an unoiled control, was later oiled and then was abandoned in 1973 because it was completely changed by storm action.

Additional observations detailed below were carried out only at the locations and times listed: (1) At Stations 1 and 3, barnacles (*Balanus balanoides*) were counted and photographed within a marked area. (2) At Station 6 until 1973, 1 m^2 of sediment from M.L.W. (mean low water) was dug to 25 cm deep and screened through 4 mm mesh. Benthos were identified, counted and their fresh weight biomass determined. A second sample of 0.1 m^2 was dug in eelgrass (*Zostera marina*) bed lying 4 cm vertically below E.L.W. (extreme low water) level. From this sample, eelgrass shoots were counted and examined. (3) At Station 7, a 0.1 m^2 sample was dug at mean tide level in a soft shell clam (*Mya arenaria*) bed. Clams collected were measured individually and the total fresh weight biomass recorded. To the side of the transect, samples of clams were dug from four marked locations. These clams were examined and classified as normal, moribund (alive but unreactive), newly dead, (body remains present) or dead (jointed shells without contents). At each site surface oil cover was measured and the visible penetration of oil into the sediment was observed. Wherever samples were dug, care was taken to avoid previous sample sites on subsequent occasions. The large sample taken at Station 6 quickly exhausted available benthos necessitating abandonment.

Photographs proved to be of particular value in the comparison of sites at different dates. In general no quantitative measurements of the quantity of oil in sediments or biota have been carried out due to lack of laboratory facilities. However, cores were taken at Station 7 and biota collected at other stations. These are deep frozen and will be analysed later, if possible. Oil was extracted from some clam samples from Station 7 using methylene chloride in a Soxhlet apparatus. The extract was weighed but not analysed.

Results

Attrition of Surface Oil. As the data on surface oil cover is mainly descriptive, no attempt is being made to analyse it in detail. However, since the amount of surface area of shore free of adherent oil may be critical for the settlement of the larvae or spores of local species and since the rate of disappearance of visible oil under varying climatic conditions is of general interest, general trends in the results are presented below. Observations show that oil did not adhere to shores in the lower 27-33% of the tidal range (mean range in Chedabucto Bay is 135 cm.). Above this, the percentage of surface covered by oil increased to M.H.W. (mean high water) then rapidly declined. Most stations showed 100% oil cover at M.H.W. in 1970. Following initial oiling, surface oil cover declined at a logarithmic rate, attrition taking place fastest at lower tidal levels on shores exposed to heavy wave (and ice) action and slowest at higher tidal levels in sheltered locations. These observations are illustrated in Figure 2, which shows attrition rates at two tidal levels, M.H.W. and 35-40 cm above M.L.W. Not all stations are represented in this figure because a complete series of observations was not always possible due to natural or man made changes to the shore. At Stations 6 and 7 the lower level did not exist due to a bar at the mouth of the lagoon which prevented complete emptying at low tide. The faster attrition at lower levels is clear and surface oil persisted only until 1973 there. Based on the situation of stations in relation to wave action, Stations 1 and 2 are exposed, Stations 3 and 4 moderately exposed and Stations 6 and 7, sheltered. The figure also demonstrates the faster disappearance of oil in proportion to exposure. At the most sheltered locations, surface oil still persists at M.H.W.

Effects on Rocky Shore Biota. Most eastern Canadian rocky shores exhibit a narrow zone at high water dominated by the small fucoid seaweed, *Fucus spiralis*. Such a zone was present in Chedabucto Bay at the time of the oil spill but by the fall of 1970 had disappeared at all stations. Since that time, sporelings of fucoid algae have repeatedly settled in this zone but have never survived to a size where they could be identified. For instance sporelings have been observed to settle each year from 1972 on at mean high water at Station 4. During this period mean oil cover at this level has declined from 15 to 5%.

The shore below the *F. spiralis* zone is dominated in the Chedabucto Bay area by a close relative, *Fucus vesiculosus*. Following the oil spill the upper limit for this species was depressed to lower tide levels and has slowly returned to successively higher levels, as surface oil cover decreased. These data are presented in Table 2. Suitable habitat for this species only occurs at Stations 1 - 4; at Station 2 a band of sand prevents colonisation of the upper shore.

TABLE 2. Upper Limits of the Intertidal Distribution of *Fucus vesiculosus* at Stations 1-4 from 1970-1975. The limit is expressed as a percentage of the mean tidal range above mean low water.

Station	1970	1971	1972	1973	1974	1975
1	34	41	42	40	52	62
2	48	48	52	52	52	52
3	60	54	64	78	72	85
4	94	68	74	99	102	106

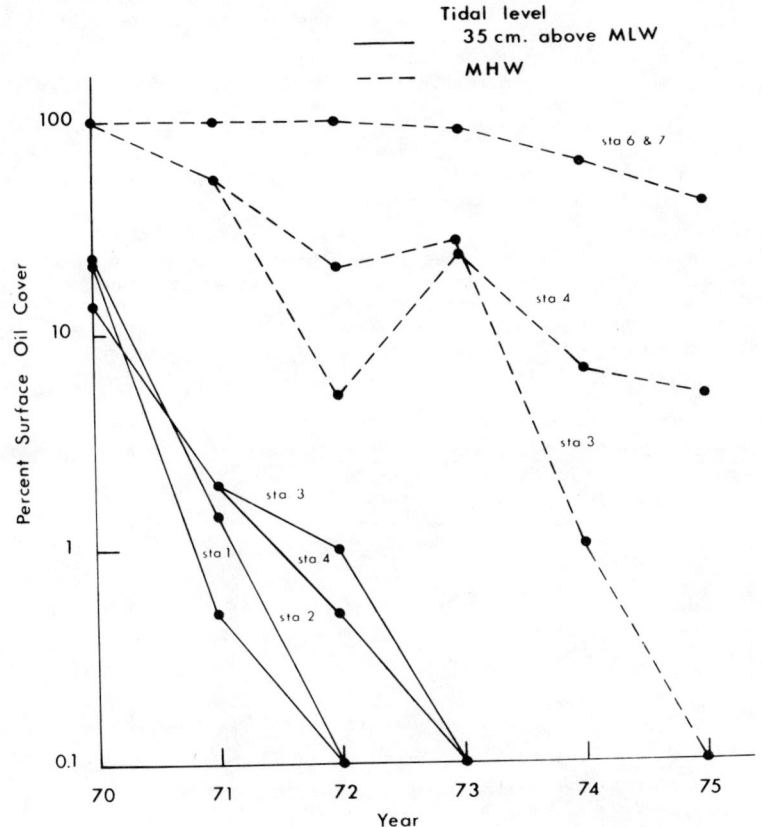

Fig. 25.2 Surface oil cover at two tidal levels at Chedabucto Bay stations from 1970-1975. Abbreviations are: MHW, mean high water; MLW, mean low water.

The extension of range varies from 28% of the tidal range at Station 1 to 38% at Station 4. The range now occupied is now within the normal variation for this region (Thomas, 1974, 1975).

Lack of prior knowledge of intertidal animal populations has prevented meaningful observations on most species. Among the commonest species, barnacles, *Balanus balanoides* showed no unusual mortalities and settlement of larvae followed by normal growth occurred in 1970 and each year thereafter as shown in Table 3. The common periwinkles (*Littorina saxatilis*, *L. littorea*, *L. obtusata*) remained abundant and their intertidal zonation was normal. A minor contraction of the range of *L. obtusata* was attributed to the disappearance of its algal habitat in the upper part of its range.

TABLE 3. Abundance per 100 cm^2 of Newly Settled and One Year Old *Balanus balanoides*, at study sites at Stations 1 and 3 in summer from 1970-1975.

Year	Station 1		Station 2	
	Young of Year	1 Year old	Young of Year	1 Year old
1970	32	40	103	95
1971	137	2	130	34
1972	153	29	0	106
1973	41	95	153	0
1974	58	0	176	93
1975	97	1	130	99

Observations on Sedimentary Shore Biota

On sedimentary shores, most of which occur in lagoons, oil attrition, as shown in Fig. 2 has been very slow. In most locations the soft shell clam (*Mya arenaria*) is abundant below mean tide level. Mortalities of this species were observed as soon as the ice melted in 1970 and have been sporadic but slowly declining since then. In plots of graded oiliness, mortality has been proportional to surface oil cover. Results are summarized in Table 4. This table also gives data on the oil content of clams from these plots in 1970.

At a fifth study plot at Station 7 which was located below the level of persistent surface oil cover, the abundance of soft shell clams has declined steadily since 1970. Data are presented in Table 5 together with information on the weight of oil extract until 1973. Length frequency diagrams of these collections show that recruitment has occurred in most years but survival has been low.

TABLE 4. Percentage Surface Oil Cover and Total Mortality of (*Mya arenaria*), in four experimental plots at Station 7 for the periods 1970-1972 and 1973-1975. The weight of methylene chloride extracted oil is given for 1970.

	Plot Number			
	1	2	3	4
Total Mortality 1970-1972. %	52.4	64.7	73.4	19.8
Mean % Oil Cover 1970-1972	51.6	74.0	85.2	1.5
mg Oil Extracted/g dry clam tissue 1970	59	100	175	57
Additional Mortality 1973-1975. %	15.9	24.2	5.6	17.8
Mean % Oil Cover 1973-1975	1.5	30.0	0.0	0.0

TABLE 5. Abundance of Soft Shell Clams (*Mya arenaria*) per m^2 in a Sample Plot at Mean Tide Level at Station 7 from 1970-1975. The weight of methylene chloride oil extract is given for 1970-1973.

	Year					
	1970	1971	1972	1973	1974	1975
Abundance/m^2	990	680	420	460	305	250
mg oil extract/g clam tissue	55	44	50	39		

Above mean tide level, the salt marsh cord grass is generally abundant. It also suffered mortalities following the oil spill but in a different pattern from other species. Data for Station 6 are presented in Table 6.

TABLE 6. Abundance of Salt Marsh Cord Grass (*Spartina alterniflora*) Clumps and the maximum height of plants at Station 6, at 105 cm above mean low water, from 1970-1975.

Year	*Spartina alterniflora* Clumps/m^2	Maximum height, cm
1970	102	48
1971	29	31
1972	1	8
1973	29	36
1974	48	25
1975	111	40

Abundance declined steadily along with plant height to 1972 and thereafter steadily recovered. The only other abundance plant on sedimentary shore transects was eelgrass (*Zostera marina*). Populations of this plant which grows below low tide level showed no changes following the spill, although oil still persists in the sediment.

Discussion

There is now a large body of literature on effects of oil on intertidal biota (Butler and Berkes 1972; Moore and Dwyer 1974; Moulder and Varley 1975) but little work on long term effects has been published. Most work on mortalities associated with oil contamination suggests that this is generally associated with more volatile fractions that are rapidly lost in weathering (Butler and Berkes 1972; Moore and Dwyer 1974; Nelson-Smith 1973; Percy and Mullin 1975). Among the few studies on long term effects, other agents such as detergents often complicate the situation. (Baker 1971a,b; Nelson-Smith 1973). Nevertheless persistent oil has been shown to affect behavior in some species, to be toxic to others and to alter the natural environment (Baker 1971a,b; Dunning and Major 1974; Gilfillan 1972; Hargrave and Newcombe 1973; Moore and Dwyer 1974). The situation in Chedabucto Bay is unique in that serious long-term effects from a single spill of Bunker oil have been observed in a cold climatic regime.

Shortly after the spill occurred, large quantities of seaweeds, mixed with oil, were deposited at the strand line. Oil had been observed to adhere tenaciously to seaweeds from the upper intertidal zone and it was assumed that extra floatation and reduced flexibility and streamlining resulting from the oil, rendered plants vulnerable to mechanical removal by waves. Stranded seaweed appeared normal and was broken off above the holdfast. This process was clearly implicated in the removal of high level seaweeds but the fact that further mortalities occurred during the summer of 1970, suggests that some toxic effect may also have contributed to the eradication of fucoid algae above mean tide level. This assumption is strengthened by observations on fucoid sporelings that settled in this zone every year, but failed to survive more than a few months. The fact that seaweeds re-colonised higher shore areas as surface oil disappeared could be related to the availability of clean spore settlement areas. However, oil normally covered a minority of the shore surface and spores did settle in even the most heavily contaminated zone. Persistent toxicity of oil seems most likely.

The situation involving salt marsh cord grass (*S. alterniflora*) in lagoons is evidently different. The pattern of mortality was peculiar in that it occurred over several years following the spill, was incomplete and in that recovery then proceeded steadily. Baker (1971) has shown that oiling of this species reduces needed oxygen diffusion through the plant to the roots and that if oiling was repeated mortalities occurred. In Chedabucto Bay the fact that the oil spill occurred in winter, when this species is dormant and without aerial portions, prevented early contact between oil and living tissues. Penetration of oil into the sediment would not occur as the latter would be frozen and the oil extremely viscous. During 1970 the plants grew, shoots penetrated the oil and the foliage appeared normal, not exhibiting typical chlorotic symptoms described by Baker (1971a). However, during that summer, the foliage did contact oil that re-mobilized in warm summer conditions usually on several occasions. The following year, 1971, plants were reduced in number and exhibited chlorosis. The effect could simply be one of delayed toxicity due to cold weather conditions. However, it is also possible that the continuous thick oil also resulted in abnormal sedimentary conditions, and that this also affected the plants' growth. In any case the recovery of the population proceeded rapidly when oil cover started to decline at this station in 1972 (Fig. 2).

Thomas (1973) described causes of early mortalities in the soft shell clam *Mya arenaria*. Many clams left their burrows when these filled with oil and either died or were eaten on the sediment surface. However, this phenomenon was short lived and clams started to die within their burrows. Dead clams were visibly contaminated with oil and mortalities were proportional to surface oil cover. Tests carried out in 1970 also showed more oil in the tissues of living clams from more heavily surface oiled areas. Mortalities are therefore clearly reltaed to oil but it is impossible to determine whether they were a result of direct toxicity or some other effect. It has been demonstrated that oil may have marked metabolic effects on various intertidal biota. Metabolic rates may increase or decrease and in filter feeding species such as soft shell clams, filtration rates may also change (Anderson, 1972; Avolizi and Nuwayhid, 1974; Dunning and Major, 1974; Gilfillan, 1972, 1973, 1975; Hargrave and Newcombe, 1973). Gilfillan (1972, 1973) showed that in two intertidal mussels there was a decreased net carbon balance and an increased energy demand correlated with oiling. In Chedabucto Bay all clam populations are dense and of small size, which suggests that food may well normally be in short supply. If this is true, the synergistic effect of low food availability and oil toxicity may have contributed to extensive mortalities. It will be interesting to see if populations return with time to their original levels or stabilise at lower ones. Much depends on the effects of weathered bunker C oil on clams, a subject on which there is no information.

The study has documented long term deleterious effects of bunker C oil on dominant organisms in several intertidal communities in Chedabucto Bay. That mortalities were not observed in less common species, can almost certainly be attributed to lack of control situation, rather than the true state of affairs. During 1976 a thorough study of oiled and non oiled communities in Chedabucto Bay was undertaken; results will be available later. However, much fuller studies of the long term behavior, physiological and ecological effects of long term oil pollution are urgently needed.

Acknowledgements

This work was supported partly by a National Research Council of Canada grant (A6389) and partly by the Environment Canada, Marine Ecology Laboratory, Dartmouth, Nova Scotia, Canada. I am most grateful to Mary Lou Harley who assisted in field and laboratory work. Thanks are also due to many others who helped in the field, laboratory or with advice.

References

Anderson, G. E., The effects of oil on the gill filtration rate of *Mya arenaria*. Va. J. Sci. 23, 45-47 (1972).

Anon. Report of the task force - operation oil. Vol. II Can. Minis. Transp. 104 p. (1971).

Avolizi, R. J. and M. Nuwayhid, Effects of crude oil and dispersants on bivalves. Mar. Poll. Bull. 5, 149-152 (1974).

Baker, J. M., Successive spillages. In. Colwell, E. B. Ed. The Ecological Effects of Oil Pollution on Littoral Communities. Inst. of Petroleum, London 1971, 21-23 (1971a).

Baker, J. M., Oil and salt marsh soil. *Ibid*. 62-71 (1971b).

Butler, M. J. A. and F. Berkes, Biological aspects of oil pollution in the marine environment: A review. Marine Sciences Center M.S. Rept. 22, McGill University. 118 p. (1972).

Dunning, A. and C. W. Major, The effects of cold seawater extracts of oil fractions upon the blue mussel, *Mytilus edulis*. In Pollution and physiology of marine organisims. F. J. Vernberg and W. B. Vernberg Eds. Academic Press, N.Y. 346-366 (1974).

Galtsoff, P. S., The American oyster, *Crassostrea virginica* Gmelin. U.S. Fish and Wildlife Serv., Bureau of Comm. Fish. Fishery Bull. 64, 480 p. (1964).

Gilfillan, E. S., Effects of water soluble fractions of crude oil on feeding and respiration in *Mytilus edulis*. Amer. Soc. Limnol. Oceanogr. Spec. Meeting Aug. 27-Sept. 1, 1972. (Abs. only) (1972).

Gilfillan, E. S., Effects of seawater extracts of crude oil on carbon budgets in two species of mussels. In <u>American Petroleum Inst. Proc. joint conf. on prevention and control of oil spills</u>, Washington, D. C. March 13-15, 1973: 691-695 (1973).

Gilfillan, E. S., Decrease of net carbon flux in two species of mussels caused by extracts of crude oil. <u>Mar. Biol.</u> <u>29</u>, 53-57 (1975).

Hargrave, B. T. and P. Newcombe, Crawling and respiration as indices of sublethal effects of oil and a dispersant on an intertidal snail *Littorina littorea*. <u>J. Fish. Res. Bd. Canada.</u> <u>31</u>, 1789-1792 (1973).

Moore, S. F. and R. L. Dwyer, Effects of oil on marine organisms: a critical assessment of published data. <u>Wat. Res.</u> <u>8</u>, 819-827 (1974).

Moulder, D. S. and A. Varley, A bibliography on marine and estuarine oil pollution. <u>Supplement I Mar. Pollut. Inf. Center. Mar. Biol. Assoc. U.K. Plymouth</u>, 152 p. (1975).

Nelson-Smith, A., <u>Oil pollution and marine ecology</u>. Plenum Press, N.Y. 260 p. (1973).

Owens, E. G., The restoration of beaches contaminated by oil. Chedabucto Bay, Nova Scotia. Mar. Sci. Branch Dept. Energy Mines Resour. Manscr. Ser. #19, 75 p. (1971).

Percy, J. A. and T. C. Mullin, Effects of crude oils on arctic marine invertebrates. <u>Beaufort Sea Tech. Report</u>. #11. Beaufort Sea Project, Victoria, B.C., 167 p. (1975).

Thomas, M. L. H., Effects of bunker C oil on intertidal and lagoonal buota in Chedabucto Bay, Nova Scotia. <u>J. Fish. Res. Bd., Canada</u> <u>30</u>, 83-90 (1973).

Thomas, M. L. H., <u>Introducing the Sea</u>. University of New Brunswick, Saint John, N.B., 140 p. (1974).

Thomas, M. L. H., Intertidal zonation on western Bay of Fundy rocky shores. <u>Trans. Altlantic Chapter Can. Soc. Envir. Biol</u>. 102-111 (1975).

CHAPTER 26

BIOLOGICAL SURVEY OF INTERTIDAL AREAS IN THE STRAITS OF MAGELLEN IN
JANUARY, 1975, FIVE MONTHS AFTER THE METULA OIL SPILL

Dale Straughan

Institute of Marine and Coastal Studies
University of Southern California
Los Angeles, California 90007

Abstract

Field sampling was conducted in areas in the Straits of Magellen that were oiled and unoiled by oil spilled from the tanker Metula in August 1974. Statistical analysis of physical parameters such as beach slope and intertidal height did not show a significant difference between the group of oiled and the group of unoiled sites. However, comparison of the grain size of the oiled and unoiled group of sites showed a statistical difference.

Marsh plants had started to grow through oil in the oiled areas. High levels of petroleum hydrocarbons were recorded in mussels in the oiled areas. The presence of byssus threads alone suggested recent loss of mussels in part of the heavily oiled area.

Biological, physical, and chemical data were analyzed using techniques of ordination, classification, and discrimination. These analyses indicate a negative relationship between the biota and presence of petroleum. While grain size was important in governing the distribution and abundance of species, the visible presence of petroleum was the most significant factor.

The Kuwait crude oil spilled in the Straits of Magellen is similar to that spilled from the Torrey Canyon. In both instances there was large scale mousse formation. It is suggested that it is this physical impact that is the most significant factor and not the cold water conditions.

Key words: Chocolate mousse, Kuwait crude, Metula, mussels, oil in sediments, Straits of Magellen.

Introduction

On August 9, 1974 the _Metula_ grounded at Satellite Patch just west of the First Narrows in the Straits of Magellen. The vessel was not refloated until 25 September, 1974. 50,000 to 56,000 tons of oil were spilled during this period (Hann, 1974, 1975, Baker 1974, Baker, et al., 1975). Most of this was light Arabian crude oil but 3,000 to 4,000 tons of Bunker C were lost during the last few days of the grounding. The light Arabian crude oil spilled was similar to the Kuwait crude oil used as an API reference oil (Warner, 1975) and to that spilled after the _Torrey Canyon_ oil spill (Warner, pers. comm.).

In January 1975, at the request of the National Oceanographic and Atmospheric Administration (NOAA), a field survey was conducted in intertidal areas of the Strait of Magellen. Since no detailed background data were available, the sampling program was designed to account for abiotic variables as well as graded amounts of petroleum from the _Metula_.
The abiotic variables studied are those that are known to naturally influence the distribution and abundance of intertidal organisms e.g. intertidal height, grain size of sediment (Straughan and Patterson, 1975), moisture content of sediments. This approach was initiated to eliminate abiotic gradients which may parellel the dosage of oil.
Analysis of the data by correlatory techniques should then reveal any significant relationship between species distribution and abundance, and the presence of petroleum.

The distribution of oil in the intertidal zone was documented in August 1974, (Hann, 1974, September-October 1974, Baker, 1974), and January-February 1975, (Baker, et al., 1975, Hann, 1975). Most of the oil was ashore between Punta Remo and Punta Anegado (Fig. 1). The highest concentrations of oil were observed in the Puerto Espora area. Isolated patches of oil were found in high intertidal areas as far east as Bahia San Felipe.

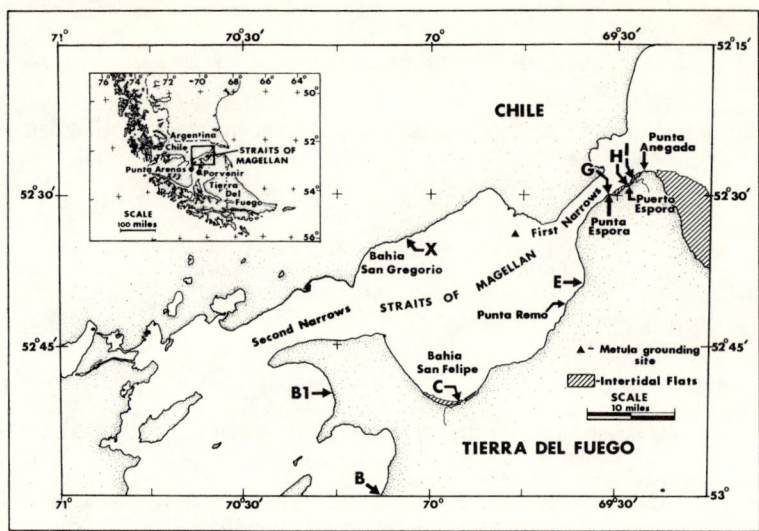

Fig. 26.1. Map of Straits of Magellan to show survey sites. Site A is at Porvenir - see inset.

In January 1975, most forms of oil were still present in at least one of the areas sampled. This included dry oil in patches (Bahia San Felipe), wet chocolate mousse on the surface of sediments, wet chocolate mousse incorporated in sediments, oil/sediment quicksand, brown surface at the waters edge due to presence of oil, buried layers of oil, oil sheen leaching from sediments, layer of wet black oil on sediments, pools of oil, oil sediment mixtures that resemble asphalt paving.

All data are not included herein because this paper is limited in length. A report on this research was submitted to NOAA in 1975 (Gunnerson and Peter, 1976). Since submission of that report, several samples which were believed lost in shipment, were recovered and analyzed. A report containing all data was subsequently prepared Straughan, (1976).

Materials and Methods

Site Selection
Site selection was initially based on the distribution of petroleum reported by (Baker 1974 and Hann, 1974) and later modified by on-site inspection. Four sites (E,G,H,I) were surveyed in the area where oil was initially most heavily deposited and where oil was still visually heavily deposited. Site C was chosen to represent a lightly oiled area. This lightly oiled category assignment was based on Baker's reports and on visual observations of oil in a dry state remaining only in upper intertidal areas. Four sites (A,B,B1,X) were initially selected as possible control areas. Sites B and B1 were visited but were not surveyed. They were areas of large cobbles and were completely dissimilar to any of the oiled sites and were regarded as ecologically unsuitable control sites.

Site A is at Porvenir. While it suffers from the disadvantage of being in an inlet adjacent to a town and thus could be exposed to other sources of man-made pollution, it is similar to some of the oiled sites (e.g., I-5) in that it has both coarse and fine sediments in the lower intertidal areas. Site X, on the northern side of the Straits of Magellen, likewise has mixed intertidal sediments with surface cobbles and intervening sand and gravel.

Sampling Techniques
Quadrats were sampled at each site to account for all visible physical and chemical variables. These included: Intertidal height, Substrate, Presence of Oil, and Presence of Kelp.

Sites I and C both included areas on the open coastline and areas on the edge of a creek draining a marsh. The marsh at Site I was very heavily oiled while that at Site C only had a few scattered patches of oil in the upper intertidal area.

In unstable substrates, namely sediments that are readily mobile and/or less than 3 inches in diameter, the quadrats sampled were 30 feet by 30 feet ($\approx$ 6 x 6 m). Ten points, selected from a table of random numbers, were sampled using a core 3 inches ($\approx$ 7.5 cm) in diameter. Where possible, three cores, 8 inches (20 cms) deep were collected at each of these points. However, much of this area is very coarse and hard and it was impossible to obtain 8 inch cores. The sediments were then sieved through screens (1.5 mm mesh). The organisms were removed from the remaining coarse sediments and preserved in 100% ethanol for later identification. Animals were sorted and stored in 70% ethanol on return to the laboratory. In each quadrat, a sediment sample was collected for grain size analyses and sediment samples were collected for petroleum analysis. The latter samples were collected directly into chemically cleaned 4 oz. aluminum containers provided by Dr. J. Scott Warner, Battelle Columbus, Ohio. Samples were kept cool (air temperatures did not exceed 40°F) until dry ice was obtained. The maximum period between collection and freezing was 2 weeks. Samples were shipped to the United States of America on dry ice and then stored at − 80°C. They were later shipped in dry ice to Dr. J. Scott Warner, Battelle Columbus, Ohio for chemical analysis (Warner, 1975).

In areas of stable substrates, namely large rocks, smaller quadrats (1m square) were surveyed. Subsamples (10 x 10 cms) were recorded and/or collected at ten points selected using a table of random numbers. Samples were only collected if it were not possible to either see and/or identify all species in the field.

All quadrats were related to each other, and, as far as possible to permanent bench marks, by recording intertidal profiles at each site using Emery sticks Emery, (1961).

Species that were not in the areas subsampled but were within either type of quadrat, were recorded. Special attention was paid to either dead or sick animals. Empty shells in good condition were counted. These were separated from old and battered shells in an attempt to obtain a record of recent mortality.

Supplementary samples of the common mussel, *Mytilus edulis chiliensis*, were collected at both oiled and unoiled sites for chemical analyses of tissues for petroleum hydrocabrons. Animals were placed in chemically cleaned glassware provided by Dr. J. Scott Warner. They were kept as cool as possible in the field (maximum 1 week), frozen in a domestic freezer for 3 days, transported to the United States of America on dry ice, and then stored at −80°C. Prior to shipment to Dr. Warner for chemical analysis, the animals were shucked into chemically cleaned glassware. Care was taken to avoid contamination of the tissues from the outside of the shell. The length of all shells was measured and the total tissue weight in each sample was later weighed by Dr. Warner prior to chemical analysis.

Oiled sediments were washed in acetone to remove the large quantities of petroleum prior to grain size analysis. In most cases, the petroleum was removed after 24 hours of soaking in 6 parts acetone to 1 part sediments. However, two samples were soaked for a second twenty-four hour period. This was necessary because in many instances so much petroleum in the form of 'chocolate mousse' was present, the sediment stuck together in a single mass. However the washing process probably removed some fine grain fractions from oiled samples.

Data Analysis
The Kolmogorov-Smirnov test (Siegel, 1956: 127) was used to determine if the normal physical characteristics (e.g. grain size) of the oiled and non-oiled groups of quadrats were significantly different. This test was chosen because it allowed a comparison between samples of unequal size groups. The error of the chi-square approximation with small samples is always in the 'safe' direction so that if the null hypothesis is rejected, one can view this decision with confidence.

An agglomerative, polythetic method of classification (Williams, 1971) was used to define groups of sites with a similar species' composition. The species were also classified into groups to show similar patterns of occurrences at the sites. Specifically, a Bray-Curtis distance (Bray and Curtis, 1957) was used to quantify the relationships between all pairs of sites (for the site analysis) and all pairs of species (in the species analysis). Flexible sorting strategy (Lance and Williams, 1967) was used to build the dendrograms which display the groups and their hierarchical relationship. Two-way coincidence tables (Stephenson, Williams and Cook, 1972) were employed to show the relationships between the defined site and species groups.

Principle components analysis (Seal, 1964) using the Q matrix form of (Orloci, 1966) and variance-covariance coeffecient was used to analyze the sites according to their abiotic characteristics. Essentially this analysis provides a rank order of sites along the axis created by the analysis to account for the maximum variance in the data.

If the biotic site groups obtained using classificatory techniques coincides with the abiotic site groups obtained by principle components analysis, it would indicate that the species distribution and abundance is related to the abiotic parameters. Spearman Rank Correlation Co-efficients (Siegel, 1956) were calculated to compare the rank order of sites obtained by principle components analysis of the biotic data with that obtained using the abiotic data to test for possible cause and effect relationships.

However, the cause of the biotic grouping or gradients could be one of the less distinctive abiotic gradients which could be overshadowed by a larger but less significant gradient as far as the distribution of these species are concerned. Hence, the data were further analyzed by multiple-discriminant analysis (Smith, 1976). This method is used to study direct relationships between predetermined groups of entities (e.g. the biotic site groups formed by the classificatory analyses) and a given set of attributes or variables measured on or at the entities (e.g. the abiotic parameters of the quadrats).

This method has an advantage over the principal components analysis because the biotic data are directly related to the abiotic data. It has the advantage over the use of single means and standard deviations of individual abiotic parameters because frequently combinations of abiotic variables rather than individual abiotic variables best explain group differences.

Rare species were eliminated from these statistical analyses by considering only species in which more than five individuals were recorded. Algae were not considered in these statistical analyses because species identification is incomplete. Species abundance data were transformed into a relative scale with the largest occurrence equal to 100% and the other occurrances rated as a percentage of this. Petroleum content was determined by the percentage of total carbon tetrachloride organic extractables in the sediments followed by gas chromatography to determine whether there was Metula oil present.

Results

Water temperatures during these surveys ranged from 8 to 10°C while air temperatures averaged 8°C. Ocean salinities ranged from 28 to 31 parts per thousand. The salinity in the estuary at I-4 and I-6 was 30 parts per thousand and the salinity was 18 parts per thousand at C-5.

The intertidal range varied between sites from 22 feet at sites E and G to 6 feet at site A (Table 1). The intertidal range presented here is that recorded at the time of the survey. Hence, the differences are due not only to the difference in tidal range between locations at the First Narrows and those further west, but also to the tides on the actual of the survey.

TABLE 1. Physical Characteristics at Coastal Sites

Site	Intertidal Range (feet)	Profile Slope	Length (feet)	Kelp	Dry	Wet	Mousse	Seep	Quick Sand	Asphalt
A	6.0	1:32	190							
*C U	10.5	1:10	105	+						
L	2.5	1:426	1,115							
T	13.0	1:94	1,220	+						
E	22.0	1:10	220	+	+	+	+	+		+
G	22.0	1:10	220	+	+		+			
H	Adjacent and similar to Site I									
*I U	9.0	1:10	90	+	+	+	+			+
L	3.5	1:286	1,000		+	+				
T	12.5	1:87	1,090	+	+	+	+			+
X	15.5	1:19	300	+						

* = Marsh areas excluded.
U = Upper, L = Lower, T = Total.

Sites E and G are both relatively short steep sloping beaches. Both have relatively coarse sediments with a high (more than 50%) percentage of gravel. Gravel is defined as sediment with a diameter of 2 mm or greater.

Sites C and I, in contrast, tend to have some gravel and cobbles in low intertidal areas but they also have finer sediments in these areas. This may take the form of a layer a few inches thick on the coarser sediments or be in the form of discrete patches. This was also observed at A-1 and at site H. The lower intertidal areas at sites C,H,I are long and relatively flat (over 1000 feet wide and with a total elevation of less than 4 feet). The upper intertidal areas at these sites are short and steep with a similar gradient to that found at sites E and G (1:10).

Two sites, G and C, were profiled on two dates a few days apart. There was little overall difference in the profile during that period.

Site X is also a short relatively steep beach (gradient 1:19). It bears more cobble than the other sites but still has gravel, coarse sand, and finer sediments between the cobble.

Kelp was stranded in upper intertidal areas at all sites except sites A and C. However, at sites E and I the kelp was covered in petroleum but it appeared clean of petroleum at site G and had no visible petroleum at site X.

Twelve of the sediment samples that were heavily oiled were washed in acetone 1 to 2 days (Table 2). This means that these samples may have lost finer sediments. Unfortunately there is no way of determining what fraction of the sample was lost. However, ∅ sizes as fine as 4.00 were recorded from non-acetone washed samples (5 samples) while the finest ∅ size from acetone washed samples was 3.75 (1 sample).

TABLE 2. Characteristics of Sediment Samples

Sample Designation	Acetone Washed (Days)	% Content Gravel	Sand	Mean ∅ (Duplicates)
A - 1 No Mud	0	65	35	0.58, 0.71
A - 1 Mud	0	60	40	2.50
A - 2	0	55	45	1.15, 1.77
C - 1	0	20	80	1.15
C - 2	0	65	35	1.51, 1.14
C - 3	0	35	65	0.83, 0.92
C - 4	0	55	45	1.85, 2.12
C - 5	0	30	70	0.65, 0.79
C - 6	0	28	72	0.79, 1.18
C - 7	0	20	80	1.29, 1.30
C - 8	0	22	78	0.98, 1.06
C - 9	0	65	35	1.63, 1.67
E - 2	1	15	85	0.41, 0.28
E - 3	1	66	34	0.19, 0.14
E - 4	1	80	20	0.83, 0.30
E - 5	1	34	66	0.30, 0.12
G - 5	0	92	8	0.22, 0.19
G - 6	0	65	35	0.60, 0.62
G - 7	0	80	20	0.89, 0.83
G - 8	0	30	70	1.28
G - 9	0	0	100	1.35, 1.32
H - 4	1	55	45	0.94, 1.12
I - 1	1	55	45	0.12, 0.35
I - 2	1	70	30	0.43, 0.36
I - 3	1	45	55	0.29, 0.39
I - 4	2	30	70	1.21, 1.12
I - 5	1	80	20	1.36, 1.27
I - 6	2	80	20	0.44, 0.41
X - 3	0	35	65	1.87, 2.30
X - 4	0	35	65	2.20, 2.14

Comparison of similar intertidal areas, for example sites C-1, C-2, C-3, and C-4 which are on a low flat unoiled intertidal area and whose samples were not acetone washed, with samples from sites I-1 and I-5 which were both heavily oiled and acetone washed, showed that the unoiled, non-acetone washed samples contained a slightly lower percentage of gravel (20, 65, 35, and 55% respectively) than the oiled, acetone washed samples (55 and 80% respectively) but that the finest sediments recorded in both areas were within a similar $\emptyset$ range (3.00 to 4.00). The differences in gravel percentage and $\emptyset$ size could easily be due to acetone washing but even if they were not, do not suggest a biologically significant change in sediment size between C-1 and C-3 and I-1 and I-5. The mean $\emptyset$ at these four quadrats from site C fell between the 0.12 level recorded at I-1 and the 1.36 level recorded at I-5 (Table 2). Therefore, it is doubtful that any differences in species composition at these quadrats would be related to different sediment size parameters.

The Kolmogorov-Smirnov test was used to compare the grain size, intertidal height, and beach slope of the group of quadrats that were oiled with the same parameter in the group of quadrats that were not oiled. $X^2 = 1.29$ for beach slope and 0.822 for intertidal height. Neither of these values were significant at 0.01 level. However, $X^2 = 10.04$ for the grain size comparison. This is significant at the 0.01 level for a one-tailed test. This indicates that the grain size of sediment analyzed from the oiled sites was significantly coarser than that analyzed from the unoiled sites. How much, if any of this difference exists in the field, is difficult to determine because the oiled sediment samples were washed with acetone.

At low tide, sand in the upper intertidal areas dried and was blown by the wind. This occurred in all upper intertidal quadrats on the open coastline except at site X where these quadrats were dominated by cobble and in quadrats where the sand was bound by oil. Complete binding by oil was observed at I-2, I-3, E-1, E-2; partial binding by oil was observed at C-8 and E-5; while no binding by oil was observed at C-6, C-7, E-4, G-9, G-8, and G-7. Quadrats below mean water level did not dry sufficiently to blow in the wind. The sand was bound in several ways:

1) by wet oil and/or mousse on or mixed with the sand (E-1, E-2)

2) by a layer of dry oil on top of the sand (C-8, I-3)

3) by an 'asphaltic' formation of sand and oil (I-2)

4) by a quicksand effect (E-2)

Oil was observed seeping out of the sand at G (G-5, G-6) and was also mixed into the water as indicated by the brown waves breaking on the beach at sites E and G.

At one site, site G, oil in the upper intertidal zone was already being eroded away by wave action.

Petroleum was recorded in the sediments at sites C,E,G,H, and I, and not at the two control sites (A,X, Table 3). Petroleum was only recorded both visibly and analytically in two of the quadrats at site C. These data illustrate the patchy distribution of petroleum in the environment unless it is present in massive amounts such as in the main area of the oil spill. The samples from C-5 illustrate this in detail. Sample C-5A was collected from one of a series of a row of dry black deposits of tar at the upper intertidal level of the quadrats; samples C-5B was collected from one of a series of a row of black tar deposits that were brown and wet internally and were fifteen feet closer to the ocean than C-5A; sample C-5C was collected fifteen feet closer to the ocean than C-5B and where there was no visible tar. The tar at C-5B appeared fresher and to have been deposited on a later and lower high tide than at C-5A. The data presented in Table 3 support these observations. Both C-5A and C-5B contained petroleum while C-5C did not contain petroleum. C-5B, the sample with the wet petroleum, had a higher water content (26%) than either the dry petroleum sample (C-5A) (1%) or the sediment sample without any petroleum (C-5C) (3%). Sample C-5C was actually slightly lower intertidally than C-5B and the sediments at C-5C would normally be expected to contain more water than similar sediments at C-5B.

Site C can then be characterized as a site that was initially exposed to a small amount of Metula oil in comparison with Sites E,G,H, and I. Since all of the oil was confined to defined horizontal patches in upper intertidal areas in January 1975, the lower intertidal quadrats were not being re-exposed to this petroleum at the time of the survey and may never have been exposed to this oil. If there were continued re-exposure of lower intertidal areas, one would have expected traces of petroleum in the sediments. The petroleum probably contained lighter and more volatile compounds during the period of initial exposure, so that while the lower intertidal areas may never have been exposed to the Metula oil that stranded, these quadrats could have been exposed to some of the lighter soluble compounds.

At five quadrats, E-1, E-2, I-4, I-5, and I-6, where thick layers of petroleum were present on the sediment surface, two samples were collected - one on the surface and one at a depth of 15 to 20 cms. In all instances there was four to six times more petroleum in the surface samples than in the subsurface samples. This was mainly because the surface sample was more oil than sediment. For example at I-4, I-5, and I-6, the field notes record that the oil was still present in a layer up to 2 inches thick on the surface of the sediment. The high quantity of oil at a depth of 15 to 20 cms indicates that the oil had penetrated much deeper into the sediments.

The petroleum formed "chocolate mousse" - a brown oil and water mixture. The ratio of this mixture can be as high as 20% oil to 80% water Berridge, et al., (1968). Such a mixture has a significant increase in the volume of the spill and can increase the area of exposure to the pollutant.

TABLE 3. Analysis of Sediment Samples from the Metula Oil Spill by J. Scott Warner

Sample Designation	Density[a]	Sediment,[b] %	Water,[c] %	CCl_4 Extractables, % by Given Method		Petroleum,[e] Contamination
				IR	Gravimetric	
A-1 No Mud[d]	2.14	89.3	10.7	0.0024	0.0059	No
A-1 Mud[d]	1.64	58.9	41.1	0.031	0.049	No
A-2[d]	1.84	95.4	4.6	0.0050	0.0046	No
C-1[d]	2.27	89.9	10.1	0.013	0.012	No
C-2[d]	2.17	89.2	10.7	0.063	0.046	No
C-3[d]	1.85	69.4	30.6	0.0230	0.034	No
C-4	2.10	88.6	11.4	0.0010	0.0019	No
C-5A	1.85	93.7	1.4	5.4	4.9	Yes
C-5B[d]	1.53	53.1	26.1	25.4	20.8	Yes
C-5C[d]	1.78	96.9	3.1	0.0003	0.0009	No
C-6	1.76	96.8	3.2	0.0002	0.0010	No
C-7	1.64	95.1	4.9	0.0002	0.0010	No
C-8[d]	1.78	93.4	3.0	4.4	3.6	Yes
C-9[d]	1.76	70.5	29.5	0.010	0.010	No
E-1 Surface	1.59	58.2	32.6	12.3	9.2	Yes
E-1 Below Surface	1.89	92.1	6.3	2.3	1.6	Yes
E-2 Surface	1.96	76.5	16.5	7.3	7.0	Yes
E-2 Below Surface	1.71	93.1	5.9	1.2	1.0	Yes
E-3	1.87	95.6	4.4	0.0037	0.0045	Yes
E-4[d]	1.79	98.7	1.3	0.0027	0.0031	Yes
E-5[d]	1.84	97.0	3.0	0.016	0.017	Yes
G-2[d]	2.02	82.4	17.6	0.020	0.025	Yes
G-3	1.07	14.6	51.5	37.9	33.9	Yes
G-4 Nearby	1.45	91.2	3.1	6.21	5.7	Yes
G-5	1.86	97.1	2.8	0.050	0.044	Yes
G-6	1.93	86.4	13.6	0.046	0.061	Yes
G-7[d]	1.57	96.2	3.7	0.052	0.060	Yes
G-8	1.59	95.6	4.4	0.028	0.026	Yes
G-9	1.54	93.4	6.6	0.015	0.015	Yes
H-1	2.13	84.6	15.4	0.072	0.082	Yes
H-3[d]	1.52	53.9	29.9	18.4	16.2	Yes
H-4	1.96	75.5	24.1	0.31	0.35	Yes
I-1	2.04	80.5	12.7	7.60	6.8	Yes
I-2	1.65	63.4	25.1	13.3	11.5	Yes
I-3 Loose Sand	1.64	97.4	2.0	0.60	0.59	Yes
I-3 Asphaltic Sand	1.83	85.8	8.5	6.2	5.7	Yes
I-4 Surface	1.04	5.8	56.3	44.7	37.9	Yes
I-4 Below Surface	1.95	79.1	14.6	7.3	6.3	Yes
I-5 Surface	1.86	69.5	20.8	11.3	9.7	Yes
I-5 Below Surface	2.16	85.6	10.0	4.9	4.4	Yes
I-6 Surface	0.99	2.6	64.9	36.3	32.5	Yes
I-6 Below Surface	1.99	77.9	14.9	7.6	7.2	Yes
X-1[d]	1.97	93.6	6.4	0.0030	0.0033	No
X-2[d]	2.83	97.2	2.8	0.0037	0.0044	No
X-3[d]	2.04	94.3	5.7	0.0016	0.0032	No
X-4[d]	2.29	83.0	17.0	0.0012	0.0015	No
X-4[d]	2.08	81.6	18.4	0.0027	0.0031	No

a = Density of wet sample as received; b = Extracted with carbon tetrachloride and dried; C = By difference [100- (% sediment + % CCl_4 extractables determined gravimetrically 10^{-1})]. d = The CCl_4 extract contains greater than 5% carbonyl containing compounds as indicated by absorption in the 1750-1650 cm^{-1} range. e = Assessment is determined by gas chromatographic analysis.

Water in the sediment samples analyzed in Table 3, will be due to several factors. One, of course, is the oil-water content of the mousse. However, intertidal sediments also contain water which will be dependent on the sediment type, intertidal height, and amount of time the sediments have drained between exposure by the tide and collection of sediments. The difference in water content due to sediment type can be best illustrated by two samples from A-1. The water content of the coarse gravel sample was 10.7% and water content in the finer sample was 41.1%. These two samples were collected only two feet apart at the same intertidal level and the same time.

However, in spite of the variations due to factors other than oil, it is obvious that in some instances, namely when thick layers of oil were still present, that these were associated with a high water content. For example, the surface sample at E-1 was light brown fresh appearing mousse. This was in a gravel type of sediment in the upper intertidal area where normally the water content would be low (less than 5%). The water content was 32.6% and it is suggested that the oil:water ratio was in the order of 1:2. At Site I where the oil was darker, the oil water ratio, based on similar assumptions is in the order of 1:1. At site G the ratio varies. Sample G-3 was mainly oil and little sediment so that most of the water was associated with the oil (ratio in the order of 1:1). The high water content at G-2 and G-6 was related to water seepage from the sand in these areas. This seepage contained oil in the form of a sheen. In the other samples, the amount of water appeared related to physical variables other than oil.

Forty-six species of marine invertebrates were recorded on this survey (Table 4). Twenty-eight of these species were polychaetes. This is not a complete species list for the entire Straits of Magellan but a list of the invertebrates found in the study quadrats. Thirty species were recorded in 13 unoiled quadrats (A-1, A-2, C-1, C-2, C-3, C-4, C-6, C-7, C-9, X-1, X-2, X-3, X-4) while 14 species were recorded in 25 oiled quadrats (C-5, C-8, E-1, E-2, E-3, E-4, E-5, G-1, G-2, G-3, G-4, G-5, G-6, G-7, G-8, G-9, H-1, H-2, H-3, I-1, I-2, I-3, I-4, I-5, I-6). Quadrat H-4 was in the edge of a shallow channel that always contained running seawater. Unlike the surrounding black oiled areas, H-4 did not contain visible oil, (Table 5) and it did contain biota that appeared to be healthy. Nearby H-3 contained 16-18% total C Cl_4 extractables while H-4 contained 0.3% total C Cl_4 extractables. H-4 contained 15 invertebrate species compared with 5,4, 1 found on H-1, H-2, H-3 respectively. The 26 oiled quadrats contained a total of 25 species.

TABLE 4. List of Living Invertebrates in Quadrats Sampled

CRUSTACEA
  Atyloella sp.
  Edotea tuberculata
  Eurypodius latreillei
  Exosphaeroma gigas
  Halicarcinus planatus
  Macrochiridothea michaelseni
  Serolis
  Valvifera
 	
NEMERTEA
 	
NEMATODA
 	
SIPUNCULOIDEA
 	
INSECTA
  Larvae
 	
MOLLUSCA
  Mytilus edulis chiliensis
  Pareuthria plumbea
  Patinigera magellanica
  Sphenia hatcheri
	ANNELIDA
	  Oligochaetes
	  Arabellidae
	  Brania sp.
	  Boccardia cf. polybranchia
	  Capitellidae
	  Ceratocephale crosslandi n.s. sp.
	  Chaetozone sp.
	  Cirratulidae
	  Cirratulis cf. cirratus
	  Eteone rubella
	  Euzonus fucifera
	  Exogone (2 sps.)
	  Hauchiella sp.
	  Isocirrus sp.
	  Lagisca cf. lamillifera
	  Langerhansia anops
	  Lumbrinereis latreilla
	  Lumbrinereis sp.
	  Nereis eugeniae
	  *Nothria sp.
	  Notocirrus chilensis
	  Notomastus sp.
	  Onuphidae
	  Phyllodocidae
	  Rhynchospio sp.
	  Terebellidae
	  Thelepsus setosus
	  Travisia cf. gigas
	  Typosyllis ?

* This species has been described by F. Piltz in a manuscript sumitted to J. Linn. Soc. in 1976.
n.s. sp. = new subspecies.

Comparison of the number of invertebrate species in visibly oiled and visibly unoiled quadrats in January 1975, showed that there were more species in the unoiled than in the oiled quadrats in all instances except site E (Table 5). The two lower intertidal quadrats at site E contained very coarse sand and thus probably never any marine invertebrates. The observations at H-4, where species appear healthy and abundant within a matter of feet of the oiled areas, would suggest that the continued effects on the distribution and abundance of species are possibly related to the physical presence of the mousse in that area and not to continue leaching of soluble components. Likewise, the most abundant Mytilus population was found in H-2, a quadrat with no visible oil but surrounded by visibly oiled areas.

TABLE 5. Number of Living Invertebrate Species in Quadrats With and Without Petroleum Visible, January 1975.

	A	C	E	G	H	I	X
Oiled		0	1		0	6	2
Not Oiled	9	19	0	7	17		14

The rare occurrences of barnacle scars, a dead crab, and a dead murid would not in themselves be regarded as important. However, the more common dead limpets (Patinigera magellanica), and observations on the mussels Mytilus edulis chiliensis are of importance. Mussels were found attached to rocks at all sites except site E where there were no rocks.

However, all attached M. edulis at site I were dead and some of the attached M. edulis at sites H were dead. At sites I-2 and I-3 there was a stranded layer of mussel shells in good condition suggesting that the animals had died recently. At G-4 there was a zone where mussel byssus threads remained, suggesting that there had been a recent mortality among these mussels.

Analysis of the biological data by classificatory techniques resulted in the formation of seven distinct site groups (Fig. 2). Site group 1 contained sites at which oligochaetes were recorded. The sites were not confined to any section of the intertidal zone. All sites except one (E-1) had no petroleum present. While site group 2 contained sites with oligochaetes at some sites, it is characterized by the presence of Mytilus edulis chiliensis shells. Empty limpet (Patinigera magellanica) shells were also found at some of these quadrats. These shells were all in good condition and showed no signs of deterioration. The quadrats in site group 2 were all heavily oiled.

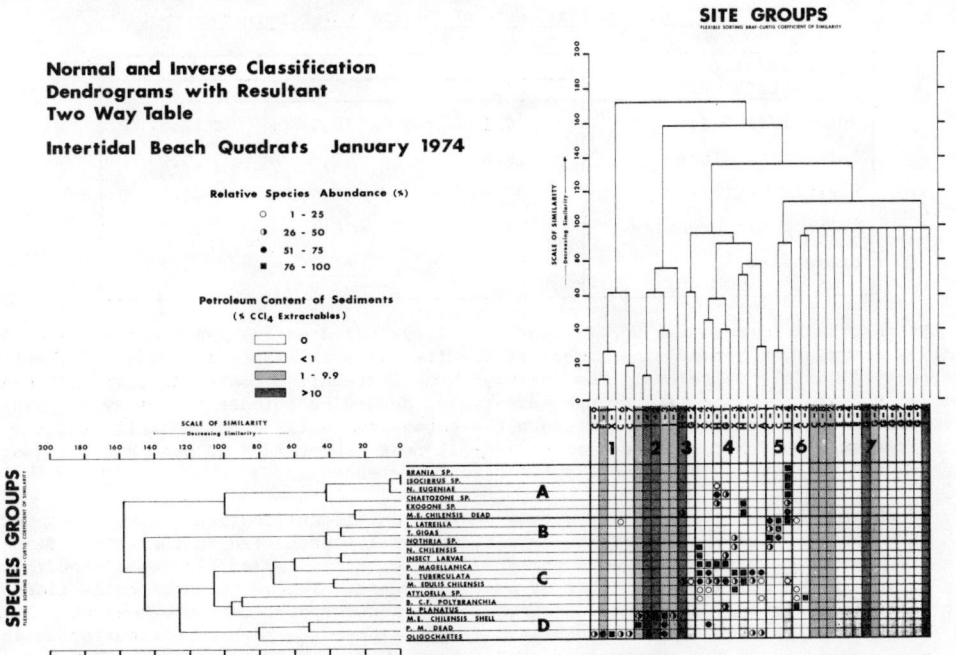

Fig. 26.2 Normal and Inverse Classification Dendrograms with Resultant Two-way Table. Petroleum content of sediments was determined on the basis of % total C Cl$_4$ extractables and gas chromatography to determine if the Metula oil was present.

Quadrats in site groups 3 and 4 were all in middle or low intertidal areas. All quadrats bore living M. e. chilensis. Both quadrats in site group 3 were oiled. No other invertebrates were recorded above the rare species category at these two quadrats. Site group 4 contained both oiled and unoiled sites but all oiled sites were only lightly oiled. Invertebrates in addition to M. e. chilensis were recorded in these quadrats.

Site group 5 contained only quadrats found in low intertidal areas. The polychaete, Lumbrinereis latreilla, was the dominant species in this group. Both quadrats at site C were not oiled when surveyed while the quadrat at site H was the lightly oiled quadrat in the midst of a heavily oiled area.

Site group 6 contained two quadrats that were not oiled and were dominated by the polychaete Boccardia c.f. polybranchia.

Site group 7 contained a large number of oiled quadrats in which no animals were collected. There is no consistent trend in intertidal height, beach slope, or sediment size between these sites.

Species groups A,B, and C were found in the middle and low intertidal areas. The oligochaetes, the only living invertebrates in species group D, were not confined to any intertidal level. None of the species groups were restricted to either oiled or unoiled sites. However, two most abundant polychaetes Nothria sp. and Boccardia c.f. polybranchia, and the amphipod, Atyloella sp., were not recorded at the oiled quadrats.

Principal components analysis of the abiotic data indicate that intertidal height and visible petroleum are the most important factors, in that order, operating on axis 1 (eigenvalue 1.87, % total variance 37.56) and that total carbon tetrachloride extractables and visible petroleum are the most important factors operating on axis 2 (eigenvalue 1.74, % total variance 34.94) in that order. Slope and grain size are the most important factors on axis 3 (eigenvalue 0.84, % total variance 16.91). Comparison of the rank order along these axis with the rank order of sites obtained by ordination techniques using the biological data and the Spearman Rank Correlation Coefficient revealed no significant correlations.

Multiple-discriminant analysis of the data indicates visible petroleum is the most important abiotic factor (coefficient of separate determination= 40.6). Grain size (coefficient of separate determination = 64.9) is the dominant factor on the second axis and total carbon tetrachloride extractables (coefficient of separate determination = 46.9) is the dominant factor on the third axis (Table 6).

TABLE 6. Coefficients of Separate Determination

Abiotic Attribute	Axis		
	1	2	3
Intertidal Height	7.2	14.2	13.8
Intertidal Slope	18.5	2.5	29.9
Visible Petroleum	40.6	11.9	2.6
Extractable Organics	17.6	6.4	46.6
Grain Size	16.1	64.9	6.7

The Spearman Rank Correlation Coefficient of - 1.09 indicated a significant (α = 0.01) negative correlation between the number of species found in each site group and the mean frequency of visible petroleum. The Spearman Rank Correlation Coefficient was not significant (α = 0.05) when the site groups were ranked according to mean frequency of visible petroleum and mean grain size. Therefore these two gradients do not coincide and the negative correlation between species numbers and oil must be regarded as real and not just a coincidental reflection of sediment grain size differences.

Many of the algae collected on this survey await final identification. However, genera such as Porphyra, Ulva, Enteromorpha, Iridea, are well represented in the area. Since no data are available in the areas on a pre-spill basis, it is impossible to say how the abundance of the algae compared with that of previous years. Tentatively the collection has been separated into over 20 genera. However, bleached Porphyra were observed at H-1, H-2, and I-2. This algae could have been bleached when exposed to high temperatures due to the oil in the area and/or the exposure to high (about 80°F) temperatures the week before the survey. At G-4 Enteromorpha was already growing over a thin layer of oil on rocks. However, it was not growing over the mussel byssus threads on the rocks suggesting that the loss of these mussels had occurred only a short time prior to the survey.

Several species of plants were oiled in upper-intertidal areas at sites C and I. The perennial Salicornia ambigua had reshot through the oil at both sites. All oiled specimens at C-5 had shot while 12 out of 32 oiled specimens at I-4 had shot. The speciments at I-4 which had not shot appeared to be dead. The grass, Elymus arenarius, was also oiled. Six oiled plants at I-4 had all shot. The woody plant Lepidophyllum cupressiforme was heavily oiled at Site I. A high percentage of these plants had already sprouted even though at low tide they remained in pools of oil and water about a foot deep.

Table 7 shows the relative size and hydrocarbon content of M. e. chiliensis tissues. Most of the animals collected in the oiled areas were relatively thinner than those from the un-oiled areas. This is possibly indicative of stress from the very high concentrations of hydrocarbons recorded (1000 to 5000 ug/g) in the tissues. The background levels from the control sites were in the order of 10 to 15 ug/g.

TABLE 7. Tissue Analysis in Mytilus Edulis Chiliensis

Sample	No. Animals	Wet Weight (g) Total	Wet Weight (g) Average	Length (mm) Max.	Length (mm) Min.	Length (mm) Mean	Hydrocarbon ($\mu g/g$) Fraction 1	Hydrocarbon ($\mu g/g$) Fraction 2	Hydrocarbon ($\mu g/g$) Fraction 3
A-1	6	16.6	2.77	54	36	44.1	7	3	1
G-1	9	15.3	1.7	53	36	46.1	940	1100	100
G-3	14	13.0	0.9	49	27	37.3	1600	1300	140
G-4	8	8.0	1.0	46	31	37.0	2700	2000	280
H-2	21	15.2	0.7	42	28	34.1	500	710	110
H-3	5	10.0	2.0	46	34	41.6	490	460	60
X-2	10	15.3	1.5	44	26	36.9	10	5	1

Discussion

The survey of the Straits of Magellan in January 1975, showed that large areas of the inter-tidal zone were still covered from oil from the Metula oil spill. This is detailed by (Hann, 1975). The present study on the biological effects showed that in the areas most heavily oiled in the First Narrows, there is evidence of continued detrimental effects of the oil spill. These include the changing of sediment characteristics by binding sand that formerly blew in the wind at low tide into an asphaltic type of bed. This will undoubtedly influence the species which will recolonize the area.

Data comparisons on unstable intertidal substrates are generally more difficult than on stable rock intertidal substrates because of the lower number of species in these unstable areas. For example, the infaunal comparison of marsh species only revealed one specimen of a oligochaete species at I-4. This species was also recorded at other oiled and unoiled high intertidal quadrats outside the marshes.

In general, more species are found in lower intertidal than higher intertidal areas (Straughan, 1973, 1975; Straughan and Patterson, 1975, Patterson, 1974). This trend is also found in the Straits of Magellan. Hence, comparison of data on the low intertidal flats at sites C,H and I should provide data that is more readily interpreted than comparison of data from upper intertidal areas. Likewise, comparison of the fauna in the similar marsh qua-rats reveals little information since only one invertebrate species was recorded.

Comparison of similar quadrats from the low flat intertidal areas at sites C and I revealed little difference in physical parameters but a marked contrast between the number of species in the oiled and unoiled areas. There were no living animals collected in I-1 and I-5 while there were 9,5,6, and 1 species recorded at C-1, C-2, C-3, C-4 respectively. It should be noted that while only one species of tube dwelling polychaete Boccardia, was recorded at C-4, over 500 of the specimens were collected in the quadrat samples.

The physical presence of the oil on and in the sediments appears a very important factor. In other words, any soluble components may not be having a significant effect on the biota after 5 months. This was supported both by visual observations and discriminant analyses. H-4 had no oil visibly present but is surrounded by visibly oiled areas (e.g., H-3). Fif-teen species were recorded in H-4 and only one species was recorded in H-3. The number of species in H-4 also belies any suggestion that the sites where the oiled quadrats were located contained fewer organisms prior to the oil spill than the sites where the non-oiled quadrats were located.

The data reveal fewer organisms in the heavily oiled area, large numbers of empty shells in good condition, and evidence that mussel communities were once more extensive. While this does not conclusively demonstrate cause and effect, because the mortalities were recent, because the oil had been present in large amounts for five months, because species distribution and abundance was more closely correlated to the presence of petroleum than natural physical parameters such as sediment grain size, one can only conclude that the continued presence of spilled oil was responsible for reducing the biota.

These findings have been contrasted to those reported by Baker 1974, in the months immediately following the spill. Baker concluded that there was little damage from the oil spill. The present report shows that in the five months after the spill there must have been mortality because the differences cannot be related to physical parameters. The differences between Baker's conclusions and the observations reported herein, are probably partially a result of the age old question, when does one count the dead bodies? Species such as mussels are able to survive adverse conditions for weeks and/or could be counted as living in early surveys but be covered by oil with the shells closed, and really be dead. Other differences such as differences in infauna between the two surveys are related to the differences in the two surveys. Baker surveyed a large number of sites in a month and directed her report to almost the entire ecosystem. In the present survey, the time was shorter but more detailed observations were made on a few specific areas confined to the intertidal zone.

Recovery of the vegetation had commenced in the marsh areas as indicated by the regrowth of plants through the oil. However, this optimistic note must be viewed with caution because studies after the Arrow spill in Canada have shown that maximum mortality in at least one marsh plant, Spartina alterniflora, may be recorded 2 years after the oil spill (Thomas, 1977).

This spill differs from most of the documented large oil spills in several respects:

1) oil was deposited intertidally in the form of chocolate mousse
2) oil remained intertidally for a long period
3) lower, as well as higher intertidal areas, were heavily contaminated by this chocolate mousse for at least 5 months.

It is difficult to compare the suspected mortality with that of other crude oil spills because of the lack of documentation of similar habitats. However, data recorded from unstable areas 6 months after the oil spill in the Santa Barbara Channel did not reveal any effects on the biota from the oil spill (Trask, 1971, Straughan, 1973). Differences between these two crude oil spills include chemical composition of oil, physical behavior of oil, and time of exposure to oil. Sandy beaches in the Santa Barbara Channel were not exposed to prolonged dosages of oil of the magnitude recorded in the Straits of Magellan, and there are no records of the extensive type of mousse formation that occurred in the Straits of Magellan.

The only other large spill of similar oil to the Metula spill, was that from the Torrey Canyon. Kuwait crude oil was spilled in the latter case. (Warner, 1975) reports that the Metula oil is very similar to the API standard Kuwait crude oil. In both cases there was extensive mousse formation. However, the shores of Great Britain were subjected to a rigorous cleanup program which included the use of detergents that are now considered too toxic for general use. The available biological data reports mainly on communities associated with more stable rocky shores in Britain, rather than those less stable substrates found on the south shores of the First Narrows.

It is interesting to compare the oil content of sand examined after the Torrey Canyon with that reported herein. For example, the highest oil content of sand reported by (Smith, 1968) was 11% from very heavily polluted sand at Sennen approximately 3 months after the oil spill. This is considerably lower than the oil content of the most polluted sites in January 1975 (30%-40%). This suggests a lower level of contamination of British than Chilean shore lines.

Therefore, in addition to being one of the largest oil spills to date, the polluted intertidal areas of Chili have been subjected to the longest and highest dosage of this type of crude oil. As the oil has been in a mousse form it has not dried out to even form recolonizable substrates but has remained in a sticky form which is not at all similar to any normal intertidal physical habitat. Exposure to these physical parameters doubtlessly had a significant influence on the biota without even considering the chemical toxicity of the oil.

Acknowledgements

I wish to acknowledge the assistance of many hard working personnel in the United States and Chili. I am also grateful for the opportunity to discuss the studies with Jenifer Baker, Brian Dicks, and Wardley Smith in Great Britain.

The field survey was conducted with assistance from Jean and Bill Texera, from the Peace Corps, Claudio Venegas and Italo Campodonico of the Instituto de la Patagonia, Charles Gunnerson of the MESA program for NOAA, Roy Hann from Texas A & M University and Ken Adams of the Environmental Protection Agency. Species identification was with assistance from Leonardo Guzman, Instituto de la Patagonia, Valarie Anderson and Robert Setzer from the Allan Hancock Foundation, University of Southern California (tentative algal identification); Fred Piltz, Allan Hancock Foundation, University of Southern California (Polychaeta); Mary Wicksten, Allan Hancock Foundation, University of Southern California and Jim Wilkins, Texas A & M University (Crustaceans); Joe Franz, Allan Hancock Foundation, University of Southern California (Molluscs).

Photography was by Charles Gunnerson, Dale Straughan, and Bill Texera and illustrations we were by Tom Licari. Kevin Lee Brown and Joe Martin analyzed the sediments for grain size J. Scott Warner of Battelle, Columbus, Ohio, analyzed sediments and tissues for petroleum content. Beverly Allen and Pam Smith typed the manuscript. Robert Kanter and Lou Fredkis conducted the computer analysis.

I am also grateful for the hospitality and assistance shown me by the Director, Instituto de la Patagonia and Contra Almirante Edwardo Allen, Commander of the Chilean Third Naval Zone and his staff.

References

Baker, J. M., Grounding of "Metula". Magellan Straits Ecological Survey, 9 September to 4 October, 1974. Report of the Oil Pollution Research Unit. Orielton Field Center (1974).

Baker, J., I. Campodonico, L. Guzman, J. J. Texera, B. Texera, C. Venegas, and A. Sanhuega, An Oil Spill in the Straits of Magella. Proc. Conference on Marine Ecology and Oil Pollution. Aviemore, Scotland: 441-471 (1975).

Berridge, S. A., M. T. Thew, and A. G. Louiston-Clarke, The Formation and Stability of Emulsions of Water in Crude Petroleum and Similar Stocks. In Scientific Aspects of Pollution of the Sea by Oil. Pub. Institute of Petroleum, London (1968).

Bray, J. R., and J. T. Curtis, An Ordination of the Upland Forest Communities of Southern Wisconsin. Ecol. Monogr. 27:325-349 (1957).

Emery, K. O., A Single Method of Measuring Beach Profiles. Limnol. Oceanog., 1: 90-93 (1961).

Gunnerson, C. G. and G. Peter, The Metula Oil Spill NOAA Special Report, U. S. Department of Commerce. Boulder, Colorado 37 pp. (1976).

Hann, R. W., VLCC "Metula" Oil Spill. Report to the U. S. Coast Guard Research and Development Program. NTIS #AD/A-003 805/wp (1974).

Hann, R. W., Follow-up Field Survey of the Oil Pollution from the Tanker "Metual". Report to the U. S. Coast Guard Research and Development Program, 57 pp (1975).

Lance, G. N., and W. T. Williams, Mixed-Data Classificatory Programs. I Agglomerative Systems. Aust. Compt. J. 1:15-20 (1967).

Orloci, L., Data Centering: A Review and Evaluation with Reference to Component Analysis. Systematic Zoology: 208-212 (1966).

Patterson, M. M., Intertidal Macrobiology of Selected Sandy Beaches in Southern California. Sea Grant Publication, University of Southern California USC-SG-9-74, 41 pp (1974).

Seal, H. L., Multivariate Statistical Analysis for Biologists. Methuen, London 207 pp (1964).

Siegel, S., Nonparametric Statistics for the Behavioral Sciences. McGraw-Hill, New York, 312 pp (1956).

Smith, J. E., "Torrey Canyon" Pollution and Marine Life. Pub. Mar. Biol. Ass U.K. Cambridge, 196 pp (1968).

Smith, R. W., Numerical Analysis of Ecological Survey Data. Ph.D. Dissertation, Biology Department, University of Southern California, Los Angeles, 401 pp (1976).

Stephenson, W., W. T. Williams, and S. Cook, Computer Analysis of Peterson's Original Data on Bottom Communities. Ecol. Monogr. 42 (4): 387-415 (1972).

Straughan, D., The Influence of the Santa Barbara Oil Spill (January-February, 1969) on the Intertidal Distribution of Marine Organisms. Report presented to the Western Oil and Gas Association. June, 1973, 66 pp (1973).

Straughan, D., Intertidal Sandy Beach Macrofauna at Los Angles - Long Beach Harbor Part 2. Marine Studies of San Pedro Bay, California. Part 8. Pub. University of Southern California: 89-107 (1975).

Straughan, D., Intertidal Biological Studies of the Metula Oil Spill in the Straits of Magellan, January 1975. Sea Grant Publication, University of Southern California USC-SG-76 (1976).

Straughan, D. and M. Patterson, Intertidal Sandy Beach Macrofauna at Los Angles - Long Beach Harbor Part 1. Marine Studies of San Pedro Bay, California. Part 8. Pub. University of Southern California: 75-88 (1975).

Thomas, M. L. H., Long Term Effects of Bunker C Oil in the Intertidal Zone. Proc. NOAA-EPA Symposium on Fate and Effects of Petroleum Hysrocarbons, Seattle, November 1976: (1977).

Warner, J. S., Determination of Petroleum Components in Samples from the Metula Oil Spill. Report to Marine Ecosystems Analysis Program National Oceanic and Atmospheric Administration (Contract No. 03-5-022-47) from Battelle Columbus, Ohio. 15 pp., 2 Appendices (1975).

Williams, W. T., Principles of Clustering. Ann. Rev. Ecol. Syst. 2:302-326 (1971).

CHAPTER 27

STUDIES ON PETROLEUM BIODEGRADATION IN THE ARCTIC

Ronald M. Atlas

Department of Biology
University of Louisville
Louisville, Kentucky 40208

Abstract

Microorganisms capable of biodegrading petroleum were found to be widely distributed in the Beaufort and Chukchi Seas, but to constitute only a low percentage of the indigenous heterotrophic microbial populations. Concentrations of hydrocarbon utilizing microorganisms were lower in ice than in water or sediment. Hydrocarbon biodegradation potential was also lower in ice than in water or sediment. Natural rates of degradation were slow; maximal losses from experimental oil spills were less than 50% during the Arctic summer due to combined abiotic and biodegradative losses. Rates of biodegradation were found to be limited by temperature and concentrations of available nitrogen and phosphorus. Residual oil had similar percentages of hydrocarbon classes as fresh oil; biodegradation of all oil component classes, including paraffinic and aromatic fractions, apparently proceeded at similar rates.

Key words: Heterotrophic microorganisms, Prudhoe Crude, Beaufort Sea, nitrogen, phosphorus

Introduction

This paper will report on several studies conducted during the past four years aimed at determining the fate of petroleum hydrocarbons in Arctic marine and nearshore ecosystems. The expanding exploration for and exploitation of Arctic petroleum reserves, the engineering hazards associated with the drilling and transport of oil in the frozen Arctic, and the logistic impracticality of physically removing oil spillages from ice laden Arctic waters, all point to the fact that Arctic microorganisms will someday soon face the task of biodegrading unwanted oil pollutants. The ability of microorganisms to biodegrade and effectively remove these petroleum hydrocarbons from pristine Arctic waters will be critical to maintaining the ecologic balance of the region.

What is the potential for petroleum biodegradation by the indigenous microorganisms of Arctic marine and coastal ecosystems? What biotic and abiotic factors influence microbial degradation of petroleum hydrocarbons? What seasonal and spatial variation exists in petroleum biodegradation potentials? At what rates are the myriad of different hydrocarbons found in petroleum transformed, and to what products? What will be the composition and properties of the residual oil? All of these are critical questions in assessing the probable fate of petroleum hydrocarbons in the Arctic.

Biodegradation of oil requires that microorganisms be present with the enzymatic capability of transforming hydrocarbons. The abundance and natural distribution of such hydrocarbon utilizing microorganisms is one area of study reported in this paper. Even when high numbers of microorganisms capable of hydrocarbon biodegradation are present, they may not be metabolically active. Abiotic parameters, such as temperature and limiting concentrations of essential nutrients, may restrict oil biodegrading activity. The biodegradation potential and limiting factors of oil biodegradation in a variety of Arctic ecosystems are other areas of study reported in this paper. Within a crude oil, some components will be more susceptible to biodegradation than others. Results of studies on the relative rates of hydrocarbon biodegradation and composition of residual oil are also presented in this paper.

Methods

Enumeration of Microorganisms

Hydrocarbon utilizing and heterotrophic microorganisms were enumerated from a variety of nearshore and offshore sites in the Chukchi and Beaufort Seas. A map showing the general area and some specific sites that were sampled is shown in Fig. 1. Sampling was done from a variety of small craft and aircraft. Surface water samples were collected in sterile

Niskin water or whirlpak bags. Sediment samples were collected by divers or with a Mud grabber (Kahl Scientific) and transferred to sterile containers. Ice samples were collected by divers or with an ice corer and transferred to sterile containers. Temperature and salinity of samples were generally measured. Samples were maintained on ice during transport to the laboratory. Samples were processed for enumeration within a few hours of collection.

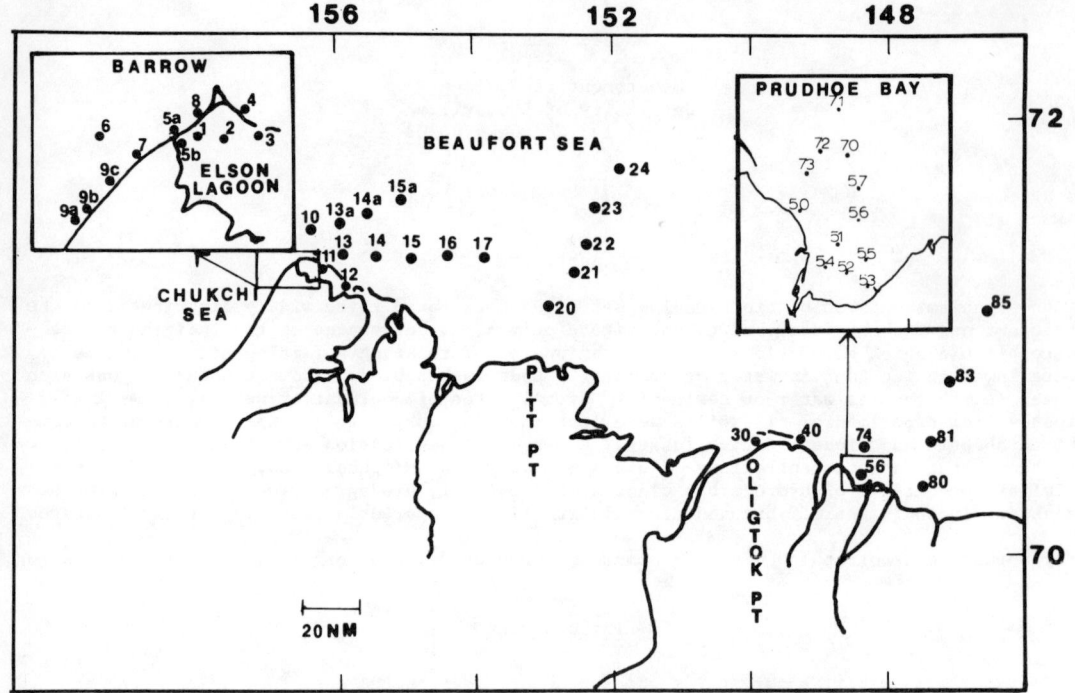

Fig. 27.1. Map showing sampling locations.

The viable heterotrophic microorganisms were enumerated by surface spreading serial dilutions onto marine agar 2216 (Difco). Hydrocarbon utilizing microorganisms were enumerated by surface spreading or plating filtered samples onto Bushnell Haas agar with 1% Prudhoe crude oil as sole carbon source. Counts on oil agar were corrected by subtracting counts obtained on Bushnell Haas agar without added oil. Plates were incubated at 5°C for 21 days to enumerate psychrophilic and psychrotrophic populations or at 20°C for 10 days to enumerate mesophilic populations.

Colonies were selected at random from the enumeration plates for isolation. Isolated microorganisms were tested for their ability to utilize various pure hydrocarbons as sole carbon source, e.g. hexadecane, pristane, pentadecylcyclohexane, methylnaphthalene and phenyldecane. In some cases, the isolated organisms were further classified according to classical or numerical taxonomy.

Hydrocarbon Biodegradation Potential
Several methods were used to determine the hydrocarbon biodegradation potential of a variety of samples. These methods involve exposing a sample to petroleum hydrocarbons, either <u>in vitro</u> or <u>in situ</u>, and measuring either oxygen consumption or carbon dioxide production, both metabolic processes. In different experiments Prudhoe crude oil or Prudhoe crude oil spiked with radiolabelled hexadecane were used.

Oxygen consumption was measured in a Gilson respirometer. Five g samples were placed in the respirometer with 0.1% hydrocarbon. In some experiments the samples were oiled <u>in situ</u> and periodically subsamples were collected and the rate of oxygen consumption measured.

To measure CO_2 production, 100 g samples were incubated with 0.5% hydrocarbon in Biometer flasks for 42 days at 5°C. The CO_2 was trapped in alkali and quantitated by titration. CO_2 production was also measured using ^{14}C radiolabelled spiked crude oil. Fifty g samples were incubated with 0.1% oil in sealed 100 ml vials. Following 2 months of incubation at 5°C, the atmosphere in the vials was flushed through hyamine hydroxide to trap the CO_2. The $^{14}CO_2$ was quantitated by liquid scintillation counting.

Biodegradative Changes in Oil Composition

Several types of experiments were conducted to measure the changes in oil resulting from biodegradation. In these experiments Prudhoe crude was exposed to indigenous microbial communities and periodically the oil was recovered from replicate oil samples for analysis. Oil was recovered by 24 hr liquid-liquid extraction with diethyl ether. Poisoned controls were used to correct for non-biological losses. For in vitro and flow through experiments oil was added at 1% wt/vol. For in situ experiments oil was added at 1 g/50 cm^2 surface area contained within plexiglas cyclinders. In vitro experiments were conducted with water collected in Prudhoe Bay and Elson Lagoon. Flow through experiments were conducted with water pumped from off Barrow, Alaska. In situ open water exposure was conducted during July and August, 1973, in Prudhoe Bay. In situ under and over ice exposure was conducted during May and June, 1976 in Elson Lagoon. Under ice experiments were conducted by taking ice cores, stoppering the bottom of a plexiglas cyclinder, adding the ice core over oil and replacing the core in place. Also, under ice exposure was accomplished by adding oil directly to the underside of the ice using scuba divers.

Oil recovered following exposure was analysed in various ways in these different experiments. The weight of recovered oil was measured as an overall determination of degradative losses. Gas chromatography was used to analyse the fate of major individual hydrocarbon components below molecular weight 450. Gas chromatographic separation was accomplished on 2 m 10% Apiezon columns and detected by flame ionization (Atlas, 1975). The weight percents of major classes of hydrocarbon components in the recovered oil was determined by column chromatographic separation on silica gel with differential solvent polarity elution (Atlas, 1975). Detailed class compositional analyses were performed with computerized gas chromatography-mass spectrometry by Continental Oil Co.

Results

Distribution of Hydrocarbon Utilizing Microorganisms

Winter samples generally had a collection temperature of -1.8°C. Winter water samples had an average salinity of 25°/oo and ice samples about 5°/oo. Summer samples had an average temperature of 2°C and salinity of 20°/oo.

Hydrocarbon utilizing microorganisms were found to comprise only a small percentage of the total heterotrophic population (Table 1). The percent hydrocarbon utilizers was calculated by dividing the number of organisms enumerated on oil-agar by the heterotrophs enumerated on marine agar. Generally the percent hydrocarbon utilizers was estimated at less than 1%, but in some cases reached as high as 10%.

The highest number of hydrocarbon utilizing bacteria was found in a nearshore water sample collected at Barrow. Other water samples had much lower numbers of hydrocarbon utilizing microorganisms. Sediment samples generally had higher counts, but not necessarily higher percentages of hydrocarbon utilizers. The summer counts of hydrocarbon utilizers in sediment were typically two orders of magnitude higher than comparable winter counts. In summer the number and percent hydrocarbon utilizers found in sediment was higher than in the overlying surface water. In winter, however, while the numbers of hydrocarbon utilizers were higher in sediment than in the overlying water and ice, the percent hydrocarbon utilizers was usually higher in the ice and water.

Concentrations of hydrocarbon utilizing microorganisms and percent hydrocarbon utilizers in surface ice were generally an order of magnitude lower than in the water underlying the ice. The concentrations and percent hydrocarbon utilizers in ice collected near Prudhoe Bay (Stations 56, 74) were higher than in ice from other regions. In water collected in winter the percent hydrocarbon utilizers was higher near Barrow than in other areas, but no similar pattern of geographic distribution was apparent in summer.

There was a high degree of variation in both numbers and percent hydrocarbon utilizers capable of growth at 4 and 20°C at a given location. In surface ice there were occasional instances of very high percent hydrocarbon utilizers capable of growth at 20°C which may be due to non-indigenous sources of these organisms.

An alternate estimate of the percent hydrocarbon utilizers was obtained by testing the ability of dominant heterotrophs, randomly isolated from summer samples, to utilize pure hydrocarbons. These tests showed that 1% of all organisms isolated from water and 2% of all organisms isolated from sediment at 4°C incubation temperature and 5% of organisms isolated from water and 3% of organisms isolated from sediment at 20°C incubation temperature, were capable of utilizing some hydrocarbon. This estimate is somewhat higher than the one obtained from the enumeration on oil-agar.

When oil was experimentally added to water in Prudhoe Bay or near Barrow, there was an increase of several orders of magnitude in the numbers of oil degrading microorganisms. In

Prudhoe Bay this increase was mostly due to a Pseudomonas spp. (Atlas and Schofield, 1975). In flow through experiments at Barrow, the percent of the population capable of utilizing hydrocarbons rose to 100% after several weeks of exposure (Horowitz and Atlas, 1976).

TABLE 1. Enumeration of Microorganisms (#/ml or #/g)

Site	Heterotrophs 4C	*HC Utilizers 4C	%HC Utilizers 4C	Heterotrophs 20C	HC Utilizers 20C	%HC Utilizers 20C
WINTER						
Ice						
2[a]	16.0	0.02	0.10	0.5	0.06	12.00
3[a]	0.2	<0.01	<5.00	1.0	<0.01	1.00
3[b]	10.0	<0.10	<1.00	30.0	<0.10	0.30
7[b,d]	5900.0	0.20	<0.01	4700.0	0.10	<0.01
13[a]	16.0	0.02	0.10	80.0	0.02	0.02
17[a]	1.6	0.02	1.00	<0.1	0.02	-
24[a]	63.0	0.02	0.03	5.2	1.90	36.00
74[a]	550.0	12.20	2.20	3100.0	0.70	0.02
56[a]	290.0	1.10	3.80	200.0	0.36	0.10
Water						
2	1400.0	3.10	2.20	2600.0	0.10	<0.01
3	10.0	<0.10	<1.00	25.0	<0.10	<0.40
13	47.0	0.20	4.20	2.0	0.05	2.50
17	3.3	0.32	9.70	0.2	0.02	10.00
24	9100.0	0.04	<0.01	16000.0	-	-
74	62.0	0.66	1.00	32000.0	0.50	0.01
Sediment						
2	640000.0	3.00	<0.01	170000.0	1.70	<0.01
13	280000.0	-	-	120000.0	0.10	<0.01
17	27000.0	4.70	0.01	2000.0	0.10	<0.01
74	63000.0	1.60	<0.01	39000.0	1.70	<0.01
56	54000.0	11.00	0.02	130000.0	0.90	<0.01
SUMMER						
Water						
2	2500.0	<1.00	0.04	2000.0	3.00	0.10
3	5000.0	1.00	0.02	1000.0	5.00	0.50
7[c]	1700.0	0.80	0.05	310.0	0.70	0.20
7[d]	100000.0	10000.00	10.00	7500.0	3000.00	40.00
12	10000.0	1.00	0.01	10000.0	50.00	0.50
71	9000.0	3.30	0.04	9000.0	1.00	0.01
56	14000.0	11.00	0.07	16000.0	2.00	0.01
Sediment						
2	120000.0	500.00	0.40	80000.0	500.00	0.60
3	190000.0	500.00	0.20	100000.0	200.00	0.20
12	680000.0	5000.00	0.70	770000.0	340.00	0.04
71	50000.0	100.00	0.20	100000.0	40.00	0.04

*HC= Hydrocarbon
a= surface
b= bottomside
c - offshore
d - nearshore

Oil Biodegradation Potential

The respiratory response to exposure to crude oil as measured by rate of oxygen consumption is shown in Table 2. Rates of oxygen consumption were significantly different between types of samples and times of collection, but not between samples exposed to oil and control samples. The only changes detected were an increase in oxygen consumption in underside ice and a decrease in oxygen consumption in sediment collected in summer after exposure to Prudhoe crude oil for two weeks. Sediment showed the greatest rates of natural oxygen consumption, especially in summer. The lowest rates of natural oxygen consumption were found in surface ice.

Carbon dioxide production from Prudhoe crude oil was greater in summer water samples than in winter (Table 3). All summer water samples were found to produce significant amounts of carbon dioxide. The highest rate of CO_2 production was found in the Prudhoe Bay sample (Station 53). The CO_2 produced by this sample accounts for mineralization of 25% of the oil, i.e. 25% of the maximal CO_2 released by chemical oxidation of the oil with acid dichromate. In the winter samples, CO_2 production was similar in underside ice and water samples. CO_2 produced from these winter ice and under ice water samples accounted for mineralization of only 3% of the oil.

TABLE 2. Rates of O_2 Consumption of Control and Oil Exposed Samples Collected at Station 3.

Treatment	$\mu l\ O_2$ Consumed/h/g
WINTER	
Surface ice - control	<0.1
Surface ice + Prudhoe crude	<0.1
Surface ice - after 2 wk exposure to Prudhoe crude	<0.1
Underside ice - control	1.7
Underside ice + Prudhoe crude	1.8
Underside ice - after 2 wk exposure to Prudhoe crude	4.2
Water under ice - control	1.3
Water under ice + Prudhoe crude	1.2
Water under ice - after 2 wk exposure to Prudhoe crude	1.1
Sediment - control	5.8
Sediment + Prudhoe crude	6.1
SUMMER	
Water - control	0.3
Water + Prudhoe crude	0.3
Sediment - control	200
Sediment - 2 wk exposure to Prudhoe crude	180
Sediment - 2 wk exposure to Arctic diesel	150

TABLE 3. CO_2 Production from Biodegradation of Prudhoe Crude During 42 Days at 5°C.

Treatment	$\mu moles\ CO_2$ Produced
Summer-Water-Station 7	4000
Summer-Water-Station 3	2400
Summer-Water-Station 53	8300
Winter-Under Ice Water-Station 3	100
Winter-Surface Ice-Station 3	80

TABLE 4. Hydrocarbon Biodegradation Potential of Winter Samples by $^{14}CO_2$ Release Method

Station	CPM $^{14}CO_2$ Produced During 50 Days Incubation at 5°C		
	Ice	Water	Sediment
2	2,400	1,050	9,500
3	2,150	2,500	1,600
13	1,250	13,200	2,550
15	2,150	13,650	2,200
17	350	2,800	1,550
24	400	1,600	-
56	-	-	2,550
74	1,200	6,800	5,300
83	145	4,350	6,650

The relative biodegradation potential of winter samples measured by $^{14}CO_2$ release is shown in Table 4. Ice samples had the lowest average biodegradation potential; water samples had the highest. Only minimal $^{14}CO_2$ release was detected in ice samples collected off Pitt Pt. and offshore near Prudhoe Bay. The greatest biodegradation potential was found in water collected along a transect off Pt. Barrow. Other samples collected along this transect (not shown in Table 4) also showed similarly high biodegradation potential. The biodegradation potential of surface ice and sediment in this area though, was not higher than other areas. Relatively high biodegradation potentials were found in both sediment and water samples collected near Prudhoe Bay.

Oil Biodegradation Experiments

Actual changes in oil exposed to indigenous microbial communities were used as direct measures of biodegradation. Table 5 shows the weight and percent weight losses from several miniature experimental oil spillages. In summer experiments, conducted *in situ* in Prudhoe Bay during summer, 30% of the oil was lost due to abiotic factors, mainly evaporation. The losses were doubled when the oil was exposed to the indigenous microorganisms, i.e. 30% of the oil was lost abiotically and 30% was lost due to biodegradation. It was possible to further increase the biodegradative losses by supplementing the oil slicks with nitrogen and phosphorus nutrients (Atlas and Schofield, 1975). These nutrients were in limiting concentrations in the water column. Gas chromatographic analysis of the residual oil from the poisoned control lost all paraffinic components below C_{12}, but paraffins above this molecular weight showed no measureable losses. The residual oil following biodegradation showed that the normal paraffins had been completely degraded and that the branched paraffins had been extensively degraded. The average temperature during this experiment was 11°C.

Experiments conducted in a subsequent summer 1975 at Barrow using a flow through chamber for oil exposure, showed only a minimal weight percent loss. The actual weight loss, shown in Table 5, is misleading because of the larger size of the oil spill used in this experiment. The summer in which this experiment was conducted was especially cold, average temperature 2°C, and there was shore ice during the entire experiment; the water was pumped through holes in the ice. Poisoned controls showed similar weight losses and the percent weight loss probably reflects largely abiotic losses with minimal biodegradation. Some biodegradation was probably occurring because there was a shift in the microbial population underlying the oil, from 35 to 75% capable of utilizing hydrocarbons at 5°C (Horowitz and Atlas, 1976).

Gas chromatographic analysis of the residual oil showed that the losses were restricted to the low molecular weight hydrocarbons (Harowitz and Atlas, 1976). As in the Prudhoe Bay experiments, it was possible to increase the biodegradative losses by addition of nitrogen and phosphorus nutrients. Even with nutrient addition, the losses were only 32% less than half that found in the Prudhoe Bay experiments. Gas chromatographic analysis of the residual oil recovered after nutrient stimulated biodegradation did show that some paraffins had been degraded. Column chromatographic analysis of the residual oil showed that biodegradation was not restricted to the paraffinic fraction, but that the aromatic fraction was also degraded. Detailed gas chromatographic-mass spectral computerized analysis showed that the residual oil had similar concentrations of all component classes as fresh oil (Table 6). This was true regardless of the extent of degradation, i.e. for nutrient supplemented and non-supplemented oil. The fact that the relative concentrations remained the same in degraded oil indicates that all component classes were being degraded at similar relative rates.

The experiments conducted in winter, over and under ice in Elson Lagoon, showed only minimal losses (Table 5). These losses were almost identical to poisoned controls and chromatographic analyses showed that only light weight compounds had been lost from the oil. It should be noted that the oil in these ice experiments was exposed for only about half the time as in the summer water experiments.

Discussion

Arctic marine and nearshore ecosystems were found to posses the potential for petroleum biodegradation. Petroleum degrading bacteria were found to be widely distributed but to constitute only a low percentage of the indigenous heterotrophic microbial populations. The numbers of hydrocarbon utilizing microorganisms present in an ecosystem determines in part the ability of that ecosystem to degrade petroleum pollutants. The low numbers of hydrocarbon utilizing microorganisms found in winter ice and some water samples may indicate a restricted ability to degrade petroleum in those samples.

TABLE 5. Degradative Losses from Prudhoe Crude Oil Exposed to the Indigenous Microorganisms in Different Ecosystems

Exposure	Time (days)	Wt.Loss (mg)	% Loss
Summer-Water-Poisoned control-Prudhoe Bay-In situ	42	300	30
Summer-Water-Prudhoe Bay-In situ	42	600	60
Summer-Water-Barrow-Flow through	51	20000	15
Winter-Surface ice-Elson Lagoon	28	120	12
Winter-Underside ice-Elson Lagoon	21	90	9

TABLE 6. Mass Spectral Analysis of Fresh and Exposed Prudhoe Crude.

Treatment	\multicolumn{14}{c}{Concentrations (%) of Oil Fractions}
	1	2	3	4	5	6	7	8	9	10	11	12	13	14
Fresh Oil	14.4	11.7	8.3	6.4	4.9	0.3	0.2	8.4	6.2	5.1	8.2	10.2	7.0	6.5
Residual Oil (51 day exposure)	14.5	11.9	8.2	6.2	4.8	0.3	0.2	8.8	6.0	5.2	8.1	10.5	7.1	6.2

Fractions
1. Paraffins
2. Monocycloparaffins
3. Dicycloparaffins
4. Tricycloparaffins
5. Tetracycloparaffins
6. Pentacycloparaffins
7. Hexacycloparaffins
8. Alkylbenzenes/C_NH_{2N-22} cpds.
9. C_NH_{2N-8}/C_NH_{2N-22} aromatics, e.g. tetralin
10. $C_NH_{2N-10}/C_NH_{2N-24}$ aromatics, e.g. chrysene
11. $C_NH_{2N-12}/C_NH_{2N-26}$ aromatics, e.g. napththalene
12. C_NH_{2N-14} aromatics, e.g. acenaphthene
13. C_NH_{2N-16} aromatics, e.g. fluorene
14. C_NH_{2N-18} aromatics, e.g. anthracene
15. C_NH_{2N-20} aromatics, e.g. phenylnaphthalene

The numbers of hydrocarbon utilizing microorganisms in an ecosystem have been found to correspond to the presence of petroleum. Levels of hydrocarbon utilizing microorganisms can be used as an indicator of the presence of petroleum pollutants (Atlas and Bartha, 1973). Walker and Colwell (1976) found that the percentage of the heterotrophic population in Chesapeake Bay capable of hydrocarbon degradation correlated with the amount of extractable hydrocarbon.

Both numbers of hydrocarbon utilizers and the percent of the heterotrophic populations capable of hydrocarbon metabolism were reported in this paper. The higher percent hydrocarbon utilizers found in summer in the Prudhoe Bay area and in winter off Pt. Barrow, may indicate that these areas have previously been exposed to hydrocarbons. The generally low percent hydrocarbon utilizers is indicative of a pristine environment.

The biodegradation potential measurements showed that water and sediment samples had the ability to metabolize oil, but that activity in ice was lower. The fact that there was no rapid increase in the rate of oxygen consumption when oil was added indicates that these samples were not actively degrading hydrocarbons when collected and that the samples were not capable of immediately beginning to degrade petroleum pollutants. The CO_2 production experiments showed that given adequate time, samples, other than ice, were capable of mineralization of petroleum, i.e. conversion of hydrocarbons to CO_2 and H_2O or complete removal of petroleum hydrocarbon pollutants. The ice samples may also have been capable of oil mineralization if given a longer incubation period. The $^{14}CO_2$ release experiments also showed that samples collected in winter were capable of mineralization of petroleum hydrocarbons if given enough time. Only a few surface ice samples failed to produce significant amounts of $^{14}CO_2$. The $^{14}CO_2$ experiments showed that Prudhoe Bay and an area off Pt. Barrow had high biodegredation potentials. The water in the area off Pt. Barrow, which had the greatest release of $^{14}CO_2$, was also an area characterized by having a high percent of the heterotrophic microbial population capable of metabolizing hydrocarbons. Sediment samples were found to have fairly high biodegradation potentials.

The oil biodegradation experiments confirmed that oil can actually undergo biodegradation in situ in the Arctic. The weight losses showed, though, that abiotic parameters can greatly influence the rates of oil biodegradation. During an especially cold summer only minimal biodegradation occurred near Barrow despite the very high numbers of petroleum degrading microorganisms found there. During a warmer summer in Prudhoe Bay extensive biodegradative losses could be measured by weight loss. As predicted from the biodegradation potential measurement, biodegradative losses could not be detected during the several weeks after oil was spilled over or under sea ice. Biodegradation in the Barrow and Prudhoe Bay summer water experiments was severely limited by available concentrations of nitrogen and phosphorus. The addition of these nutrients was found to enhance biodegradation.

Abiotic losses were found to be restricted to low molecular weight compounds. Biodegradation resulted in some cases in the complete degradation of detectable n-paraffins and extensive degradation of resolvable branched paraffins. Aromatic compounds, including polynuclear aromatic compounds. were also found to undergo biodegradation at rates similar to the paraffins. Alicyclic hydrocarbons were also biodegraded at similar rates. These findings, as shown by detailed chromatographic-mass spectral analyses, do not support the hypothesis that n-paraffins are degraded initially and most rapidly and that other petroleum components are only degraded later at much lower rates (Kallio, 1975). It is possible that cometabolism of petroleum hydrocarbons at low biodegradation rates in low temperature Arctic waters accounts for the surprisingly equal rates of biodegradation of the different compositional classes of hydrobarbons in crude oil. In spite of the fact that biodegradation of all hydrocarbon classes was detected, there was still residual oil left after several months of exposure. Complete biodegradation of a crude oil has never been found.

In summary, Arctic waters possess the potential for oil biodegradation. There is spatial and seasonal variation in the oil biodegradation potential, dependent in part on variations in the levels of hydrobarbon degrading microorganisms and in part on abiotic factors. The natural rates of oil biodegradation were found to be slow. Less than 50% of the hydrocarbons in an Arctic marine oil spill would be expected to be lost during one summer due to combined abiotic and biodegradative losses. Rates of biodegradation during the winter on or under ice would be very low. Surprisingly, these studies indicate that residual oil, after degradation for a year, would be expected to possess the same relative chemical class composition as fresh oil.

Acknowledgements

This work was supported by contracts with the Office of Naval Research and the National Oceanic and Atmospheric Administration. Logistic support was supplied by the Naval Arctic Research Laboratory. Contributions to the work were made by Mr. George Roubal, Mr. Amikam Horowitz, Mr. Michael Busdosh, Dr. Tatsuo Kaneko and Dr. Micah Krichevsky. Mass spectral analyses were generously performed by Dr. Harrell Ford, Continental Oil Co.

References

Atlas, R. M., Effects of temperature and crude oil composition on petroleum biodegradation. Appl. Microbiol., 30, 396 (1975).

Atlas, R. M., and R. Bartha, Abundance, distribution and oil biodegradation potential of microorganisms in Raritan Bay, Environ. Pollut., 4, 29 (1973).

Atlas, R. M., and E. A. Schofield, Petroleum degradation in the Arctic, p. 183, In: A. Bourquin, D. G. Ahearn and S. P. Meyers (eds.), Impact of the Use of Microorganisms on the Aquatic Environment, EPA 660/3-75-001, Environmental Protection Agency, Corvallis, OR., 1975.

Horowitz, A., and R. M. Atlas, Continuous open flow-through system as a model for oil degradation in the Arctic Ocean, Appl. Environ. Microbiol., in press, (1976).

Kallio, R. E., Microbial degradation of petroleum, Paper presented 3rd International Biodegradation Symposium, Kingston, R. I., August 1975.

Walker, J. D., and R. R. Colwell, Enumeration of petroleum degrading microorganisms, Appl. Environ. Microbiol., 31, 198 (1976).

CHAPTER 28

ARCTIC HYDROCARBON BIODEGRADATION

S. D. Arhelger, B. R. Robertson, and D. K. Button

University of Alaska, Fairbanks, Alaska 99701

Abstract

Supplemented sterile sea water was inoculated with small sample volumes of raw sea water and examined for oil related microbial activity. A ubiquitous Alaskan water microbial population of 10^3 to $10^5/l$ was indicated. In situ ^{14}C-dodecane oxidation rates based on $^{14}CO_2$ recovery were: Port Valdez, 0.7 g/l day; Chukchi Sea, 0.05 g/l day; and Arctic Ocean, 0.001 g/l day. Computations show that population estimates and oxidation rate measurements are compatible with each other.

Key words: Biodegradation, Microbial Populations, ^{14}C-dodecane, Port Valdez, Arctic Ocean

Introduction

Because of the potential for oil spills in Alaskan waters, the indigenous hydrocarbon metabolism has attracted considerable interest. Yet few reliable field measurements of hydrocarbon oxidation in the marine environment exist (Wilson et al 1975). We inferred hydrocarbon biodegradation in Cook Inlet from oil input rates and the lack of measurable hydrocarbon accumulation in the water column (Kinney et al 1969). Hydrocarbon oxidizing microorganisms were demonstrated throughout Port Valdez by conventional plating methods (Robertson et al 1973). Atlas (1973) observed bacterial build up associated with small contained oil slicks on Alaska's Prudhoe Bay, yet little information exists that would support hydrocarbon oxidation as a normal marine process in the arctic. In fact as these experiments were designed and initial data collected (1969-1973), it was questionable whether the process was significant in the open ocean. Low solubility restricts dissolved hydrocarbon metabolism velocity for kinetic reasons because the concentrations of most hydrocarbons cannot build to sufficient levels to satisfy the capacity of organisms to accumulate substrate. Oil phase metabolism is also restricted due to what must be low oil droplet inoculation frequency at typical populations (Robertson et al 1973) and by restricted mineral nutrient penetration into a semi-solid oil mass from dilute surroundings (Atlas and Bartha 1972b).

Our objective was to test the hydrocarbon oxidation capacity and determine the abundance and activity of hydrocarbon oxidizing organisms in arctic sea water samples. Hydrocarbon oxidation activity of the indigenous flora was monitored by collecting $^{14}CO_2$ from ^{14}C-dodecane in situ incubations. Population estimates were based on the smallest sample volume that, when diluted into a supplemental sea water medium, produced cultures that would generate colonies on agar plates, became turbid, formed ATP, or disrupted added oil slicks. While a preliminary account of the Port Valdez oxidation rates has appeared (Robertson et al), a digest of these data is included for comparison.

Materials and Methods

Critical Volume for Oil Slick Disruption
A series of sample volume sizes: 0.01, 0.1, 1.0, 10 and 100 ml; was used to inoculate 250 m screw cap bottles containing enriched sea water. The sea water was supplemented with the following additions per liter: $(NH_4)_2SO_4$, 100 mg; Na_2HPO_4, 4 mg; and EDTA, to 10 µM. Preliminary experiments, where each supplementary nutrient was varied independently showed thes particular concentrations to produce maximal response. Following sterilization, this enriched sea water medium was equilibrated with Prudhoe Bay crude oil at room temperature resulting in the addition of approximately 1 mg/l of primarily aqueous phase low boiling hydro carbons. Ten days following inoculation, 0.1 ml of sterile crude was added to form a contir uous oil phase. Bottles containing the small but developing cultures were incubated in the dark at 10 C with weekly agitation. They were checked monthly for presence of microbial pop ulations (turbidity, the ability to produce colonies on spread plates, ATP content, and visible disruption of the oil slick). The latter was quite apparent - the slick became discontinuous upon agitation, probably due to the production of surfactants.

In Situ Oxidation Rates

Either a uniformly labeled [^{14}C]amino acid mixture or [1-^{14}C]dodecane was added to 1 liter sea water samples. These were incubated in light-tight bottles returned to the depth from which they had been filled. The amino acid addition to each bottle was approximately 2.5 µCi in a total of 550 ng carbon (see Fig. 2), specific activity 57 mCi/matom carbon (Amersham Searle Corp., Arlington Heights, Ill.). A typical dodecane addition was 825 nCi in 75.5 µg (see Fig. 2), specific activity 27.4 mCi/m mole (ICN Tracerlab, Waltham, Mass.). Control incubations were poisoned with 1×10^{-6} M $HgCl_2$ (we have since found it necessary to raise this concentration to 10^{-5} M to stop all oxidative activity). After incubations of from 0 to 35 days, selected bottles were raised. The reaction was stopped with acid, the CO_2 liberated, collected and counted (Robertson et al 1973).

Sampling

Small samples involved use of an "ecological water sampler" (Cm^2 Inc., Mountain View, Calif.). Large samples were drawn into sterile 2 liter vacuum flasks through weighted Tygon tubing. In situ incubation bottles were filled with sea water pumped up through polypropylene hosing with a polyvinyl chloride lift (bilge) pump. The apparatus was soaked in methanol, then flushed with several hundred volumes of sea water from the appropriate depth. Valdez samples were taken aboard ship. Chukchi Sea samples were collected from a small boat among the ice pack in August, those in May through borings in the ice. Mid-winter Arctic Ocean samples were taken through a hole cut through the Ice Island ice. Sample locations are shown in Fig. 1.

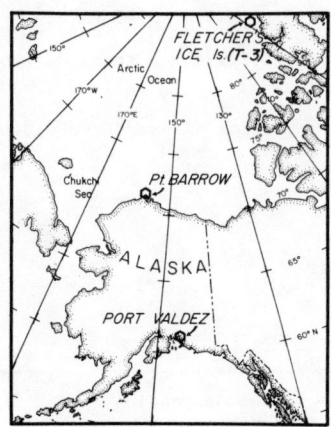

Fig. 28.1 Study area locations.

Results

The activity generated in 250 ml bottles of sterile sea water media by Chukchi Sea samples as a function of added raw sea water volume is shown in Table 1. Inoculation occurred upon adding 1 ml of raw sea water; however, it was only 25% successful at the 0.1 ml level. The size of final populations in sub-culture bottles was increased by oil addition to these bottles, as shown by the final ATP concentrations. In experiments not shown, three of the other medium supplements (nitrogen, phosphate and EDTA) all increased response at a given inoculation level; the fourth supplement (soluble crude oil component added before inoculation) was without noticeable effect; thus the sub-culture medium alone was deemed adequate. While population development could not be detected from spread plates at the 0.01 ml inoculation level (Table 1), ATP was found, the amount of which increased upon oil addition. The reason for this observation is unclear; however many workers believe that substantial numbers of marine bacteria do not respond to these cultivation techniques in a conventional way. Contamination is of constant concern in field experiments. Although subjected to repeated openings for sampling, 90% of the controls where only the inoculation step was omitted remained sterile throughout the 5 month incubation period. Furthermore, a sign test comparing the response between the two replicate samples (at the 0.1 ml and 1.0 ml inoculum level of Table 1) showed the samples to be indistinguishable. Thus we believe that response observed was generated by indigenous organisms. The smallest volume of sea water from the three sample locations consistently generating these various responses is shown in Table 2. Population estimates based on these responses are 10^3 to 10^5/l.

In Situ Oxidation Rates

Samples charged with radioactive substrate and resuspended began to liberate $^{14}CO_2$ almost immediately. Typical isotope distribution data are shown in Table 3. Absence of recovered radioactivity from zero time incubations demonstrated minimal unreacted substrate carry-over. Rate reduction in response to mercuric ion, found to prevent Corynebacterium growth at 10^{-6} M in sea water medium during preliminary experiments, demonstrated that the response was biological. Amino acid oxidation rates are provided for comparison. Particulate radioactivity in the case of amino acids represents accumulation of the approximately 1 µg/l supplied into cell mass. In the case of dodecane, particulate radioactivity includes some unreacted hydrocarbon. Oxidation rates during 74 separate incubations of this type are shown in Fig. 2. Reactors inoculated with very small mixed populations ($10^{0.3}$ to 10^2/ml calculated from Table 2) can be expected to grow after a somewhat variable "lag" period. The fact that metabolism does proceed was indicated by the quantity of substrate-derived CO_2 produced. This quantity was statistically larger than the zero time incubations or the poisoned controls according to linear regression analysis of oxidation rate data with respect to time. A more meaningful numerical indication of an indigenous flora with hydrocarbon oxidation capacity is probably the initial CO_2 production rate. This is the rate established before extensive population build up or substrate depletion. Although subject to considerable error, these rates, taken from the initial (1 to 4 day) slopes of Fig. 2, are shown in Table 4.

TABLE 1. Activity Generated by Inoculation with Chukchi Sea Samples.[1]

Volume taken[2]	No oil added		Oil added[2]		
	Population generated[3]	ATP µg/l	Population generated[3]	ATP µg/l	Slick disrupted[4]
0.00	0 of 2	0.0	0 of 2	0.0	0 of 4
0.01	0 of 2	0.2	0 of 4	0.6	0 of 3
0.1	1 of 2	0.1	1 of 3	0.3	1 of 5
1.0	1 of 2	0.4	4 of 4	6.5	4 of 4
10.0	1 of 2	0.8	4 of 4	3.0	4 of 4
100.0	0 of 2	0.0	4 of 4	1.4	4 of 4
100.0[5]	1 of 1	0.0	1 of 2	0.0	0 of 2

[1] Observations of duplicate samples from duplicate 3 m casts taken from among the pack ice in August 3 km north of the Barrow sand spit.
[2] Duplicate bottles were incubated from each of the duplicate samples giving a total of 4 observation sequences at each dilution.
[3] As determined from spread plates (Buck and Cleverdon, 1960).
[4] The number of samples in which the oil slick present during incubation was rendered discontinuous.
[5] Sea water only; no carbon or mineral supplements.

TABLE 2. Minimum Viable Inoculation Volume.

Location	Date	Depth m	Minimum Volume Generating Response ml		
			Colony production	Slick disruption	ATP
Arctic Ocean	January	3	0.1	1.0	0.1
Port Valdez	August	10	0.1	0.1	0.1
Chukchi Sea	April	1	1.0	1.0	0.01

Discussion

These experiments demonstrated alkane metabolism is a wide spread capacity of the arctic near-surface marine microflora. While dodecane was the only hydrocarbon added here, various observations of mixed substrate consumption by pure culture isolates, including other alkanes cycloparaffins (unpublished, this laboratory) and polynuclear aromatics (Gibson and Yeh, 1973), suggest that hydrocarbon metabolism by mixed cultures would extend to most common soluble crude oil components. Our preliminary experiments show that the metabolism of various aromatics in sea water is even faster than those reported here.

Because hydrocarbon oxidation is considered to be an inducible process (Grund et al 1975) and because we only infrequently located hydrocarbon oxidizers by incubation of membrane filters on nutrient sources, we were surprised to observe *in situ* dodecane oxidation in the arctic. However, Schwartz et al (1974) demonstrated deep water hydrocarbon oxidizers and the process appears to be common to many organisms (Duncan and Ulrich 1973). Lee et al (1975), using similar $^{14}CO_2$ recovery techniques, observed 16 µg/l day octadecane oxidation rates in crude oil-containing systems, but highly variable rates in crude oil-free systems. We regard the variability observed here as normal for processes initiated with very small

TABLE 3. Typical Material Distributions During *In Situ* Incubations.[1]

Substrate[2]	Days incubation	ng C/l particulate	ng C/l dissolved	ng C/l oxidized[3]
Amino acids	0	0.04	594	0.09
Amino acids, $HgCl_2$	4	0.57	577	0.45
Amino acids	4	253	46.6	33.3
Dodecane	0	7170[4]	2.1	1.04
Dodecane, $HgCl_2$	9	16016[4]	3.6	22.8
Dodecane	9	5648[4]	4.1	222.7

[1] Data were obtained from below the Chukchi Sea shore ice off Point Barrow in May.
[2] The initial concentration of mixed amino acids was 614 ng C/l, that of dodecane 383 µg/l, approximately 1.8 µg/l of which was dissolved.
[3] Calculated from recovered $BaCO_3$ specific activity.
[4] Large values result from retention of accommodated oil by filter.

numbers of organisms from relatively pristine systems. With better solubility information now available, *in situ* oxidation measurements can be improved by avoiding the presence of an oil phase. We had no difficulty in demonstrating dodecane metabolism by single species in continuous culture when supplied at only 1.78 µg/l (Button 1976).

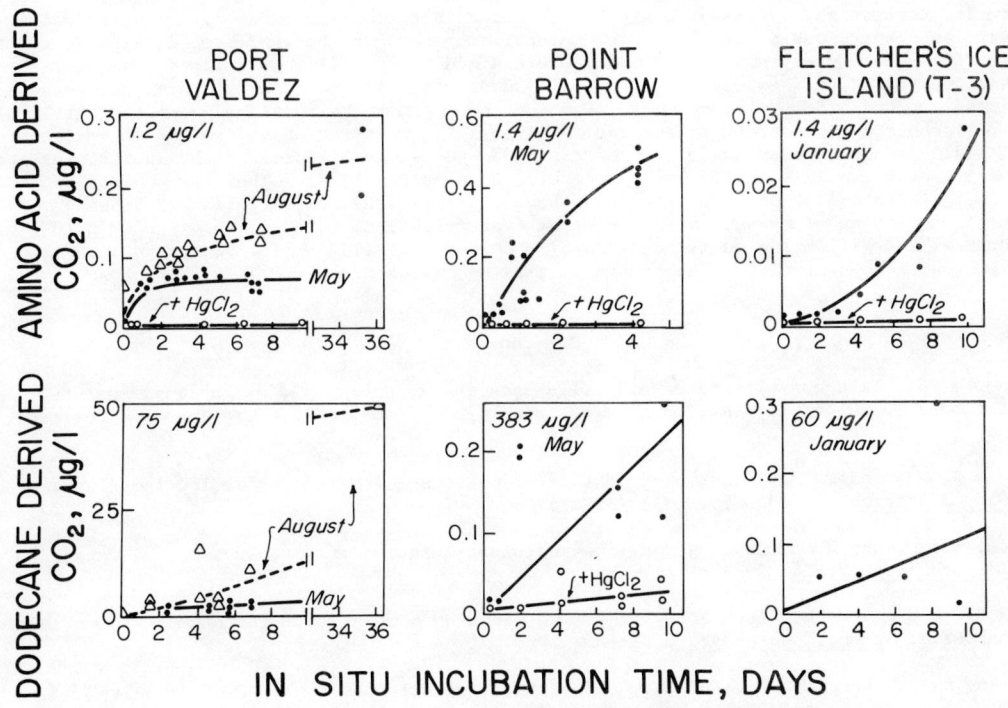

Fig. 28.2 Production of CO_2 from *in situ* incubation of an amino acid mixture and from dodecane. Initial substrate concentrations are also shown. Open symbols show 1 µM $HgCl_2$ poisoned control rates. Each datum point represents total $^{14}CO_2$ collection from a 1 m dark bottle after incubation for the time shown. Sample and incubation conditions were: for Port Valdez, 10 m depth, 3 C, salinity 30°/₀₀, April 1972; Point Barrow, 4 m depth through the shore ice, 2 C, salinity 28.3°/₀₀, April 1973; T-3 Ice Island, 3 m depth through the ice, -1.7 C, salinity 30.1°/₀₀, January 1973.

TABLE 4. *In Situ* Oxidation Rate Estimates from Initial CO_2 Production Rates.

	Oxidation rate, µg/l day		T-3
	Port Valdez	Point Barrow	Ice Island
Amino acids	0.05	0.02	0.001
Dodecane	0.7	0.05	0.001

The oil slick disruption technique for hydrocarbon oxidizer detection was surprisingly definitive. Although we have been unable to demonstrate measurable aqueous surface tension effects, introduction of an oil phase to culture fluid from any of our eight arctic hydrocarbon oxidizer isolates (Robertson *et al* 1973) produces visible effects. Population measurement precision can be increased by observing larger numbers of samples. We have recently done this without finding significant changes in the results.

It is often concluded that low temperatures (Atlas and Bartha 1972a) and low nutrient concentrations (Atlas and Bartha 1972b) severely restrict oil biodegradation. However, the typical 0.5 to 2.0 µM phosphate concentration near Point Barrow (Schell 1975) exceeds typical Michaelis constants for growth (Button *et al* 1973) by 10^2. Furthermore, temperature has little influence on diffusivity, an important term in dissolved hydrocarbon metabolism where solubility is rate determining for all but the very low molecular weight representatives. Compare, for example, the dodecane solubility of 1.78 µg/l (Button 1976) with the apparent Michaelis constant for methane uptake by lake water of 75.2 µg/l (Rudd and Hamilton 1975). Hence, low temperatures and nutrient availability are severely restrictive only in oil phase containing oil-water mixtures.

A question often raised is whether population measurements by relatively conventional techniques such as the "most probable number" based tests used here accurately reflect the real bacterial population. One way to check this value is to compare metabolic activity with population. Where substrate is dilute, metabolism rate v can be expressed in kinetic terms by the second order expression $v = a_\mu XA$, where X and A are cell population (g/l) and substrate concentration, respectively. a_μ is a substrate-organism affinity term analogous to the usual second order rate constant. We find this term to be relatively constant at 10^6 l/g cells hour among most organisms and substrates (in preparation). The one exception has been a single observation of *Corynebacterium* 198 - dodecane affinity which, in the presence of arginine, was significantly lower (computed from Button 1976). Thus the population, from the Port Valdez oxidation rate data, the above affinity, the concentration of dodecane at saturation, and at an average mass/bacterium (wet weight) of 10^{-13} g, is $v/a_\mu A$ (2 x 10^{-9} g dodecane/1 hour)/(10^6 l/g cells hour) (1.87 g dodecane/l) (10^{-13} g/bacterium) = 1.5 x 10^5 cells/l which is compatible with the population estimates made.

References

Atlas, R. M., Fate and effects of oil pollutants in extremely cold marine environments. National Technical Information Service, report AD-769 895. U.S. Department of Commerce, 1973.

Atlas, R. M., and R. Bartha, Biodegradation of petroleum in seawater at low temperature. Can. J. Microbiol. **18**, 1851-1855 (1972a).

Atlas, R. M., and R. Bartha, Degradation and mineralization of petroleum in sea water. Biotech. Bioeng. **14**, 297-318 (1972b).

Buck, J. D., and R. C. Cleverdon, The spread plate as a method for the enumeration of marine bacteria. Limnol. Oceanogr. **5**, 78-80 (1960).

Button, D. K., The influence of clay and bacteria on the concentration of dissolved hydrocarbon in saline solution. Geochim. Cosmochim. Acta **40**, 435-440 (1976).

Button, D. K., S. S. Dunker, and M. L. Morse, Continuous culture of Rhodotorula rubra. Kinetics of phosphate-arsenate uptake, inhibition and phosphate-limited growth. J. Bacteriol. **113**, 599-611 (1973).

Duncan, J., and J. Ulrich, Assimilation of hydrocarbons by Pseudomonas strains isolated from human clinical specimens. Appl. Microbiol. **26**, 894-898 (1973).

Gibson, D. T., and W. K. Yeh, Microbial degradation of aromatic hydrocarbons, p. 33-38. In: D. G. Ahearn and S. P. Meyers (eds.), Microbial degradation of oil pollutants. Center for Wetland Resources, Louisiana State University, 1973.

Grund, A., J. Shapiro, M. Finnewald, P. Pacha, J. Leahy, K. Markbreiter, M. Nieder, and M. Topfer, Regulation of alkane oxidation in Pseudomonas putida. J. Bacteriol. 123, 546-556 (1975).

Kinney, P. J., D. K. Button, and D. M. Schell, Kinetics of dissipation and biodegradation of crude oil in Alaska's Cook Inlet, p. 333-340. In: Proceedings Joint Conference on Prevention and Control of Oil Spills. American Petroleum Inst., Washington, D. C., 1969.

Lee, R. F., M. Takahashi, J. R. Beers, W. H. Thomas, D. L. R. Seibert, P. Koeller, and D. R. Green, Controlled ecosystems: their use in the study of the effects of petroleum hydrocarbons on plankton. In: Symposium proceedings. Pollution and physiology of marine organisms. Milford, Connecticut, 1975.

Robertson, B., S. Arhelger, P. J. Kinney, and D. K. Button, Hydrocarbon biodegradation in Alaskan waters, p. 171-184. In: D. G. Ahearn and S. P. Meyers (eds.), Microbial degradation of oil pollutants. Center for Wetland Resources, Louisiana State University, 1973.

Rudd, W. M., and R. D. Hamilton, Factors controlling rates of methane oxidation and the distribution of the methane oxidizers in a small stratified lake. Arch. Hydrobiol. 75, 522-538 (1975).

Schell, D. M., Seasonal variation in the nutrient chemistry and conservative constituents in coastal Alaskan Beaufort Sea waters. In: Environmental studies of an arctic estuarine system. EPA final report 660/3-75-026. Corvallis, Oregon, 1975.

Schwartz, J. R., J. D. Walker, and R. R. Colwell, Growth of deep-sea bacteria on hydrocarbons at ambient and in situ pressure. Dev. Ind. Microbiol. 15, 239-249 (1974).

Wilson, E. B., et al, Petroleum in the marine environment. National Academy of Sciences Printing Office, Washington, D. C., 107 p., 1975.

CHAPTER 29

BIOAVAILABILITY OF SEDIMENT-SORBED NAPHTHALENES
TO THE SIPUNCULID WORM, *Phascolosoma agassizii*

J. W. Anderson, L. J. Moore, J. W. Blaylock,
D. L. Woodruff, and S. L. Kiesser

Battelle, Pacific Northwest Division
Marine Research Laboratory
Sequim, Washington 98382

Abstract

The peanut worm (Sipunculida: *Phascolosoma agassizii*) was exposed to petroleum hydrocarbons from Prudhoe Bay crude (PBC) oil. Uptake and release of naphthalene and alkylnaphthalenes were compared for worms exposed to hydrocarbons in solution, oil on the surface of sediments and oil mixed in sediment. Spiunculids exposed for 24 hr to a water-soluble fraction of PBC contained from 2 to 10 times the concentration of naphthalenes initially in the water. After living for two weeks in sediment, either oiled on the surface or mixed thoroughly with oil, animals contained from 2 to 4 ppm of total naphthalenes. The tissue content of hydrocarbons in organisms exposed to sediment-bound hydrocarbons compared closely with hydrocarbon content of the oil-contaminated sediments and may therefore represent contaminated sediment within the gut. Depuration of naphthalenes was rapid when worms were transferred to clean water and/or sediment, perhaps indicating egestion. After two weeks of depuration, both water- and sediment-exposed worms released naphthalenes to background levels. From these results it does not appear that significant bioaccumulation of naphthalenes occurs from hydrocarbon fractions bound to sediment.

Key words: Sipunculids, bioavailability, sediment-sorption, naphthalenes

Introduction

Over the past two to three years, a considerable amount of information has been generated regarding the availability of petroleum hydrocarbons in water to marine organisms and the effects of these compounds (for reviews see Anderson, 1976; Neff *et al.*, 1976; Anderson, 1975). Research in this field has been greatly enhanced by the development of sensitive analytical techniques to determine the content of specific petroleum hydrocarbons in water, tissue and sediment (Warner, 1975; 1976). From such extensive analyses of numerous animal tissues, exposed in the laboratory or field experiments, it became evident that naphthalenes (naphthalene plus alkylnaphthalenes) could be singled out as being (1) among the most toxic of petroleum hydrocarbons tested, (2) accumulated in tissues to a significant extent on exposure to oil-water-dispersions (OWD), seawater extracts (WSF) of several oils, and solutions of the specific compound(s), and (3) retained as long or longer than other hydrocarbons tested, when animals were allowed to release (depurate) contamination in clean seawater.

The results of various laboratory studies, involving relatively short exposures to hydrocarbons in solution and/or dispersion form, have produced the majority of the information on the uptake, release and toxicity of the naphthalenes (Anderson, 1975). Organisms studied thus far include oysters, clams, grass shrimp, penaeid shrimp, mysids, polychaetes and several species of fish. Taking the results of these studies into consideration, there are still a few important questions to be answered: (1) What are the levels of these and other petroleum hydrocarbons that significantly effect populations of marine organisms chronically exposed one or more months?   (2) Are the turnover rates (depuration via whatever means) significantly increased by lengthening exposure periods to months?   (3) Are ingested hydrocarbons, adsorbed to particulate matter or incorporated into food-organisms, released by animals at slower rates?

The subject of this paper relates to one aspect of the last question concerning bio-availability of sediment-bound naphthalenes. In an earlier investigation, Cox *et al.* (1975) found the accumulation and release of these compounds by postlarval brown shrimp *(Penaeus azteca)* and clams *(Rangia cuneata)* followed the appearance and loss of the hydrocarbons in the water of an experimental shrimp pond. At the time that water levels dropped to background concentrations, these latter animals had released the majority of the accumulated naphthalenes. At this same time interval, oysters *(Crassostrea virginica)* still retained relatively high levels approximating those associated with the fine clay sediment. Investigating the bioavailability of naphthalenes from sediment and detritus, Rossi (1976) exposed the polychaete, *Neanthes arenaceodentata*, to sediment containing 9 ppm of total naphthalenes (naphthalene, N + methylnaphthalenes, MN + dimethylnaphthalenes, DMN) for 28 days. The animals, sampled at several time intervals, never contained levels above background (0.1 ppm). In additional tests with a contaminated artificial detritus (powdered alfalfa) containing about 10 ppm ^{14}C-2-methylnaphthalene, Rossi found accumulated C^{14}-activity was voided with the feces during a 24-hr period with clean food and water. This lack of accumulation was exhibited even when feeding on contaminated detritus was continued for 16 days.

In light of this accumulation of particulate-bound hydrocarbons by filter-feeding oysters (Cox et al., 1975) and mussels (Fossato and Canzonier, 1976) and the lack of this process in the benthic detritus-feeder *(Neanthes)*, it seemed essential to examine additional organisms for the potential of bioaccumulation from particulates.

The objective of this study was to determine the accumulation and release parameters of naphthalenes for the burrowing peanut worm, (Sipunculida) *Phascolosoma agassizii*. So that uptake from sediment could be compared to that from contaminated water, the latter was also briefly investigated. Sediment-oil mixtures were produced and used in two ways to examine the variability that might result from different types of actual oil spills.

Materials and Methods

Hydrocarbon Analyses

The majority of tissue and sediment samples were analyzed for content of naphthalenes by the method of Neff and Anderson (1975, and all water was analyzed by this technique. Spectro-grade hexane was used to extract naphthalenes from water with brief shaking, from sediment by shaking for 24 hr in sealed flasks (Na_2SO_4 added to remove water), and from tissue by homogenizing in hexane and then centrifuging (8000 rpm for 10 min). Hexane extracts from tissue and sediment were treated with activated Florisil for several hours to overnight to remove polar compounds that may absorb at the wavelengths between 220 and 235 μm. Water extracts and some tissue extracts produce a baseline on the ultraviolet spectrophotometer (u.v.) that is very near zero. However, sediment extracts and tissues exposed to oil-water-dispersions such as those adsorbing to the contaminated sediments often produce baselines that are significantly above zero. Therefore, the baseline optical density (O.D.) value at 235 μm was subtracted from all u.v. scans before the concentrations of specific naphthalenes were calculated from the O.D.s at 221, 224 and 228 μm (see Neff and Anderson, 1975, for more explanation of the technique).

Hexane extracts of tissues of u.v. analysis were carefully evaporated to 0.5 ml under a nitrogen stream and injected into a flame ionization gas chromatograph (GC) for verification of u.v. determinations. The GC conditions were: column - 1/8 in x 6 ft 3% Dexsil 300 on 100/120 mesh Supelcoport; temperatures - injector 250°C, detector 300°C, column 80° to 250°C at 6°/min.

A third technique was utilized to determine the levels of naphthalenes in sipunclids exposed to oil contaminated sediments. When silica gel chromatography and GC failed to show anything but trace amounts of naphthalenes, extracts of the worms were analyzed by high-pressure liquid chromatography (HPLC) combined with florescence spectrophotometry (HPLC-FS) (Warner, in press). Sediment samples were normally analyzed by u.v. for naphthalenes and infrared spectrophotometry (i.r.) for total hydrocarbons (API, 1958), but in a few cases gravimetric determinations (total hydrocarbons) and GC analyses were conducted to provide additional specific information.

Exposure Systems

Oil in water. Regardless of the exposure system, the ambient seawater conditions from Feb. 1, 1976, to Sept. 1, 1976, ranged between 29 and 31 parts per thousand salinity and 7 to 12°C. Temperature never varied more than 1°C during any exposure. The same batch of Prudhoe Crude oil was used in all of the experiments described in this paper. A seawater extract was obtained by mixing 1 part oil over 9 parts water for 20 hr in a glass bottle (see Anderson et al., 1974, for details of method and analyses of other oils). While one cannot say what portion of the hydrocarbons in this extract are in true solution as opposed to fine dispersed droplets, no oil film nor droplets could be observed; therefore, the water accommodating these hydrocarbons will be referred to as the water-soluble fraction (WSF). One hr after removal of the water phase (WSF) from beneath the oil, this 100% extract was added to several finger bowls containing sipunculids. Samples of the water extract taken at time 0 and at termination (24 hr) were analyzed by u.v. for content of total naphthalenes. Tissues were analyzed immediately after exposure (24 hr) and after 2 weeks in clean water (depuration).

Oil in sediment. Sediments were placed in partitioned polyvinyl chloride (PVC) trays (30x50x10 cm) which were situated in fiberglass exposure aquaria. Glass partitions were inserted to provide 3 uniform sections 8 cm deep within each tray. The trays were fitted with fiberglass-mesh window screen (1.6 mm openings) over the bottom plate which had 9 perforations (38 mm dia) and the top surfaces (1500 cm^2) of the boxes were open.

The exposure aquaria which contains the PVC trays and received influent seawater at 1 l./min were commercially built, gel-coated fiberglass tanks with inside dimensions of 36x46x81 cm. In order to accommodate simulated tide cycles, each aquarium was equipped with 2 discharge locations. The discharge that maintained a "high tide" level in the aquarium was a constant overflow glass tube which was mounted in the end of the aquarium opposite the influent seawater delivery. The water depth in the aquarium at "high tide" was 30 cm. At predetermined time intervals, "low tide" level was obtained by opening a valve on a glass tube discharge situated directly below the "high tide" discharge. Although some variability occurred, "low tide" for all exposure aquaria was accomplished within a 2-hr interval, followed by a 1-hr "low slack" tide. Likewise, the return to "high tide" took about 1 hr, resulting in a total cycle time of approximately 4 hr. Two complete cycles were accomplished for each day of the experiment (4 weeks).

Sediment for this experiment was collected from the lower intertidal zone at the mouth of Sequim Bay. The sediment texture was coarse sand and fine gravel, and was sieved through 12.5 mm screen mesh before being placed in PVC trays. Each tray was filled to within 2 cm of the top with sieved sediment, and subsequently placed in respective exposure aquaria for oil treatment. Sediment volume for each tray was 12,000 cm^3, and trays were held off the bottom for drainage by use of cement bricks.

Introduction of oil in the 4 treated trays was affected by layering oil during a "low tide" cycle on the thin water surface above the sediment. In order to obtain complete surface coverage, 480 ml of oil were required for each tray, which resulted in a 0.32 cm surface layer of oil. After the oil was introduced, the "low tide" cycle was continued, affording about 2 hr for the oil to contact the sediment directly. Subsequent tide cycles were effected at previously described time intervals. Oil which was lifted from the sediment by "high tide" was discarded via the constant flow "high tide" discharge tube.

During exposure, sediment core samples were collected from partitioned sections of the trays for hydrocarbon analysis. Three core samples were taken from appropriate sections of each tray for each of three sample dates. Core samplers were made from sections of PVC pipe 3.7 cm inside dia by 9.8 cm in length. Each sample tube was washed in hot soapy water, rinsed with clean hot water, and immersed in isopropyl alcohol prior to use. Sampling was accomplished by inserting coded sample tubes into sediment, inserting a #8 neoprene stopper tightly in the top of each tube, and removing tubes to coded ziploc plastic bags for immediate freezing. Samples remained frozen until analysis was initiated.

Thirteen days after oil treatment was initiated, peanut worms (*Phascolosoma agassizii*) were introduced into sediments. The worms were collected from the intertidal zone in the vicinity of the sediment collection site. Ten worms were placed on the sediment surface in each of 4 trays at "high tide." After 2 weeks of exposure, one-third of the worms were randomly removed from each tray and frozen for subsequent u.v. spectrophotometric analysis. The remaining organisms after rinsing and blotting were transferred to respective clean glass containers with clean sediment for depuration. After either 6 or 14 days of depuration, the remaining animals were frozen for analyses.

Oil in sediment. To simulate a spill in which oil was thoroughly mixed with intertidal sediment, a different exposure system was designed. Preliminary results of the oil-on-sediment research led to the conclusion that hydrocarbon levels in a heterogeneous sediment were highly variable. Sediment in this system was, therefore, sieved through a 1.6 mm screen before oil was introduced. To avoid loss of the fine organically-rich particles, water was not added to the sieved sediment. Mixing of the wet, freshly-collected sediment was accomplished by using a small cement mixer, which was first internally coated with fiberglass resin. While the 18 l. of sediment were turning in the mixer, an oil-water emulsion was added. The emulsion was prepared in a blender by adding 36 ml of oil containing ^{14}C-2-methylnaphthalene to 400 ml seawater. A second 400 ml of seawater was then blended to remove the remainder of the oil from the vessel and was transferred to the sediment. Complete mixing was checked by liquid scintillation counting of the several small sediment samples selected at random.

After approximately 1 hr of mixing, the sediment analysis (C^{14}) indicated homogeneity and the moist sand was transferred to Pyrex glass dishes (22x11x6 cm) containing 10 sipunculids each. The animals were lying on fiberglass screen (1.6 mm) which covered two 25mm holes in the bottom of the dish. All animals were therefore required to burrow up through the sediment to reach the surface (5 cm deep). In addition to the 12 containers with animals and exposed sediment, there were 4 dishes with animals and clean sediment, and 3 dishes with only control sediment. All containers were placed in the previously described fiberglass aquaria receiving a flow of 1 l./min. Three aquaria contained the 12 exposed dishes (4 each), and the remaining 7 were divided into two others. Circulation of water through the glass dishes was provided by elevating them on bricks such that the holes in the bottom would allow exchange with water, but no tidal fluxes were used over the 2-week holding period. Sampling of tissue and sediment took place at various intervals by removing individual dishes, and analyses were as described earlier.

Results

Oil in Water

When *Phascolosoma* was exposed to the 100% water-soluble extract of Prudhoe Crude oil, the uptake was approximately a factor of 2 over the initial content of naphthalenes in the water (Table 1). In another experiment utilizing ^{14}C-2-methylnaphthalene in a constant flow system, the magnification factor was approximately 10, illustrating the effect of a constant depletion of naphthalenes in static systems (Table 1). It should be noted that there is approximately equal distribution of N, MN and DMN in the WSF, and after 24 hr of exposure, both naphthalene and alkylnaphthalenes are well represented in the tissues of the animals.

Animals which were transferred to clean sediment for 2 weeks following the 24-hr exposure period released naphthalenes to a level which is within the range of background absorbence for control animals (0.2 - 0.3 ppm).

TABLE 1. Naphthalenes (ppm) in Exposure Water and Tissues of Sipunculids (*Phascolosoma*)

	Time	N	MN	DMN	TN (Mean±S.D.)		n
100% WSF Prudhoe Crude	0	.56	.47	.43	1.45	.08	2
	24 hr	.024	.023	.020	.07	.004	2
					(95% decrease)		
Phascolosoma	24 hr	.938	.403	.933	2.27	.47	4
	24 hr + 2 wk depuration	.108	.080	.060	.24	.15	3
					(89% decrease)		

Oil on Sediment

Analysis of the core samples taken from the sediment exposed to a layer of oil showed that the total hydrocarbon (i.r. analyses) values of the surface layer varied on the same day from about 900 to 4900 ppm. Such variation was anticipated, since this substrate was a heterogeneous mix of gravel, sand, and fine detritus particles. There was a sharp drop in contamination levels of 1 to 2 orders of magnitude as one progressed from the top 2 cm of the sediment cores to the lower 4 cm (center and bottom regions). The correlation between the i.r. method of analysis and that of the gravimetric method was quite close. On examination of the n-paraffin levels in the sediment cores, the same large variations were observed, and a peak in concentration occurred in the C15 - C18 range (Table 2). Some contamination remained after 29 days, even at the lower regions of the cores, but in general these lower sections contained 2 orders of magnitude less than the surface layer.

Determinations of naphthalenes in the surface sediments also showed considerable variations, but one can see a general decrease in levels between days 14 and 29 (Table 3). The extremely high concentrations obtained in 2 samples (4 and 42) produced significantly higher means and standard deviations than would have been exhibited if they had been omitted. High variability was not the result of analytical error, but due to the heterogeneity of the substrate. Analyses showed that naphthalene was not retained, while methyl- and dimethylnaphthalenes were still present at reasonably high concentrations after 29 days. Where analyses were available (J. S. Warner, Battelle, Columbus), trimethylnaphthalenes (TMN) were also found to be retained on sediments for the duration of the experiment. A control sediment sample also analyzed by GC (Warner) possessed no N, MN, DMN or TMN at levels above 0.04 ppm. Analyses of the lower 4 cm of the sediment cores by GC also showed very low levels of contamination by naphthalenes (<0.1 ppm), regardless of the time after exposure.

TABLE 2. Content of n-Paraffins (µg/g or ppm) in Samples from Oil-On-Sediment Experiment

No. carbon atoms	6 Days						29 Days					
	Sample #4			Sample #16			Sample #85			Sample #103		
	T**	C	B	T	C	B	T	C	B	T	C	B
11	17.8	0.01	0.02	2.2	*	*	1.4	*	0.01	2.1	0.01	0.01
12	17.9	0.02	0.03	2.8	*	*	1.8	*	0.01	2.3	0.02	0.01
13	17.9	0.03	0.03	3.2	*	*	1.9	*	0.01	2.4	0.03	0.02
14	20.1	0.04	0.05	3.9	*	*	2.3	0.01	0.03	2.8	0.05	0.03
15	22.6	0.05	0.06	4.3	*	*	3.3	0.01	0.04	3.2	0.07	0.05
16	25.5	0.07	0.09	5.0	*	*	3.0	0.02	0.06	3.6	0.08	0.04
17	26.8	0.07	0.09	4.9	*	*	3.1	0.02	0.06	4.0	0.08	0.05
18	22.5	0.07	0.09	4.9	*	*	2.6	0.02	0.06	3.7	0.06	0.03
19	19.6	0.06	0.08	4.4	*	*	2.3	0.02	0.05	3.4	0.05	0.03
20	16.3	0.05	0.06	3.2	*	*	2.0	0.01	0.04	2.3	0.05	0.03
21	13.0	0.04	0.05	2.6	*	*	1.6	0.01	0.04	1.9	0.04	0.02
22	12.9	0.04	0.05	2.5	*	*	1.6	0.01	0.04	1.7	0.04	0.02
23	12.3	0.04	0.05	2.4	*	*	1.5	0.01	0.03	1.7	0.04	0.02
24	11.9	0.04	0.05	2.3	*	*	1.4	0.01	0.03	1.7	0.03	0.02
25	10.3	0.03	0.04	2.0	*	*	1.2	0.01	0.03	1.5	0.03	0.02
26	9.1	0.03	0.04	1.8	*	*	1.1	0.01	0.03	1.4	0.03	0.02
27	5.9	0.02	0.03	1.2	*	*	0.8	0.01	0.02	0.9	0.02	0.02
28	4.9	0.02	0.02	1.0	*	*	0.7	0.01	0.02	0.8	0.02	0.01
29	3.8	0.01	0.02	0.8	*	*	0.6	0.01	0.01	0.7	0.02	0.01
30	3.5	0.01	0.01	0.6	*	*	0.3	0.01	0.01	0.5	0.01	0.01
31	2.7	0.01	0.01	0.6	*	*	0.2	*	0.01	0.5	0.01	0.01

*Less than 0.01 µg/g.

**T,C,B denote top, center and bottom 2 cm portions of cores.

Note: Control sample is not included, since analyses showed all n-paraffins were less than 0.01 ppm (µg/g).

TABLE 3. Concentrations of Naphthalenes in the Top 2 cm of Sediment Receiving Oil on the Surface

Sample	Days after application	µg/g wet			
		MN	DMN	TMN	TN
3	6	.43	.69		1.12
4		15.7	19.8	7.3	42.8
9		.67	1.33		2.00
16		1.8	2.7	1.2	5.7
18		.74	1.23		1.97
				Mean	10.72 ± 18.02 S.D.
39	14	.32	.90		1.22
42		13.98	25.99		39.97
45		2.09	3.64		5.73
54		5.84	9.51		15.35
				Mean	15.57 ± 17.30 S.D.
77	29	1.77	3.04		4.81
78		3.09	5.17		8.26
85		.9	1.8	0.8	3.5
86		2.26	4.52		6.78
87		.23	.55		.79
88		.67	1.49		2.17
89		.29	.40		.70
103		.5	1.1	0.5	2.1
				Mean	3.64 ± 2.77 S.D.

It should be observed that the WSF of Prudhoe Crude oil and organisms exposed to it for 24 hr contained reasonably high levels of naphthalene, while oiled sediments in a flowing system apparently did not retain this compound. If one were to disregard the 3 samples containing very high contamination (4, 42, 54), it would appear that the level of naphthalenes in the sediments changed very little over the 29 days in a system receiving 2 tidal flushes per day and constant flowing seawater.

Animals exposed to the contaminated sediment were analyzed for tissue naphthalenes immediately after the 2-week exposure, after 6 days of depuration, and after 13 days depuration in clean sediment. The results shown in Table 4 illustrate the extensive variability between organisms. When trimethylnaphthalenes are included, the maximum concentrations of total naphthalenes were 1.35 to 2.64 ppm in animals which were removed directly after the 2 weeks exposure. Even under these conditions, the remainder of the organisms contained relatively low levels of naphthalenes (0.02 - 0.44 ppm), especially when one considers the fact that the surface sediments contained a mean of 3.6 ppm of total naphthalenes at 29 days. Of course, there is no way of knowing how much time or feeding activity was spent in the upper layer, and the lower 4 cm contained negligible amounts of contamination.

During the two depuration periods, the worms apparently lost a significant portion of the naphthalenes, as shown in Table 4. Again, the variability between animals makes summarization difficult. Detailed and sensitive analyses by HPLC-FS showed that 1 of 6 animals depurated for 6 days still possessed detectable contamination (0.78 ppm total naphthalenes). After 13 days, 4 of 12 worms contained levels which ranged from .18 to .46 ppm after subtraction of 0.30 ppm for background absorbence of control tissue. Analyses of 14 control animals produced a mean of 0.204 and standard deviations of 0.096, which were used to correct the levels of naphthalenes in Table 4. It is interesting to note that, as in the determinations on sediment, tissue analyses resulted in levels of naphthalene which were below background and dimethylnaphthalenes were generally higher than the monomethyl-isomers.

Oil in Sediment

By first sieving the collected sediment through 12.5 mm and then 1.6 mm mesh screens, relatively uniform material was produced. The mixing of this more homogeneous sediment with an oil-water emulsion of relatively small volume (800 ml oil-water + 18 l. substrate) produced an evenly mixed contaminated sediment. Initial counts of ^{14}C-2-methylnaphthalene from 8 small (2-4 g) substrate samples collected randomly gave readings of 162 ± 40 counts/min/g sediment. Additional evidence of the uniformity of oil and sand is available in Fig. 1 where the standard deviations are shown by vertical bars. Both i.r. and u.v. analyses of samples taken at 0, 16 hr, 40 hr, 88 hr, 1 week and 2 weeks demonstrate that total hydrocarbon and naphthalenes concentrations remained quite constant over the period from 16 hr to 2 weeks. There was a sharp drop in the content of total hydrocarbons from time 0 to 16 hr, as might be expected when the sediment was flushed for this period with flowing water. Presumably, this oil which was removed represented emulsion droplets which resided in the interstitial spaces or on the surface of the sediment. Perhaps due to the sorption characteristics of the alkylnaphthalenes, the level of contamination from these compounds changed very little from time 0 to 2 weeks after mixing.

Tissues of sipunculids exposed to the oil-in-sediment mixture for 2 weeks were analyzed by both u.v. and GC. The results of these determinations are shown in Fig. 2. At most time intervals, there is quite good correspondence between the means produced from the two analytical methods. Variations between organisms are still rather large as can be seen from the standard deviations for both types of analysis, but those from GC determinations were generally smaller and quite consistent. There was a trend for the peak tissue concentrations to occur at 40 hr of exposure, but the large standard deviations at this interval overlap all other ranges except those of the final interval. These latter termination samples analyzed by u.v. showed levels significantly below those of the 1 week exposure animals, and GC determinations indicated the 2 week animals possessed even lower concentrations (less than 0.03 ppm of both MN and DMN).

Sipunculids which were transferred to clean sediment after exposure to mixed oil and sediment for 40 hr showed a decrease in tissue concentration of between 63 (u.v.) or 80% (GC) in 2 days. At the 2 week termination date, 5 animals were found on the bottom of the aquarium and had therefore removed themselves from the contamination at some point during the exposure. The levels of naphthalenes in these animals (0.17 ppm) were very near those of control animals sampled at the 2-week interval (0.10 ppm total naphthalenes). Four sipunculids which were still in the contaminated substrate after 2 weeks were transferred to glass beakers, with no sediment, and placed in the flowing seawater. After 2 weeks in clean water, without sediment, these animals contained only 0.02 ± .01 ppm of total naphthalenes. If we assume these worms originally contained the same amount of naphthalenes as others exposed for 2 weeks, then 2 weeks of depuration produced a decrease of about 2 orders of magnitude.

TABLE 4. Levels of Naphthalenes in *Phascolosoma* Exposed to Surface Contaminated Sediment

Condition	Analytical Method	Concentration μg/g wet (ppm)			
		MN	DMN	TMN	TN[1]
Exposed for 2 weeks	u.v.	.15	.60		.45
	u.v.	.23	.46		.39
	HPLC-FS	1.00	.90	.74	2.64
	HPLC-FS	.45	.34	.56	1.35
	u.v.	.10	.26		.06
	u.v.	.30	.44		.44
	u.v.	.11	.21		.02
	u.v.	.10	.07	.05	.22
	HPLC-FS				<.04(1)[2]
	u.v.				<.30(5)[3]
Depurated for 6 days	HPLC-FS	.15	.36	.27	.78
	HPLC-FS				<.04(5)
Depurated for 13 days	u.v.	.33	.43		.46
	u.v.	.17	.31		.18
	u.v.	.22	.28		.20
	u.v.	.22	.53		.45
	u.v.				<.30(8)

[1] TN is total naphthalenes, where u.v. analyses were capable of only determining methyl- and dimethyl-isomers, but the HPLC-FS method also produced values for the trimethylnaphthalenes.

[2] Parenthetical numbers indicate the number of samples below the level of detection which was .04 for HPLC and .30 for u.v. ([3]). This latter value (.30) has been subtracted from all tissue samples analyzed by u.v.

Discussion

The brief experiment conducted to determine the extent of accumulation of naphthalenes from a WSF of Prudhoe Crude oil showed that tissue magnification was only somewhat less than a factor of 2. Of course, this type of calculation is based on the initial content in the exposure water (1.5 ppm TN) and cannot take into account the gradual decrease in concentration with time, which resulted in a 95% loss in 24 hr. Under similar exposure conditions (24 hr), Rossi (1976) showed the polychaete *Neanthes* accumulated 14 ppm from a 0.6 ppm TN concentration of WSF of No. 2 Fuel oil, and this represented a magnification factor of 23. Differences between uptake rates of various species could be the subject of an entire paper, but the factors involved certainly include characteristics of the body wall, surface-to-volume ratios, activity level (metabolic and/or water circulation), and detoxification-degradation capabilities of the organisms.

Sipunculids, similar to other species tested, accumulated naphthalene and alkylnaphthalenes to approximately the same extent, when exposed for a short period to a water extract of oil. This uniformity in the uptake rates of the various compounds was the reason for originally expressing the findings in terms of total naphthalenes. It should be emphasized that the analyses of sediments and tissues, exposed to the oil-contaminated sediments for several days to 1 month, showed that naphthalene was absent. Mono-, di- and trimethyl isomers were well represented in both sediment and tissue samples. Several earlier studies, reviewed by Anderson (1975) and Neff *et al.* (1976), demonstrated the longer term retention of methyl- and dimethylnaphthalenes, as compared to the parent compound. From these earlier findings, it was apparent that the more water-soluble and more volatile naphthalene was released from tissues more rapidly and was probably degraded at a faster rate. Presumably, these factors possibly combined with differential sorption characteristics contributed to the more rapid "wash-out" of naphthalene from the substrate in both oil-on-top and oil-in-sediment experiments. These findings should be utilized in monitoring programs which regard the detection of petroleum hydrocarbons resulting from spills or effluents. The relatively long association of alkylnaphthalenes with sediment particles makes them excellent candidates for detecting petroleum hydrocarbon input, while naphthalene would be a poor choice.

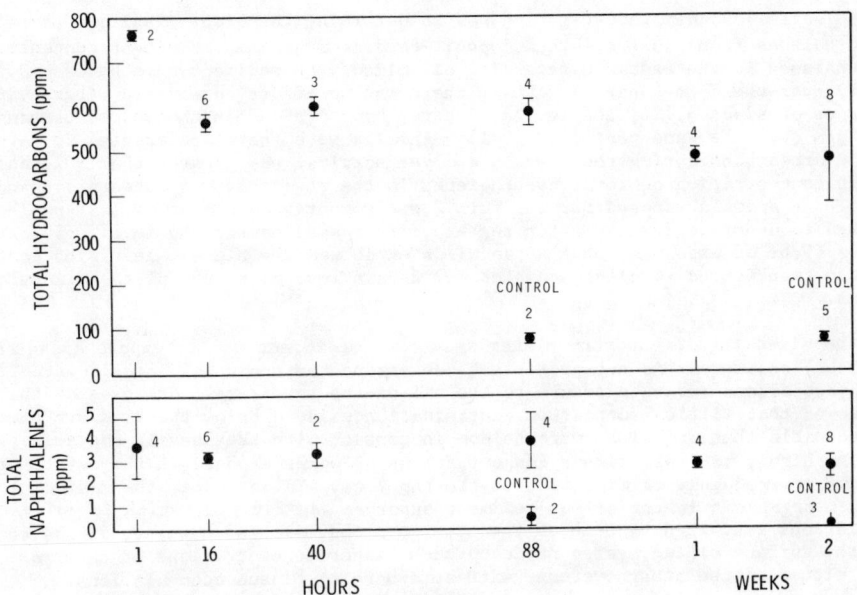

Fig. 29.1. Total hydrocarbons and total naphthalenes in oil-in-sediment exposure system over the 2-week period. Number of samples analyzed, means, and standard deviations >10% (vertical bars) are shown for each time interval.

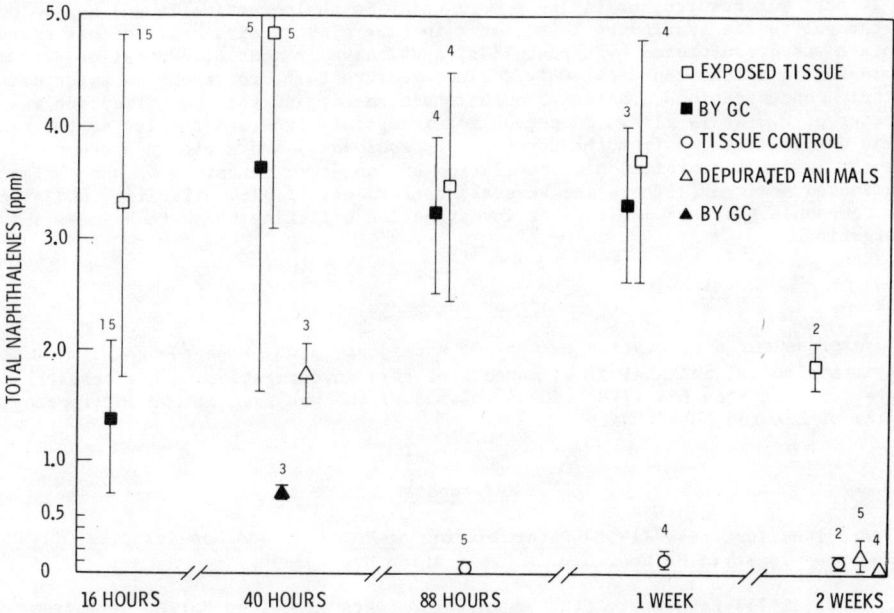

Fig. 29.2. Content of methyl- and dimethylnaphthalenes in sipunculids (*Phascolosoma*) exposed to oil-in-sediments. Means, standard deviations (>10%), and the number of samples are shown for exposed (squares), depurated (triangles), and control (circles) tissues. Solid symbols designate GC analyses, while all others were determined by u.v. Depuration of the 40-hr exposed animals was for 48 hr, while that of the 2-week exposed groups was either 2 weeks (4 samples) or undetermined (5 samples).

The major objective of this investigation was to determine the bioavailability of sediment sorbed naphthalenes from crude oil to a deposit-feeding organism. Sediment concentrations of total naphthalenes in the research regarding oil mixed with sediment were between 3.7 and 2.6 ppm (Fig. 1) over the 2-week period. While there was no effort to measure other hydrocarbons in the tissues of sipunculids, the levels of total hydrocarbons in the sediments ranged from 765 to 475 ppm over the same period (Fig. 1). Animals were therefore exposed to rather high amounts of hydrocarbons during the 2 weeks and yet survival was greater than 90%, and the highest mean concentration of total naphthalenes in the organisms was between 3.6 and 4.8 ppm (Fig. 2). Those animals exposed for the full 2 weeks contained a mean of 1.9 ppm TN, and those allowed to depurate 2 weeks after the exposure showed background levels of contamination. After 40 hr of exposure, when sipunculids exhibited the highest levels of contamination, animals transferred to clean seawater for 2 days lost 63 to 80% of the accumulated naphthalenes.

We assume the mixed-in-oil exposure system is a more efficient way to expose organisms to oiled sediment, because it is not possible to determine the amount of actual contact with petroleum hydrocarbons during exposure to the oil-on-top substrate. Analyses of this latter sediment showed that little hydrocarbon contamination existed below the 2 cm surface section, and it is possible that organisms were seldom in contact with the heavily contaminated surface. The highly variable tissue concentrations of worms exposed in this system may be partially due to frequency of migration to the top 2 cm. In any case, the maximum level obtained from organisms in the oil-on-sediment exposure was 2.6 ppm, which is quite similar to concentrations exhibited by both of the other systems. It is, however, significant to note that the surface oiling system produced much higher concentrations of substrate naphthalenes than either of the other systems, without enhancing tissue accumulations.

While numerous investigations have reported on the availability of water-born hydrocarbons to several species of marine organisms, very few have concerned transfer from particulates. Fossato and Canzonier (1976) showed that hydrocarbons adsorbed to particles were significantly and constantly accumulated by filter-feeding mussels. However, to our knowledge, the only other report on hydrocarbon uptake by a benthic deposit-feeder is that of Rossi (1976). The results of the latter study were described earlier, but in summary, naphthalenes that entered the gut via either sediment or detritus were not accumulated but voided with the feces in 24 hr. Our results, utilizing a worm which is phylogenetically and morphologically close to the polychaete studied by Rossi, were in general very similar. In most cases, little uptake was demonstrated by sipunculids, and 2 days or more of depuration produced sharp decreases in tissue content. Only during exposure to hydrocarbons in water did tissue levels attain concentrations greater than the environment, and this magnification was less than a factor of 2. It is always dangerous to extrapolate from one species to the next, but two benthic deposit-feeders from the northern (*Phascolosoma*) and southern extremes of the Pacific coast have not exhibited bioaccumulation or long-term retention of naphthalenes from oil-contaminated sediment. There are several other facets of bioavailability and other petroleum compounds that require further investigation utilizing these techniques and other benthic organisms.

Acknowledgements

We wish to express our appreciation to Drs. G. Roesijadi, J. S. Warner, and R. M. Bean for their aid in biological and analytical aspects of this investigation. This research was funded by a contract from EPA (77BCF-EHA540-M2.B.3.a) and was part of the Interagency Pass-thru Program of EPA and DBER/ERDA.

References

American Petroleum Institute (1958) Determination of Volatile and Non-Volatile Oily Material: Infrared Spectrometric Method, A.P.I. Publication No. 733-48.

Anderson, J. W. (1975) Laboratory Studies on the Effects of Oil on Marine Organisms: An Overview, A.P.I. Publication No. 4249.

Anderson, J. W., Effects of petroleum hydrocarbons on the growth of marine organisms, in: Petroleum Hydrocarbons in the Marine Environment, McIntyre, A. D., and K. Whittle, ed., Rapp. P.-v. Reun., Cons. int. Explor. Mer, 17. (in press).

Anderson, J. W., J. M. Neff, B. A. Cox, H. E. Tatem and G. M. Hightower, Characteristics of dispersions and water-soluble extracts of crude and refined oils and their toxicity on estuarine crustaceans and fish, Mar. Biol. 27, 75 (1974).

Cox, B. A., J. W. Anderson and J. C. Parker (1975) An experimental oil spill: The distribution of aromatic hydrocarbons in the water, sediment and animal tissues within a shrimp pond, in: Proceedings of 1975 Conference on Prevention and Control of Oil Pollution, A.P.I., E.P.A., and U.S.C.G.

Fossato, V. U., and W. J. Canzonier, Hydrocarbons uptake and loss by the mussel Mytilus edulis, Mar. Biol. 36, 243 (1976).

Neff, J. M., and J. W. Anderson, An ultraviolet spectrophotometric method for the determination of naphthalene and alkylnaphthalenes in the tissues of oil-contaminated marine animals, Bulletin of Environmental Contamination and Toxicology 14, 122 (1975).

Neff, J. M., J. W. Anderson, B. A. Cox, R. B. Laughlin, Jr., S. S. Rossi and H. E. Tatem, Effects of petroleum on survival, respiration and growth of marine animals, in: Proceedings of Symposium on Sources, Effects and Sinks of Petroleum, sponsored by American Institute of Biological Sciences, Washington, D.C., August, 1976 (in press).

Rossi, S. S. (1976) Interactions Between Petroleum Hydrocarbons and the Polychaetous Annelid, Neanthes arenaceodentata: Effects on Growth and Reproduction; Fate of Diaromatic Hydrocarbons Accumulated from Solution or Sediments, Texas A&M University, Ph.D. Dissertation.

Warner, J. S. (1975) Determination of sulfur-containing petroleum components in marine samples, in: Proceedings of 1975 Conference on Prevention and Control of Oil Pollution, A.P.I., E.P.A., and U.S.C.G.

Warner, J. S., Determination of aliphatic and aromatic hydrocarbons in marine organisms, Analytical Chemistry 48, 578 (1976).

CHAPTER 30

FACTORS AFFECTING THE RETENTION OF A PETROLEUM HYDROCARBON
BY MARINE PLANKTONIC COPEPODS

R. P. Harris, V. Berdugo*, E. D. S. Corner,
C. C. Kilvington and S. C. M. O'Hara

Marine Biological Association of the United Kingdom, The Laboratory,
Citadel Hill, Plymouth PL1 2PB, England

*Present address: Israel Oceanographic and Limnological Research Ltd.,
Haifa Laboratories, P.O. Box 1793, Haifa, Israel.

Abstract

^{14}C-1-Naphthalene was used as a model compound to study the retention of an aromatic hydrocarbon by marine planktonic copepods during 24-hour exposure experiments. Seven species were investigated, including representative estuarine, neritic and oceanic forms. Naphthalene concentrations varied from 0.2 to 1000 µg/l, a range including those that might occur temporarily under an oil spill.

Significant positive correlations were demonstrated between naphthalene retention and copepod size measured as dry weight and total lipid content; but a negative correlation was observed with temperature, and retention was diminished in animals starved for progressively longer periods. Amounts of the hydrocarbon absorbed on the surfaces of the animals appeared to be only a small fraction of the totals accumulated.

Supplementing the quantity of ^{14}C-1-naphthalene in solution with a relatively small amount as suspended food led to a marked increase in radioactivity in the animals. In addition, studies on the fate of naphthalene ingested by male and female Calanus helgolandicus during feeding for 24 hr on a plant diet showed that, compared with bulk constituents of normal foodstuffs, the hydrocarbon was more readily assimilated. About half the assimilated fraction was released in soluble form during feeding, either as unchanged hydrocarbon or metabolites, and the other half retained. There was no evidence that the size of the portion retained varied with the sex of the animal.

Upon transfer of the animals to clean seawater following exposure exponential depuration was observed, but in the case of Eurytemora affinis radioactivity accumulated by nauplius I was still detectable in the resultant adults 34 days later.

Key Word Index: Petroleum hydrocarbon; ^{14}C-1-naphthalene; zooplankton; Calanus; Eurytemora; uptake; retention; depuration.

Introduction

Ideally, experiments involving a petroleum hydrocarbon should be carried out using concentrations that occur naturally. However, concerning the bi-cyclic aromatic hydrocarbon used in the present work, naphthalene, as far as we are aware, no date for the amounts dissolved in the sea have been published. Brown, Searl, Elliot, Phillips, Brandon and Monaghan (1973) give values for certain oceanic waters showing that the total amounts of dissolved aromatic hydrocarbons averaged 1.0 µg/l, with naphthalenes accounting for 4.3%, i.e. 0.043 µg/l. On the other hand, Barbier, Joly, Saliot and Tourres (1973) found that total hydrocarbons dissolved in seawater from an inshore area near Brest amounted to 137 µg/l, of which 3.5% was bi-cyclic aromatics, which were therefore at a concentration of 4.8 µg/l. By contrast, in areas subject to chronic oil pollution (e.g. Milford Haven) offshore effluent discharges contain oil at an average concentration of 25 mg/l. (C.U.E.P. Pollution Paper No. 6, 1976) and the data of Anderson, Neff, Cox Tatem and Hightower (1974: Table 4) can be used to show that, for No. 2 fuel oil, this would give rise to levels of soluble bi-cyclic aromatics of approximately 100 µg/l.

In the present work it was not possible to achieve meaningful results using the very low concentrations calculated from the work of Brown et al.(1973): however, many of the data were obtained using levels of naphthalene within the range 0.2 - 100 µg/l and, if it is assumed that this hydrocarbon is typical of bi-cyclic aromatics, our data could relate to a wide range of natural conditions.

Copepods are animals of central importance in the marine foodweb and one of the main foods of planktivorous fish and the levels of hydrocarbons retained and the extent to which these components persist in the animals are matters of considerable interest that have already received a certain amount of study (Lee, 1975; Corner, Harris, Kilvington and O'Hara, 1976). As part of their earlier investigation, Corner et al., (1976a) measured hydrocarbon retention by the copepod Calanus helgolandicus after its immersion for 24 hr in seawater containing ^{14}C-1-naphthalene concentrations ranging from 0.2 to 125 µg/l. This type of study has now been extended in the present work to include similar experiments with 7 species of copepod, representing oceanic, neritic and estuarine forms, the main purpose being to investigate the relationships between hydrocarbon retention and dry weight, ash-free dry weight and total lipid content with a view to assessing the possible use of these parameters in predicting hydrocarbon retention by populations of copepods briefly exposed to the effects of an oil spill. In addition, this objective has been widened to include further experiments with particular species, concerned with the effects on hydrocarbon retention of two important environmental factors, namely seawater temperature and the nutritional state of the animals.

Direct uptake from solution in seawater is only one of the ways in which copepods can accumulate hydrocarbons: another route of entry is by way of the diet. In the previous work with Calanus, Corner et al., (1976a) found that, in terms of providing the same level of hydrocarbon retention in the animals, the quantity needed in solution was much higher than that required in the form of particulate food, a result supporting the view that in the sea the dietary route of uptake is far more important quantitatively. In the present work, more direct and detailed evidence of this has been obtained in experiments in which levels of radioactivity were compared in animals simultaneously allowed to accumulate ^{14}C-1-naphthalene over 24 hr from solution alone and from solution supplemented by a known quantity as suspended food. In addition, because of the quantitative importance of the dietary route of uptake, further experiments were carried out to study the fate of the hydrocarbon in animals accumulating it in this way; that is to say, the relative amounts of ingested naphthalene assimilated and subsequently retained or released in soluble form, either unchanged or as metabolites. Investigations of zooplankton production have included many such studies, but with natural foodstuffs such as the bulk dietary constituents carbon, nitrogen and phosphorus. It therefore seemed of interest, in the context of hydrocarbon retention, to obtain comparative data for naphthalene, an unnatural food component (or "xenobiotic").

Frankenfeld (1973: Fig. 4, Table 10) found that under laboratory conditions simulating weathering, the amounts and types of organic materials that leached into the seawater under a slick of fresh No. 2 fuel oil were such that total dissolved hydrocarbons after 3 days amounted to 5.4 mg/l, of which naphthalene compounds accounted for 16.3%, being therefore present at a concentration of 0.86 mg/l. In the depuration experiments forming part of the present work, copepods were exposed for 24 hr to a wide range of naphthalene concentrations in seawater, including some close to 1.0 mg/l, and the hydrocarbon levels in the animals were subsequently measured over several weeks. Such experiments are therefore relevant to the important question of what happens to the hydrocarbons in an animal that accumulates them during a brief period spent in the immediate vicinity of an oil spill before removing to clean seawater.

Methods

Animals
Experimental animals were collected by townet from coastal waters within a radius of 10 miles of Plymouth breakwater with the exception of the estuarine species, Eurytemora affinis, which was collected from the Tamar river estuary. All animals were returned to the laboratory within 2 hr of capture, and were subsequently sorted and sexed. Animals were maintained at 10°C before the experiment, and fed the following unialgal phytoplankton cultures: Isochrysis galbana (Eurytemora affinis, Acartia clausi and Euterpina acutifrons), Peridinium trochoideum and Prorocentrum micans (Centropages typicus, Temora longicornis and Anomalocera patersoni), Peridinium trochoideum and Biddulphia sinensis (Calanus helgolandicus). All phytoplankton species were cultured using an enriched seawater medium (Guillard and Ryther, 1962).

Seawater Solutions of Naphthalene
Concentrated solutions of ^{14}C-1-naphthalene were prepared using methods described by Corner et al. (1976a). Aliquots were added by sterile pipette to 0.2 µm membrane-filtered seawater to prepare the experimental concentrations. In all experiments a sample was taken from each container at the start and end of each 24 hr period to measure the naphthalene concentration in solution. In the closed systems losses of radioactivity were typically about 8% in 24 hr, naphthalene presumably being adsorbed to glass surfaces. Experimental concentrations were calculated as the geometric mean of the initial and final levels.

24 hour retention experiments

Uptake experiments were performed in 300 ml wide-necked Erlenmeyer flasks. The number of copepods per flask ranged from 10 for large species and high naphthalene concentrations to 50 for small species and lower concentrations. After addition of the requisite number of animals to each flask, containing 0.2 μm membrane filtered seawater, a predetermined aliquot of naphthalene stock was added, together with additional seawater to fill each flask completely. After gentle stirring a 10 ml water sample was taken from each flask for determination of the initial concentration; the neck of the flask was then sealed with a glass plate and the flasks were maintained in a 10°C constant temperature room for the 24 hr exposure period under a 12 hr light: 12 hr dark low intensity illumination regime. At the end of 24 hr, 10 ml water samples were taken from each flask to determine the final naphthalene concentration. The copepods were then collected on a nylon mesh, recounted, washed by immersion for 15 seconds in three successive rinses of 500 ml of clean seawater to remove traces of the experimental naphthalene solution, and finally collected by filtration on a nylon mesh disk. The animals were immediately transferred to a scintillation vial containing 1 ml Soluene, and after subsequent digestion, addition of Fluor, and dark-equilibration, the samples were counted in a Packard 20003 liquid scintillation counter. The count was converted to naphthalene equivalents per animal, and the retention per copepod during 24 hr.

The effect of temperature on retention was investigated by maintaining animals in flasks immersed in constant temperature baths over a range from 5.0 to 16.0°C, representing approximately the annual extreme seawater temperatures off Plymouth. The seawater and naphthalene solutions were equilibrated, and the experimental animals acclimated, for 48 hr before the start of the experiments.

Experiments in which Calanus helgolandicus females were exposed to naphthalene solutions during feeding were conducted with the dinoflagellate Peridinium trochoideum. Cells cultured in uncontaminated medium were added at the same time as the naphthalene at the start of the 24 hr exposure period. In this system the naphthalene could enter the copepod via two routes, either directly from solution, or via the diet after uptake by Peridinium cells. A standard food concentration of approximately 100 μg algal carbon/l was used in all experiments. Cell concentrations were measured with a model ZB Coulter Counter, and converted to algal carbon using the factor determined by Mullin, Sloan and Eppley (1966).

In certain experiments retention was studied either in animals that had been previously starved, or during such a period of starvation over a number of days. The starving animals were maintained in 0.2 μm membrane-filtered seawater which was changed daily. Control animals, which were allowed to feed during the pretreatment period, were maintained in a suspension of Peridinium trochoideum, the concentration of which was adjusted daily to give a mean of 100 μg algal carbon/l.

Accumulation experiments

These experiments were carried out at 10°C in the dark using the rotating columns described in an earlier paper (Corner, Head and Kilvington, 1972). The food was the diatom Biddulphia sinensis. Adult female Calanus were first starved for 3 days, during the last of which they were treated with the antibiotics streptomycin sulphate and neomycin sulphate, each at a concentration of 50 mg/l; batches of 50 were then added to each of 3 glass columns containing seawater solutions of naphthalene at a known concentration; other samples of 50 animals were added to 3 more columns, each containing a similar concentration of naphthalene in the column. After a further period of 24 hr techniques described earlier (Corner et al. 1976a) were used to measure the quantities of radioactivity in animals that had taken up hydrocarbon from solution alone and in those that had accumulated it from both solution and food.

In order to express the importance of the dietary route quantitatively, the increase in level of radioactivity attained by animals accumulating the hydrocarbon from both solution and food was calculated in terms of the additional quantity of ^{14}C-1-naphthalene in solution alone that would have produced the same increase (C-A in Table 3: see Results section). This extra amount was then compared with the quantity as suspended food (B in Table 3) that gave the same increase in level in the animals. To estimate B, two further columns were used in each experiment, in which Biddulphia cells were suspended in seawater solutions of ^{14}C-1-naphthalene for 24 and 48 hr at 10°C in the dark and the levels of radioactivity per cell measured after both intervals. The average level during the period of grazing by the Calanus (24 - 48 hr) was then estimated as the logarithmic mean (a).

To facilitate accurate determination of the levels of radioactivity in the animal samples, batches of 50 Calanus had to be used, for which high concentrations of plant cells were needed in order to provide each animal with its maximum daily ration. Previous work (Corner et al. (1972) had shown that this was approximately 2000 Biddulphia cells if the food level were maintained above 6000 cells/l: accordingly, it was expected that the 50 animals present in each column would remove 100,000 cells during grazing. In fact the average daily ration removed by each Calanus was slightly higher (2062 cells), the number of cells in the columns at the start of the experiment, 109,500 being reduced to 6,400 at the end. The average concentration of cells in the columns during grazing, calculated from the logarithmic means of the levels at the start and end of each experiment, was 26,500, equivalent to 20,000 cells/l (b). The average level of hydrocarbon represented by suspended food during the grazing period in each experiment (B in Table 3) was given by a x b.

Adult male Calanus were used in one of the experiments and as preliminary studies had indicated that, compared with females, these animals removed a much smaller daily ration of Biddulphia cells, the food concentration in the columns at the start of the experiment was halved so as to allow a more accurate measurement of the small number of cells removed during grazing. The daily ration removed was unexpectedly low, only 246 cells per animal, and so the average cell concentration in the columns during grazing, 33,600/l, was considerably higher than that observed in the experiments using females (20,000/l). This means that the value for B in the experiment with males is also considerably higher (1.11 compared with 0.66 ng/l, which substantially reduces the size of the ratio (C-A):B for males in Table 3.

Assimilation and retention experiments
The animals, having been separated as to sex, were usually starved for three days and treated with antibiotics in the way described earlier. However, in one experiment the animals were maintained on a diet of Biddulphia cells until only 5 hr before the grazing experiment was started, this being done in order to increase the lipid levels of the females. Fifty females were then added to each of three glass columns and fifty males to each of three more, all the columns containing a seawater suspension of Biddulphia cells that had previously been immersed for 24 hr in a known concentration of ^{14}C-1-naphthalene alone and then transferred to fresh seawater. Two further columns, containing the seawater suspension of previously treated Biddulphia cells but no animals, were used as controls to determine losses of radioactivity from the cells during the 24 hr period of the experiment. The logarithmic mean of the levels of radioactivity found in these cells at the start and end of the 24 hr period of grazing was taken as the average quantity during the experiment; and this value, when multiplied by the number of cells grazed by each animal, gave the daily ration of radioactivity removed. In all the experiments the cells, when transferred to hydrocarbon-free seawater, lost a substantial amount of the radioactivity they had originally taken up, an observation similar to that made by Soto, Hellebust and Hutchinson (1975) in studies of the effects of naphthalene on the flagellate Chlamydomonas angulosa.

After 24 hr the quantity of radioactivity in each bath of animals was determined, giving the amount retained at the end of the grazing period. In addition, all the plant cells and faecal pellets left in each column after grazing were collected and counted and levels of radioactivity then measured in samples containing known numbers of each. The number of plant cells alone, when multiplied by the amount of radioactivity per cell at the end of the experiment, gave the quantity present in each sample as cells; and this value, when subtracted from the total radioactivity in the sample, gave the quantity accounted for as faecal pellets. The amount of radioactivity assimilated by a Calanus was then estimated as the difference between the quantity assimilated during grazing and the amount still retained in the animals at the end of the experiment.

Depuration experiments
Loss of hydrocarbon by Eurytemora affinis was studied after exposing the animals (either adult females or nauplius stage I) to naphthalene solutions for 24 hr periods. Subsequently the animals were washed, and transferred to uncontaminated seawater either in the presence or absence of food (Isochrysis galbana) and allowed to depurate for a number of days. At intervals samples of animals were taken to determine the level of radioactivity remaining. In the case of exposure of nauplius I, the long term retention was studied by rearing the exposed nauplii to adults which were then examined for radioactivity. Culturing techniques were essentially those described by Heinle (1969). Animals were reared in 5 l. beakers at 10 and 15°C on a diet of Isochrysis galbana.

Dry weight and lipid content
Dry weight determinations were made, using a Cahn 4150 Electrobalance, on groups of animals dried for 24 hr at 60°C. No distilled water rinse was used as this has been observed to result in loss of lipid droplets from some species of copepod. Ignition in a muffle furnace for 4 hr at 500°C, followed by re-weighing, enabled an estimate to be made of the ash-free dry weight of each species. Total lipid analyses were performed by the method of Zollner and Kirsch (1962), as applied to marine animals by Barnes and Blackstock (1973).

Results and Discussion

<u>Naphthalene retention by different species in relation to body size and lipid content</u>
For all the species investigated there was a highly significant correlation between the retention of naphthalene during 24 hr exposure experiments and the concentration in solution when plotted on a log:log scale (Fig. 1). Details of the regression equations are given in Table 1.

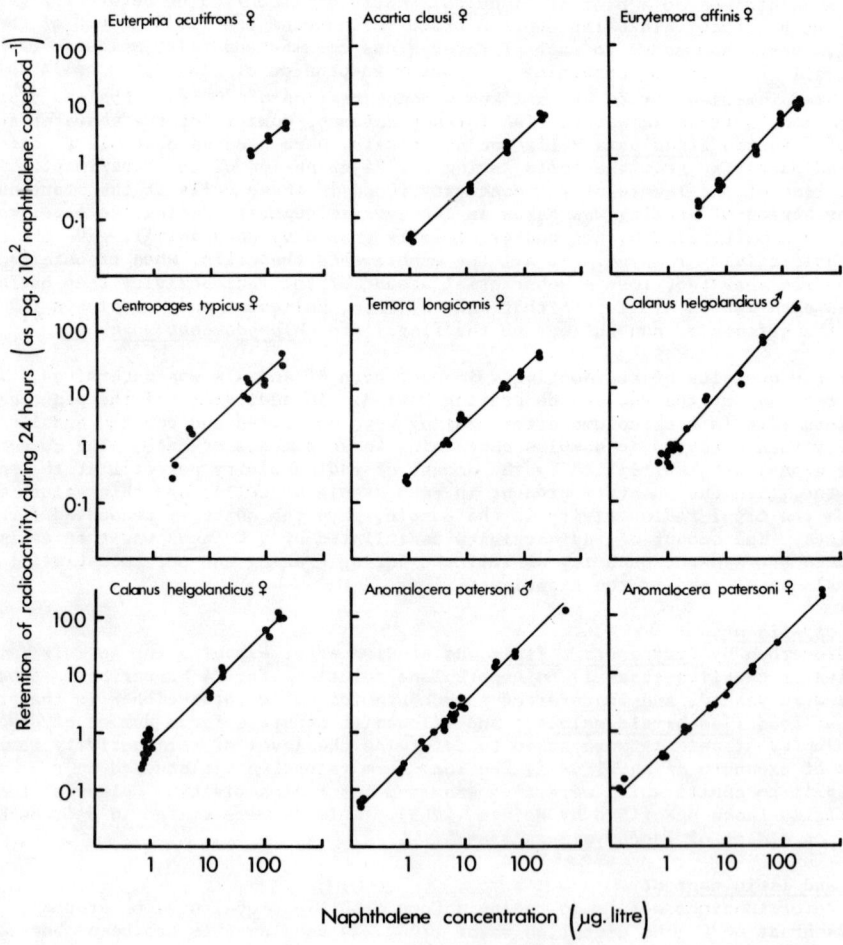

Fig. 30.1. Retention of radioactivity by different species of marine copepod immersed in seawater containing ^{14}C-1-naphthalene at different concentrations for 24 hr (no food present). Regression equations are given in Table 1.

TABLE 1. Regression Equations Relating 24 hr Retention by Marine Copepods to Concentration of ^{14}C-1-Naphthalene in Solution

Species		Equation		
Euterpina acutifrons	♀	Log y = 0.8334 log x + 0.6674;	r = 0.9813;	n = 6
Acartia clausi	♀	Log y = 0.9076 log x + 0.6528;	r = 0.9982;	n = 12
Eurytemora affinis	♀	Log y = 1.0491 log x + 0.6089;	r = 0.9947;	n = 18
Centropages typicus	♀	Log y = 0.9208 log x + 1.4040;	r = 0.9692;	n = 12
Temora longicornis	♀	Log y = 0.9516 log x + 1.4546;	r = 0.9929;	n = 16
Calanus helgolandicus	♂	Log y = 1.1622 log x + 1.8055;	r = 0.9886;	n = 20
Calanus helgolandicus	♀	Log y = 0.9794 log x + 1.7584;	r = 0.9888;	n = 26
Anomalocera patersoni	♂	Log y = 0.9745 log x + 1.5254;	r = 0.9940;	n = 25
Anomalocera patersoni	♀	Log y = 0.9625 log x + 1.6998;	r = 0.9971;	n = 18

y = level of radioactivity after 24 hours (as pg naphthalene, copepod^{-1})
x = naphthalene concentration in solution (µg liter^{-1})

This finding confirms that reported for female Calanus helgolandicus by Corner et al. (1976a) there being no significant difference between the regression line calculated in the present work and that described in the previous study despite the slightly different experimental conditions under which they were determined. Similarly a straight line relationship between retention by the American oyster, Crassostrea virginica, and hydrocarbon concentration in solution up to 450 µg/l has been reported by Stegeman and Teal (1973).

There is no significant difference between the regression coefficients given in Table 1 when tested at the 5% level, the regression lines being parallel with a pooled slope of 1.0107. However, the height of the lines (the y intercept of the regression equations) shows significant differences between the species. This value, which is an estimate of the retention at 1 µg/l, enables comparison to be made between the copepods studied. From Table 1 it can be seen that naphthalene retention varied from 4.1 pg/copepod for Eurytemora affinis to nearly 16 times that level, 63.9 pg/copepod, for male Calanus helgolandicus. The species of copepod investigated varied considerably both in weight and body lipid as is illustrated in Table 2, the latter being of particular interest because of the high solubility of many hydrocarbons in lipid. The question of how closely the naphthalene retention could be correlated with size and lipid content was examined by regression analysis. The linear regressions of the intercept values for each species from Table 1 (24 hr retention at 1 µg/l) against body lipid, dry weight, and ash-free dry weight are illustrated in Fig. 2.

TABLE 2. Weight and Lipid Content of the Copepod Species Investigated

Species		Total lipid (µg)			Dry Weight (µg)			Ash-free dry weight (µg)		
		$\bar{x}$	s	n	$\bar{x}$	s	n	$\bar{x}$	s	n
Euterpina acutifrons	♀	0.5	–	1	2.8	0.6	2	1.9	0.7	2
Acartia clausi	♀	1.4	0.4	2	19.3	1.8	4	14.4	1.2	4
Eurytemora affinis	♀	1.1	0.1	3	7.7	0.7	5	6.7	0.7	5
Centropages typicus	♀	7.8	1.3	4	59.6	7.5	7	44.4	6.0	7
Temora longicornis	♀	4.9	0.9	5	38.3	1.8	4	34.2	1.9	4
Calanus helgolandicus	♂	33.9	8.3	11	185.5	22.9	11	147.8	20.8	11
Calanus helgolandicus	♀	10.5	6.4	14	138.0	28.5	14	114.8	24.9	14
Anomalocera patersoni	♂	9.0	1.9	5	186.4	17.4	5	159.6	15.0	5
Anomalocera patersoni	♀	12.9	2.1	5	249.9	39.4	5	213.7	38.8	5

(mean, $\bar{x}$; standard deviation, s; and number of determinations, n)

The best correlation was observed with total lipid content, the correlation coefficient of 0.9503 being significantly higher than those for dry weight and ash-free dry weight at the 5% level. Thus it appears that for the species investigated, including oceanic, neritic and estuarine forms, the total lipid content is a good indicator of their relative retention of naphthalene from seawater solution during short-term exposure. However, body size and total organic content as estimated by dry weight and ash-free dry weight respectively also exhibit a high degree of correlation with naphthalene retention, and therefore lipid content may not be the only factor involved. Clearly body size and lipid content are themselves interrelated, and both will in turn be correlated with surface area of the copepod which may be of some importance in regulating adsorptive processes.

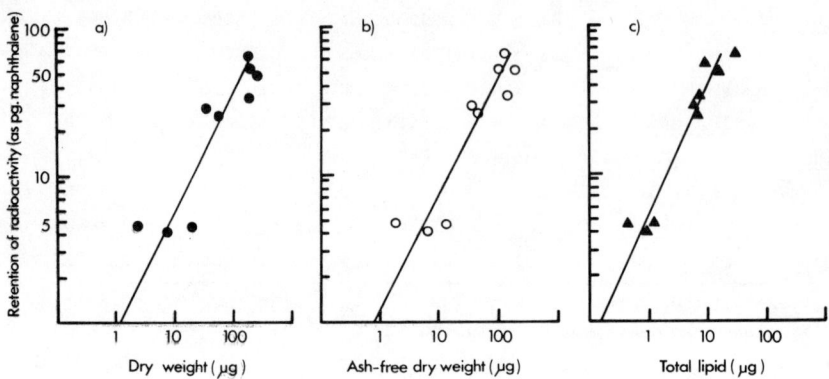

Fig. 30.2. Relationship between 24 hr retention of radioactivity (calculated for a concentration of 1 µg/l. naphthalene from the regression equations in Table 1) and dry weight (–●——●–), ash-free dry weight (–○——○–) and total lipid (–▲——▲–) content for different species of marine copepod. Details of regression equations: a) Log y = 1.2426 log x + 0.0705, r = 0.9125, n = 9; b) Log y = 1.2814 log x - 0.0763, r = 0.9106, n = 9; c) Log y = 1.1230 log x - 0.7692, r = 0.9503, n = 9.

A measure of the quantitative importance of surface adsorption can be obtained from data presented in Fig. 5 where retention over a range of concentrations is illustrated for female Calanus exposed for only 30 sec to the naphthalene solution. The amount retained, taken to by a measure of surface uptake, is about 5% of that retained by starving females during a 24 hr period: accordingly, it appears that rapid surface adsorption may not be a major factor in retention of naphthalene. These points will be discussed further later. However, the present results indicate that knowledge of total body lipid can be used, together with an empirical relationship, in order to predict retention of the hydrocarbon naphthalene by marine copepods during short-term exposure. Expressing all the retention data in Fig. 1 in terms of the animal's lipid content enables one to calculate a common regression equation for all the species investigated. The equation relating retention after 24 hr as pg naphthalene/(µg copepod body lipid), ($\underline{y}$), to the concentration in seawater solution as µg/l, ($\underline{x}$), is : log $\underline{y}$ = 0.9739 log $\underline{x}$ = + 0.6109, r = 0.9798, n = 153.

The role of size of animal as a factor affecting the amount of hydrocarbon retained by marine copepods has been mentioned by Lee (1975). During 24 hr exposure to ^{3}H-benzpyrene in seawater solution the large copepods Calanus hyperboreus and Calanus plumchrus (2.2 and 1.0 mg dry weight respectively) took up and retained significantly more of the hydrocarbon than the much smaller C. helgolandicus (0.2 mg dry weight). However, interpretation of these results is complicated by the fact that uptake for each species was studied at a different temperature (14°C for C. plumchrus, 3°C for C. hyperboreus, and 16°C for C. helgolandicus). All the data illustrated in Fig. 1 were determined at 10°C, which is close to the mean annual seawater temperature off Plymouth. To investigate the effect of temperature on retention, female Calanus helgolandicus were exposed to naphthalene in seawater solution at different temperatures covering the seasonal range in the sea off Plymouth. The results indicate (Fig. 3) that there is an inverse relationship between retention and temperature, the retention per copepod in 24 hr decreasing by about 39 pg, or by 3.23 pg/(µg copepod lipid), per 10°C rise in temperature. This suggests that the retention calculated in Figs. 1 and 2 at 10°C would be about 44% higher at winter temperatures (6°C) and about 50% lower in summer (16°C). Although the present results were obtained with only one species, and work with many other species would need to be done before any generalization could be made, they do have some relevance to the problem of hydrocarbon retention by arctic and tropical zooplankton. It is assumed that the temperature effects observed in the present results are due to the rate of hydrocarbon metabolism increasing more rapidly than the rate of uptake as temperature increases. Both Corner et al. (1976a) and Lee (1975) have shown that copepods have the ability to metabolize hydrocarbons including naphthalene.

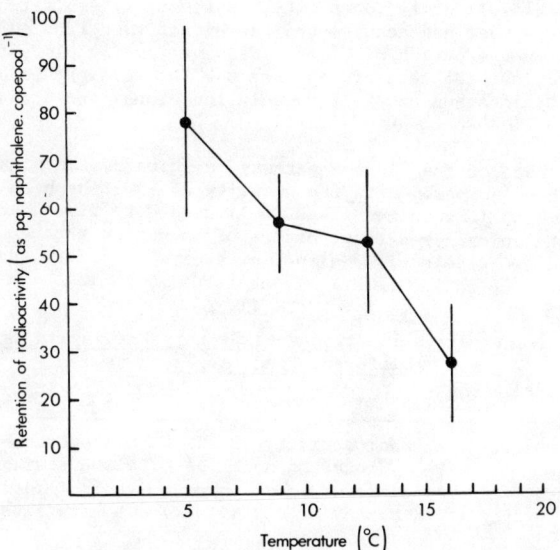

Fig. 30.3. The retention of radioactivity at different temperatures by female Calanus helgolandicus immersed in sea water solutions of naphthalene (concentration 1 µg/l.) for 24 hr in the absence of food. Each point is the mean of four observations; vertical bars represent ± standard deviation.

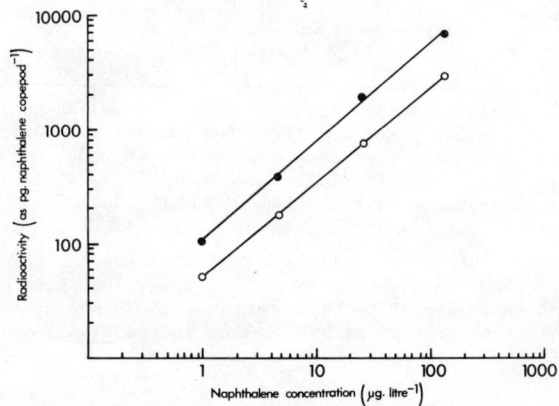

Fig. 30.4. Levels of radioactivity in female Calanus helgolandicus immersed in sea water containing ^{14}C-1-naphthalene at different concentrations, when starving (—o——o—), and feeding on Biddulphia sinensis cells (—●——●—),: regression line for starving animals is log y = 0.832 log x + 1.712, r = 0.993, n = 12; for feeding animals log y = 0.854 log x + 2.04, r = 0.986, n = 12. Each point represents the mean of three determinations at the same concentration. Other details as in text.

Accumulation from solution and diet

Shown in Fig. 4 are the levels of radioactivity, expressed as equivalent weights of naphthalene, detected in Calanus that had accumulated the hydrocarbon from solution alone and from both solution and diet over a period of 24 hr. Using a log-log plot, a linear relationship is observed with each of the two sets of data and the two straight lines are nearly parallel with slopes of 0.823 (hydrocarbon uptake from solution alone) and 0.854 (hydrocarbon uptake by both routes).

The quantitative importance of the dietary pathway is shown by the high values for the ratio (C-A):B in Table 3. Thus, compared with the quantity of ^{14}C-1-naphthalene present as suspended food (B), the amount in solution alone (C-A) needed to give the same increase in hydrocarbon level in the animals was three orders of magnitude greater in the experiments with female Calanus and two orders of magnitude greater when males were used, even though these latter animals removed only a relatively small daily ration.

TABLE 3. Quantitative Importance of Dietary Pathway in Calanus at Different Food Levels

Hydrocarbon concentration		Radioactivity as pg hydrocarbon animal^{-1}		Hydrocarbon in sol^n alone equivalent to (sol^n + food) level (µg/liter^{-1})	Ratio (C-A:B)
In sol^n (µg liter^{-1}) A	As food (ng liter^{-1}) B	(Sol^n alone)	(Sol^n + food)	C	
Experiments with females					
0.96	0.66	53	106	2.40	2186
4.69	2.38	179	386	11.55	2892
25.52	122.4	758	4538	230.80	1677
26.00	29.0	749	1830	76.54	1743
134.10	61.4	2997	6774	375.40	3930
Experiments with males					
0.93	1.10	75	88	1.24	282

Hydrocarbon in solution alone giving 'solution + food' level in animals calculated from straight line log y = 0.823 log x + 1.712 for females and log y = 1.153 x 1.817 for males, obtained in immersion experiments using the columns. Average daily ration captured is 2062 Biddulphia cells female^{-1} and 246 male^{-1}; average cell concentration is 20,000 liter^{-1} in experiments with females and 33,600 liter^{-1} in experiment with males).

Values for the ratio (C-A):B vary considerably and show no obvious trend with levels of dissolved hydrocarbons throughout the range 0.96 to 34.1 µg/l. Nor did the amount of hydrocarbon present in the plant cells appear to have any effect. In one of the experiments using 25 µg ^{14}C-1-naphthalene/l the quantity of hydrocarbon in the Biddulphia cells was roughly four times that found in the other, yet the two values for the ratio (C-A):B - 1677 and 1743 - were nearly the same.

One factor that does affect the size of the ratio, however, is the level of available food. As mentioned earlier, the design of the experiments was such that during grazing the average level of plant cells was 20,000/1. Natural concentrations of Biddulphia cells found in the sea, however, are much lower: for example, Bainbridge (1957) gives the highest value in offshore waters of the U.K. as 2620/1. Data obtained in the present work, combined with those reported earlier by Corner et al. (1972) for daily rations removed by female Calanus at different food levels, can be used to estimate the value of the ratio (C-A):B when the hydrocarbon is present at a low concentration (0.96 µg/l) and the food level is more typical of those found in the sea (1000 cells/1). Thus, in the present study it was found that an adult female Calanus feeding on Biddulphia cells at a concentration of 20,000/1 in seawater containing 0.96 µg ^{14}C-1-naphthalene/l removed a maximum daily ration of 2062 cells, which increased the level of hydrocarbon in the animal from 82 pg, accumulated from solution alone, to 104 pg accumulated from both solution and food (see Table 3). If it is assumed that the increase due to feeding is directly proportional to the number of cells removed, it follows that an animal capturing only 500 cells - which previous work showed to be the daily ration at a food level of 1000 cells/1 (Corner et al., 1972) - would show an increase in level of hydrocarbon by 5.3 pg to give a total of 87.3 pg. The quantity of hydrocarbon in solution alone that produces this level in the animals can be calculated from Fig. 4 as 1.35 µg/l, representing an increase of 0.39 µg/l. The average level of hydrocarbon in a Biddulphia cell

TABLE 4. Fate of Naphthalene in Female Calanus

Hydrocarbon level in soln for previously treating plant cells (μg liter^{-1})	Daily ration per animal		Daily loss in faeces per animal			Hydrocarbon assimilated per animal		Hydrocarbon retained per animal		Hydrocarbon released per animal		Retained Assimilated (%)	Lipid as % ash-free dry weight
	(Cells)	(pg H-C)	(Pellets)	(pg H-C pellet^{-1})	(Total as pg H-C)	(pg)	(% Ration)	(pg)	(% Ration)	(pg)	(% Ration)		
1.37	1782	321.9	90.1	1.71	154.1	167.8	52.1	71.6	22.2	96.2	29.9	42.7	4.17
1.37	1773	320.0	87.6	2.11	184.8	135.3	42.3	66.3	20.7	69.0	21.6	49.0	4.17
1.37	1722	311.1	115.0	1.56	179.4	131.7	42.3	68.4	22.0	63.3	20.3	51.9	4.17
5.56	2160	282.1	110.6	0.772	85.4	196.7	69.7	145.1	51.4	51.6	18.1	73.8	9.92
5.56	2101	274.4	99.4	0.951	94.5	179.9	65.5	134.8	49.1	45.1	16.4	74.9	9.92
5.56	2106	275.0	105.6	0.858	90.6	184.4	66.9	122.6	44.6	61.8	22.3	66.5	9.92
127.7	2234	1553	84.8	6.82	578	975	62.8	370.0	23.8	605	39.0	37.9	17.4
127.7	2234	1553	81.6	6.94	566	987	63.5	388.2	25.0	599	38.6	39.3	17.4
127.7	2214	1539	86.4	7.51	648	891	57.9	338.9	22.0	552	35.9	38.0	17.4

(All levels of radioactivity expressed as equivalent weight of hydrocarbon (i.e. pg H-C)

TABLE 5. FATE OF NAPHTHALENE IN MALE CALANUS

Hydrocarbon level in sol^n for previously treating plant cells (µg liter^{-1})	Daily ration per animal		Daily loss in faeces per animal			Hydrocarbon assimilated per animal		Hydrocarbon retained per animal		Hydrocarbon released per animal		Retained / Assimilated (%)	Lipid as % ash-free dry weight
	(Cells)	(pg H-C)	(Pellets)	(pg H-C pellet^{-1})	(Total as pg H-C)	(pg)	(% Ration)	(pg)	(% Ration)	(pg)	(% Ration)		
1.37	677	122.3	86.8	0.461	40.0	82.3	67.3	28.4	23.2	53.9	44.1	34.5	27.8
1.37	572	103.3	82.4	0.357	29.4	73.9	71.5	30.7	29.7	43.2	41.8	41.5	27.8
1.37	611	110.5	79.6	0.576	45.9	64.6	58.5	31.2	28.3	33.4	30.2	48.4	27.8
5.56	785	102.5	92.5	0.400	37.0	65.5	63.9	33.2	32.4	32.3	31.5	50.7	30.1
5.56	654	85.4	79.2	0.501	39.7	45.7	53.5	32.5	38.1	13.2	15.5	71.1	30.1
5.56	688	89.9	88.4	0.472	41.7	48.2	53.6	34.9	38.8	13.3	14.8	72.4	30.1
127.7	682	490	102.8	1.59	163	327	66.7	87.8	17.9	239	48.8	26.8	17.1
127.7	727	522	114.4	1.93	221	301	57.7	84.9	16.3	216	41.4	28.2	17.1
127.7	821	589	98.0	2.80	274	315	53.5	97.4	16.5	218	37.0	30.9	17.1

(All levels of radioactivity expressed as equivalent weight of hydrocarbon (i.e. pg H-C)

during this experiment was 32.9 x 10^{-3} pg so that, at a food concentration of 1000 cells/l, the amount present as suspended food would have been 24.1 pg, giving the values of the ratio (C-A):B as 11,680, an order of magnitude greater than that found using the food level of 20,000 cells/l. (see Table 3)

Corner, Harris, Whittle and Mackie (1976) have summarized the few quantitative estimates that have been made of the total amounts of dissolved and particulate hydrocarbons in samples from the same sea-area, 'dissolved' material being that passing through and particulate that retained by a membrane filter: it has generally been found that the concentration of dissolved hydrocarbons is higher than that represented by suspended material (Jeffrey, 1970; Levy, 1971; Zsolnay, 1971). What is not known from these studies is the extent to which the analyzed particulate material could have been used by zooplankton as a food. Assuming, however, that it consisted mainly of phytoplankton, which was certainly the case in the study by Jeffrey (1970), then the present findings indicate that although such material may contain relatively small amounts of hydrocarbon in comparison with those dissolved in the surrounding seawater, it could still be of greater quantitative importance in terms of hydrocarbon accumulation by an animal such as Calanus.

Ingestion, Assimilation and Retention by Female and Male Calanus
Data from the experiments using females are shown in Table 4 and from those using males in Table 5. All levels of radioactivity are expressed as an equivalent weight of naphthalene.

Under the conditions used there was no indication that the amounts of ingested hydrocarbon inhibited feeding. Thus, females receiving either more than 1500 pg or less than 300 pg hydrocarbon in the diet removed similar daily rations of plant cells: moreover, the daily number of cells captured were slightly higher than the maximum of 1800 found by Corner et al. (1972) for female Calanus feeding on uncontaminated Biddulphia cells.

Although the average numbers of faecal pellets released by females and males were very similar (95.7 and 91.6 per day respectively), male faecal pellets were markedly smaller and contained much less hydrocarbon, the average value being 1.0. pg per pellet compared with 3.25 pg for those from females. The lower levels of hydrocarbon in the male faecal pellets doubtless reflect the smaller number of plant cells removed by these animals and hence the smaller amounts of hydrocarbon ingested.

Freegarde, Hatchard and Parker (1971) and Conover (1971) have shown that zooplankton, feeding in the presence of a fine suspension of crude oil, can ingest this material and release it in faecel pellets, which being slightly heavier than seawater and containing bacteria, could be responsible for immobilizing a substantial fraction of an oil-spill. The present findings show that the soluble components of crude oil, such as naphthalene, could also be immobilized in the faecal pellets of zooplankton by become adsorbed on particulate foods such as plant cells which are subsequently ingested by the animals.

It is clear from the data in Tables 4 and 5 that both the total quantity of hydrocarbon released as faecal pellets and that assimilated by the animals, whether females or males, vary according to the size of the ration captured. However, no such variation with size of ration is observed when either value is expressed as a percentage of the ration captured. For example, the percentage assimilated by female Calanus is similar whether the animals capture less than 300 pg or more than 1500 pg hydrocarbon daily; likewise, with male Calanus, the proportion assimilated does not significantly change when the daily amount of hydrocarbon ingested varies from 100 to 500 pg. Further data in Tables 4 and 5 show that neither with females nor males is any correlation found between the size of ration taken and the percentages retained by the animals or released in soluble form. All the data for assimilation, retention and soluble release obtained with females on the one hand and with males on the other have therefore been combined to give average values, together with standard errors in Table 6.

The data in Table 6 show that although males removed a much smaller ration of plant cells, and hence a much smaller daily quantity of hydrocarbon, the proportions of ingested naphthalene actually assimilated by males and females were very similar, as were those either retained by the animals or released in soluble form. Thus, with either males or females, about 60% of the captured food was assimilated, about half of this being retained and the other half released in soluble form. Concerning the latter fraction it should be noted that, in the light of previous work with copepods (Lee, 1975; Corner et al. 1976a), this is likely to be a mixture of unchanged hydrocarbon and its water-soluble metabolites. In their study using female Calanus grazing Biddulphia cells Corner et al, (1972) obtained data for the assimilation, retention and release in soluble form of the natural dietary constituents nitrogen and phosphorus. These data are also included in Table 6. It is noteworthy that, compared with either dietary nitrogen or phosphorus, the hydrocarbon is much more readily assimilated.

TABLE 6. Summarized Data for Calanus Feeding on Biddulphia Cells

Dietary constituent	Sex	% Ration				Retained Assimilated (%)	Soluble release Assimilated (%)
		Faeces	Assimilated	Retained	Soluble Release		
Naphthalene	♀	41.9(3.4)	58.1(3.4)	31.2(4.4)	26.9(3.0)	53.7(5.1)	46.3(5.1)
Naphthalene	♂	39.3(2.2)	60.7(2.3)	26.8(2.9)	33.9(5.8)	44.1(5.8)	55.9(5.8)
Nitrogen	♀	65.9	34.1	7.5	26.6	22.0	78.0
Phosphorus	♀	59.6	40.4	0	40.4	0	100

(Figures in brackets are standard errors. Data for nitrogen and phosphorus taken from Corner et al, (1972).

Compared with the species of diatom used as foods for Calanus by Marshall and Orr (1955) Biddulphia is less readily digested. Nevertheless, naphthalene being highly lipid-soluble one would expect its assimilation from the gut to be almost complete, with no substantial loss as faecal pellets. A possible explanation is that some of the hydrocarbon may have already been converted into water-soluble metabolites by the plant diet; and although such a view is not supported by evidence so far available in the literature (Soto et al., 1975; unpublished observations reported by Lee, 1975), a detailed study of this interesting problem has not yet been described, and so the question should still be regarded as open (Vandermuelen and Ahern, 1976).

Metabolism and Lipid Content as Factors Affecting Naphthalene Retention
Earlier data have shown that the respiration of starved Calanus is less than that of fed (Corner, Cowey and Marshall, 1965; Conover and Corner, 1968) and as the metabolism of naphthalene by marine crustaceans, including zooplankton, has been shown to involve oxidative processes (Corner, Kilvington and O'Hara, 1973; Lee, 1975; Burns, 1976) it seemed that excretion of the hydrocarbon in the form of metabolites would be less likely to occur with starved animals than with fed, the starved thereby retaining greater amounts. However, studies of the effects of starvation on the retention of naphthalene, using male Calanus, showed the reverse, the levels of hydrocarbon accumulated over 24 hr by direct uptake from solution progressively decreasing when the test animals had been previously starved for longer and longer periods. In addition, data included in Figure 5 show that, when immersed for 24 hr in seawater containing various concentrations of ^{14}C-1-naphthalene, compared with living Calanus, animals that had been killed (by being kept at 37°C for 5 mins) retained substantially smaller amounts of the hydrocarbon: yet these dead animals, being unable to metabolize the compound, should have accumulated more than the living if metabolism was the key factor in hydrocarbon retention. These findings led to the view that the lipid content of the animals might be more important than metabolism in determining naphthalene retention, prolonged starvation causing a fall in lipid level in the animal, which would thus have less chance of retaining a highly lipid-soluble substance such as an aromatic hydrocarbon.

Some evidence of the importance of lipid in this respect had already been obtained: thus, as mentioned earlier, the levels of hydrocarbon retained by several species of zooplankton immersed in seawater containing 1.0 µg ^{14}C-1-naphthalene/1 show a better correlation with lipid content than with either dry weight or ash-free dry weight (see Fig. 2). More detailed experiments, involving measurements of the levels of naphthalene retained after 24 hr by animals accumulating the hydrocarbon directly from solution in seawater, were therefore carried out using both male and female Calanus which had either been starved for 5 days or fed on Biddulphia cells before the uptake experiment was begun. The data (Table 7) obtained with either males or females showed that, compared with the fed animals, the starved retained a significantly smaller amount of the hydrocarbon and that both total lipid and lipid as % dry weight were significantly lower for the starved animals than for the fed.

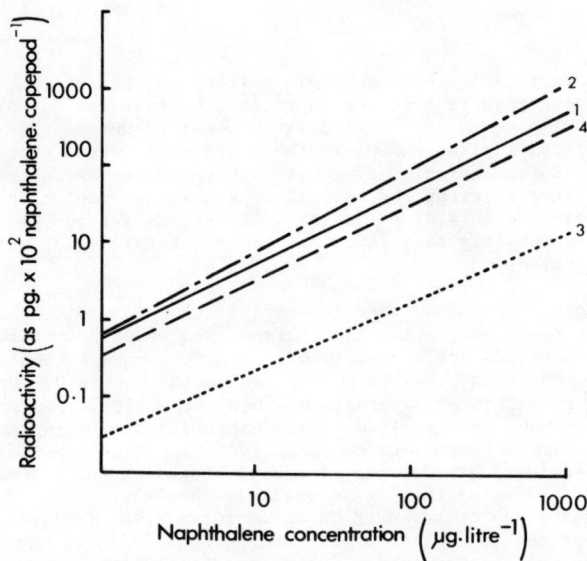

Fig. 30.5. Retention of radioactivity by female <u>Calanus helgolandicus</u> immersed in sea water containing ^{14}C-1-naphthalene at different concentrations under various conditions. 1) (————) Starving, 24 hr exposure, log y = 0.9794 log x + 1.7584, r = 0.9888, n = 26; 2) (—·—·—) Feeding on the dinoflagellate, <u>Peridinium trochoideum</u>, at a food concentration of about 100 μg algal carbon/l., log y = 1.003 log x + 1.7698, r = 0.9453, n = 26; 3) (· · · · · ·) Starving, exposed for 30 seconds only, log y = 0.8833 log x + 0.4778, r = 0.9975, n = 12; 4) (— — —) Dead animals (heat treated at 37°C for 5 mins), log y = 0.9916 log x + 1.5315, r = 0.9971, n = 12). For further details see text.

Nevertheless, although the data obtained with either males or females are consistent with the view that hydrocarbon retention is correlated with lipid level, no such correlation is found when the data for males are compared with those for females: thus, fed males and females retained similar amounts of naphthalene (87.45 and 86.60 μg/animal respectively), but the males had roughly thrice the amount of lipid (33.12 compared with 12.12 μg/animal); likewise, starved males and females retained nearly the same level of hydrocarbon (57.84 and 50.48 pg/animal respectively), but once again the males had nearly thrice the level of lipid (16.13 compared with 5.04 μg/animal).

A possible explanation is that male and female <u>Calanus</u> contain lipids differing in chemical composition and it is not so much the total lipid present, but the level of a particular lipid fraction that is important in determining hydrocarbon retention: in testing which, further studies would be needed that included qualitative as well as quantitative analyses of lipid content.

So far, the role of lipid in hydrocarbon retention has been discussed in terms of the data obtained in experiments in which animals accumulated naphthalene from solution alone. However, experiments were also carried out in which the hydrocarbon entered the animals exclusively by way of the diet; and the data from these experiments (Tables 4, 5 and 6), provide no evidence that lipid content is an important factor. Thus, the results summarized in Table 4 show that the average proportion of assimilated hydrocarbon retained by females (53.7%) was close to that found using males (44.1%): yet those in Tables 4 and 5 can be used to show that the average level of total lipid in the males (25.0% ash-free dry weight) was substantially higher than that in the females (10.5%).

These findings were similar to those obtained with males and females in the studies of hydrocarbon uptake from solution alone. However, differing from the results of these latter studies are further data, in Table 4, showing that the percentage of assimilated hydrocarbon retained by females with a lipid content of only 4.17% ash-free dry weight was higher than that found for animals of the same sex which, having been maintained on <u>Biddulphia</u> cells until only a few hours before the experiment was started, had a lipid content as high as 17.4%.

Possibly, the lack of correlation between lipid content and hydrocarbon retention found in these feeding experiments results from the fact that, whether the animals were fed or starved beforehand, during the actual period of hydrocarbon uptake they were voraciously feeding on Biddulphia; and lipids present in the plant cells could therefore have been rapidly incorporated into the animals, together with the hydrocarbon. Indeed, it is known that female Calanus, after starving for several days, rapidly replenish their levels of triglyceride when feeding on this diet (Sargent, Gatten, Corner and Kilvington: unpublished observations) and it is possibly this fraction, and not total lipid, which is important in determing hydrocarbon retention.

Retention of naphthalene during long-term depuration
The experiments so far described, using naphthalene as a model hydrocarbon, show that after a short period of exposure (24 hr) a variety of marine copepods accumulate significant amounts of the hydrocarbon from seawater concentrations as low as 0.2 µg/l: and certain factors affecting the retention of naphthalene - body size, lipid content, nutritional state and temperature - have been studied under conditions simulating those close to an oil spill where levels of naphthalene could exceed 1 µg/l (see Introduction), but also where volatilization, photo-oxidation and rapid microbial breakdown cause removal of the hydrocarbons (Lee and Ryan, 1975), so that it is unlikely that this level of naphthalene would persist for longer than a day or so, even in an enclosed area, following a single discrete input of oil. These processes, together with hydrographic effects, mean that zooplankton, after encountering the hydrocarbon for only a comparatively short-time would then be in the presence of much lower levels, or even "clean" seawater: in recognition of which, the experiments were carried out over a relatively short period of time (24 hr). Also recognized, however, was the need to know what happens to hydrocarbons retained by zooplankton when, after a preliminary short exposure to these compounds in an oil-polluted area, the animals move into uncontaminated water.

To examine this aspect of the problem two types of experiment were carried out with the estuarine copepod Eurytemora affinis. Firstly, adult females exposed to the hydrocarbon in solution under conditions identical to those used in the 24 hr retention experiments, were subsequently rinsed and transferred to clean seawater, depuration of the hydrocarbon then being followed by determining the levels of radioactivity in samples of the animals at regular intervals, some of them being fed and others starved. Secondly, recently hatched nauplius stage I were exposed to the hydrocarbon under similar conditions, and then reared to adulthood over several weeks in clean seawater, samples of the early copepodites (CI/CII) and, subsequently, of the adults being examined for levels of radioactivity. In both types of experiment high concentrations of naphthalene were used (192-1090 µg/l), in order to facilitate detection of radioactivity in the animals after depuration: nevertheless, such levels could still relate to those prevailing immediately beneath an oil-slick (see Introduction).

The levels of radioactivity detected during these experiments are shown in Fig. 6, from which it is clear that a rapid loss of radioactivity from adult females occurs upon their transfer to clean seawater, the rate of loss decreasing as depuration continues. This type of depuration curve has been previously obtained in experiments involving a variety of marine copepods (Lee, 1975; Corner et al. 1976a and b). An important question, however, is whether depuration continues until all the hydrocarbon is completely lost from the animal, or whether a significant fraction still persists even after a long period. In the work of Corner et al. (1976 a and b) traces of radioactivity were found in the copepods (adult male and female Calanus helgolandicus and female Eurytemora affinis) after depuration periods of 7 - 10 days, further studies with Calanus showing that the rate of depuration was slower after the animals had accumulated the hydrocarbon from the diet rather than from solution. Longer experiments were carried out by Lee (1975), using a variety of labelled hydrocarbons and several species of zooplankton, in which small amounts of hydrocarbon could always be detected, even after a depuration period of 28 days. Similar results were obtained in the present work (Fig. 6) in which detectable amounts of radioactivity could still be measured in adult Eurytemora raised from nauplius I - which had been immersed in a seawater solution of naphthalene for 24 hr and then transferred to clean seawater - after a period of 34 days. The residue in the adult, on average, accounted for as much as 10% of the initial level accumulated by the nauplius; which is considerably greater than that found in previous studies, and may have been caused by the fact that all the earlier work was done with non-growing adults, whereas the animals reared from nauplius I in the present work were actively feeding and growing throughout the experiment, thereby retaining the hydrocarbon whilst increasing the levels of natural body constituents.

TABLE 7. Comparison of Naphthalene Retention by Calanus

			Naphthalene concentration in solution (μg liter^{-1})	Level of radio-activity in copepod after 24 hr exposure (as pg naphthalene copepod^{-1})	Copepod Ash-free dry weight (μg)	Copepod Dry Weight (μg)	Copepod Total lipid (μg)	Lipid % Ash-free dry weight
Male	Fed	$\bar{x}$	1.08	87.40	147.80	185.48	33.12	22.4%
		s	0.16	13.50	20.84	22.95	5.48	
		n	5	4	5	5	5	
	Starved	$\bar{x}$	1.08	57.84	101.86	138.34	16.13	15.8%
		s	0.16	13.58	24.31	24.49	3.10	
		n	5	5	5	5	3	
		t		3.26*	3.21*	3.14*	5.02**	
Female	Fed	$\bar{x}$	0.97	86.60	120.31	148.08	12.12	10.1%
		s	0.03	17.97	27.94	28.75	3.43	
		n	5	5	10	10	5	
	Starved	$\bar{x}$	0.97	50.48	87.67	111.30	5.04	5.8%
		s	0.03	8.17	16.29	10.52	1.56	
		n	5	5	10	10	5	
		t		4.09**	3.19**	3.80**	4.20**	

(Values of t for comparisons between fed and starved animals; * significant at 5% level; ** significant at 1%)

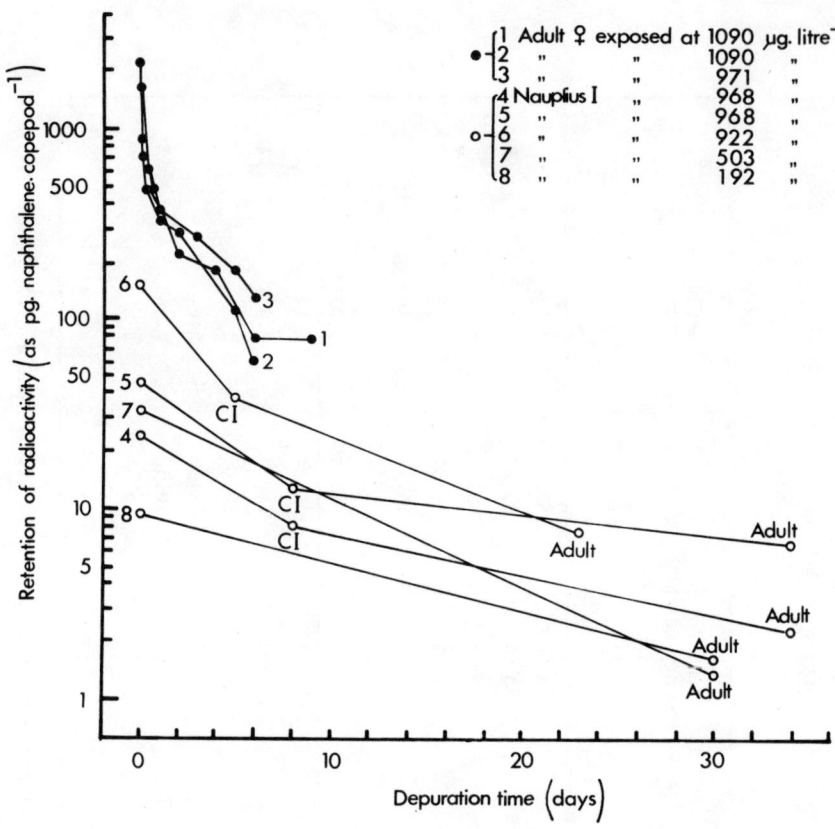

Fig. 30.6. Levels of radioactivity in <u>Eurytemora affinis</u> transferred to uncontaminated conditions following 24 hr exposure to ^{14}C-1-naphthalene at various concentrations.

Lee (1975) tentatively identified some of the radioactivity retained in his animals after lengthy depuration as hydrocarbon metabolites, and it would be interesting to know whether the same holds for the radioactivity measured in the present work with <u>Eurytemora</u>, although analyzing the very small amounts of test material would present considerable technical problems. More important, however, is the finding that, whether the hydrocarbon is present, unchanged or in the form of metabolites, it can persist for 34 days in the animal, at any time during which it is available to predators. This implies that although natural processes may rapidly reduce the levels of hydrocarbons in the sea after an oil spill, weeks later the transfer of aromatic components like naphthalene to a higher trophic level could still be taking place.

Acknowledgements

We are indebted to Skipper Hutchings and Skipper Knott and the crews of R. L. "Sepia" and R. L. "Gammarus" for regular supplies of research material. The work of one of us (V. B.) was supported by a U.N.E.S.C.O. Fellowship and by funds from the Marine Biological Association of the United Kingdom and the Israel Oceanographic and Limnological Research Institute.

References

Anderson, J. W., J. M. Neff, B. A., Cox, H. E. Tatem and G. M. Hightower, The effects of oil on estuarine animals: toxicity, uptake and depuration, respiration, pp. 285-310, In: F. J. Vernberg and W. B. Vernberg (eds.) Pollution and physiology of marine organisms. Academic Press, London, 1974.

Bainbridge, R., The size, shape and density of marine phytoplankton concentrations. Biol. Rev. 32, 91-116 (1957).

Barbier, M., D. Joly, A. Saliot and D. Tourres, Hydrocarbons from sea water, Deep-Sea Res. 20, 305-314 (1973).

Barnes, H. and J. Blackstock, Estimation of lipids in marine animals and tissues: detailed investigation of the sulphophosphovanillin method for "total" lipids, J. exp. mar. Biol. Ecol. 12, 103-118 (1973).

Brown, R. A., T. D. Searl, J. J. Elliot, B. G. Phillips, D. C. Brandon and P. H. Monaghan, Distribution of heavy hydrocarbons in some Atlantic Ocean waters. pp. 505-519. In: Proc. Joint Conf. on Prevention and Control of Oil Spills. American Petroleum Institute, Washington, D.C. (1973).

Burns, K.A., Hydrocarbon metabolism in the intertidal fiddler crab Uca pugnax, Mar. Biol. 36, 5-11 (1976).

Conover, R. J., Some relations between zooplankton and Bunker C oil in Chedabucto Bay following the wreck of the tanker Arrow, J. Fish. Res. Bd. Can. 28, 1327-30 (1971).

Conover, R. J. and E. D. S. Corner, Respiration and nitrogen excretion by some marine zooplankton in relation to their life cycles, J. mar. biol. Ass. U.K., 48, 49-75 (1968).

Corner, E. D. S., C. B. Cowey and S. M. Marshall, On the nutrition and metabolism of zooplankton, III. Nitrogen excretion by Calanus, J. mar. biol. Ass. U.K., 45, 429-442 (1965).

Corner, E. D. S., R. P. Harris, C. C. Kilvington and S. C. M. O'Hara, Petroleum compounds in the marine food-web; short-term experiments on the fate of naphthalene in Calanus, J. mar. biol. Ass. U.K., 56, 121-133 (1976a).

Corner, E. D. S., R. P. Harris, K. J. Whittle and P. R. Mackie, Hydrocarbons in marine zooplankton and fish. pp. 71-105. In: A. P. M. Lockwood (ed.) Effects of pollutants on aquatic organisms, Society for Experimental Biology Seminar Series, Vol. 2, Cambridge University Press (1976).

Corner, E. D. S., R. N. Head and C. C. Kilvington, On the nutrition and metabolism of zooplankton. VIII. The grazing of Biddulphia cells by Calanus helgolandicus, J. mar. biol. Ass. U.K., 52, 847-861 (1972).

Corner, E. D. S., C. C. Kilvington and S. C. M. O'Hara, Qualitative studies on the metabolism of naphthalene in Maia squinado (Herbst), J. mar. biol. Ass. U.K., 53, 819-32 (1973).

Department of Environmental Central Unit on Environmental Pollution, The separation of oil from water for North Sea oil operations, 34 pp. London, H.M.S.O. (Pollution Paper No. 6) (1976).

Frankenfeld, J. W., Factors governing the fate of oil at sea; variations in the amounts and types of dissolved or dispersed materials during the weathering process, pp. 485-495, In: Proc. Joint Conf. on Prevention and Control of Oil Spills. American Petroleum Institute, Washington, D.C. (1973).

Freegarde, M., C. G. Hatchard and C. A. Parker, Oil at sea; its identification, determination and ultimate fate, Laboratory Practice 20, 35-40 (1971).

Guillard, R. R. L. and J. H. Ryther, Studies of marine planktonic diatoms. 1. Cyclotella nana Hustedt, and Detonula confervacea (Cleve) Gran, Canad. J. Microbiol. 8, 229-239 (1962).

Heinle, D. R., Culture of calanoid copepods in synthetic seawater, J. Fish. Res. Bd. Can. 26, 150-153 (1969).

Jeffrey, L. M., Lipids of marine waters, pp. 55-75. In: D. W. Hood (ed.) Symposium on Organic Matter in Natural Waters, University of Alaska Institute of Marine Science Occasional Publication No. 1. (1970).

Lee, R. F., Fate of petroleum hydrocarbons in marine zooplankton, pp. 549-553. In: Proc. 1975 Conf. Prevention and Control of Oil Spills, American Petroleum Institure, Washington, D.C. (1975).

Lee, R. F. and C. Ryan, Biodegradation of petroleum hydrocarbons by marine microbes, pp. 119-125. In: J. M. Sharpley and A. M. Kaplan (eds.) Proceedings of the Third International Biodegradation Symposium, Applied Science Publishers, London (1976).

Levy, E. M., The presence of petroleum residues off the east coast of Nova Scotia, in the Gulf of St. Lawrence, and the St. Lawrence river, Water Research 5, 723-733 (1971).

Marshall, S. M. and A. P. Orr, On the biology of Calanus finmarchicus, VIII. Food uptake, assimilation and excretion in adult and stage V Calanus, J. mar. biol. Ass. U.K., 34, 495-529 (1955).

Mullin, M. M., P. R. Sloan and R. W. Eppley, Relationship between carbon content, cell volume and area in phytoplankton, Limnol. & Oceanogr. 11, 307-311 (1966).

Soto, C., J. A. Hellebust and T. C. Hutchinson, Effect of naphthalene and aqueous crude oil extracts on the green flagellate Chlamydomonas angulosa. II. Photosynthesis and the uptake and release of naphthalene, Canad. J. Bot. 53, 118-126 (1975).

Stegeman, J. J. and J. M. Teal, Accumulation, release and retention of petroleum hydrocarbons by the oyster Crassostrea virginica, Mar. Biol. 22, 37-44 (1973).

Vandermuelen, J. H. and T. P. Ahern, Effect of petroleum hydrocarbons on algal physiology: review and progress report. pp. 107-125. In: A. P. M. Lockwood (ed.) Effects of pollutants on acquatic organisms, Society for Experimental Biology Seminar Series, Vol. 2, Cambridge University Press (1976).

Zollner, N. and K. Kirsch, Über die quantitative Bestimmung von Lipoiden (Micromethode) mittels der vielen naturlichen Lipoiden (allen bekannten Plasmalipoiden) gemeinsamen Sulphophosphovanillin-Reaktion, Z. ges. exp. Med. 135, 545-561 (1962).

Zsolnay, A., Preliminary study of the dissolved hydrocarbons and hydrocarbons on particulate material in the Gotland Deep of the Baltic, Kieler Meeresforschungen, 27, 129-134, (1971).

CHAPTER 31

EFFECTS OF TEMPERATURE AND SALINITY OF NAPHTHALENES UPTAKE IN THE
TEMPERATURE CLAM, RANGIA CUNEATA AND THE BOREAL CLAM, PROTOTHACA STAMINEA

Kenneth W. Fucik and Jerry M. Neff

Department of Biology, Texas A&M University
College Station, Texas 77843

Abstract

The temperate clam, Rangia cuneata, and the boreal clam, Protothaca staminea, were exposed to a 25% WSF (water soluble fraction) of Southern Louisiana crude oil for three days under varying temperature and salinity regimes. The R. cuneata were exposed under temperature and salinity combinations of 15, 20, 25 and 30°C and 0, 10, 20 and 30 °/oo. Temperature-salinity combinations of 5, 10 and 15°C and 25, 30 and 35 °/oo were used in P. staminea exposures. Clams were allowed to depurate for three days under these same temperature-salinity combinations.

In three uptake experiments, two using R. cuneata and one using P. staminea, the greatest naphthalenes concentrations were measured in those clams exposed at the lowest temperatures. Statistical analysis confirmed that naphthalenes uptake in the different temperature-salinity groups in each experiment was significantly different. Temperature was shown to have the greatest effect on this difference. Salinity had a slight effect in only one of the R. cuneata uptake experiments. Tissue naphthalenes concentrations in P. staminea were less than those measured in R. cuneata after the same exposure period. Temperature and salinity had no effect on the release of naphthalenes in either R. cuneata or P. staminea.

Key words: Rangia cuneata, Protothaca staminea, temperature, salinity, naphthalene.

Introduction

The ability of bivalve molluscs to accumulate hydrocarbons has been demonstrated under both laboratory (Anderson et al. (1974), Clark and Finley (1975), Fossato and Canzonier (1976), Neff et al. (1976), among others) and field conditions (Farrington and Quinn (1973), Stegeman and Teal (1973), Cox et al. (1975), DiSalvo et al. (1975), Shaw et al (1976), Fucik et al. (1977). For the most part, however, these authors have failed to discuss the patterns of uptake observed in these molluscs in relation to the environmental conditions in which they live. Pringle et al. (1968) have suggested that the hydroclimate (particularly temperature and salinity), dosage and duration of exposure, and the physiological condition of the organism are some of the factors affecting heavy metal uptake in molluscs. Slowey and Neff (1976) have concluded that (1) heavy metals are often present at higher concentrations in the tissues of animals from low-salinity environments that in those from sea water; (2) seasonal variations in tissue heavy metals concentrations in benthic invertebrates may be related to ambient salinity and temperatures; and (3) the accumulation potential of a metal may be affected by the salinity or the ambient temperature. In view of this evidence, it would appear that the rate of hydrocarbon uptake by bivalves could vary with the environmental conditions. Such effects could be especially important in the estuarine system where a wide range of constantly changing environmental parameters is encountered.

Materials and Methods

Two sets of clams representing both a temperate and a boreal species were used in these experiments. The temperate clam, Rangia cuneata, was collected from San Antonio Bay, Texas. Hopkins et al. (1972) considers Rangia cuneata to be the dominant benthic macroinvertebrate in the 0-15 °/oo zone of the Gulf coast estuaries. He also notes that this species cannot maintain a population outside of a salinity range of 1-15 °/oo although it is found along the Texas coast in bays where the salinities range from 0-40 °/oo.

The boreal clam Protothaca staminea, is considered to be first in order of importance in the lower reaches of the intertidal region (Kozloff 1973). Specimens of this clam were supplied by Dr. Jack W. Anderson of the Battelle Northwest Marine Laboratories in Sequim,

Washington. Water conditions along the Strait of Juan de Fuca, where these clams were collected, are more uniform than those encountered along the Texas coast and more closely approximate open ocean conditions. Salinities in Puget Sound and the Strait of Juan de Fuca area range from 27-30 $^o/oo$ with water temperatures ranging from 8-14°C (Weiser (1959), Lie (1963), Anderson personal communication).

Uptake and Depuration

In the Rangia experiments, clams were exposed in rectangular glass aquaria to a 25% WSF (water soluble fraction) of So. Louisiana crude oil in a total volume of 3 l. The exposure solutions were changed daily. Temperature and salinity combinations of 15, 20, 25 and 30°C and 0, 10, 20 and 30 $^o/oo$ were used. Six clams in each temperature-salinity regime were exposed for three days. After the exposure to the WSF was terminated, three clams were removed for analysis of naphthalenes uptake and the remaining three were moved to clean water under the same conditions in which they had been exposed and allowed to depurate for three additional days.

Two experiments utilizing the Rangia were completed, differing only in the amount of time the clams were acclimated to the new temperature and salinity regimes. Prior to the experiment the clams had been maintained in seawater at 10 $^o/oo$ and 20°C. In the first experiment, clams were moved directly to their experimental conditions and allowed only a 24 hr. acclimation period before exposure began. In the second experiment, clams were gradually taken to the new salinities over a two week period with temperature acclimation occurring over a 3 day period before exposure to the WSF.

Protothaca staminea was maintained at 10°C and 30 $^o/oo$ until it was determined that the clams had stabilized in the lab. These clams were then moved to the experimental chambers in temperature and salinity combinations of 5, 10 and 15°C and 25, 30 and 35 $^o/oo$. Five clams at each combination were exposed and allowed to depurate in the same manner as were the Rangia. Three clams were used for naphthalenes uptake and two for depuration experiments.

All clams were analyzed spectrophotometrically for naphthalenes content by the method of Neff and Anderson (1975). Naphthalenes content is reported as the total naphthalenes (naphthalene, methylnaphthalenes and dimethylnaphthalenes) concentration in the tissues determined on the basis of the wet weight of the clams.

Filtration Rates

Filtration rates were measured in Rangia using neutral red, a vital stain. Individual clams were placed in 600 ml glass jars after first being cleansed with a wire brush. Neutral red was added to the water at a concentration of approximately 5 µg/ml in a 300 ml volume. Filtration experiments were conducted at the same temperature and salinity combinations used in the naphthalene uptake experiments. Optical density readings were made at 520 mµ on a Bausch and Lomb Spectronic 20 spectrophotometer at the beginning of the experiment, every hour for the next three hours, and finally after 19 hours when the experiment was terminated. In a control chamber having an empty Rangia shell, there was little change in O.D. readings during the duration of the experiment. The disappearance of neutral red in the solution was converted to filtering rates by the formula:

$$m = \frac{M \log(conc_0/conc_t)}{t \log e}$$ (Ward and Aiello 1973),

where m is the filtration rate, M is the total volume of water present, and $conc_o$ and $conc_t$ are the concentrations of indicator at time zero and time t. It is assumed that the efficiency of removal of the neutral red by the clams is 100%.

Statistical Analysis

Each set of experimental data was subjected to ANOVA (analysis of variance) techniques using SAS-76 (Barr et al 1976). An Amdahl 370/6 computer on the Texas A&M University campus was used in constructing the ANOVA tables.

Results

Naphthalenes Uptake by Rangia cuneata

Figure 1 shows the pattern of naphthalenes uptake observed in the first Rangia experiment in which the clams were allowed only a 24 hr. acclimation period before exposure began. Uptake was generally quite variable within and between temperature-salinity groups, but certain consistent patterns were evident. Overall, the greatest uptake was measured in those clams exposed at 15°C. Intermediate values were obtained in the 20°C and 25°C exposed clams with the lowest tissue naphthalenes concentration being measured at 30°C. Within each temperature group, the lowest naphthalenes concentrations were consistently measured in those clams exposed at 30 o/oo. The highest concentrations were noted in those clams exposed at 15°C and 30°C in the 20 °/oo water and at 20°C and 25°C in the 0 °/oo water. Because of mortalities, depuration experiments were not conducted in this group.

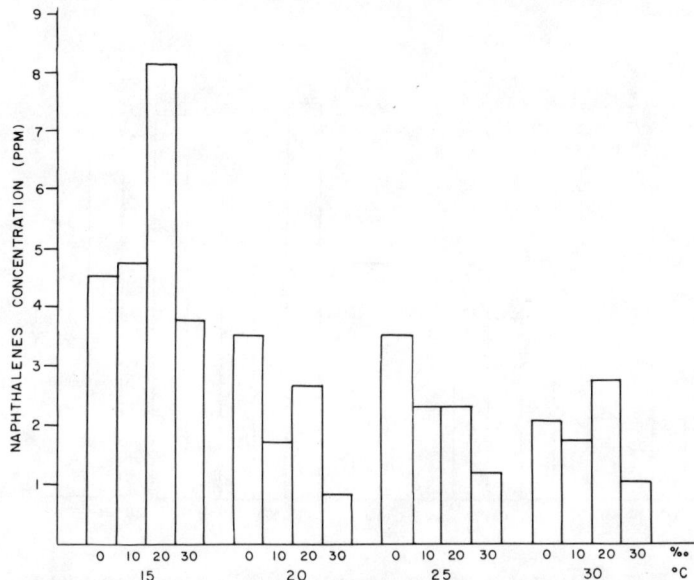

Fig. 31.1. Uptake of naphthalenes by clams exposed to a 25% WSF of So. Louisiana crude oil after a 24 hr. acclimation period to the different salinity-temperature regimes.

In the second experiment, uptake was again variable within and between temperature-salinity groups although trends were similar to those observed in the previous experiment. Figure 2 shows the uptake and depuration patterns of the clams exposed in this experiment. The greatest uptake was noted in those clams exposed at 15°C. Unlike the previous experiment, however, lowest values were measured in clams at 20°C and 25°C with intermediate levels being found at 30°C. In all groups, naphthalenes uptake was again lowest at the highest salinity (30 °/oo) and variable at the other salinities. Depuration was quite variable in Rangia cuneata as measured by naphthalenes concentrations in the clam tissues after 3 days in clean water. Levels ranging from 15-83% of the accumulated hydrocarbons remained in the clams at this time.

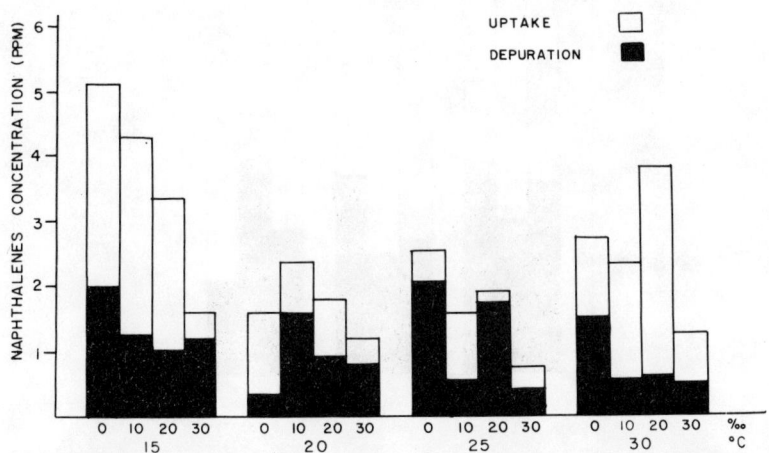

Fig. 31.2. Uptake and depuration of naphthalenes by clams exposed to a 25% WSF of So. La. crude oil. These clams were acclimated to the varying conditions over a 14 day period.

Filtration Rates

The results of the filtration experiments are presented in Fig. 3. Filtering rates showed a close linear relationship with temperature as the lowest rates were measured at 15°C and the highest at 30°C. Salinity appeared to have no effect on these rates.

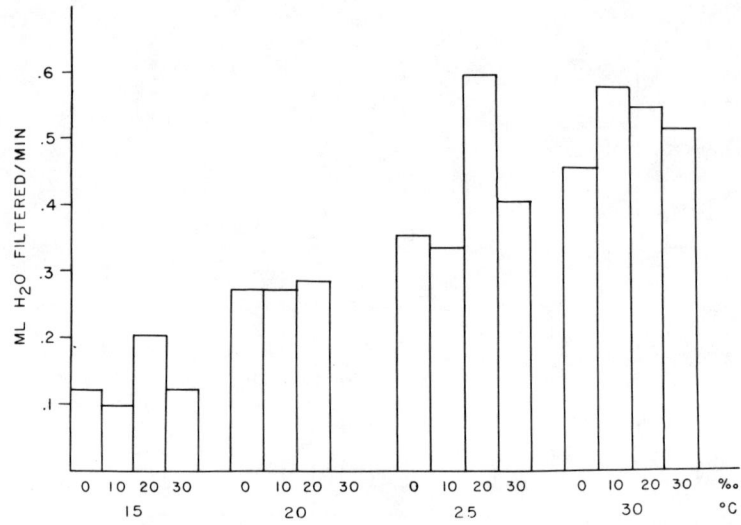

Fig. 31.3. Filtration rates of Rangia cuneata at various temperatures and salinities. Each point is the average rate of three clams measured over 19 hours. No measurements were made at 20°C, 30 °/oo because of mortalities.

Naphthalenes Uptake by Protothaca staminea

The results of the filtration experiments are presented in Fig. 3. Filtering rates of the So. Louisiana crude oil were much lower than those measured in the Rangia cuneata in the previous experiments (Fig. 4). Naphthalenes uptake in these clams was less than 2 ppm at all temperature and salinity combinations.

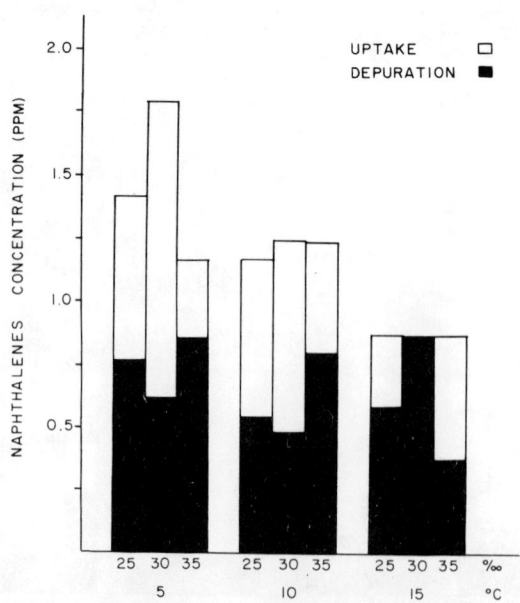

Fig. 31.4. Naphthalenes uptake and depuration by the clam, P. staminea, exposed for three days to a 25% WSF of So. La. crude oil. The depuration period lasted for 3 days.

As in earlier experiments with Rangia, naphthalenes uptake appeared to be somewhat higher at the lower temperature (5°C) with the least uptake measured at 15°C. Salinity appeared to have less of an effect than was observed in the Rangia uptake experiments. Protothaca staminea was also quite variable in its ability to depurate the accumulated hydrocarbons (Fig. 4). From 37-100% of the naphthalenes remained in the clam tissues after 3 days in clean seawater.

Statistical Analysis

In the first R. cuneata experiment, a very highly significant difference ($p<0.0001$) existed in the tissue naphthalenes concentrations in the different temperature-salinity groups. A similar level of significant differences was also calculated in the second Rangia experimental group ($p<0.0001$) and in the Protothaca uptake experiment ($p<0.0005$).

Those factors which contributed most to these differences are presented in Table 1. In the first Rangia experiment, only the linear and quadratic temperature terms met the criterion for significantly contributing to the uptake differences. The two temperature terms again proved to be the most significant sources of difference in naphthalenes concentrations in the second Rangia uptake experiment. However, slight differences were due to the quadratic salinity term as well as the temperature-salinity interaction. Protothaca uptake was influenced only by temperature.

TABLE 1. The Statistical Significant of Temperature and Salinity in Contributing to Naphthalenes Uptake in the Clams, R. cuneata and P. staminea

R. cuneata uptake experiment no. 1	
Source	F Value
Temperature	13.31***
Salinity	1.11
Salinity x Salinity	2.60**
Temperature x Temperature	9.78**
Temperature x Salinity	0.09

R. cuneata uptake experiment no. 2	
Source	F Value
Temperature	17.23***
Salinity	0.37
Salinity x Salinity	4.95*
Temperature x Temperature	14.18***
Temperature x Salinity	3.08*

P. staminea uptake	
Source	F Value
Temperature	21.02***
Salinity	0.33

*** very highly significant
** highly significant
* significant

ANOVA tables generated for depuration experiments showed that there were more significant differences in naphthalenes release due to temperature or salinity. These results were consistent in both the Rangia and Protothaca experiments.

Discussion

Temperature obviously had an effect on naphthalenes uptake in both the Rangia and Protothaca clams although explanations as to how temperature is affecting the uptake mechanisms remains speculative. It is unlikely that the higher tissue naphthalenes concentrations at the lower temperatures is a result of a longer naphthalenes residence time in the colder waters. Initial naphthalenes concentrations in the water and after 24 hours were determined in both the Rangia and Protothaca experiments. These results are presented in Table 2. Both initial and final concentrations were similar at all temperatures and salinities in both experiments suggesting a similar availability of naphthalenes to all of the clams.

TABLE 2. Naphthalenes Levels in Water After 24 hrs. The Original Mixture was a 25% WSF of So. Louisiana Crude Oil. Only Naphthalenes Levels in 30 $^o/oo$ Water were Determined in the Rangia Exposure

Rangia Exposure			
Temp. ^oC		Original Concentration (ppm)	Final Concentration (ppm)
15°	(a)	0.103	0.018
	(b)	0.101	0.016
20°	(a)	0.098	0.012
	(b)	0.101	0.011
25°	(a)	0.099	0.011
	(b)	0.098	0.009
30°	(a)	0.096	0.013
	(b)	0.098	0.012
Protothaca Exposure			
Temp. ^oC	Salinity (o/oo)	Original Concentration (ppm)	Final Concentration (ppm)
5°	25	0.14	0.06
	30	0.13	0.06
	35	0.12	0.05
10°	25	0.13	0.03
	30	0.14	0.04
	35	0.14	0.05
15°	25	0.17	0.05
	30	0.16	0.05
	35	0.14	0.04

Cunningham and Tripp (1975) have suggested that the increased metabolism that results with higher temperature could account for a more rapid turnover of all tissue constituents, including contaminants, in bivalves. However, this explanation is complicated by the fact that temperature in the present experiment did not significantly affect the release of naphthalenes by the clams. Furthermore, since there is no evidence that molluscs are able to metabolize hydrocarbons, it is likely that some other mechanism independent of temperature controls the release of hydrocarbons. Stegeman and Teal (1973) proposed that a portion of the accumulated hydrocarbons enters a stable tissue compartment where it is retained for an extended period. This stable fraction is then released very gradually. Di Salvo et al. (1975) and Neff (1975) support this view.

The patterns of uptake observed are also difficult to explain in view of the measured filtration rates in Rangia. Theoretically, much higher naphthalenes levels should have been expected in those clams exposed at the higher temperatures where filtration rates were greater. This would be especially true since depuration was shown to be independent of temperature. The fact that the lowest naphthalenes uptake was observed at the higher temperatures points out the necessity of further studying the mechanisms involved in the hydrocarbon uptake process.

In the second Rangia exposure, no mortalities were recorded in the clams during the two week acclimation period. However, one clam died in the 25°C, 30 °/oo group 24 hours after the addition of the WSF. Two other clam deaths were noted in the 30°C, 30 °/oo group 48 and 72 hours after the exposure to the WSF began. Although filatration rates were measured for Rangia, the clams were not exposed to the WSF at the time the measurements were made. Therefore, it cannot be ruled out that the high temperatures and salinities and the WSF exposures acted synergistically to decrease clam filtration rates at the higher temperatures and salinities. This could then account for the lower uptake in these groups of exposed Rangia. However, the Protothaca were, for the most part, constantly filtering during the exposure period so that this would not appear to explain the patterns of naphthalenes uptake observed in this species of clams.

Stegeman and Teal (1973) have indicated that the lipid content of the animal influences the level to which hydrocarbons are accumulated during the exposure period. Although no comparisons of lipid content in either the Rangia or the Protothaca were made, differences in lipid concentrations could account for the much higher naphthalenes uptake in the Rangia as compared to the Protothaca.

The present research suggests, then, that seasonal variations in naphthalenes uptake in bivalves exist. The magnitude of the seasonal effects on naphthalene uptake in the natural environment may be considerably different than that encountered in the present research since uptake was determined in clams acclimated to simulated conditions. In the natural environment, the tissue composition of the animals can be expected to be different from that found in lab acclimated clams and to vary seasonally.

References

Anderson, J. W., J. M. Neff, B. A. Cox, H. E. Tatem, and G. M. Hightower, The effects of oil on estuarine animals: Toxicity, uptake and depuration, respiration. In F. J. Vernberg and W. B. Vernberg (eds.), Pollution and Physiology of Marine Organisms, Academic Press, New York, 1974.

Barr, A. J., J. H. Goodnight, J. P. Sall, and J. T. Helwig, A User's Guide to SAS 76, SAS Institute, Raleigh, N. C., 1976.

Clark, R. C., Jr., and J. S. Finley, Uptake and loss of petroleum hydrocarbons by the mussel, Mytilus edulis, in laboratory experiments, Fish. Bull. 73, 508 (1975).

Cox, B. A., J. W. Anderson, and J. C. Parker, An experimental oil spill: The distribution of aromatic hydrocarbons in the water, sediment, and animal tissues within a shrimp pond, Proc. Joint Conference on Prevention and Control of Oil Pollution, American Petroleum Institute, Washington, D. C., 607 (1975).

Cunningham, P. A., and M. R. Tripp, Factors affecting the accumulation and removal of mercury from tissues of the American oyster Crassostrea virginica, Mar. Biol. 31, 311 (1975).

Di Salvo, L. H., H. E. Guard, and L. Hunter, Tissue hydrocarbon burden of mussels as potential monitor of environmental hydrocarbon insult, Environ. Sci. Technol. 9, 247 (1975).

Farrington, J. W., and J. G. Quinn, Petroleum hydrocarbons in Narragansett Bay. I. Survey of hydrocarbons in sediments and clams (Mercenaria mercenaria), Estuarine Coastal Mar. Sci. 1, 71 (1973).

Fossato, V. U., and W. J. Canzonier, Hydrocarbon uptake and loss by the mussel Mytilus edulis, Mar. Biol. 36, 243 (1976).

Frazier, J. M., The dynamics of metals in the American oyster, Crassostrea virginica. I. Seasonal effects, Chesapeake Sci. 16, 162 (1975).

Fucik, K. W., H. W. Armstrong, and J. M. Neff, Naphthalenes uptake by the clam, Rangia cuneata, in the vicinity of an oil separator platform in Trinity Bay, Texas, Proc. Joint Conference on Prevention and Control of Oil Pollution (1977, in press).

Hopkins, S. H., J. W. Anderson, and K. Horvath, The brackish water clam Rangia cuneata as indicator of ecological effects of salinity changes in coastal waters, A Report to the Office of the Chief of Engineers, Corps of Engineers, U. S. Army, 1972.

Huggett, R. J., M. E. Bender, and H. D. Slone, Utilizing metal concentration relationships in the eastern oyster (Crassostrea virginica) to detect heavy metal pollution, Water Res. 7, 451 (1973).

Kozloff, E. N., Seashore Life of Puget Sound, the Strait of Georgia, and the San Juan Archipelago, University of Washington Press, Seattle, 1973.

Lie, U., A quantitative study of benthic infauna in Puget Sound, Fiskeridir. Skr. Ser. Havunders. 14, 229 (1968).

Neff, J. M., and J. W. Anderson, An ultravi-let spectrophotometric method for the determination of naphthalene and alkylnaphthalenes in the tissues of oil-contaminated marine animals, Bull. Environ. Contam. Toxicol. 14, (1975).

Neff, J. M., B. A. Cox, D. Dixit, and J. W. Anderson, Accumulation and release of petroleum derived aromatic hydrocarbons by four species of marine animals. Mar. Biol. (1976, in press).

Pringle, B. H., D. E. Hissong, E. L. Katz, and S. T. Mulawka, Trace metal accumulation by estuarine molluscs, J. Sanit. Eng. Div. Proc. Am. Soc. Civ. Eng. 94, 455 (1968).

Shaw, D. G., A. J. Paul, L. M. Cheek, and H. M. Feder, Macoma balthica: An indicator of oil pollution, Mar. Pollut. Bull. 7, 29 (1976).

Slowey, F., and J. M. Neff, Research study to determine the availability of sediment adsorbed heavy metals to benthos with particular emphasis on deposit feeding infauna, Technical Report submitted to the U. S. Army Corps of Engineers, Vicksburg, Ms. (1976, in press).

Stegeman, J. J., and J. M. Teal, Accumulation, release and retention of petroleum hydrocarbons by the oyster Crassostrea virginica, Mar. Biol. 22, 37 (1973).

Ward, M. E., and E. Aiello, Water pumping, particle filtration, and neutral red absorption in the bivalve mollusc Mytilus edulis, Physiol. Zool. 46, 157 (1973).

Wieser, W., Free-Living Nematodes and Other Small Invertebrates of Puget Sound Beaches, University of Washington Press, Seattle, 1959.

CHAPTER 32

THE ACCUMULATION AND DEPURATION OF NO. 2 FUEL OIL BY THE
SOFT SHELL CLAM, Mya arenaria L.

Dennis Stainken

Oil & Hazardous Materials Spills Branch
Industrial Environmental Research Laboratory-Ci
U.S. Environmental Protection Agency
Edison, New Jersey 08817

Abstract

Young soft shell clams, Mya arenaria, were exposed to subacute concentrations of No. 2 fuel oil-in-water emulsions under simulated winter ($4^{o}C$) spill conditions. A pattern of accumulation and discharge of petroleum constituents, an experimental depuration time (biological half-life, TB_{50}), and a potential transport mechanism of aromatic compounds from the fuel oil to the clams were experimentally determined. The clams were exposed to single dose concentrations of 10, 50, and 100 ppm of No. 2 fuel oil-in-water emulsion for 28 days. Clams accumulated the greatest amount of hydrocarbons within one week after the initial exposure. The accumulated hydrocarbons decreased each week as the hydrocarbon content of the water decreased. Mass spectrometric analysis determined that the principle compounds accumulated and retained after three weeks of oil exposure were monomethyl, dimethyl, and trimethylnaphthalene isomers. Depletion of oil from the water column and accumulation and discharge of fuel oil constituents appeared to involve a mucus-oil complex formation by the clam.

The depuration period was determined when the clams were transferred to an uncontaminated system for 14 days subsequent to the 28 day oil exposure. Accumulated hydrocarbons were rapidly, although incompletely discharged. At the end of the depuration period, mass spectrometric analysis revealed that many of the hydrocarbons present in the clams were dimethyl and trimethylnaphthalene isomers. The biological half-lifes (TB_{50}) calculated were: 10 ppm (50 days); 50 ppm (11 days); 100 ppm (13.5 days).

Key words: Mya arenaria, oil accumulation, depuration, retention, No. 2 fuel oil, aromatic hydrocarbons, oil emulsion

Introduction

Chemical analyses have shown that various components of petroleum oils are accumulated within marine invertebrates. Blumer et al (1970), Ehrhardt (1972), Clark and Finley (1973) and Farrington and Quinn (1973) have reported the occurrence of petroleum oils in bivalve molluscs sampled from the environment. These molluscs were sampled from areas which had either been exposed to oil spills or from areas considered industrially contaminated. Laboratory studies of the accumulation of petroleum oils and oil fractions by mussels or oysters were investigated by Lee et al (1972), Blaylock et al (1973) and Stegeman and Teal (1973). Their work indicated that various fractions of petroleum oils can be accumulated and retained by mussels and oysters.

Oil spillage continues to be a serious problem in coastal and estuarine waters. Field studies of spills and their effects on Mya arenaria were documented by Thomas (1973) and Dow and Hurst (1975), and the occurrence of oil in Mya arenaria subsequent to environmental contamination was reported by Zitko (1971) and Scarratt and Zitko (1972).

Soft shell clams, Mya arenaria, occur frequently in areas receiving acute and chronic oil exposures and are often harvested for human consumption. This study was therefore performed to determine experimentally, a pattern of accumulation and discharge of petroleum constituents, a depuration time (biological half-life, TB_{50}), and a potential transport mechanism of aromatic constituents of the oil in soft shell clams exposed to No. 2 fuel oil-in-water emulsions. Few studies of oil accumulation and discharge have been performed at winter temperature conditions and experimentation was therefore performed at $4^{o}C$ to simulate the winter state during which time spills and discharges were more likely to occur resulting from winter storms.

A number two fuel oil was chosen for study because it is commonly shipped in coastal waters, used in coastal industrial installations, and has already been involved in a well documented spill (Blumer, et al, 1970). The oils were added in an emulsified form to simulate a potential naturally occurring condition. Forrester (1971), Gordon et al, (1973) and Kanter (1974) have reported the formation of oil emulsions in sea water by various mechanisms. In the event of an oil spill during the colder months, it is probable that much of the oil would be dispersed and emulsified in the water column through turbulent wave action. Emulsions tend to be relatively stable and a mechanism for the accumulation of emulsified oil by soft shell clams has been reported by Stainken (1975).

Materials and Methods

The No. 2 fuel oil was supplied by the U.S. Environmental Protection Agency, Industrial Environmental Research Laboratory-Ci, Edison, N.J. The specific gravity of the oil was 2.40 centistokes. The oil was composed of 14% aromatics and 86% nonaromatics according to ASTM method No. D2549-68. Oil-in-water emulsions were ultrasonically prepared according to a procedure developed by Gruenfeld and Frederick (1977).

Clams for the experiments were collected from Sequine Point, Staten Island, N.Y. Young clams with a mean shell length of 25 mm (Newcombe, 1936; Pfitzenmeyer, 1965) were utilized because young bivalves tend to have greater respiration and filtration rates than those of older bivalves (Prosser and Brown, 1961; Walne, 1972).

An exposure period of 28 days to No. 2 fuel oil emulsions having concentrations of 10, 50, and 100 ppm was utilized. Four 20 gallon covered aquaria containing 60 liters of filtered seawater per aquaria (Salinity = 20%) were employed. The seawater was collected from Sandy Hook Bay and filtered through a coarse plankton net to remove macro debris. One aquarium served as a control. Each of the remaining aquaria received a sufficient volume of a stock oil emulsion to attain an initial concentration of either 10, 50 or 100 ppm of emulsified oil termed Time 0. The water was continuously aerated and the temperature was maintained at $4^{\circ}C$. The clams were acclimated to the experimental conditions for a duration of 6 days before the stock emulsions were added. Sampling for hydrocarbon contents of water and clams was performed every 7 days after Time 0. The controls were sampled at Time 0 and every 7 days subsequent to Time 0.

Chemical confirmation of the oil being bound or adsorbed to mucus was provided by extracting mucus secreted by the clams exposed to 100 ppm emulsified No. 2 oil. The mucus was collected after the four weeks oil exposure. It had a gray-green flocculent appearance and formed flocculent clumps on the surface and in the water column. The methodology used for hydrocarbon analysis has been described by Stainken (1975).

The depuration of the clams was determined for 14 days after the 28 day oil exposure. Each experimental group of clams was removed from their aquaria and rapidly placed in clean 20 gallon aquaria containing 60 liters of fresh coarse filtered Sandy Hook Bay sea water maintained at the same temperature and salinity. As clams were transferred, the valves were wiped clean to remove deposited pseudofaeces, etc. Samples for chemical analysis were obtained on day 7 and 14.

The hydrocarbon contents of the Sandy Hook Bay sea water and the experimental aquaria water were determined using a Freon extraction technique (Gruenfeld, 1973; Zeller, 1974). Each week, 400 ml samples were removed from the center of each tank and at the same depth. Prior to the experiments, the hydrocarbon content of the sea water was analyzed to ensure external contamination did not occur.

The extraction and column chromatograph method used for chemical analysis of tissue was a modification of that employed by Blaylock, et al, (1973). All solvents were of distilled spectral grade (Burdick & Jackson, Muskegon, Mich.). All glassware was prerinsed with hexane. Once a week, 5 clams from each experimental exposure were shucked and the water from the mantle cavities was drained. Soft tissues (5-8 gm tissue weight) were placed in a round bottom flask (500 ml) having a standard taper ground glass neck, and mixed with 95% ethanol (150 ml) and KOH (10 gm). The shells were not extracted. The mixtures were then refluxed at approximately $65-70^{\circ}C$ on a heating mantle and under a Friedrich reflux condenser for 1½ hours. After cooling to ambient temperatures, the interior of each condenser was washed with 5 ml hexane. The washing was collected in the receiver flask.

The digested material was transferred to a Teflon stoppered separatory funnel (1 liter) using distilled water (80 ml) and two portions (50 ml each) of hexane. The mixture was then shaken by hand for one minute. The two phases were then drained into separate flasks and the aqueous phase was returned to the separatory funnel using a hexane (50 ml) wash. The extraction and separation was repeated for a total of three times.

The combined extracts were washed (minimum 3 times) with aliquots (500 ml each) of distilled water and transferred to an Erlenmeyer flask (300 ml) using additional hexane. Water was removed from the extract by adding 2-3 gm of anhydrous Na_2SO_4. The extracts were then concentrated to approximately 40 ml under a stream of filtered nitrogen in a warm water bath (50°C) and transferred to a conical flask (50 ml) where concentration was continued to a final volume of 2 ml.

The concentrated extract was placed on a glass column (25 cm length x 1.0 cm I.D.). The column was packed with 6.5 cm of 5% water deactivated silica gel (100-200 mesh, Grade 923, Davison Chemical, Baltimore, Md.). Another 6.5 cm of 5% water deactivated alumina (Neutral, 90-20 mesh, Brockman Activity 1: Fisher Chemical Co.) was layered over the silica. Both the silica gel and alumina were activated at 110°C for 24 hours and then deactivited prior to column packing. Sand was layered over the alumina (0.5 cm). When packed, the column was rinsed with one column volume of hexane.

The residual concentrate was charged onto the column and then elutriated with 50 ml hexane. Subsequent analysis demonstrated that all petroleum hydrocarbons in the sample had passed through. The column was rinsed with 50 ml of benzene but few hydrocarbons were found in this fraction and it was later discarded. The eluate was concentrated for gas chromatographic analysis to a volume of 0.1 ml under nitrogen.

The identification of sample hydrocarbons was achieved by comparison with gas chromatograms of hydrocarbon standards and of No. 2 fuel oil in hexane. A Perkin-Elmer (Model 900) gas chromatograph was employed, equipped with a flame ionization detector and a 6 ft stainless steel column packed with 8% Dexsil on 80/100 mesh of Chromosorb W. Standard operating conditions were as follows: carrier gas: N_2, 2 cc/min; detector: H_2, 20 cc/min; air: 40 cc/min; injector temperature: 200°C; detector temperature: 300°C; manifold temperature: 300°C; column temperature: 70°C with a two minute hold, and programmed at an increase of 8°C per minute to 300°C; injection sample: 0.1 microliter. Further identifications of compounds were accomplished by mass spectrometric analysis provided by the U.S. EPA (IERL-Ci), Edison, N.J.

Hydrocarbons were quantified by cutting and weighing the chromatograms. The total areas under the peaks were obtained by subtraction of the baseline area of the first week control measurement. It was assumed that the area under the first week control measurement would be most representative of background levels and therefore any material above this baseline were considered accumulated petroleum hydrocarbons. The sample peak areas were compared to that of an internal standard (octacosane, C_{28}) having a known concentration (2 µg/gm clam tissue).

Hydrocarbon quantification results were analyzed for regression. The regression analysis used the No. 1987A/ST3 package from the 700 series Standard Statistical Program, Wang Laboratories, on a taped program. The significance of the regressions were further analyzed by a t-test (Zar, 1974).

Results

Water quality appeared excellent throughout the 28 day period. The dissolved oxygen and ph remained optimum in all tanks and bacterial contamination was not apparent. Control animals were reactive to tactile stimuli and appeared to be feeding. Very little mortality occurred at all oil concentrations. The actual mortalities were: Controls - 0.0%, 10 ppm - 0.5%, 50 ppm - 2.0%, 100 ppm - 3.0%.

During the 14 day depuration period, the 10 ppm clams seemed to recover their tactile irritability, in comparison to the controls. The 50 and 100 ppm clams were sluggish and after shucking many smelled foul. Their visceral mass was less firm than controls and their color had paled from the normal gold-brown. Mortalities occurred in the clams previously exposed to the concentration of 50 and 100 ppm. The actual mortality at the end of the depuration period was: Control 0.0%, 10 ppm - 0.0%, 50 ppm - 18.3% and 100 ppm - 13.6%.

The sea water before all tests and the controls during the tests did not have detectable hydrocarbon contents. Within the oiled tanks, some of the oil emulsion had broken within two hours after addition. Much of the oil emulsion and water soluble fraction remained in the water column as indicated in Table 1. One week later, the detectable hydrocarbon concentrations in the water column of the oiled tanks were similar. Much of the oil was bound in mucus which adhered to the glass cooling coils and tank walls or formed floating organic flocculent conglomerates. The IR spectra also showed the emergence of an apparent aromatic band at 3030 cm^{-1}. The hydrocarbon peaks at 2930 cm^{-1} often masks the aromatics. Apparently, the water soluble aromatics of the fuel oil were becoming unmasked as more of the non-aromatic hydrocarbons became dissipated from the water column. The Week 2 data were similar to those of Week 1 and a peak emerged at ca. 3000 cm^{-1}.

TABLE 1. Hydrocarbon concentration (ppm) in the water column during the 28 day exposure period

Aquaria	Time 0*	Week 1	Week 2	Week 3	Week 4
Control	0	0	0	0	0
10 ppm	4.5	1.31	0.56	0.37	0
50 ppm	43.7	1.04	0.71	0.37	0.29
100 ppm	60.7	1.52	0.78	0.32	0.47

*The Time 0 sample measurement was made two hours after addition of the emulsified oil.

By Week 3, the measurable hydrocarbon content had dropped further and in the 10 and 50 ppm tank samples the possible aromatic peak became less apparent. The 100 ppm tank sample still had a strong peak at 3030 cm^{-1}. At the end of the fourth week, the hydrocarbon content in the tanks had fallen to almost trace concentrations and peaks were becoming indistinguishable.

The tanks used in the oil exposure experiments were drained after the fourth week. There was an extremely strong odor of No. 2 fuel oil, especially in the 100 ppm tank. As the mucus was wiped off the sides and cooling oils, the odor became readily apparent in all the oiled tanks. This indicates that much of the oil was bound by the mucus which had settled on the cooling coils and sides of the tanks. The mucus-oil complex formation may have aided in the dissipation of hydrocarbons from the water column. The mechanism of the binding of oil in mucus by the clams has been documented (Stainken, 1975). Mass spectrometric analysis demonstrated that the mucus contained predominantly dimethyl and trimethyl-naphthalenes with paraffins, mainly in the C-14 and C-15 regions. Figure 1 illustrates the gas chromatograms of No. 2 fuel oil and an extract of hydrocarbons from a mucus sample. A close correlation of peak heights is readily noticeable. The chromatogram pattern of the mucus is similar to No. 2 fuel oil but some of the lower boiling compounds are missing.

During the depuration period, measurable amounts of hydrocarbons in the aquaria water were not detected. There was an emergence of a peak at 3000 cm^{-1} during the second week in the experimental tanks.

Gas chromatograms for the control clams are contained in Fig. 2. The basic pattern for the normal control clams remained unchanged during the experiment. The chromatogram for Week 1 was chosen as a baseline for the exposed animals to compare the amount of No. 2 fuel oil hydrocarbons accumulated. The chromatograms for the clams exposed to 10 ppm No. 2 fuel oil are illustrated in Fig. 3. The chromatograms A-C show a characteristic well defined pattern of peaks. Mass spectrometric analysis of the third weeks sample (chromatogram C) revealed that the bulk of the chromatogram consisted of monomethyl, dimethyl and trimethylnaphthalenes. By the fourth week, the resolution of the peaks had diminished and the chromatograms of the clams placed in fresh sea water would superficially indicate there were few hydrocarbons present. However, mass spectrometric analysis revealed that the sample (Chromatogram F, Fig. 3) contained dimethyl and trimethylnaphthalenes at the end of the depuration period.

Figure 4 depicts the chromatograms from the clams exposed to 50 ppm of No. 2 fuel oil emulsion. The same pattern of peaks as those of the 10 ppm is present. The peaks have much higher resolution and are still apparent even after the depuration period. A higher attenuation (10x64) had to be used because the samples contained a much higher hydrocarbon concentration. Mass spectrometric analysis of the Week 3 sample revealed that most of the sample was comprised of monomethyl, dimethyl and trimethylnaphthalenes. Mass spectrometric analysis of the clam sample at the end of depuration (Chromatogram F, Fig. 4) showed that the sample was composed of alkyl benzenes, monomethyl, dimethyl and trimethyl-naphthalenes. The chromatogram of Week 1 control was drawn under chromatograms E and F, Fig. 4, to illustrate the quantity of petroleum hydrocarbons retained by the clams.

Chromatograms of clams exposed to a concentration of 100 ppm No. 2 fuel oil are illustrated in Fig. 5. The same recurrent pattern is readily visible and the peaks have sharper resolution. Mass spectrometric analysis of the week 3 sample showed it to consist of naphthalene, monomethylnaphthalene, a mixture of dimethyl and trimethylnaphthalene isomers, phenanthrene, and C_{12} and C_{14} paraffins. Mass spectrometric analysis of the clam sample after the two week depuration (Chromatogram F, Fig. 5) confirmed that the sample consisted of naphthalene, a mixture of dimethyl and trimethylnaphthalene isomers and no paraffins.

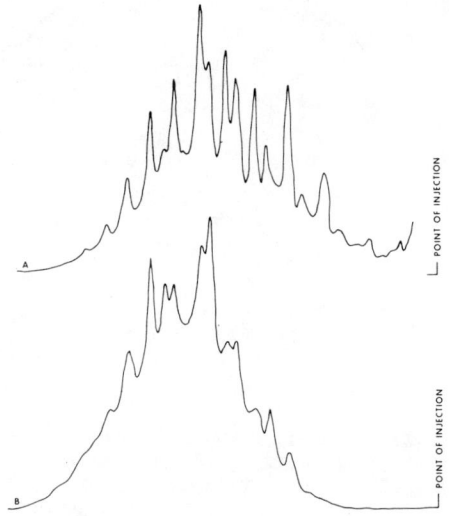

Fig. 32.1 Gas chromatograms
(A) 8 ppm #2 FUEL OIL IN HEXANE, ATTENUATION 1x128
(B) CARBON TETRACHLORIDE EXTRACT OF CLAM MUCUS, ATTENUATION 10x1024

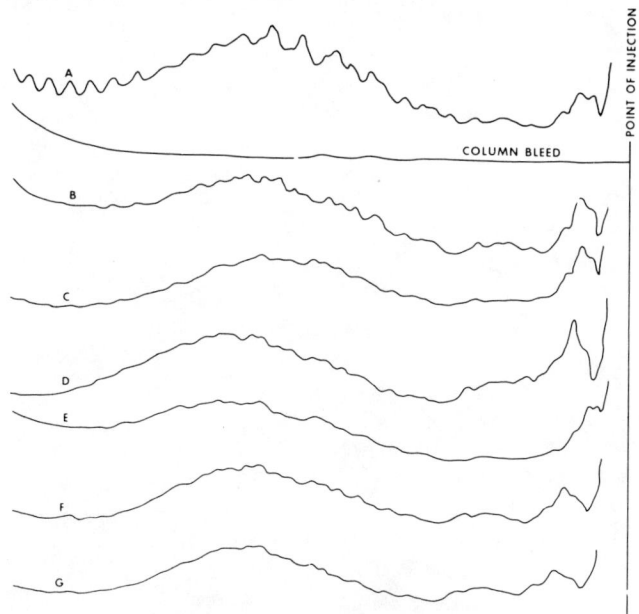

Fig. 32.2. Control animal gas chromatograms.
(A) TIME 0, WITH COLUMN BLEED DRAWN BELOW; (B) WEEK 1; (C) WEEK 2; (D) WEEK 3;
(E) WEEK 4; (F) CLAMS IN FRESH SEA WATER, WEEK 1; (G) CLAMS IN FRESH SEA WATER, WEEK 2
ALL CHROMATOGRAMS AT 1x128 ATTENUATION

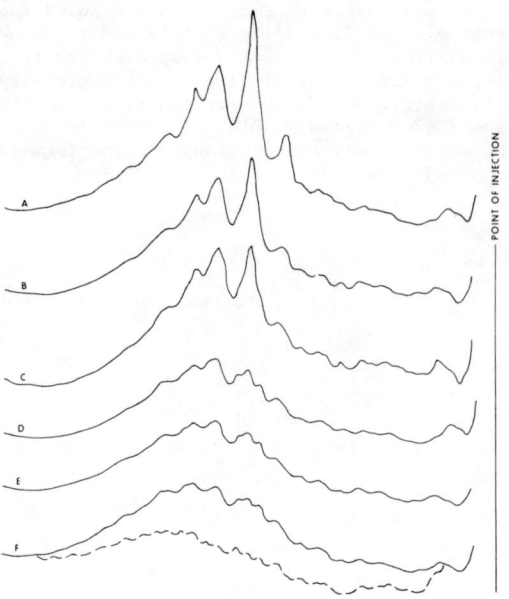

Fig. 32.3. Gas chromatograms clams exposed to 10 ppm #2 Fuel Oil.
(A) WEEK 1; (B) WEEK 2; (C) WEEK 3; (D) WEEK 4; (E) CLAMS PLACED IN FRESH SEA WATER, WEEK 1;
(F) CLAMS PLACED IN FRESH SEA WATER, WEEK 2, WITH CONTROL WEEK 1 DRAWN AS BASE LINE
ALL CHROMATOGRAMS AT 1x128 ATTENUATION

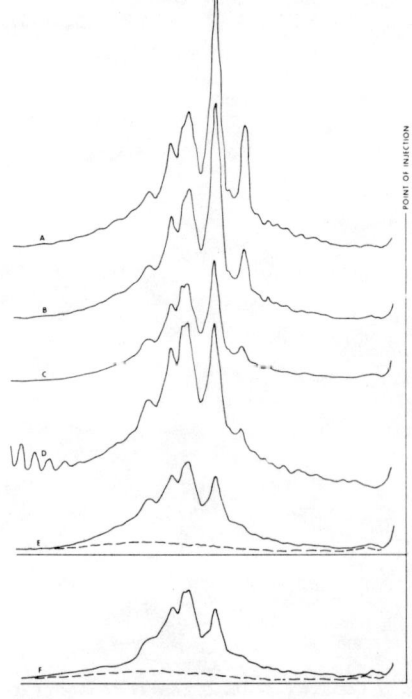

Fig. 32.4. Gas chromatograms clams exposed to 50 ppm #2 Fuel Oil.
(A) WEEK 1; (B) WEEK 2; (C) WEEK 3; (D) WEEK 4; (E) CLAMS PLACED IN FRESH SEA WATER,
WEEK 1, WITH CONTROL WEEK 1 AS BASELINE; (F) CLAMS IN FRESH SEA WATER, WEEK 2,
WITH CONTROL WEEK 1 AS BASELINE
ALL CHROMATOGRAMS AT 10x64 ATTENUATION

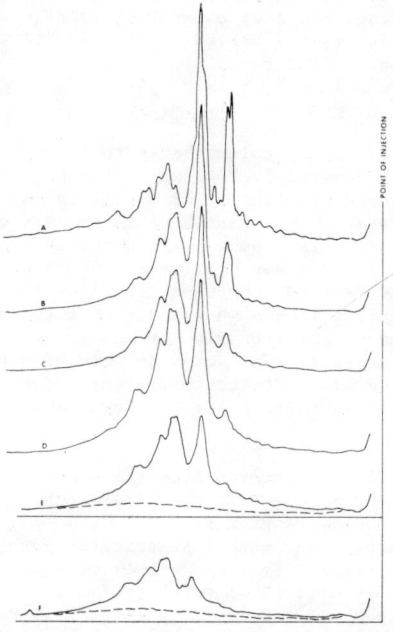

Fig. 32.5. Gas chromatograms clams exposed to 100 ppm #2 Fuel Oil.
(A) WEEK 1; (B) WEEK 2; (C) WEEK 3; (D) WEEK 4; (E) CLAMS PLACED IN FRESH SEA WATER, WEEK 1, WITH CONTROL WEEK 1 AS BASELINE (F) CLAMS IN FRESH SEA WATER, WEEK 2, CONTROL WEEK 1 AS BASELINE
ALL CHROMATOGRAMS AT 10x64 ATTENUATION

The results of quantification of the gas chromatograms are listed in Table 2. Results are expressed as micrograms of petroleum hydrocarbons/gram clam sample. After one weeks exposure to oil, the values diminish with time. The sudden upsurges in Week 3 at 10 ppm and in Week 4, of the 50 and 100 ppm are unexplained.

Experimental concentration factors were calculated for the first week of exposure from the data in Table 1 and 2. The hydrocarbon concentration of the water column, Week 1, was subtracted from the Time 0 value. This net hydrocarbon concentration in the water column was then divided into the hydrocarbon value accumulated by the clam after the first week of oil exposure. The computed concentration factors were: 10 ppm - 8.5, 50 ppm - 3.3, 100 ppm - 1.8.

Table 2. Quantification of petroleum hydrocarbons (micrograms per gram of clam tissue).

	10 ppm	50 ppm	100 ppm
Week 1	27.3	143.1	109.3
Week 2	22.9	120.5	99.2
Week 3	31.6	97.0	90.7
Week 4	17.3	146.8	117.5
Week 5 (depuration)	12.9	66.8	83.6
Week 6 (depuration)	14.9	55.3	57.0

The biological half lifes (TB_{50}) or residence times were calculated using a regression analysis of the depuration period and plotting on semilog paper. The fourth weeks data were calculated as Time 0 for the depuration period. The calculated TB_{50} were: 10 ppm - 50 days; 50 ppm - 11 days; 100 ppm - 13.5 days.

Discussion

The concentrations of oil in the water column began to decrease several hours after the addition of the oil emulsion. Several factors were probably responsible for this. Much of the oil was apparently removed from the water column by the mucociliary feeding and ejection mechanisms of the clams. Large masses of mucus were ejected from the clams and were accumulated on the cooling coils. Subsequent chemical analysis revealed a large content of oil in the mucus. It is presumed that as the clams filtered water, they were exposed primarily to naphthalene derived compounds. Earlier investigators reported that fuel oil can form droplets and micelles in sea water, and that the major water soluble components were the aromatic compounds, particularly the naphthalenes (Boylan & Tripp, 1971; Boehm & Quinn, 1974). The accumulation of large amounts of oil and aromatic molecules in the mucus may be a potential mechanism of concentrating and disseminating petroleum hydrocarbons in the environment. The mechanism for the formation of an oil-mucus complex was discussed by Stainken (1975).

After Time zero, the oil rapidly disappeared from the water column. The amount of dissolved compounds attained a peak measured in the first week, and then decreased to an equilibrium value for several weeks ranging from 0.5-0.29 ppm. Analysis of mucus and clam tissue indicated that naphthalenes and methyl substituted naphthalenes were the apparent predominant components in the water. This is further substantiated by other reports (Frankenfeld, 1973; Anderson, et al., 1974) which indicate that the water soluble fractions of oils often contain aromatic compounds, particularly naphthalenes.

The peak accumulation of petroleum derived hydrocarbons by the clams was reached one week after exposure, followed by a gradual loss in accumulated hydrocarbons. The bulk of the material accumulated were naphthalenes and methyl substituted naphthalene isomers. Other aromatics were detected and accumulated but the naphthalenes were the most persistent. The results of this study are comparable to those of Vaughan (1973). He reported that oysters exposed to No. 2 fuel oil often accumulated methyl and dimethyl naphthalenes. Other studies also indicate that aromatic hydrocarbons may be accumulated by bivalves (Ehrhardt, 1972; DiSalvo, et al, 1973.

Paraffins or aliphatics were not detected in clam tissue in this study, except in the higher oil exposures. The aromatic compounds constituted a much greater percentage of the total oil extracted from the clams. The extracted concentrations of aromatics from the clams were also much greater than the original diluted concentrations when added at Time 0 as part of the fuel oil emulsion. Similar results of Stegeman and Teal (1973) demonstrate that a high aromatic content in the organism need not be the result of a high aromatic content in the contaminating oil. This phenomena is probably due to the greater solubility of the aromatic compounds (Anderson, et al., 1974; Boehm and Quinn, 1974).

The greater uptake of hydrocarbons by clams exposed to 50 ppm may be due to a dose dependent narcosis. Galtsoff, et al., (1935) reported that water soluble substances of crude oil could produce anaesthetic effects on the ciliated epithelium of the gills. The 100 ppm oil concentration may have reduced the filtration activity of the clams below that of the clams exposed to 50 ppm. The reduced filtration rate may account for the animals exposed to the higher concentration (100 ppm) accumulating less oil than the animals exposed to the lower concentration (50 ppm).

Depuration of accumulated petroleum derived hydrocarbons actually began as the hydrocarbons decreased in the water column. When the clams were removed to fresh sea water, depuration proceeded rapidly for the first week. The calculated half-time for depuration ranged from 11-50 days. The retention of hydrocarbons at the end of the depuration period indicated that after oil exposure, depuration will proceed rapidly to a low concentration after which depuration proceeds slowly. Other investigators have reported similar findings in other bivalve species (Lee, et al., 1972; Stegeman & Teal, 1973; Vaughan, 1973). However, the actual retention of oil by Mya arenaria in this study is further substantiated by the observations of Thomas (1973) in which clams were observed to eject oil for at least one month after an oil spill.

Oil retention can have several deleterious effects. Long term low level petroleum hydrocarbon contamination may interfere or weaken the ability of the clam to withstand further environmental stresses such as those of temperature, salinity, spawning, disease and insult from other contaminants. Furthermore, oil retained in the sediment and clams may also be passed on to predators, including man.

Summary

1. Clams accumulate large amounts of petroleum derived hydrocarbons after initial exposure to petroleum (No. 2 fuel oil). The amount and the type of petroleum hydrocarbon accumulated are related to the initial dose, time following initial exposure and the relative "solubilities" of the various hydrocarbon components. Naphthalene and methyl substituted naphthalene isomers (mono, di, tri) are the most common compounds concentrated by the clams. Occasionally, some alkyl benzenes, phenanthrenes and paraffins in the C-12 and C-14 region are accumulated.

Depuration by soft-shell clams of accumulated petroleum hydrocarbons may actually begin as soon as hydrocarbon levels drop in the water column. When clams are placed in fresh hydrocarbon free sea water, depuration proceeds rapidly within the first week after which the rate slows. Dimethyl and trimethylnapthalenes were still present in the clams after two weeks in fresh sea water.

2. Chemical analysis of clam mucus extract demonstrated that clams can concentrate and ingest or release to the environment extremely large amounts of petroleum hydrocarbons by means of their mucus - oil binding mechanism. Analysis of petroleum compounds present in the mucus revealed them to be methyl substituted naphthalenes and paraffins, all compounds similar to those which the clams accumulate. Types of chemical compounds accumulated or bound in the mucus are probably related to differential solubilities of petroleum constituents.

References

Anderson, J. W., J. M. Neff, B. A. Cox, H. E. Tatem, and G. M. Hightower., Characteristics of dispersions and water-soluble extracts of crude and refined oils and their toxicity to estuarine crustaceans and fish. Mar. Biol. 27, 75-88 (1974).

Blaylock, J. W., P. W. O'Keefe, J. N. Roehm, and R. E. Wildung., Determination of n-alkane amd methylnaphthalene compounds in shellfish. Proc. of Joint Conf. on Preyent. & Contr. of Oil Spills, Wash., D.C., A.P.I., 173-178 (1973).

Blumer, M., J. Sass, G. Souza, H. Sanders, F. Grassle, and G. Hampson. (1970) The West Falmouth Oil Spill. Tech. Rept. No. 70-44, Woods Hole Oceanographic Institution.

Boehm, P. D. and J. G. Quinn., The solubility behavior of No. 2 fuel oil in sea water. Mar. Pollut. Bull. 5(7), 101-105 (1974).

Boylan, D. B. and B. W. Tripp., Determination of hydrocarbons in sea water extracts of crude oil and crude oil fractions. Nature 230, 44-47 (1971).

Clark, R. C. Jr., and J. S. Finley., Paraffin hydrocarbon patterns in petroleum polluted mussels. Mar. Pollut. Bull. 4(11), 172-176 (1973).

DiSalvo, L. H., H. E. Guard, L. Hunter, and B. A. Cobet., Hydrocarbons of suspected pollution origin in aquatic organisms of San Francisco Bay: Methods and preliminary results. In: The Microbial Degradation of Oil Pollutants, La. State Univ. Public. No. LSU-SG-73-01, 205-220 (1973).

Dow, R. L., and J. W. Hurst, Jr., The ecological, chemical and histopathological evaluation of an oil spill site. Part 1. Ecological Studies. Mar. Pollut. Bull. 6(11), 164-166 (1975).

Ehrhardt, M., Petroleum hydrocarbons in oysters from Galveston Bay. Environ. Pollut. 3, 257-271 (1972).

Forrester, W. D., Distribution of suspended oil particles following the grounding of the tanker Arrow. J. Mar. Res. 29(2), 151-170 (1971).

Frankenfeld, J. W., Weathering of oil at sea. Final Rept. No. CG-D-7-75, Dept. of Transportation, U.S.C.G., Off. of R & D, Wash., D.C. (1973).

Galtsoff, P. S., H. F. Prytherch, R. O. Smith, and J. Koehring., Effects of crude oil pollution on oysters in Louisiana waters. Bull Bur. Fish. Wash., D.C. 18, 143-210 (1935).

Gordon, D. C., Jr., P. D. Keizer, and N. J. Prouse., Laboratory studies of the accomodation of some crude and residual fuel oils in sea water. J. Fish. Res. Bd. Can. 30, 1611-1618 (1973).

Gruenfeld, M., Extraction of dispersed oils from water for quantitative analysis by Infrared Spectrophotometry. Environ. Sci. & Tech. 7(7), 636-639 (1973).

Gruenfeld, M. and R. Frederick., The ultrasonic dispersion, source identification, and quantitative analysis of petroleum oils in water. Rapp. P.-v. Reun. Con. Int. Explor. Mer. 171, 33-38 (1977).

Kanter, R., Susceptability to crude oil with respect to size, season and geographic location in Mytilus californianus (Bivalvia). U. So. Cal., Sea Grant Prog. Publ. No. USC-SG-4-74, (1974).

Lee, R. F., R. Saurheber, and A. A. Benson., Petroleum hydrocarbons: uptake and discharge by the marine mussel Mytilus edulis. Science 177, 344-346 (1972).

Newcombe, C. L., Validity of concentric rings of Mya arenaria for determining age. Nature 137(3457), 191-192 (1936).

Pfitzenmeyer, H. T., Annual cycle of gametogenesis of the soft shell clam, Mya arenaria, at Solomon, Maryland. Chesap. Sci. 6(1), 52-59 (1965).

Prosser, C. L. and F. A. Brown, Jr. (1961) Comparative Animal Physiology. 2nd ed., W. B. Saunders, Phila.

Scarratt, D. J. and V. Zitko., Bunker C oil in sediments and benthic animals from shallow depths in Chedabucto Bay, N.S. J. Fish. Res. Bd. Can. 29, 1347-1350 (1972).

Stainken, D. M., Preliminary observations on the mode of accumulation of ho. 2 fuel oil by the soft shell clam, Mya arenaria. Proc. Conf. on Prevent. & Contr. Oil Pollution, San Francisco, Calif., A.P.I., 463-468 (1975).

Stegeman, J. J. and J. M. Teal., Accumulation, release and retention of petroleum hydrocarbons by the oyster, Crassostrea virginica. Mar. Biol. 22(1), 37-44 (1973).

Thomas, M. L., Effects of Bunker C oil on interstitial life and lagoonal biota in Chedabucto Bay, Nova Scotia. J. Fish. Res. Bd. Can. 30, 83-90 (1973).

Vaughan, B. E. (ed.)., Effects of oil and chemically dispersed oil on selected marine biota - a laboratory study. A.P.I. Pub. No. 4191, (1973).

Walne, P. R., The influence of current speed, body size and water temperature on the filtration rate of five species of bivalves. J. Mar. Biol. Assoc., U.K. 52, 345-374 (1972).

Zar, J. J. (1974)., Biostatistical Analysis. Prentice Hall, Englewood Cliffs, N.J.

Zeller, M. V., Oil in water: use of non-toxic solvent and importance of acidification. Perkin Elmer, Infrared Bull. 41 (1974).

Zitko, V., Determination of residual fuel oil contamination of aquatic animals. Bull. Environ. Contam. & Toxicol. 5(6), 559-564 (1971).

CHAPTER 33

EFFECTS OF CHLORINATED BIPHENYLS AND PETROLEUM HYDROCARBONS
ON THE ACTIVITY OF HEPATIC ARYL HYDROCARBON HYDROXYLASE
OF COHO SALMON (*Oncorhynchus kisutch*) AND CHINOOK SALMON (*O. tshawytscha*)

Edward H. Gruger, Jr., Marleen M. Wekell, and Paul A. Robisch

Northwest and Alaska Fisheries Center
National Marine Fisheries Service
National Oceanic and Atmospheric Administration
2725 Montlake Boulevard East
Seattle, Washington 98112 U.S.A.

Abstract

Saltwater-adapted coho (*Oncorhynchus kisutch*) and chinook (*O. tshawytscha*) salmon were fed two mixtures of test compounds: one composed of chlorobiphenyls and the other of petroleum hydrocarbons. The chlorobiphenyl mixture consisted of biphenyl, 2-chloro-, 2,2'- and 2,4'-dichloro-, 2,5,2'-trichloro-, and 2,5,2',5'-tetrachlorobiphenyl. The hydrocarbon mixture consisted of *n*-pentadecane, 2,3-benzothiophene, 2,6-dimethyl- and 2,3,6-trimethylnaphthalene, fluorene, phenanthrene, 1-phenyldodecane, and heptadecylcyclohexane. The mixtures were fed separately in Oregon moist pellets at concentrations of 1 ppm and 5 ppm and together at 2 ppm and 10 ppm. Induction of aryl hydrocarbon (benzo[a]pyrene) hydroxylase activity occurred in hepatic microsomes from coho salmon within the first two weeks of hydrocarbon exposure. This induction was potentiated by the presence of chlorobiphenyls; however, no effect on enzyme activity was found with chlorobiphenyls alone. The results indicate that chlorobiphenyls act synergistically with hydrocarbons to induce the enzyme system in coho salmon. In chinook salmon, the activity of aryl hydrocarbon hydroxylase in hepatic microsomes was depressed by the chlorobiphenyls and hydrocarbons, administered both separately and together. The results provide evidence indicating that benzo[a]pyrene hydroxylase responds differently in different species of salmonids to chlorobiphenyls and petroleum hydrocarbons in food. It is concluded that chlorobiphenyls in marine environments may alter the activity of hydroxylases regulating the accumulation and discharge of petroleum hydrocarbons in organisms.

Key words: PCB, Aryl hydrocarbon hydroxylase, petroleum, salmon, chlorobiphenyl, synergism

Introduction

Recent studies have shown that petroleum hydrocarbons cause increases in the activity of microsomal drug metabolizing enzymes, such as aryl hydrocarbon hydroxylase (AHH), in marine organisms (Pedersen et al 1974, Payne and Penrose 1975). Increased AHH activity in fishes has been associated with petroleum contamination of aquatic environments (Payne 1976, Gruger et al 1977), and measurement of this enzyme system was suggested as a useful index of recent or long-term petroleum exposures (Payne and Penrose 1975). Recently, PCBs were found to induce this enzyme system in freshwater rainbow trout (*Salmo gairdnerii*) (Lidman et al 1976) and in freshwater coho salmon (*Oncorhynchus kisutch*) (Gruger et al 1976, Gruger et al 1977). In addition, the hepatic AHH was induced by a mixture of PCBs and petroleum fed to the coho salmon; however, no induction occurred when only petroleum was fed (Gruger et al 1977).

Comparable studies on the combined effects of dietary petroleum and PCBs on the induction of AHH in saltwater-adapted fishes are important because the ubiquitous chlorinated biphenyls are likely to coexist with petroleum hydrocarbons in contaminated marine environments (Gutsell 1921, Holcomb 1969, Risebrough et al 1968, Stalling and Mayer 1972, Ocean Affairs Board 1975). Accordingly, the present study was initiated to determine the influences of dietary chlorobiphenyls and petroleum hydrocarbons on the hepatic microsomal AHH of saltwater-adapted coho and chinook salmon (*O. tshawytscha*).

Significant differences in AHH activities were found between coho and chinook salmon in relation to whether chlorinated biphenyls or petroleum hydrocarbons were fed separately or together. Alterations found in enzyme activities suggest that the retention and discharge of petroleum hydrocarbons in salmonids may be changed by the presence of chlorinated biphenyls from marine diets.

Methods

Selection of Experimental Fish

Approximately 2,400 coho salmon were obtained from Dom Sea Farms, Gorst, Washington. Groups of 20-30 fish were anesthetized with 50 mg/l tricaine methanesulfonate (MS-222) in seawater, and lengths and weights were measured. The fish were maintained in outdoor nylon net pens in seawater located at the aquaculture facility of the Northwest and Alaska Fisheries Center, near Manchester, Washington. Each pen (4 ft x 7 ft x 6 ft deep with 1.5 ft above water), covered with framed nylon netting, contained 150 fish. There were 14 pens of fish for duplicate exposure experiments. The initial range of mean weight $\pm$ S.D. for the 14 groups was 184 ± 41 to 204 ± 41 g.

About 600 chinook salmon, obtained from the aquaculture facility, were treated in the same manner as the coho salmon and allocated to four pens in seawater. Their weights ranged from 71.6 ± 13.9 g to 76.3 ± 15.6 g at the start of the experiment.

Preparation of Control and Test Diets

The basal control diet for all fish was Oregon moist pellets (Hublou 1963), which were prepared from ingredients obtained from Moore-Clark Co., La Connor, Washington. In order to minimize extraneous organic contaminants such as PCBs in the diets, a fish protein concentrate (25.2% wt/wt) was substituted for herring meal and extra food-grade soybean oil was added to the preparations. Test diets were prepared similar to the control diets, except that the chlorobiphenyls and hydrocarbons were added as separate mixtures in soybean oil solutions. The mixture of chlorobiphenyls consisted of 17.8% (wt/wt) biphenyl, 11.6% 2-chlorobiphenyl, 19.8% 2,2'-dichlorobiphenyl, 9.6% 2,4'-dichlorobiphenyl, 23.8% 2,5,2'-trichlorobiphenyl, and 17.3% 2,5,2',5',-tetrachlorobiphenyl. The hydrocarbon mixture consisted of 12.1% 2,3-benzothiophene, 9.8% n-pentadecane, 12.2% 2,6-dimethylnaphthalene, 12.6% 2,3,6-trimethylnaphthalene, 12.6% fluorene, 12.7% 1-phenyldodecane, 12.4% phenanthrene, and 15.5% heptadecylcyclohexane. The chlorobiphenyls (Analabs Inc., North Haven, Conn.; RFR Corp., Hope, Rhode Island) and the hydrocarbons (Chemical Samples Co., Columbus, Ohio; Chem. Service, Inc., Westchester, Penn.; Aldrich Chemical Co., Inc., Milwaukee, Wis.) were obtained from commercial sources and used without further purifications.

The pelleted diets were prepared in lots of 4.5 kg, using aliquots (10 g and 50 g) of stock solutions of mixtures of test compounds in food-grade soybean oil. The test diets consisted of 1 and 5 ppm (wt/wt) of chlorobiphenyls, 1 and 5 ppm of hydrocarbons, and 2 and 10 ppm of 50-50 mixtures of the two types of compounds. The pellets were stored at -15°C until used.

Feeding Samples

The fish were acclimated for three weeks before the initiation of the exposures. Coho salmon were fed 1 ppm and 5 ppm hydrocarbons and chlorobiphenyls alone. In addition, coho salmon designated to receive the 2 and 10 ppm of the 50-50 mixtures of xenocompounds were fed the diets containing 1 and 5 ppm of chlorobiphenyls, respectively, for one week before beginning the feeding of the mixtures. The coho were fed three and four times each day at a daily rate of 0.75% of the biomass, which was a maintenance ration. Daily biomasses were adjusted for the weights of fish removed from the pens. Experiments were carried out from February to April 1976, when ambient seawater temperature was low (7°C). Test coho salmon were fed their respective diets for four weeks, followed by control diets for four weeks.

The four individual groups of chinook salmon were fed the control diet, and diets containing 5 ppm chlorobiphenyls, 5 ppm hydrocarbons, and 10 ppm of the 50-50 mixtures, respectively. The experiments with the chinook salmon were performed during July and August 1976, when ambient seawater temperature was slightly greater than during the experiments with coho salmon. The chinook salmon were fed twice each day at a daily rate of 2.5% of their biomass; that rate provided a minimum growth of the fish at that time of year. Test fish were fed diets containing the xenocompounds for four weeks, followed by control diets for the remainder of the experiments.

Handling of Samples

Ten coho salmon were taken from half of duplicate exposure groups at weekly intervals for AHH analyses, alternating between duplicates every other time. Also, ten chinook salmon from each group were taken daily and weekly. Fish were sacrificed by cervical dislocation, and their lengths and weights were measured. At the same time, examinations were made for pathological abnormalities of the fish, and hematocrits were measured with blood taken from the caudal vein. Finally, livers were excised.

The livers were rinsed with chilled 0.25 M-sucrose, placed immediately into screw-capped vials, and frozen in liquid nitrogen. In the case of chinook salmon, however, the livers were frozen over dry ice. The livers were then transferred to the laboratory on dry ice and stored at -60°C until preparation of microsomes and subsequent measurements of AHH activities. Livers and microsomes suspended in 0.25 M-sucrose were stored for periods up to five months at 60°C without significant loss of AHH activity.

Analyses of Enzyme Activities in Livers

The procedure of DePierre et al (1975) for assaying AHH activities for mammalian tissues was examined and modified for application to salmon. To establish optimum conditions for AHH activities in salmon hepatic microsomes, the following variables of the assay procedure were examined: temperature, pH, required cofactors, and characteristics of the substrate--^{3}H-benzo[a]pyrene. ^{3}H-benzo[a]pyrene was purified according to DePierre et al (1975), except the hexane solution of substrate was dried over sodium sulfate prior to making an acetone stock solution. The specific activity of the acetone solution was measured weekly. Purified benzo[a]pyrene was stored at -20°C under nitrogen for up to four weeks.

For optimum specific activity of AHH, reaction mixtures contained the following: 0.8-1.0 mg microsomal protein, 3 mM-$MgCl_2$, 1.1 mM-NADPH, 62.5 mM-tris hydrochloride buffer, and 66 μM-tritiated-benzo[a]pyrene (12.5 mCi/mmole). The final volume was 1 ml. The pH of 7.5 was involved in reactions for coho salmon, while pH 7.8 was for chinook salmon AHH reactions but changed to pH 7.5 for analyses of reactions at the seventh day and thereafter (cf., Results). The reaction mixtures were shaken for 10 min at 25 $\pm$ 0.2°C before initiation of the reaction by the addition of 20 μl of acetone containing the benzo[a]pyrene. Duplicate reactions were shaken at 25°C in 15 x 125 mm open culture tubes in subdued light. After 20 min, reactions were stopped, and reaction products were extracted and analyzed by the method of DePierre et al (1975). Radioactivities of 0.3 ml aliquots of neutralized aqueous phase in 15 ml "PCS" (Amersham/Searle Corp., Arlington Heights, Ill.) were measured in a scintillation spectrometer (Packard Tri-Carb Model 3003), with corrections determined for background and quenching. Nonenzymatic background values were obtained with either boiled microsomes or reactions that contained a stop solution before addition of the substrate (DePierre et al 1975). Microsomal protein was determined by an automated system adapted to the method of Lowry et al (1951), using bovine serum albumin as a standard (Technicon Instruments Corp., Tarrytown, N. Y.). AHH activities were calculated as nmoles products/mg microsomal protein/20 min. All activity values were determined as means $\pm$ S.D.: statistical significances were determined by Student's t test.

Results

Method of AHH Analysis

By applying the method of DePierre et al (1975) to salmon hepatic AHH analysis, we found that the rate of the enzymic reaction was linear to 20 min. Results indicated an optimum reaction rate at 25°C (Table 1), which agrees with other studies with fish (Adamson 1967).

TABLE 1. AHH Activity of Coho Salmon Hepatic Microsomes as a Function of Temperature of the Enzyme Analysis Reaction[a]

Temperature °C	AHH Activity nmoles product formed/mg protein/20 min
10	1.29
15	2.14
20	2.29
25	2.23
40	0.63

[a] Incubation mixture contained 3 μmoles $MgCl_2$. 1.1 μmoles NADPH (reduced nicotinamide adenine dinucleotide phosphate), 62.5μmoles tris hydrochloride, pH 7.5, 66 nmoles ^{3}H-benzo[a]pyrene (12.5 mCi/mmole), and 0.8 mg microsomal protein (from 100,000 g pellet resuspended in 4 ml of 0.25 M-sucrose/g liver), in 1 ml, in air.

Employing the reduced form of nicotinamide adenine dinucleotide phosphate (NADPH), rather than a NADPH-generating system in situ, analyses resulted in higher specific activities of AHH. Also, optimum rates of AHH reaction with coho salmon microsomes occurred at pH 7.5 (in tris buffer), which is the same as for AHH reactions involving mammalian tissues (Nebert and Gelboin 1968, DePierre et al 1975). The same reactions for chinook salmon microsomes were initially optimal at pH 7.8, but the optimum changed to pH 7.5 after the fish were exposed to hydrocarbons for one week (cf. Table 2). This pH shift is discussed below.

Magnesium ion was required for the AHH reaction for salmon; AHH activity was increased 48% when 3 mM-MgCl$_2$ was included. Manganese ion, however, had no effect on the results, in deference to results for mammals (DePierre et al 1975).

There was essentially no loss in AHH activity when salmonid hepatic microsomes were stored in suspensions of 0.25 M-sucrose, at -60°C, for up to five months. This information was important because samples of suspended microsomes were held frozen for two to four months before they were analyzed for enzyme activities. Also, repeated thawing and refreezing of suspended microsomes caused substantial losses in AHH activities; therefore, AHH analyses were performed only for once-frozen microsomes.

TABLE 2. Exposures to Chlorobiphenyls and Hydrocarbons: Influence on Optimum pH of AHH Activities for Chinook Salmon Hepatic Microsomes

Treatment[a]	AHH activity nmoles products/mg protein/20 min	Observed pH[b]	Optimum pH
Control, #1; zero time	0.44	7.29	7.9
	0.48	7.75	
	0.66	7.81	
	0.70	7.96	
	0.56	8.42	
Control, #2; zero time	1.02	7.32	7.8
	1.38	7.74	
	1.75	7.82	
	1.44	8.01	
5 ppm chlorobiphenyls; 3 days	0.44 ± 0.24 (n=4)	7.5	7.8
	0.64 ± 0.25 (n=6)	7.8	
5 ppm chlorobiphenyls; 7 days	0.43	7.23	7.8
	0.46	7.43	
	0.56	7.76	
	0.26	7.96	
5 ppm chlorobiphenyls; 14 days	0.30 ± 0.18 (n=5)	7.5	7.8
	0.43 ± 0.34 (n=5)	7.8	
5 ppm hydrocarbons; 3 days	0.18 ± 0.18 (n=6)	7.5	7.8
	0.54 ± 0.29 (n=5)	7.8	
5 ppm hydrocarbons; 7 days	0.25	7.27	7.5
	0.26	7.50	
	0.07	7.84	
	0.06	8.05	
	0.02	8.53	
5 ppm hydrocarbons; 14 days	0.42 ± 0.22 (n=10)	7.5	7.5
	0.30 ± 0.19 (n=10)	7.8	
10 ppb mixed chlorobiphenyls and hydrocarbons; 14 days	0.78 ± 0.42 (n=5)	7.5	7.8
	0.91 ± 0.54 (n=5)	7.8	

[a] Chinook salmon fed various diets and sampled at 0, 3, 7, and 14 days, as described in the text.
[b] pH measurements made with a research pH meter (Radiometer-Copenhagen, Type PHM-25, London Co., Westlake, Ohio).

Other studies of the method of DePierre et al (1975) revealed that storage conditions and handling of the enzyme substrate, ^{3}H-benzo[a]pyrene, were important to the analytical procedure. Analyses showed that acetone solutions of benzo[a]pyrene undergo moderate oxidation when not protected from air with an inert gas. During 65 days of storage, oxidation products increased from 0.4 to 2.3 nmoles in blank AHH reactions, when a single solution of ^{3}H-benzo[a]pyrene (0.01 µCi/nmole) was opened to air each day for analysis and stored at -20°C in between times. Protection with a nitrogen atmosphere did not entirely prevent oxidation of benzo[a]pyrene in solution; however, reaction products were reduced by approximately 50% in blank reactions employing substrate that had been protected for 46 days with nitrogen. It was important to work with fresh benzo[a]pyrene solution that was used in 10 to 14 days, so as to avoid undesirably large blank values.

Conditions of Experimental Fish

Coho salmon. The general health of experimental animals was considered to be important; therefore, fish were monitored for signs of pathology. By the fourth week of exposures to the dietary test compounds, it appeared that the entire stock of coho salmon was developing health problems. Hence, only fish in experiments for the first three weeks were considered in studying AHH activities.

The body weights of coho salmon were similar among the various groups. In fact, no overall growth occurred among control or test fish, which was consistent with the feeding plan (cf. Methods).

Some mortalities were observed throughout the experiments, but these were not attributed to the composition of the diets. At the beginning of the study, all coho salmon appeared healthy. However, all test coho exhibited random and progressive signs of fin lesions, kidney lesions, enlarged spleens and orange pigmentations of livers during the study. Control salmon appeared generally healthier than fish in the test groups. Hematocrits, measured during the fourth and fifth weeks only, ranged from 4 to 50% with mean $\pm$ S.D. values of $34 \pm 12\%$ and $32 \pm 14\%$, respectively.

Chinook Salmon. All chinook salmon were in excellent condition at the start of experimentation; however, lesions (e.g., fin erosion) were evident after one week of feeding the test compounds. After four weeks of exposure to the test compounds, kidney lesions were observed. In general, control chinook salmon were healthier than experimental animals for a longer time. As with coho salmon, only data from the first three weeks of study are considered in evaluating AHH activities of chinook salmon.

Measurements of hematocrits for chinook salmon were made for the entire time that AHH activity analyses were performed, because of the poor condition of test fish in the study of coho salmon. Hematocrits ranged from $25 \pm 11\%$ to $40 \pm 10\%$.

Comparisons of mean body weights of chinook salmon over a 49-day period indicated slow growth for all groups. The growth was according to plan (cf. Methods) and did not reflect effects due to dietary xenocompounds.

Specific Activities of Hepatic AHH in Relation to Dietary Xenocompounds

Coho salmon. The specific activities of AHH (benzo[a]pyrene hydroxylase) for various groups of coho salmon are presented in Table 3. Control coho salmon exhibited a gradual increase

TABLE 3. Specific Activities of Hepatic Microsomal Aryl Hydrocarbon Hydoxylase (AHH) in Coho Salmon Fed a Control Diet and Diets Containing Test Compounds[a]

Time Days	AHH activity (nmoles products/mg microsomal proteins/20 min						
	Control	Chlorobiphenyls		Hydrocarbons		Mixed chlorobiphenyls and hydrocarbons	
		1 ppm	5 ppm	1 ppm	5 ppm	2 ppm	10 ppm
7	0.58±0.35	0.52±0.29	0.90±0.54	0.79±0.48	0.60±0.41	0.73±0.42	1.09±0.25[b]
14	0.61±0.32	0.72±0.29	0.74±0.19	0.70±0.55	1.09±0.69[c]	1.06±0.45[d]	0.73±0.26
21	0.72±0.31	0.76±0.37	0.43±0.29[d]	0.80±0.50	0.71±0.30	0.81±0.54	0.61±0.41

[a] Values represent means ± S.D. of 10 fish.
[b] Statistically different from controls ($P<0.01$).
[c] Statistically different from controls ($P<0.10$).
[d] Statistically different from controls ($P<0.05$).

in mean values of AHH activities, but there are no significant differences among the values.

The coho salmon fed 1 ppm chlorobiphenyls exhibited no significant differences in AHH activities compared to controls. At the 5 ppm concentration, the chlorobiphenyls produced a significant ($P<0.05$) decrease in AHH activities after three weeks, but there were no differences compared to controls at other times.

Coho salmon fed 1 ppm hydrocarbons exhibited no differences in hepatic AHH activities compared to controls during the first three weeks. The 5 ppm hydrocarbon feeding resulted in significant ($P<0.10$) increases in hepatic AHH activities for coho salmon only at the second week of exposure.

Data in Table 3 indicate that coho salmon fed the mixtures of chlorobiphenyls and hydrocarbons together in 1:1 proportions (wt/wt) at 2 and 10 ppm concentrations, respectively, exhibited induction of AHH in one to two weeks. Those inductions were significant ($P<0.01$ and $P<0.05$); however, no other significant differences were found for AHH activities in relation to the chlorobiphenyl-hydrocarbon diets at other times.

Chinook salmon. Because results with test compounds in coho salmon indicated AHH induction within one to two weeks, experiments with chinook salmon included more frequent analyses of AHH in the first week. Only the effects of xenocompounds at 5 ppm concentration in diets were studied for chinook salmon (Table 4).

Control chinook salmon exhibited no significant change in AHH activities in three weeks of feeding. Table 4 includes data for chinook salmon before and after exposure to the anesthetic MS-222; no effect of anesthesia on hepatic AHH activity is indicated. The anesthetic was necessary for weighing and sorting fish, so its potential effect on the enzymic system was important.

Significant decreases were found in chinook salmon hepatic AHH activities in relation to the three test diets (Table 4). Significant differences compared to controls were generally consistent but varied in degree of confidence ($P<0.001$ to $P<0.05$). No induction of hepatic AHH was observed in the chinook as was found with coho salmon.

TABLE 4. Specific Activities of Hepatic Microsomal Aryl Hydrocarbon Hydroxylase (AHH) in Chinook Salmon Fed a Control Diet and Diets Containing Test Compounds[a]

Time Days	AHH activity (nmoles product/mg microsomal proteins/20 min)			
	Control	Chlorobiphenyls 5 ppm	Hydrocarbons 5 ppm	Chlorobiphenyls and hydrocarbons 10 ppm
0	1.09 ± 0.45[b]	-	-	-
0	0.96 ± 0.56	-	-	-
1	(1.22 ± 0.55)[c]	0.76 ± 0.24[d]	0.82 ± 0.4	0.67 ± 0.23[d]
2	(1.22 ± 0.55)[c]	0.96 ± 0.22[d]	0.77 ± 0.35[d]	0.85 ± 0.38
3	(1.22 ± 0.55)[c]	0.72 ± 0.37[d]	0.86 ± 0.48	0.62 ± 0.47[e]
4	(1.22 ± 0.55)[c]	0.44 ± 0.28[f]	0.50 ± 0.27[e]	0.91 ± 0.48
7	1.39 ± 0.63	0.83 ± 0.49[d]	0.63 ± 0.36[e]	0.52 ± 0.34[e]
14	1.08 ± 0.25	0.96 ± 0.45	0.65 ± 0.32[e]	1.27 ± 0.66
21	1.18 ± 0.56	0.60 ± 0.35[d]	0.73 ± 0.42[d]	0.65 ± 0.45[d]

[a] Values represent means $\pm$ S.D. of 10 fish.
[b] Activity for fish before exposures to the anesthetic MS-222.
[c] Average value of controls at 0 and 7 days.
[d] Significantly less ($P<0.05$) than corresponding control.
[e] Significantly less ($P<0.01$) than corresponding control.
[f] Significantly less ($P<0.001$) than corresponding control.

The data show that the hepatic AHH activity was statistically similar on the second and fourteenth day of exposures to that of controls when the 10 ppm combination of chlorobiphenyls and hydrocarbons was fed. However, at the same times, 5 ppm hydrocarbons resulted in a statistically significant depression of hepatic AHH, and 5 ppm chlorobiphenyls had no effect on the AHH, compared to controls. These findings suggest that chlorobiphenyls antagonize the effect on AHH by the hydrocarbons, when the two types of xenocompounds are present together in chinook salmon. The net result is a counteracting effect with the AHH activity no longer depressed.

Discussion

Significance of Hemotology

In the studies on the effects of xenobiotics on the hepatic AHH, every effort was made to obtain data from healthy animals. Although hemotocrit values were determined, such values were found to be doubtful indices of health because high values were not consistently associated with only healthy animals. Many factors affect hemotocrits in fish, such as water temperature, state of maturity, spawning cycle, metabolic activity, and sampling techniques (Katz 1949, Barnhart 1969, Wedemeyer and Chatterton 1971). More hematocrit information about healthy salmon is needed, especially as it relates to experimental salmon used in xenobiotic studies.

Hepatic AHH Activities

Coho salmon. Evidence was presented to show that hepatic microsomal aryl hydrocarbon (benzo[a]pyrene) hydroxylase is induced in saltwater-reared coho salmon by petroleum hydrocarbons administered separately and together with chlorobiphenyls in diets. The highest activities of hepatic AHH were observed when the hydrocarbons were fed to the coho salmon; the activities ranged from 1.5 to 1.9 times the activities for corresponding control fish. The effects observed for the hydrocarbons were probably caused by the polycylic aromatic fraction, which made up most of the mixture.

Chlorobiphenyls alone did not cause increased AHH activity in coho salmon; however, two circumstances suggest that these compounds influenced the hepatic enzyme system when administered together with hydrocarbons. First, fish treated with 10 ppm of the mixture of combined chlorobiphenyls and hydrocarbons exhibited higher AHH activity after one week than did fish that were treated with 5 ppm hydrocarbons. Second, induction of AHH activity occurred when fish were fed 2 ppm of the mixture but not when either chlorobiphenyls or hydrocarbons alone were fed at 1 ppm. The findings imply that chlorobiphenyls act synergistically with petroleum hydrocarbons inducing the hepatic AHH in coho salmon.

Chinook salmon. Studies with chinook salmon in seawater showed that the enzyme activity was *depressed* by the chlorobiphenyls and hydrocarbons administered alone or in combination. Results obtained with chinook salmon, under certain conditions, suggest that the chlorobiphenyls may be antagonistic to the tendency of hydrocarbons to depress the hepatic AHH; however, such a conclusion requires further verification because evidence for antagonism occurred intermittently during the experiments.

The shift in optimum pH of the reaction of AHH for chinook salmon from 7.8 to 7.5, occurring after one week of hydrocarbon exposures, is of interest. The shift, resulting in significantly different AHH activities, may result from the presence of multiple enzyme forms (isozymes) of AHH or from altered enzyme properties due to hydrocarbon interactions with enzyme structure.

A brief review of previous findings may place the results with chinook salmon in perspective. Many xenocompounds may not induce AHH activity in animals, or if so, the extent to which aromatic hydroxylation occurs with different compounds varies considerably among species (Williams 1971). Benzo[a]pyrene hydroxylase in mammals is not especially responsive to many inducers of drug-metabolizing enzymes, including PCB's and certain chlorine-containing pesticides (Fouts 1973). Some compounds exert an opposite effect of inducers. For example, pike exposed to domestic and industrial wastes exhibited lower AHH activity than controls (Ahokas et al 1976). In trout, continued administration of inducers of xenobiotic metabolizing enzymes leads to a marked diminution of enzyme activity, which is attributed to collapse and regression of the endoplasmic reticulum (Scarpelli 1974). Alterations in the endoplasmic reticulum have been observed in mammels fed chlorobiphenyls (Fishbein 1974).

Comparative Implications

Our results are consistent with other studies showing that activities of xenobiotic metabolizing enzyme systems are often lower in fish compared to higher animals (Adamson 1967, Buhler and Rasmusson 1968). In addition, while the microsomal xenobiotic metabolizing enzyme systems of fish appear to be similar in many respects to those of mammals (Stanton and Khan 1975), there are differences such as the lack of mixed function oxidase induction by phenobarbital in marine fishes (Philpot et al 1976). There is no reason to assume that the effects of xenocompounds on fish AHH will parallel changes brought about by these compounds in mammals: the pattern and extent of AHH induction within a given species of fish can vary among different strains and can be related to their geographic origin (Pedersen et al 1976). Also, induction of AHH in fish is inversely affected by both water temperature (Dewaide 1970) and salinity (Gruger et al 1977).

The specific activities of the AHH system from fish hepatic microsomes are expected to be useful indices of the capability of fish to initiate the biotransformation of aromatic hydrocarbons (Scarpelli 1974, Payne 1976). The enzyme system in fish is believed essential either to detoxify or discharge aromatic hydrocarbons and other xcnocompounds. Many compounds other than hydrocarbons affect the activity of AHH in animal systems (Conney 1971, Gelboin 1971), thereby suggesting that the ability of fish to metabolize aromatic hydrocarbons may be significantly altered by exposure to a variety of xenocompounds. The metabolic fate of petroleum hydrocarbons in fish may depend on whether other ubiquitous contaminants (e.g., PCB's) are synergistic, antagonistic, or just additive in the alteration of AHH activities (Gruger et al 1977). Alteration in AHH would presumably affect the storage of aromatic hydrocarbons and metabolic products in tissues.

Conclusions

The pronounced differences for the coho and chinook salmon in terms of the responses of the hepatic AHH systems to petroleum hydrocarbons and chlorobiphenyls imply that chlorinated hydrocarbons from the environment may enhance or retard the normal biotransformations of petroleum in exposed marine organisms. These xenobiotic effects may be influential factors in marine environments.

Acknowledgements

The authors thank Neil Stewart and Philip Schwartz for assistance in the chemistry; and Conrad Mahnken for use of the aquaculture facility; and Earl Prentice for technical assistance. Work was supported under Interagency Agreement EPA-IAG-E693 between the National Oceanic and Atmospheric Administration and the Environmental Protection Agency.

References

Adamson, R. H., Drug metabolism in marine vertebrates, Fed. Proc. 26, 1047-1055 (1967).

Ahokas, J. T., N. T. Karki, A. Oikari, and A. Soivio, Mixed function monooxygenase of fish as an indicator of pollution of aquatic environment by industrial effluent, Bull. Environ. Contam. Toxicol. 16, 270-274 (1976).

Barnhart, R. A., Effects of certain variables on hematological characteristics of rainbow trout, Trans. Am. Fish. Soc. 98, 411-418 (1969).

Buhler, D. R. and M. E. Rasmusson, The oxidation of drugs by fishes, Comp. Biochem. Physiol 25, 223-239 (1968).

Conney, A. H., Environmental factors influencing drug metabolism, p. 253-278, In: B. N. LaDu, H. G. Mandel, and E. L. Way (eds.), Fundamentals of Drug Metabolism and Drug Disposition, Williams and Wilkins, Baltimore, 1971.

DePierre, J. W., M. S. Moron, K. A. M. Johannesen, and L. Ernster, A reliable, sensitive, and concenient radioactive assay for benzopyrene monooxygenase, Anal. Biochem. 63, 470-484 (1975).

Dewaide, J. G., Species differences in hepatic drug oxidation in mammals and fishes in relation to thermal acclimation, Comp. Gen. Pharmacol. 1, 375-384 (1970).

Fishbein, L., Toxicity of chlorinated biphenyls, Annu. Rev. Pharmacol. 14, 139-156 (1974).

Fouts, J. R., Some selected studies on hepatic microsomal drug-metabolizing enzymes - environment interaction, Drug. Metab. Disposition 1, 380-385 (1973).

Gelboin, H. V., Mechanisms of induction of drug metabolism enzymes, p. 279-307. In B. N. LaDu, H. G. Mandel, and E. L. Ways (eds.), Fundamentals of Drug Metabolism and Drug Disposition, Williams and Wilkins, Baltimore, 1971.

Gruger, E. H. Jr., T. Hruby, and N. L. Karrick, Sublethal effects of structurally related tetrachloro-, pentachloro-, and hexachlorobiphenyl on juvenile coho salmon, Environ. Sci. Technol. 10, 1033-1037 (1976).

Gruger, E. H., Jr., M. M. Wekell, P. T. Numoto, and D. R. Craddock, Induction of hepatic aryl hydrocarbon hydroxylase in salmon exposed to petroleum dissolved in seawater and to petroleum and polychlorinated biphenyls, separate and together, in food, Bull. Environ. Contam. Toxicol., In press (1977).

Gutsell, J. S., Danger to fisheries from oil and tar pollution of waters, p. 3-10. In: Appendix VII to the Report of the U.S. Commissioner of Fisheries for 1921, Bureau of Fisheries Document No. 910, Government Printing Office, Washington, D.C., 1921.

Holcomb, R. W., Oil in the ecosystem, Science 166, 204-206 (1969).

Hublou, W. F., Oregon pellets, Progr. Fish-Cult. 25, 175-180 (1963).

Katz, M., The hematology of the silver salmon, Oncorhynchus kisutch (Walbaum), Ph.D. Thesis, University of Washington, Seattle, 1949.

Lidman, U., L. Forlin, O. Molander, and G. Axelson, Induction of the drug metabolizing system in rainbow trout (Salmo gairdnerii) liver by polychlorinated biphenyls (PCBs), Acta Pharmacol. Toxicol. 39, 262-272 (1976).

Lowry, O. H., N. J. Rosebrough, A. L. Farr, and R. J. Randall, Protein measurement with the Folin phenol reagent, J. Biol. Chem. 193, 265-275 (1951).

Nebert, S. W. and H. V. Gelboin, Substrate-inducible microsomal aryl hydroxylase in mammalian cell culture, J. Biol. Chem. 243, 6242-6249 (1968).

Ocean Affairs Board, Inputs, p. 1-18. In: Petroleum in the Marine Environment, Natl. Acad. Sci., Washington, D.C., 1975.

Payne, J. F., Field evaluation of benzopyrene jydroxylase induction as a monitor for marine petroleum pollution, Science 191, 945 (1976).

Payne, J. F. and W. R. Penrose, Induction of aryl hydrocarbon (Benzo(a)pyrene) hydroxylase in fish by petroleum, Bull. Environ. Contam. Toxicol. 14, 112-116 (1975).

Pedersen, M. G., W. K. Hershberger, and M. R. Juchau, Metabolism of 3,4-benzpyrene in rainbow trout (Salmo gairdnerii), Bull. Environ. Contam. Toxicol. 12, 481-486 (1974).

Pedersen, M. G., W. K. Hershberger, P. K. Zachariah, and M. R. Juchau, Hepatic biotransformation of environmental xenobiotics in six strains of rainbow trout (Salmo gairdnerii), J. Fish. Res. Board Can. 33, 666-675 (1976).

Philpot, R. M., M. O. James, and J. R. Bend, Metabolism of benzo(a)pyrene and other xenobiotics by microsomal mixed-function oxidases in marine species. In: Sources, Effects and Fates of Petroleum in the Marine Environment, Symposium sponsored by Am. Inst. Biol. Sci. Washington, D.C., August 1976, In press.

Risebrough, R. W., P. Rieche, D. B. Peakall, S. G. Herman, and M. N. Kirven, Polychlorinated biphenyls in the global ecosystem, Nature 220, 1098-1102 (1968).

Scarpelli, G., Hepatic function in fish, p. 191-198. In: Marine Bioassay Workshop Proceedings, Am. Petrol. Inst., Environ. Protection Agency, and Mar. Technol. Soc., Washington, D.C., 1974.

Stalling, D. L. and F. L. Mayer, Jr., Toxicities of PCBs to fish and environmental residues, 159-164. In: Environmental Health Perspectives, Experimental Issue No. 1, Natl. Inst. of Environ. Health Sci., Bethesda, Maryland, 1972.

Stanton, R. H. and M. A. Q. Khan, Components of the mixed-function oxidase of hepatic microsomes of freshwater fishes, Gen. Pharmacol. 6, 289-294 (1975).

Wedemeyer, G. and K. Chatterton, Some blood chemistry values for the juvenile coho salmon (Oncorhyunchus kisutch), J. Fish. Res. Board Can. 28, 606-608 (1971).

Williams, R. T., Species variations in drug biotransformation, p. 187-205. In: B. N. LaDu, H. G. Mandel, and E. L. Way (eds.), Fundamentals of Drug Metabolism and Drug Disposition, Williams and Wilkins, Baltimore, 1971.

CHAPTER 34

THE FATE OF PETROLEUM HYDROCARBONS FROM A NO. 2
FUEL OIL SPILL IN A SEMINATURAL ESTUARINE ENVIRONMENT

Rudolf H. Bieri and Vassilios C. Stamoudis

Virginia Institute of Marine Science, Gloucester Point, VA 23062

Abstract

Oysters, Crassostrea virginica, and clams, Mercenaria mercenaria, have been exposed to surface spills of #2 fuel oil in an estuarine area under seminatural conditions. The fate of oil components as a function of time was investigated in detail by gas chromatography and a computerized GC-MS system in extracts from water, sediments, oysters and clams. The natural background in the different sample types was established by the collection of additional samples outside the spill area.

The results show that accommodated petroleum hydrocarbons in water reach their maximum concentration a few hours after the spills and then disperse very rapidly. Most hydrocarbons are below detection 25 hours after the spills. In oysters, while substantial hydrocarbon concentrations are present within 6 hours after the spill, uptake continues in some cases past 100 hours for both branched aliphatic and aromatic hydrocarbons, and appears to be related to molecular weight. Normal alkanes up to about heptadecane, while clearly present in the 6 hour samples, have either decreased to very low levels or are below background 25 hours after the spill. Mass spectra of the unresolved envelope, typically present in the aliphatic fraction of petroleum, indicate increasing unsaturation with time. After reaching a maximum, all hydrocarbon concentrations decrease until they eventually disappear in the background. It is shown that all observations can be understood only if uptake past 6 hours is via ingestion of adsorbed hydrocarbons and if modifications in the composition of oil occur external to the oyster. Clams acquired hydrocarbons at much lower concentrations than oysters. Natural stress and differences in feeding habit may have contributed to this situation. No spill-related hydrocarbons could be detected in extracts from sediments. However, traces of some polycyclic aromatics were present in all, including control and prespill samples. Aliphatic fractions from sediment samples reflected the odd-even distribution typical of plant origin.

Key words: No. 2 fuel oil, oyster, clam, fate, aliphatic fraction, aromatic fraction, biomodification, depuration, uptake, chromatography, computerized GC-MS.

Introduction

During the summer of 1975, experimental surface spills of #2 fuel oil were investigated in the Penniman Spit area of Cheatham Annex (U. S. Navy) in Yorktown, VA. To contain the surface film, two separate structures of marine plywood were erected, enclosing the spill area on three sides each while the fourth side was a natural intertidal border. The total area enclosed was approximately 24 x 24 meters (80 x 80 ft.) each. An aerial view is presented in Fig. 1. Water within the structures could communicate with the surrounding water through gaps in the walls at a subsurface level, while the movement of sediment and detritus was hindered. The spill area was further protected by a boom stretched out along the walls. Oysters (Crassostrea virginica) and clams (Mercenaria mercenaria) were kept in trays suspended from posts within the spill area and were located close to the sediment-water interface. These trays were found to be necessary after an initial transplant of animals revealed that the silty, soft sediments could not provide the support needed to keep the shellfish close to the interface.

In designing the chemical research for the fate study, we attempted to derive as much detail as we possibly could within a given frame of time and costs. Much emphasis was therefore placed on the use of a computerized GC-MS system during the analytical work.

Experiment

The Spill

Different amounts of a #2 fuel oil were spilled on July 7, 1975 within the two structures: structure I received 85 liters (22.5 gallons) and structure II 28 liters (7.5 gallons).

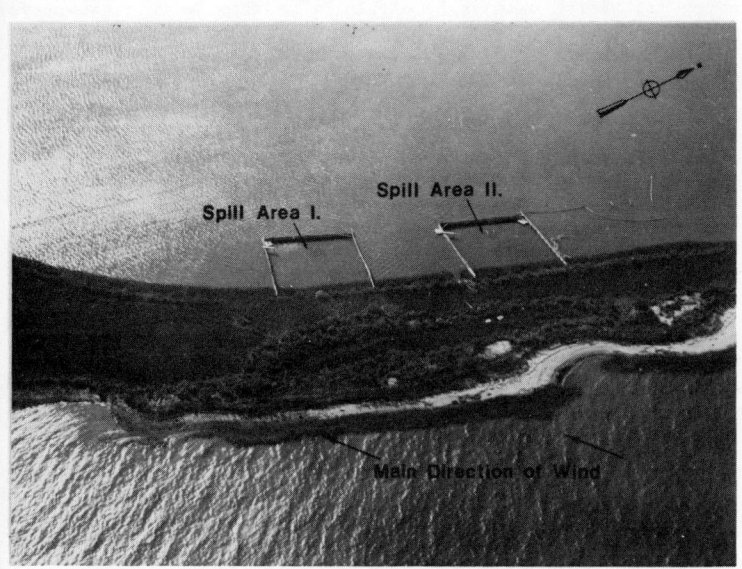

Fig. 34.1 Aerial view of the two spill-areas at Penniman Spit, Cheatham Annex, Yorktown, VA. At the time this picture was taken, the booms had not yet been emplaced. (Photo: Courtesy of NASA).

Spilling began approximately one hour past high slack. There was light rain. The air temperature was 26.9°C, that of the water 26.8°C. Spreading of the oil was not consistent, as light winds out of the N.E. appeared to impart considerable drag to the surface film. In both spills, all the oil ended up close to the S.W. corner of the boom, piling up in a layer about 1/2" thick. Within a few minutes of the spill, an oil film began to appear on the outside of the boom. While some small amount of oil seemed to be leaking through seams holding individual lengths of boom together, the remainder surmounted the barrier in ways that could not be determined.

Ten to fifteen minutes after the spill, frothy masses of organic debris began to show up at the water surface—both inside and outside of the boom—forming irregular patches of linear appearance. No debris had been observed before the spill. Where the oil was thick, large amounts of this debris were collected in it and remained there for hours. Most of the debris had disappeared one day after the spills. At this time, the water also appeared to be more transparent than it was before the spill.

The following hypothesis is offered to explain these observations: during the summer months, the water in York River estuaries is loaded with plankton, giving it a dark, greenish color and limiting visibility to a few centimeters. Much of the plankton residing near the surface may have been killed by the spilled oil, thereby creating the organic debris that was observed. The collection of debris in the oil patches would indicate that this material had hydrophobic properties. What is less readily interpreted, however, is its disappearance from the surface after one day. Since an increase of density is necessary to make this material sink, attachment of mineral-solids (through deposition on silt during low tide or pickup of suspended particles, for example) would have to occur. Except for the tidal renewal of water and a subsequent influx of new fauna, an increase in transparency through first the killing, concentration in the oil, and finally removal, would be expected. Such an environment can be assumed to be conducive to rapid biodegradation.

Sampling (Bierl 1977)

Four types of samples were collected each time from spill area I, spill area II and from a control area located about 50 meters north of area II: water, sediment, oyster and clam. In all cases special care was exercised to prevent the presence of surface film in the samples.

Water was sampled with a bottle collector of special design about 30 cm from the bottom.

Sediments were retrieved with the help of a grab sampler. The grab did not give a uniform type of sample, especially not in the presence of a hydrocarbon gradient near the sediment water interface. Also, since the uppermost part of the sediment was not consolidated, an unknown fraction of unconsolidated sediment may have been blown away upon impact of the grab.

Oysters and clams were raised in their trays to a level approximately 10" below the water surface in order to prevent the contamination with surface film of animals left in the trays for continued exposure. Collected animals, however, had to be brought through the interface.

Sample Pretreatment and Extraction

Water samples (3 liters) were extracted in their collection jugs, using essentially the methods of Hites and Biemann (1972). An internal standard consisting of $\underline{n}$-hexacosane and pyrene in acetone was added prior to the extraction. Hydrocarbons were partitioned twice into 40-60 ml of CH_2Cl_2, combined, and dried with anhydrous sodium sulfate.

Sediment samples were collected in Mason jars with Teflon® lined covers and kept in an icebox until return to the lab. Then ca. 50g aliquots were placed (after good mixing) in coarse fritted glass Soxhlet thimbles and an internal standard was added. This was followed by Soxhlet extraction with ca. 170 ml of methanol for 48 hours at 3 cycles per hour. To the methanol extract ca. 450 ml of water and ca. 15g of sodium chloride were added and the mixture was extracted three times with 80 ml pentane. The combined pentane extracts were then washed three times with 450 ml of water and dried with anhydrous sodium sulfate for 16 hours.

Oysters and Clams were collected in plastic bags and placed in an ice chest until return to the lab. The outsides of the shells were then thoroughly brushed in warm tap water containing Alconox®. This was followed by rinsing with plenty of tap water and distilled water. Shucking was done on a Teflon® covered board using stainless steel knives. Care was taken to transfer both juice and tissue to a Mason jar with Teflon® lined cover.

The extraction method used was a modification of the $KOH-CH_3OH-H_2O$ digestion and pentane back-extraction method (Farrington and Medeiros, 1975). Approximately 50g of animal tissue was mixed in a flask (equipped with a Snyder column) with internal standard, ca. 5.5g KOH pellets, 60g CH_3OH and 20 ml water and then digested for 2 hours at 83-87°C.

To the oyster digestate, ca. 250 ml of water and 6-12g of sodium chloride were added and the mixture extracted 3 times with 80 ml pentane. The combined pentane extracts were then washed 3 times with 300 ml water and dried with anhydrous sodium sulfate for 16 hours.

The clam digestate was transferred to a Soxhlet extractor by filtration through a coarse fritted glass thimble and extracted with an additional ca. 120 ml methanol for 18 hours at 2 cycles per hour. This extract was then treated the same way as the methanol extract from sediments.

Concentration and Column Chromatography

The dried pentane or methylene chloride extract was transferred to a Kuderna-Danish type evaporator (with 10 ml Kontes tube) and 6-8 ml of hexane were added. The mixture was then concentrated on a hot water bath until the volume was reduced to 3-7 ml. The sample was further concentrated in a Kontes evaporator to about 1 ml volume.

For the chromatography, the samples were placed in a (17.5 x 1) cm wet packed (with hexane) silica gel column (Bio-Sil®, 100-200 mesh, activated at 235°C for 16 hours). Fractions H_1=5 ml and H_2=13 ml were eluted with hexane. Fractions Hb_1=8 ml and Hb_2=25 ml were eluted with 70/30 (V/V) hexane/benzene. Fractions H_2 (aliphatic) and Hb_2 (aromatic) were then further concentrated to 0.4-0.5 ml, using the Kontes evaporator. The samples were sealed in 1 ml vials with Teflon^R covered silicone rubber seals and stored at -20°C.

Solvents and Reagents

Pentane, hexane (UV), benzene, methanol and methylene chloride, all "distilled in glass", were purchased from Burdick and Jackson. Sodium chloride and sodium sulfate, both "analytical reagent" grade, were Soxhlet extracted with hexane and dried at 135°C for 16 hours. Potassium hydroxide (analytical reagent) was washed with pentane and then dried over nitrogen. Water (pretreated by reverse osmosis, charcoal treated and ion exchanged in a central system) before use was extracted with hexane. All glassware used was first brushed

with soapy (Alconox®) warm water, sonicated if needed, rinsed with plenty of tap water and
acetone, placed in dichromate cleaning solution, rinsed with plenty of tap water, distilled
water, acetone (pesticide grade) and dried at 135°C. Before use, each clean item was
rinsed with pentane.

Analysis
The concentrated, separated fractions were then injected into a gas chromatograph to derive
a "fingerprint". A packed column (6 1/2 ft. long x 1/8" O.D., Chromosorb G-AW-DMCS, coated
with 1.5% OV 17) and detection by flame ionization was used. In view of the wide boiling
range of the sample, temperature programming (50-290°C at 12°C/min) was employed.

The use of an internal standard for the derivation of quantitative information is not without problems. Ideally, to have good control over fractionation introduced during extraction, separation and concentration, it is desirable to add as many standard compounds
(covering a wide range of volatilities and polarities) as possible and at positions in the
chromatogram which are devoid of sample compounds. For petroleum-related samples, this is
clearly impossible. A compromise resulted in an internal standard containing n-hexacosane
and pyrene only.

Other problems concerning the use of an internal standard are well known: while it
certainly is the best method for homogenous samples (water), it is of disputable value
(Warner 1976) for inhomogenous sample types (sediment, tissue). However, faced with
reality, there is some justification for the use of the internal standard method in the
latter case. For example, while an internal standard is of little value in sediment
samples where the standard and the compounds to be extracted reside in radically different
sites (and for this reason may be extracted with different yields), this almost certainly
will not be so if one is interested in petroleum hydrocarbons that originate from a recent
spill. It is more reasonable to expect that in this particular case both, the petroleum
hydrocarbons and the internal standard, are adsorbed on similar sites and with similar
energies of adsorption, from where they also can be extracted with similar yields.

Most petroleum hydrocarbons may be absorbed (that is, located inside the cell) while the added standard may be adsorbed on the cell walls. In this case, by effectively disrupting the
cells before extraction, one would expect to find little fractionaction in the extraction
yields and under these conditions find the addition of internal standard justifiable.
Although we know of no direct study which would conclusively prove that digestion in methanolic KOH achieves this goal, it seems reasonable to assume that very few cells would be able
to withstand such harsh treatment. Digestive extraction experiments by Farrington and
Medeiros (1975) on clams lend some support to the above assumption, although the extraction
yields are influenced in addition by other factors.

The quantitation of a sample by chromatography requires:

a. the identification of the compound or compounds present in the peak; and
b. that this peak can be accurately related, either by its height or by its area, to the
internal standard. This seems fairly straightforward, but in reality again there are severe
limitations imposed by the presence of a complex unresolved envelope, the coincidence of
clearly identifiable (by M.S.) compounds with valleys in the chromatogram and shifts in the
retention of fused peaks due to compositional changes. For these reasons best judgment must
often be used where clearcut decisions are not possible.

The identification of compounds in quantifiable peaks in all cases is based on mass
spectrometry and retention data. Due to excessive superimposition present in most
chromatogram peaks, computer-search procedures (Hites and Biemann 1968 and 1970, Hertz,
Hites and Biemann 1971) could not be employed, but had to be replaced with individual
inspection of the mass spectra. All mass spectra were recorded on a Du Pont 21-492B
spectrometer (interfaced with a G.C. via jet separator, and a data system), using 70 volt
electrons for the ionization and a repetitive scan for a mass to charge range of 27-600 at
a resolution of about 700.

Compound identifications were based on several criteria:

a. mass spectra by comparision with published reference spectra (Stenhagen, McLafferty and
Abrahamsson, 1974; API-44, Mass spectral data).
b. mass spectra by comparison with special information derived from chemical standards.
c. mass spectra and retention data from chemical standards.

d. addition of a compound previously identified by a, b, or c to the sample and observation of qualitative changes in the mass spectrum and quantitative changes in the chromatogram.

Identification based on mass spectrum and retention is the most reliable method available. In cases where a good and characteristic mass spectrum can be obtained, identification also is reliable if correlation with reference specta, either from the literature or from spectra derived by injection of a standard sample, can be achieved. Where a, b, and c fail to allow positive identification, criterion d should be chosen.

Precision and Errors
In the presence of peak-superimposition (caused by the extreme complexity of the sample and insufficient resolving power of the G.C. column) and dramatic compositional changes with time, it is almost meaningless to talk about precision and error, as there is no rational way to evaluate these factors in an exact fashion. On the other hand, the display of results requires some "feel" for the magnitude of uncertainty present. To bridge this gap, we present the following estimates:
a. the precision for aliphatic or dominant aromatic compounds within one sample is of the order of a few percent in the average, but about 10-20 per cent for mixtures leading to fused peaks of changing appearance (due to sample modification with time), as is common for the aromatic fraction.
b. the error between specific compounds in different samples may be up to 50 percent, or more where an unresolved hump in the chromatogram is much larger than a quantifiable peak.

The distinction between "precision" for listed compounds within one sample and "error" for the concentration of a particular compound in different samples is arbitrary. By repeat injection of samples into the G.C., we have not been able to substantiate any systematic error between different components of the same sample and thus label such deviations as precision. All mechanisms that result in a proportional discrepancy for all compounds within one sample are labeled as errors simply for reasons of distinction.

Results and Discussion

Petroleum Hydrocarbons Accumulated by Oysters
The data presented in Tables 1 and 2 are the aliphatic and aromatic fractions from oysters exposed to the 85 liter and 28 liter #2 fuel oil spills. Chromatograms from pre-spill samples and from the control area indicate a possible contribution from background in amounts up to 0.03 ppm at retention times closely approaching those of some compounds listed in Tables 1 and 2. This background, however, was quite variable and the concentrations were too small to derive a usable mass spectrum. Since the weather conditions shortly after the spill were rather calm, concentration levels in the 6 hour sample were expected to be near background. The substantial concentrations observed in these samples were thus surprising. Even more surprising were the dramatic changes taking place between +6h and +25h, especially in the aliphatic fraction (Fig. 2). In a first attempt to correlate G.C.-fingerprints from samples collected at different times after the spill, we were confused by a complete change in the appearance of the chromatograms. Only by mass spectrometry could we confirm what we suspected by a closer inspection of the chromatograms: that most of the n-alkanes had disappeared to a level where they--with few exceptions--ceased to be measurable. Where peaks at the correct retention times for n-alkanes remained ($>$n-C_{18}), the mass spectra indicated the presence of considerable branching. While it is well known that some microbes preferentially degrade n-alkanes (Zobell 1969, Blumer et al. 1970, Floodgate 1972, Stegeman and Teal 1973) the almost complete absence of normal alkanes after 25 hours and an already drastic reduction of n-hexadecane (the most abundant n-alkane in the #2 fuel oil) after 6 hours is unusual. Of the branched alkanes, only two could be identified with certainty, pristane and phytane, while those corresponding to peaks "a" (not identifiable), "b" (tentatively identified as methyl-tetradecane), "c" (mixture consisting of branched alkanes and olefins,) "d" (tentatively identified as methyl-pentadecane) and "f" (tentatively identified as methyl-hexadecane) produced mass spectra that indicated the presence of several compounds. Since characteristic ions of branched alkanes in general are not very abundant, superimposition of several compounds makes a derivation of structure from mass spectra nearly impossible. In addition, if the mass spectra of compounds in the unresolved envelope at identical positions in the G.C. are followed as a function of time after the spill, increasing unsaturation (as judged from increases in the abundance of fragments at masses 41, 55, 69, 83, etc.) is observed and complicate the identification even more.

Most of the branched aliphatic (and possibly alicyclic hydrocarbons) in samples from spill area I reach their maximum concentrations in oysters 100h after the spill, but pristane and phytane are still at the 0.5 ppm level in the 242 hour sample. In the 509 hour oyster sample, most chromatographic peaks except "d" have disappeared or have completely

TABLE 1. Concentration of Aliphatic Hydrocarbons in Oysters (Crassostrea virginica) Exposed to Spills of No. 2 Fuel Oil (in ppm wet weight)

Compound*	Area I - 85 liter spill					Area II - 28 liter spill				
	Time After Spill					Time After Spill				
	+6h	+25h	+100h	+242h	+509h	+6h	+25h	+100h	+242h	+509h
a	0.09	0.22	0.27	0.08	---	---	---	---	---	---
$\underline{n}$-$C_{13}H_{28}$	0.19	---	---	---	---	<0.21†	---	---	---	---
b: methyl-tetradecane	0.14	0.26	0.57	0.16	---	---	0.07	0.11	0.04	---
$\underline{n}$-$C_{14}H_{30}$	0.25	---	---	---	---	<0.25†	---	---	---	---
c	---	0.24	0.31	0.20	---	---	0.07	0.16	0.05	---
d: methyl-pentadecane	0.13	0.29	0.60	0.40	---	0.14	0.16	0.17	0.17	0.16
$\underline{n}$-$C_{15}H_{32}$	0.17	<0.03†	---	---	---	<0.21†	---	---	---	---
$\underline{n}$-$C_{16}H_{34}$	0.13	<0.005†	---	---	---	0.14	---	---	---	---
f: methyl-hexadecane	0.12	0.20	0.30	0.17	---	0.10	0.14	0.09	0.08	0.07
Pristane	0.22	0.44	0.65	0.45	---	0.18	0.30	0.16	0.14	0.08
$\underline{n}$-$C_{17}H_{36}$	0.19	---	---	---	---	0.14	---	---	---	---
Phytane + $\underline{n}$-$C_{18}H_{38}$	0.29	0.40	0.65	0.47	---	0.19	0.27	0.16	0.14	0.07
R = 1900**	0.18	0.08	0.10	0.11	---	0.13	0.06	0.09	0.05	---
R = 2000**	0.13	0.05	0.04	0.04	---	0.11	0.03	0.07	0.02	---
Total unres. env.††	8	23	24	24	12	5	11	40	10	14

* For position of unidentified compounds, see chromatogram Fig. 2.
** Relative retention is given because mass spectrum indicates presence of branching.
† Where <sign is used, peak at correct retention is present but superimposition of peak nearby allows determination of upper limit only.
†† Semiquantitative estimate from unresolved area.

Note: Where dash (-) is used, compound either is absent or cannot be identified with certainty.

unrecognizable mass spectra. This sample could therefore not be quantified. The same general trends apply to samples from spill area II, except that methyl-hexadecane, pristane and phytane have reached their maximum concentrations already at +25h. Observed maximum concentrations in animals exposed to the smaller amount of oil spilled to the surface appear similar to those exposed to the larger amount of oil, but it must be remembered that the actual maximum may occur between sample collections.

Except for an unexplained increase of olefinic, aliphatic or alicyclic hydrocarbons in the unresolved envelope with time of exposure, no increase of distinct compounds not originally present in the oil to indicate metabolism was found. However, the extraction and separation methods employed exclude more polar compounds from samples admitted to the G.C. and so this may not be surprising. All hydrocarbons in Table 1 are major compounds in #2 fuel oil, their absolute increase with time is a function of uptake and their increase relative to $\underline{n}$-alkanes is probably due to microbial degradation. This is in accord with Lee et al. (1972) who found no evidence for metabolism in oysters. Whether the observed degradation of $\underline{n}$-alkanes occurs inside the oysters by guest species or on their source of food cannot be determined from the results on aliphatic hydrocarbons, but will be a subject of further inquiry in the discussion of aromatic hydrocarbons.

The results for the aromatic fraction are presented in Table 2. The compound numbers used in this table can be related to compound names by Table 3. Again one notes the presence of substantial concentration levels in oyster tissue in the 6 hour samples, reflecting rapid transfer of hydrocarbons from the surface to the sediment-water interface and subsequent uptake by the oysters. Tissue concentrations in samples from spill area I continue to increase until 25 hours after the spill for substituted naphthalenes with up to 5 additional carbons, biphenyls with up to 2 additional carbons and fluorenes with up to one methyl group attached. For naphthalenes, biphenyls and fluorenes of higher substitution, dibenzothiophenes and penanthrenes, the maximum concentration is reached 100 hours after the spill. At the same time, many of the compounds peaking near 25 hours have either disappeared, are at a substantially reduced level or can no longer be recognized with certainty.

Most aromatic compounds in spill area II reached maximum concentrations after 25 or 100 hours of exposure. This maximum shifted to 242 hours only for the last few compounds of the table (C_2- and C_3-phenanthrenes, C_3-dibenzothiophenes). Again, the maximum near 25 hours appeared to be quite sharp, as one can see from the low concentrations encountered at 100 hours.

TABLE 2. Concentrations of Aromatic Hydrocarbons in Oysters (Crassostrea virginica) Exposed to Spills of No. 2 Fuel Oil (in ppm wet weight)

Compound*	Area I - 85 liter spill Time After Spill:					Area II - 28 liter spill Time After Spill:				
	+6h	+25h	+100h	+242h	+509h	+6h	+25h	+100h	+242h	+509h
109	0.13	0.15	---	---	---	---	0.10	---	---	---
111	0.11	0.12	---	---	---	---	0.09	---	---	---
116 + 117	0.20	0.28	---	---	---	0.04	0.34	0.01	---	---
118	0.25	0.35	---	---	---	0.06	0.43	0.02	---	---
119 + 120	0.07	0.13	---	---	---	0.02	0.16	0.01	0.01	---
123 + 124	0.10	0.16	---	---	---	0.05	0.25	0.02	0.01	---
125	0.18	0.31	0.16	---	---	0.11	0.50	0.06	---	---
126	0.14	0.17	0.09	0.03	---	0.10	0.40	0.03	---	---
127 + 128	0.13	0.23	0.07	---	---	0.06	0.23	0.05	0.01	---
129 - 131	0.07	0.19	---	---	---	0.08	0.33	0.06	---	---
132 - 134	0.09	0.18	---	---	---	0.08	0.29	0.04	0.01	---
135	0.04	0.09	0.13	0.05	---	0.04	0.12	0.05	0.03	---
136 + 137	0.05	0.13	0.37	0.13	---	0.06	0.17	0.07	0.05	---
138 + 139	0.16	0.40	0.47	---	---	0.20	0.65	0.18	0.08	---
140 + 141	0.08	0.14	}0.44	0.20	---	0.11	0.34	0.12	0.13	---
142 + 143	0.05	0.11								
145 - 148	0.20**	0.71	1.41	0.56	---	0.31	1.16	0.37	0.28	---
149	n.d.	0.18	0.32	0.17	---	0.05	0.29	0.10	0.11	---
150 - 152	0.07	0.33	0.82	0.27	---	0.11	0.41	0.23	0.07	---
153 + 154	0.05	0.17	0.34	0.17	---	0.06	0.22	0.16	0.08	---
155	0.02	0.06	0.18	0.13	---	0.03	0.07	0.09	0.07	---
156 + 157	0.03	0.09	0.35	0.20	---	0.04	0.11	0.17	0.12	---
158 + 159	0.10	0.27	0.77	0.48	---	0.12	0.25	0.35	0.28	---
162	0.03	0.04	0.21	0.15	---	0.03	0.06	0.09	0.10	---
Total unres. envelope†	5	9	24	12	4	3	17	5	8	4

* For compound names, see Table 3.
** Methyl-dibenzothiophene is superimposed on peak from contaminant.
† Semiquantitative estimate from unresolved area.

Note: Where dash (-) is used, compound either is absent, below background, or cannot be identified.

Concentration of Hydrocarbons in water

The concentrations of aromatic hydrocarbons in post-spill water samples are given in Table 4. No concentrations could be calculated for aliphatic fractions, as they were below detectability. In the aromatic fraction, only the methyl- and some of the dimethyl-naphthalenes reach levels around 10 ppb six hours after the spill in both spill areas, all other compounds are at the ppb or 0.1 ppb level. Twenty five hours after the spill all aromatics are at the 0.1 ppb level or below. Beyond 25 hours, no petroleum hydrocarbons could be detected. These results indicate very rapid dispersion of accomodated hydrocarbons in the water column. While it may appear at first glance that the enhancement of aromatics over aliphatics is due to solubility (McAuliffe 1966, Boylan and Tripp 1971, Bieri et al. 1974), the relatively high concentrations of some polycyclic aromatics (mainly fluorene, methyl-fluorenes, dimethyl-fluorenes, phenanthrene and methyl-phenanthrenes) indicate that this fractionation has a more complex cause.

Bioaccumulation

In comparing the results on oyster tissue with the hydrocarbon concentrations in water, some rather puzzling questions become apparent. For instance, if the hydrocarbon concentrations in water reach their maximum shortly after the spill and then decay rapidly, why do the concentrations in oysters continue to increase until up to 10 days after the spill? Furthermore, if the disappearance of the lower molecular weight aromatic hydrocarbons in oyster after 25 hours is interpreted as depuration (R. Anderson 1973) how can they continue to accumulate the heavier molecules? Finally, if they are not capable of metabolizing hydrocarbons (Lee et al. 1972) why does the "unresolved envelope" increase until 4-10 days after the spill and why does the apparent unsaturation in this envelope increase? In order to find answers to these questions, previous work is briefly recapitulated.

Uptake and discharge of hydrocarbons by Mytilus edulis has been discussed first by Lee et al. (1972), in oysters (Crassostrea virginica) by Anderson (1973), Anderson el al. (1974), Stegeman and Teal (1973) and for chlorinated hydrocarbons in a variety of aquatic animals by Hamelink et al. (1971). Stainken (1975) investigated the uptake of dyed #2 fuel oil emulsion in softshell clams (Mya arenaria) from a biological point of view.

The chemical studies all indicate the following trends: molluscs exposed to constant levels of hydrocarbons are capable of accumulating such compounds as a function of time and concentration, with the rate of uptake being highest at the beginning of exposure and eventually decreasing to zero (where hydrocarbons reach saturation levels) after some time. The concentrations of hydrocarbons in tissue at saturation are from 2 to 3 orders of magnitude higher than those in water (bioaccumulation) and have been found to correlate with total pentane-extractable lipid (Stegeman and Teal 1973). Such uptake had previously also been observed in natural spill situations (Blumer et al. 1970, Ehrhardt 1972, Ehrhardt and Heinemann 1974).

After transfer of contaminated molluscs to "clean" water (quotation marks to indicate that this does not necessarily guarantee absence of hydrocarbons) tissue concentrations of hydrocarbons were found to decrease rapidly at first and then approach a residual level, about 10% of the original concentration at saturation. Prolonged exposure to clean water caused this residual amount to decay further. The observation of decreasing hydrocarbon concentrations upon exposure to clean water is variably called "discharge" (Lee et al. 1972) or "depuration" (R. Anderson 1973).

While the general trends of the present research effort do not contradict the conclusions of these laboratory studies, there are several new aspects that must be included:

1. the apparent discharge of petroleum hydrocarbons by molluscs from surface spills is shown to take place under natural conditions.
2. very rapid compositional changes are shown to occur in the hydrocarbon

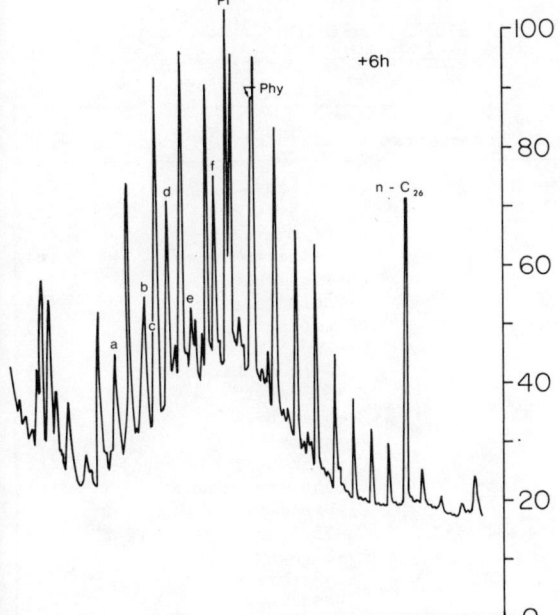
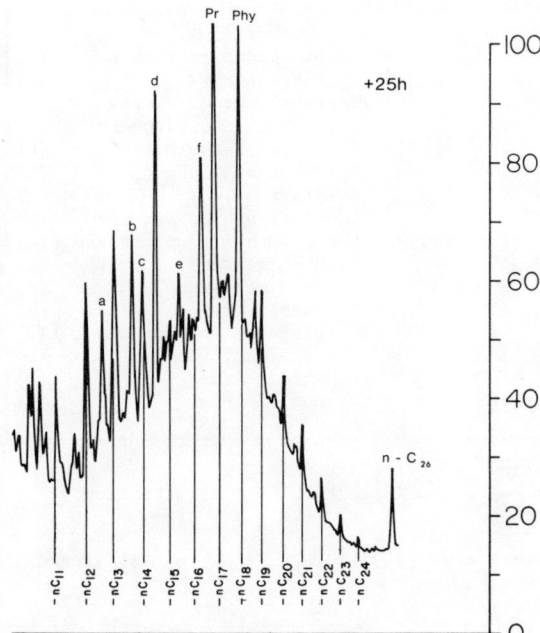

Fig. 34.2 Aliphatic fractions of post-spill extracts from oyster, demonstrating the dramatic changes in composition taking place between +6 and +25 hours. In the lower chromatogram, the approximate position of the n-alkanes is indicated by vertical lines.

mixtures, mainly in the aliphatic fractions, from oyster tissue extracts.
3. the hydrocarbon concentrations in oyster tissue continue to increase long after the concentration in water has decayed to unmeasurably low levels.
4. apparent uptake and depuration, as judged from increasing or decreasing tissue concentrations, are observed to occur simultaneously.
5. there are indications of increasing unsaturation in the chromatographically unresolvable part of the aliphatic fraction from oyster extracts with time of exposure.

TABLE 3. Listing of Identified Major Aromatic Compounds in No. 2 Fuel Oil and Post-Spill Extracts from Tissue

Peak No. in Chromatogram	Compound Name		Peak No. in Chromatogram	Compound Name	
101	Naphthalene	(a)	140	C_2-Biphenyl	(c)
109	2-Methyl-naphthalene	(a)	141	C_5-Naphthalene	(c)
111	1-Methyl-naphthalene	(a)		C_2-Biphenyl or	
116	Biphenyl	(b)		C_2-Acenaphthene	(c)
117	2,6-Dimethyl-naphthalene*	(a)		C_3-Biphenyl	(c)
	2-Methyl-biphenyl	(a)		Methyl-fluorene	(b)
118	1,3-Dimethyl-naphthalene*	(a)	142	C_6-Naphthalene	(c)
119	1,5-Dimethyl-naphthalene*	(a)		C_2-Biphenyl or	
120	2,3-Dimethyl-naphthalene*	(a)		C_2-Acenaphthene	(c)
123	3-Methyl-biphenyl	(a)		C_3-Biphenyl	(c)
	C_3-Naphthalene	(b)	143	C_2-Biphenyl	(c)
124	4-Methyl-biphenyl	(a)		C_3-Biphenyl	(c)
125	C_3-Naphthalene	(b)		C_4-Biphenyl	(c)
126	C_3-Naphthalene	(b)	145	Dibenzo-thiophene	(b)
127	2,3,5-Trimethyl-			C_4-Biphenyl	(c)
	naphthalene*	(a)	146	C_2-Fluorene	(b)
128	C_4-Naphthalene	(b)		C_4-Biphenyl	(c)
129	C_2-Biphenyl	(b)	147	Phenanthrene	(a)
	C_3-Naphthalene	(b)		C_2-Fluorene	(b)
130	C_4-Naphthalene	(b)	148	C_2-Fluorene	(b)
131	C_4-Naphthalene	(b)		C_4-Biphenyl	(c)
132	Fluorene	(b)	149	Methyl-dibenzothiophene	(b)
	C_4-Naphthalene	(b)		(C_4-Biphenyl)**	(c)
	Methyl-acenaphthene	(c)	150	Methyl-dibenzothiophene	(b)
133	C_4-Naphthalene	(b)		C_3-Fluorene	(c)
	Methyl-acenaphthene	(c)	151	2-Methyl-phenanthrene	(a)
134	C_4-Naphthalene	(b)	152	Methyl-dibenzothiophene	(b)
	C_2-Biphenyl	(c)	153	1-Methyl-phenanthrene	(b)
135	C_4-Naphthalene	(b)	154	C_2-Dibenzothiophene	(b)
	C_2-Biphenyl	(c)		C_3-Fluorene	(c)
136	C_4-Naphthalene	(b)	155	C_2-Dibenzothiophene	(b)
	C_5-Naphthalene	(c)	156	3,6-Dimethyl-phenanthrene	(a)
137	C_5-Naphthalene	(c)	157	C_2-Phenanthrene	(b)
138	C_5-Naphthalene	(c)	158	C_2-Phenanthrene	(b)
	C_2-Biphenyl	(c)	159	C_2-Phenanthrene	(b)
	Methyl-fluorene	(b)		C_3-Dibenzothiophene	(c)
139	C_2-Biphenyl	(c)	161	Fluoranthene	(b)
	Methyl-fluorene	(b)	162	C_3-Phenanthrene	(c)
140	C_5-Naphthalene	(c)	163	Pyrene (Standard)	(a)

(a) identification by mass spectrum and retention time from standard.
(b) identification by published mass spectrum and approximate knowledge of retention.
(c) identified by mass spectrum alone.

* Other isomers of substituted naphthalenes with very similar retention time may also be present.
** Not present in tissue extracts.

The results of this study, in addition, show non-selective changes occurring in the aromatic fraction, that mainly appear to be related to molecular weight. Such changes can be suspected as being caused by biodegradation as well as metabolism, but the selective, quick removal of n-alkanes and the delayed disappearance of the lighter polycyclic aromatics make the former a more likely candidate.

Thus, a continued increase in the concentrations of hydrocarbons in tissue, after they have practically disappeared from water, may reflect the presence of a different hydrocarbon source available to the oyster. Such a source could be found in the organic detritus formed by the dead plankton mentioned at the end of chapter II, paragraph 1. The hydrocarbons in this case would enter the oyster through a different interface--the gut--as distinguished from a direct water-lipid interface through which water-accommodated hydrocarbons could be transferred to the oyster. This is in agreement with Lee et al. (1972) and Stegeman and Teal (1973) who also proposed the two principal paths of uptake.

TABLE 4. Concentration of aromatic hydrocarbons in subsurface water samples from spills of #2 fuel oil (in ppb).

	85 liter spill			28 liter spill	
	Time After Spill			Time After Spill	
Compound*	+6h	+25h	+100h	+6h	+25h
101	5	---	---	3	---
109	11	0.2	---	8	---
111	9	0.2	---	7	---
116 + 117	8	0.2	---	5	---
118	11	0.3	---	8	---
119 + 120	3	---	---	2	---
123 + 124	2	---	---	1	---
125	4	---	---	2	---
126	3	---	---	2	---
127 + 128	2	---	---	1	---
129 - 131	1	---	---	1	---
132 - 134	2	---	---	2	---
135	0.4	---	---	---	---
136 + 137	0.5	---	---	---	---
138 + 139	2	---	---	1	---
140 + 141	---	---	---	---	---
142 + 143	---	---	---	---	---
144 - 148	2	---	---	2	---
149	0.3	---	---	---	---
150 - 152	1	---	---	1	---
153 + 154	1	---	---	---	---
155	0.1	---	---	---	---
156 + 157	0.2	---	---	---	---
158 + 159	0.4	---	---	---	---
162	---	---	---	---	---

* For compound names, see Table 3.

Note: Where dash (-) is used, compound is absent, below background, or cannot be identified.

d. Kinetic considerations

The apparent simultaneous uptake and discharge of hydrocarbons from oyster tissue can be interpreted in several ways. Before continuing the discussion, however, it is necessary to first remove any ambiguities about the phrase "apparent uptake and discharge (or depuration)". Uptake in most experiments is identified by an increase of hydrocarbon concentrations extracted from oyster tissue with time of exposure, and depuration or discharge is characterized by decreasing concentrations. Starting from a hydrocarbon mixture of constant composition and assuming that oysters acquire their hydrocarbons basically by partitioning (Hamelink et al. 1971), the hydrocarbon concentrations in tissue, C_z^i, can be described in the simplest case by:

$$C_z^i = C^i K^i (1 - e^{-\frac{t}{\tau^i}})$$

where C_z^i is the concentration of a hydrocarbon in the water or in the digested material, K^i is the partition coefficient for this hydrocarbon between water or ingestate and interfacial lipid, t is the time of exposure, τ^i is the characteristic time for uptake or discharge, probably mainly related to diffusion and thus dependent on molecular structure and weight of the hydrocarbon, and the microstructure of the cell membrane. For two aromatic compounds of molecular weight m^l and m^h, $m^h > m^l$, at concentrations C^l and C^h, characteristic times $\tau^l < \tau^h$ and partition coefficients $K^l = K^h = K$, the concentrations in tissue thus would be

$$C_z^l = C^l K (1 - e^{-\frac{t}{\tau^l}})$$

$$C_z^h = C^h K (1 - e^{-\frac{t}{\tau^h}})$$

Both concentrations in tissue increase (with different time constants until equilibrium is reached at $\frac{t}{\tau} \gg 1$. Upon transfer to clean water at time t_1, the flux of hydrocarbons is reversed and the concentration of hydrocarbons in tissue is given by:

$$C_z^i = K^i \{ [C^i (1 - e^{-\frac{t_1}{\tau^i}}) - C_o^i] e^{-\frac{t - t_1}{\tau^i}} + C_o^i \} \ ; \ t \geq t_1$$

(C_o^i = concentration of hydrocarbon in clean water).

For the same pair of hydrocarbons as above:

$$c_z^l = K\{[C^l(1-e^{-\frac{t_1}{\tau^l}}) - C_o^l]e^{-\frac{t-t_1}{\tau^l}} + C_o^l\}$$

$$c_z^h = K\{[C^h(1-e^{-\frac{t_1}{\tau^h}}) - C_o^h]e^{-\frac{t-t_1}{\tau^h}} + C_o^h\}$$

Both hydrocarbons now decrease with time constant τ^l and τ^h respectively until a lower equilibrium is reached that depends on the residual concentrations C_o^l and C_o^h only.

It is thus impossible to find decreasing and increasing hydrocarbon concentrations simultaneously under these conditions and observation 4 is not satisfied.

In order to satisfy 4, the assumption of constant composition must be dropped and a mechanism for the modification of the hydrocarbon composition must be found. This modification in principle can occur external or internal to the oyster.

For external modification, the exposure concentration C^i itself becomes a function of time $C^i(t)$ and in the simplest case--if $\tau^i \ll t$-- becomes the parameter that determines the change of hydrocarbon concentration in tissue with time:

$$c_z^i = C^i(t)K^i \quad ; \quad \tau^i \ll t$$

Thus, C^i simply reflects the concentration C^i of hydrocarbon i to which the oyster is exposed. Concentration changes are no longer dependent on biological residence times of hydrocarbons, but are a function of environmental residence times determined by dispersion, solubility, evaporation, emulsification, photo-and biochemical oxydation etc. Under such circumstances, the terms "uptake" and "depuration" loose their meaning as they relate to the transient states determined by the biological residence times. Clearly, this is an attractive possibility that is capable of explaining observation 2-5 if a mechanism for the compositional changes can be found. We suggest that the rapid disappearance of some n-alkanes, the slower decay of the lower molecular weight polycyclic aromatics (although their relatively high volatility and solubility may also contribute to these losses), and the increasing unsaturation in the unresolved envelope all point to biodegradation. With absolute amounts of oil adsorbed on spill related organic detritus being small and microbial populations at the detritus-water interface large, the effects of biodegradation would be expected to be very pronounced.

Internal modification, both by microbes and by metabolism, is a less likely contender. For biodegradation, it can be shown that unless the rate of biodegradation is much faster than the rate of uptake, one still should find n-alkanes in tissue extracts. As for metabolism, a more selective removal of hydrocarbons of one or more specific types, but not of others, would be expected. Since this is not observed, we tend to agree with Lee et al. (1972) that molluscs do not metabolize hydrocarbons.

e. Petroleum hydrocarbons accumulated by clams

Aromatic hydrocarbons found in clam tissue (Mercenaria mercenaria) in general are lower by about a factor of 10 than those in Crassostrea virginica. Most concentrations are below 0.1 ppm for individual petroleum hydrocarbons in G.C. integratable peaks. An unusual amount of extraction problems were encountered (information of stable emulsions, and of gels during concentration, probably due to very high lipid levels and extraction yields--especially those for the aliphatic fraction--were found to be low in some cases (10-30%). However, extractions with high yields gave equally low concentrations. There is a distinct possibility that the clams, being suspended in trays instead of buried in sediment, were under stress (an indefinable but intuitively meaningful term borrowed from biology) and did not feed, since all clams, including those in the control, died after 509 hours. Since a similar situation was encountered in another large scale experiment, this may indicate that clams are not well suited study objects. All aliphatic hydrocarbons were below the level necessary for quantification (0.01 ppm).

f. Sediments

Sediment extracts presented special problems insofar as the chromatograms indicated the presence of numerous peaks in background and pre-spill samples. In the aliphatic fraction, they are superimposed on a large unresolved envelope. While there is much variation present in background and post-spill samples (probably due to sample inhomogeneity) they both contain the same main peaks and qualitatively are difficult to distinguish from each other. All aliphatic fractions show a preference of odd over even n-alkanes in the C_{20}-C_{32} carbon range that is characteristic of natural production (Eglinton et al. 1963).

In view of what has been said about hydrocarbons in oyster tissue, one may wonder why the oysters from the control or from pre-spill collections do not reflect the composition of (mainly natural) hydrocarbons found in sediment samples. Two explanations are offered. First, the oysters clearly would have to be able to ingest the sediment. Since they were suspended above the sediment in trays, they may not have been able to reach this material. Second, as mentioned, the grab sampler on impact "blows" away most of the interface material. Thus, the sediment sample from which the hydrocarbons were extracted may not be representative of interfacial material. The preservation of hydrocarbons, including n-alkanes, in sediments separated from exchange with overlying water is well known.

The aromatic fraction of sediment extracts from control-, pre- and post-spill samples tend to support the second explanation. Quantitatively, and within the variability mentioned before, they are very similar and reveal the presence of some polycyclic aromatic compounds (substituted naphthalenes, substituted biphenyls, phenanthrene and substituted phenanthrenes, and fluoranthene) that cannot be related to the #2 fuel oil spill. On the other hand, isomeric distributions of C_3-, C_4- and maybe C_5-naphthalenes, C_2-biphenyls, methyl-phenanthrene and C_3-phenanthrenes are similar to those in oils. Since a fuel pier of the U. S. Navy is located about 1/2 mile upstream, a chronic contamination may exist. In addition, this area may receive some runoff from a nearby highway. Levels of individual aromatic compounds are below 0.1 ppm (dry sediment). The most abundant single compound is fluoranthene. Many of the compounds eluting with aromatic hydrocarbons appear to contain oxygen and have high molecular weight ($\sim$300-400). They cannot be identified without further detailed study.

Summary

Oysters exposed to surface spills of #2 fuel oil in a seminatural estuarine environment had rapid access to petroleum hydrocarbons and accumulated them in their tissues. Drastic quantitative changes in the aliphatic fraction of oyster extracts within 25 hours are indicative of biodegradation. These changes and a continued increase of hydrocarbon concentrations in tissue, long after water-accommodated hydrocarbons had disappeared, suggested that the oysters acquired their hydrocarbons from oil-contaminated organic detritus via the digestive tract. Concentrations of individual hydrocarbons reach a maximum 25 to 242 hours after the spill (depending on molecular weight) and then decreased again to levels beyond quantitative assessment. An unresolvable envelope of increasing complexity followed similar trends, but disappeared more slowly. Differences in the interpretation of laboratory-exposure studies and the present experiment are discussed with the help of a simple model. It is shown that while the laboratory studies describe changes in tissue concentrations that relate to the biological residence time of hydrocarbons in the animal, the changes in a natural spill situation are determined mainly by the residence times of the different hydrocarbons in the environment. Clams accumulated considerably smaller amounts of petroleum hydrocarbons. The fact that all clams (including the controls) died 500 hours after the spill raises serious questions about their suitability as experimental animals under these conditions. Because of a high natural hydrocarbon background, the presence of fuel related aliphatic hydrocarbons in sediments could not be quantitatively assessed. Polycyclic aromatic hydrocarbons found in sediment extracts appear to be unrelated to #2 fuel oil.

Acknowledgements

During all phases of the work, we were assisted by Alice Chang, Jerry Losser and Michu Tcheng. The construction of the enclosures was supervised by Harold Slone. We are grateful to Robert J. Huggett and Dr. Michael Bender for many discussions during the planning stages of the project and for their support, and to Susan Lofurno for typing the manuscript. Financial support was provided by contract No. 68-03-2111 with EPA and project No. 210-75T of API. VIMS Contribution No. 796.

References

Anderson, J. W., J. M. Neff, G. A. Cox, H. E. Tatem and G. M. Hightower, Characteristics of dispersions and water-soluble extracts of crude and refined oils and their toxicity to estuarine crustaceans and fish, Marine Biology, 27, 75-84 (1974).

Anderson, R. D., Effects of petroleum hydrocarbons on the physiology of the american oyster Crassostrea Virginica Gmelin, Dissertation, Texas A&M University, 1973.

American Petroleum Institute, Research project 44, Selected Mass Spectral Data, Thermodynamic research center, Texas A&M University, 1972.

Bieri, R. H., Report to E.P.A., Contract 68-03-2111 (1977). See this report for detailed description of the experiment and analytical methods.

Bieri, R. H., A. L. Walker, B. W. Lewis, G. Losser, and R. J. Huggett, Identification of hydrocarbons in an extract from estuarine water accomodated No. 2 fuel oil. pp 149-153. In: NBS Spec. Publ. 409, Marine Pollution Monitoring (Petroleum), Proceedings of a Symposium and Workshop at NBS, Gaithersburg, Maryland, May 13-17, 1974 (Issued Dec. 1974).

Blumer, M., G. Souza, and J. Sass, Hydrocarbon pollution of edible shellfish by an oil spill, Marine Biology, 5, 195-202 (1970).

Boylan, D. B. and B. W. Tripp, Determination of hydrocarbons in seawater extracts of crude oil and crude oil fractions, Nature (London), 230, 44-47 (1971).

Englinton, G. and R. J. Hamilton, The distribution of alkanes. pp 187-217. In: Chemical Plant Taxonomy, Ed., T. Swain, Academic Press, London and New York, 1963.

Ehrhardt, M., Petroleum hydrocarbons in oysters from Galveston bay. pp 257-271. In: Environmental Pollution (3), Applied Science Publishers Ltd, England, 1972.

Ehrhardt, M. and J. Heinemann, Hydrocarbons in blue mussels from the Kiel bight. pp 221-225. In: NBS Spec. Publ. 409, Marine Pollution Monitoring (Petroleum), Proceedings of a Symposium and Workshop at NBS, Gaithersburg, Maryland, May 13-17, 1974 (Issued Dec. 1974).

Farrington, J. W. and G. C. Medeiros, Evaluation of some methods of analysis for Petroleum hydrocarbons in marine organisms. pp 115-121. In: Proceeding of the Conference of Prevention and Control of Oil Pollution, American Petroleum Institute, 1975.

Floodgate, G. D., Biodegradation of hydrocarbons in the sea. pp 153-171. In: R. Mitchell (ed.) Water Pollution Microbiology, Wiley Interscience, New York, 1972.

Hamelink, J. L., R. C. Waybrant, and R. C. Ball, A proposal: Exchange equilibria control the degree chlorinated hydrocarbons are biologically magnified in lentic environments, Trans. Amer. Fish. Soc., 100, 207-214 (1971).

Hertz, H. S., R. A. Hites, and K. Biemann, Identification of mass spectra by computer-searching a file of known spectra, Anal. Chem., 43, 681-691 (1971).

Hites, R. A. and K. Biemann, Mass spectrometer-computer system particularly suited for gas chromatography of complex mixtures, Anal. Chem., 40, 1217-1221 (1968).

Hites, R. A. and K. Biemann, Computer evaluation of continuously scanned mass spectra of gas chromatographic effluents, Anal. Chem., 42, 855-860 (1970).

Hites, R. A. and K. Biemann, Water Pollution: Organic compounds in the Charles river, Boston, Science, 177, 158-161 (1972).

Lee, R. F., R. Sauerheber, and G. H. Dobbs, Uptake, metabolism and discharge of polyclic aromatic hydrocarbons by marine fish, Marine Biology, 17, 201-208 (1972).

McAuliffe, C., Solubility in water of paraffin, cycloparaffin, olefin, acetylene, cycloolefin and aromatic hydrocarbons, J. Phys. Chem., 70, 1267-1275 (1966).

Stainken, D. M., Preliminary observations on the mode of accumulation of No. 2 fuel oil by the soft shell clam, Mya arenaria. pp 463-468. In: Proceedings of the Conference of Prevention and Control of Oil Pollution, American Petroleum Institute, 1975.

Stegeman, J. J. and J. M. Teal, Accumulation, release and retention of petroleum hydrocarbons by the oyster Crassostrea virginica, Marine Biology, 22, 37-44 (1973).

Stenhagen, E., S. Abrahamsson, and F. W. McLafferty, Registry of mass spectral data, Wiley-Interscience, New York, 1974.

Warner, J. S., Determination of aliphatic and aromatic hydrocarbons in marine organisms, Anal. Chem., 48, 579-583 (1976).

Zobell, C. E., Microbial modification of crude oil in the sea. pp 317-326. In: Proceedings of the Conference of the Prevention and Control of Oil Spills, American Petroleum Institute, Publ. No. 4040, 1969.

CHAPTER 35

INTERLABORATORY CALIBRATION FOR THE ANALYSIS
OF PETROLEUM LEVELS IN SEDIMENT

S. A. Wise, S. N. Chesler, B. H. Gump,
H. S. Hertz and W. E. May

Trace Organic Analysis Group, Bioorganic Standards Section
Analytical Chemistry Division
National Bureau of Standards, Washington, D.C. 20234

Abstract

The large number of environmental analyses to be performed in the future necessitates the existence of a common basis for comparing the data. Otherwise data obtained from different laboratories would be of limited usefulness. Furthermore, unless the data can be put on an equivalent basis, environmental standards can neither be set nor enforced.

A sample, split between NBS and one other laboratory, was analyzed in order to determine the suitability of Katalla River sediment for a more extensive intercalibration exercise. The results of this limited intercomparison are discussed. The results encouraged us to initiate an enlarged intercalibration exercise which is now in progress. Comparison of the interlaboratory data should provide a measure of the analytical variability among the participating laboratories.

Key words: Gas chromatography; gas chromatography-mass spectroscopy; hydrocarbons; intercalibration; petroleum analysis; trace analysis.

Introduction

The large number of environmental analyses to be performed in the future necessitates the existence of a common basis for comparing the data. Otherwise data from different laboratories would be of limited usefulness. Furthermore, unless the data can be put on an equivalent basis, environmental standards can neither be set nor enforced.

In setting up intercalibration exercises, three types of materials of increasing scientific value and increasing sophistication are possible:

1) sample splits on field samples;
2) interim calibration materials;
3) reference material.

Sample splits are of limited value, since they usually consist of intercomparison of only a few (often only two) laboratories. Sample splits are, however, easy to initiate and of value if only as a means of beginning intercalibration. Interim calibration materials could be field samples which have been homogenized as well as possible and collected in sufficient quantity to permit wide-scale distribution. Although results obtained on such an intercalibration experiment would not yield any data on accuracy of analysis, they would at least provide an indication of interlaboratory comparability or precision. An intercalibration exercise of this form is now in progress.

A limited interlaboratory sample split between NBS and Dr. John A. Calder at Florida State University was conducted first in order to determine the suitability of Katalla River sediment for the more extensive intercalibration exercise. The Katalla River is located in the Northeastern Gulf of Alaska near a natural oil seep.

For this sample split we were interested in comparing the following data:

1) total hydrocarbons in the GC elution range (roughly $C_{10}-C_{25}$) with each laboratory specifying the exact range reported.

2) total extractable hydrocarbons

3) pristane/phytane ratio

4) % water

5) three most abundant compounds and their amounts.

Methods

The Katalla sediment samples were obtained from a 3 kg. pooled sample which was homogenized by mixing in a clean battery jar for 30 minutes. The samples (∼250 g. each) were then packed into bottles. Except for the brief period of homogenization, the samples were kept frozen from the time of collection. Two bottles of frozen sediment were shipped to the collaborative laboratory and five bottles were analyzed by NBS.

Of the five bottles analyzed by NBS, four were analyzed by headspace sampling and GC/GC-MS. The remaining bottle was utilized for Soxhlet extraction of the hydrocarbons to obtain the total extractable hydrocarbon value and the water content of the sediment. The headspace sampling procedure for the analysis of hydrocarbons in sediment has been described previously (Chesler et al. 1976, May et al. 1975). This method involves dynamic headspace sampling at room temperature for 2 hours and then at 70°C for 2 more hours and the trapping of the volatile components on a TENAX-GC packed pre-column. The trapped components are subsequently analyzed by GC or GC-MS using a glass SE-30 coated, SCOT analytical column (100 m x 0.65 mm i.d.). An internal standard of ∼ 1 µg each of 2-methylundecane, 5-methyltetradecane, 7-methylhexadecane, and 2-methyloctadecane was added prior to headspace sampling to facilitate quantitation by GC.

One of the NBS sediment samples was Soxhlet-extracted for 48 hours with diethyl ether. The ether extract was passed through anhydrous Na_2SO_4 and reduced to ∼2 ml, chromatographed on a µBondapak NH_2 column and again reduced to a small volume for GC analysis, and finally reduced to dryness for weighing. Normal phase liquid chromatography using a non-polar mobile phase with a µBondapak NH_2 column has been shown to be effective in the removal of polar organic compounds of biological origin from the sediment extract (May et al. 1976). These columns also yield more efficient class separations of the various compounds of interest (aliphatics, aromatic hydrocarbons, and polynuclear aromatic hydrocarbons) than do the silica columns generally employed for such separations. The internal standard was added to the sediment extract immediately prior to analysis by GC.

Dr. Calder's analytical procedure involved multiple extractions of the sediment with methanol and then chloroform in an ultrasonic bath, followed by saponification of the lipids. The non-saponifiable extracts were subsequently chromatographed on alumina:silica gel, and reduced first to a small volume for GC and then to dryness for weighing. The GC analysis was performed using a 3.2 mm O. D. x 1.8 m stainless steel column packed with 4% FFAP on Gas Chrom Z (80/100).

Results and Discussion

The results of this interlaboratory study are summarized in Table 1.

In comparing the quantitative results obtained from gas chromatography, several comments may be made. The overall agreement between the results (0.97 µg/g vs 3.5 µg/g) is quite good considering the present state-of-the art of hydrocarbon analyses in sediment samples. It is evident from the list of the four most abundant compounds and their concentrations, that the extraction methods appear to emphasize the recovery of the aliphatic hydrocarbons at the expense of the substituted two-condensed ring aromatic hydrocarbons. The substituted naphthalenes, of interest due to their high toxicity to marine life, elute chromatographically in the $\underline{n}-C_{11}$ - $\underline{n}-C_{13}$ range. An examination of Calder's and the NBS Soxhlet extraction-GC data indicates some losses of compounds up to ∼$\underline{n}-C_{15}$ (presumably during the extract-concentration step). The four largest peaks from Calder's aromatic fraction listed in Table 1, while unidentified, elute in the range of the substituted naphthalenes and would be subject to losses during the concentration step.

The differences in GC elution profiles from solvent-extracted and headspace-sampled sediment samples are made more evident by comparing histograms of concentration vs time corresponding to the respective gas chromatograms (Fig. 1). While the $\underline{n}$-aliphatics are more prominent in the extracted sediment (Fig. 1A), the substituted naphthalenes are more prominent in the headspace-sampled sediment. The GC-MS total ion chromatogram and m/e 43 and m/e 142, 156 and 170 single-ion records (Fig. 2) confirm the latter. It appears, then, that while both the solvent extraction and headspace sampling methods yield essentially the same value for low level hydrocarbon contamination in a sediment sample, they emphasize different aspects of that contamination. Solvent extraction methods primarily provide information about the

TABLE 1. Results of an Interlaboratory Study of a Katalla River Sediment

Laboratory	Total Hydrocarbons (μg/g dry weight basis)		Pristane/Phytane Ratio	Sample Size (dry weight)	Percent water	Four most abundant compounds/amounts (μg/kg dry weight)
	Gravimetry	GC				
J.A. Calder	33.0	3.5 (C_{12}–C_{25} range)	3.37 ± 0.44* (n=2)	207 g	20	pristane/270, $\underline{n}$-C_{19}/178; $\underline{n}$-C_{20}/178; $\underline{n}$-C_{21}/170 **four largest peaks from aromatic fraction (retention time/amount) 3.09 min/100; 2.61 min/82; 3.92 min/74; 3.33 min/59
NBS-Headspace sampling GC/GC-MS	----	0.97 ± 0.07*	2.34 ± 0.08* (n=8) (C_{11}–C_{22} range)	25–35 g		CH_3-naphthalene/53; $(CH_3)_2$-naphthalene/37; $\underline{n}$-C_{14}/36; $(CH_3)_2$-naphthalene/34
NBS-Soxhlet extraction	85.5	3.5 (C_{11}–C_{22} range)	1.56 (n=1)	190 g	24	$\underline{n}$-C_{14}+ $(CH_3)_2$-naphthalene/325; $\underline{n}$-C_{13}/138; $\underline{n}$-C_{17}/101; $\underline{n}$-C_{15}/89.

* Standard deviation of replicate values from the mean of n replicate values.

** Identity of peaks not provided with data; see text for further discussion.

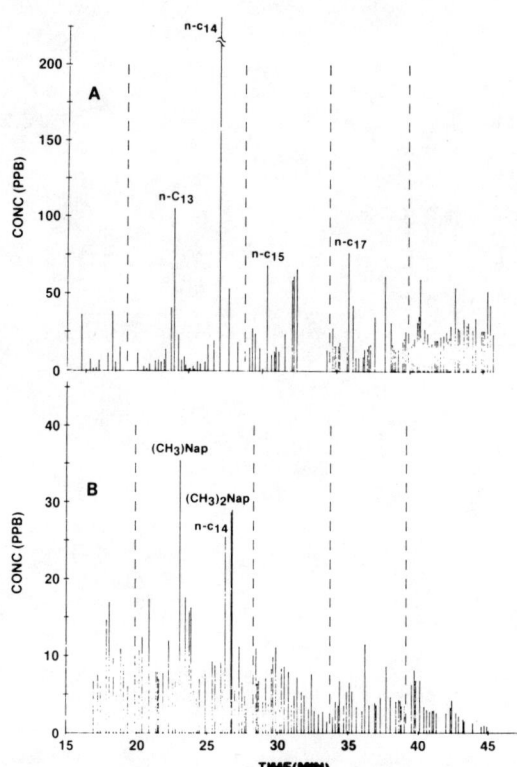

Fig. 35.1 Concentration vs time histograms of (A) Soxhlet extracted and (B) headspace-sampled Katalla River sediment. Peak heights from the respective gas chromatograms have been reduced to the internal standard peak heights (positions noted by vertical lines) and plotted as single species concentrations based on wet weight.

aliphatic hydrocarbons, in the $\underline{n}$-C_{12} - $\underline{n}$-C_{30} molecular weight range; whereas the headspace sampling method provides additional information about the toxic substituted naphthalenes.

Although the results obtained from this intercomparison were quite limited, they did encourage us to initiate a more extensive intercalibration exercise using Katalla River sediment.

For this purpose, large quantities of intertidal sediment were collected from the Katalla River area and from a relatively pristine region, Hinchinbrook Island. After homogenization, samples of the sediment from each site were distributed to a number of laboratories including the various NOAA contracting laboratories. In addition to the information compared in the preliminary sample split, we have requested the participants of this more extensive intercalibration exercise to respond with information regarding the three most abundant aromatic hydrocarbons and total polynuclear aromatic hydrocarbons (four rings and larger). Comparison of this new interlaboratory data should provide a measure of the analytical variability among the participating laboratories.

Acknowledgement

The authors wish to express their appreciation to Dr. John A. Calder for his participation in this study.

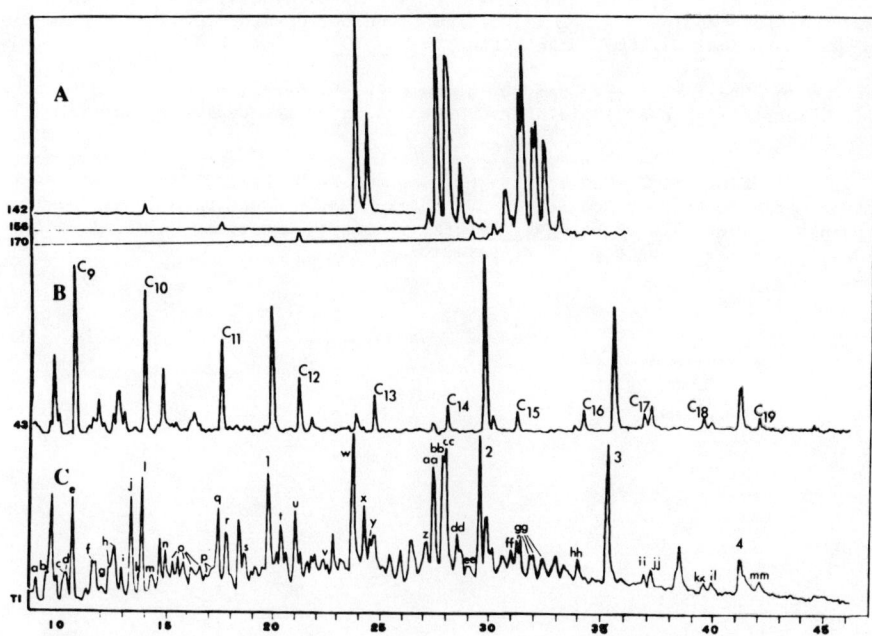

Fig. 35.2 GC-MS analysis of Katalla River sediment: (A) composite m/e 142, 156 and 170 single ion records indicating presence of C_1, C_2, and C_3-naphthalenes, respectively, (B) m/e 43 single ion record, (C) total ion chromatogram. C_x=alkane containing x carbon atoms. C_x-∅=benzene substituted with x carbon atoms (e.g. C_3-∅ could be trimethyl-, propyl-, isopropylbenzene, etc.) Peaks labeled 1,2,3, 4 are the internal standards methyl-C_{11}, methyl-C_{14}, methyl-C_{16}, and methyl-C_{18}, respectively. Identifications followed by "?" are not definite due to incompletely resolved spectra.

a	C_2-cyclohexane	t	C_2-decalin
b	C_2-∅	u	n-C_{12}
c	C_2-∅	v	$\overline{C}_6$-cyclohexane
d	C_3-thiophene	w	C_1-naphthalene
e	n-C_9	x	C_1-naphthalene
f	$\overline{C}_3$-cyclohexane	y	n-C_{13}
g	propyl-∅	z	ethyl-naphthalene
h	C_3-∅	aa	C_2-naphthalene
i	C_3-∅	bb	n-C_{14} & C_2-naphthalene
j	C_3-∅ & C_4-thiophene	cc	$\overline{C}_2$-naphthalene
k	C_4-thiophene	dd	C_2-naphthalene
l	n-C_{10}	ee	ethyl-naphthalene
m	$\overline{C}_3$-∅	ff	n-C_{15}
n	C_4-cyclohexane	gg	$\overline{C}_3$-naphthalene
o	C_4-∅	hh	n-C_{16}
p	C_5-thiophene ?	ii	$\overline{n}$-C_{17}
q	n-C_{11}	jj	pristane
r	$\overline{C}_4$-∅	kk	n-C_{18}
s	C_5-cyclohexane	ll	phytane
		mm	n-C_{19}

References

Chesler, S. N., B. H. Gump, H. S. Hertz, W. E. May, S. M. Dyszel, and D. P. Enagonio, Trace Hydrocarbon Analysis: Prince William Sound/Northeastern Gulf of Alaska Baseline Study, NBS Technical No. 889, Washington, D. C., 1976.

May, W. E., S. N. Chesler, S. P. Cram, B. H. Gump, H. S. Hertz, D. P. Enagonio and S. M. Dyszel, Chromatographic Analysis of Hydrocarbons in Marine Sediments and Seawater, J. Chromatog. Sci. 13, 535 (1975).

May, W. E., S. A. Wise, S. N. Chesler, B. H. Gump and H. S. Hertz, State-of-the-Art Chromatographic Techniques Applied to the Analysis of Hydrocarbons in Marine Tissue, Proceedings of the Division of Environmental Chemistry, American Chemical Society Meeting, Fall 1976, manuscript in preparation.

CHAPTER 36

DETERMINATION OF THE LEEWAY OF OIL SLICKS

Craig L. Smith

Virginia Institute of Marine Science
Gloucester Point, Virginia 23062

Abstract

The leeway of oil slicks was determined as a function of wind velocity in the range 5-25 knots to enable more precise forecasting of the trajectory of oil spills, and thus aid effective containment and cleanup operations. Leeway was calculated by measurement of the separation of oil slicks from a dyed patch of surface water at sea, using time-sequenced nadir aerial photography. Five oil types, Nos. 2, 4, and 6 fuel oils, and light and heavy crude oils, were found to exhibit similar leeway as a function of wind speed. Oil spill volume had no measureable effect on leeway, and slicks moved in the direction of the wind. The leeway increases with sea state and obeys a linear relationship with wind velocity in the wind ranges studied. Oil spill leeway, μ_o, may be calculated for this wind range from the equation:

$$\mu_o = 0.0179 U_{10} + 0.0196, \text{ where } U_{10}$$

is the windspeed measured in knots at 10 meter elevation. Wind drift factors, obtained by summing oil slick leeway and water surface drift velocities, agree well with values calculated from accidental spills and from laboratory scale tests. The mean wind drift factor found from these experiments is 3.64% $\pm$ 0.51%.

Key Words: oil spills, leeway, wind drift factor, trajectory.

Introduction

The movement of spilled oil on the sea is mainly dependent on two factors, the current velocity of the water supporting the oil slick, and the velocity of the local wind. The primary objective of this project is to measure the leeway of various oil types on the sea surface as a function of the wind velocity. Leeway is defined as the movement of the oil slick through (over) the water due to the action of the wind. Previous investigations have shown that oil slicks do move in the general direction of the wind, and with a speed related to the wind velocity. This information is the result of observations of actual oil spills (Smith, 1968; Tomczak, 1964; Stroop, 1927; Munday, 1970; Kolpack, 1969), of laboratory simulated oil spills (Schwartzberg, 1970; Teeson, et al., 1970), and by theoretical inference from the wind-induced drift of the uppermost layers (see Teeson et al., 1970 and Schwartzberg, 1970, and references cited therein). These various approaches suggest oil slick leeway velocities ranging from Munday's smallest value of 0.0% to that of Tomczak, 4.3% of the wind velocity.

It is known that the wind-induced current in the uppermost layers of water diminishes rapidly as a function of depth; the drift velocity 2 cm below the surface may be as little as half that at the surface (see Teeson et al., 1970; Kenyon, 1969; and Neumann and Pierson, 1966). All previous measurements of oil slick drift have assumed little or no motion of the oil slick relative to that of the very uppermost water layer (surface microlayer). The leeway as defined in this project refers to motion of the oil slick relative to the underlying water mass, whose currents may be measured by current meters, surface drifters, etc.

Measurements and interpretation of oil slick leeway must include consideration of the following factors:

1) The effect of oil type on slick leeway
2) The effect of spill volume on leeway
3) The effect of wind speed on leeway
4) The effect of wave climate on leeway
5) The deviation of leeway from wind direction due to Coriolis forces.

Under most conditions at sea, the type of oil spilled would not be expected to affect slick leeway. Laboratory experiments by Schwartzberg (1970) showed only very minor differences between oil types. However, various physical properties of oils such as density, viscosity, and the tendency to form water-in-oil emulsions of the "chocolate mousse" variety (Berridge et al., 1968) may produce observable differences at sea.

There is no a priori justification for the assumption that spill volume will not affect leeway. Again, Schwartzberg (1970) found no effect in tank tests, but these results may not apply to the open sea.

Although one might expect leeway of oil slicks to be directly proportional to wind velocity, there is some evidence that the relationship is more complex, as Munday (1970) suggests a threshold velocity of 7 knots before the oil slick begins to move, and Thorade (1914) claims that the surface microlayer current is represented by two different functions, one for wind speeds less than 6 m/sec., and one for those greater.

Wave action is expected to have some effect on oil slick leeway, but its contribution is in general difficult to separate from that of wind speed, due to the causal relationship of the two. The drift current of the surface microlayer which would carry an oil slick generated by the wind on the water is a combination of the frictionally induced current and the wave produced "Stokes drift", first postulated by Stokes (1847). Although these effects may be approximated mathematically, exact solutions under transient wind and sea conditions are of overwhelming difficulty (see discussion by Kenyon, 1969; Schwartzberg, 1970; and Neumann and Pierson, 1966).

There is disagreement over the magnitude of deviation from wind direction to be expected for oil slick leeway. James (1966) predicts a drift angle of wind driven currents of 20°-22° to the right of true wind direction for mid-latitudes in the northern hemisphere. However, the angle diminishes in shallow water, where frictional forces balance the Coriolis forces. Oil slick drift observed by Munday (1970) was either directly downwind, or to the left of the wind, while the Torrey Canyon slick drifted dead downwind. Conversely, Teeson et al. (1970) claim to have observed drift angles for polyethylene sheets ranging from 0-22° to the right of the wind.

Oil slick leeway in this project was measured using time-sequenced nadir aerial photographic recording of separation distances of small volume oil slicks from a dyed patch of water. A smoke plume generated from the support vessel indicated wind direction in each photograph, and scale was determined from the known length of the support vessel. Wind speed, direction, and sea state were determined from the support vessel during each experiment.

Methods

The site chosen for experimental determination of oil slick leeway was the immediate vicinity of the U. S. Coast Guard Chesapeake Light Station, a tower located about 13 nautical miles off the Virginia coast at $75^\circ 42.8'$ W longitude and $36^\circ 54.02'$ N latitude. This site was selected because it is not as subject to irregular wind and current conditions as a near-shore location, and because of the low probability of inconvenience or damage caused by released oil to public and private property and to the aquatic environment.

Photographs of experiments being conducted at the sea surface were obtained at regular intervals (usually 5 or 10 minutes), with three exposures per pass of the aircraft over the experiment to ensure that at least one frame would record the entire experimental setup. Duration of experiments ranged from 40-150 minutes, depending on surface conditions. Photographs were obtained with a nadir-viewing, motor-driven, single lens reflex camera equipped with a prism viewfinder mounted in a single-engine light aircraft. A 50 mm, f/4, wide angle (75°) lens was used at a shutter speed of 1/500 sec. Ordinary 70 mm daylight color slide film, exposed through a yellow filter provided good definition of all parts of the experimental setup and sufficient resolution at the 3000 ft. altitude flown by the aircraft.

Five different oil types were examined in this project. A sufficient supply of each oil was obtained so that experiments were conducted with aliquots from the same batch. The No. 2 fuel oil was dyed prior to release with Nitro Fast Red B (Sandoz), an oil-soluble red dye, at a concentration of about 1 lb/50 gallons to improve its detectability from the air.

Oil spill experiments were initiated from either a 42 ft. workboat or a 90 ft. research vessel. Oil, usually in 40 liter volumes, was dumped from a steel container off the leeward side of the vessel, which was running at dead slow speed at right angles to the local wind. Usually, two slicks of different oil types were deployed, approximately 150 meters apart. A surface marker, either a dye cannister or about 8 liters of concentrated dye

TABLE 1. Physical Characteristics of Oil Types

Oil	API Gravity, 77°F	Kinematic Viscosity (centistokes)
No. 2 Fuel Oil	33.8° (a)	3.1 (a)
No. 4 Fuel Oil	24° (b)	1000 (b)
No. 6 Fuel Oil	18.9° (a)	2800 (a)
Light Crude Oil	44.2° (c)	2.4 (c, d)
Heavy Crude Oil	14.0° (c)	2600 (a)

[a] Values reported by Schatzberg and Nagy (1971), at 77°F

[b] Values estimated, at 77°F

[c] Values furnished by suppliers

[d] At 100°F

solution (Uranine, fluorescein sodium salt, Dow Chemical) was then deployed upwind of the two slicks. A smoke plume, generated by injection of 30 SAE motor oil into the exhaust of a small gasoline engine, was maintained from the support vessel during each experiment. The support vessel was positioned close to the oil slicks, usually downwind or to the side to prevent interference with the wind flow over the slicks, so that dye patch, oil slicks, support vessel, and smoke plume could all be recorded in a single photograph. Experiments were usually terminated after 2 hours, or whenever the separation distance between the slicks and dye patch became too great for inclusion in a single photograph. Fig. 1 shows the typical experimental layout.

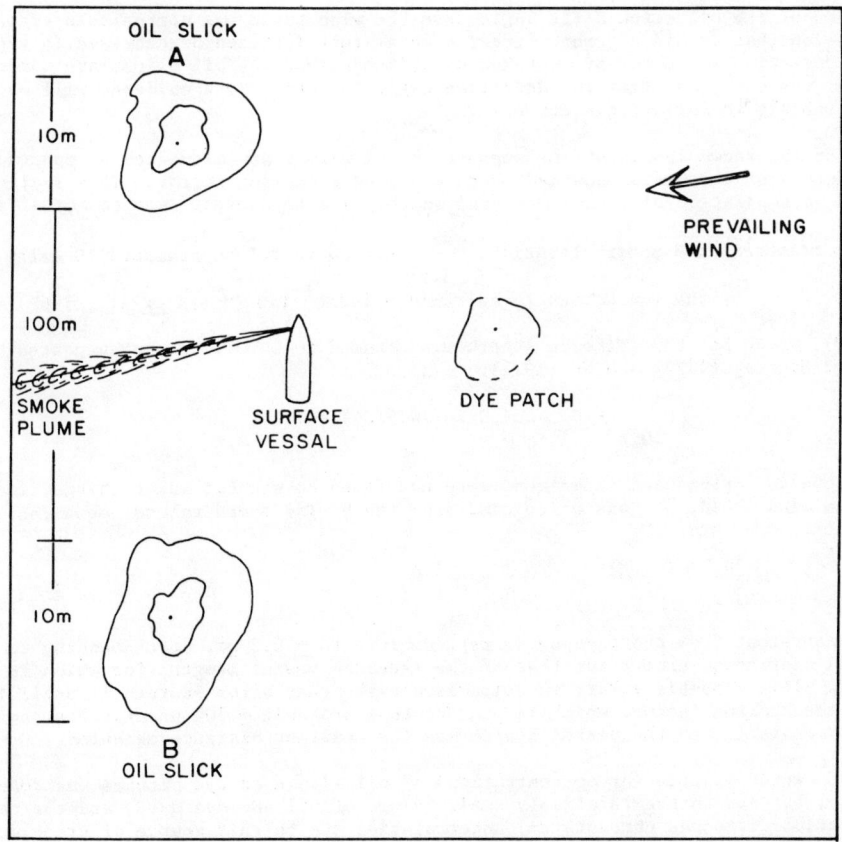

Fig. 36.1 Experimental Layout.

Wind speed was recorded manually aboard the support vessel, using a cup anemometer at an elevation of either 3 or 10 meters. Estimated mean wind speed over a one minute interval was recorded every five minutes. Wind direction was determined from the bearing of the smoke plume.

Film strips developed by a commercial processer were measured using a binocular microscope body with a vernier micrometer X-Y stage. The eye pieces were removed and the objective replaced with a vertically mounted measuring magnifier with a cross-hair reticle. Film was pressed between thin glass plates, and the stage raised until the magnifier reticle almost touched the cover glass. The X-Y coordinate of any given position on the film can then be recorded from the stage micrometer scales. Coordinates of the visually estimated center of the dye patch or dye marker, of the center of mass of each oil slick, the bow and stern of the support vessel, and any two points along the smoke plume were then recorded for each photograph. Use of a more sophisticated technique for determination of centers of slicks and dye was found unnecessary, as visual estimates proved highly accurate and reproducible.

Separation distances between the center of the dyed water (or dye marker) and the center of mass of each oil slick were measured and the components parallel to ($\vec{S_t}$) and normal to ($\vec{S_t'}$) the axis of the wind were calculated using vector algebra. The magnitude of the leeway vector, $\mu_{o,t}$, of each oil slick was calculated for each photographic interval using equation (1).

$$\mu_{o,t} = \frac{\vec{S_t} - \vec{S_o}}{t} \qquad (1)$$

The mean leeway vector, $\overline{\mu_o}$, for the experiment was obtained from equation (2), for n intervals.

$$\overline{\mu_o} = \frac{1}{n} \sum_t \mu_{o,t} \qquad (2)$$

The deviation of the oil slick drift angle from the wind angle was not calculated explicitly. The observation that $\vec{S'}$ did not show either a consistent increase or decrease in experiments where wind direction was constant is taken as evidence that the oil slicks are moving in the direction of the wind, and that the deviation angle is zero. More evidence will be presented on this subject in following sections.

Comparison of the known length of the support vessel with that calculated by subtraction of position vectors of the vessel bow and stern produced a scaling factor. This scaling factor was applied to separation distances measured on the film to convert data to actual scale.

Wind speeds measured at 3 meter elevation, $U_{3,t}$ were converted to standard 10 meter elevation values, $U_{10,t}$ by the von Karman logarithmic velocity law (Myers et al., 1969) shown in equation (3), where Zo, the friction length was assumed to be 0.1 cm, as suggested by Sivadier and Mikolaj (1973) and Wu (1971).

$$U_{10,t} = U_{3,t} \ln(10/Zo)/\ln(3/Zo) = 1.15 U_{3,t} \qquad (3)$$

Since wind speeds during each experiment were not found to exhibit major changes in magnitude, a mean wind speed, $\overline{U}_{10}$ was calculated from the N wind speed values, equation (4).

$$\overline{U}_{10} = \frac{1}{N} \sum_t U_{10,t} \qquad (4)$$

Distance measurement from photographs is reproducible to ± 0.2 mm, corresponding to about ± 1% for all distances, except for that of the research vessel length, for which it is about ± 10%. This probable error, in comparison with other error sources is negligible, except for the scaling factor, which is subject to a probable ± 10% error. The scaling factor error is large, as the vessel length was the smallest distance measured.

Selection of center of mass (or concentration) of oil slicks or dye patches was done subjectively, but due to the relatively small volume of oil and dye used, and the relatively large separation distances encountered, uncertainties due to this source of error are less than ± 10%.

An additional source of error in determination of distances traveled by a slick in the direction of the wind is due to instantaneous deviation of the wind from its mean direction. In some experiments, periodic, frequent wind shifts were observed at the sea surface, with angular deviation of $\pm$ 15-20° from the mean direction. As a result, slicks traveled farther in terms of actual distances over the water than is estimated by the measurement techniques employing the mean wind direction. The maximum error, present in those experiments with variable winds, is on the order of 5% underestimation of the leeway vector.

Therefore, maximum probable error, due to experimental design and measurement techniques, is expected to be on the order of $\pm$ 25%.

Results

The values of the experimentally determined mean leeway, μ_o for the five oil types and the associated mean wind speed values $\bar{U}_{10}$, are summarized in Table 2. Also tabulated are the standard deviation from the means, as each mean is the result of multiple determinations during a single experiment. The mean for each experiment is used in the following statistical treatment rather than the group of μ_o to avoid bias, since the number of determinations in each experiment was far greater for low wind speeds than for the higher ones. For the most part, standard deviations of $\bar{\mu}_o$ do not exceed $\pm$ 25%, and some of the variation must be ascribed to deviations of U_{10} about the mean. It appears, therefore, that variance of measured leeway is of the same order of magnitude as the estimated probable error in measurement from the preceding section.

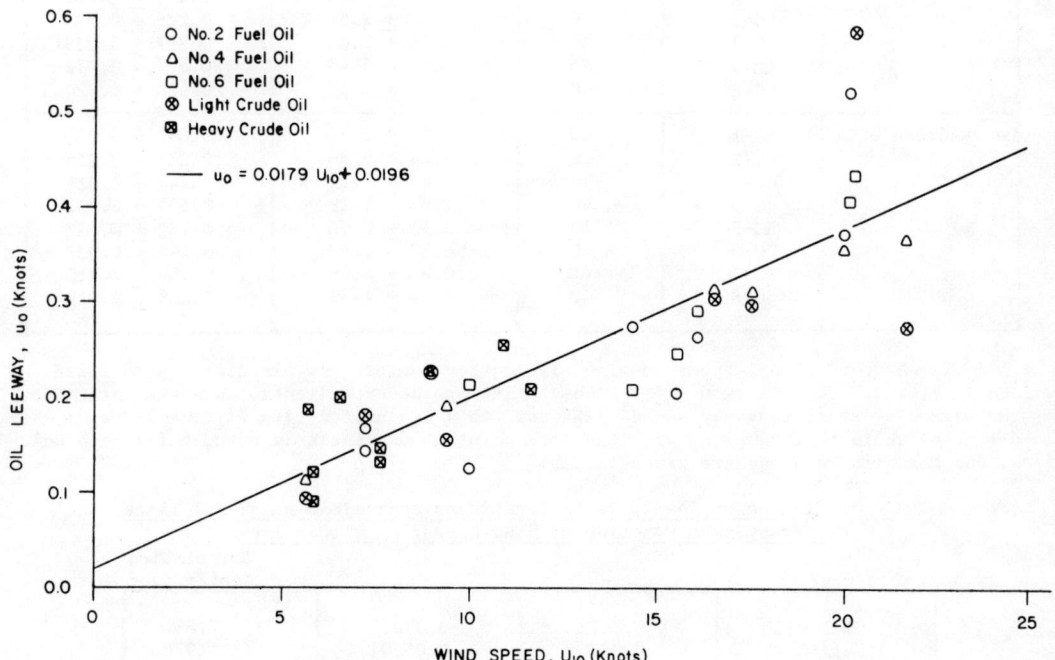

Fig. 36.2 Oil slick leeway as a function of wind speed.

TABLE 2. Measured leeway of oil slicks

Oil type	Experiment	Volume (liters)	Mean wind speed, $\overline{U}_{10}$ (knots) $\pm$ SD	Mean oil slick leeway, $\overline{\mu}_o$ (knots) $\pm$ SD
No. 2 Fuel Oil	2-1	40	20.13 $\pm$ 3.18	0.523 $\pm$ 0.020
	2-2	40	16.14 $\pm$ 2.11	0.264 $\pm$ 0.025
	2-3	40	14.37 $\pm$ 0.62	0.275 $\pm$ 0.029
	2-4	40	10.05 $\pm$ 1.20	0.125 $\pm$ 0.048
	2-5	40	15.55 $\pm$ 1.03	0.205 $\pm$ 0.016
	2-6	40	19.98 $\pm$ 0.78	0.372 $\pm$ 0.010
	2-7	80	7.24 $\pm$ 1.17	0.141 $\pm$ 0.012
	2-8	8	7.24 $\pm$ 1.17	0.166 $\pm$ 0.014
No. 4 Fuel Oil	4-1	60	5.68 $\pm$ 0.94	0.114 $\pm$ 0.028
	4-2	40	17.55 $\pm$ 3.12	0.314 $\pm$ 0.028
	4-3	40	21.67 $\pm$ 2.88	0.369 $\pm$ 0.075
	4-4	40	9.40 $\pm$ 0.88	0.101 $\pm$ 0.022
	4-5	40	8.98 $\pm$ 2.59	0.226 $\pm$ 0.012
	4-6	40	16.54 $\pm$ 0.40	0.314 $\pm$ 0.019
	4-7	40	19.98 $\pm$ 0.78	0.359 $\pm$ 0.038
No. 6 Fuel Oil	6-1	40	20.13 $\pm$ 3.18	0.409 $\pm$ 0.025
	6-2	40	16.14 $\pm$ 2.11	0.292 $\pm$ 0.018
	6-3	40	14.39 $\pm$ 0.62	0.208 $\pm$ 0.106
	6-4	40	10.05 $\pm$ 1.20	0.223 $\pm$ 0.006
	6-5	40	15.55 $\pm$ 1.03	0.247 $\pm$ 0.007
	6-6	40	20.25 $\pm$ 5.49	0.435 $\pm$ 0.119
Light Crude Oil	LC-1	60	5.68 $\pm$ 0.94	0.093 $\pm$ 0.034
	LC-2	40	17.55 $\pm$ 3.12	0.298 $\pm$ 0.030
	LC-3	40	21.67 $\pm$ 2.88	0.275 $\pm$ 0.079
	LC-4	40	9.40 $\pm$ 0.88	0.155 $\pm$ 0.011
	LC-5	40	8.98 $\pm$ 2.59	0.226 $\pm$ 0.039
	LC-6	40	16.54 $\pm$ 0.40	0.307 $\pm$ 0.023
	LC-7	40	20.25 $\pm$ 5.49	0.587 $\pm$ 0.072
	LC-8	40	7.24 $\pm$ 1.17	0.180 $\pm$ 0.020
Heavy Crude Oil	HC-1	40	5.84 $\pm$ 0.86	0.091 $\pm$ 0.039
	HC-2	40	5.84 $\pm$ 0.86	0.119 $\pm$ 0.025
	HC-3	40	7.64 $\pm$ 1.29	0.147 $\pm$ 0.020
	HC-4	40	7.64 $\pm$ 1.29	0.131 $\pm$ 0.021
	HC-5	40	5.75 $\pm$ 1.76	0.187 $\pm$ 0.032
	HC-6	40	6.57 $\pm$ 0.98	0.199 $\pm$ 0.027
	HC-7	40	10.86 $\pm$ 1.46	0.256 $\pm$ 0.015
	HC-8	40	11.64 $\pm$ 1.86	0.208 $\pm$ 0.019

The mean leeway for all oil types combined is plotted against the associated wind speed values in Fig. 2. The function $\mu_o(U_{10})$ best fitting the experimental data was determined by the method of least squares, and is expressed as a linear function of type I: $\mu_o = aU_{10} + b$, which is plotted along with the data points. Coefficients a and b for each oil type, and all oils combined are given in Table 3.

TABLE 3. Values of Coefficients for Linear Approximation of Oil Slick Leeway as a Function of Wind Speed: $\mu_o = aU_{10} + b$

Oil type	a	b(knots)	Correlation Coefficient
No. 2 Fuel Oil	0.0223	-0.0502	0.860
No. 4 Fuel Oil	0.0150	0.0550	0.978
No. 6 Fuel Oil	0.0225	-0.0602	0.887
Light Crude Oil	0.0183	0.0206	0.763
Heavy Crude Oil	0.0160	0.0434	0.681
Common	0.0179	0.0196	0.853

F-tests (Snedecor, 1956) demonstrated that significant departure from linearity was not present. Differences in the regression functions $\mu_o(U_{10})$ among oil types, and between oil types and combined oils were found to be statistically non-significant, using the method of convariance. Since these differences do not follow any systematic relationship with physical characteristics of oil type such as viscosity or density, it is concluded that they are produced by variations in the conditions of the individual experiments, such as wave climate, wind velocity constancy, and perhaps temperature. Thus, leeway of oil slicks of any of the 5 types of oil examined, and probably of any petroleum-based oil, may be predicted by a single equation:

$$\mu_o(\text{knots}) = 0.0179\ U_{10}(\text{knots}) + 0.0196 \tag{5}$$

It is of interest to note that the coefficient b in equation (5) is not zero, implying that oil slicks exhibit leeway in the absence of wind. This result would appear to be in contradiction with the definition of leeway, which may be construed to imply the boundary condition that $\mu_o(0) = 0$. Data for wind speeds in the range of 0-5 knots were not collected. Since values of b for the several oil types showed considerable scatter, with some of them negative, it is concluded that oil slicks do not exhibit leeway in the absence of wind, and that extrapolation of equation (5) to wind speeds of less than 5 knots is not valid.

A function which fulfills the boundary condition is the type IA linear expression: $\mu_o = a'U_{10}$. Values of a' calculated from the experimental data for the five oil types and all oils combined are listed in Table 4. Oil slick leeway values calculated using this function are subject to a slightly greater standard error than those from equation (5) in the wind range of 5-25 knots, but this function is appropriate for winds in the range of 0-5 knots. There is no experimental basis for preference of either for wind speeds greater than 25 knots.

TABLE 4. Values of Coefficients for Linear Approximation of Oil Slick Leeway as a Function of Wind Speed: $\mu_o = a'U_{10}$

Oil Type	a'	95% Confidence Limits
No. 2 Fuel Oil	0.0185	± 0.0038
No. 4 Fuel Oil	0.0196	± 0.0025
No. 6 Fuel Oil	0.0188	± 0.0033
Light Crude Oil	0.0200	± 0.0047
Heavy Crude Oil	0.0221	± 0.0052
Common	0.0199	± 0.0016

Oil slicks were found to move along the same axis as the wind, as determined by the constancy of the distance of the oil slick from a line parallel to the wind passing through the center of the dye patch. This result is further substantiated by observation of a hitherto unreported phenomenon which will be referred to as "tailing". Aerial photographs of oil slicks moving under the influence of the wind show that after a period of time, a circular oil slick becomes distorted into a characteristic "teardrop" profile, with the thicker portions of the oil concentrated near the downwind end (head) of the teardrop, and a long tail of increasingly thin oil extending upwind toward the point of release. This behavior is shown schematically in Fig. 3. Although there is some uncertainty in defining the upwind edge of the thinnest film, it appears to coincide with the point of initial oil release on the water's surface. This phenomenon was observed for all oil types in all wind conditions, but is more noticeable in low to mid-range conditions, as the tail is more rapidly dispersed in rough, whitecapping seas. The tails are observed to be straight in winds of constant direction, and parallel to the wind direction, showing graphically that there is no deviation of oil slick leeway direction from that of the wind.

A detailed analysis of the effect of waves on oil slick leeway is not possible due to the lack of precise sea state data. However, effects were noted, and may be described in a qualitative manner. For most experiments, a fully developed sea was not present, due either to lack of fetch or duration of wind intensity. In two instances, however, experiments HC-5 and HC-6, measurements were made in rather light winds and high seas developed during the previous night. Leeway values were considerably larger than those predicted from equation (5), or from leeway measured under comparable wind speeds (experiments 4-1, LC-1, HC-1, and HC-2). These differences are larger than would be expected from experimental error, and are presumably due to wave action. In other experiments, sea conditions were of varying stages of development, and the effect on oil slick leeway is inseparable from other contributiosn. It is concluded that an increase in sea state will cause an increase in oil slick leeway when wave motion is aligned with the wind.

The effect of oil spill volume was investigated in experiments 2-7 and 2-8 under light wind conditions where spill volumes of 80 and 8 liters of No. 2 fuel oil respectively were tested. No significant differences in slick leeway were noted. However, the phenomenon of tailing, discussed previously, suggests that thick oil slicks exhibit greater leeway than do very thin oil slicks. In the case of very large oil spills where a significant period of time is required for the oil to spread to an equilibrium area, the volume of oil spilled may prove to be a significant variable. This cannot be demonstrated by the small spill volumes investigated in this project.

AT RELEASE

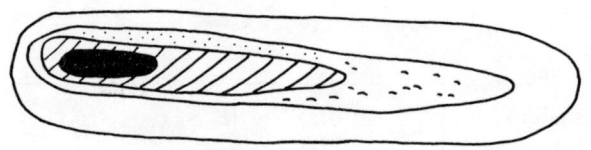

AFTER DRIFTING

Fig. 36.3 Tailing phenomenon.

Discussion

The drift of oil slicks produced by the action of the wind is frequently described in terms of a wind factor: slick drift rate as a percentage of wind speed. The wind factor from data in these experiments is readily obtainable using the value of a' for all oil types from Table 5: 1.99%. This value is somewhat smaller than wind factors observed for the Torrey Canyon slick, 3.4%, and the Gerd Maersk slick, 4.3%, (Tomczak, 1966). Wind factors for these two spills were deduced from changes in geographical position of the drifting oil with time, and are the sum of drift rates due to oil slick leeway and wind-driven surface current. Data from this project may be modified by consideration of surface current velocities for direct comparison with the accidental spill figures. An estimate of the wind-driven surface current may be obtained from the following equation proposed by Thorade (1914), where ϕ is the geographical latitude (37° in these experiments).

$$\mu_s = \frac{0.0361 \sqrt{U_{10}}}{\sqrt{\sin\phi}}, \text{ for } U_{10} < 11.64 \text{ kt,}$$

$$\mu_s = \frac{0.0126 \ U_{10}}{\sqrt{\sin\phi}}, \text{ for } U_{10} > 11.64 \text{ kt.} \tag{6}$$

The modified wind factors calculated by the express $100 \ (\mu_o + \mu_s)/U_{10}$ are plotted in Fig. 4, with the wind factors observed for the accidental spills cited above for comparison. The mean wind factor calculated in this manner is 3.64% ± 0.51% in excellent agreement with the other findings. The wind factor is almost identical to that found by Schwartzberg (1970) from experiments in a small scale test basin, 3.66% ± 0.17%. The large positive deviations of wind factor at a wind speed of about 6 knots were observed in experiments HC-5 and HC-6, where sea state was considerably higher than that to be expected from such a low wind speed.

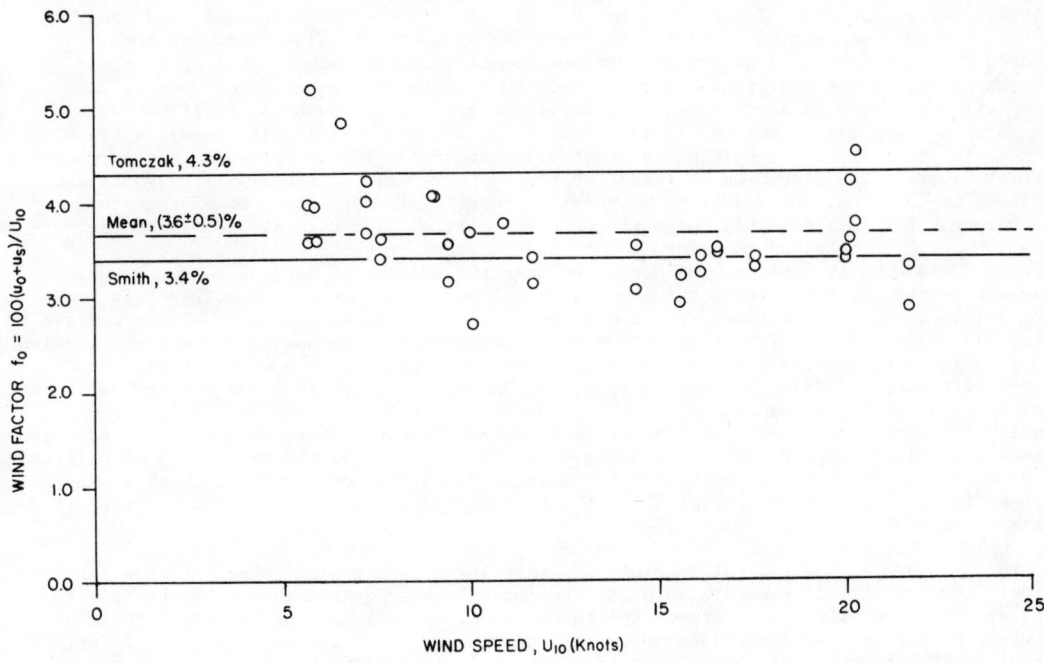

Fig. 36.4 Wind factor as a function of wind speed.

As noted in the previous section, the difference found for leeway among the five oil types lack statistical significance. The apparent order, arranged in decreasing leeway as a function of wind speed is: heavy crude > light crude ≈ No. 4 fuel oil > No. 6 fuel oil ≈ No. 2 fuel oil. For comparison, the oils, in terms of decreasing viscosity (and density) are ordered: heavy crude ≈ No. 6 fuel oil > No. 4 fuel oil > No. 2 fuel oil ≈ light crude. Thus, there is no apparent systematic correlation of either viscosity or density with oil slick leeway.

Deviations of oil slick drift from the observed direction of the surface wind were not observed in the experiments, although Ekman spiral drift theory, based on the action of Coriolis forces on moving bodies, predicts drift angles of up to 22° to the right of the wind for mid-latitudes in the northern hemisphere (James, 1966). There are several possible explanations for these observations. First, the deflections of surface currents from wind direction are known to attenuate in shallow water due to frictional coupling with the bottom (Doebler, 1966; Hela, 1952; Mandelbaum, 1965). Water depths for the experiments described in this project average near 20 meters. The frictional layer calculated according to Sverdrup et al. (1942) may be expected to reach the bottom for wind speeds in the range of this study. Surface current drift angles analyzed by Doebler (1966) in waters of similar depth at comparable latitude ranged between 0° and 10° to the right of the wind. Second, frictional coupling between the surface of the water and the oil tends to resist the Coriolis force. Finally, the oil slick leeway velocities measured in this study are all relative to surface water, and do not consider velocities of the geostrophic, tidal, or other currents in the water column carrying the slick. Thus, the velocities of the slicks relative to the earth, i.e. the velocities which give rise to the Coriolis force, are largely unknown for these experiments.

The most unexpected result of this project was the phenomenon of tailing, described in the previous section. It appears that there is a substantial difference in leeway between very thin oil slicks and thicker ones. Since the amount of oil left behind in the "tail" is always quite small compared with that in the thicker "head" portion, the effects of tailing do not significantly affect the determination of oil slick leeway. However, the phenomenon is of considerable theoretical interest, and suggests some possible limitations on extrapolation of data determined from small oil spills to very large ones.

The thin oil films are chemically different from the thicker oil, as evaporation of volatile constituents proceeds more rapidly for thinner films (Smith and MacIntyre, 1971; Kreider, 1971), causing them to be enriched in the less volatile non-hydrocarbon constituents of oil, many of which exhibit surface activity (Garrett and Barger, 1970). The hydrophilic portion of these surface-active molecules becomes associated with the water surface, and tends to spread to form a monomolecular film. The behavior of surface-active films has been studied in somewhat more detail than that of oil slicks. The main feature of interest in the change in surface energy of water which is covered by such a monolayer, which, among other effects, tends to dampen the small capillary waves. On one hand, Phillips (1966) suggests that damping of the capillary waves impinging on a surface film causes a transfer of momentum to the film, resulting in a net velocity of the film relative to the surrounding surface water in the direction of travel of the short waves. As the capillary waves are produced by the wind, the resultant velocity is normally in the direction of the wind. Furthermore, Fitzgerald (1964), measuring the velocity of talc particles on the water surface, observed an increase in velocity for water covered by surface-active material relative to a clean water surface. He suggested that the damping of capillary waves caused a decrease in the friction length which produced an increased transfer of energy from the wind to the surface. A similar increase in monolayer film speed over that of the surface water was also observed by MacArthur (1962), although Fitzgerald suggested that his results may be subject to errors from failure to consider the spreading rate of the film. In contrast, Keulegan (1951), Van Dorn (1953), and Vines (1962) all observed similar values of surface velocity for clean and slick covered water. Garrett and Barger (1970) report equilibrium spreading pressures of a series of surface-active chemicals sufficient to oppose forces from winds ranging from 11-18 knots. The observations of the phenomenon of tailing suggest that there is no net velocity of the thinnest portion of the tail, which should be richest in surface-active material, relative to the dyed surface water. This apparently contradicts some of the previous findings.

The thicker portions of the oil do exhibit a positive velocity with respect to the surface water. The thicker oil presents a larger drag profile than does a thin film, because the drag profile, the difference between the thickness of the oil lens or clump and the submergence of the bottom of the lens below the undisturbed water surface, is directly proportional to the thickness of the oil. Also, in an oil slick which has begun to spread, the wind must first pass over the thinner portion of the slick, which is damping the capillary waves, before it can act upon the thicker oil. As described above, this leads to a reduction in the friction length of the slick covered surface, and an increase in wind speed near the slick surface over that of the clean water surface. Thus, having a larger drag profile, and feeling the greater wind speed, the thick oil moves with greater velocity than the thin film. It is proposed that thin films do not move with velocities much in excess of the surface water because the interaction of the hydrophilic portion of surface-active molecules in the thin films produces a sufficient coefficient of friction to resist both the wind drag and the momentum transfer from damped capillary waves. In the case of an oil containing large quantities of surface-active materials, a component due to spreading of the surface-active monolayer against the wind may also be present.

Acknowledgement

I wish to thank Dr. Bill MacIntyre for his assistance with the aerial photography and for comments on the manuscript. This work was supported by Contract DOT-GG-33, 183A with the U. S. Coast Guard Research and Development Center.

References

Berridge, S. A., M. T. Thew, and A. G. Loriston-Clarke. The formation and stability of emulsions of water in crude petroleum and similar stocks. pp. 35-59. <u>Scientific Aspects of Pollution of the Sea by Oil</u>, P. Hepple, Editor, Institute of Petroleum, London, (1968).

Doebler, H. J., A study of shallow water drift currents at two stations off the East Coast of the United States, U. S. Navy Underwater Sound Laboratory, USL Report 755, New London, (1966).

Fitzgerald, L. M., The effect of wave-damping on the surface velocity of water in a wind tunnel. <u>Austral. J. Phys.</u>, <u>17</u>, 184-188, (1964).

Garrett, W. D. and W. R. Barger, Factors affecting the use of monomolecular surface films to control oil pollution on water. <u>Environ. Sci. and Tech.</u> 4(2), 123-127, (1970).

Hela, I., Drift currents and permanent flow. <u>Commentaciones Physio-Mathematicae, Societas Scientiarum Fennica</u>, Vol. XVI (14), 1-28 (1952).

James, R., Ocean Thermal Structure Forecasting, U.S. Naval Oceanographic Office, Washington, D.C., (1966).

Kenyon, K. E., Stokes drift for random gravity waves. J. Geophys. Res., 74, 6991-6994, (1969).

Keulegan, G. H., J. Res. Nat. Bur. Stand., 46, 358, (1951).

Kolpack, R., Movement of oil and water in Santa Barbara Channel, California. Presented at the Joint Conference on Prevention and Control of Oil Spills, American Petroleum Institute, New York, (1969).

Kreider, R. E., Identification of oil leaks and spills, pp. 119-124. Proceedings of the Joint Conference on Prevention and Control of Oil Spills, American Petroleum Institute, Washington, D.C., (1971).

MacArthur, I. K. H., Research, 15, 230, (1962).

Mandelbaum, H., Wind generated ocean currenta at Amrum Bank Lightship. Trans. Am. Geophys. Un., 36, 77-86, (1955).

Munday, J. C., W. Harrison, and W. G. MacIntyre, Oil slick motion near Chesapeake Bay entrance, Water Res. Bull., 6, 879-984, (1970).

Myers, J. J., C. H. Holm, and R. F. McAllister, Handbook of Ocean and Underwater Engineering, p. 12-6ff, McGraw-Hill, New York, (1967).

Neumann, G. and W. J. Pierson, Jr., Principles of Physical Oceanography. Prentiss-Hall, Inc. Englewood Cliffs, N.J., (1966).

Phillips, O. M., The Dynamics of the Upper Ocean, P. 40, Cambridge University Press, Cambridge, (1966).

Schatzberg, P. and K. V. Nagy, Sorbents for oil spill recovery, pp. 221-233, Proceedings of the Joint Conference on Prevention and Control of Oil Spills, American Petroleum Institute, Washington, D.C., (1971).

Schwartzberg, H. G., Spreading and movement of oil spills. Water Pollution control Research Series, Environmental Protection Agency. Water Quality Office, Report 15080 EPL, (1970).

Sivadier, H. O. and P. G. Mikolaj, Measurement of evaporation rates from oil slicks on the open sea, pp. 475-484. Proceedings of the Joint Conference on Prevention and Control of Oil Spills, American Petroleum Institute, Washington, D.C., (1973).

Smith, C. L. and W. G. MacIntyre, Initial aging of fuel oil films on seawater, pp. 457-461, Proceedings of the Joint Conference on Prevention and Control of Oil Spills. American Petroleum Institute, Washington, D.C., (1971).

Smith, J. E., Editor, Torrey Canyon: Pollution and Marine Life, pp. 150-162, Cambridge University Press, Cambridge, (1968).

Snedecor, G. W., Statistical Methods. 5th Edition, Iowa State College Press, Ames, (1958).

Sverdrup, H. U., M. W. Johnson, and R. H. Fleming, The Oceans, p. 497. Prentiss-Hall Inc., Englewood Cliffs, N.J., (1942).

Stokes, G. G., On the theory of oscillatory waves, Trans. Camb. Phil. Soc. 8, 441-455, (1847).

Stroop, D. V., Report on oil pollution experiments: behavior of fuel oil on the surface of the sea. Bureau of Standards, Department of Commerce, Washington, D.C., (1927).

Teeson, D., F. M. White, and H. Schenck. Simulation of drifting oil by polyethylene sheets. Ocean Engineering, 2, 1-11, (1970).

Thorade, H. Die Geschwindigkeit der Triftstromungen und die Ekman'sche theorie, Ann. D. Hydr. U. Marit. Meteorol., 42, 379, (1914).

Thorade, H. Die Geschwindigkeit der Triftstromungen. Wiss. Beilag, Realschule zu Eilbeck, Hamburg, (1914).

Tomczak, G., Ozeanographie 10, 129-130, (1964).

Tomczak, G., Investigations with drift cards to determine the influence of wind on surface currents, <u>Studies on Oceanography</u>, p. 129-139, Tokyo, University of Tokyo Press, (1964).

Van Dorn, W. G., Wind stress on an artificial pond. <u>J. Mar. Res.</u>, <u>12</u>, 249-276, (1953).

Vines, R. G., Evaporation control: a method of treating large water stores, pp. 137-160, <u>Retardation of Evaporation by Monolayers: Transport Processes</u>. V. K. LaMer, Ed., Academic Press, New York, (1962).

Wu, J., Evaporation retardation by monolayers: another mechanism. <u>Science</u>, <u>174</u>, 283-285, (1971).

CHAPTER 37

EVAPORATION AND SOLUTION OF C_2 TO C_{10} HYDROCARBONS
FROM CRUDE OILS ON THE SEA SURFACE

Clayton D. McAuliffe

Chevron Oil Field Research Company
La Habra, California 90631

Abstract

Evaporation and solution of C_2 to C_{10} hydrocarbons were measured from four ocean spills of two crude oils (24° and 39° API gravities).

Loss rates for these light hydrocarbons were in accordance with their vapor pressures. The trimethylbenzenes (C_9), the slowest of these to weather, were gone from the oil in 4 to 8 hrs.

C_2 to C_{10} hydrocarbons were found in the water under the oil slicks only during the first 30 min, in concentrations of 2 to 60 µg/l. Their relative concentrations indicated that they were residual hydrocarbons in dispersed oil droplets in the water column, and not in solution at the time of collection.

Oil discharged on a water surface is a nonequilibrium condition with respect to evaporation and solution. Hydrocarbons (alkanes, cycloalkanes, and aromatics) should dissolve from an oil slick in amounts related to their mole fractions in oil, and inversely proportional to their molecular weights. This was not observed, indicating that dissolved hydrocarbons quickly evaporated from the near-surface waters.

The rates of loss of benzene and cyclohexane from the oils were the same. These hydrocarbons have similar vapor pressures, but very different solubilities in water. This also suggests that loss by solution is minor, compared with evaporation.

These studies indicate that solution of hydrocarbons into the water column from crude oil slicks, followed by evaporation, resulted in immeasurably low concentrations (<1 µg/l) of dissolved C_2 to C_{10} hydrocarbons 15 min after each spill.

Key words: oil spills, light hydrocarbons, evaporation, dissolution, crude oils, weathering.

Introduction

Experimental data are presented on the evaporation and solution of low-molecular-weight hydrocarbons from four ocean spills of two different crude oils.

If adverse biological effects (immediate toxicity) result from oil spills, they are thought to be produced principally by the more soluble low-molecular-weight hydrocarbons (principally aromatics). Of importance, therefore, are the concentrations of dissolved hydrocarbons and the duration of organism exposure to them. Several studies (Harrison et al 1975, Kreider 1971, Sivadier and Mikolaj 1973, Smith and MacIntyre 1971) have shown the rapid loss of hydrocarbons with up to about 12 carbon atoms in the molecule in a few hours. Other studies (Anderson et al 1974, Rice et al 1976) have established the equilibrium concentrations of hydrocarbons dissolved into water from an added excess of oil, for laboratory bioassay experiments. These concentrations were obtained by preventing evaporation during 22 hr or longer equilibration.

The above investigations demonstrate both the rapid loss of C_2 to C_{10} hydrocarbons from oil on a water surface, and the maximum dissolved hydrocarbons obtainable in water without permitting evaporation. Little information is available as to what happens when the two processes operate simultaneously. Theory (Harrison et al 1975) and laboratory studies (McAuliffe 1971, 1974) predict that evaporation dominates.

Harrison et al (1975) attempted to quantify the relative rates of evaporation and dissolution from oil discharged on a sea surface, but their analytical method of determining dissolved hydrocarbons was not sufficiently sensitive. Their results did suggest that only a small percentage of the low-molecular-weight hydrocarbons entered the water column.

The only previous measurements of dissolved hydrocarbons under open water conditions were reported by McAuliffe et al (1975) in the oil-in-water emulsion plume produced by spraying chemical dispersants (surfactants) during a Gulf of Mexico spill. Low-molecular-weight hydrocarbons were not found under surface oil slicks that had been on the water for more than 1 hr.

Reported here are measurements made of C_2 to C_{10} hydrocarbons in water under four controlled oil spills, along with the rates of loss of these water-soluble hydrocarbons from the surface oil with time. Details of the spills (rate of spreading, movement, and total crude oil in the water column) will be reported elsewhere (Johanson et al).

Experimental

The API conducted 4 separate spills (10.5 bbl each) through a contractor. The spills were made on open ocean water, two with a 39.0° API gravity Murban crude oil (Abu Dhabi) and two with a 23.9° API gravity La Rosa crude oil (Venezuela). These were selected to bracket the densities and viscosities of many crude oils produced in offshore waters, or transported by tankers in world trade.

Each spill was made from a 500-gal tank through two 3-in rubber hoses. The ends of the hoses were on floats and discharged the oil horizontally on the water surface. This minimized evaporation losses due to discharge above the water, and minimized vertical descent of the oil into the water column. The less viscous Murban oil discharged in 3 min, the La Rosa in 6 min. Table 1 summarizes the wind, wave, and temperature conditions.

Sixty-eight water samples were collected with time at 5 and 10 ft (1.5 and 3.0 m) under the oil slick, by either (a) opening and closing 350 ml glass bottles fitted with spring-loaded teflon stoppers and mounted on an outrigger on the research vessel, or (b) by submersible pump through polypropylene tubing to the deck of the vessel. All sampling devices were

TABLE 1. Wind, Wave, and Temperature Conditions at Time of Crude Oil Spills

	Crude Oil Spills			
	La Rosa 1	Murban 1	La Rosa 2	Murban 2
Wave height, ft	1-2	3-5	3-5	2-3
Whitecaps	Occasional	Frequent	Frequent	Occasional
Wind, knots	10-18	14-20	12-28	8-15
Water temperature, °C	14	14	11	11
Air temperature, °C	17	16	12	12

lowered into and removed from the water outside the observed oil slick areas, to avoid surface oil contamination. Drogues were deployed to measure movement of surface currents relative to surface oil slicks. From the positions of drogue and slick, water samples were collected from locations that should have maximized the oil concentrations in water.

The crown-top glass sample bottles were new, and had been flushed with reactor-grade helium and sealed with crimp-type (crown) caps for storage and shipment prior to sample collection. The bottles were uncapped immediately before use. After sample collection, a 0.25 to 0.5 in gas space was left in each bottle. Solid mercuric chloride was added (100 mg/l) to prevent bacterial destruction of hydrocarbons prior to analysis. The bottles were sealed with plastic-lined crown caps.

A minimum sample depth of 5 ft was selected, to prevent sample contamination by surface oil. Wave heights were generally from 3 to 5 ft (Table 1). Background water samples were collected prior to each oil spill.

Samples of oil were collected in a time sequence from the visually observed thickest parts of each oil slick. Collections were made principally with a small special skimmer developed by JBF Scientific Corporation and towed by a boom. Some oil samples were collected by repeatedly dipping a galvanized steel bucket into the slick, and pouring into a separatory funnel. Excess water was drained from the funnel. Oil collected by the skimmer represented tows through both thick and thin patches. It must be assumed that despite the effort to achieve a spatially composite sample, the thicker oil patches were collected at a faster volumetric rate. Therefore samples were probably predominantly oil from the thick, less weathered lenses. Samples collected by the bucket-separatory funnel method were all taken in the lenses, because very little oil could be collected in the thinner slick areas. Oil samples were transferred to aluminum-lined screw-cap bottles, and mercuric chloride was added to samples with a separate water phase, to prevent possible bacterial degradation of the oil prior to analysis.

Each water sample was analyzed for low-molecular-weight hydrocarbons by a gas chromatographic analysis method of repeated helium equilibrations with water (McAuliffe 1971). The method measures individual C_1 to C_{10} alkanes, cycloalkanes, benzene, and mono-, di-, and trimethyl-substituted benzenes, with sub-µg/l sensitivity.

Oil samples were analyzed for remaining low-molecular-weight hydrocarbons by equilibrating 10 to 20 ml of the original oil or surface-collected oils with 70 ml of sea water collected outside the spill area. The oil and water were hand shaken gently and periodically for 24 hrs or more, to establish equilibrium. Mercuric chloride added to water samples at time of collection prevented possible biodegradation of hydrocarbons during the oil-water equilibration. Twenty-five ml of this water was analyzed by the gas chromatographic gas equilibration technique, as for the under-slick water samples. The water was first transferred from one syringe to another through a 0.45 µm filter, to remove separate-phase oil that may have been dispersed in the water during oil-water mixing. Water samples collected under the slicks were not filtered.

Results and Discussion

Measurement of the concentrations of soluble hydrocarbons as a function of time determines the exposure of organisms to the water-soluble constituents of oil in the water column. The rate of loss of these low-molecular-weight hydrocarbons from surface oil slicks determines the maximum extent to which soluble hydrocarbons can be contributed to underlying waters. Rate of loss of these low-molecular-weight hydrocarbons also determines if they will remain in oil sufficiently long to be transported to shorelines, or possibly to bottom sediments.

Rate of Loss of C_2 to C_{10} Hydrocarbons from Oil Slicks

When oil is discharged on an open sea surface, the slick varies in thickness. Thicker lenses are surrounded by thinner films. The configurations of slicks in this study appeared to depend on the type of oil and the various combinations of wind velocity, wind direction, and currents.

Dissolved hydrocarbons found in water equilibrated with the original La Rosa crude oil and in samples collected with time following the first La Rosa spill are presented in Table 2. The underscored values are the percents remaining in the surface-collected oil samples.

Seawater equilibrated with unweathered La Rosa crude oil (from the spill tank) contained hydrocarbons dissolved in proportion to their individual solubilities and amounts in the oil. The C_2 to C_4 alkane concentrations are high, but benzene and toluene concentrations stand out compared with C_6 and C_7 alkane and cycloalkane hydrocarbons. This hydrocarbon signature is typical for water equilibrated with crude oils [Kuwait, South Louisiana (Anderson et al 1974); Cook Inlet, Prudhoe Bay (Rice et al 1976); and Murban, (reported here)].

Table 2 also shows hydrocarbons weathering from the surface oil rapidly and in accordance with their vapor pressures. Figure 1 is a plot of the losses of benzene, toluene, dimethylbenzenes, and trimethylbenzenes from the surface samples collected with time (Table 2). The length of the bar represents the time the skimmer was towed through the slick. Percentage loss lines have been drawn through the midpoint of the towing time. The loss of these aromatic hydrocarbons may not be representative of the weathering of all the discharged oil, because of nonuniform sample collection. However, the curves show the maximum time for the oil to become depleted of these constituents.

The loss of C_2 to C_{10} hydrocarbons with time from the surface oil for the three other spills is shown in Tables 3-5. All low-molecular-weight hydrocarbons were gone in 4 to 8 hrs. The times required for complete loss of trimethylbenzenes (the last to weather) were similar for the four spills. The second La Rosa spill weathered more quickly than the first, due to higher wind and sea states, even though the water and air temperatures were respectively 3°C and 5°C lower. The first Murban spill weathered more rapidly than the second, because of greater slick agitation. The second La Rosa and first Murban, although different oils, weathered similarly, because of compensating effects of winds, waves, and temperature vs the different oil characteristics, such as viscosity. Other studies (Harrison et al 1975, Smith and MacIntyre 1971) have shown similar results.

Harrison et al (1975) found that trimethylbenzene (cumene) was completely lost in 40 to 80 min (as related to wind and wave conditions) from 5 spills of a South Louisiana crude oil discharged onto 24°C water. Air temperatures ranged from 20.5° to 24.1°C. Normal C_9 was lost at a slightly lower rate (40 to 90 min). Cumene and n-C_9 have similar vapor pressures. By association, these investigators inferred that naphthalene, which has the same boiling point as n-C_{12}, should have disappeared in from 3 to 8 hrs.

TABLE 2. Hydrocarbons Dissolved in Sea Water Equilibrated with Oil Samples, and Percentage Remaining in Surface Slicks (First La Rosa Spill)

Time After Spill	0	18–73 Min		3.3–5.2 Hrs		5.3–6.5 Hrs (b)	
Source of Oil	Tank	Skimmer		Skimmer		Skimmer	
Hydrocarbon		Concentrations in μg/l (ppb)					
Ethane	2011	8.0	0.4 (a)				
Propane	3630	2.0	0.1				
Isobutane	760	12.5	1.6				
n-Butane	1880	6.9	0.4				
Isopentane	600	11.7	2.0				
n-Pentane	600	7.4	1.2				
Hexanes	500	5.8	1.2				
n-Hexane	150	3.2	2.1				
Methylcyclopentane	275	8.5	3.1				
Benzene	3300	80	2.4	0.63	0.02		
Cyclohexane	190	4.0	2.1				
n-Heptane	105	5.9	5.6				
Methylcyclohexane	165	13.7	8.3	0.17	0.10		
Toluene	2800	300	11	2.60	0.09	0.4	0.01
Ethylbenzene	275	110	40	3.20	1.2	0.9	0.3
m-, p-Xylene	840	315	38	10.1	1.2	3.4	0.4
o-Xylene	350	150	43	7.1	2.0	2.4	0.7
Trimethylbenzenes	300	170	57	23.8	7.9	12.8	4.3
Total Saturates	11200	90	0.8	0.2			
Total Aromatics	7860	1125	14	47.0	0.6	20	0.25
Total Hydrocarbons	19000	1210	6.4	47	0.25	20	0.11

(a) __ Underscored value is percent of hydrocarbon remaining in surface-collected oil.
(b) Samples collected 22 and 24 hrs after spill contained no detectable C_2 to C_{10} hydrocarbons.

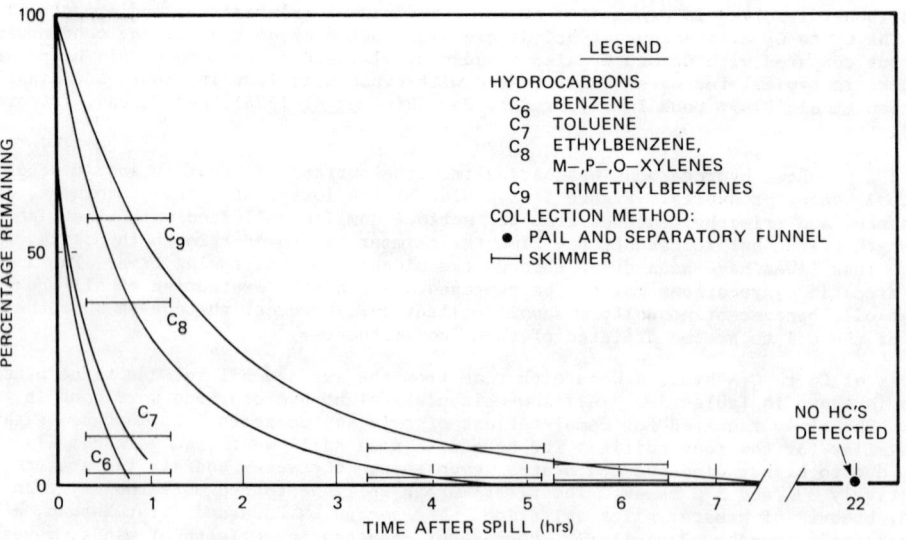

Fig. 37.1 Percent of aromatic hydrocarbons remaining in surface oil slick – first La Rosa spill.

Smith and MacIntyre (1971) measured the complete loss of n-C_{10} in 7 hrs from a 4.8-bbl No. 2 fuel oil spill 15 miles offshore. The air and water temperatures were about 5°C. The wind and waves at the time of release were calm, but the wind increased to 15-18 knots and seas built up to whitecaps after 4-5 hrs.

TABLE 3. Hydrocarbons Dissolved in Sea Water Equilibrated
With Oil Samples (Second La Rosa Spill)

Time After Spill	0	84 Min	9-89 Min	3.7 Hrs[a]
Source of Oil	Tank	Sep. Funnel	Skimmer	Sep. Funnel
Hydrocarbon	Concentrations in µg/l (ppb)			
Ethane	2011			
Propane	3630		0.16	
Isobutane	760		0.37	
n-Butane	1880		0.48	
Isopentane	600		0.42	
n-Pentane	600		0.30	
Hexanes	500		0.27	
n-Hexane	150		0.21	
Methylcyclopentane	275		0.51	0.10
Benzene	3300		3.1	1.20
Cyclohexane	190			0.05
n-Heptane	105			
Methylcyclohexane	165	0.34	0.87	0.34
Toluene	2800	1.40	14.0	4.4
Ethylbenzene	275	1.9	13.5	5.7
m-, p-Xylene	840	6.4	42.0	18.3
o-Xylene	350	3.6	24.3	11.6
Trimethylbenzenes	300	17.3	54	36
Total Saturates	11200	0.34	3.6	
Total Aromatics	7860	30.6	150	77
Total Hydrocarbons	19000	31	155	77

[a] Sample collected by separatory funnel 4.6 hrs after spill contained no detectable C_2 to C_{10} hydrocarbons.

TABLE 4. Hydrocarbons Dissolved in Sea Water Equilibrated
with Oil Samples (First Murban Spill)

Time After Spill	0	72-110 Min	110 Min[a]
Source of Oil	Tank	Skimmer	Sep. Funnel
Hydrocarbon	Concentrations in µg/l (ppb)		
Ethane	230		
Propane	2150		
Isobutane	800		
n-Butane	2880		
Isopentane	1030		
n-Pentane	1340		
Hexanes	850		
n-Hexane	500		
Methylcyclopentane	355		
Benzene	6080		4
Cyclohexane	410		
n-Heptane	330		0.2
Methylcyclohexane	235		0.4
Toluene	6160	25	50
Ethylbenzene	825	80	115
m-, p-Xylene	1940	270	375
o-Xylene	1010	175	240
Trimethylbenzenes	750	270	375
Total Saturates	11100		
Total Aromatics	16800	820	1160
Total Hydrocarbons	27900	820	1160

[a] Sample collected with separatory funnel 4.5 hrs after spill contained no detectable C_2 to C_{10} hydrocarbons.

Low-Molecular-Weight Hydrocarbons in the Water Column under Oil Slicks

Of the 68 water samples collected with time under the four separate oil slicks, C_2 to C_{10} hydrocarbons were found in only the first five samples. These were collected 15 to 20 min after each spill, the time required for the research vessel to maneuver into sampling position after discharging the oil. These hydrocarbons were not found in water samples collected 30 min or later after each spill. Hydrocarbons are not reported for the first La Rosa spill because the first samples were collected at 39 min (10 ft) and 48 min (5 ft) after oil discharge.

The observed concentrations of individual low-molecular-weight hydrocarbons in the water samples are shown in Table 6. The highest observed total concentration was 60 µg/l (ppb), in the 5 ft water sample collected 20 min after the first Murban spill. A 10 ft sample was not collected. Total hydrocarbon concentrations in the other four samples ranged from 2 to 16 ppb. The higher winds at the time of the first Murban spill, compared with those during the second Murban spill, account for the higher values in the water column.

In all samples the dimethyl- and trimethylbenzenes predominated over benzene and toluene. The relative concentrations of hydrocarbons in these samples decreased with decrease in molecular size, and closely resembled the relative concentrations for these hydrocarbons remaining in oil samples collected from the water surface (Tables 2-5). This suggests that the measured dissolved hydrocarbons are from separate-phase oil (droplets) dispersed in the water column. The measured water-soluble hydrocarbons in these samples probably resulted from equilibration of water with these dispersed droplets after sample collection. This will be discussed later.

TABLE 5. Hydrocarbons Dissolved in Sea Water Equilibrated With Oil Samples (Second Murban Spill)

Time After Spill	0	5-35 Min	1.7 Hrs	2.5-2.75 Hrs	4.3-4.7 Hrs	4.75 Hrs (b)
Source of Oil	Tank	Skimmer	Sep. Funnel	Skimmer	Skimmer	Sep. Funnel
Hydrocarbon		Concentrations in µg/l (ppb)				
Ethane	230	0.07	.03 (a)			
Propane	2150	0.38	.02			
Isobutane	800	0.26	.03			
n-Butane	2880	1.25	.04			
Isopentane	1030	1.80	.17			
n-Pentane	1340	3.60	.27			
2,2-Dimethylbutane	15	0.11	.73			
Hexanes	850	7.3	.86			
n-Hexane	500	15	3.0			
Methylcyclohexane	235	20	8.5			
Benzene	6080	515	8.5			
Cyclohexane	410	40	9.8			
n-Heptane	330	40	12.1			
Methycyclohexane	235	55	23	0.26		
Toluene	6160	3130	51	26	0.81	0.80	1.6
Ethylbenzene	825	705	85	70		1.3	4.3
m-, p-Xylene	1940	1825	94	240	2.8	7.3	16.5
o-Xylene	1010	890	88	170	1.6	5.1	14.1
Trimethylbenzenes	750	715	95	360		42	65
Total Saturates	11100	185					
Total Aromatics	16800	7780	870	5	56	100	
Total Hydrocarbons	27900	7960	870	5	56	100	

(a) Underscored value is percent of hydrocarbon remaining in surface-collected oil.

(b) Samples collected 5.5 and 6.5 hrs after spill contained no detectable C_2 to C_{10} hydrocarbons.

In light of the possible contribution of dissolved hydrocarbons from dispersed oil, the Table 6 values should be considered the maximum concentrations and times that organisms would be exposed at 5 and 10 ft under the oil slicks to these hydrocarbons. These concentrations are lower than those known to cause adverse effects on marine organisms, especially at the very short exposure times of 30 min or less.

Harrison et al (1975), as reported above, spiked a Southern Louisiana crude oil with 4.2% isopropylbenzene (cumene) and made 5 separate 6.6-bbl spills. Four water samples were collected under two oil slicks with a Van Dorn sampler. In two of the water samples, cumene was detected at a concentration of about 0.3 mg/l (300 µg/l), but the method of analysis was not quantitative below 1 mg/l. The authors suggested that the samples may have also been contaminated during collection. They suggested that dissolution of cumene into water is two orders of magnitude slower than evaporation.

During discharge of 1500 bbl/day of 34°API gravity South Louisiana crude oil, about 30% was emulsified (McAuliffe et al 1975). The sum of C_2 to C_{10} hydrocarbons in the water column of the dispersed oil plume ranged from 20 to 200 µg/l near the source, decreasing to 2 µg/l at 1 mile. The relative concentrations of the individual hydrocarbons were similar to those of the present study. Again, dispersed oil in the sample probably contributed the dissolved hydrocarbons after collection and prior to analysis.

Evaporation vs Solution of C_2 to C_{10} Hydrocarbons

The concentrations of benzene, toluene, dimethylbenzenes, and trimethylbenzenes in the two crude oils as determined by mass spectrometry (Johanson et al), are shown in Table 7. Although the relative proportions of these hydrocarbons are similar, the total C_6-C_9 aromatic content of Murban crude oil is 2.3%, compared with 0.96% for La Rosa crude oil. Seawater equilibrated with an excess of these oils (without evaporation) will contain aromatic hydrocarbons in concentrations reflecting their mole fractions in the oils and their individual solubilities (McAuliffe 1966). These concentrations are shown in the first numerical column of Tables 3-5, along with the concentrations of alkane and cycloalkane hydrocarbons. Ethane-through-pentane hydrocarbons are present in relatively high concentrations, reflecting their compositions in the crude oils and their intermediate solubilities. Note that the amounts of the aromatic hydrocarbons benzene and toluene (high solubilities) are much higher (~6 mg/l each in Murban, and 3 mg/l each in La Rosa) compared with values of 0.3 to 0.5 mg/l (Murban) and 0.1 to 0.3 mg/l (La Rosa) for C_6 and C_7 saturate hydrocarbons (n-hexane, n-heptane, methylcyclopentane, cyclohexane, and methylcyclohexane). Amounts of benzene and toluene are each approximately 10 times higher than the individual C_8 aromatics (ethylbenzene, m-, p-, o-xylene), and approximately 10 times higher than the sum of the trimethylbenzenes.

Under nonequilibrium conditions and without evaporation, the predominance of lower-molecular-weight hydrocarbons dissolved in water would be futher accentuated over that observed for equilibrium conditions. For a given class of hydrocarbons, the smaller the molecule, the more rapidly it will enter the water phase. For example, if water were mixed with oil until the trimethylbenzenes reached, say, 1% of their equilibrium values, the lower-molecular-weight hydrocarbons would be progressively closer to their equilibrium values as molecular weight decreased. Thus, for aromatic hydrocarbons, benzene would be in solution in higher concentrations than toluene, as contrasted with approximately equal concentrations for Murban and La Rosa crude oils under equilibrium conditions. Similar nonequilibrium relative concentrations would be found for the alkane and cycloalkane classes of hydrocarbons. Measured concentrations in water for all hydrocarbons would become progressively lower as the degree of departure from equilibrium increased.

Oil discharged on the sea surface is a nonequilibrium condition with reference to evaporation and solution of hydrocarbons. Wind is constantly renewing the air over the slick, and waves and water currents renew the water under the slick (at a lower rate). The relative concentrations of the hydrocarbons dissolved in water should be as postulated above for nonequilibrium conditions unless evaporation completely dominates. Namely, the smaller the molecule, the higher the concentration in water for each class of hydrocarbon (alkane, cycloalkane, and aromatic).

This, however, was not observed, as shown by the analysis of the five water samples in Table 6. The compositions of hydrocarbons in Table 6 are very similar to the composition of water equilibrated against oil samples collected from the sea surface. It is known that water under the oil slicks contained dispersed oil, as measured by solvent extraction and infrared analysis (Johanson et al). Dispersed oil was not measured in the same sample collection bottle as the C_2 to C_{10} hydrocarbons. Therefore a direct correlation between total oil and C_2 to C_{10} hydrocarbons is not possible. Three samples for total oil analysis by IR were collected at approximately the same times and depths as for three of the analyses shown in Table 6 (Murban 1, 20 min; Murban 2, 15 min; and La Rosa 2, 19 min). These samples contained respectively 1500, 6000, and 330 µg/l of total oil. Based upon the original composition of the oil (Table 7) and the percent of weathering of hydrocarbons (for example from Table 2), these quantities of oil approximate the amounts and relative concentrations required to give the observed C_2 to C_{10} dissolved hydrocarbon compositions and amounts shown in Table 6.

TABLE 6. Hydrocarbons Dissolved in Water Samples under Oil Slicks

Oil Spill	Murban 1	La Rosa 2	Murban 2	Murban 2	Murban 2
Time After Spill, Min	20	19	15	15	18
Depth, Ft	5	5	5	10	5
Hydrocarbon	Concentrations in µg/l (ppb)				
Ethane	(a)				
Propane					
Isobutane	0.01	0.02			
n-Butane	0.02	0.04			
Isopentane	0.01	0.05			
n-Pentane	0.04	0.04			
Hexanes	0.07				
n-Hexane	0.24				
Methylcyclopentane	0.20	0.10	0.05	0.04	
Benzene	1.58	0.50	0.51	0.25	0.12
Cyclohexane	(b)	(b)	0.68		
n-Heptane	2.30	0.54			
Methylcyclohexane	7.60	0.80	0.30		
Toluene	6.20	0.61	1.22	0.61	0.41
Ethylbenzene	5.30	0.55	1.66	0.34	0.50
m-, p-Xylene	11.4	1.43	3.57	0.71	1.07
o-Xylene	12.2	0.79	3.95	0.39	0.79
Trimethylbenzenes	12.9	0.23	3.87		
Total Saturates	10.5	1.59	1.03	0.04	
Total Aromatics	49.6	4.11	14.8	2.40	2.89
Total Hydrocarbons	60	5.7	16	2.4	2.9

(a) No value, not detected.
(b) Present, but not resolved by GC integrator from benzene.

TABLE 7. Aromatic Hydrocarbons in La Rosa and Murban Crude Oils

Hydrocarbon	La Rosa	Murban
	Weight Percent	
Benzene	0.07	0.13
Toluene	0.22	0.49
Dimethylbenzenes	0.37	0.87
Trimethylbenzenes	0.30	0.74
Total	0.96	2.23

The low-molecular-weight hydrocarbons remaining in dispersed oil collected 15 to 20 min after the spill might be expected to resemble surface oil collected at 1 to 3 hrs, because the small droplets with large specific surfaces would allow more rapid weathering of the C_2 to C_{10} hydrocarbons than would occur from the thicker parts of the oil slick on the water surface.

The above discussion indicates that the measured C_2 to C_{10} dissolved hydrocarbons in the water samples are probably from the dispersed oil, and were not in solution at the time of sample collection, because nonequilibrium conditions would favor the highest concentrations of dissolved hydrocarbons for those hydrocarbons with the lowest molecular weights. These observations indicate that hydrocarbons that dissolve in water quickly evaporate. A surface slick, as mentioned above, is nonuniform in thickness, and the oil film is being continually broken by waves. The surface slick often moves relative to the underlying water. Thus, dissolved hydrocarbons can evaporate. Harrison et al (1975), in their model of evaporation and solution, predict that the ratio of evaporation to solution will be 100 to 1 for aromatic hydrocarbons, and 10,000 to 1 for alkanes. Although there is less tendency for higher-molecular-weight hydrocarbons to dissolve into water from a surface oil slick, those that do will also subsequently evaporate. The partitioning of each class of hydrocarbons between water and air is approximately a constant (McAuliffe 1974). As molecular weights of a class of hydrocarbons increase, vapor pressures and solubilities decrease by about the same percentages. Actually vapor pressures decrease slightly less rapidly than solubilities, thereby favoring evaporation over solution with increase in molecular weight. The data and discussion above indicate that the concentrations for all dissolved hydrocarbons under oil slicks is extremely low, less than 1 µg/l after 15 min. Similar observations were made previously (McAuliffe et al 1975).

Additional evidence that solution is a minor process compared with evaporation is found in the analysis of surface oils collected following the spills. Both processes are operating simultaneously. If solution were important, it should be evident by comparing the loss of hydrocarbons from oil with the same or similar vapor pressures, but with very different solubilities in water (Harrison et al 1975). Such a comparison is benzene (VP, 95.5 mm Hg; solubility, 1780 mg/l), and cyclohexane (VP, 97.8 mm Hg; solubility, 55 mg/l). The third numerical columns of Tables 2 and 5 show the percents of hydrocarbons remaining in the first oil samples collected from the water surface. The percentages remaining are benzene (2.4) and cyclohexane (2.1) for the first La Rosa spill (Table 2), and benzene (8.5) and cyclohexane (9.8) for the second Murban spill (Table 5). The La Rosa spill suggests a preferential loss of cyclohexane - Murban, benzene. However, the values are close and considered in light of possible intermixing of variously weathered samples during collection (skimmer) and errors in analysis, no preferential loss is shown. This indicates that solution is a minor process compared with evaporation even for aromatic hydrocarbons. These data confirm the prediction (Harrison et al 1975) and other experimental evidence for the loss of low-molecular-weight hydrocarbons from oil and water (McAuliffe 1971, 1974).

Acknowledgement

The hydrocarbon measurements were performed under a contract from the American Petroleum Institute.

References

Anderson, J. W., J. M. Neff, B. A. Cox, H. E. Tatem, and G. M. Hightower, Characteristics of dispersions and water-soluble extracts of crude and refined oils and their toxicity to estuarine crustaceans and fish, Mar. Biol. 27, 75-88 (1974).

Harrison, W., M. A. Winnik, P. T. Y. Kwong, and D. Mackay, Crude oil spills. Disappearance of aromatic and aliphatic components from small sea-surface slicks, Environ. Sci. Technol. 9, 231-234 (1975).

Johanson, E. E., J. C. Johnson, C. D. McAuliffe, and R. A. Brown, Physical and chemical weathering of crude oil slicks on the ocean, (Manuscript in preparation).

Kreider, R. E., Identification of oil leaks and spills, Proceedings Joint Conference on Prevention and Control of Oil Spills, American Petroleum Institute, Washington, 119-124 (1971).

McAuliffe, C. D., Solubility in water of paraffin, cycloparaffin, olefin, acetylene, cycloolefin, and aromatic hydrocarbons, J. Phys. Chem. 70, 1267-1275 (1966).

McAuliffe, C. D., GC determination of solutes by multiple phase equilibrium, Chem. Technol. 1, 46-51 (1971).

McAuliffe, C. D., Determination of C_1-C_{10} hydrocarbons in water, Marine Pollution Monitor-(Petroleum), NBS Spec. Publ. 409, U.S. Government Printing Office, Washington, 121-125 (1974).

McAuliffe, C. D., A. E. Smalley, R. D. Groover, W. M. Welsh, W. S. Pickle, and G. E. Jones, The Chevron Main Pass block 41 oil spill: chemical and biological investigations, Proceedings 1975 Conference on Prevention and Control of Oil Pollution, American Petroleum Institute, Washington, 555-566 (1975).

Rice, S. D., J. W. Short, C. C. Brodersen, T. A. Mecklenburg, D. A. Moles, C. J. Misch, D. L. Cheatham, and J. F. Karinen, Acute toxicity and uptake-depuration studies with Cook Inlet crude oil, Prudhoe Bay crude oil, No. 2 fuel oil and several subarctic marine organisms, Northwest Fisheries Center Auke Bay Fisheries Laboratory, Processed Report, May (1976).

Sivadier, H. O. and P. G. Mikolaj, Measurement of evaporation rates from oil slicks on the open sea, <u>Proceedings Joint Conference on Prevention and Control of Oil Spills</u>, American Petroleum Institute, Washington, 475-484 (1973).

Smith, C. L. and W. G. MacIntyre, Initial aging of fuel oil films on sea water, <u>Proceedings Joint Conference on Prevention and Control of Oil Spills</u>, American Petroleum Institute, Washington, 457-461 (1971).

CHAPTER 38

INPUT OF LOW-MOLECULAR-WEIGHT HYDROCARBONS FROM PETROLEUM OPERATIONS INTO THE GULF OF MEXICO

James M. Brooks, Bernie B. Bernard and William M. Sackett

Department of Oceanography
Texas A&M University
College Station, Texas 77843

Abstract

Dissolved C_1 to C_4 hydrocarbon patterns measured during the last 6 years in the Gulf of Mexico indicate that underwater venting of waste gases and brine discharges, both associated with offshore platforms, are the major sources of non-methane light hydrocarbons to upper Gulf coastal waters. These sources are apparently responsible for the two orders of magnitude increase in Louisiana Shelf waters over open ocean levels of the light hydrocarbons with average concentrations of 3100, 31, and 22 nanoliters per liter of methane, ethane, and propane, respectively.

Analyses of the hydrocarbon composition of vented gases and brines and estimates of their annual discharge rates indicate that up to 450 metric tons of C_5 to C_{10} hydrocarbons are being added to Louisiana Shelf waters each year.

Although the C_1 to C_4 hydrocarbons *per se* are apparently not toxic to marine organisms, they nevertheless are proving to be highly sensitive indicators of the more toxic components of petroleum which are being introduced to the sea by man's activities.

Key words: methane, ethane, propane, butane, Gulf of Mexico, gas venting

Introduction

Dissolved C_1 to C_4 hydrocarbon patterns in surface waters of the Gulf of Mexico have been measured for the past six years during cruises of the Texas A&M University research vessels. Thousands of analyses made on these cruises have identified ports and estuaries with their associated commercial and petrochemical activities, offshore petroleum operations, and shipping activities, as the major man-derived sources of low-molecular-weight hydrocarbons (Brooks and Sackett, 1973, Brooks *et al.*, 1973). The water column for at least one example of each of these three types of inputs has shown hydrocarbon concentrations several orders of magnitude higher than the water column in the open Gulf of Mexico. Low-molecular-weight hydrocarbon concentrations are proving to be highly sensitive indicators of man-derived petroleum pollution. This paper will focus on the major man-derived source of light hydrocarbons to the Gulf of Mexico--offshore drilling and production operations.

Surface concentrations of light hydrocarbons are determined by hydrocarbon "sniffing" which involves the continuous extraction of dissolved gases from solution followed by gas chromatographic analysis (Brooks and Sackett, 1973). The "sniffer" data given in this paper are reported in relative hydrocarbon units which are multiples of open ocean concentrations. The open ocean concentration of methane is approximately 50 nanoliters of gas per liter of sea water (nl/L), ethane 0.5 nl/L, propane 0.35 nl/L and the butanes 0.05 nl/L (Swinnerton and Lamontagne, 1974). Thus a relative methane concentration unit of 10 is equal to 500 nl/L (10 x 50 nl/L).

Light Hydrocarbon Patterns

Louisiana and Upper Texas Shelf
Offshore petroleum operations are the major source of light hydrocarbons to the Texas-Louisiana Shelf. The production of petroleum from over 1200 OCS blocks leased in this region results in the highest hydrocarbon concentrations observed in the Gulf of Mexico. There are also significant contributions from onshore petrochemical industries and refineries, and runoff from the Mississippi River. These factors result in a very dramatic and complex temporal and spatial distribution of hydrocarbons on the Louisiana Shelf.

Figures 1 and 2 show relative C_1 to C_3 hydrocarbon concentrations across portions of the Louisiana Shelf. Most methane concentrations range from 2 to 200 times open ocean concen-

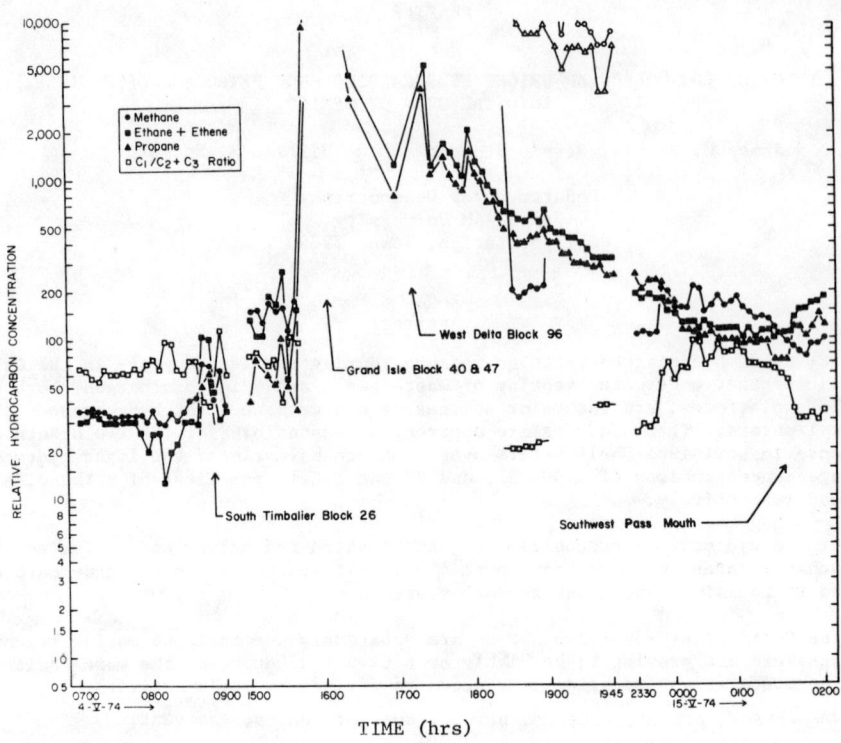

Fig. 38.1 Relative light hydrocarbon concentrations on the Louisiana Shelf --- Cruise 74-G-8 (May 1974).

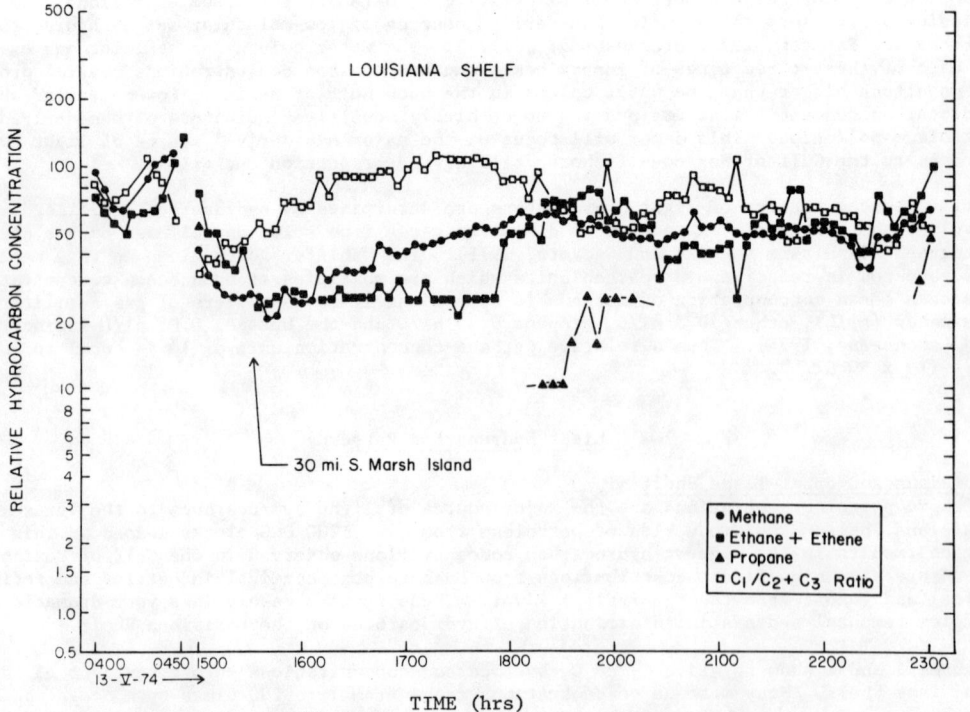

Fig. 38.2 Relative light hydrocarbon concentrations on the Louisiana Shelf --- Cruise 74-G-8 (May 1974).

trations (100 to 10,000 nanoliters of CH_4/liter of sea water). The exceptions are in the immediate vicinity of some platforms where concentrations may climb into the range of milliliters of CH_4/liter (1600 hrs on 14-V-74, Fig. 1) and on the outer continental shelf where open ocean values are observed. Assuming the approximately 1600 individual analyses of methane performed during Cruises 74-G-5 (March 1974) and 74-G-8 (May 1974) on the Louisiana Shelf provide a representative sampling of this area, the average methane concentration in the surface water is 62 times open ocean concentrations or 3100 nl/L.

The 74-G-5 and 74-G-8 cruise data (examples shown in Fig. 1 and 2) indicate that relative methane, ethane, and propane concentrations closely parallel one another over much of the Louisiana Shelf. Thus, the ethane and propane concentrations are approximately 1.0 and 0.7%, respectively, of the methane concentration. The data indicate that these hydrocarbons originate from similar sources, although sufficient deviations exist between their relative concentrations to suggest that there are differences in the hydrocarbon compositions of the various point sources. Assuming that the relative average ethane and propane concentrations over the Louisiana Shelf are similar to methane, ethane averages 31 nanoliters/liter (62 x 0.5 nl/L) and propane 22 nanoliters/liter (62 x 0.35 nl/L).

Also plotted on Fig. 1 and 2 are values of the mole ratio of methane to ethane plus propane, $C_1/(C_2+C_3)$, which has been suggested as an indicator of hydrocarbon origin (Frank et al., 1970). Ninety-five percent of the light hydrocarbon compositions of oil and gas from wells in Texas and Louisiana have $C_1/(C_2+C_3)$ ratios between 5 and 50 (Moore et al., 1966). Bernard et al. (1976) reported $C_1/(C_2+C_3)$ ratios from several underwater vents between 7 and 36 and brines have similar values (Table 1). There are many regions on the Louisiana

TABLE 1. Light Hydrocarbon Concentrations in Produced Brines

Sample[1]	C_1	C_2	C_3	$i-C_4$	$n-C_4$	$i-C_5$	$n-C_5$	Benzene[2]	Toluene	$C_1/(C_2+C_3)$
Brine 1[3]	13,000	970	300	75	85	31	17	9.8	3.3	10
Brine 2	160	30	24	9.1	4.0	4.2	-	3.3	0.29	3
Brine 3	171	8	0.3	0.02	<0.01	-	-	-	-	21
Brine 4	966	48	5.8	0.14	<0.01	-	-	-	-	18
Brine 5	105	7.9	0.4	<0.01	<0.01	-	-	-	-	13
Brine 6	36	2.0	0.1	<0.01	<0.01	-	-	-	-	17

[1]C_1 to C_5 hydrocarbon concentrations expressed as 10^{-6} liters of gas/liter of brine.

[2]Benzene and Toluene concentrations expressed as mg/L by weight (ppm).

[3]Brine 1 is the only sample stored in glass, other brine samples were weathered for 2 to 3 weeks by storage in 1-liter plastic containers with an air space above the brine.

Shelf where this dissolved gas ratio ranges from 15 to 40 indicating a petrogenic origin. There are, however, other regions on the Louisiana shelf that have ratios in the 50 to 100 range. These higher values may indicate contributions from biological sources producing primarily methane and/or natural seepage of gas having a high $C_1/(C_2+C_3)$ ratio. To complicate the system, it is also quite likely that the ratio is altered as the light hydrocarbons are dispersed and degraded in surface water.

The spatial surface distribution of low-molecular-weight hydrocarbons on the Louisiana Shelf is extremely variable and is largely controlled by the proximity of production platforms. The highest concentrations are generally found on the mid-Louisiana Shelf region (Ship Shoal, Eugene Island and South Marsh Island Areas) and in the Mississippi Delta region (Grand Island, West Delta, and South Pass areas). Although both of these regions have large numbers of production platforms, it is difficult to accurately estimate the relative contributions of man and nature to the average value of 3100 nl/L methane because of the lack of early baseline measurements. In contrast, much of the southwestern part of the Louisiana Shelf still has low, open ocean concentrations of methane in areas not yet developed by oil companies.

Most of the upper Texas Shelf, subject to relatively little production, is "clean" with respect to light hydrocarbons. However, there are indications of an increase in light hydrocarbon inputs into parts of the upper Texas Shelf over the last few years, as illustrated by the comparison of data collected during transects from Galveston Harbor to Claypile Bank (approximately 60 miles SE of Galveston) on 12 October 1971 and 16 March 1974.

The values of 16 March 1974 are considerably higher, averaging 3 times open ocean concentrations of methane over much of the transect. Values from 12 October 1971 only average 1.3 times open ocean concentrations over the majority of the cruise track, with several anomalies up to 2 times open ocean concentrations. This increase over the years could be attributed to a seasonal variation; *e.g.*, possibly a greater intrusion of Louisiana Shelf water into this region during the 74-G-5 cruise, but is most likely the result of an actual increase in light hydrocarbon inputs into the upper Texas Shelf region as a result of expanded offshore operations during the time interval between the cruises. Some of the hydrocarbon anomalies along the Galveston to Claypile Bank transect can easily be attributed to the proximity of oil fields.

Lower Texas Shelf
The lower Texas Shelf is almost free of oil production, and has characteristic, open ocean concentrations of light hydrocarbons except in regions near coastal inputs; *e.g.*, rivers and bays. Figure 3 shows station locations and surface methane concentrations along the lower part of this shelf. Table 2 lists concentrations with depth of C_1 to C_3 hydrocarbons

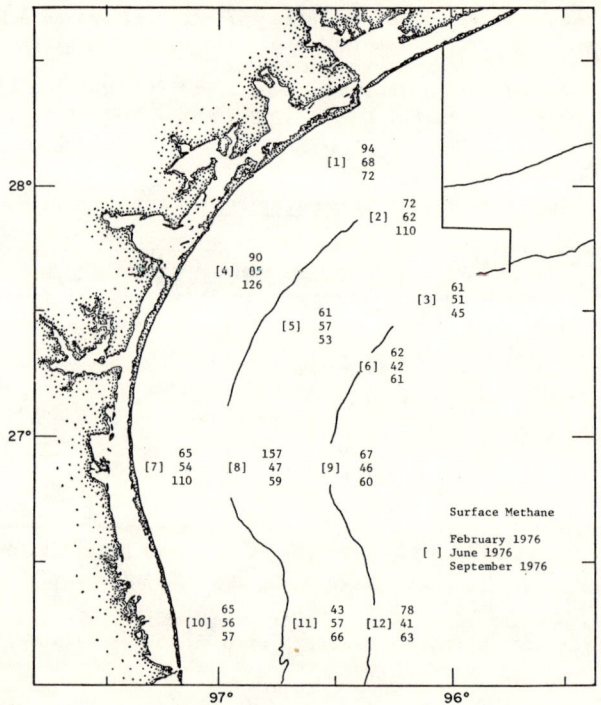

Fig. 38.3 Surface methane concentrations on South Texas Shelf (units = nl/L).

at six of these stations during June 1976. Sampling at depth is most easily accomplished by taking discrete water samples in Nansen bottles, usually at 10 meter intervals. All of the stations contain natural levels of hydrocarbons, in contrast to the high levels shown by Fig. 1 and 2 on the Louisiana Shelf. The increases observed in near-bottom samples are most likely due to methane seepage from the shelf sediments, as gas seepage is common in these areas. Light hydrocarbon concentrations are now being monitored monthly at the locations shown in Fig. 3 to determine baseline concentrations of the light hydrocarbons in South Texas Shelf waters. As the petroleum industry develops the South Texas OCS region, light hydrocarbon concentrations can be expected to rise, and the measured baseline concentrations will become valuable in assessing relative contributions of man and nature.

Platforms, Venting, and Brine Discharges

As demonstrated earlier, the distribution of light hydrocarbons on the Louisiana shelf can largely be attributed to hydrocarbon inputs associated with offshore platforms. In waters in the immediate vicinity of production platforms, dissolved light hydrocarbon concentrations are not as predictable, being controlled by several factors. For example, waters adjacent to platform groups farthest offshore (presumably the newest) contain light hydro-

TABLE 2. C_1 to C_3 Hydrocarbon Concentrations in South Texas Coastal Waters

Station	Depth	Methane (nl/L)	Ethene (nl/L)	Ethane (nl/L)	Propene (nl/L)	Propane (nl/L)
[1]	0 m	68	7.9	0.35	1.86	0.49
	9 m	67	4.4	-	-	-
	18 m	370	6.4	0.58	1.52	0.61
[2]	0 m	62	6.7	0.41	1.86	0.24
	10 m	58	4.7	-	-	-
	20 m	71	3.4	-	-	-
	30 m	138	2.8	-	-	-
	40 m	135	3.6	0.50	0.56	0.42
[3]	0 m	51	12.2	0.29	1.86	0.42
	15 m	48	8.1	-	-	-
	28 m	60	11.0	-	-	-
	40 m	57	9.6	-	-	-
	50 m	58	5.3	-	-	-
	60 m	61	3.4	-	-	-
	70 m	58	2.7	-	-	-
	80 m	96	2.0	-	-	-
	90 m	103	1.8	-	-	-
	100 m	100	1.6	-	-	-
	110 m	124	1.4	-	-	-
	120 m	129	1.3	-	-	-
	130 m	166	1.3	0.38	0.22	0.42
[10]	0 m	85	7.8	0.44	1.60	0.49
	7 m	83	4.2	-	-	-
	15 m	99	4.7	0.41	1.12	0.36
[11]	0 m	56	5.8	0.35	1.80	0.52
	10 m	56	4.1	-	-	-
	21 m	91	3.8	-	-	-
	30 m	101	2.6	-	-	-
	40 m	99	2.8	-	-	-
	45 m	121	2.9	0.46	0.59	0.49
[12]	0 m	42	13.1	0.35	1.63	0.41
	10 m	54	7.1	-	-	-
	21 m	58	8.4	-	-	-
	30 m	56	8.8	-	-	-
	40 m	61	9.7	-	-	-
	50 m	52	7.8	-	-	-
	60 m	55	2.9	-	-	-
	70 m	57	2.0	-	-	-
	80 m	95	2.7	-	-	-
	90 m	106	2.0	-	-	-
	100 m	122	1.6	-	-	-
	110 m	115	1.4	-	-	-
	120 m	84	1.0	-	-	-
	130 m	74	0.7	0.37	0.28	0.52

carbon levels two to three orders of magnitude above levels found in open Gulf surface waters. Water around platforms nearer the coast, presumably in place for a longer period of time and having a more highly developed pipeline system than the platforms cited above, had much lower levels but still one to two orders of magnitude higher than open Gulf surface waters. More important than stage of development, however, are the two principal inputs that appear to control the magnitude of hydrocarbon levels around platforms: underwater venting of gas and discharge of oil field brines.

Hydrocarbon Venting

Offshore production platforms are built to provide a working facility to separate water, liquid and gaseous hydrocarbons. Separation is required offshore because of the extremely large pressure drops which would be experienced due to two-phase flow of liquids and gases being transported together by pipeline (Gunn and Gueringer, 1970). The oil would also require a larger storage system for barge or tanker transport if the dissolved gases were not removed. Separation of crude oil and gas then, is made offshore to allow single-phase transmission of crude oil and gas through separate pipelines. Depending on the economics of transportation, the separated gas may either be transported for sale or disposed of in

TABLE 3. Molecular and Isotopic Composition of Underwater Vent

Compound	Composition (molar)
Methane	84±1%
Ethane	7.0±0.1%
Propane	4.8±0.1%
Isobutane	0.9±0.1%
n-Butane	2.0±0.1%
Isopentane	0.6±0.1%
n-Pentane	0.6±0.1%
$\delta^{13}C$†(methane)	$-42\pm1°/_{\circ\circ}$

†$\delta^{13}C = [(R/R_{std})-1] \times 1000$ where R and R_{std} represent the ratio of ^{13}C and ^{12}C in sample and standard, respectively, the standard is PDB_1---the Chicago belemnite standard.

one of several ways. Waste gas may be released by venting either below or above water, by flaring, or by compressing the gas and pumping it back into the reservoir to provide gas lift. The first two options account for the majority of the natural gas disposal in the Gulf of Mexico. Although some gas is flared, it appears that most waste gases are vented underwater. The petroleum industry considers underwater venting preferable because it eliminates possible ship-pipe collisions or hurricane damage associated with above-water flaring. Since platform operations involve the separation of volatile and flammable gases and liquids, underwater venting eliminates the problems associated with burning flares in the vicinity of producing platforms. Underwater venting also permits the safe disposal of larger amounts of gas than can be eliminated easily by flares.

Produced natural gas can originate from the operations of oilfields as a by-product (associated gas) or from reservoirs producing only gas (non-associated gas). Presumably all gas being flared or vented in the Gulf is associated gas. All oils at reservoir pressures contain associated gas ranging in volume from a few cubic feet to over 1000 cubic feet per barrel. The associated gas at reservoir pressures is dissolved in the oil, but as the pressure is reduced during production the gas solubility in the oil drastically decreases and the gas separates from solution. Formation-volume factors (volume of reservoir oil necessary to yield one barrel of stocktank oil) for oils generally range between 1.14 and 1.60 (Levorsen, 1967). Heavier crude oils have a much greater capacity to hold gas in solution both at reservoir and atmospheric pressures. The associated gas is separated from the oil in gas/oil separators mainly by physical density separation.

The associated natural gas, unlike nonassociated gas, generally contains appreciable amounts of condensates. Since the associated gas has been dissolved in the liquid hydrocarbon phase it contains at equilibrium a complex mixture of both aliphatic and aromatic hydrocarbons. Thus a wide range of C_1 to C_{12} and greater hydrocarbons are found in associated gas. Depending on the operating procedures, various amounts of the liquid condensates are removed by the oil/gas separator before disposal of the gases.

Our analyses of the water surrounding approximately 10 underwater vents and the vent gas itself indicate that there are large variations in the C_5 and higher hydrocarbon content of vented gases in the Gulf of Mexico, ranging from light gases with only small amounts of greater than C_5 to gases whose liquid hydrocarbon compositions produce surface hydrocarbon films in the vicinity of the vent. Table 3 shows the chemical and carbon isotope composition of the gas collected from one of these vents. The table indicates the vent has appreciable quantities of all the light gaseous hydrocarbons.

In addition, the vent gas contains quantities of the light liquid hydrocarbons. Figure 4 is a mass spectrum of this vent gas showing m/e^+ peaks as high as 169, corresponding to a 12 carbon compound. The limits of detection of the CEC110B analytical mass spectrometer were approximately 10 ppm relative to the total gas sample. More sensitive analyses by gas chromatography revealed compounds as high as C_{15}. Figure 5 shows the gas chromatogram of the vent gas collected at the surface after bubbling through the water column. It shows a complex spectrum of peaks up to C_{15}. This spectrum differs considerably from the chromatogram of a produced brine (Fig. 6) which shows that greater than 90% of the C_5 to C_{10} compounds in the sample are aromatic. This can be explained by selective solution into the water column of the aromatic compounds in the gases coming from the vent. The aliphatic hydrocarbons have a much greater partition coefficient into the gas phase due to much smaller solubilities than the aromatics.

Figure 7 shows station locations from a detailed investigation around an underwater vent in East Cameron lease block 272 (vent is located at Sta. 7). Dissolved hydrocarbon concen-

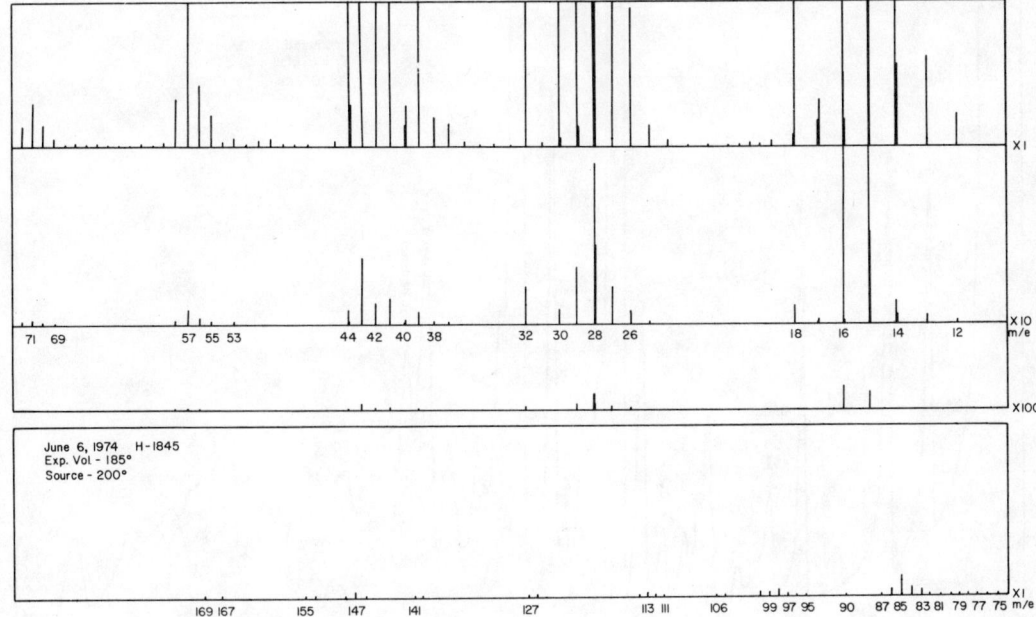

Fig. 38.4 Mass spectrum of gas collected from underwater vent.

trations and hydrographic parameters are shown in Table 4. The underwater vents have only a slight influence on the salinity, temperature, dissolved oxygen and nutrient structure of the water column. One observed influence, however, apparent in XBT traces from these sta-

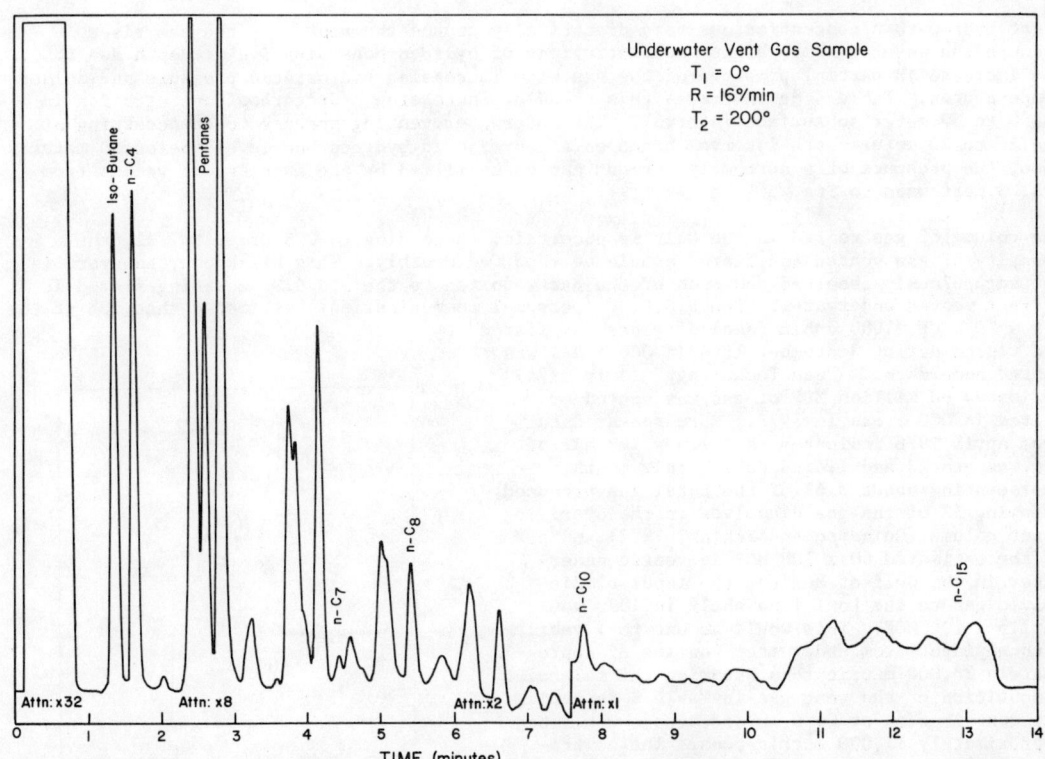

Fig. 38.5 Gas chromatogram of gas collected from an underwater vent ---same gas as Fig. 4.

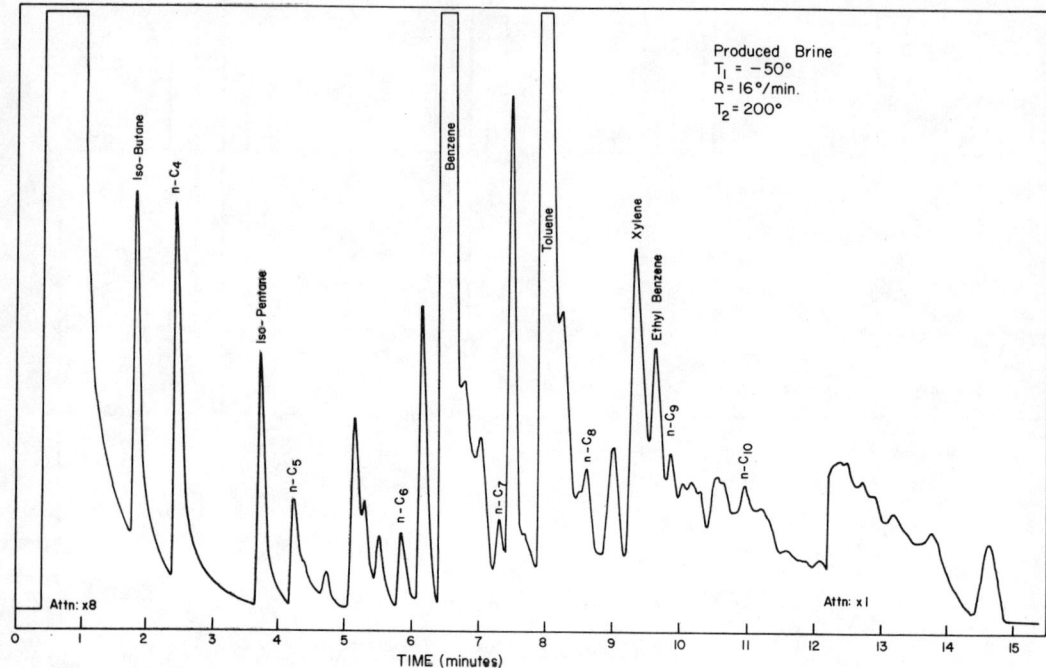

Fig. 38.6 Gas chromatogram of hydrocarbons in a produced brine in the Gulf of Mexico.

tions, is a slight temperature drop at ∼10 meters for stations in close proximity to the vents. This results from deeper, colder water pumped to the surface by the vent, sinking immediately to ∼10 meters and spreading laterally.

Light hydrocarbon concentrations vary dramatically around the vent. As the gas rises through the water column, higher concentrations of hydrocarbons dissolve at depth due to the increase in partial pressure of the gas with increasing hydrostatic pressure and colder temperatures. Table 4 demonstrates this trend of increasing hydrocarbon concentration in the 0 to 30 meter subsurface interval. The underwater venting appears to be occurring at the 20 to 30 meter depth interval based on a decrease in hydrocarbon levels below 30 meters. Also, the presence of a northerly current can be confirmed by the increase in values from Sta. 3 northward to Sta. 5 (Fig. 7).

The volume of gas vented in the Gulf is uncertain. According to OCS Order No. 11, the quantity of gas vented and flared should be reported monthly. This breakdown, however, is not scrupulously observed and much of the gas reported to the U.S.G.S. as being flared is in fact vented underwater. The U.S.G.S. (personal communications) estimated that 70% of the 5.9×10^6 MCF (1000 cubic feet) of waste gas flared and vented during September 1974 in OCS areas was vented underwater. "Sea Technology" (July 1974) estimates 60 million MCF of gas was vented or flared in OCS areas in 1973. More recent data from April 1976 indicates that 1.3×10^6 MCF of gas was vented and flared during this month, representing about 3.6% of the total gas produced. Assuming 5% of the gas dissolves in the overlying water column (Guinasso and Schink, 1973) and 50% of the estimated 60×10^6 MCF is vented underwater in the Gulf of Mexico, the input of dissolved gas to the Louisiana shelf in 1973 would be 1.5×10^6 MCF. This would amount to a yearly methane input from underwater venting of approximately 25,000 metric tons assuming the molecular composition of the vent gas in Table 3 is typical. The contribution of C_2-C_4 hydrocarbons would be approximately 11,000 metric tons. The contribution of light liquid hydrocarbons (C_5-C_{10}) is more conjectural because of the limited compositional data available on these vents. Assuming

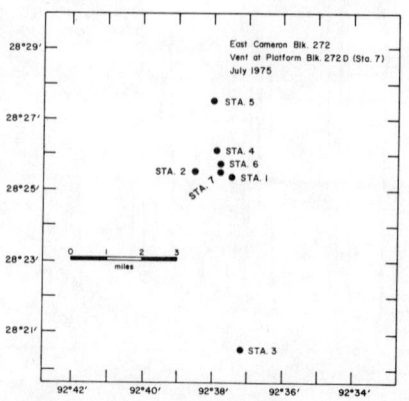

Fig. 38.7 Station locations.

TABLE 4. Hydrocarbon Concentrations and Hydrographic Parameters Near an Underwater Vent in East Cameron Blk. 272.

Station	Depth	Salinity (°/₀₀)	Temp. (°C)	O_2 (ml/L)	Methane (nl/L)	Ethane (nl/L)	Propane (nl/L)
Sta. 1	0 m	29.305	28.94	4.88	10,700	335	180
	10 m	32.957	29.17	4.61	41,300	1,380	380
	20 m	35.051	26.42	4.83	20,000	1,680	570
	30 m	35.828	22.94	4.69	4,730	140	30
	40 m	35.998	21.44	4.41	1,630	36	15
	50 m	36.048	21.14	3.59	1,000	<30	<15
Sta. 2	0 m	–	28.2	–	73,300	2,620	850
	10 m	–	28.1	–	373,000	–	18,500
	20 m	–	28.0	–	93,300	4,400	880
	30 m	–	24.0	–	11,700	–	–
	40 m	–	22.0	–	3,300	60	–
	50 m	–	21.2	–	1,730	19	–
Sta. 3	0 m	29.838	29.33	5.03	633	40	<20
	10 m	31.031	28.58	4.84	270	7	<20
	20 m	34.728	26.72	3.58	1,900	45	<20
	30 m	35.842	22.65	3.61	8,900	40	<20
	40 m	35.984	21.45	3.30	2,000	7	<20
	50 m	36.039	20.86	1.87	1,030	<5	<20
Sta. 4	0 m	28.357	29.4	5.35	380	22	<20
	10 m	31.709	29.0	4.95	38,700	1,300	380
	20 m	34.354	28.5	4.75	46,700	2,300	700
	30 m	35.652	23.8	4.79	11,300	36	<20
	40 m	35.947	22.2	4.67	1,930	11	<20
	50 m	36.017	21.5	3.71	1,430	7	<20
Sta. 5	0 m	29.122	29.2	–	400	15	14
	10 m	32.714	28.6	–	15,900	550	160
	20 m	34.386	29.0	–	3,600	200	60
	30 m	35.680	25.0	–	17,900	400	70
	40 m	36.022	22.0	–	2,110	<10	14
	50 m	36.033	21.4	–	1,570	<10	<10
Sta. 6	0 m	28.968	28.8	5.05	3,700	140	43
	10 m	31.505	28.5	4.80	268,000	8,400	2,360
	20 m	34.144	28.0	4.77	101,000	3,400	1,090
	30 m	35.621	23.0	4.72	7,730	35	50
	40 m	35.967	21.3	4.50	1,710	7	<40
	50 m	36.017	21.0	3.79	1,560	<10	<40
Sta. 7	0 m	25.895	28.8	5.27	566	22	160
	10 m	32.635	28.2	4.73	241,000	8,400	2,300
	20 m	34.300	26.6	4.76	296,000	10,200	2,700
	30 m	35.596	23.0	4.65	21,000	70	<40
	40 m	35.783	21.9	4.24	1,750	15	<40
	50 m	35.745	21.7	3.61	5,460	7	<40

C_5-C_{10} hydrocarbons only constitute 0.1% of the vented gas and the average molecular weight is 86 (corresponding to C_6), the annual input of liquid hydrocarbons from vents is 160 metric tons. This estimate does not take into account the higher dissolution rates of the aromatic components during ascent through the water column as compared to the gaseous hydrocarbons.

The environmental impact of underwater venting in OCS areas does not appear to have drawn the concern of either government or industry. The impact of these vents is largely unknown. Governmental regulations prevent the dishcarge into the ocean of produced brine containing over 50 ppm oil, but allow, albeit on a semicontrolled basis, the venting of huge amounts of hydrocarbons that contain uncontrolled quantities of the immediately toxic light aromatic hydrocarbons. It is difficult to follow the rationale of regulating the hydrocarbon content of brines and leaving the spectrum of hydrocarbons in gas vents largely up to the discretion of the individual producer. In addition to the input of hydrocarbons, the effects to marine ecosystems of induced turnover of the water column due to the large, rising column of gas bubbles have not been fully evaluated.

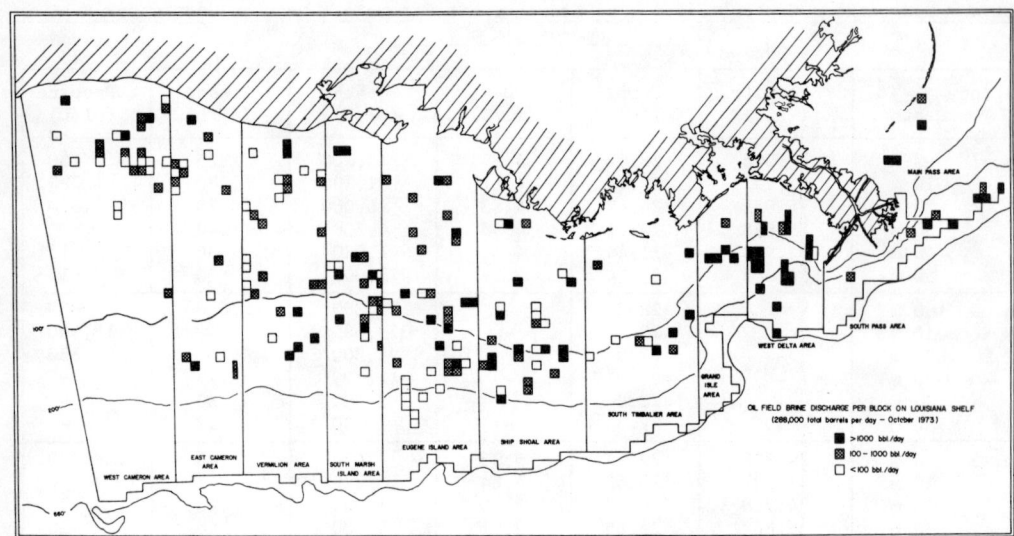

Fig. 38.8 Discharge locations of oil field brines in 1973 on the Louisiana Shelf outside of the three-mile limit (U.S.G.S., personal communications).

Oil Field Brines

Oil field brines appear to be a major source of light hydrocarbons to the upper Texas and Louisiana Shelf. These brines, produced along with oil and gas, are usually discharged into the sea after passing through oil/water separators. To meet federal regulations the brines are treated to reduce oil levels down to less than 50 ppm. The treatment does not, however, remove the large concentrations of dissolved light hydrocarbons that are present in the brines, whose discharge may be responsible for some of the high hydrocarbon levels on the Louisiana Shelf.

The production of brines in many fields can be higher than the production of oil. An example of this is Shell Oil Company's approximately 700 wells in South Pass Block 24 and 27 fields, which in 1969 were producing 90,000 barrels of oil per day along with 200,000 barrels of water (Sport, 1969). These brines, treated onshore and discharged into Southwest Pass of the Mississippi River, may be responsible for the higher light hydrocarbon concentrations reported in Southwest Pass as compared to South Pass (Brooks and Sackett, 1973). The offshore discharge of brines from OCS lease blocks appears to be increasing annually. In 1972 there were approximately 180,000 bbl/day discharged into the Gulf of Mexico (U.S. Dept. Interior, 1972). As of October 1973, the U.S.G.S. reports 290,000 bbl/day being discharged into OCS lease areas, and by October 1974 over 320,000 bbl/day were reported.

The locations where offshore disposal of brines is occurring in the Gulf of Mexico are shown in Fig. 8. This figure indicates that the brine discharges are quite scattered across the entire Louisiana Shelf, although there is a high concentration in West Delta area. Most of the 300,000 + bbl/day are discharged on the Louisiana shelf, as less than 10,000 bbl/day were discharged into Texas waters in 1973, mainly in the High Island Area. These discharges only include disposal in OCS lease areas. By 1974, an additional 300,000 bbl/day of brine were being treated onshore and disposed of in coastal waters inside the 3 mile limit.

The concentration of dissolved light hydrocarbons in produced brines varies widely from brine to brine. Gaseous hydrocarbons and light aromatics are enriched in brines collected near oil and gas reservoirs (Buckley et al., 1958; Zarrella et al., 1967; McAuliffe, 1969). The brines produced in the Gulf of Mexico are probably in close equilibrium with the crude oil and therefore will have high hydrocarbon concentrations. Assuming Dalton's law of partial pressures is valid, subsurface brines in equilibrium with crude oils should have dissolved in them individual hydrocarbons in proportion to their concentration in the crude oil (McAuliffe, 1969).

The spectrum of hydrocarbons in brines shows a similar aliphatic hydrocarbon pattern as observed in collected vent gases but a much larger concentration of aromatic hydrocarbons. Table 1 shows the light hydrocarbon composition of a number of brine samples. The majority of these brines were weathered because they were shipped and stored several weeks in plastic

TABLE 5. Estimated Annual Inputs to the Gulf of Mexico of C_1-C_{10} Hydrocarbons Via Brines

Components	Assumed Average Conc. in Brine	Input per Annum (metric tons)
Methane	$1,000 \times 10^{-6}$ L/L	18
C_2 - C_4	100×10^{-6} L/L	4
Benzene & Toluene	10 ppm (by weight)	270
C_5 - C_{10} (aliphatic)	1 ppm (by weight)	27

containers with an air space above the liquid. Brine 1 is the only properly preserved brine sample which probably accounts for its higher concentrations. Figure 6 shows a gas chromatogram of this brine. It contains appreciable amounts of hydrocarbons up to C_{12}. Benzene, toluene, and xylene are by far the major C_5 to C_{12} constituents.

By using the above mentioned 1974 U.S.G.S. data on daily brine discharges into the northwest Gulf of Mexico, various hydrocarbon contributions can be estimated (Table 5).

Minor Sources
The spillage of crude oil from platform operations is apparently not a significant source of light hydrocarbons to the Gulf of Mexico. The U.S. Dept. of Interior (1976) reported only 300,000 barrels of oil were lost through major offshore incidents (greater than 50 bbls) in the Gulf of Mexico during the period from 1964-1976. This figure does not include small spills. Sea Technology in 1974 has estimated 2000 to 30,000 bbls/yr of oil are spilled in OCS lease areas from platforms. Even considering the additional introduction from major spillage (e.g., blowouts, ruptures of gathering lines and similar offshore accidents) the total contribution of light hydrocarbons from oil losses is only a very small part of the total light hydrocarbons addition to the Gulf. They could be a major source, however, on a local basis following a major spill.

Conclusions

Six years of surveying and thousands of analyses indicate that offshore petroleum operations are contaminating much of the Gulf of Mexico coastal waters with low-molecular-weight hydrocarbons. The chief input from offshore operations in Louisiana waters is the underwater venting of gases. This input, over 25,000 metric tons per annum, appears to be three orders of magnitude higher than the input from brine discharges. However, the brines now being discharged into the Gulf are the major contributors of aromatics such as benzene and toluene.

The input of light hydrocarbons from underwater vents is particularly significant because the hydrocarbons are distributed throughout the entire water column as opposed to brines which are discharged chiefly at the surface where most of the dissolved hydrocarbons are rapidly lost to the atmosphere. The solution of hydrocarbons from a vent is greatest near the vent outlet where maximum water pressure and minimum temperature solubilizes the greatest amount of light hydrocarbons.

The biological effects of the hydrocarbons dissolved in the Gulf of Mexico from the vents and brine discharges are largely undefined. If the vents and brines only contained the dissolved light gaseous hydrocarbons, there would be little problem with effects on ecosystems. However, this is not the case, as these inputs contain appreciable amounts of the C_5 to C_{10} aliphatics which inhibit primary productivity (Brooks, 1975). They also contain large amounts of the low-boiling aromatics which are considered to be the most immediately toxic of all the water-soluble fractions of petroleum (Blumer, 1969, 1971; Holcomb, 1969). Thus the estimated input of $\sim$450 tons per year of light liquid hydrocarbons at specific point sources in the Gulf of Mexico cannot be totally ignored.

Acknowledgments

This study was supported by NSF Grant No. GX-37344.

References

Bernard, B. B., J. M. Brooks, and W. M. Sackett, Natural gas seepage in the Gulf of Mexico, Earth and Planetary Science Letters 31, 48-54 (1976).

Blumer, M., Oil pollution of the ocean. In: D. P. Hoult (ed.) Oil on the Sea. Plenum Press, New York, 1969.

Blumer, M., Scientific aspects of the oil spill problem, Environ. Affairs 1, 54-73 (1971).

Brooks, J. M. and W. M. Sackett, Sources, sinks and concentrations of light hydrocarbons in the Gulf of Mexico, J. Geophys. Res. 78, 5248-5258 (1973).

Brooks, J. M., A. D. Fredericks, W. M. Sackett and J. W. Swinnerton, Baseline concentrations of light hydrocarbons in the Gulf of Mexico, Environ. Sci. & Technol. 7, 639-642 (1973).

Brooks, J. M., Sources, sinks, concentrations and sub-lethal effects of light aliphatic and aromatic hydrocarbons in the Gulf of Mexico. Ph.D. Dissertation, Texas A&M University, May 1975.

Buckley, S. E., C. R. Hocott and M. S. Taggart, Jr., Distribution of dissolved hydrocarbons in subsurface waters. In: L. G. Weeks (ed.) Habitat of Oil. Am. Assoc. Petrol. Geol., Tulsa, 1958.

Frank, D. J., W. M. Sackett, R. Hall and A. D. Fredericks, Methane, ethane, and propane concentrations in the Gulf of Mexico, Am. Assoc. Petrol. Geol. Bull. 54, 1933 (1970).

Guinasso, N. L. and D. R. Schink, A simple physiochemical acoustical model of methane bubbles rising in the sea, Unpub. Rept. 73-15T, Texas A&M University (1973).

Gunn, B. G. and S. L. Gueringer, Gas disposal systems for offshore platforms, Second Annual Offshore Technology Conf., April 22-24, Paper #OTC 1191, (1970).

Holcomb, R. W., Oil in the ecosystem, Science 166, 204-206 (1969).

Levorsen, A. I., Geology of Petroleum, Freeman and Co., San Francisco, 1967.

McAuliffe, C., Determination of dissolved hydrocarbons in subsurface brines, Chem. Geol. 4, 225-233 (1969).

Moore, B. J., R. D. Miller, and R. D. Shrewsbury, Analysis of natural gases of the United States, 1964, U. S. Bureau of Mines, Inf. Circ. 8302 (1966).

Sea Technology, Interior to collect royalties on vented and flared natural gas, Sea Technology 16, 10 (1974).

Sport, M. C., Design and operation of gas floatation equipment for the treatment of oil-field produced brines. First Annual Offshore Technology Conf., Houston, May 18-21, Paper #OTC 1015, (1969).

Swinnerton, J. W. and R. A. Lamontagne, Distribution of low-molecular-weight hydrocarbons: Baseline measurements, Environ. Sci. & Technol. 8, 657-663 (1974).

U. S. Dept. of Interior, Draft environmental statement of the proposed outer continental shelf, oil and gas general lease sale--offshore Louisiana, Washington, D. C. (1976).

Zarrella, W. M., R. J. Mousseau, N. D. Coggeshall, M. S. Norriss and G. J. Schroyer, Analysis and significance of hydrocarbons in subsurface brines, Geochim. Cosmochim. Acta 31, 1155-1166 (1967).

CHAPTER 39

INTERTIDAL SEDIMENT HYDROCARBON LEVELS AT TWO SITES
ON THE STRAIT OF JUAN DE FUCA

W. D. MacLeod, Jr., D. W. Brown, R. G. Jenkins, and L. S. Ramos

Northwest and Alaska Fisheries Center
National Marine Fisheries Service
National Oceanic and Atmospheric Administration
2725 Montlake Boulevard East
Seattle, Washington 98112   U.S.A.

Abstract

Hydrocarbon baseline data are needed to assess the potential impact of oil pollution from tanker traffic in the Strait of Juan de Fuca. Initial studies were directed to intertidal sediments from two physically similar areas along the shipping lanes: Port Angeles, WA and Dungeness Bay, WA. Latest analytical techniques such as solvent/slurry extraction, silica gel chromatography, and glass capillary GC were adapted for analyzing large numbers of intertidal sediments efficiently and expeditiously. The method used applies to alkanes in the $C_{14}-C_{32}$ range, or to arenes ranging from substituted benzenes through benzpyrenes. Recoveries of hydrocarbons averaged better than 90%. Individual hydrocarbons were determined within 20% standard error at the ng/g level.

Lowest hydrocarbon levels were found at Dungeness Bay, a relatively pristine area. The GC profile of the alkanes suggests that the Dungeness hydrocarbons have a biogenic origin. Hydrocarbon levels were substantially greater at Port Angeles harbor, especially adjacent to Peabody Creek. The even distribution of $C_{14}-C_{22}$ n-alkanes from creekside sediment is consistent with known fuel oil seepage upstream. Above C_{22}, the n-alkane pattern appears more biogenic, although the levels remain high. Greatest individual divergences between the two areas were found among the arenes: phenanthrene, fluoranthene, and pyrene. This suggests that arenes be monitored as indicators of oil pollution; over ten arenes are proposed for baseline analyses and monitoring.

Key words: Intertidal - Petroleum - Hydrocarbons - Baseline - Capillary GC

Introduction

Until recently, most of the crude oil for the U. S. Pacific Northwest has been supplied by pipeline from Canada. As a result, the greater Puget Sound region has experienced few difficulties from massive oil spills or low-level chronic contamination due to refining. However, with dwindling supplies now available by pipeline, tanker traffic in the Strait of Juan de Fuca (Fig. 1) has risen sharply, as has the risk of acute and chronic oil pollution in the marine environment. Predictably, these trends could continue with the opening of the Alaskan pipeline (Clark 1976), especially if the greater Puget Sound region becomes a pipeline terminal for the Midwest.

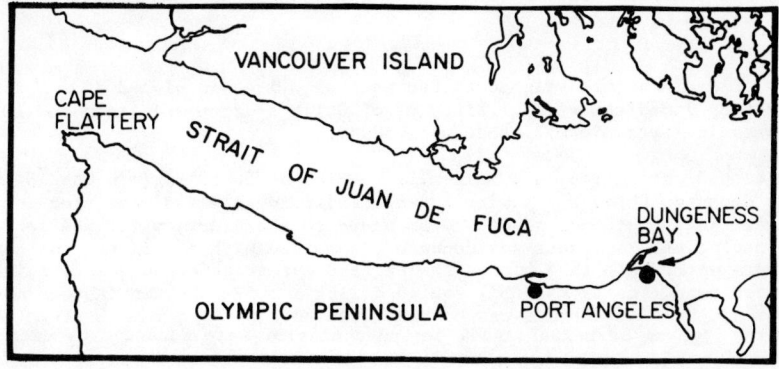

Fig. 39.1   The Strait of Juan de Fuca and the Olympic Peninsula, Washington.

Knowledge of the present distribution and concentrations of hydrocarbons along the Strait of Juan de Fuca is necessary in order to establish a baseline for measuring the future impact of petroleum pollution. This requires highly-sensitive and highly-resolving methods of organic chemical analysis, such as those reviewed in Analytical Chemistry (Bradley 1975) and in various symposia (Farrington et al. 1972; Natl. Acad. Sci. 1975; Am. Petrol. Inst. 1975). Recent studies by Farrington and Tripp (1975), Rohrback and Reed (1975), Chesler et al. (1976), and Warner (1976) have contributed significantly to this field.

The primary objective of this study was to test analytical techniques for their utility in processing large numbers of marine sediment samples. Techniques developed in the research cited above, particularly that of Rohrback and Reed (1975) and Warner (1976), were adapted for our use (MacLeod et al. 1976). To test our procedure for speed and efficiency in determining the abundance and variation of hydrocarbons in marine sediments, intertidal sediments were analyzed from two physically-similar sites on the Strait of Juan de Fuca. The sites were selected for their contrast in probable exposure to petroleum contamination. This analysis of samples at both ends of the contamination scale was valuable in demonstrating the scope of the analytical procedure.

Methods and Materials

Sample Collection
Samples of sediment were collected from two intertidal, physically-similar sites twenty miles apart on the Strait of Juan de Fuca (Fig. 1):
1. Port Angeles Harbor - the site of a proposed tanker pipeline terminal (already exposed to petroleum hydrocarbons); and
2. Dungeness Bay - next to a wildlife refuge (believed to be relatively uncontaminated).

Figures 1 and 2 show the locations of sediment sampling on the beaches near Peabody Creek, Port Angeles, and near Three Crabs restaurant, Dungeness Bay, respectively. Composite samples were obtained from multiple cores, using 3 cm deep by 7.5 cm dia tin cans previously cleaned with detergent, 96% sulfuric acid, and organic solvents. The subsample cores were collected at the corners of a 1 m square, and deposited in a clean aluminum pail. The resultant composite sample was thoroughly mixed with a hand trowel, and a portion of the composite was placed in either a clean glass jar or in washed aluminum foil. Packaged samples were stored in a cold chest until returned to the laboratory freezer within 6 hr.

Materials
Only clean glass, Teflon, or metal materials, or residue-free solvents and reagents contacted the samples. Glassware was washed in hot laboratory detergent, rinsed with water, dried, and rinsed with acetone and methylene chloride. Teflon and metal sheeting, and metal implements were rinsed with acetone and methylene chloride. Hydrochloric acid, anhydrous sodium sulfate, coarse sand, sodium hydroxide, silica gel, and glass wool were extracted with methylene chloride before use. Solvents (Burdick and Jackson* distilled in glass, or Mallinckrodt nanograde) gave no measurable residues in procedural blank analyses.

Dry Weight
Pebbles were removed from thawed sediment by spatula or sieve, and the sediment was thoroughly mixed. Approximately 20 g were weighed to the nearest 0.1 g in a tared aluminum dish. After covering loosely with aluminum foil, the sediment was dried in an oven at 120°C for 24 hr. It was then cooled for 30 min in a desiccator, reweighed, and dry weight percent was calculated.

Extraction
A 100 g aliquot of sediment was weighed to the nearest 0.1 g and placed in a 1 liter bottle fitted with a Teflon-lined screw cap. Fifty ml of 0.1 N hydrochloric acid and 100 ml of ethanol-free, peroxide-free diethyl ether were added.

The bottle was capped and rolled on a ball-mill tumbler at 30-60 rev/min for 18 hr. The ether phase was decanted through a powder funnel containing a glass-wool plug into a 500 ml erlenmeyer flask. Another 100 ml of ether was added to the slurry which was tumbled again for 1 hr. The second ether extract was decanted, combined with the first, and concentrated to 15 ml in a warm water bath in a hood. The extract was transferred quantitatively to a 25 ml concentrator tube (Kontes K570050) equipped with a reflux column (Kontes K569351) and an ebullator was added. The extract was concentrated to 2 ml in a tube heater (Kontes K720001, K720003). Two ml of hexane and a second ebullator were added. The extract was then reconcentrated to 1 ml.

---

* Reference to a company or a product does not imply endorsement by the U. S. Department of Commerce to the exclusion of others that may be suitable.

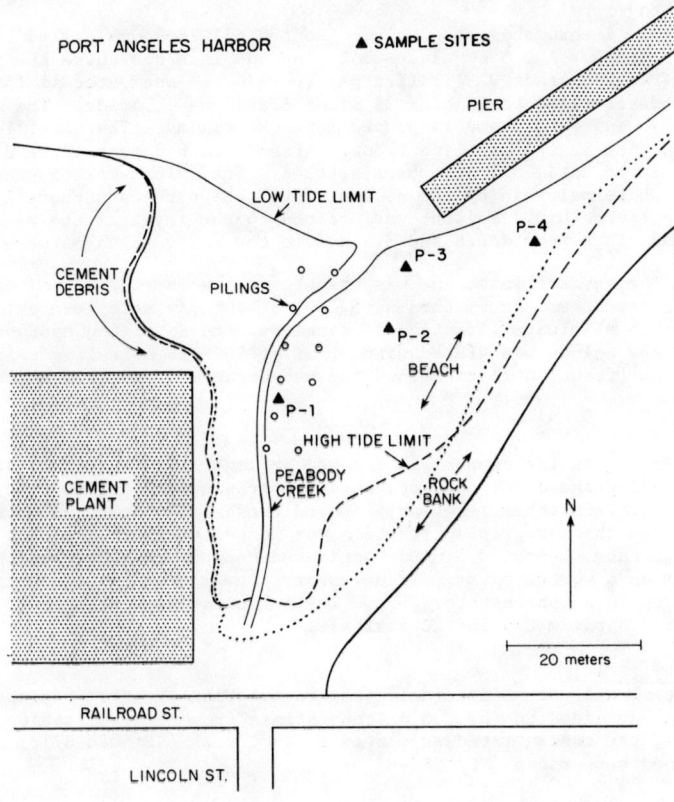

Fig. 39.2 Peabody Creek intertidal sampling area at Port Angeles, Washington.

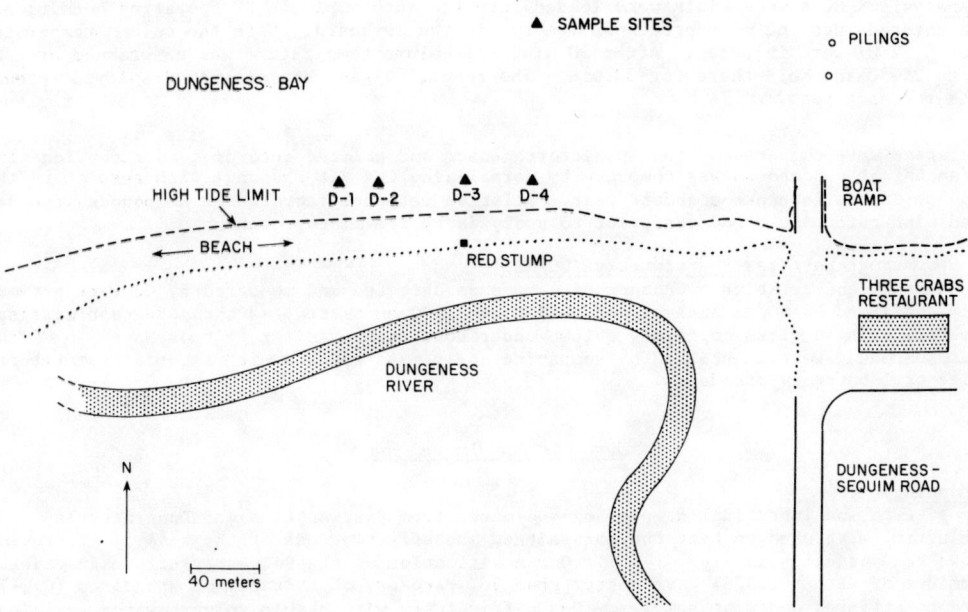

Fig. 39.3 Dungeness Bay intertidal sampling area at Dungeness, Washington.

Silica Gel Chromatography

Column Preparation. A chromatography column (Kontes K420280) was filled to the reservoir with methylene chloride. A 0.5 cm glass-wool plug was inserted above the stopcock. Seven g of 100-200 mesh (Davison grade 923) silica gel (previously activated at 150°C) were placed in a 250 ml erlenmeyer flask containing 25 ml of methylene chloride. The mixture was swirled vigorously, and then promptly poured into the column. The particles settled quickly with little turbulence at the settling front. After 1 cm had settled, 1-2 drops per second of solvent were eluted until all adsorbent settled. The column was compacted by fully opening the stopcock while maintaining the solvent level above the adsorbent. A 1 cm layer of sand was added to the column. Solvent was drained to the level of the sand; 2 ml and then 40 ml of petroleum ether were added and drained to the sand. Eluates were discarded.

Sample Elution. The extract in hexane was transferred to the column and eluted at 2-4 ml/min with 15 ml of petroleum ether, then with 3 ml of 20% v/v methylene chloride in petroleum ether. This 18 ml eluate (fraction 1) contained the saturated hydrocarbons. Receivers were changed and the column was eluted with 25 ml of 40% v/v methylene chloride in petroleum ether. This eluate (fraction 2) contained the unsaturated hydrocarbons, including the arenes.

Desulfurization

The surface of fine, granular copper was cleaned by immersion in concentrated hydrochloric acid. After decanting the acid, the activated copper was washed five times with acetone and five times with petroleum ether. It was prepared fresh daily and stored under petroleum ether. A silica gel chromatographic fraction was concentrated to 1 ml and placed in a 40 ml conical centrifuge tube. About 0.5 ml of activated copper was added and the mixture was stirred for 2 min on a vortex mixer. If necessary, the mixture was centrifuged and transferred with washing to a concentrator tube. The solution was concentrated to 0.5 ml in the tube heater for microgravimetry and GC analysis.

Gas Chromatography (GC)

GC Sample. A 1.0 ml aliquot of internal standard solution (4 ng/µl hexamethylbenzene in carbon disulfide) was added to the 0.5 ml concentrate from desulfurization. After reconcentration to 0.5 ml, the concentrate was placed in a GC vial (HP 5080-8712) which was sealed with a Teflon-lined septum cap (HP 5080-8703).

GC Apparatus and Modifications. GC samples were analyzed on a Hewlett-Packard automatic gas chromatograph (HP 5840A) equipped with an autosampler, a J & W Scientific glass capillary column (30 m x 0.25 mm i.d., coated with SE 30), and an FID. The GC sample injection port was modified to split the carrier gas as shown in Fig.   . Helium carrier gas pressure was adjusted to 2 ml/min flow through the column at 60°C with a bubble flow-meter. The needle valve was adjusted for 20 ml/min bypass flow to give a 10:1 split ratio. A charcoal trap ahead of the needle valve absorbed compounds from the split stream.

GC Analysis. GC sample vials were loaded into the autosampler. GC operating conditions were entered into the microprocessor memory via the keyboard. With the column temperature at 60°C, 2 µl were injected. After 10 min, the column temperature was programmed at 2°C/min to 250°C and held there for 30 min. The resulting gas chromatogram displayed retention times for each peak.

Peak areas were computed by the GC microprocessor and printed according to retention times. The amount of a compound was computed by normalizing its FID response with respect to the response of the internal standard peak. Relative response factors for compounds were determined with reference standards prior to analysis by the microprocessor.

Gas Chromatography/Mass Spectrometry (GC/MS)

The identity and relative abundance of compounds detected and measured by GC were periodically confirmed by GC/MS analysis on a Finnigan 3200 automated gas chromatograph/mass spectrometer, using a glass capillary column under conditions similar to those used in GC analysis. Compounds were identified by comparing their mass spectra with spectra from reference tables or laboratory standards.

Results and Discussion

Methodology
Composite marine intertidal samples of sediments from Port Angeles and Dungeness Bay, Washington, were used to test the convenience and effectiveness of the analytical procedure (shown schematically in Fig.   ). Our modification of the solvent/slurry extraction technique of Warner (1976) gave better than 90% recovery of hydrocarbon standards ($C_{14}+$) added to stripped sediment samples. No difficulties with stable solvent-water emulsions were encountered. The saturated hydrocarbons were readily separated from alkylated aromatic hydrocarbons by silica gel adsorption chromatography. When the column preparation procedure was followed carefully, there was a solvent volume of 2-3 ml between the last measurable

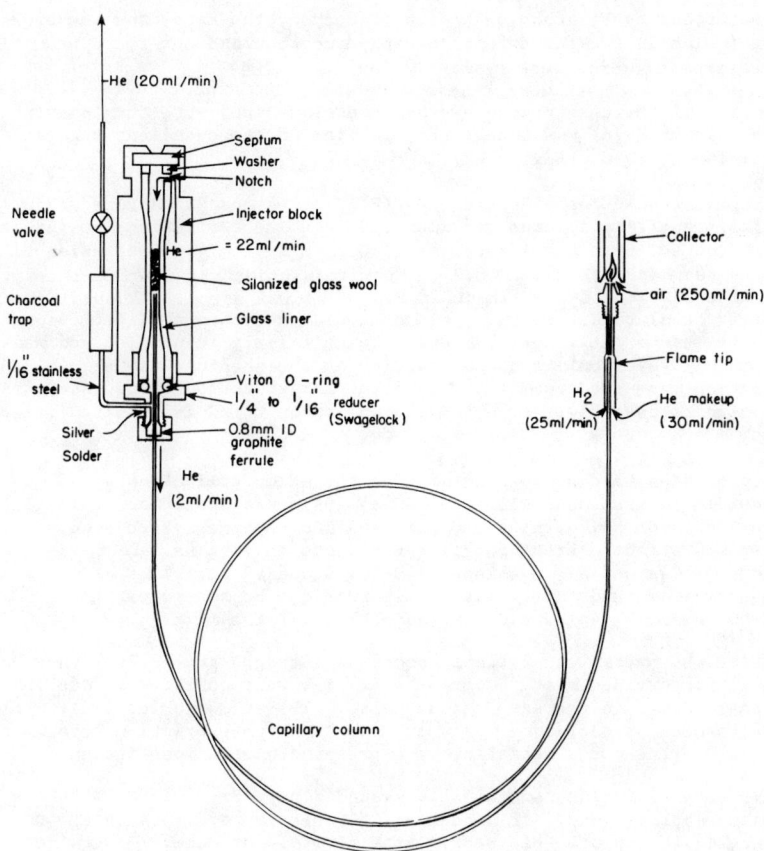

Fig. 39.4 Schematic details of the GC sample train: injector, column, and detector.

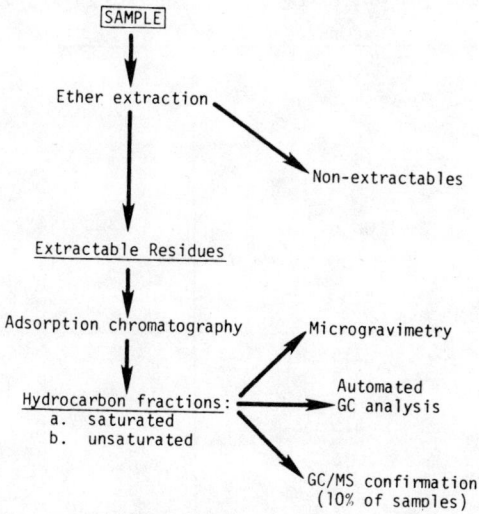

Fig. 39.5 Schematic of sediment analysis.

elution of a saturated hydrocarbon such as heptadecylcyclohexane, and the elution front of phenyldodecane (which is two-thirds saturated hydrocarbon and one-third aromatic hydrocarbon). Other adsorbents which were evaluated including other silica gels, failed this test. Adsorbents finer than 200 mesh were excessively slow, even under pneumatic pressure. Customary subdivision of the unsaturated hydrocarbon class into more than one fraction containing the arenes (Warner 1976) was unnecessary in view of the excellent separation of the arenes by glass capillary GC (Fig.   ).

Gas chromatographic resolution of the arenes (Fig.   ) and the alkanes (Fig.   ) with wall-coated glass capillary columns was improved over the resolution obtainable with packed or even support-coated, open tubular columns. Our GC system gave retention times for compounds generally reproducible within 0.2%. Such resolution and reproducibility made it possible to determine that the C_{28} n-alkane in Fig. ___ was actually a shoulder on a larger peak. With less resolution, the C_{28} n-alkane would have merged with the larger peak. This could have easily led to the larger component being falsely identified and measured as C_{28} n-alkane. GC/MS analysis confirmed the identity of the shoulder as C_{28} n-alkane. The electron ionization mass spectrum of the adjacent larger compound suggested that it was a branched or cyclic alkane; however, chemical ionization GC/MS is required for positive identification.

Reproducibility studies were conducted first on the GC determination alone, and then on the overall procedure. The reproducibility of GC analysis was tested on the n-alkanes from C_{14} to C_{31}, and on the isoprenoidal alkanes, pristane and phytane. Precision was excellent, averaging under 15% standard error for five replicate injections. Poor quantitation was observed only for the C_{30} and C_{31} n-alkanes (40-50% standard error). The difficulty with measuring these two compounds appears to arise from the broadness of the GC peaks toward the end of the chromatogram, coupled with the complexity of the mixture.

The precision for the overall analytical procedure averaged around 20% standard error at the 2 ng/g level for dry sediment. Composite samples were analyzed in duplicate as denoted by the lower case letter in the sample code P-1a, P-1b, etc. (Tables 1-3). Inspection of the individual results of duplicate analyses, carried from extraction through the GC determination (Tables 2,3), shows a satisfactory correspondence in most instances.

The lower limit of practical sensitivity is around 2 ng/g dry weight, which is acceptable for many baseline applications. If necessary, further improvements in the GC sample introduction hardware could improve this sensitivity fivefold or more. Otherwise a larger sample must be taken to increase the sensitivity.

TABLE 1. Hydrocarbons from Intertidal Sediment, Port Angeles Harbor (P) and Dungeness Bay (D), determined by Microgravimetry

Sample	Silica gel chromatography fraction, µg/g dry sediment	
	1. (saturated)	2. (unsaturated)
P-1 a	950	360
b	770	350
P-2 a	76	95
b	78	94
P-3 a	1000	430
b	1100	400
P-4 a	170	160
b	130	190
D-1 a	2.5	34
b	2.7	37
D-2 a	5.5	29
b	3.7	27
D-3 a	2.8	23
b	1.5	24
D-4 a	3.9	36
b	4.2	33

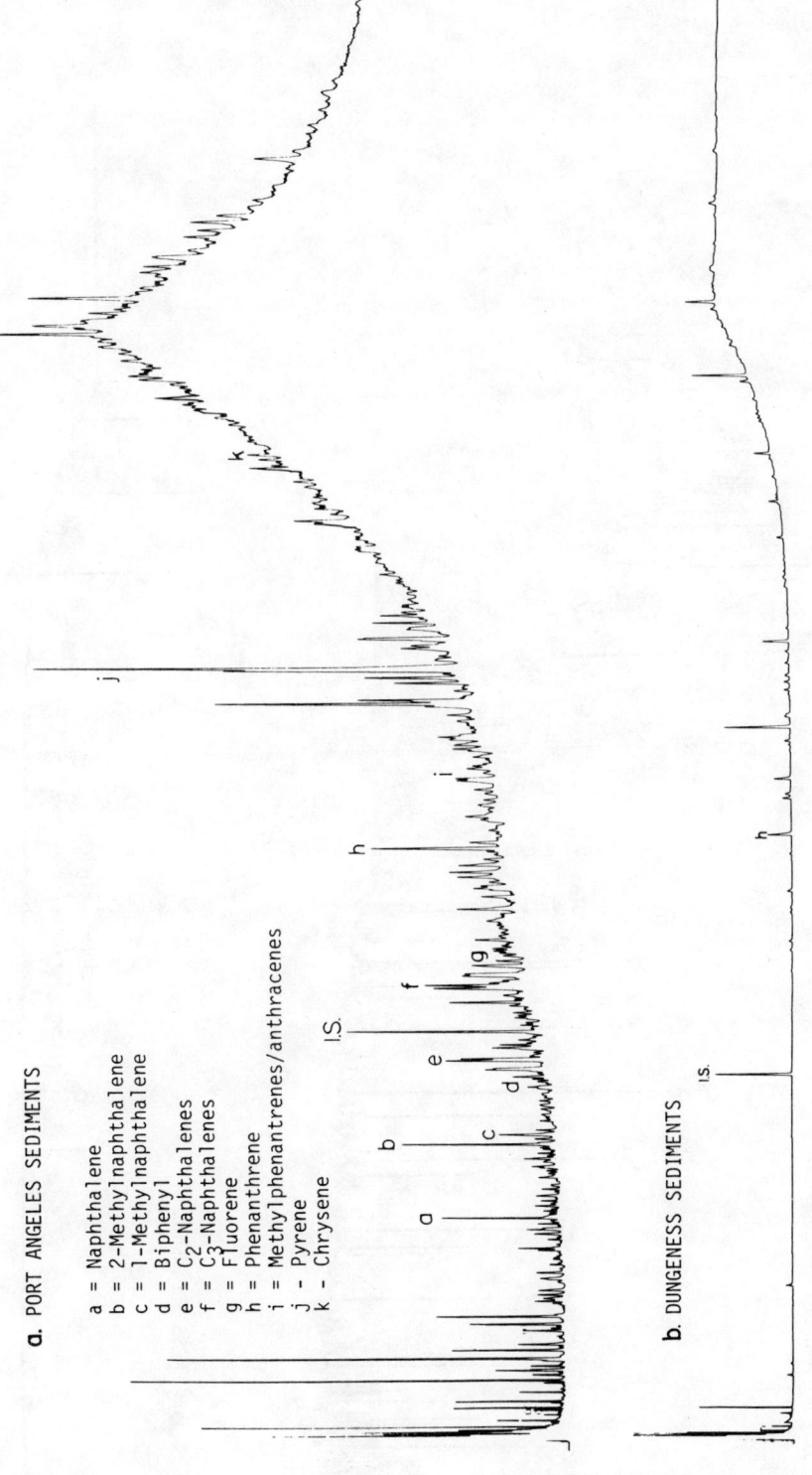

Fig. 39.6 Glass capillary gas chromatograms of unsaturated hydrocarbons from (a) Port Angeles sediments and (b) Dungeness sediments.

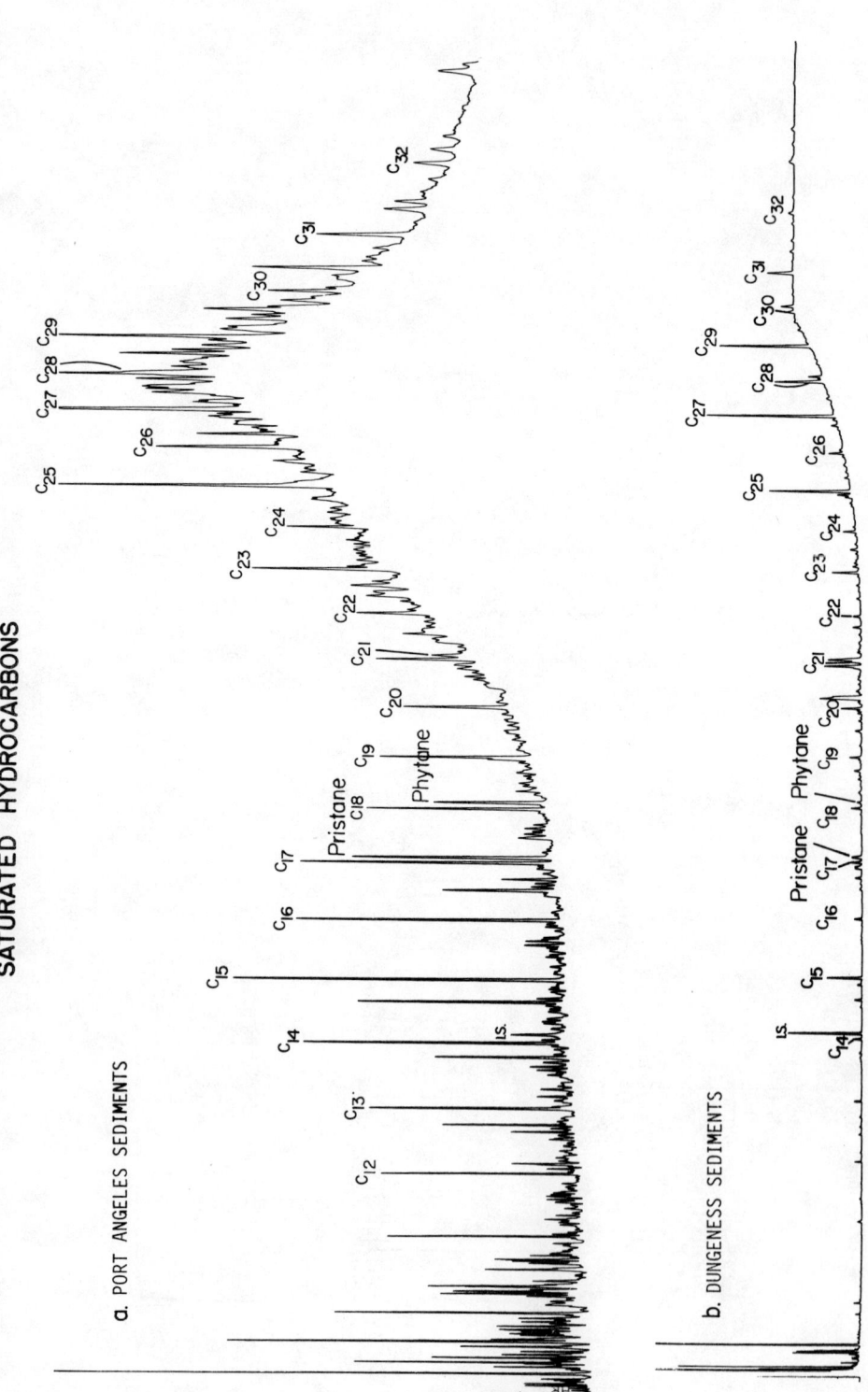

Fig. 39.7 Glass capillary gas chromatograms of saturated hydrocarbons extracted from (a) Port Angeles sediments and (b) Dungeness sediments.

TABLE 2. Alkanes from Intertidal Sediment, Port Angeles Harbor (P) and Dungeness Bay (B), determined by GC

Sample		C_{14}	C_{15}	C_{16}	C_{17}	C_{18}	C_{19}	C_{20}	C_{21}	C_{22}	C_{23}	C_{24}	C_{25}	C_{26}	C_{27}	C_{28}	C_{29}	C_{30}	C_{31}	Pristane	Phytane
P-1	a	290	370	300	420	290	370	230	220	180	450	270	1700	630	1700	320	870	94	1100	400	290
	b	210	380	300	410	300	370	250	250	210	480	290	1400	670	1800	350	750	240	1600	370	280
P-2	a	<2	19	15	25	17	13	18	28	22	250	9.9	1200	120	2300	270	1400	150	870	47	24
	b	8.6	20	15	24	16	10	18	30	23	150	9.5	1000	230	1400	150	1400	150	820	47	25
P-3	a	170	200	180	240	130	34	110	32	74	170	120	630	210	930	290	470	88	430	240	94
	b	160	190	170	240	120	32	110	32	60	170	92	570	170	870	240	450	88	360	230	99
P-4	a	30	43	38	64	33	<2	28	22	16	14	3.4	140	78	250	160	150	<2	160	77	43
	b	25	35	32	50	31	<2	26	2	2	21	6.6	130	69	240	180	170	<2	180	74	42
D-1	a	9.1	42	11	26	11	19	11	15	8.0	33	13	130	13	160	56	110	4.8	44	33	12
	b	9.0	45	11	30	12	20	10	16	9.8	34	14	120	13	160	19	100	2.5	41	36	15
D-2	a	12	87	18	31	16	27	15	21	12	42	15	110	13	190	48	150	13	62	45	11
	b	10	73	15	29	14	26	12	21	13	43	17	140	26	220	23	70	14	88	41	10
D-3	a	4.8	31	8.5	13	11	17	30c	10	10	42	23	130	25	230	100c	180	12	110	16	5.2
	b	5.7	31	7.7	11	6.5	13	5.7	11	6.5	30	16	110	18	160	25	120	8.8	66	16	4.5
D-4	a	12	71	16	32	15	31	16	26	18	76	25	220	8.9	350	21	68	1.9	82	59	16
	b	11	72	17	34	16	36	19	33	32	100	50	300	21	450	40	71	2.6	57	62	16

* $\underline{n}$ - alkane denoted where chain length given as C_{14}, C_{15}, etc.

c denotes contaminated peak

TABLE 3. Selected Arenes from Intertidal Sediment, Port Angeles Harbor (P) and Dungeness Bay (D), determined by GC

Sample		ng/g dry sediment		
		Phenanthrene	Fluoranthene	Pyrene
P-1	a	320	730	450
	b	240	1000	480
P-2	a	43	120	110
	b	50	110	100
P-3	a	190	660	340
	b	230	1100	510
P-4	a	320	600	390
	b	110	370	300
D-1	a	11	17	13
	b	6.7	18	17
D-2	a	11	24	24
	b	5.4	13	13
D-3	a	3.3	7.5	13
	b	3.1	8.0	15
D-4	a	11	22	11
	b	-	-	-

Hydrocarbon Levels: Port Angeles vs. Dungeness Sediments

Relatively high levels of the saturated hydrocarbon fraction (Table 1) were found in Port Angeles sediments compared to those found in Dungeness sediments. Higher levels at Port Angeles were also evident among the individual alkanes (Table 2). The levels of the unsaturated hydrocarbon fraction at Port Angeles and Dungeness (Table 1) show analogous differences in abundance, although not as great as the selected arenes (aromatic hydrocarbons) in this Fraction (Table 3).

The microgravimetric data from Port Angeles (Table 1) also revealed an upward gradient in hydrocarbon levels in the direction of Peabody Creek (Fig.   ). For example, sediment samples P-1 and P-3 taken adjacent to the stream bed contained up to tenfold higher levels of the saturated hydrocarbon fraction than did samples P-2 and P-4 farther away. The unsaturated hydrocarbon fraction (containing the arenes) reflected this trend, though to a lesser degree. Individual alkanes (Table 2) and the arenes (Table 3) showed similar gradients on the beach near Peabody Creek, with the highest values at creekside (samples P-1 and P-3). These gradients strongly indicate that the hydrocarbon levels there are related to the known contamination of Peabody Creek by long-standing oil seepage upstream.

The n-alkanes from all sediment samples showed an odd-carbon predominance from C_{23} to C_{31} (Table 2 and Fig.   ). This is believed to result from biogenic input, presumably land-derived. Below C_{23} the comparison changed: the Dungeness n-alkanes still displayed a marked odd-carbon predominance characteristic of a biogenic origin (Bray and Evans 1961) but the Port Angeles n-alkanes did not. Little or no odd-carbon n-alkane predominance is a characteristic of oil-derived hydrocarbons; therefore, in view of the known fuel oil seepage upstream at Peabody Creek, it is not surprising that the Port Angeles C_{14} to C_{22} n-alkanes showed much less odd-carbon predominance. In comparing Figs.   and   , the extensive hump of unresolved compounds and the much more complex alkane profile in the Port Angeles sediment provides further evidence of petroleum-related contamination (Farrington and Tripp 1975).

Chromatograms of the unsaturated hydrocarbons were also significantly more complex at Peabody Creek, Port Angeles (Fig.   ) than at Dungeness (Fig.   ). The divergence of the data in Table 3 between Port Angeles and Dungeness suggests that phenanthrene, fluoranthene, and pyrene could be useful indicators of oil-related contamination. In addition to these, Table 4 lists a number of other compounds which were identified by GC/MS from sediments sampled adjacent to Peabody Creek (Fig.   ). Most are arenes commonly occurring in petroleum and its products. The others (pinene and dichlorobenzene) presumably resulted from the dumping of industrial solvents.

The arenes denoted by an asterisk in Table 4 are potentially useful indicators of petroleum contamination. These compounds were selected because: (a) they are found in crude and refined petroleum; (b) they are recovered efficiently by our procedures; and (c) they cover a wide range of arenes from 1-5 rings, yet can be determined in a single gas chromatogram.

TABLE 4. Unsaturated Compounds Identified in Peabody Creek Sediment by GC/MS

toluene	C_4-naphthalene
xylene	phenathrene*
pinene	anthracene*
C_3-benzene*	methyl fluorene
C_4-benzene	methyl phenanthrene* and/or
C_5-benzene	methyl anthracene*
naphthalene*	C_2-(phenanthrene and/or anthracene)
C_6-benzene	fluoranthene*
dichlorobenzene	pyrene*
methylnaphthalene*	benzanthracene*
C_2-naphthalene	chrysene*
C_7-benzene	benzofluoranthene
C_3-naphthalene	benzpyrene*
Fluorene	perylene*

* Good prospect for petroleum monitoring.

The C_3-benzenes and naphthalene are about 60% and 70% recoverable, respectively, by our procedure. The arenes beyond these compounds in Table 4 are over 80% recoverable. The benzenes and naphthalenes are the most abundant arenes in crude oil, as well as the most water-soluble, volatile, and acutely toxic (Malins 1977); hence, they deserve prominent consideration in baseline surveys. Beyond the C_2-substituted benzenes, the GC pattern becomes extremely complex due to the many isomeric forms, so the list must be limited to arenes which are prominent in pollution and well separated by GC. Fig.   shows numerous substituted benzenes preceding naphthalene (a). The four-ring arenes and larger are evident beyond chrysene (k). Among these, the benzanthracenes, benzpyrenes, and perylenes occur near the top of the hump and beyond. These polynuclear arenes, though less abundant and less water-soluble, are important for their potential chronic biological effects (Malins 1977).

Acknowledgements

We thank Victor D. Henry for laboratory assistance and Robert C. Clark, Jr., for helpful discussions. This study was sponsored by the NOAA Marine Ecosystems Analysis Program with funding from the U. S. Environmental Protection Agency.

References

Am. Petrol. Inst., Proceedings of 1975 Conference on Prevention & Control of Oil Pollution. 612 p. American Petroleum Institute, Washington, D. C. (1975); see also Proceedings of 1973, 1971 Conferences, ibid.

Bray, E. E. and E. D. Evans, Distribution of n-Paraffins as a Clue to Recognition of Source Beds, Geochim. Cosmochim. Acta 22, 2-9 (1961).

Bradley, M. P. T., Hydrocarbons. p. 189R-199R. In: J. M. Fraser (ed.). Petroleum. Anal. Chem. 47, 169R-232R (1975); see also previous reviews listed therein (Ref. 1A-11A).

Clark, R. C., Jr., Impact of the Transportation of Petroleum on the Waters of the Northeast Pacific, Mar. Fish. Rev. 38, 20-26 (1976).

Farrington, J. W., C. S. Giam, G. R. Harvey, P. Parker and J. Teal, Analytical Techniques or Selected Organic Compounds: Petroleum. p. 154-176. In: E. D. Goldberg (ed.) Marine Pollution Monitoring: Strategies for a National Program. Allan Hancock Foundation (Santa Catalina Mar. Biol. Lab.), University of Southern California, Los Angeles, California. (October 25-28), 1972).

Farrington, J. W., and B. W. Tripp, A Comparison of Analysis Methods for Hydrocarbons in Surface Sediments. p. 267-284. In: T. M. Church (ed.). Marine Chemistry in the Coastal Environment. Am. Chem. Soc. Symposium Series No. 18, American Chemical Society, Washington, D. C. (1975).

MacLeod, W. D., Jr., D. W. Brown, R. G. Jenkins, L. S. Ramos, and V. D. Henry, <u>A Pilot Study on the Design of a Petroleum Hydrocarbon Baseline Investigation for Northern Puget Sound and Strait of Juan de Fuca</u>. 18 p. Natl. Ocean. Atmos. Admin. Tech. Memo. No. ERL-MESA 8, Environmental Research Laboratory, Boulder, CO. (1976).

Malins, D. C. (ed.), <u>Effects of Petroleum on Arctic and Subarctic Marine Environments and Organisms</u>. Vol. I & II. Academic Press, New York, In press.

Natl. Acad. Sci., <u>Petroleum in the Marine Environment</u>. 107 p. National Academy of Science, Washington, D. C. (1975).

Rohrback, B. G. and W. E. Reed, <u>Evaluation of Extraction Techniques for Hydrocarbons in Marine Sediments</u>. 23 p. Inst. of Geophys. and Planet. Phys., Publ. No. 1537, University of California at Los Angeles, Los Angeles, CA. (1975).

Warner, J. S., Private Communication. Battelle Columbus Lab., Columbus, OH. (1976).

CHAPTER 40

CHARACTERIZATION OF VOLATILE HYDROCARBONS
IN FLOWING SEAWATER SUSPENSIONS OF NUMBER 2 FUEL OIL

R. M. Bean and J. W. Blaylock

Battelle, Pacific Northwest Laboratories
Richland, WA 99352

Abstract

Dilute suspensions of No. 2 fuel oil in seawater (less than one milligram per liter) have been analyzed for monocyclic aromatic hydrocarbons using helium partition gas chromatography. The methodology permits routine quantitation of volatile aromatic constituents at concentrations well below one microgram per liter. Gas chromatograms obtained from helium extracts of the dilute oil in seawater suspensions exhibited an unresolved envelope upon which were superimposed aromatic component peaks identified by gas chromatography/mass spectrometry (GC/MS). The envelope was not removed by repeated helium extractions of an oil in seawater suspension with helium, but was removed if the sample was filtered prior to extraction. The unresolved envelope has been shown by GC/MS to consist largely of saturate hydrocarbons, presumably extracted directly into the helium from an insoluble or accommodated oil phase. These results are consistent with the existence of soluble and insoluble oil phases in suspensions of No. 2 fuel oil in seawater and with the presence of the monocycle aromatic components largely in true solution.

Key words: No. 2 Fuel Oil, Gas chromatography, Water-soluble fraction, Oil-Water Suspensions

Introduction

Current studies being conducted at the Battelle-Northwest Marine Research Laboratory are concerned with the impact of seawater suspensions of petroleum hydrocarbons on population distributions of marine biological communities (Vanderhorst, et. al., 1976a). These experiments are being carried out in continuous-flow bioassay apparatus (Vanderhorst, et. al., 1976b) which provide oil in seawater at concentrations considerably less than one part per million. To conduct these experiments, it is important to repeatedly characterize the exposure system for hydrocarbon concentration and composition to maintain adequate experimental control and to provide a statistically valid description of the treatment. Infrared spectrophotometry (IR) is frequently used to monitor total oil concentrations in continuous-flow petroleum bioassay experiments (Vaughan, 1973; Bean, et al., 1973); however, the concentrations of hydrocarbons being used for the current biological experiments require an inconveniently large water sample for an accurage IR analysis. IR methodology does not provide information about the distribution of hydrocarbon types.

Several investigators have reported on the hydrocarbon types in seawater suspensions of petroleum and petroleum fractions. Boylan and Tripp (1971) identified thirty mono- and dicyclic aromatic compounds extracted into seawater from kerosene. Studies of the composition of water soluble oil fractions in flowing (Bean, et. al., 1973) and static (Anderson, et. al. 1974) bioassay systems have shown that the concentrations of monocyclic aromatic hydrocarbons are very high with respect to those of other component types. For example, benzene, toluene, ethylbenzene, and xylenes were found to comprise about one-half the total measured hydrocarbons and over 90% of the aromatic hydrocarbons in a water soluble fraction of South Louisiana crude oil (Anderson, et. al., 1974). Although the concentrations of volatile aromatic hydrocarbons can be relatively high, their contribution to the infrared absorbance at the analytical wavelength used for oil in water determinations is small compared to saturate hydrocarbons (Bean, 1974). Thus, alternative analytical methods are desirable for the routine analysis of the volatile aromatic hydrocarbons in water.

Several analytical techniques have been employed to study the hydrocarbon type distribution in aqueous suspensions of petroleum hydrocarbons. Methods involving solvent extraction followed by gas chromatography (Boylan and Tripp, 1971; Boehm and Quinn, 1974; Bean, 1974) require evaporation of solvent, with concurrent losses of volatile hydrocarbons, to obtain sufficient analytical sensitivity. Extraction of volatile hydrocarbons using a helium purge followed by concentration of absorbents (May, et. al., 1975; Swinnerton and Lamontagne, 1974) provides sufficient analytical sensitivity but quantitative recovery of hydrocarbons is difficult (May, et. al., 1975). We have found the gas partitioning methodology developed by McAuliffe (1971) to be adaptable for the analysis of monocyclic aromatic hydrocarbons in continuous flow petroleum bioassay experiments.

Materials and Methods

Suspensions of No. 2 Fuel Oil in Seawater

The fuel oil employed for this work is an American Petroleum Institute reference oil. Much of the methodological development was conducted using oil-in-seawater suspensions generated in the continuous-flow apparatus (Vanderhorst, et. al., 1976b). However, the No. 2 fuel oil suspensions discussed here were generated by vigorously shaking one part of fuel oil with ten parts of saline water (35 $^0/_{00}$ NaCl) allowing phase separation for at least 12 hours, determining the oil concentration in a subsample of the aqueous phase, and carefully diluting to the desired concentration with saltwater. Care was taken during subsampling and dilution to avoid loss of volatiles to the atmosphere by avoiding the creating of large airwater interface areas. Suction pipeting of samples was avoided.

Sampling of Fuel Oil Suspensions

A 100 ml glass syringe fitted with a luer-lok® Kel-F® valve, was used to sample the fuel oil suspensions. The syringe was filled and expressed several times to avoid contamination by oily surface film. The syringe was then filled with 50 ml of sample water and placed in a water-bath for 30 min at 25°C. After removal from the bath, the syringe was fitted to the source of ultra high purity helium gas (Fig. 1) and a volume of helium equal to the volume of water sample admitted to the syringe. The syringe was shaken vigorously for two min and the helium sample introduced into the gas sample loop.

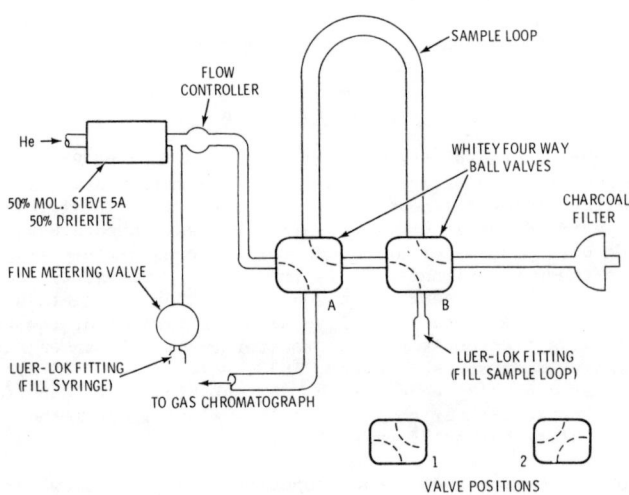

Fig. 40.1 Gas sampling loop.

Gas Sample Loop

The gas sample loop used for the introduction of helium containing the partitioned hydrocarbons is shown in Fig. 1. The four-way ball valves (Whitey Company) are an important feature of the loop. The amount of dead-volume is small and surfaces exposed to the sample are largely stainless steel with a minimum of exposure to the teflon seas. Operation of the loop is as follows: the sample syringe is locked to the "fill sample loop" fitting and valve B turned to position 2. Approximately 10 ml of helium is expressed into the system through the charcoal filter. Both valve A and B are turned to position 1 and helium placed into the sample loop. More than the amount of helium required is used to insure complete filling of the sample loop. Turning valve B to position 2 seals the loop, and turning valve A to position 2 allows carrier gas from the gas chromatograph to sweep the sample from the loop onto the chromatographic column. We have found that filling and emptying the sample loop as described above results in reproducible, contamination-free sampling, and permits acceptable aromatic component separation.

Gas Chromatography

The gas chromatograph employed was a Varian 2400 series instrument equipped with linear temperature programmer, flame ionization detectors, and cryogenics. Two matched 10 ft x 1/8 in. stainless steel columns containing bonded carbowax 400 on Porosil F were used to obtain the chromatograms. Instrument sensitivities were set at either 4 or 8 x 10^{-12} amps full scale, which required matched columns for good baseline stability. Ultra high purity helium was used as carrier gas at 30 mls/min. Initial column temperature was $20^{\circ}C$ with temperature programming at $6^{\circ}C$/min. to $190^{\circ}C$. Areas of component peaks were obtained using an Autolab ® System IV Computer/calculator. The signal from the detector amplifiers was attenuated at the recorder when necessary but not the signal to the Autolab®. Instrument calibration was obtained by direct injection of pure hydrocarbon components in hexane solvent onto the chromatographic column. Calculation of component concentrations followed the method developed by McAuliffe (1971). At least two successive equilibrations of the same sample are necessary to calculate component concentrations. Caution: Direct injection of solvent onto the column used for the separation of fuel oil hydrocarbons tends to cause deterioration of the separation efficiency of the carbowax column.

Identification of Component Peaks

A Finnigan Model 3200 GC/MS controlled by a System Industry Data System 250 was used to identify separated aromatic peaks. The sample loop previously described was used for sample introduction. GC conditions were essentially the same as described above. Because of sensitivity limitations, saltwater suspensions containing about 10 parts per million (as measured by IR) were required. Chromatograms obtained from these concentrated samples were sufficiently similar to those obtained on the more dilute samples to permit cross comparisons to be made. Baseline subtraction methods were used to obtain spectra as free from contamination as possible. Where possible, identifications of compounds were made by a computer search system similar to the MSSS system developed by Battelle Memorial Institute for the EPA. Contamination of aromatic component peaks by saturate compounds complicated the specific identification of some tri- and tetramethyl benzene isomers. GC/MS analysis of fuel oil suspensions that had previously been filtered through 0.45 µ Millipore® filters reduced the saturate contamination and aided in component identification by visual inspection. Where possible, component peaks tentatively identified by GC/MS were confirmed by comparing the retention times of hydrocarbon standards. n-Paraffins were identified by retention time.

Filtration of Fuel Oil Suspension in Water

One hundred milliliters of aqueous oil suspension was taken into the sample syringe as previously described. The syringe was then attached through a luer-lok® fitting to a filtration device containing a 13 mm 0.45 µ Millipore® filter (Water Associates, Inc.). Fifty milliliters of the aqueous sample was forced through the filter into another 100 ml syringe attached to the other side of the filter. Filtered and unfiltered samples were analyzed using the helium partitioning technique.

Determination of Fuel Oil Concentration by IR

Acidified samples of aqueous fuel oil suspensions were extracted into carbon tetrachloride (Burdick and Jackson) and the absorbance at 2930 cm^{-1} determined on a Beckman Acculab® infrared spectrophotometer. Methodological details have been described elsewhere (Bean, 1974).

Results and Discussion

We found the technique of gas partitioning well suited to the routine analysis of volatile hydrocarbons in flowing seawater systems. Analysis of the sample (two successive partitionings) can be accomplished in about one hour. Chromatographic peaks are sufficiently well separated to permit integration by the data processor even though introduction of sample to the column can require several min. A calibration curve prepared from n-hexane in distilled water (Fig. 2) shows that the method is both accurate and sensitive. Levels of detectability are about 5 ng/l for saturate compounds and 15 ng/l for aromatic compounds using a 25 ml sample loop.

The chromatograms shown in Fig. 3A and 3B are obtained from two successive partitionings of an unfiltered fuel oil suspension (0.15 ppm oil by IR). Component peak designations given in Fig. 3 were tentatively identified by GC/MS and confirmed by GC retention time, where possible. Differences can be observed when comparing the two partitionings. The components seen early in the chromatogram of the first extraction (ret. time 2-10 min) are largely absent in the chromatogram of the second extraction, with the exception of benzene. McAuliffe (1969) noted that saturates are almost completely extracted into the gas phase on the first partitioning; thus, the behavior of these light components in the fuel oil suspension is diagnostic of saturate hydrocarbons. The peaks attributable to aromatic components are largely retained in subsequent partitionings (McAuliffe, 1971).

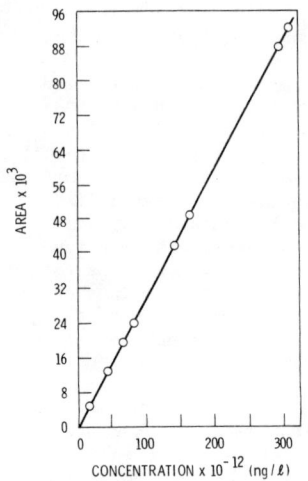

Fig. 40.2 n-Hexane calibration using 25 ml sample loop. Instrument conditions as in text.

Another feature of the chromatograms from the unfiltered suspension is an envelope appearing under the component peaks between the retention times of 16 to 24 min. During the mass spectrometric examination of the component peaks, material from this envelope interfered with the cracking patterns obtained and made positive identification of components difficult using computer algorithms. Since most of the interfering spectra consisted of a series of fragments separated by 14 mass units, we conclude that the envelope consists largely of saturate hydrocarbons. Although the envelope is substantially reduced in the chromatogram of the second helium extraction, the mass spectra obtained in this region continued to indicate saturate contamination.

The chromatogram from the same fuel oil/water suspension after filtration through 0.45 μ Millipore ® filters (Fig. 3C) contrasts with the sample obtained before filtration. Although the light saturate components are still evident, the saturate components comprising the envelope observed in the unfiltered sample are absent. This was also confirmed from the mass spectra; filtered samples had the least saturate spectral interference with the aromatic component spectra.

The results obtained from the filtered and unfiltered samples suggest that the suspended fuel oil hydrocarbons exist in the water soluble and the insoluble form even after standing for 24 hr. Boehm and Quinn (1974) reached a similar conclusion from examination of the heavier hydrocarbon compositions of filtered and unfiltered fuel oil suspensions. Chromatograms obtained from samples of bioassay water generated in a continuous flow-through apparatus (Vanderhorst, et. al., (1976b), containing less than 1 ppm fuel oil (IR), have much larger saturate envelopes relative to the aromatic component peak areas than do the laboratory prepared fuel oil suspensions. This would indicate a larger proportion of particulate oil to soluble oil in the flow-through apparatus, where the oil is in contact with seawater only a few hours prior to sampling.

The presence of an insoluble oil envelope does not appear to interfere with the analysis of aromatic components by the gas partitioning technique. Fig. 4 shows the plots of peak area versus partition number of several monocyclic aromatic components determined in an aqueous suspension of fuel oil. The data follows the expected semilogarithmic behavior (McAuliffe, 1971). The slopes obtained for individual components varied from sample to sample during repeated analysis of the flowing seawater system under constant conditions of temperature and salinity. This variation in slope can be explained from a consideration of the equilibria involved in the partitioning of a fuel oil suspension containing both soluble and insoluble oil. A sample to sample variation in the quantity of insoluble oil would be expected to result in changes in equilibrium distribution of volatile hydrocarbons between aqueous and vapor phases and therefore in a change of partitioning slope. Large variations in insoluble oil have in fact been observed in the flow-through bioassay apparatus at very low oil concentrations (Vanderhorst et al., 1976a). At least two helium partitionings are therefore important for each analysis, even if temperature and salinity (two other variables affecting partitioning slope) are constant.

Quantitative data for several major components in unfiltered and filtered fuel oil suspensions are presented in Table 1. Some component peaks could not be satisfactorily quantitated because of incomplete resolution from other components. A comparison of GC retention times with tentative identifications made by GC/MS revealed that more than one major aromatic component was present in several of the chromatographic peaks. For example, the principal peak under the component mixture labeled C_4-benzenes (Fig. 3) was tentatively identified as 1, 2-diethy benzene by GC/MS; however, 1, 3-dimethyl-5-ethylbenzene has the identical retention time.

Recoveries of aromatic components after filtration were high for the more volatile aromatics through the xylenes, ranging from 80 to 89%. Recoveries for the last three components in Table 1 were considerably lower, (40 to 59%). The difference in ability of aromatic components to pass through the filter is interpreted to be a function of their equilibrium distribution between soluble and insoluble oil. The more highly substituted aromatic molecules, being less hydrophyllic, tend to associate with water insoluble oil droplets and thus are filtered out of the aqueous phase to a greater extent than the more water soluble aromatic components.

Filtration of the fuel oil suspension removed 26% of the monocyclic aromatic components reported. However, both filtered and unfiltered samples were found to contain substantially more hydrocarbon than could be accounted for by IR analysis of the unfiltered sample (Table 1). Discrepancies of similar magnitude between IR and GC analyses of the soluble components of crude oils have been previously reported (Bean, 1974), and can be attributed to a much lower contribution to the IR absorbance of the water soluble aromatic hydrocarbons than the less soluble saturate hydrocarbons. This suggests that petroleum bioassay studies involving water soluble oil constituents should not rely solely on IR analysis to define exposure levels.

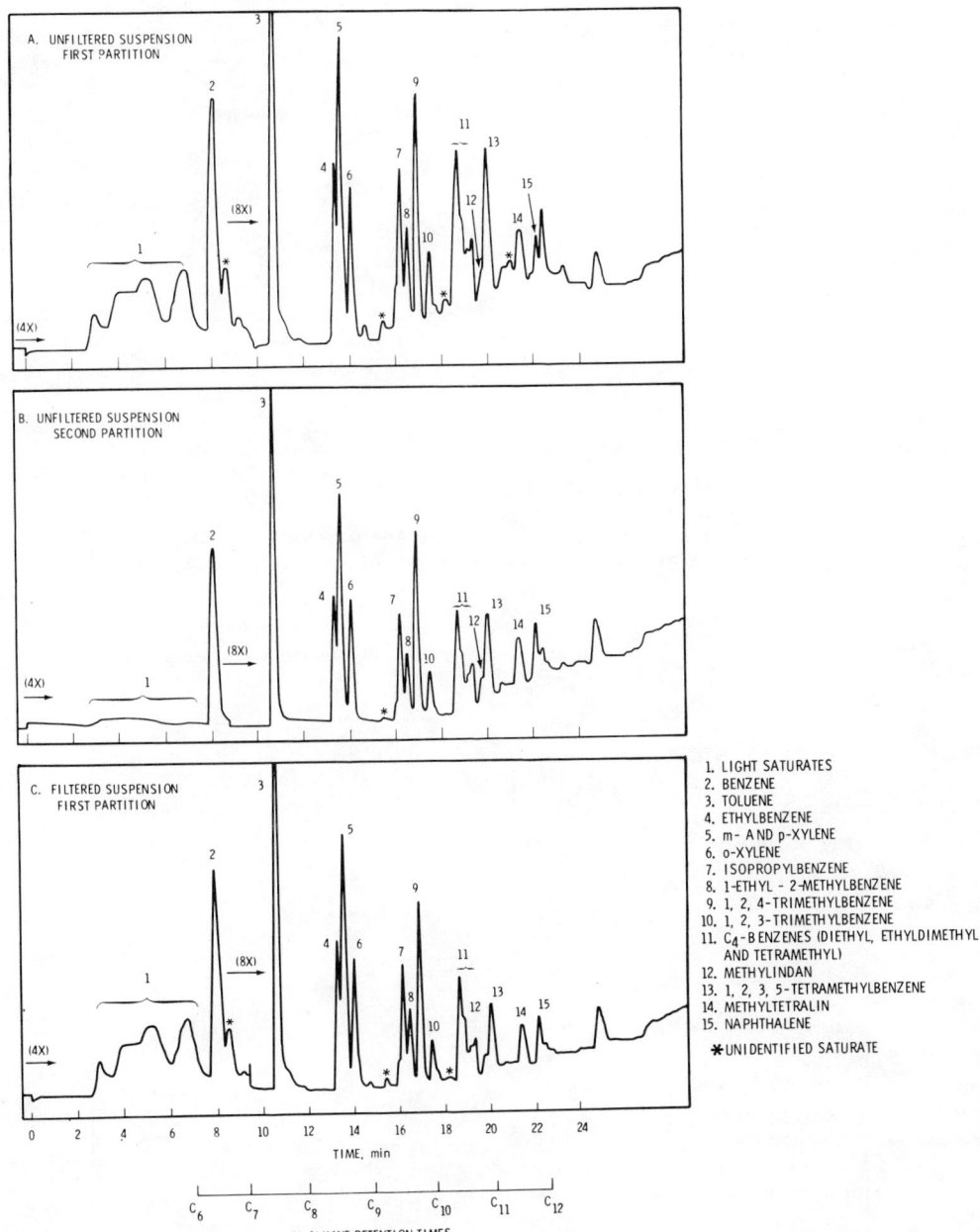

Fig. 40.3 Gas chromatograms of seawater suspension of No. 2 fuel oil.

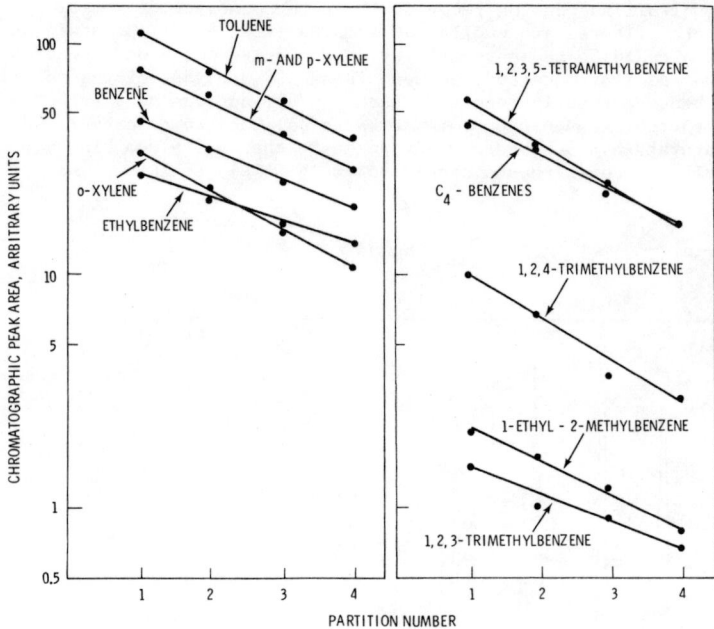

Fig. 40.4 Partitioning of monocyclic aromatics from a seawater suspension of No. (0.15 ppm by IR).

Table 1. Concentrations of aromatic components determined by helium partitioning in filtered and unfiltered samples of seawater containing No. 2 fuel oil (unfiltered sample analyzed for 396 µg oil per liter of seawater by IR)

	Concentration In Unfiltered Sample	Concentration In Filtered Sample	Recovery After Filtration
	----µg/l----	----µg/l----	------%-------
Benzene	120	100	83
Toluene	185	162	88
Ethylbenzene	89	71	80
m + p-Xylene	229	186	81
o-Xylene	143	127	89
1-Ethyl-2-Methylbenzene	109	86	79
1,2,4-Trimethylbenzene*	480	386	80
1,3,3-Trimethylbenzene	76	45	59
1,3-Dimethyl-5-Ethylbenzene*	138	80	58
1,2,3,5-Tetramethylbenzene	258	104	40
TOTAL	1827	1347	74

* Mass spectrometric and gas chromatographic evidence for more than one major aromatic component present.

Acknowledgments

This work was sponsored by the United States Energy Research and Development Administration. We are grateful to C. Bruce Koons and David L. Johnson of Exxon Production Research Company, Houston, Texas for advice and assistance regarding gas sampling techniques. GC/MS spectra were obtained by Denis C. K. Lin of Battelle Memorial Institute, Columbus, Ohio.

The registered trademarks and companies are referenced for reader convenience in replicating experiments and do not represent endorsement by Battelle, Pacific Northwest Laboratories.

References

Anderson, J. W., J. M. Neff, B. A. Cox, H. E. Tatem and G. M. Hightower, Characteristics of dispersions and water-soluble extracts of crude and refined oils and their toxicity to extuarine crustaceans and fish, Marine Biology 27, 75 (1974).

Bean, R. M. (1974) Suspensions of Crude Oil in Seawater: Rapid Methods of Characterizing Light Hydrocarbon Solutes, in: NBS Special Publication 409, Marine Pollution Monitoring (Petroleum), Proceedings of a Symposium and Workshop held at NBS, Gaithersburg, MD, May.

Bean, R. M., J. R. Vanderhorst, and P. Wilkinson (1973) Interdisciplinary Study of Toxicity of Petroleum to Marine Organisms, Battelle, Pacific Northwest Laboratories, Richland, WA.

Boehm, P. D. and J. G. Quinn, The solubility behavior of No. 2 fuel oil in seawater, Marine Pollution Bulletin 5, 101 (1974).

Boylan, D. B., and B. W. Tripp, Determination of hydrocarbons in seawater extracts of crude oil fractions, Nature 320, 44 (1971).

May, W. E., S. N. Chester, S. P. Cram, B. H. Gump, H. S. Hertz, D. P. Enagonio and S. M. Dyszell, Chromatographic analysis of hydrocarbons in marine sediments and seawater, J. Chromatographic Sci. 13, 535 (1975).

McAuliffe, C., GC determination of solutes by multiple phase equilibration, Chem Tech 1, 46 (1971).

McAuliffe, C., Solubility in water of normal C_2 and C_{10} alkane hydrocarbons, Science 158, 478 (1969).

Swinnerton, J. W. and R. A. Lamontagne, Oceanic distribution of low-molecular-weight hydrocarbons, Environmental Science and Technology 8, 657 (1974).

Vanderhorst, J. R., R. M. Bean, L. J. Moore, P. Wilkinson, C. I. Gibson, and J. W. Blaylock (1976a), Effects of a continuous low level No. 2 fuel dispersion on laboratory held intertidal colonies, accepted for publication in the Proceedings, 1977 Conference on Prevention and Control of Oil Pollution.

Vanderhorst, J. R., C. I. Gibson, L. J. Moore and P. Wilkinson (1976b), Continuous-flow apparatus for use in petroleum bioassay, Accepted for publication by Bull. Env. Contam. and Toxicol.

Vaughan, B. E., editor, (1973), Effects of Oil and Chemically Dispersed Oil on Selected Marine Biota, Battelle, Pacific Northwest Laboratores, Richland, WA.

CHAPTER 41

SEDIMENT HYDROCARBONS AS ENVIRONMENTAL INDICATORS IN THE NORTHEAST GULF OF MEXICO

Julia S. Lytle and Thomas F. Lytle

Gulf Coast Research Laboratory
Ocean Springs, Mississippi 39564

Abstract

Surface sediment samples were collected in 1975-76 from forty-five locations in the Gulf of Mexico along the continental shelf from Pascagoula, Mississippi to Fort Myers, Florida. Hydrocarbons were analyzed in the sediments for the Bureau of Land Management in an effort to survey hydrocarbons in the northeast Gulf and detect man-induced and seasonal effects in hydrocarbon profiles. Clearly, there are three zones in this area as distinguished by their aliphatic hydrocarbon distributions. The Florida shelf contains aliphatics which consist primarily of a branched-cyclic-unsaturated complex at Kovats Index 2-75-2150, with lesser components at 1640-2500 and 3400 on FFAP. The small, $\underline{n}$-alkane fraction is dominated by C_{17} and resembles algal hydrocarbons.

The sediments from the Mississippi-west Alabama shelf and from the outermost edge of the Florida shelf are characterized by an abundance of high molecular weight $\underline{n}$-alkanes of pronounced odd/even preference typical of terrestrial hydrocarbons. Occurring in lesser amounts, is a suite of low molecular weight $\underline{n}$-alkanes with a uniform distribution characteristic of degraded crude oil. The west Florida and east Alabama shelf contain hydrocarbons intermediate in nature between these two extremes. These characteristics have remained virtually unchanged through several seasons and suggest that seasonal effects upon hydrocarbons are minimal in this area.

Key words: sediments, hydrocarbons, alkanes, pollution, olefins, organic geochemistry.

Introduction

The offshore oil leases in the northeastern Gulf of Mexico in 1974 have resulted in an extensive program of scientific activity on the continental shelf of Mississippi-Alabama-Florida (MAFLA). The 1974 program included a baseline hydrocarbon survey of sediments of the inner continental shelf extending from Tampa, Florida to the Mississippi River delta. Though intended as an environmental impact study, the efforts of investigators in this region have complemented previous studies concentrated in the deep basin (Aizenshtat et al., 1973; Newman, 1973; Rogers and Koon, 1970), barrier islands (Palacas et al., 1976) and estuarine systems (Palacas et al., 1972; Swanson et al., 1968; Johnson and Calder, 1973) of the eastern Gulf.

Results of the initial study disclosed two distinct hydrocarbon provinces in the northeast Gulf (Gearing et al., 1976). The Florida shelf, rich in carbonate materials, was characterized by very complex mixtures of hydrocarbons which were apparently of marine origin and were dominated by a group of C_{25} branched-unsaturated compounds. Sediments of the Mississippi and west Alabama shelf, chiefly comprised of silt and clay materials, yielded hydrocarbons with a very distinct terrestrial signature of high molecular weight $\underline{n}$-alkanes of high odd/even preference. Also in evidence was a suite of petroleum-like hydrocarbons indicating possible pollution on the Mississippi-Alabama shelf. The east Alabama-west Florida shelf acted as a transition zone containing pronounced contributions of terrestrial marine and petroleum-like hydrocarbons.

In 1975-76 the monitoring phase of the study expanded the 1974 sample program to include deep water sites on the outer continental shelf, sites further south on the Florida shelf and collections made during more than one season to detect short-term or seasonal changes in hydrocarbon profiles.

Experimental

Forty-five box corings were made along six transects in the northeastern Gulf of Mexico continental shelf during June, 1975. Twenty-one of the station sites were resampled in January, 1976. Samples consisted of approximately 3000g of sediment from the top 20 cm of each core. Sediment was placed in prewashed (hot soapy water followed by methanol rinsing) tin cans and stored at -18°C until analyzed. All precautions were taken to eliminate possible contamination aboard ship.

Extraction and analysis of the sediments followed the procedure of Sever et al., (1972) and Gearing et al. (1976). Following methanol-chloroform extraction, elemental sulfur was removed, and the purified lipid was separated by elutions from silica gel-alumina into an 'aliphatic' (hexane) and 'aromatic' (benzene) fraction. Each fraction was weighed and analyzed by gas chromatography using external standards for quantification. Further aids in identification employed were hydrogenation (after the method of Lytle and Sever, 1973), urea adduction and mass spectrometry.

Percent carbonate materials were measured as loss upon acidification; the resulting residue was used for gasometric determination of organic carbon.

Study Area

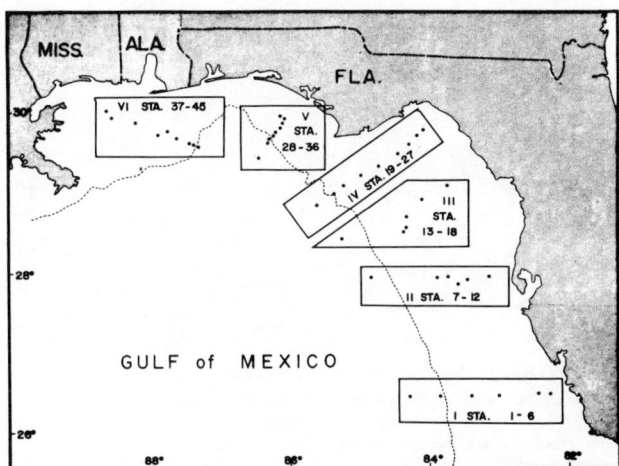

Fig. 41.1 Sample stations for 1975-1976. Sation Nos. at all Transects run consecutively seaward except Transect 4 stations which run shoreward. The Dotted line represents the 50 fathom contour.

A map depicting the sampling sites for 1975-76 is shown in Fig. 1. Ludwich (1974) has described the sediments in this area, and Griffin (1962) has described the factors affecting water movements on the eastern Gulf shelf. A detailed geophysical and mineralogical survey of the area is also available (Final Reports, BLM 1974, 1975). Transects 2, 3, 5 and 6 include areas sampled in 1974 (Gearing et al., 1976). Transect 1 contains sites further south than the 1974 sampling program and extends east-west from Fort Myers, Florida in water depths of 11 to 165m. Transect 2 sites lie off Tampa, Florida in 20 to 190m depths. Transect 3 begins at a point midway between Tampa and St. Marks, Florida in 20m of water and extends southwest to 170m depths. In a west-southwest direction from St. Marks, Florida, Transect 4 covers stations at 10-170m depths. Transect 5 stations lie southwest of Panama City, Florida at 38-180m depths. Stations on Transect 6 extend east-southeasterly from Pascagoula, Mississippi from 20m depths out to 110m depths. At all but the deep-water stations of Transects 1-5, sediments are composed of variable mixtures of carbonate-rich sands and shell-hashes. Fine-grained materials are enriched in all deep-water stations and stations along Transect 6.

Results & Discussion

TABLE 1. Gravimetric Data of Sediments

Parameter	Transect 1	Transect 2* Area I	Transect 3 Area II	Transect 4	Transect 5 Area III	Transect 6 Areas IV&V
Percent Carbonate	81.6	81.9 86.4	68.4 85.9	48.9	75.2 71.0	31.4 44.2 15.8
Percent Organic Carbon (Carbonate-free basis)	3.47 (3.19)**	3.17 (2.58) 3.60 (1.87)	1.32 (1.42) 2.09 (0.98)	0.771 (1.03)	1.70 (1.53) 1.42 (1.07)	1.16 (1.22) 0.77 0.58 (0.92)(0.20)
Lipid/Total Sediment, ppm	66.8 (30.3)	104. (33.4) 158. (79.)	75.2 (75.3) 79. (131.)	82.8 (100.)	95.1 (146.) 107. (109.)	89.9 (77.9) 44. 232. (39.)(122.)
Hydrocarbons/Total Sediment, ppm	1.97 (0.840)	2.66 (0.695) 4.10 (1.53)	1.96 (1.05) 2.40 (0.70)	2.08 (4.21)	2.44 (2.12) 2.30 (0.80)	5.00 (4.29) 1.50 11.7 (1.40)(6.4)
Aliphatic HC/Lipid X 100	1.44 (0.423)	1.34 (0.821) 1.14 (0.29)	1.34 (0.608) 1.22 (0.29)	2.25 (1.31)	1.52 (0.890) 1.43 (0.68)	3.22 (1.51) 2.18 3.21 (1.74)(1.67)
Aromatic HC/Lipid X 100	1.14 (0.371)	1.47 (0.523) 1.55 (0.84)	1.92 (1.17) 1.92 (0.66)	2.07 (0.864)	1.71 (1.62) 1.78 (0.94)	2.77 (1.05) 1.91 2.56 (1.67)(1.11)

*Transect (1975) and Area (1974) data means are tabulated with corresponding locales combined, Transect above line and Area below line.

**Nos. in parentheses represent standard deviations.

Comparison of 1974 and 1975. Table 1 contains values of certain gravimetric hydrocarbon data of the 1975 sampling period; for comparison, values reported by Gearing et al. (1976) for 1974 areas of the same locations are included. The data are quite consistent indicating good agreement between the two years. Percent carbonate is quite high in all transects especially Transects 1-5, reflecting the shell-hash composition of the sediments on the Florida shelf. Organic carbon on a carbonate-free basis is somewhat enriched in Transects 1-2 though in the total sediment organic carbon values are very similar throughout the northeastern Gulf. Hydrocarbons are generally higher in total sediments of Transect 6 samples, but in the acid-insoluble fraction (carbonate-free) both lipid and hydrocarbons are decidedly enriched in Transects 1-5 vs 6. The reverse is true of aliphatic and aromatic hydrocarbons when expressed as a percent of the lipid fraction; both occur in higher amounts in Transect 6 vs 1-5. The data tend to confirm that all Florida-Gulf shelf sediments (Transects 1-5) are quite similar in their major chemical constituents and are distinguishable from Mississippi shelf sediments (Transect 6). The transition from Florida carbonate-rich sediments to Mississippi silt and clay occurs at the eastern end of Transect 6 at a location corresponding to 1974 Area IV sites.

The distribution of aliphatic hydrocarbons in a typical Florida sediment collected in the summers of 1974 and 1975 is shown in Fig. 2. In both chromatograms n-alkanes occur as minor components somewhat dominated by n-C_{17}, a likely reflection of organic detritus from the marine community (Blumer et al., 1971; Brown et al., 1972; Clark and Blumer, 1967; Gelpi et al., 1968; Mackie et al., 1974; Lee and Loeblich, 1971; Oro et al., 1967; Parker et al., 1972; Winters et al., 1969; Youngblood et al., 1971). The most pronounced

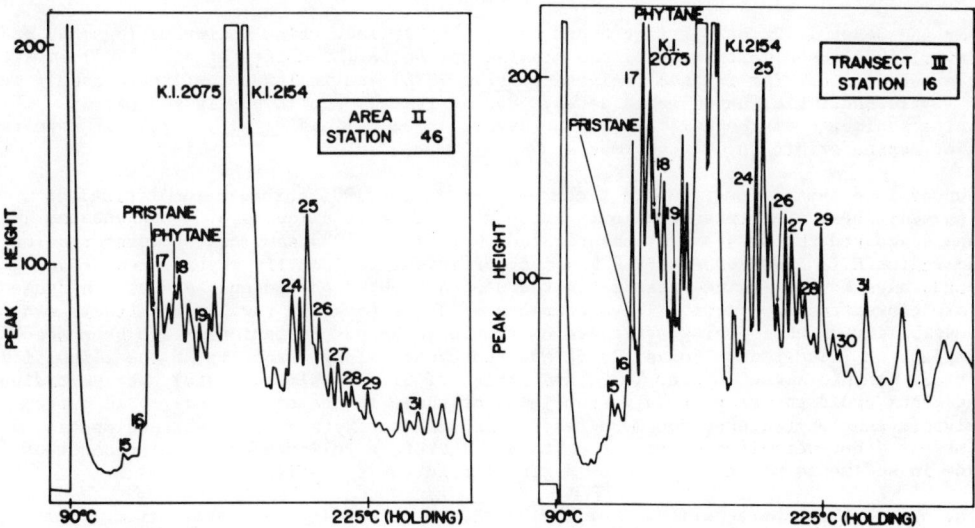

Fig. 41.2 Aliphatic hydrocarbons of Florida shelf sediments in 1974 and 1975. The chromatograms are from the station designated Area II-46 in 1974 and Transect III-16 in 1975. The 'x' refers to a group of phytadiene peaks.

feature is the peak conglomerate in the Kovats Index range ca. 2070-2075. The peak consists of a mixture of isomeric $C_{25}H_{48}$ compounds (Gearing et al., 1976). Further clarification of this mixture was effected in 1975 by $AgNO_3$-impregnated silica gel separation of aliphatics and repeated preparative GLC and capillary GC-MS. One typical component yields a mass spectrum seen in Fig. 3. The most consistent structures are those of

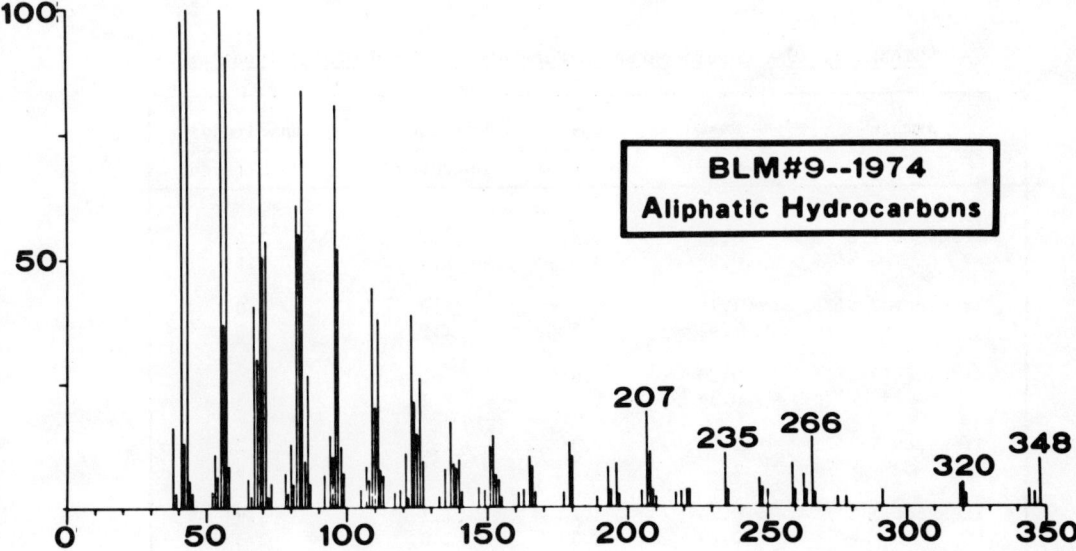

Fig. 41.3 Mass spectrum of a Kovats Index 2070-2075 component of Florida shelf aliphatics.

di-methyl branched C_{25} non-conjugated diolefins (Hoering, T., unpublished data). As yet no precursor hydrocarbon has been found among phyto- and zooplankton, water, neuston, particulate material, epifauna or epiflora of the NE Gulf (Calder, 1975; Meyers, 1975; Lytle and Lytle, 1975). Compounds eluting at or very near this Kovats Index (ca. 2070) and possibly of similar structure can be seen in hydrocarbon chromatograms of bay sediments

(Blumer and Sass, 1972a,b; Farrington and Quinn, 1973), lacustrine sediments (Wakeham and Carpenter, 1976), sediments west of the Mississippi Delta (McAuliffe et al., 1975), quartz sands of NE Gulf barrier islands (Palacas et al., 1976) and in Texas continental shelf sediments (Lytle and Lytle, unpublished data). Due to its absence in purely terrestrial-estuarine sediments of the Gulf of Mexico (Sever, 1970; Palacas et al., 1972), it appears to be of marine origin in Gulf sediments.

A group of four phytadienes labeled X can be seen in the 1975 chromatogram typical of Florida sediments. The intense magnification of phytane in aliquots of hydrogeneated hydrocarbons suggested the presence of unsaturated phytanes in 1974 samples. However resolution in the region K.I. 1900-2000 on FFAP was not sufficient to identify phytadienes. The compounds have Kovats Indices that match those of authentic phytadienes and upon hydrogenation are converted to phytane. The occurrence of these peaks in raw lipid extracts excludes the possibility of their being artifacts of sample work-up. Phytadienes have been reported in zooplankton (Blumer and Thomas, 1965) and represent a large fraction of the aliphatics of certain benthic algae (Lytle, T. F. and Lytle, J. S., unpublished data); the phytadienes in sediments could therefore result from input of these marine organisms. It is quite conceivable that phytadienes could be readily reduced to phytane in recent sediments and account for a non-petroleum source of phytane. Therefore reliance upon the presence of phytane in sediments may not per se be a strong argument for pollution.

All the hydrocarbon characteristics of the Florida shelf indicate an area strictly of deposition of marine organic matter. A related study of fatty acids by Walker (1976) offers further proof of this suggestion by demonstrating the abundance of fatty acids characteristic of marine organisms in Florida shelf sediments but not those from the Mississippi shelf.

The fact that little if any temporal variation exists in hydrocarbon distribution of the Florida shelf is apparent in the chromatograms of Fig. 2 but is further documented by observing some gas chromatographic parameters from all Florida Transects (1-5) summarized in Table 2. These parameters, which include some used to indicate petroleum pollution, exhibit only minor changes more as a result of natural variability than any true seasonal effect.

TABLE 2. Gas Chromatographic Parameters of Sediment Aliphatics

Sample	$\dfrac{C_n \leq 20}{C_n \geq 21}$	$\dfrac{C_{17}}{\text{pristane}}$	$\dfrac{C_n \leq 20 \text{(odd)}}{C_n \leq 20 \text{(even)}}$	$\dfrac{C_n \geq 21 \text{(odd)}}{C_n \geq 21 \text{(even)}}$
Transect 1 S'75*	0.36	2.5	2.7	3.4
W'76	0.28	3.2	2.4	2.4
Transect 2 S'75	1.1	1.5	1.8	3.0
W'76	1.4	1.3	1.9	2.5
Transect 3 S'75	0.20	1.6	–	–
W'76	0.34	1.7	2.1	2.5
Transect 4 S'75	1.0	2.5	1.2	3.2
W'76	1.1	3.3	1.5	2.3
Transect 5 S'75	0.24	3.1	1.1	3.9
W'76	0.20	2.0	1.0	3.3
Transect 6 S'75	0.10	1.5	1.2	3.6
W'76	0.18	1.3	1.3	3.2

* Values represent means of the 2 or 3 samples from the summer of 1975 (S'75) resampled in the winter of 1975-76 (W'76) except Transect 6 where 6 stations were resampled.

Figure 4 contains a comparison of 1974 and 1975 hydrocarbon analysis of sediments from the Mississippi continental shelf and represents most samples of Transect 6. This area is distinct from the Florida shelf in terms of hydrocarbon distribution. Both

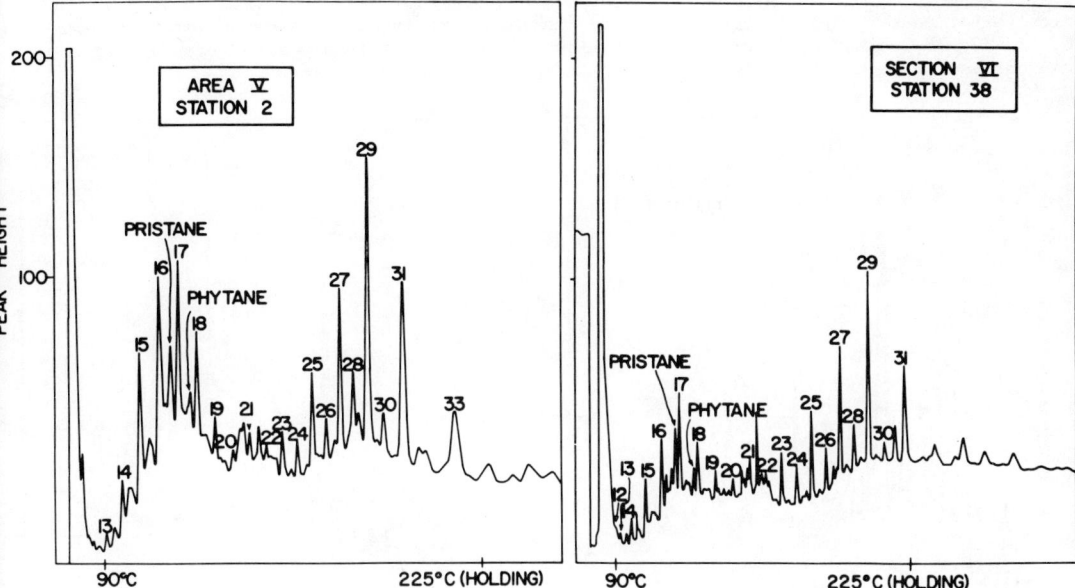

Fig. 41.4 Aliphatic hydrocarbons of Mississippi shelf sediments in 1974 and 1975. The chromatograms are from the station designated Area V-2 in 1974 and Transect VI-38 in 1975.

chromatograms of Fig. 4 display a suite of high molecular weight n-alkanes primarily of odd carbon number. This distribution is unique for terrestrial plants (Clark and Blumer, 1967; Eglington and Hamilton, 1967) and reflects input of terrestrial material from the Mississippi River, Mississippi Sound and Mobile Bay. The diminished quantity of the K.I. 2070 complex in this area probably indicates the minor contribution of marine lipids. (Most samples in the western portion of Transect 6 contain considerably less of this complex than samples 41-45.) The abundance of pristane and phytane, the even distribution of n-alkanes between $C_{15}-C_{20}$, and the unresolved envelope in this range were used to implicate a petroleum source for hydrocarbons in Mississippi sediments in 1974 (Gearing et al., 1976). Similar characteristics have been recorded in other sediments known to be polluted with petroleum hydrocarbons (Blumer and Sass, 1972a,b; Giger et al., 1974; Lytle, 1975; Farrington and Quinn, 1973; Blumer et al., 1970). The striking change that has occurred in the year between collections is the relative decrease in the n-alkane suite $C_{15}-C_{20}$. Preferential loss of n-alkanes by bacterial activity or other processes can account for the rapid loss of the petroleum alkanes in sediments (ZoBell, 1969, 1973; Miget et al., 1969). The C_{17}/pristane ratio dropped from 1974-1975 in the Mississippi sediments, another indication of loss of low molecular weight n-alkanes. Shown in Table 2, the C_{17}/pristane ratio shows further reduction in the winter 1975-1976 collection along Transect 6, attesting to a progressive decline of the petroleum n-alkane signature since the summer of 1974. At some stations along Transect 6, hydrocarbons have approached 'background' levels with little remaining evidence of pollution. Though a petroleum-like distribution of the low molecular alkanes is a necessary condition for detecting petroleum hydrocarbons it may not always be sufficient to distinguish indigenous from petroleum hydrocarbons. An approximate condition of steady state exists in the deposition of sedimentary hydrocarbons, thereby maintaining a relatively constant hydrocarbon profile. To suggest that the $C_{15}-C_{20}$ alkanes as seen in 1974 are indigenous to the Mississippi shelf sediments, in view of their decline, is to suggest that the biological community in the Mississippi River drainage area has dramatically changed from 1974 to 1975. This supposition would be much more difficult to accept than a sporadic addition of pollutant hydrocarbons from any number of sources in the area.

An intermediate or mixing zone of hydrocarbons in the northeastern Gulf coincides with the transition zone of carbonate-rich to silt-clay composition at the eastern portion of Transect 6. Here the marine and terrestrial components are at about equal abundance with even lesser amounts of the petroleum hydrocarbons.

Outer Shelf Sediments. Perhaps the most intriguing results in the 1975 samples involve the outer continental shelf along the northeastern Gulf. Here sediments on the Florida coast lose some of the shell hash-sand appearance of inner shelf samples and are composed of the higher quantities of fine grained materials like those found along Transect 6 on the Mississippi coast. Hydrocarbon levels are generally higher at these deeper stations (Nos. 6, 12, 13, 27, 36 and 45) than at the shallower stations. All deep water stations give a gas chromatographic pattern typified by that shown at Station 13 in Fig. 5.

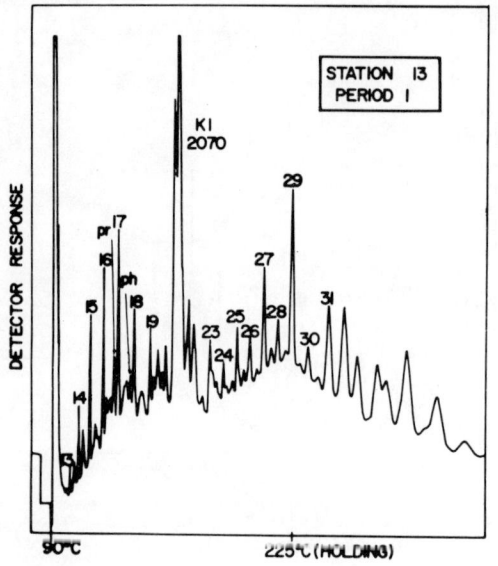
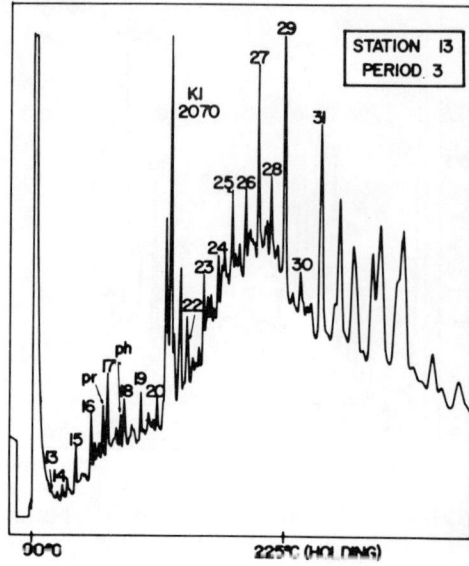

Fig. 41.5 Aliphatic hydrocarbons of outer continental shelf of northeastern Gulf of Mexico. Period 1 represents sampling of June, 1975 and period 3, January, 1976.

Some contribution of marine materials is evidenced by the presence of considerable amounts of the K.I. 2070 compounds, but what is surprising is the obvious presence of terrestrial and petroleum-like hydrocarbons. It appears that sediments of composition similar to those of the Mississippi shelf are being transported as far south as the outer shelf off Ft. Myers, Florida. If that is the source, then the migration of riverine sediments and any associated pollutants may be more extensive than was previously thought. Carbon isotope ratios do not reflect any major contribution of terrestrial organic material to these sediments (Parker, Patrick L., unpublished data). Evidently the hydrocarbons, representing a very small fraction of the total organic material, must be transported with very fine-grained material capable of extensive migration. Indeed correlation coefficients of aliphatic and aromatic hydrocarbons vs % clay reveal significant correlations at the 98.5% confidence level. Hedges and Parker (1976) have shown that the bulk of river-supported terrestrial organic material is transported very short distances from shore, so that it may come as no surprise that decidedly terrestrial hydrocarbons may exist in a marine-organic matrix. It has been suggested that very finely divided material may be transported along the outer continental shelf in a layer close to the bottom (Proceedings - ERDA, 1976).

The chromatrograms in Fig. 5 are another example of the only prominent short-term effects seen in sediment hydrocarbon patterns of the northeastern Gulf. Only those stations exhibiting traces of petroleum-like hydrocarbons reveal a discernible change that being a steady decrease in low molecular weight n-alkanes with time. Even in samples of Transect 5 just east of the track of Hurricane Eloise (September, 1975) temporal effects were of very small order.

Conclusions

Florida inner shelf sediments presently appear to be reservoirs of marine-derived hydrocarbons consisting of olefinic isomers of the formula $C_{25}H_{48}$ with lesser quantities of phytadienes and an n-alkane assemblage dominated by C_{17}. The hydrocarbons are distributed in patterns that are consistent from one year to the next and from season to season. This invariability could be perturbed with an input of pollutants as it has been in the sediments of the Mississippi-W. Alabama shelf. These Mississippi-W. Alabama sediments contain primarily odd-numbered high molecular weight n-alkanes of a decided terrestrial origin. The imprint of these hydrocarbons does not reflect a significant effect of time. On the other hand these sediments also contain a suite of petroleum-like hydrocarbons that are susceptible to noticeable loss in relatively short periods of time. Therefore in using n-alkanes to monitor sediments for petroleum pollution, frequency of sampling must be such that the pollutants can be readily detected.

An interesting feature of the outer continental shelf of the northeastern Gulf of Mexico is the occurrence of hydrocarbons that bear little resemblance to any of the inner shelf samples except those of the Mississippi shelf. This band on the outer continental shelf though containing some marine hydrocarbons is principally an area of deposition of terrestrial and petroleum-like hydrocarbons presumably of the same source as Mississippi shelf sediments. An explanation of this observation is not readily apparent but should prove intriguing to those interested in pollutant transport phenomena.

Acknowledgements

The authors gratefully acknowledge the technical and interpretive assistance of Dr. Patrick J. Gearing and Dr. Juanita N. Gearing. We thank Dr. Thomas C. Hoering for mass-spectral analysis and interpretations. Mrs. Sheron Bond, Miss Rebecca Cirlot and Miss Donna Bond were invaluable in tabulating data.

Financial support from the Bureau of Land Management (Contract Numbers 08550-CT4-11 and 08550-CT-5-30) is readily acknowledged.

References

Aizenshtat, Q., Baedecker, M. J., and Kaplan, I. R. (1973) Distribution and diagenesis of organic compounds in JOIDES sediment from Gulf of Mexico and western Atlantic. Geochim. Cosmochim. Acta. 37, 1881-1898.

Blumer, M., Guillard, R. R. L. and Chase, T. (1971) Hydrocarbons of marine phytoplankton. Mar. Biol. 8, 183-189.

Blumer, M., Sass, J. (1972a) The west Falmouth oil spill. Woods Hole Oceanogr. Inst. Tech. Rep. 72-19, p. 60.

Blumer, M., and Sass, J. (1972b) Indigenous and petroleum derived hydrocarbons in a polluted sediment. Mar. Pollut. Bull. 3, 92-94.

Blumer, M., Souza, G. and Sass, J. (1970) Hydrocarbon pollution of edible shellfish by an oil spill. Mar. Biol. 5, 195-202.

Blumer, M., and Thomas, D. W. (1965) Phytadienes in zooplankton. Science 147, 1148-1149.

Brown, F. S., Baedecker, M. J., Nissenbaum, A. and Kaplan, I. R. (1972) Early diagenesis in a reducing fjord, Saanich Inlet, British Columbia-III. Changes in organic constituents of sediment. Geochim. Cosmochim. Acta. 36, 1185-1203.

Calder, John A., In 1975 Final Report, Baseline Environmental Survey of the Mississippi, Alabama, Florida (MAFLA) Lease Areas, BLM Contract No. 08550-CT5-30, State University System of Florida Institute of Oceanography, St. Petersburg, Florida (In preparation).

Clark, R. C., and Blumer, M. (1967) Distribution of n-paraffins in marine organisms and sediment. Limnol. Oceanog. 12, 79-87.

Eglington, G. and Hamilton, R. J. (1967) Leaf epicuticular waxes. Science 156, 1322-1334.

Farrington, J. W., and Quinn, J. C. (1973) Petroleum hydrocarbons in Narragansett Bay - I. Survey of hydrocarbons in sediment and clams (Mercenaria mercenaria). Estuarine Coastal Mar. Sci. 1, 71-79.

Final Report, In 1974 Baseline Environmental Survey of the Mississippi, Alabama, Florida (MAFLA) Lease Areas, BLM Contract No. 08550-CT4-11, State University System of Florida Institute of Oceanography, St. Petersburg, Florida.

Final Report, In 1975 Baseline Environmental Survey of the Mississippi, Alabama, Florida (MAFLA) Lease Areas, BLM Contract No. 08550-CT5-30, State University System of Florida Institute of Oceanography, St. Petersburg, Florida. (In preparation).

Gearing, Patrick J., Gearing, Juanita Newman, Lytle, Thomas F. and Lytle, Julia Sever (1976) Hydrocarbons in 60 northeast Gulf of Mexico shelf sediments: a preliminary survey. Geochim. Cosmochim. Acta. 40, 1005-1017.

Gelpi, E., Oro, J., Schneider, H. J. and Bennet, E. O. (1968) Olefins of high molecular weight in two microscopic algae. Science 161, 700-702.

Giger, W., Reinhardt, M., Schaffner, C., and Stumm, W. (1974) Petroleum-derived and indigenous hydrocarbons in recent sediments of Lake Zug, Switzerland. Environ. Sci. Technol. 8, 454-455.

Griffen, G. M. (1962) Regional clay-mineral facies - products of weathering intensities and current distribution in the northeast Gulf of Mexico. Bull. Geol. Soc. Amer. 73, 737-768.

Hedges, John I. and Parker, Patrick L. (1976) Land-derived organic matter in surface sediments from the Gulf of Mexico. Geochim. Cosmochim. Acta 40, 1019-1030.

Johnson, Robert W. and Calder, John A. (1973) Early diagenesis of fatty acids and hydrocarbons in a salt marsh environment. Geochim. Cosmochim. Acta 37, 1943-1955.

Lee, R. F., and Loeblich, A. R., Jr. (1971) Distribution of 21:6 hydrocarbon and its relationship to 22:6 fatty acid in algae. Phytochemistry 10, 593-602.

Ludwick, J. C. (1974) Sediment in the northeastern Gulf of Mexico. In Papers in Marine Geology, pp. 204-238 MacMillan.

Lytle, J. S. (1975) Fate and effects of crude oil on an estuarine pond. In Proceedings of the 1975 Conference on Prevention and Control of Oil Pollution, pp. 595-600. American Petroleum Institute.

Lytle, J. S. and Lytle, Thomas F., In Final Report 1975 Baseline Environmental Survey of the Mississippi, Alabama, Florida (MAFLA) Lease Areas, BLM Contract No. 08550-CT5-30, State University System of Florida Institute of Oceanography, St. Petersburg, Florida. (In Preparation).

Lytle, T. F. and Sever, J. R. (1973) Hydrocarbons and fatty acids of Lycopodium. Phytochemistry 12, 623-629.

Mackie, P. R., Whittle, K. J. and Hardy, R. (1974) Hydrocarbons in the marine environment. I. n-Alkanes in the Firth of Clyde. Estuarine Coastal Mar. Sci. 2, 359-374.

Miget, R. J., Oppenheimer, C. H., Kator, H. I. and LaRock, P. A. (1969) Microbial degradation of normal paraffin hydrocarbons in crude oil. In Proc. API/FWPCA Joint Conference on Prevention and Control of Oil Spills, American Petroleum Institute, New York, 327-332.

Myers, Phil, In Final Report 1975 Baseline Environmental Survey of the Mississippi, Alabama, Florida (MAFLA) Lease Areas, BLM Contract No. 08550-CT5-30, State University System of Florida Institute of Oceanography, St. Petersburg, Florida. (In Preparation)

Newman, J. W. (1973) Quarternary deep sea sediments from the Gulf of Mexico: an organic geochemical study. Ph.D. Thesis. The University of Texas at Austin.

Oro, J., Tarnabene, T. G., Nooner, D. W. and Gelpi, E. (1967) Aliphatic hydrocarbons and fatty acids of some marine and freshwater microorganisms. J. Bacteriol. 93, 1811-1818.

Palacas, J. G., Gerrild, P. M., Love, A. H. and Roberts, A. A. (1976) Baseline concentrations of hydrocarbons in barrier-island quartz sand, northeastern Gulf of Mexico. Geology 4, 81-84.

Palacas, J. G., Love, A. H. and Gerrild, P. M. (1972) Hydrocarbons in estuarine sediments of Choctawhatchee Bay, Florida, and their implications for genesis of petroleum. Bull. Amer. Assoc. Petrol. Geo. 56, 1402-1418.

Parker, P. L., Winters, J. K. and Morgan, J. (1972) A baseline study of petroleum in the Gulf of Mexico. In Baseline Studies of Pollutants in the Marine Environment, pp. 555-581. National Science Foundation, IDOE.

Proceedings of a workshop on environmental oceanography of the Gulf of Mexico, sponsored by the Energy Research and Development Administration. 15-16 March, 1976.

Rogers, M. A. and Koons, C. B. (1970) Generation of light hydrocarbons and establishment of normal paraffin preference in crude oils. In Refining of Petroleum for Chemicals. (editor R. F. Gould) American Chemical Society.

Sever, J. R. (1970) The organic geochemistry of hydrocarbons in coastal environments. Ph.D. Thesis. The University of Texas at Austin.

Sever, J. R., and Lytle, T. F. and Haug, P. (1972) Lipid geochemistry of a Mississippi coastal bog environment. Contrib. Mar. Sci, 16 149-161.

Swanson, V. E., Palacas, J. G., Love, A. H., Ging, T. G., Jr. and Gerrild, P. M. (1968) Hydrocarbons and other organic fractions in recent tidal flat and estuarine sediments, northeastern Gulf of Mexico. Talk presented at ACS Meeting, Atlantic City, N.J., Sept. 8-13.

Wakeham, Stuart G. and Carpenter, Roy (1976) Aliphatic hydrocarbons in sediment of Lake Washington, Limnol. Oceanog. 21, 711-723.

Walker, Sharon H. (1976) An environmental survey of fatty acids in the northwestern Gulf of Mexico, Master's Thesis, Louisiana State University at Baton Rouge.

Youngblood, W. W., Blumer M., Guillard, R. L., and Fiore, F. (1971) Saturated and unsaturated hydrocarbons in marine benthic algae. Marine Biology 8, 190-201.

Zobell, C. E. (1969) Microbial modification of crude oil in the sea. In Proc. API/FWPCA Joint Conference on the Prevention and Control of Oil Spills, pp. 317-326. American Petroleum Institute, New York.

Zobell, C. E. (1973) Microbial degradation of oil: present status, problems and perspectives. The Microbial Degradation of Oil Pollutants, D. G. Ahearn and S. P. Myers editors, Publication No. LSU-SG-73-01 1973. 3-16.

CHAPTER 42

THE STABILITY OF EMULSIFIED CRUDE OILS
AS AFFECTED BY SUSPENDED PARTICLES

C. P. Huang and H. A. Elliott

Department of Civil Engineering
University of Delaware
Newark, Delaware 19711

Abstract

Interaction between emulsified crude oil and naturally occurring suspended solids, such as oxides, hydroxides, carbonates, and clays can significantly affect emulsion stability, and thereby govern the distribution and the toxicity potential of oily pollutants in the aquatic environment. The stability of three crude oil emulsions, viz., Nigerian, Venezuelan, and Iranian, was studied in response to various types and concentrations of suspended solids, viz., Al_2O_3, SiO_2, and kaolinite. Interfacial reactions between emulsion and solid are influenced by the physical and chemical properties of both kinds of particles involved as well as the chemical characteristics of the aqueous phase.

The magnitude and sign of the surface charge of both the oil droplet and solid particles have been demonstrated to be important in the context of crude oil emulsion stabilization. Finely divided solid particles have been found to stabilize emulsions when both the oil droplet and solid possess small surface charge of the same or opposite sign. If the droplet and solid carry moderate to large surface charge of opposite sign, a complex destabilization phenomenon is observed. However, the solid concentration necessary to influence emulsion stability greatly exceeds mean oceanic values. Therefore, solid stabilization of crude-oil-in-water emulsions may occur in the riverine environment or off shorelines. Deep ocean drilling and estuarine dredging operation can also provide enough suspended solids for emulsion stabilization.

Key words: emulsion, stability, suspended particle, crude oil

Introduction

The continuing increase in the worldwide use of fossil fuel, especially petroleum, has resulted in a sizeable flux of the hydrocarbons into the aquatic environment. Upon introduction into a receiving water, crude oil will first spread out as a thin film and then disperse further through numerous steps of transformation. Prevailing wind and wave action aided by surface active materials tend to produce fine droplets or oil-in-water (O/W) emulsions (Pilpel, 1967).

Crude O/W emulsion is an important entity in controlling the fate of the spilt oil. Since an O/W emulsion is an ensemble of small droplets, large O/W interface is created. For instance, a fine emulsion prepared from only 10 ml of oil can produce an interface covering more than a quarter-acre (Becher, 1962). Transport processes such as bacterial degradation, zooplankton uptake (Conover, 1972), chemical decomposition, and dissolution (Zobell, 1946) of crude oil are thereby enhanced by the presence of an immense oil-water interface.

Moreover, crude oil-in-water emulsions can have deleterious effects on marine aquatic organisms (Swedmark, et al., 1973). Emulsion toxicity appears to be associated with certain high-boiling-point hydrocarbons in the crude oil (Blumer, et al., 1973). The toxicity potential of crude oil was reported to increase with increasing degrees of oil dispersion (Tarzwell, 1975).

Thermodynamically, O/W emulsions are unstable (Kitchner and Mussellwhite, 1968). Therefore, in order for the O/W emulsions to be significant in controlling the fate of crude oil and in being a hazard to aquatic life, a certain degree of stability must be imparted to the emulsions. A stable O/W emulsion requires the presence of a third component (emulsifying agent) capable of collecting at the interface between the two liquid phases. Three major classes of emulsifiers have been commonly suggested (Kitchner and Mussellwhite, 1968): (1) macromolecular compounds which prevent coalescence by forming a mechanically strong skin at the liquid-liquid interface; (2) surface-active compounds whose emulsifying ability is due to the polar-nonpolar molecular structure; (3) finely divided insoluble powders which collect at the interface and armor particles against contact and coalescence.

Much has been reported on the effect of macromolecules and surfactants on the stability of emulsion (Frieberg and Wilton, 1970). However, quantitative studies on the stabilization of emulsion by finely divided particles is seldom reported. Only recently, Thuer (1975) investigated the stability of petroleum oil in a bi-colloidal system as affected by the pH and simple electrolytes. The primary objective of this research is to investigate the effect of various inorganic solids on the stability of crude oil emulsions.

Suspended particles are ubiquitous in the aquatic environment. Natural, e.g., weathering, volcanic emanations, hydrothermal activities, biosynthesis, and/or man-influenced activities, e.g., waste discharge, dredging, agricultural, and mining operations, always produce particles which can be readily brought into natural water systems (Huang, 1976).

Materials and Methods

The emulsion-solid system was prepared either with a sodium chloride (Fisher Scientific) solution system or with a synthetic seawater (Syntha-Seas Chemical Company). Inorganic solid particles (silica, alumina, or kaolinite) in varying quantities were first introduced into 500 ml of the saline water which was being continuously agitated with a magnetic stirrer (Fisher Scientific) operating at the maximum speed. After the solids were sufficiently dispersed, 0.25 ml of crude oil (Nigerian Light, Iranian Heavy, or Venezuelan) was then added and the solution was agitated for an additional 10 minutes. While stirring, the pH was adjusted with sodium hydroxide or hydrochloric acid (Baker Chemical Company) and monitored with a Sargeant Welch pH meter (Model LS) equipped with a combination pH electrode (Fisher). Typical physical-chemical properties of the crude oils and inorganic particles are listed in Table 1 and 2, respectively.

TABLE 1. Crude Oil Properties Important in the Context of Emulsification (from Berridge, et al (1968) and Getty Oil Company)

Crude Oil	Specific Gravity at 60 F	Sulfur Content wt %	Kinematic Viscosity CentiStokes at 100 F	Vanadium Content ppm
Nigerian Light	0.867	0.19	5.16	55
Iranian Heavy	0.869	1.66	8.83	107
Venezuelan	0.896	1.3	12.0	73

TABLE 2. Typical Properties of Colloids

Colloid	Particle Diameter (μm)	pH_{zpc}
1. Oil Droplets		
Nigerian	1	3.0 ± 0.2
Iranian	3	5.0 ± 0.1
Venezuelan	5	5.8 ± 0.1
2. Solid Particles		
Cabosil (Cabot Company)	0.007-0.014	
Fisher Silica (Fisher Co.)	100	2.0 ± 0.2
Kaolinite (Egar, Fe)	7.6	2.0 ± 0.2
α-Al_2O_3 (Linde A)	0.05	
α-Al_2O_3 (Linde B)	0.3	8.8 ± 0.2
α-Al_2O_3 (Linde C)	1.0	

At the end of the stirring, the sample was transferred to a 500 ml separatory funnel and allowed to stand under quiescent conditions for a prescribed period of time. A portion of the emulsified oil-solid suspension was sampled by inserting a 25 ml pipette into the funnel at mid-depth. This portion was taken as a representative sample of the emulsified oil. When appropriate, the crude oil and solids which settled to the bottom of the funnel were withdrawn directly. This constituted the settled oil portion. Virtually all of the remaining liquid was then drained from the separatory funnel. The oil that had separated or creamed to the top of the separatory funnel either adhered to the inside walls or was retained in the funnel bottom as the liquid level dropped. This is referred to as the separated oil.

Quantitative analysis of the crude oil in the emulsified, settled, and separated portions was determined by the method proposed by Gruenfeld (1973). Freon "TF" (E. I. duPont de Nemours and Company, Incorporated) was used as the extracting solvent, with addition of sulfuric acid and sodium chloride to improve extraction efficiency. Four extractions were performed on each sample. The Freon-extracted crude oil was then placed in cylindrical silica cells of 5 or 10 cm pathlength depending on the oil concentration, for transmittance measurement at a frequency of 2930 cm^{-1} with a Beckman Acculab 4 Infrared Photometer.

Particle size measurement was made with a Coulter counter (Coulter Electronics, Incorporated) system using 48, 100 and 190 μ apertures. A Zeta Meter (Zeta Meter, Incorporated) was employed to measure the electrophoretic mobility of both the oil droplets and the solid particle.

Results and Discussion

Characterization of Oil and Solid Particles

Table 2 shows the size and pH_{zpc} of the emulsified crude oil droplets and the solid particles used in this study. The average sizes of the crude oil emulsions, as obtained by Coulter counting technique, are 1, 3, and 5μ, respectively, for the Nigerian, Iranian, and Venezuelan crude oils. The difference in particle size can be attributed primarily to the viscosity of the crude oils. From Table 1, Venezuelan crude oil displays the largest viscosity, and since the mixing energy input remains identical for all three crude oils, it should be the most difficult one to be dispersed and thereby should assume the largest particle size. The electrokinetic character of the colloidal particles is shown in Fig. 1. (oil droplets) and Fig. 2 (solids).

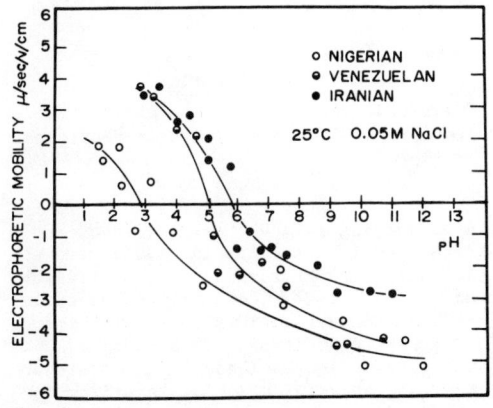

Fig. 42.1 The electrophoretic character of crude oil emulsions.

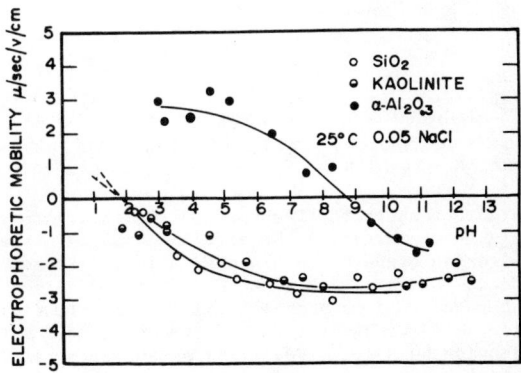

Fig. 42.2 The electrophoretic character of inorganic solid particles.

As indicated from Fig. 1, oil droplets were positively charged at low pH values. With increasing pH, the droplets became more negatively charged. The pH value at which the curves intercept the pH axis is designated as the pH of zero point of charge (pH_{zpc}). At the pH_{zpc}, the particle is considered having zero net surface charge. The pH_{zpc} was found to be 3.0 ± 0.2m 5.0 ± 0.1, and 5.8 ± 0.1 for the Nigerian, Venezuelan, and Iranian crude emulsions, respectively. The electrokinetic character of SiO_2 and kaolinite are similar, both having a pH_{zpc} of 2.0 ± 0.2. In contrast, $\alpha-Al_2O_3$ exhibits a pH_{zpc} of 8.8 ± 0.2. The results clearly indicate that at the pH conditions commonly encountered in the aquatic environment the crude oil emulsions, silica and kaolinite, will be negatively charged, while $\alpha-Al_2O_3$ will carry a positive charge.

Interaction of Similarly Charged Oil Droplets and Solid Particles

Figure 3 shows the dispersed Nigerian Crude Oil recovered after various quiescent periods as a function of the concentration of finely divided silica (Cabosil). At solid concentrations less than ca. 0.3 - 0.4 gm/l, more oil remained dispersed than when no solids were present in the system. Obviously, the solid particles enhanced the stability of the crude O/W emulsion. At $C_{S_{max}}$, the solid concentration resulting in maximum stabilization (0.1 gm/l), the quantity of oil which remained dispersed was about 40 percent more than in the absence of solids for the various quiescent periods investigated. Clearly, the solid particles must have been transported to the O/W interface to effectively armor the oil droplets against coalescence.

At solid concentration greater than $C_{S_{max}}$, a continuous decrease in the quantity of dispersed oil was noted. Figure 4 shows the quantity of oil extracted from the separated, emulsified, and settled portions for the Nigerian-Cabosil system. Concomitant with the increase in emulsified oil after $C_{S_{max}}$, an increase in settled oil occurred. High solid concentrations therefore resulted in a net transport of dispersed oil to the sediments.

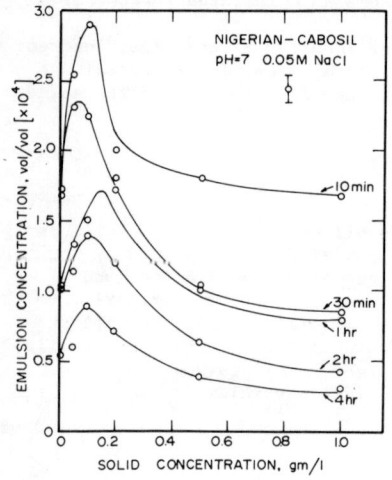
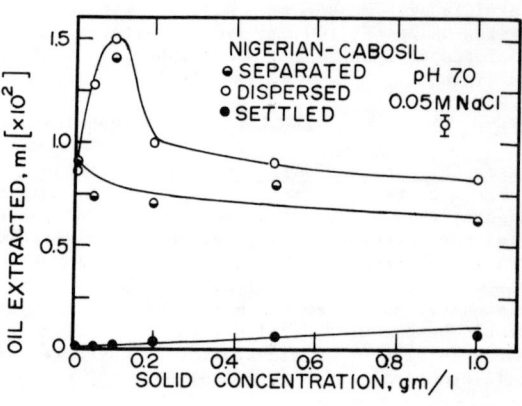

Fig. 42.3 The effect of silica on the stability of Nigerian crude oil droplets. Time (in min.) shown for various periods of quiescence following system agitation.

Fig. 42.4 The distribution of Nigerian oil droplets as affected by silica particles.

Figure 5 shows the response of the three crude oils to varying solids concentrations, and Fig. 6 demonstrates the effectiveness of three different solid particles in stabilizing the Nigerian crude O/W emulsions. The solids concentration required for maximum stabilization, $C_{S_{max}}$, as well as the degree of stabilization varied for each case. However, the general behavior remains unchanged; i.e., the stability increased to some maximum value with increasing solid concentration and then decreases with further solid addition. This typical stabilization behavior (Type I) is shown schematically in Fig. 7. Figures 1 and 2 indicate that at pH 7 both the solid particle and the oil droplet are negatively charged. According to the fundamental electrostatic principle, similarly charged particles should repell each other, and the greater the surface potential as reflected by its electrophoretic mobility, the greater this mutual repulsion. However, it is clear that the solid particles have made successful contact with the oil droplets. According to Verwey and Overbeek (1946), the electrostatic potential energy barrier must be 20 to 25 times larger than the thermal migration energy (κT) in order to prevent slow particle flocculation. This barrier for the Nigerian-Cabosil system as calculated with the equations derived by Hogg, et al., (1966) for heterocoagulation is on the order of one to two times κT. Therefore, it is clearly insufficient to prevent solid particle-oil droplet coagulation. The initial stabilization represented in Fig. 7 can consequently be attributed to solid particles collecting at the O/W interface.

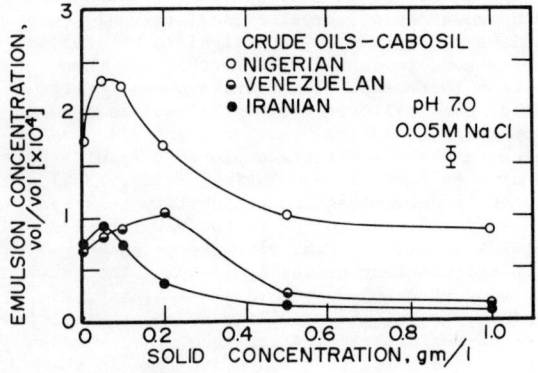

Fig. 42.5 The stability of crude oil emulsions as affected by silica particles.

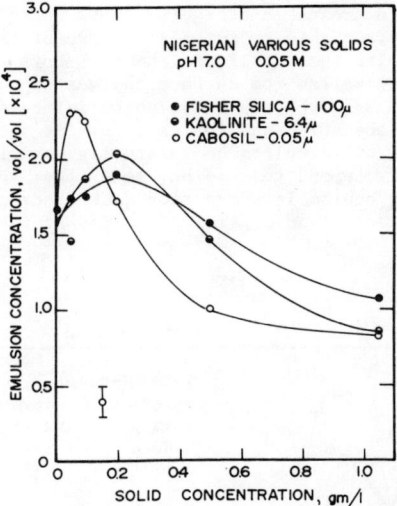

Fig. 42.6 The effect of relative particle size on the stability of Nigerian crude oil emulsions.

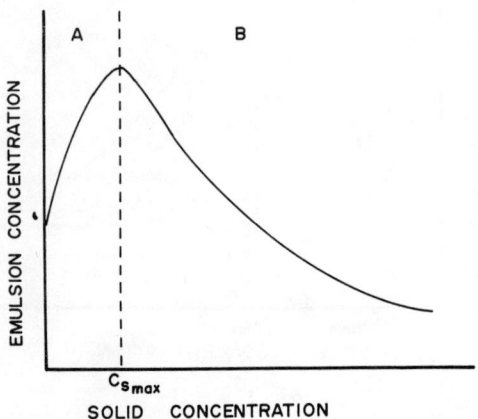

Fig. 42.7 Schematic representation of the stability of oil emulsion as affected by similarly-charged inorganic solid particles (Type I behavior).

Increasing the solid concentration beyond C_{Smax} resulted in destabilization. Introducing more solids into the system causes additional solid particles to collect on the oil droplet and the specific gravity of the aggregate will increase accordingly to eventually become greater than that of the aqueous phase. For example, for an oil droplet and solid particle with specific gravities of 0.9 and 2.6, respectively, an oil-solid aggregate of 22 percent (by weight) solids would have a specific gravity of 1.05. Moreover, at high solid concentrations, the solid-solid interparticle distance was reduced which can cause the solids to settle downward. Dispersed oil droplets would be enmeshed in a settling solid matrix and be swept out from the water column.

As shown in Fig. 5, a given solid particle may stabilize different crude oil emulsions to various degrees. Differences in stability behavior can be attributed to the physical and chemical properties of the crudes. As discussed above, the difference in viscosity may result in different droplet sizes under identical agitation intensity conditions. In principle, the smaller the size of the emulsion, the more stable it will be. Therefore, the emulsified Venezuelan and Iranian crude droplets that are larger than Nigerian droplets show the least stability. Furthermore, surface active material and crude oil can also modify O/W emulsion stability. For example, sulphonic acids and sulfonates (Pilpel, 1967) as well as vanadyl compounds (Dean, 1968) are substances with surface active properties that are frequently found in crude oils. Although no qualitative nor quantitative analyses of the impurity surfactants were conducted, it is known that the sulfur contents of all crude oils differ from each other.

Oppositely Charged Oil Droplets and Solid Particles

The fact that certain naturally occurring solids (Al_2O_3 and $CaCO_3$) are positively charged in the neutral pH range has prompted the investigation of an oppositely charged oil droplet and solid particle interaction. In the present research, aluminum oxide was used. The effect of α-Al_2O_3 on the stability of Nigerian O/W emulsions is shown in Fig. 8. Generally, the α-Al_2O_3 can collect at the O/W interface and armor the oil droplets against coalescence. However, the smallest size α-Al_2O_3, 0.05 μ in diameter, failed to show significant stabilizing effect. The 0.3 and 1.0 μ α-Al_2O_3 particles tend to destabilize further and then restabilize the emulsion as the solid concentration increases. The results shown in Fig. 5 also indicate that the smaller the solid particle, the smaller the amount of solids needed to stabilize the emulsion. This destabilization-restabilization behavior (Type II) is also depicted in Fig. 9. Transport of solid particles to the O/W interface probably results from the electrostatic attraction between solid and droplet "mutual coagulation" (Hogg, 1966) and the hydrophobic-hydrophilic nature of the solid which determines its wettability relationship with the two liquid phases. Both mechanisms attract the solids to the O/W interface and neutralize the surface charges of both colloids (Region A, Fig. 9). Charge reduction diminishes the repulsive potential between droplets; thereby gradually lowering the barrier to flocculation and coalescence. At a concentration of ca. 0.05 gm/l Al_2O_3 (Point D, Fig. 9) the oil droplets had virtually no net charge and, hence, its stability reaches the minimum level. Beyond this point, no further adsorption of solid particles can take place through purely coulombic attraction. The hydrophobicity of the solid then plays its role in the preferential adsorption of solids at the O/W interface.

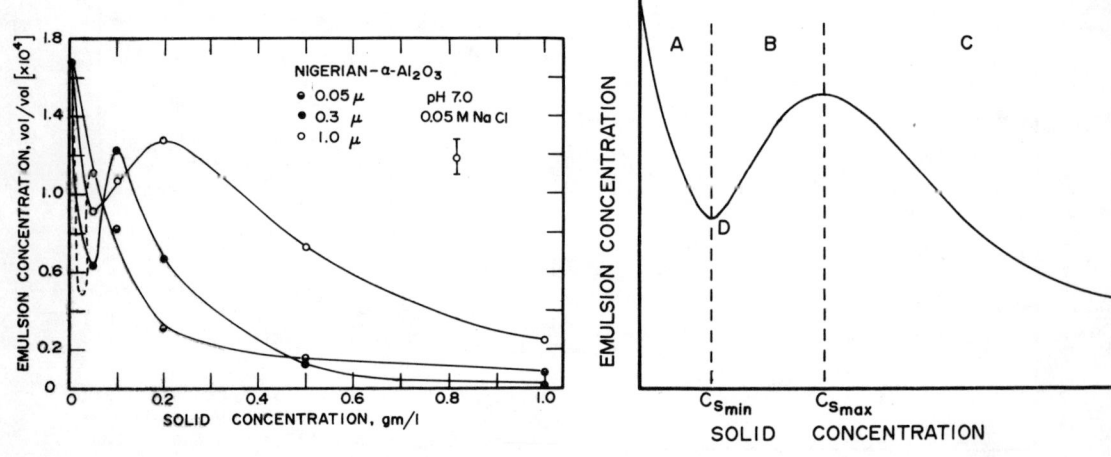

Fig. 42.8 The effect of α-Al_2O_3 on the stability of Nigerian crude oil emulsions (Type II behavior).

Fig. 42.9 Schematic representation of the stability of oil emulsions as affected by counter-charged inorganic solid particles.

At the solid concentration of the minimum stability, the surface coverage by solid particles is small, thus allowing easy accommodation of additional particles. The higher the droplet surface coverage, the greater the degree of stability enhancement. Accordingly, the stability began to increase after total charge neutralization because solid particles were still accumulating at the surface as a result of their hydrophobic nature (Region B, Fig. 9). The more solids that accumulate at the surface, the more protected the colloid is against coalescence. However, with further increase in solid concentration destabilization was again observed. Destabilization at high solids concentrations was characteristic of Type I behavior as well (Fig. 7). When the solid concentration becomes sufficiently high, the solids settle in a blanket fashion, and mechanically trap oil droplets as they subside.

Effect of Ionic Strength

Figure 10 is a plot of the electrophoretic mobility of Nigerian oil droplets as a function of NaCl concentration. Nigerian oil droplets are more negatively charged in ionic strengths typical of fresh water than in seawater. Such behavior can be interpreted as a result of electrical double layer compression. This implies that an O/W emulsion is more stable in fresh water than in seawater. For the solid-stabilized system, an analogous response to ionic strength was noted (Fig. 11). The aqueous phase electrolyte concentration used was 0.05, 0.35, and 0.7 M NaCl. Also, synthetic seawater, with a calculated ionic strength of 0.695, was employed. In general, increasing the ionic strength resulted in a decreasing ability of the Cabosil (SiO_2) particles to enhance the Nigerian emulsion stability.

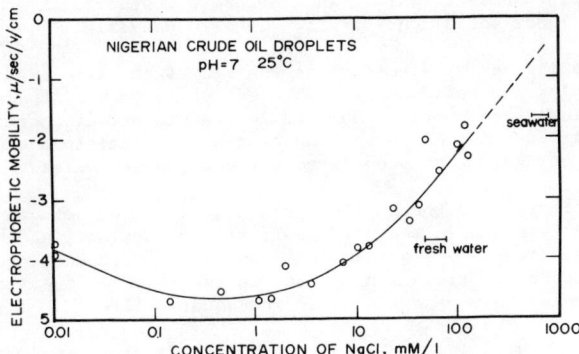

Fig. 42.10 The effect of ionic strength on the electrokinetics of Nigerian crude oil emulsions.

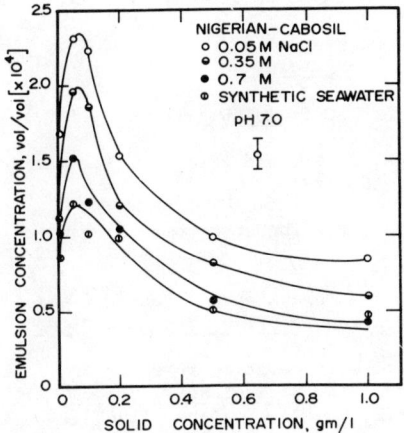

Fig. 42.11 The effect of ionic strength on the stability of the Nigerian - SiO_2 (Cabosil) system.

Summary

Predicting the environmental impact of an oil spill requires knowledge of the numerous transformations that oil undergoes when discharged into an aquatic system. Emulsification of crude oil may occur as a result of natural forces.

Finely divided suspended solids can influence the stability of crude O/W emulsions in natural aquatic systems. However, the concentration of suspended solids required to significantly improve emulsion stability was found to be about 100 to 200 ppm. Such solids concentrations greatly exceed mean oceanic values. Therefore, solid stabilization of crude O/W emulsions probably occurs only in the riverine environment or close to shorelines, particularly during and after rainstorms. Ocean drilling and estuarine dredging suspend sufficient solid material to enhance emulsion stability.

Suspended solids were observed to influence the stability of emulsified crude oil in two distinctly different ways:

1. Type I Behavior - When the oil droplet and the solid possess small surface charges of the same sign, electrostatic forces are weakly repulsive. Hence, the hydrophobicity of the solid determines the effectiveness of the solid as an emulsifier.

2. Type II Behavior - When the droplet and the solid carry moderate-to-large surface charge of opposite signs, electrostatic attraction exists, and solids will readily collect at the O/W interface. However, the solid particle adsorption results in

neutralization of droplet surface charge, and therefore a stabilization is observed. Restabilization with higher solid concentration is also possible.

If both the droplet and the solid carry large surface charges of the same sign, the electrostatic repulsion between droplet and solid may be large enough to prevent the solid from closely approaching the O/W interface where it must be located to act as an emulsifier. However, because of the prevailing neutral pH and ionic strength conditions encountered in natural waters, it is rather unlikely that such large surface charges exist.

Most colloids are negatively charged in the pH range characteristic of natural aquatic systems. Therefore, a stabilization phenomenon (Type I) will predominate under natural conditions. Hence, if the required amount of suspended solids are present, emulsified crude oil will tend to remain in the water column longer than in the absence of solids. Such stabilization of crude oil emulsions may adversely affect aquatic life.

Increasing the salinity of the aqueous phase tends to destabilize emulsions, whether they are free oil droplets or solid-stabilized droplets. Therefore, to assure the least hazard to aquatic life, oil drilling and other operations which potentially generate crude O/W emulsions may be preferentially located in high salinity waters.

Acknowledgements

This research was supported, in part, by a grant from the University of Delaware Research Foundation. The financial support of a Davis Fellowship to H. A. Elliott is also acknowledged.

References

Becher, P., The liquid-liquid interface, Official Digest, 486 (1972).

Berridge, S. A., Dean, R. A., Follows, R. G., and Fish, A., The properties of persistent oils at sea, J. Inst. Petrol. 54, 300 (1968).

Blumer, M., Ehrhardt, M., and Jones, J. H., The environmental fate of stranded crude oil, Deep Sea Res., 20, 239 (1973).

Conover, R. J., Some relations between zooplankton and bunker C oil in Chedabucto Bay following the wreck of the tanker Arrow, J. Fish. Res. Board. Can. 28, 1327 (1972).

Dean, R. A., The chemistry of crude oils in relation to their spillage on the sea, In: J. D. Carthy and D. R. arthur (eds.) The Biological Effect of Oil Pollution on Littoral Communities. Field Studies Council, Londin, England, (1968).

Friberg, S. and Wilton, I., Liquid crystals - the formula of emulsions, Amer. Perfumes and Cosmetics, 85, 12, 27 (1970).

Gruenfeld, M., Extraction of dispersed oils from water for quantitative analysis by infrared spectrophotometry, Environ. Sci. & Technol., 7, 636 (1973).

Hogg, R., Healy, T. W., and Fuerstenau, D. W., Mutual coagulation of colloidal dispersion, Trans. Faraday Soc., 62, 1638 (1966).

Huang, C. P., Solid-solution interface and its role in controlling the chemical composition of natural waters. Symposium on Solute Transport in the Marine Environment, AIChE meeting, Atlantic City, New Jersey, September, 1976.

Pilpel, N., Oil Pollution of the sea, Sci. J., 3, 73 (1967).

Swedmark, M., Granmo, A., and Kollberg, S., Effects of oil dispersants and oil emulsions on marine animals, Water Res., 7, 1649 (1973).

Tarzwell, C. M., Toxicity of oil and oil-dispersant mixtures to aquatic life. Seminar on Water Pollution by Oil, Aviemore, Scotland, 1975.

Thuer, M., Transportmechanismen von öl in naturlchen Gewässern unter besonderer Berucksichtigung von sedimentations - und flotations bedimgungen bei der Heterokoagulation von kolloidalen Oltropfchen mit mineralischen schwebstoffen, Ph.D. Thesis, Swiss Institute of Technology, Zurich, Switzerland, 1975.

Verwey, E. J. W., and Overbeck, J. Th. G., Long distance forces and colloidal particles, Trans. Faraday Soc. 42, 117 (1946).

Zobell, C. E., Action of microorganisms on hydrocarbons, Bacteriol. Rev., 10, 1 (1946).

CHAPTER 43

CHEMICAL CARCINOGENS IN THE MARINE ENVIRONMENT. BENZO(a)PYRENE IN
ECONOMICALLY-IMPORTANT BIVALVE MOLLUSKS FROM OREGON ESTUARIES*

Michael C. Mix, Ronald T. Riley, Keith I. King, Steven R. Trenholm and Randy L. Schaffer

Department of General Science
Oregon State University
Corvallis, Oregon 97331

Abstract

We have recently begun to study levels of carcinogenic polycyclic aromatic hydrocarbons that are present in bivalve mollusks from Oregon's estuaries. Because of many unique features in their life history and biology, indigenous shellfish are useful for monitoring the marine environment. In this paper, we describe benzo(α)pyrene (BAP) levels in economically-important shellfish populations from several sites in five Oregon bays. We have assayed BAP levels in clams (*Tresus capax, Saxidomus giganteus, Mya arenaria*), mussels (*Mytilus edulis*) and oysters (*Crassostrea gigas*) from Tillamook, Netarts, Yaquina, Alsea and Coos Bays. Detectable levels of BAP were present in bivalves from 38 of the 44 sampling sites. High levels (> 15 ng/g) were present in mussels collected from the Newport bayfront in Yaquina Bay and from a marina in Tillamook Bay. Significant levels (> 5 ng/g) were present in *M. arenaria* collected from an area adjacent to the shipping docks in Coos Bay.

Key words: Carcinogen, benzo(a)pyrene, mussels, clams, oysters, shellfish

Introduction

It has been suggested from analyses of epidemiological data, that 80-90 percent of all cancers that occur in human occupants of industrial societies are caused by environmental contaminants which act as primary initiators of neoplastic or malignant growths (Higgenson, 1971). To date, no definitive data are available on the proportion of cancers in humans which develop as a consequence of exposures to chemical carcinogens. However, environmental chemicals are suspected to be etiologic agents on the basis of man's proven susceptibility to some chemical carcinogens and the wide occurrence of chemicals which are known to be carcinogenic for animals or have structures similar to those of known carcinogens.

Hydrocarbons are universal components of the marine environment and originate from marine organism biosynthesis and from pollution by fossil fuels and oil products. Benzo(α)pyrene (BAP) and several other polycyclic aromatic hydrocarbons (PAH), which are both carcinogenic and noncarcinogenic, are found in petroleum, and thousands of kg of these compounds enter the sea each year by many routes, including petroleum spills, runoff from roads, sewage, effluents from industrial processes and fallout from the atmosphere.

It is quite evident that PAH levels from pollution by fossil fuels and oil by-products may be expected to increase in the biologically productive coastal regions of Oregon. Such contamination is inevitable because of the proposed use of Oregon bays for unloading oil tankers, increased ship traffic, increasing recreational demands and expanding industrialization. There are two aspects to be considered when evaluating the impact from increased loading of estuaries with this group of contaminants. First, oil and oil by-products pose a significant hazard to marine ecosystems, and much intensive research is necessary if the hazards are to be quantified and fully understood (Evans and Rice, 1974). Second, there may be potential public health problems associated with the consumption of shellfish that had been contaminated with environmental PAH.

The effects of acute, massive exposures to oil and PAH on estuarine organisms have been reasonably well documented in a variety of field and, for the latter, laboratory investigations (e.g. Stegeman and Teal, 1973; Di Salvo, Guard and Hunter, 1975; Dunn and Stich, 1976). However, little is known about the effects of chronic low-level pollution; there is a complete absence of the difficult field research which would lead to an understanding of the

---

*Supported in part by Grants R804427010 from the Environmental Protection Agency, G40 from the OSU Environmental Health Sciences Center and #04-5-158-2 from the OSU Sea Grant Program.

chronic effects of PAH at the population or community level. Too, there is virtually no information about how chronic exposure to low levels of carcinogenic PAH may affect individual marine organisms. An often overlooked possibility is that environmental carcinogens may cause proliferative disorders in plants and lower animals.

It has been stated that aromatic compounds isolated from marine sediments cause "cancerous growths" in certain seaweeds (Ishio, Yano and Nakagawa, 1972; Boney, 1972). There is also a report (Barry and Yevich, 1975) that softshell clams (*Mya arenaria*) developed "malignant gonadal tumors" after an oil spill that resulted in these organisms being exposed continuously for two years. Proliferative disorders, perhaps representing neoplasia, have been reported in oysters and clams from east coast bays (Couch, 1969; Farley, 1969a; Farley and Sparks, 1970; Barry, Yevich and Thayer, 1971; Newman, 1972; Christensen, Farley and Kern, 1974) and several reports have described such conditions in oysters, mussels and clams from Yaquina Bay, Oregon (Jones and Sparks, 1968; Farley, 1969b; Farley and Sparks, 1970; Mix, 1975a; Mix, 1975b; Mix, 1976; Mix and Riley, in press; Mix, Pribble, Riley and Tomasovic, in press). Thus, the appearance of proliferative disorders in marine invertebrate populations may serve as a warning that not only are environmental stressors affecting the environment, but they should be recognized as clues to the potential association of point source contaminators to increased incidence of human cancer in various geographical regions (Kraybill, 1976).

We have recently begun a study that has the following objectives: 1. to determine the concentration (body burdens) of selected environmental chemical carcinogens in economically-important bivalve mollusks and crustaceans from Oregon bays, estuaries and inshore areas; 2. to survey populations of bivalve mollusks, determine the prevalence of neoplastic diseases in these populations and ascertain if there is any correlation between carcinogenic body burdens and the incidence of such diseases; 3. to identify point sources of chemical carcinogens that are present in Oregon bays and estuaries utilized in this study; and 4. to ultimately assess the public health hazard, if any, from consuming contaminated shellfish.

The purpose of this paper is to present information and data about the presence, levels and distribution of BAP in several species of shellfish from 5 Oregon bays. We selected BAP for our initial studies because it is a potent carcinogen, an abundant product of petroleum combustion and other industrial processes and there are precise analytical methods available (Dunn and Stich, 1976; Dunn, in press) for quantifying low levels likely to be encountered in feral populations of shellfish.

Materials and Methods

Selection of Oregon Estuaries
Oregon has a large number of bays and estuaries (Fig. 1). We used three criteria for selecting the 5 estuaries to be utilized for these studies: they must have major commercial and/or recreational shellfisheries; there had to be a variety of bivalve species present in or on different substrates; and each bay had to reflect varying degrees of industrialization and human habitation.

Selection of Bivalve Molluscan Species
We used the following criteria for selecting the species of bivalve mollusks to be used in these studies. They had to be: economically-important; present in sufficient numbers to withstand sampling for several years; present in at least two bays so comparative studies could be made; ubiquitous, so comparisons could be made with other workers throughout the United States and Canada; representative of certain types of substrates and/or specific habitats (e.g. soft mud, low salinity); available for sampling during the entire year.

Collection and Preparation of Shellfish Samples
Clams and cockles from the various bays were dug during low, usually minus, tides while mussels were collected during the entire tidal cycle, depending on their location. Oysters were obtained from the commercial growers and simply removed from the shucking tables. At least 10 shellfish were collected from each sampling site. Immediately following collection, the shellfish from a single site were placed in labeled plastic bags, put on ice contained in coolers and transported back to our laboratory in Corvallis. Individual animals were sized, shucked (removed from the shell) and the pooled sample from each site was then weighed. Each pooled sample was then stored in a plastic bag at $-20^\circ C$ until it was analyzed for BAP.

Chemical Analysis for Benzo(α)pyrene
Prior to analysis each sample, excepting the mussels, was homogenized in an electric plastic meat grinder. Mussel samples were prepared similarly; however, because of the small size of the individual animals, no homogenization was necessary.

An aliquot of each sample was analyzed according to the method of Dunn (Dunn and Stich, 1976; Dunn, in press). A 30-40 g sample was digested by refluxing in an ethanol-KOH solution. Following digestion, the ethanol-KOH supernatant was extracted with 2,2,4-trimethylpentane (TMP) and the organic phase passed through a column of partially deactivated florisil. The PAH were eluted with benzene and, after removal of the benzene, the eluate was cleaned up by DMSO extraction in TMP.

BAP was isolated by preparative thin layer chromatography on 20% acetylated cellulose, made to volume in hexadecane and the concentration of BAP determined by spectrophotofluorimetry. Recovery of BAP by the extraction procedure was determined by spiking the original digestion mixture with an aliquot of G-^{3}HBAP and counting an aliquot of the final hexadecane solution.

Analysis of wood (piling) samples was accomplished similarly. Approximately 100 mg of the outer 0.2 cm of the piling was dried at 60°C for 24 hr and then placed in a pre-cooled porcelain mortar, liquid N_2 added, and the wood pulverized to a fine powder with a porcelain pestle. Ten mg of the powdered wood was then analyzed according to the procedures outlined previously.

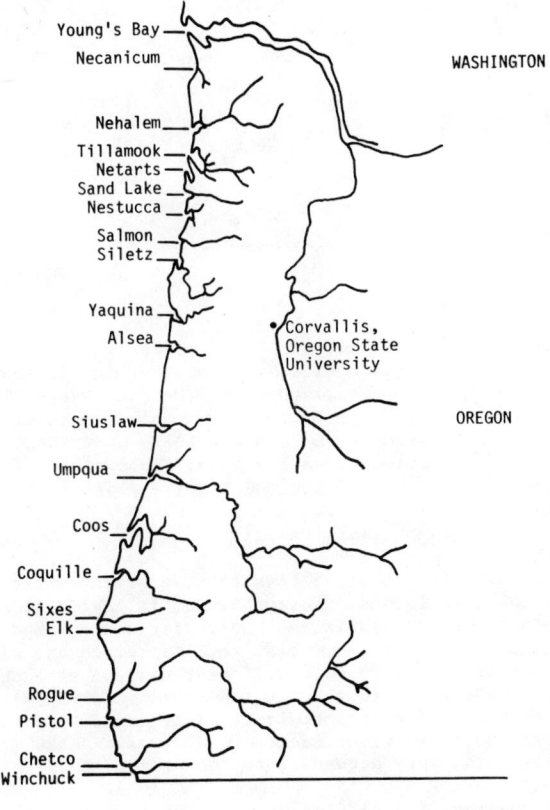

Fig. 43.1 Oregon bays and estuaries.

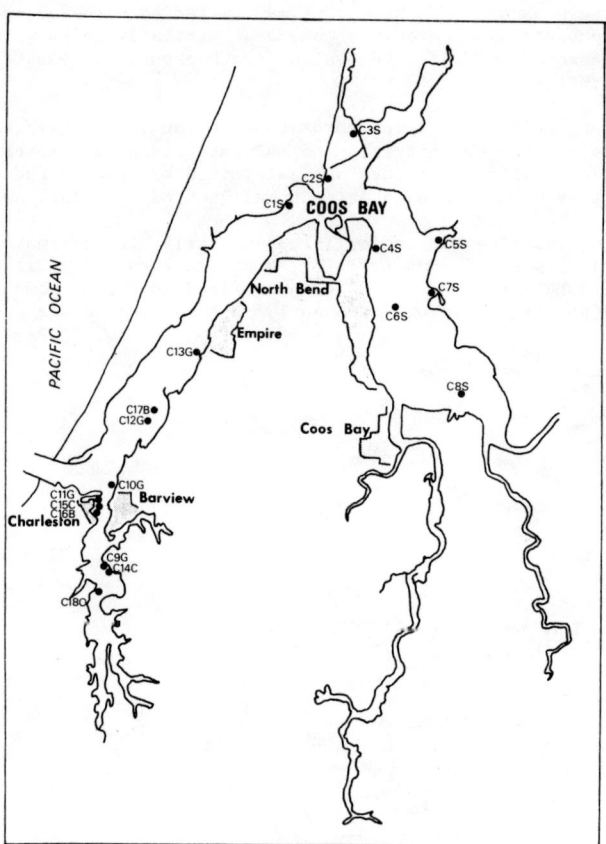

Fig. 43.2 Coos Bay, with a total surface of 10,000 acres of which 50% are tidelands, drains a basin of 605 square miles with a very high freshwater yield. Coos Bay is the most heavily industrialized of all Oregon bays. Timber, fish resources and agricultural activities are of major economic importance. Many major lumber manufacturers are located in the bay and there is a large amount of shipping traffic concentrated around Coos Bay. Two commercial oyster growers operate in South Slough and there is moderate to heavy exploitation of the clam populations.

Results

Figs. 2-6 show the sampling sites in the 5 bays that we selected for our studies, and describe some of their characteristics, (Percy, Sutterlin, Bella and Lingeman, 1974). These bays were chosen because they satisfied the criteria described previously. They are utilized for recreational and/or commercial shellfisheries and all have substantial populations of bivalve mollusks. They have different degrees of industrialization, ranging from essentially pristine (Netarts) to heavily industrialized (Coos). Their drainage basins vary widely in size from small (Netarts) to large (Tillamook, Coos) and their salinity profiles, perhaps the single most important factor in determining the distribution and abundance of shellfish populations, vary according to the size of their watersheds.

Table 1 summarizes the species of shellfish we will use in our studies along with their habitats and the bays in which they are found. All of these species are economically-important and they are found in different types of habitats and substrates. By selecting a large number of species, it is possible to study all areas and habitats of the 5 bays and we are not limited to only the high salinity areas of the lower bays.

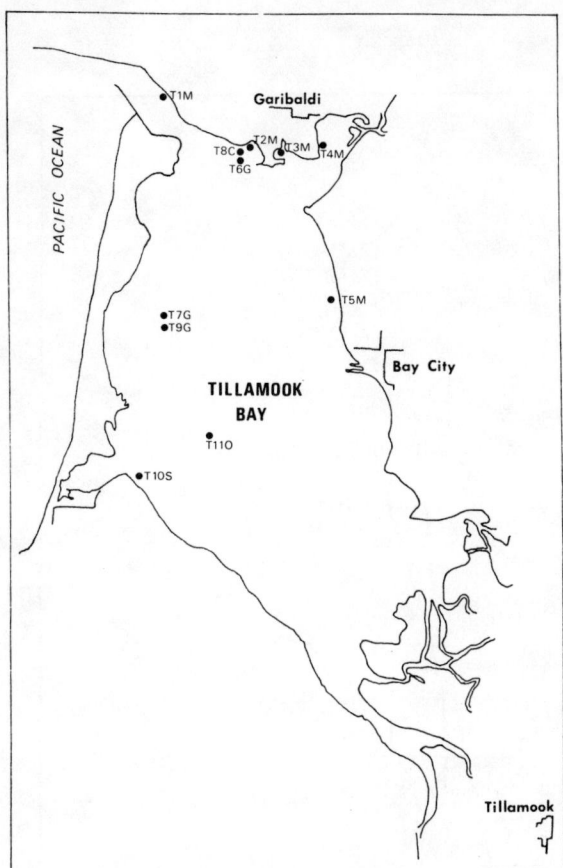

Fig. 43.3 Tillamook Bay, with a total surface area of 8,660 acres of which 50-60% are tidelands, drains a basin of 540 square miles with a high freshwater yield. Major industries: timber, agricultural products, fish and seafoods, tourism. Not considered to be highly industrialized. Three commercial oyster companies are in the bay and there is a moderate amount of recreational clamming.

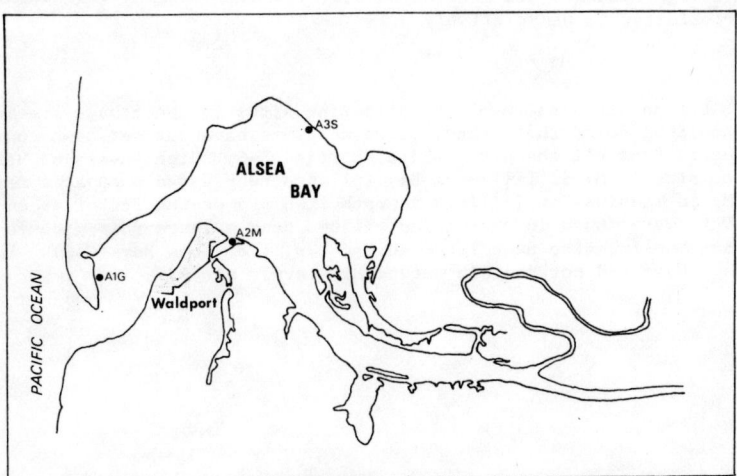

Fig. 43.4 Alsea Bay, with a total surface area of 2,140 acres of which 45-50% are tidelands, drains a basin of 474 square miles with a high freshwater yield. Lumber-related activities, tourism and agriculture are of major economic importance. Little industrial use of the bay. Clams are moderately exploited.

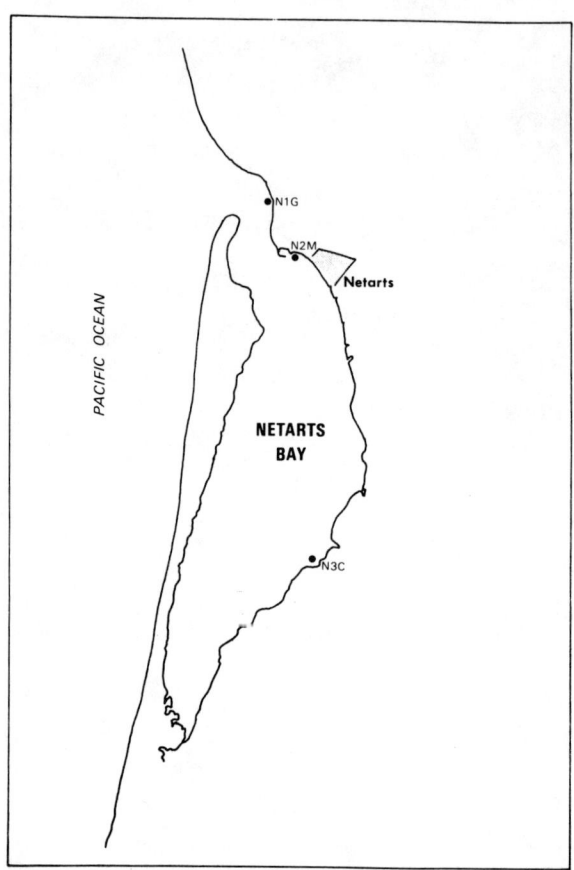

Fig. 43.5 Netarts Bay, with a total surface area of 2,200 acres of which 65-90% are tidelands, drains a basin of 14 square miles with a very low freshwater yield. Manufacturing companies are lacking completely; clam digging very popular. The bay is considered to be relatively pristine.

The levels of BAP found in bivalves from collecting sites in the 5 bays are included in Table 2. It should be noted that quantitative analyses have not yet been conducted on shellfish collected from all the sites shown in Figs. 2-6. High levels of BAP were found in mussels from one site (T3M) in Tillamook Bay (pilings near a gas pump at a marinea), and two sites (Y2M, Y4M) in Yaquina Bay (pilings beneath fish processing factories and near marinas). Significant levels were found in mussels on pilings near a large ship dock in Yaquina Bay (Y10M) and in softshells, also near large ship docks, from Coos Bay (C4S). Shellfish from Netarts and Alsea Bays did not contain detectable levels of BAP.

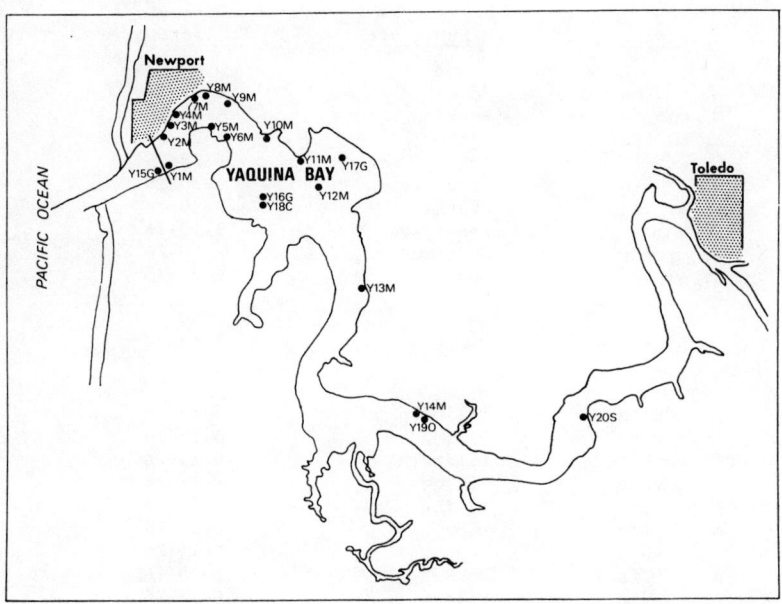

Fig. 43.6 Yaquina Bay, with a total surface area of 4,000 acres of which 35-61% are tidelands, drains a basin of 253 square miles with a medium freshwater yield. A major industrial estuary, the bay is a center for lumbering and commercial fishing activities. Toledo is the focal point of the forest industry processing facilities for the entire Mid-Coast Basin. Newport is the center for commercial fishing activities and there are numerous fish processing plants along the bayfront. Numerous marinas are scattered throughout the bay. Four commercial oyster growers are in the bay and clams are heavily dug.

TABLE 1. Clams, Mussel and Oyster Species Used in the Study

Species	Common Name	Habitat Substrate; Salinity	Bays[a]
Tresus capax	Gaper clam	Soft, sandy and rocky mud; high	T,N,Y,A,C
Saxidomus giganteus	Butter or Empire clams	Rocky mud; high	N,C,Y
Mya arenaria	Softshell clam	Soft or sandy mud; low	T,Y,A,C
Clinocardium nuttalli	Cockle	Sandy or rocky mud	T,N,Y,C
Mytilus edulis	Mussel	Pilings, rocks; low to high	T,N,Y,A,C
Crassostrea gigas	Pacific oyster	Held in trays, on sticks; moderate	T,Y,C

[a] The abbreviations used in this table: T - Tillamook Bay; N - Netarts Bay; Y - Yaquina Bay; A - Alsea Bay; C - Coos Bay

TABLE 2. Levels of Benzo(a)pyrene in Bivalve Mollusks from Oregon Estuaries

Bay Site	Species	Type of Substrate	Date of Collection	B(α)P Levels ng/g (ppb)
TILLAMOOK				
T1M	M. edulis	Rock	6-29-76	+*
T2M	M. edulis	Rock	6-30-76	0.40
T3M	M. edulis	Dock	6-28-76	16.86
T4M	M. edulis	Log	6-28-76	0.28
T5M	M. edulis	Concrete	6-30-76	0.10
T6G	T. capax	Gravel	6-30-76	0.16
T7G	T. capax	Sandy mud	8-10-76	3.21
T10S	M. arenaria	Soft mud	8-10-76	0.30
T11O	C. gigas	Mud	6-29-76	+
NETARTS				
N1G	T. capax	Rocky gravel	6-29-76	+
YAQUINA				
Y1M	M. edulis	Piling	6-15-76	0.12
Y2M	M. edulis	Piling	6-15-76	30.20
Y3M	M. edulis	Piling	6-15-76	3.15
Y4M	M. edulis	Piling	6-15-76	14.90
Y5M	M. edulis	Piling	6-15-76	0.87
Y6M	M. edulis	Dock	6-15-76	3.00
Y7M	M. edulis	Dock	6-15-76	4.13
Y8M	M. edulis	Dock	6-15-76	0.39
Y9M	M. edulis	Piling	6-15-76	0.77
Y10M	M. edulis	Piling	6-15-76	5.23
Y11M	M. edulis	Rocks	6-15-76	0.51
Y12M	M. edulis	Piling	6-16-76	0.44
Y13M	M. edulis	Iron piling	6-16-76	0.37
Y14M	M. edulis	Dock	6-17-76	4.30
Y15G	T. capax	Sand	6-15-76	0.56
Y16G	T. capax	Sandy mud	6-15-76	0.25
Y17G	T. capax	Soft mud	6-15-76	0.25
ALSEA				
A1G	T. capax	Rocky sand	7-9-76	+
A2M	M. edulis	Dock	7-9-76	+
A3S	M. arenaria	Soft mud	7-9-76	+
COOS				
C1S	M. arenaria	Sandy mud	6-28-76	0.57
C2S	M. arenaria	Sandy mud	6-28-76	0.33
C3S	M. arenaria	Soft mud	6-28-76	--**
C4S	M. arenaria	Rocky mud	7-15-76	6.66
C5S	M. arenaria	Soft mud	7-28-76	0.33
C6S	M. arenaria	Sandy mud	7-15-76	0.56
C7S	M. arenaria	Soft mud	7-28-76	0.58
C8S	M. arenaria	Soft mud	7-15-76	1.91
C9G	T. capax	Sandy mud	7-29-76	0.51
C10G	T. capax	Soft mud	6-29-76	0.44
C11G	T. capax	Sandy mud	7-29-76	0.33
C12G	T. capax	Sandy mud	6-29-76	0.14
C13G	T. capax	Sandy mud	6-29-76	0.35
C16B	S. giganteus	Sandy mud	7-29-76	1.05
C17B	S. giganteus	Sandy mud	7-15-76	0.29

*not detectable, less than 0.1 ppm   **not analyzed

Discussion

The concept of utilizing single species of bivalve mollusks to monitor the marine environment for a variety of contaminants is becoming increasingly popular (Di Salvo, Guard and Hunter, 1975; Dunn and Stich, 1976; Shaw, Paul, Cheek and Feder, 1976). We have extended this concept by using several species of shellfish to monitor estuaries. Indigenous bivalve shellfish represent ideal subjects for evaluating carcinogenic loads in the marine environment because: they are sedentary and, excepting the larval sojourn, spend their entire lives in the same location; shellfish populations inhabit waters which are polluted by domestic sewage, petroleum by-products and industrial waste products; they tend to concentrate almost all noxious substances in the environment; they are relatively easy to locate and sample; it is possible to obtain or locate spat (young shellfish that have recently metamorphosed from the free-swimming larval stage to the permanently-attached form) and thus, it is possible to monitor the rate of uptake and incorporation of various chemical carcinogens in bivalve populations starting with essentially zero-aged animals; their ubiquitous distribution permits comparisons to be made with other workers; and large numbers of shellfish can be analyzed and examined.

We feel there are two primary advantages in using multiple species in monitoring our studies. First, it is possible to monitor an entire bay which is not always possible when using one species, particularly in Oregon estuaries which receive massive volumes of freshwater during the winter. For example, softshell clams are virtually the only inhabitants of the upper bays where much of the industry is frequently located; thus, we are able to monitor such areas with that species, which is generally not found in the lower bays. Second, by using shellfish that occupy a variety of habitats, we hope to gather information about the routes of movement, rates of transfer and reservoirs of BAP in the marine environment.

In this study, we have obtained baseline information about the levels and locations of BAP in shellfish from Oregon estuaries. High levels of BAP were found in mussels from Tillamook and Yaquina Bays which were taken from pilings near marinas and fish processing factories. The sources of BAP at these sites have not been clearly established. Because of the suggestion (Dunn and Stich, 1976) that creosote may be the major source of environmental BAP in mussels, the pilings from which they were attached were analyzed. The results of those analyses suggest that creosote may not be the major source of BAP in the mussels. The pilings with the highest level of BAP (265,512 ng/g) harbored mussels with one of the lowest levels (0.22 ng/g).

Clam species have rarely been utilized in studying environmental carcinogens. Because they are buried in the substrate, it may be possible, because of food chain decay, that they will possess low levels of BAP even though they inhabit a highly contaminated environment. However, the roles of food chain transfer, sediment transfer, and microbial metabolism of BAP are not yet clearly understood (Ehrhardt, 1972; Farrington and Quinn, 1973; Cundwell and Traxler, 1973; Giger and Blumer, 1974; Mackie, Whittle and Hardy, 1974; Zitco, 1975). In future studies, we will attempt to sample two species that live in different habitats at the same location in an attempt to resolve the question about BAP levels in clams. Specifically, we will attempt to locate mussel populations at sites where clams contain significant levels of BAP (e.g. C4S in Coos Bay).

Much remains to be learned about BAP and other petroleum-related carcinogens in the estuarine environment. There is a great amount of ignorance about the sources of environmental BAP, its movement through the food web and its modification by marine organisms. The fate of BAP, once it is incorporated into shellfish tissues, is obviously extremely important when evaluating public health aspects. It has been stated (Lee, Sauerheber and Benson, 1972) that BAP is not metabolized by mussels (*M. edulis*). However, recent studies (R. Anderson, personal communication) have yielded results which suggest that oysters, *Crassostrea virginica*, can slowly metabolize BAP. Further studies are necessary to resolve these difficult questions.

Conclusions

1. Certain populations of indigenous bivalve mollusks from the industrialized Oregon bays (Tillamook, Yaquina, Coos) contain shellfish with significant levels of BAP in their tissues. Shellfish from the nonindustrialized bays (Netarts, Alsea) contain non-detectable levels of BAP.

2. By utilizing a multiple number of shellfish species, it is possible to monitor an entire estuary for BAP. It is our judgment that the single-species approach has only limited usefulness for monitoring marine ecosystems as complex as estuaries because any one species is seldom found throughout the entire system.

3. Our first data suggest that BAP is not always directly transferred from creosoted pilings to mussels growing on them. Additional analyses are required to test this finding.

Acknowledgements

We would like to express our gratitude to Drs. J. Couch, J. Harshbarger, V. Freed and Mr. W. Wick for their support and encouragement. Drs. N. Richards, J. Anderson and J. Neff provided several constructive comments and Dr. B. Dunn was most helpful in acquainting us with his analytical procedures. We thank P. Heikkila, D. Bunting, T. Gaumer, M. Henderson and C. Morehouse for their technical assistance.

References

Barry, M. and P. Yevich, The ecological, chemical and histopathological evaluation of an oil spill site. III. Histopathological studies, *Marine Poll. Bull.* 6, 164 (1975).

Barry, M., P. Yevich and N. H. Thayer, Atypical hyperplasia in the soft-shell clam *Mya arenaria*, *J. Invert. Pathol.* 17, 17 (1971).

Boney, A. D., Aromatic hydrocarbons and the growth of marine algae, *Marine Poll. Bull.* 5, 185 (1974).

Christensen, D. J., C. A. Farley and F. G. Kern, Epizootic neoplasms in the clam *Macoma balthica* (L.) from Chesapeake Bay, *J. Nat. Cancer Inst.* 52, 1739 (1974).

Couch, J. A., An unusual lesion in the mantle of the American oyster (*Crassostrea virginica*), *Nat. Cancer Inst. Monogr.* 31, 557 (1969).

Cundwell, A. M. and R. W. Traxler, Microbial degradation of petroleum at low temperature, *Marine Poll. Bull.* 4, 125 (1973).

Di Salvo, L. H., H. E. Guard and L. Hunter, Tissue hydrocarbon burden of mussels as potential monitor of environmental hydrocarbon insult, *Environm. Sci. and Technol.* 9, 247 (1975).

Dunn, B. P., Techniques for the determination of benzo(α)pyrene in marine organisms and sediments, *Environm. Sci. Technol.* (in press).

Dunn, B. P. and H. F. Stich, Release of the carcinogen benzo(α)pyrene from environmentally contaminated mussels, *Bull. Environm. Contamin. and Toxicol.* 15, 398 (1976).

Ehrhardt, M., Petroleum hydrocarbons in oysters from Galveston Bay, *Environm. Poll.* 3, 257 (1972).

Farley, C. A., Probable neoplastic disease of the hematopoietic systems in oysters (*Crassostrea virginica* and *Crassostrea gigas*), *Nat. Cancer Inst. Monogr.* 31, 541 (1969a).

Farley, C. A., Sarcomatoid proliferative disease in a wild population of edible mussels (*Mytilus edulis*), *J. Nat. Cancer Inst.* 4, 509 (1969b).

Farley, C. A. and A. K. Sparks, Proliferative diseases of hemocytes, endothelial cells, and connective tissue cells in mollusks, *Comp. Leuk. Res. Bibl. Haematol.* 36, 610 (1970).

Farrington, J. W. and J. G. Quinn, Petroleum hydrocarbons in Narragansett Bay. I. Survey of hydrocarbons in sediments and clams (*Mercenaria mercenaria*), *Estuarine and Coastal Marine Sci.* 1, 71 (1973).

Evans, D. R. and S. D. Rice, Effects of oil on marine ecosystems: a review for administrators and policy makers, *U.S. N.M.F.S. Fish. Bull.* 72, 625 (1974).

Giger, W. and M. Blumer, Polycyclic aromatic hydrocarbons in the environment: isolation and characterization by chromatography, visible, ultraviolet and mass spectometry, *Anal. Chem.* 46, 1663 (1974).

Higgenson, J., The role of geographical pathology in environmental carcinogenesis. pp. 68-69. *In: Environment and Cancer* (The Univ. of Texas, 24th Annual Symposium on Fundamental Cancer Research 1971). The Williams and Wilkens Co., Baltimore, (1971).

Ishio, S., T. Yano and H. Nakagawa, Cancerous disease of *Porphyra tenera* and its causes. *Proc. Seventh Intl. Seaweed Symp.*, 373 (1972).

Jones, E. and A. K. Sparks, An unusual histopathological condition in *Ostrea lurida* from Yaquina Bay, Oregon, *Proc. Natl. Shellfish Assn.* 59, 11 (1969).

Kraybill, H. F., The distribution of chemical carcinogens in aquatic environments. pp. 3-34. *In:* "Neoplasms in Aquatic Animals as Indicators of Environmental Carcinogens: Progress in Experimental Tumor Research" (F. Homburger, *Ed*), S. Karger, Basel, (1976).

Lee, R. F., R. Sauerheber and A. A. Benson, Petroleum hydrocarbons: uptake and discharge by the marine mussel, *Mytilus edulis, Science, 177,* 344 (1972).

Mackie, P. R., K. J. Whittle and R. Hardy, Hydrocarbons in the marine environment. I. n-alkanes in the firth of clyde, *Estuar. and Coastal Marine Sci.* 2, 359 (1974).

Mix, M. C., The neoplastic disease of Yaquina Bay bivalve molluscs. pp. 369-386. *In:* "The Cell Cycle in Malignancy and Immunity" (J. C. Hampton, Ed.), (CONF-731005), NTIS, Springfield, Va., (1975a).

Mix, M. C., Proliferative characteristics of atypical cells in native oysters (*Ostrea lurida*) from Yaquina Bay, Oregon, *J. Invert. Pathol.* 26, 289 (1975b).

Mix, M. C., A review of the cellular proliferative disorders of oysters (*Ostrea lurida*) from Yaquina Bay, Oregon. pp. 275-282. *In:* "Neoplasms in Aquatic Animals as Indicators of Environmental Carcinogens: Progress in Experimental Tumor Research" (F. Homburger, Ed.), S. Karger, Basel, (1976).

Mix, M. C. and R. T. Riley, A pericardial tumor in a nature (Olympia) oyster, *Ostrea lurida*, from Yaquina Bay, Oregon, *J. Invert. Pathol.* (in press).

Mix, M. C., H. J. Pribble, R. T. Riley, S. P. Tomasovic, Neoplastic disease in bivalve molluscs from Oregon estuaries with emphasis on research of proliferative disorders in Yaquina Bay oysters, *Ann. N.Y. Acad. Sci.* (in press).

Newman, M. W., An oyster neoplasm of apparent mesenchymal origin, *J. Nat. Cancer Inst.* 48, 237 (1972).

Percy, K. L., C. Sutterlin, D. A. Bella and P. C. Klingeman, *Description and Information Sources for Oregon Estuaries*, O.S.U. Sea Grant, Corvallis, Oregon (1974).

Shaw, D. G., A. J. Paul, L. M. Cheek and H. M. Feder, *Macoma balthica*: an indicator of oil pollution, *Marine Poll. Bull.* 7, 29 (1976).

Stegeman, J. J. and J. M. Teal, Accumulation, release and retention of petroleum hydrocarbons by the oyster, *Crassostrea virginica, Marine Biol.* 22, 37 (1973).

Zitco, V., Aromatic hydrocarbons in aquatic fauna, *Bull. Environm. Contamin. and Toxicol.* 14, 621 (1975).

CHAPTER 44

SEASONAL VARIATIONS OF HYDROCARBONS IN THE WATER COLUMN
OF THE MAFLA LEASE AREA

John A. Calder

Department of Oceanography
Florida State University
Tallahasee, FL 32306

Abstract

A series of 15 stations in the northeast Gulf of Mexico were occupied during summer, fall and winter 1975-76. Samples were collected and analyzed by gas chromatography for dissolved hydrocarbons and those associated with suspended particulate material. Average concentration of total resolved hydrocarbons was 0.4 µg/l dissolved and 0.3 µg/l particulate. Concentrations were higher near shore. Unresolved components were present in both dissolved and particulate phases, especially near the Mississippi River and Sound which may be the source of this material. Biogenic hydrocarbons, nC15, nC17, pristane and squalene in the particulate phase may be reflective of in situ biomass. A series of n-alkanes (nC21 to nC32) in both dissolved and particulate phases persisted during all seasons. Squalene was the dominant molecule in the dissolved unsaturated/aromatic fraction at most stations, but was very low in concentration at the offshore stations in the fall. Total dissolved hydrocarbons did not correlate with particulate organic carbon. Total particulate hydrocarbons did not correlate with particulate organic carbon or chlorophyll a (Chl. a).

Key words: MAFLA, dissolved hydrocarbon, particulate hydrocarbons, gas chromatography

Introduction

The sale of oil and gas leases along the entire U.S. outer-continental shelf (OCS) and heightened public awareness of the potential harmful impact of petroleum-related activities, resulted in the initiation of environmental baseline and monitoring studies in the lease areas, under the sponsorship of the U.S. Department of Interior, Bureau of Land Management. The first of these studies was the MAFLA (Mississippi-Alabama-Florida) program in the northeast Gulf of Mexico. During 1875-76, 3 sets of samples were collected from the water column in June-July 1975, September 1975 and January-February 1976. Dissolved hydrocarbons and those associated with suspended particulate matter were analyzed by my laboratory. This report represents our initial evaluation of the three data sets.

Methods

Fifteen stations in the northeast Gulf of Mexico were occupied during summer 1975, fall 1975 and winter 1976. At each station, 80 l. of water was collected from a depth of 10 m with 30 l. Niskin bottles. The Niskin bottles had been rinsed with methanol prior to use and were equipped with Teflon coated spring closures. The water was drained from the Niskin bottles through Teflon tubing into a precleaned stainless steel can of the type used to contain soft drink at soda fountains. The o-ring gasket on each can was wrapped with Teflon film. The water was immediately poisoned with $HgCl_2$ and then filtered as soon as possible on board ship. Filtration was accomplished by pressurizing the storage can with prepurified N_2 and forcing the water via Teflon tubing through a precombusted Whatman GF/F filter in a stainless steel Millipore filter holder and into a second stainless steel can. The filtrate was stored at ambient temperature until returned to the laboratory. The filters were wrapped in pre-combusted aluminum foil and frozen.

In the laboratory, the water was acidified to pH2 with concentrated HCl and then extracted with doubly distilled chloroform or methylene chloride in 2 l. separatory funnels. Each extraction consists of 1500 ml of water and 3x50 ml of solvent. The total $CHCl_3$ (or CH_2Cl_2) extract was reduced to small volume in a rotary evaporator and then transferred quantitatively to a 25 ml round bottom flask. The remaining solvent was removed under a stream of prepurified N_2. After addition of 10 ml of 0.5N KOH in methanol, the extract was saponified under reflux for at least 4 hours. Following addition of an equal volume of water, the non-saponifiable material was extracted into benzene (3x10 ml). The benzene was removed under N_2 and the residue taken up in a small volume of hexane for column chromatography.

Filter pads were placed intact into an appropriately sized round bottom flask and covered with a 1:1 mixture of benzene: 0.5 N KOH in MeOH. After a 4 hour reflux the mixture was filtered through a pre-cleaned glass fiber filter. Following addition of 25 ml of saline solution, the benzene layer was removed and the aqueous layer re-extracted with 3x25 ml of benzene. The benzene extracts were combined, reduced to dryness and taken up in hexane for column chromatography.

The non-saponifiable extracts in a small volume of hexane, were applied to a prewashed alumina overlaying silica gel column (1:3 V/V alumina to silica gel ratio, activity one) and eluted with 2 column volumes of hexane (aliphatic hydrocarbons) and 2 column volumes of benzene (unsaturated, aromatic fraction). The hexane fraction was reduced to small volume and the benzene fraction dried and taken up in a small volume of hexane for G.C. analysis.

Primary GC analysis was done with 2.2 mm ID x 2 m stainless steel columns packed with 4% FFAP on Gas Chrom Z, 80/100 mesh. Retention times were coverted to retention indices utilizing known standards of n-alkanes. Peak areas were automatically integrated and converted to weight by applying GC response factors calculated from quantitative normal and isoprenoid alkanes and aromatics. These calculations as well as calculations of peak ratios, odd-even preference, wt. % composition and concentration were done by a computer program which produced both paper and magnetic tape output for submission to a central data bank.

Glassware was washed in detergent, soaked in acid, rinsed with distilled water and oven dried. Solvents were doubly distilled. Periodic blanks were run and rejected if material with retention index greater than 1200 was present.

The use of sampling bottles which are lowered open, through the surface film has been questioned (Gordon et al. 1974). To check on possible contamination of the water samples by surface film hydrocarbons, an all glass and teflon sample bottle was attached to the hydrowire immediately below the 30 l. Niskin bottle on each hydrocast. This bottle remained closed while being lowered and was opened only after reaching the 10m sampling depth. Water from this glass bottle and an aliquot from the Niskin were analyzed for dissolved hydrocarbons by a UV-fluorescence method similar to that used by Gordon et al. (1974). Detailed results will be presented elsewhere (Byrne and Calder, unpublished). Analysis of the results shows that of the 43 sample pairs analyzed, the two sampling methods gave identical results, within experimental error, in 32 cases. The glass bottle gave a significantly higher value in six cases while the Niskin bottle gave a significantly higher value in five cases. While the problem of surface film contamination is a real one, it appears that in this case, water samples collected by Niskin bottles were not affected by surface film contamination.

Results

Water

The G.C. derived concentration of the aliphatic and unsaturated/aromatic fractions are listed in Table 1 for all 3 seasons. In summer and fall the concentrations of the unsaturated/aromatic fraction generally exceeded that of the aliphatic fraction; this situation was reversed in the winter. The fall season had the lowest average hydrocarbon concentrations and the winter the highest.

Qualitatively, the dissolved hydrocarbons displayed regional differences during each sampling season (Fig. 1). In the summer, two distinct regions were apparent. Stations 1-7 displayed a unique bimodal envelope of unresolved components in the aliphatic fraction, with the maxima centered at C-17 and C-27 (Fig. 2). Stations 8-15 had a broad envelope with no clear maximum. Both groups of samples displayed a series of n-alkanes from C-21 to C-32 with the weight ratio of total odd carbon number to total even carbon number n-alkanes averaging 1.1 ± 0.1. The unsaturated/aromatic fraction of both groups were similar and were generally dominated by a peak at RI=3060. Chromatography on a non-polar column confirmed the identity of this molecule as squalene. The concentration of squalene averaged $0.12 \pm .06$ μg/l. The concentrations of hydrocarbons in both fractions from the two groups were not significantly different (Table 2).

During the fall season, the concentration of dissolved hydrocarbons fell to about 50% of summer values (Table 1). This was true of both the aliphatic and unsaturated/aromatic fractions. Akiphatic fractions from stations 1-8 and 14 (Fig. 2) were characterized by a series of n-alkanes predominantly from C-21 to C-32. The odd/even ratio for these fractions averaged $1.09 \pm .10$. There was no detectable unresolved envelope at these stations. The remaining stations, 9-13 and 15, contained a definite envelope with a maximum near C-27 (Fig. 2). The envelope was of lesser magnitude relative to the n-alkanes at stations 9 and 10. The series of n-alkanes from C-21 and C-32 was still present and had an average odd-even ratio of $1.04 \pm .14$, excluding one value of 0.17. Samples which contained the envelope also had greater concentrations of resolved hydrocarbons in the aliphatic fraction, averaging $0.10 \pm .04$ μg/l versus $0.04 \pm .03$ μg/l for those samples without an envelope (Table 2).

TABLE 1. Concentration of Aliphatic (H), Unsaturated/Aromatic (B) and Total (T) Dissolved Hydrocarbons (μg/l)

Station	Summer			Fall			Winter		
	H	B	T	H	B	T	H	B	T
1	.11	1.01	1.12	.01	.15	.16	.69	.49	1.17
2.	.14	.22	0.36	.05	.12	.17	.45	.18	.63
3	.13	.19	0.32	.08	.06	.14	*.40	.06	.69
4	.39	.30	0.69	.04	.40	.45	.14	.10	.24
5	.14	.32	0.46	.02	.12	.13	1.08	.10	1.18
6	.08	.23	0.31	.02	.08	.10	.05	.23	.28
7	.25	.22	0.47	.10	.14	.24	.08	.04	.12
8	.05	.06	0.11	.02	.19	.21	.11	.03	.14
9	.17	.09	0.26	.11	.12	.23	.07	.35	.42
10	.25	.38	0.63	.06	.27	.32	.21	.07	.28
11	.10	.30	0.40	.09	.18	.27	.07	.08	.14
12	.17	.36	0.53	.12	.38	.50	.41	.09	.49
13	.09	.43	0.52	.06	.16	.22	.46	.15	.62
14	.13	.23	0.36	.02	.10	.12	1.14	.17	1.31
15	.06	.22	0.28	.16	.09	.25	.33	.03	.36
Avg	.15 ±.09	.30 ±.22	.45 ±.24	.06 ±.04	.17 ±.10	.23 ±.12	.38 ±.35	.14 ±.13	.54 ±.40

TABLE 2. Average Concentrations of Aliphatic (H) and Unsaturated/Aromatic (B) Dissolved Hydrocarbons (μg/l)

	Summer	H	B
Station 1-7 (bimodal envelope)		.18±.11	.36±.29
Station 8-15 (unimodal envelope)		.13±.07	.26±.13
	Fall		
Station 1-8, 14 (no envelope)		.04±.03	.15±.10
Station 9-13, 15 (envelope)		.10±.04	.20±.11
	Winter		
Station 1-5, 12-15 (envelope)		.56±.34	.15±.14
Station 6-11 (no envelope)		.10±.06	.13±.13

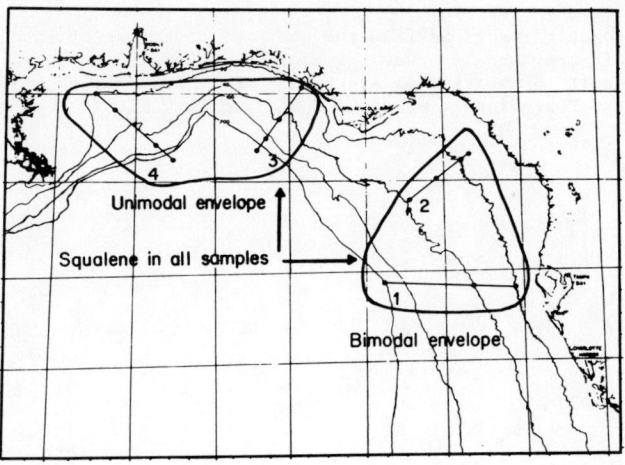
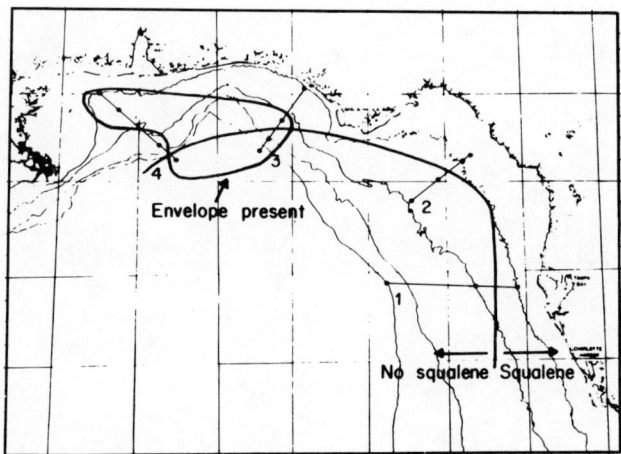
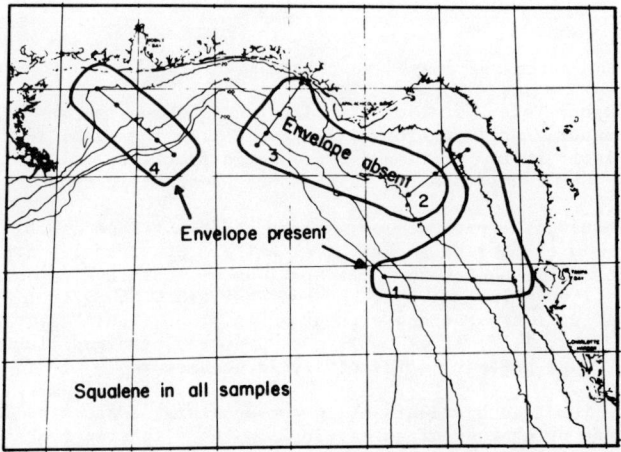

Fig. 44.1 Dissolved hydrocarbon distribution during summer (upper), fall (middle) and winter (lower) sampling seasons. Station locations are indicated by closed circles and are numbered sequentially nearshore to offshore beginning with transect 1.

The unsaturated/aromatic fractions from the shoreward stations of each transect were similar to summer samples in that squalene was the dominant molecule in this fraction. The concentration of squalene at these stations averated 0.06 ± .03 µg/l, excluding one value of .26 µg/l at station 12. The offshore stations contained very little squalene averaging 0.002 ± 0.002 µg/l.

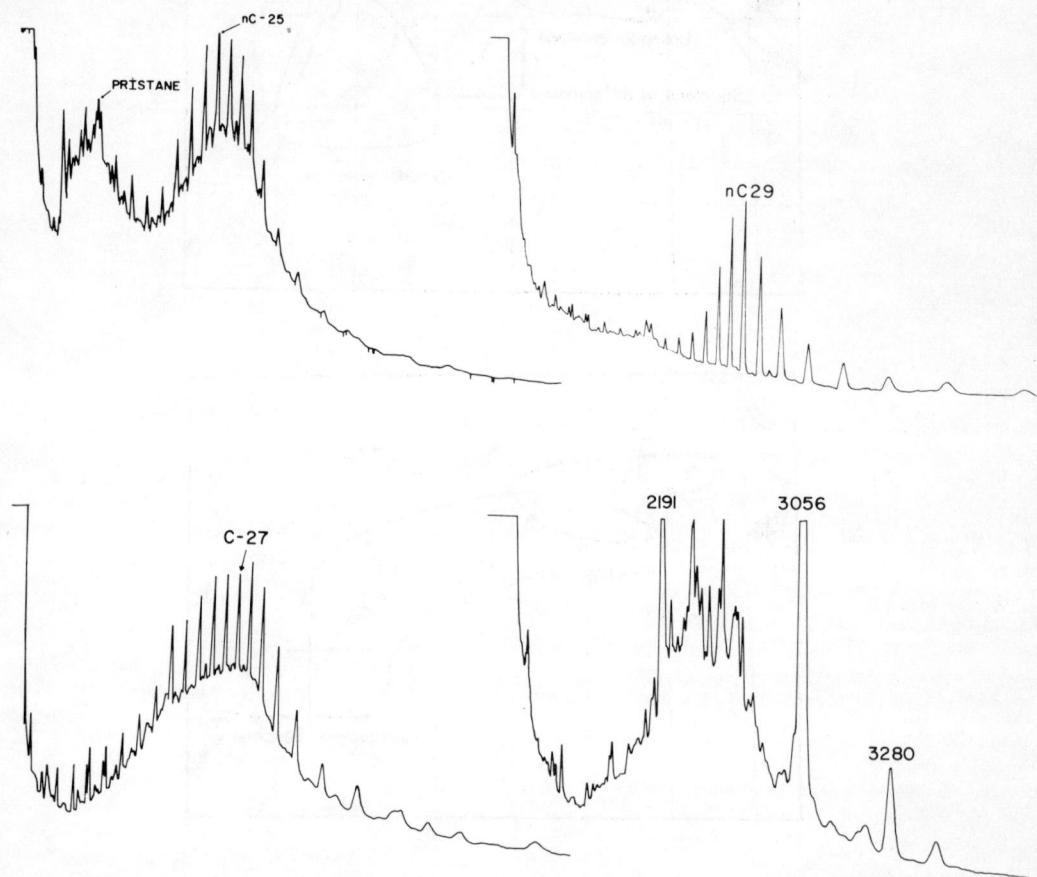

Fig. 44.2 Gas chromatograms of representative dissolved hydrocarbon samples.

upper left: station 2, aliphatic fraction, summer 1975
upper right: station 2, aliphatic fraction, fall 1975
lower left: station 15, aliphatic fraction, fall 1975
lower right: station 1, unsaturated-aromatic fraction, winter 1976.

During the winter season, the presence or absence of an envelope in the aliphatic fraction divided the stations into coherent geographical units. Stations 1-5 and 12-14 contained a large envelope with a maximum at C-25 while stations 6-11 did not contain an envelope. The concentration of resolved aliphatic hydrocarbons averaged 0.56 µg/l at stations exhibiting the envelope and 0.10 µg/l at stations without an envelope (Table 2). The odd/even ratio from both groups averaged 1.1. At all stations, a poorly resolved cluster of peaks with RI between 1600 and 1900 was present in relatively large amounts.

The concentration of resolved hydrocarbons in the unsaturated/aromatic fraction did not depend on the presence or absence of an envelope in the aliphatic fraction, averaging .15 µg/l and .13 µg/l at stations with and without the aliphatic envelope. Squalene was present in the unsaturated/aromatic fraction at all samples with an average concentration at 0.04 ± 0.03 µg/l. Many of the samples also contained an unresolved envelope in this fraction, a feature not seen in previous seasons (Fig. 2).

Particulate
The average concentration of resolved particulate hydrocarbons was .18 µg/l in the summer, with most of the material being in the aliphatic fraction (Table 3). The dominant peak in

the aliphatic fraction was nC15 with an average concentration of .044 ± .029 µg/l. Pristane and nC17 were present in all samples. An envelope in the aliphatic fractions was evident at stations 4 and 12-15 (Fig. 3). Its distribution maximum occurred around nC23. A series of n-alkanes was superimposed on the envelope.

TABLE 3. Concentration of Aliphatic (H), Unsaturated/Aromatic (B) and Total (T) Particulate Hydrocarbons (µg/l)

Station	Summer			Fall			Winter		
	H	B	T	H	B	T	H	B	T
1	.03	.04	.07	16.02*	1.36*	17.38*	.087	.027	.114
2	.66	.03	.69	.015	.002	.017	.050	.036	.086
3	.03	.01	.04	.011	0	.011	.323	.040	.363
4	.21	1.36*	1.57*	.113	.028	.141	.187	.022	.309
5	.06	.01	.07	.045	.006	.051	.151	.038	.189
6	.04	.03	.07	.050	.016	.066	.070	.024	.094
7	.18	.03	.21	.007	.004	.011	.058	.036	.094
8	.09	.05	.14	.144	.022	.167	.434	.193	.627
9	.07	.01	.08	.007	.003	.010	.073	.019	.092
10	.29	.01	.10	.014	.004	.018	.050	.020	.070
11	.09	.02	.11	.025	.010	.035	.080	.028	.108
12	.19	.21	.40	.095	.033	.128	3.341	.272	3.613
13	.13	.06	.19	.103	.052	.155	.391	.775	1.166
14	.09	.05	.14	.050	.007	.057	1.340	.046	1.386
15	.13	.01	.14	.088	.011	.099	.697	.220	.917
Avg	.14 ±.15	.04 ±.05	.18 ±.17	.055 ±.046	.014 ±.015	.069 ±.058	.49 ±.86	.12 ±.20	.62 ±.93

*omitted from average

In the unsaturated/aromatic fraction, squalene was the dominant molecule with an average concentration of 0.016 ± 0.014 µg/l. A peak at RI 2350 was also prominent (Fig. 4).

In fall, the concentration of particulate hydrocarbons fell to about 40% of summertime values and averaged 0.069 µg/l (Table 3). The dominant feature was the presence or absence of the biogenic hydrocarbons nC15, pristane, nC17 and squalene. In the aliphatic fractions, nC15 was the dominant molecule and nC17 and pristane were present at stations 1, 4, 5, 6, 8, 10-15. The concentration of nC15 averaged 0.025 ± .014 µg/l at these stations. At the remaining stations, 2, 3, 7 and 9, the biogenic hydrocarbons were essentially absent with the concentration of nC15 being 0.001 ± .001 µg/l. Stations 1, 4 and 12-15 displayed envelopes in the aliphatic fraction, with station 1 having a very high concentration of both resolved and unresolved aliphatic hydrocarbons.

The unsaturated/aromatic fractions in the fall contained squalene and in general little else. The concentration of squalene averaged .01 ± .01 µg/l and .003 ± .004 µg/l at stations having and lacking, respectively, the aliphatic biogenic hydrocarbons.

The concentration of particulate hydrocarbons was greater during the winter than the preceeding seasons, averaging .62 µg/l. However, there was a large range of .07 to 3.6 µg/l (Table 3). Aliphatic fractions at all stations contained envelopes, with these being relatively large at stations 8 and 12-15. Biogenic hydrocarbons were essentially absent at all stations except 11 and 15 where nC15 averaged 0.02 µg/l. Envelopes were also present in the unsaturated/aromatic fraction at all stations being very large at stations 12-15. Squalene was very low or absent at all stations except 15 where its concentration was 0.03 µg/l.

Discussion

The concentration of hydrocarbons in the water column of the MAFLA lease area compares well with the lower values reported in the literature for open ocean water. The overall average concentration was 0.4 µg/l dissolved hydrocarbons, and 0.3 µg/l particulate or 0.7 µg/l total resolved hydrocarbons. Brown et al. (1975) determined that total hydrocarbons in the open Atlantic and Pacific were about 1 µg/l by an IR method. In the Mediterranean the concentration ranged from 2-8 µg/l and near Bermuda the concentration was 3-6 µg/l. Levy (1971) reported values for total hydrocarbons of 2-13 µg/l in the Atlantic off Halifax by a UV-fluorescence method. Comparison of these results is made difficult because of the three different analytical methods used (G.C., IR, UVF) which are responsive to different portions of the hydrocarbons in the samples. Two reports of dissolved hydrocarbons by gravimetric analysis,

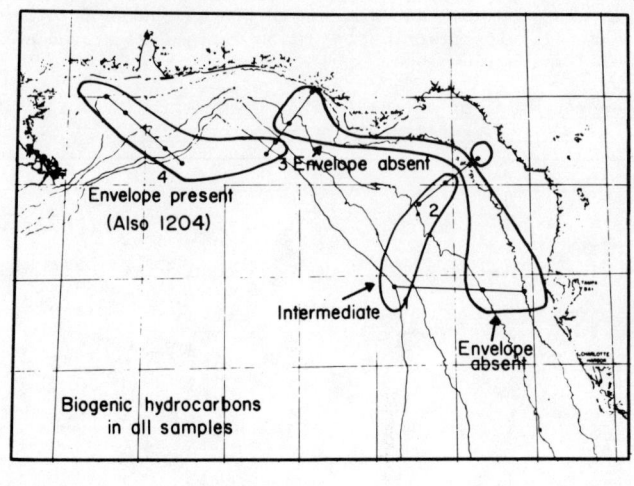
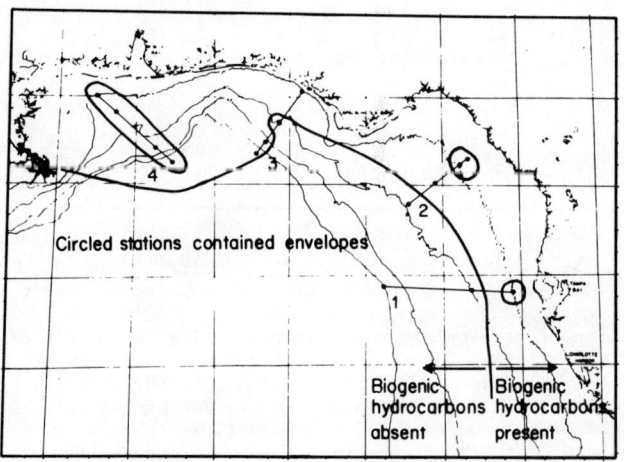
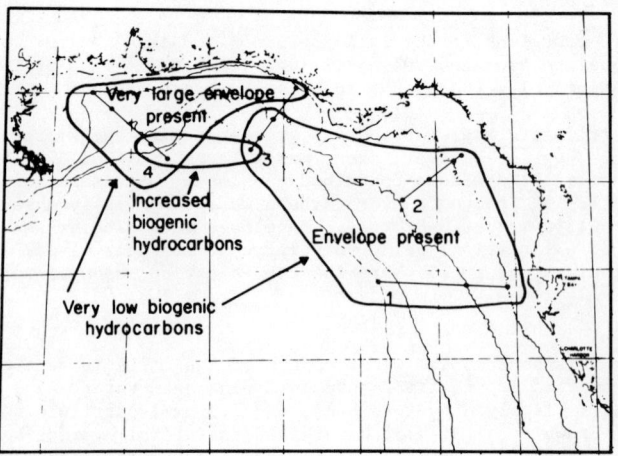

Fig. 44.3 Particulate hydrocarbon distribution during summer (upper), fall (middle) and winter (lower) sampling seasons.

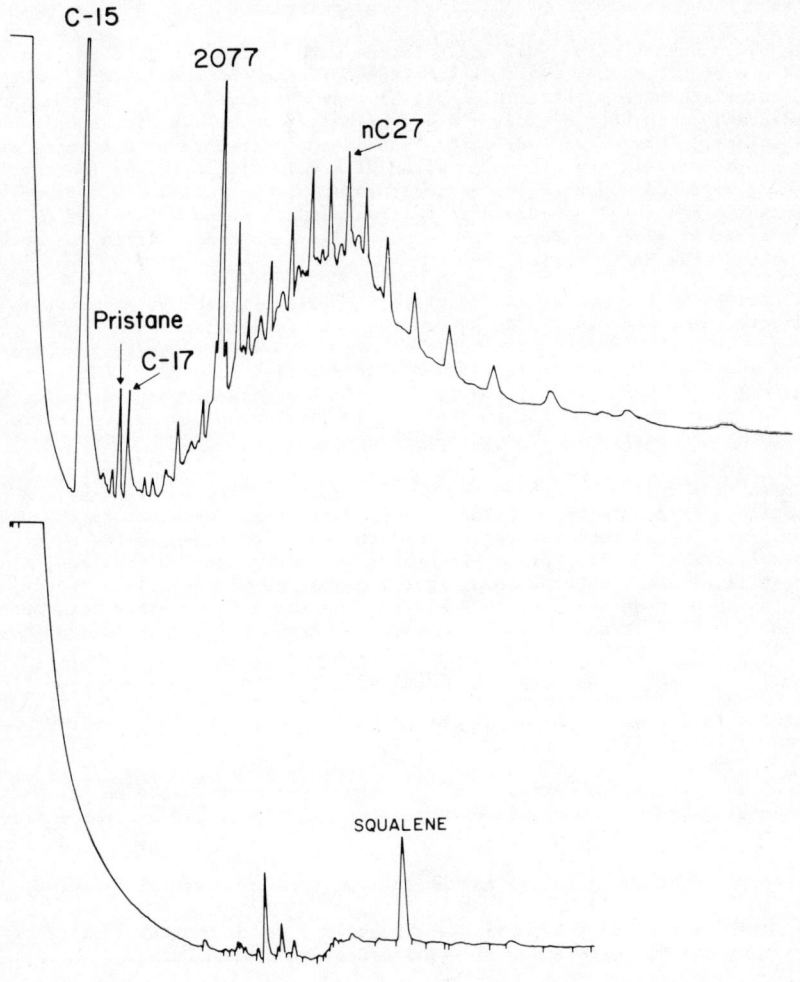

Fig. 44.4 Gas chromatograms of representative particulate hydrocarbon samples.

upper: station 4, aliphatic fraction, summer 1975
lower: station 7, unsaturated-aromatic fraction, summer 1975.

which measures all hydrocarbons, indicate concentrations greater than reported above. Iliffe and Calder (1974) reported concentrations for aliphatic hydrocarbons of 12 µg/l in the southeast Gulf of Mexico and Yucatan Straits and 47 µg/l in the Florida Straits while Barbier et al. (1973) reported values of 43 and 95 µg/l of total dissolved hydrocarbons from water collected at 50 m off the west coast of Africa. The GC-derived concentrations do not include contributions from the unresolved envelope when it is present. In those cases, total hydrocarbon may be a factor of 10 greater than reported.

There is a general trend of higher total resolved hydrocarbon concentrations near shore in both dissolved and particulate phases, although there are several exceptions to this trend.

The higher hydrocarbon concentrations near shore could be a result of direct terrestrial input or enhanced in situ production stimulated by terrestrially derived nutrients. The unresolved envelope components seem to have a terrestrial source, either Tampa Bay on transect 1, or the Mississippi River/Sound on transect 4. These unresolved components could be the remnants of highly weathered crude oil from marine sources or waste oil from terrestrial sources. Both dissolved and particulate hydrocarbons contained a series of n-alkanes from

from nC21 to nC32 with an odd/even ratio of near unity. This feature might be the result of weathered petroleum residues, but could also be derived from marine phytoplankton (Clark and Blumer, 1967). This series of alkanes was present when the lower molecular weight biogenic alkanes were absent. If they are of recent biosynthetic orgin, their stability in seawater must be greater than that of nC15, nC17 and pristane.

The biogenic hydrocarbons nC15, nC17 and pristane were dominant in the particulate aliphatic fraction and are probably the result of plankton collected on the filters. They hydrocarbons then should correlate with plankton biomass; however the remaining aliphatic and unsaturated/aromatic hydrocarbons in both dissolved and particulate phases are apparently not reflective of in situ biomass. Thus total hydrocarbon should not correlate with biomass estimators, such as Chl a. No correlation was noted with Chl a values reported by Iverson (1976) on samples taken simultaneously with our hydrocarbon samples. This differs from the correlation between Chl a and total non-aromatic hydrocarbons reported by Zsolnay (1972) for waters off the west coast of Africa. However, the upwelling region off Africa was much richer in phytoplankton than the MAFLA region.

The total dissolved hydrocarbons did correlate with dissolved organic carbon analysis of samples collected simultaneously with hydrocarbon samples (Aller, 1976). The ratio of total dissolved hydrocarbons to dissolved organic carbon was 0.4 ± 0.2 µg/mg in summer, 0.2 ± 0.2 µg/mg in fall and 0.3 ± 0.2 µg/mg in winter. The relative constancy of this ratio during each season indicates that the distribution of dissolved hydrocarbons and dissolved organic carbon are controlled by similar processes. No such relationship existed between particulate hydrocarbons and particulate organic carbon.

The high concentration of squalene in the water column is very interesting. A possible source of squalene is zooplankton (Calder, 1976). The total squalene in the average standing crop of zooplankton would be a few pg/l, while the concentrations in the water column average several tens of ng/l. For zooplankton to be the source of squalene, it must have long term stability in the water column. Yet the absence of squalene at several stations in the fall indicates that squalene is subject to degradative or other loss mechanisms. The source and dynamics of squalene in sea water deserves further investigation.

Conclusions

1. Hydrocarbons in the water column of the MAFLA area exist at low levels comparable to open ocean values.

2. The presence of weathered petroleum in dissolved and particulate phases is indicated, but not proven, by the occurrence of unresolved envelopes and n-alkanes from nC21 to nC32.

3. The unresolved envelope material may be derived from terrestrial sources.

4. Biogenic hydrocarbons in the particulate phase may be an indicator of in situ biomass, although there is no correlation of total hydrocarbon with Chl a.

5. There is a very high concentration of material with RI=3060 on FFAP and RI=2810 on SP-2100 in the water column. This material is probably squalene.

6. Total dissolved hydrocarbon correlates well with dissolved organic carbon. There is no correlation between particulate hydrocarbons and particulate organic carbon.

Acknowledgements

This research was supported by the U.S. Dept. of Interior, Bureau of Land Management under contract number 08550-CT5-30. Excellent technical assistance was provided by L. Griffin, W. Teehan, S. Horlick, C. Byrne, D. Wynne, and D. Siebert.

References

Aller, C., M. S. Thesis in progress, Florida State University, 1976.

Barbier, M., D. Joly, A. Saliot and D. Tourres, Hydrocarbons from sea water. Deep-Sea Research, 20, 305-314 (1973).

Brown, R. A., J. J. Elliot, J. M. Kelliher and T. D. Searl, Sampling and analysis of non-volatile hydrocarbons in ocean water. Adv. in Chemistry Series, No. 147, Ed. by T. R. P. Gibb, American Chemical Society, Washington, D. C. (1975).

Calder, J. A., Hydrocarbons from zooplankton of the eastern Gulf of Mexico. Proceedings of a Symposium on "Sources, Effects and Sinks of Hydrocarbons in the Aquatic Environment". American Institute of Biological Sciences, Arlington, VA. (1976).

Clark, R. C. and M. Blumer, Distribution of n-paraffins in marine sediments and organisms. Limnology and Oceanography, 12, 79-87 (1967).

Gordon, D. C., P. D. Keizer and J. Dale, Estimates using fluorescence spectroscopy of the present state of petroleum hydrocarbon contamination in the water column of the northwest Atlantic Ocean. Mar. Chem. 2, 251-261 (1974).

Iliffe, T. M. and J. A. Calder, Dissolved hydrocarbons in the eastern Gulf of Mexico Loop Current and the Caribbean Sea. Deep-Sea Research, 21, 481-488 (1974).

Iverson, R. L., Phytoplankton in the MAFLA lease area. MAFLA Final Report, 1975-76, in preparation.

Levy, E. M., The presence of petroleum residues off the east coast of Nova Scotia, in the Gulf of St. Lawrence, and the St. Lawrence River. Water Research, 5, 723-733 (1971).

Zsolnay, A., Hydrocarbon and chlorophyll: a correlation in the upwelling region off West Africa. Deep-Sea Research, 20, 923-926 (1973).

CHAPTER 45

DISTRIBUTION OF PETROLEUM HYDROCARBONS IN WESTERNPORT BAY (AUSTRALIA):
RESULTS OF CHRONIC LOW LEVEL INPUTS

Kathryn A. Burns Ph.D. and Jonathan L. Smith

Marine Chemistry Unit
5 Parliament Place
Melbourne, Victoria 3002
Australia

Ministry for Conservation
250 Victoria Parada
Melbourne, Victoria 3002
Australia

Abstract

A study of hydrocarbons in Westernport Bay was undertaken to assess the impact of the chronic low level input associated with man's use of the environment even in the absence of major oil spills. Aims of this project are 1) to measure amount of background hydrocarbon pollution associated with present land development, 2) to identify sources and levels of inputs, 3) to measure hydrocarbon accumulation from chronic input, 4) to gain a detailed picture of the partitioning of hydrocarbons in this environment which includes measurements of hydrocarbons in the water column, sediments and selected organisms, and 5) to assess the impact of chronic petroleum discharge on the structure and productivity of the ecosystem.

Analysis of solid samples included solvent extraction, saponification, column chromatography to separate saturated from unsaturated fractions and determination by gas chromatographic, fluorescence and gravimetric methods. Seawater hydrocarbons were absorbed on mixed resin columns, eluted with solvents and analyzed similarly to solid samples.

The mussel, *Mytilus edulis*, was the major indicator species used to establish problem areas and probable sources of petroleum input. Two major sources of pollution were identified, refinery and other industrial outfalls, and boating activities. Other possible sources are domestic disposal operations. Pollution levels varied from no detectable petroleum hydrocarbons to amounts close to saturation levels of body lipids in mussels (29 mg/g lipid).

The discussion includes means of relating amounts found in indicator species to levels of input, and implications on the toxicity to various components of the ecosystem.

Key words: Petroleum, hydrocarbons, chronic oil pollution, *Mytilus*, indicator species, environmental monitoring.

Introduction

Westernport Bay is a marine embayment bordering Bass Strait located 60 km southeast of Melbourne on the coast of Victoria, Australia. The Bay has an area of 680 km^2, of which about 40% is intertidal mud flat. Extensive areas are occupied by mangroves and seagrasses which are thought to be the major supporters of the aquatic food web via detritus. The marine fauna and flora of the Bay are unique assemblages of tropical and temperate representatives creating an exceedingly diverse and delicately balanced ecosystem. The Western Port region is typified as a "country atmosphere" with 68% of the 2,000 km^2 catchment area utilized for agriculture (mainly grazing), 20% for forests and only 1% for urban subdivision. Population in the catchment area is about 45,000 with only 14% served by a sewer system (Shapiro, 1974).

Development pressures to make Westernport a deep water port are mounting as the Bay provides easy access to large tankers. Three heavy industries are already located along the western shore, an oil refinery, a gas fractionating plant, and a steel mill. Both oil industries have established tanker unloading facilities. As yet, no large oil spills (greater than 150 tons) have occurred. Thus, even though the Bay is still relatively unspoiled, it suffers constant threat of oil pollution from tanker activities as well as chronic inputs from sources such as boating, refinery and other industrial outfalls, and domestic sewage.

Under the auspices of the Ministry for Conservation in Victoria, the Westernport Bay Environmental Study was undertaken to gain knowledge of the physical, chemical and biological systems in Westernport and to identify man's impact on the ecosystem with a goal of sound land/water management policies. The hydrocarbon study described in this paper is part of this program.

To predict the impact of petroleum hydrocarbon pollution in Westernport Bay, an ecosystem approach was followed. It was necessary firstly to study hydrocarbon inputs into the bay and then to establish the means by which the ecosystem dissipated these compounds.

Hydrocarbons enter the marine environment either by loss and discharge of petroleum products or by the decomposition of organisms containing biogenic hydrocarbons. The contribution from these two sources can be determined by chemical isolation of the hydrocarbon constituents and their separation into components on a gas chromatograph. Biogenics are usually much different in chemical composition than petroleum. By comparing sample chromatograms with those from known sources one can compute the contribution of each in a sample (Farrington et al., 1976). Other techniques such as fluorescence provide information on specific types of hydrocarbons and are also useful in distinguishing petroleum sources (Gordon and Keizer, 1974).

Hydrocarbons are readily accumulated in the tissues of marine organisms either by assimilation of contaminated food or by direct absorption from water through respiratory and other body surfaces (Reinert, 1969; Burns, 1976; Lee et al., 1972; etc.). Because of the high lipid solubility and low water solubility of hydrocarbons, marine organisms are expected to retain substantial quantities in their lipid stores unless they have some physiological mechanism for actively clearing their body tissues. Results of analysis of organisms taken from the environment thus depend on levels and types of hydrocarbons to which the animals were exposed, physiological mechanisms which were acting to control uptake, metabolism and excretion, and processes of exchange with clean water.

The mussel, *Mytilus edulis*, was chosen as the major monitoring species in this study because it readily concentrates hydrocarbons in its tissues even when levels are too low to be detected in the water column (DiSalvo et al., 1975). Mussels take up hydrocarbons and store them with relatively little biochemical modification (Lee et al., 1972a). Detailed uptake and discharge data are available for marine shellfish (Stegeman and Teal, 1973; Clark and Finley, 1975; Fossato and Canzonier, 1976) facilitating reasonable data interpretation. Since *Mytilus* is a cosmopolitan species we offer our results for comparison with world monitoring programs.

This is an on-going program in Victoria. Substantial progress has already been made in achieving at least our first four objectives in Westernport Bay.

Methods

Sample Collection

Mussels were collected at 18 shores and bay stations in Westernport Bay between July, 1974 and April, 1976 at approximately six-monthly intervals. Whole mussels of 4 to 7 cm length were washed in seawater and sealed in plastic bags. Surface sediments were collected from 12 intertidal shore stations (top 1 cm) and sediments from 8 subtidal bay stations were collected by divers. Precautions were taken against shipboard contamination as described by Grice et al. (1972). All apparatus used in sampling was washed with redistilled solvents before use. Samples were packaged in precleaned polyethylene bags or glass jars, transported on ice to the laboratory, and stored at -20°C until analysis.

Hydrocarbons suspended in the water column were sampled by pumping seawater through a resin column similar to that described by Harvey (1972) for the analysis of chlorinated hydrocarbons. The system consisted of a pre-filter (precombusted glass fiber type AE in a 4.5 cm stainless steel Millipore holder with Teflon seals) and a 2.5 x 30 cm stainless steel column, packed with 100 cm^3 of a 1 to 1 mixture of Amberlite XAD-2 resin (Rohm and Hass Co.) and Teflon particles of 600 to 850 microns. Resins were held in place with 60 mesh stainless steel screen. Using a vacuum pump, water was drawn from about ½m depth through copper tubing, filter and column, at a flow rate of 500 ml/min. All apparatus and resins were precleaned with redistilled solvents. After pumping 60 l of seawater, columns were capped with brass plugs, filters placed in clean glass jars, samples placed on ice and the apparatus reassembled with clean column and new filter for the next run. Filters were frozen in the laboratory until analysis as a solid sample. Hydrocarbons were eluted from the columns by running through 600 ml of redistilled methanol (60°C) at full gravity flow followed by 300 ml glass-distilled water. Eluants were collected in a separatory funnel; the hydrocarbons which partitioned into hexane were purified and analyzed as described later. This system was calibrated by pumping through seawater spiked with 10 µg/l diesel oil and subtracting background seawater levels. Comparison with direct solvent extraction of spiked water in a separatory funnel showed an overall recovery of 103%.

Hydrocarbon extraction

Solid samples were defrosted and prepared for solvent extraction in a Soxhlet apparatus. Mussels were shucked, liquid discarded, weighed, and about 80 gm of whole wet tissue placed in preextracted cellulose thimbles. Solvent flasks were filled with 300 ml methanol and samples extracted for 20 hrs. After cooling, 20 ml of 0.7 N KOH, 30 ml glass-distilled water and 30 ml benzene were added to solvent flasks. Reflux was continued for 2 hrs to effect saponification of lipids. After cooling, solvent was changed and the extraction and saponification repeated for another 48 hrs. Sediments were sorted to remove large plant particles and animals, weighed and transfered to extraction thimbles. Procedures were similar to those for mussels except that a 1 to 1 mixture of methanol/benzene was used in the second extraction.

The two solvent extract of each sample were combined in a separatory funnel and partitioned into hexane. Hexane extracts were concentrated on a rotary evaporator, dried with anhydrous Na_2SO_4, and filtered. Methanol/water phases were acidified, extracted with hexane, dried and concentrated. The sum of the gravimetric determinations of the nonsaponifiable plus saponifiable fractions defined the total extractable lipid per sample. After extraction samples were dried to constant weight at 60°C. Residual dry weight plus total extractable lipid weight constituted the total dry weight of samples. Sulphur compounds were removed from sediment extracts by trickling over a column of activated copper. Hydrocarbons were fractionated by charging concentrated lipid extracts to a column of silica gel and alumina which had been previously activated at 200°C for 1 hr and then partially deactivated with 5% water. Adsorbant to lipid weight was always greater than 100/1. Hydrocarbons were eluted with 3 column volumes of hexane. Some samples were fractionated to obtain more detailed chemical data. Saturated hydrocarbons were eluted with 1 column volume hexane (fraction A). Aromatics and unsaturates were eluted with 1 volume 10% benzene/hexane (fraction B) and 1 volume 20% benzene/hexane (fraction C).

Extraction efficiencies were calibrated by spiking methanol with diesel oil and Gippsland crude oil (characterized by Jackson *et al.*, 1975). Recoveries were 82 to 94% for diesel and 88% for crude. We have intercalibrated our procedures by analyzing the I.D.O.E. 5 sample described by Farrington *et al.*, (1976a). Results are uncorrected for % recovery.

Hydrocarbon Analyses
Total hydrocarbon extracts or fractions A & B were analyzed by gas chromatography using dual 15 ft x 1/8 in stainless steel columns packed with 3% silicon OV 101 on Gas Chrom Q. Temperatures were programmed from 150 to 340° or from 100 to 290°C at 5°/min using nitrogen as the carrier gas and dual flame ionization detectors. N-alkanes were identified by comparison of retention times with those of standards and by coinjection of sample and standards. Components were quantitated by measuring peak areas on chromatograms and comparing to areas generated by external standards in the C-12 to C-32 boiling range. These techniques detect hydrocarbons quantitatively in the C-14 to C-34 range.

Biogenic hydrocarbons were distinguished by criteria similar to those given in Burns and Teal (1973) and Farrington *et al.*, (1976b). Biogenics were subtracted from total hydrocarbons to give an estimate of the quantity of petroleum in samples. Sources of contamination were distinguished by assessing possible inputs from boating and refinery effluents in an area and testing these types of petroleum through our procedures. When estimating sources of petroleum contamination in our samples, allowance was made for qualitative changes in environmental samples due to weathering and microbial degradation (Blumer *et al.*, 1973).

Fraction C was analyzed qualitatively on an Aminco-Bowman spectrophotofluorometer modified so that emission and excitation speeds could be independently controlled. This produced a "coupled emission-excitation spectrum" as described by Gordon and Keizer (1974). Crude and diesel contamination could be distinguished qualitatively in samples on the basis of the fluorescence spectrum of this fraction.

Results

Hydrocarbon analyses of mussels showed petroleum contamination in areas of Westernport Bay close to oil industry and other industrial outfalls, boating activities and some domestically settled sites. Levels of contamination reflected presumed amounts of discharge from the various sources (Table 1). Mussels taken from areas influenced by refinery outfalls showed high (600 to 1,200 µg/gm dry wt) persistent values for petroleum hydrocarbons. Areas influenced by boating activities showed erratic high values with residual levels of 100 to 200 µg/g. Lower concentrations were found in mussels taken progressively further away from input sites until no petroleum was found at farthest sites. Based on levels of petroleum hydrocarbons in mussel tissues, we mapped out Bay areas subject to chronic oil pollution and the approximate area of influence on the water column of presumed sources (Fig. 1).

Detailed qualitative assessment of gas chromatograms and fluorescence spectra was required to substantiate possible sources and determine their relative importance in Westernport Bay. Figure 2 shows typical gas chromatograms of total hydrocarbon extracts of mussels taken from various locations in the Bay. It illustrates the variety of petroleum products discharged into even a sparsely populated and industrialized area. Comparison of these chromatograms with those of source oils with allowance for environmental degradation and weathering, clearly shows at least two separate sources. Mussels close to refinery outfalls showed g.c. patterns similar to a light weathered crude oil. Mussels from boating sites occasionally showed fresh undegraded fuel oil in high concentrations, although most boating areas showed varying proportions of diesel oil and a heavier lube oil similar to that used in outboard engine fuel mixtures. The relative height of the unresolved complex mixture (UCM) (typical of degraded petroleum) in the high (C-28) and low (C-17) boiling range was used to estimate the relative amounts of diesel and outboard fuels in samples, and is expressed in Table 1 as (UCM C17/C28). The N-alkanes in the C-21 to C-27 range were useful as markers in identifying the presence of crude oil in mussels.

Confirmation of suspected sources was achieved by comparing the fluorescence properties of mussel extracts with standard oils. Mussel extracts which showed petroleum contamination by gas chromatography also showed fluorescence characteristics of aromatic hydrocarbons. Spectra from mussels could be qualitatively matched to one of the standard oils (Fig. 3). For example, mussels at boating sites (stations 16, 17, 18) containing diesel oil showed fluorescence of mainly double ring aromatic derivatives (naphthalenes). Samples collected near petroleum industries (stations 3, 4, 5, 6, 8) showed the heavier ring characteristics of light crude oil with fluorescence maxima near 355 nm and 415 nm.

TABLE 1. Mussels

Petroleum hydrocarbons in mussels, *Mytilus edulis*, collected from stations in Westernport Bay. (% unsaturated calculated as weight of column fractions (B/A+B)x100). Results were calculated from gas chromatograms of hydrocarbons in the C14 to C34 boiling range and expressed in amounts of hydrocarbons per gram tissue on both a dry and lipid weight basis. (UCM is unresolved complex mixture.)

STATION	DATE SAMPLED	PETROLEUM HYDROCARBONS		UCM C17/C28	QUALITATIVE DESCRIPTION BASED ON G.C. & FLUORESCENCE PATTERNS
		ug/g dry	mg/g lipid		
Clean sites:					
14	15.10.75	0	0.0	1.0	
	15.10.75	13	0.1	1.0	Biogenic hydrocarbons mostly plant waxes and algal components. No fluorescence. UCM 3% to 5%. Unsaturates. 26% to 41%.
	8. 6.76	23	0.2	1.0	
1	30. 6.75	0	0.0	1.0	
	31. 1.75	0	0.0	1.0	
13	1. 7.75	0	0.0	0.3	
	27.11.75	16	0.1	0.4	
Refinery and industrial outfall influenced sites:					
4	4. 7.75	735	4.4	8.0	
	12.12.75	689	4.5	5.3	
	12. 2.76	570	3.8	3.3	High chronic levels of light crude sometimes mixed with a heavier crude type oil. Some paraffins useful as markers. Biogenics usually not visible. UCM 50% to 90% of g.c. Unsaturates 30% to 40%.
	28. 4.76	685	3.5	3.4	
	28. 4.76	1186	6.4	2.9	
5	4. 7.75	447	3.5	6.0	
	12.12.75	123	1.0	1.0	
	12. 2.76	133	1.1	1.5	
8	3. 7.75	342	2.0	4.8	
	28.11.75	370	3.4	1.8	
	10. 2.76	659	4.7	3.1	
6	15.10.75	472	3.8	3.4	
	3. 2.76	228	1.7	1.8	
9	16.10.75	468	4.6	0.6	

(Cont'd)

TABLE 1 (Cont'd)

STATION	DATE SAMPLED	PETROLEUM HYDROCARBONS		UCM C17/C28	QUALITATIVE DESCRIPTION BASED ON G.C. & FLUORESCENCE PATTERNS
		ug/g dry	mg/g lipid		
Boating sites:					
12	21. 7.74	160	1.1	0.8	
	30. 1.75	100	0.7	1.3	
	1. 7.75	162	0.9	1.1	Biogenics visible over
	27.11.75	126	1.1	0.5	degraded diesel and outboard fuel residues. UCM 20% to
15	30. 6.75	98	0.6	2.7	80% of g.c. Fluorescence
	25.11.75	119	0.8	1.2	shows varying amounts of
16	24. 7.74	75	0.9	5.3	diesel contamination.
	31. 1.75	130	1.0	2.5	Unsaturates @ 30%.
	30. 6.75	630	3.3	6.7	
	25.11.75	283	2.4	1.2	
	9. 2.76	156	1.1	1.7	
18	25. 2.76	454	3.3	2.5	
11	17. 4.75	20	0.1	2.0	
	1. 7.75	12	0.1	1.0	
	27.11.75	21	0.2	0.6	
17	15.10.75	62	0.6	2.0	
	3. 2.76	60	0.3	0.6	
Mixed influence sites:					
2	15.10.75	72	0.6	1.8	
	3. 2.76	110	0.7	1.7	
3	3. 7.74	190	1.6	5.0	Degraded diesel oil
	29. 1.74	215	1.1	2.9	Diesel/crude residues
	4. 7.75	843	9.7	3.4	Crude residue
	15.10.75	204	1.9	3.0	Degraded fuel oil
	28.11.75	113	1.1	2.0	Crude residue
	12. 2.76	3220	29.4	7.5	Fresh fuel oil
7	15.10.75	55	0.7	1.1	Crude under biogenics
	3. 2.76	49	0.4	0.6	
10	15.10.75	27	0.2	1.0	Crude under biogenics
	12. 6.75	0	0.0	0.5	
	5. 2.76	0	0.0	0.4	

Sediments

Surface sediments taken from intertidal and shallow subtidal stations in the Bay showed petroleum hydrocarbon contamination only in areas of high chronic input and small isolated oil spills (Table 2). In most areas of the Bay there appears to be little build up in the sediments.

Areas close to refinery and other industry outfalls and boat docks show low levels of oil contamination in sediments. Gas chromatograms generally showed degraded oil residues (UCM) under the biogenic alkanes and alkenes characteristic of seagrass waxes and algal hydrocarbons. The unresolved mixture was generally higher boiling than that found in mussels from the same area. Fluorescence spectra showed relative enrichment of the heavier ring structures in sediments compared to mussels. Farrington and Quinn (1973) reported similar results for sediments and clams collected from Narragansett Bay.

We sampled intertidal sediments from an area subjected to a small accidental spill of Gippsland crude oil (station I). Initial samples showed very high levels (5mg/g dry wt) of the waxy crude. Analysis of sediments collected three months later indicated the oil had dissipated to a degraded residual level of about 50 µg/g. A similar persistence was noted at another site (station H) contaminated with crude oil. The mechanism of removal of oil from these sediments or a mass balance was not determined but presumably dissipation was due primarily to the fast rate of tidal dilution in Westernport Bay.

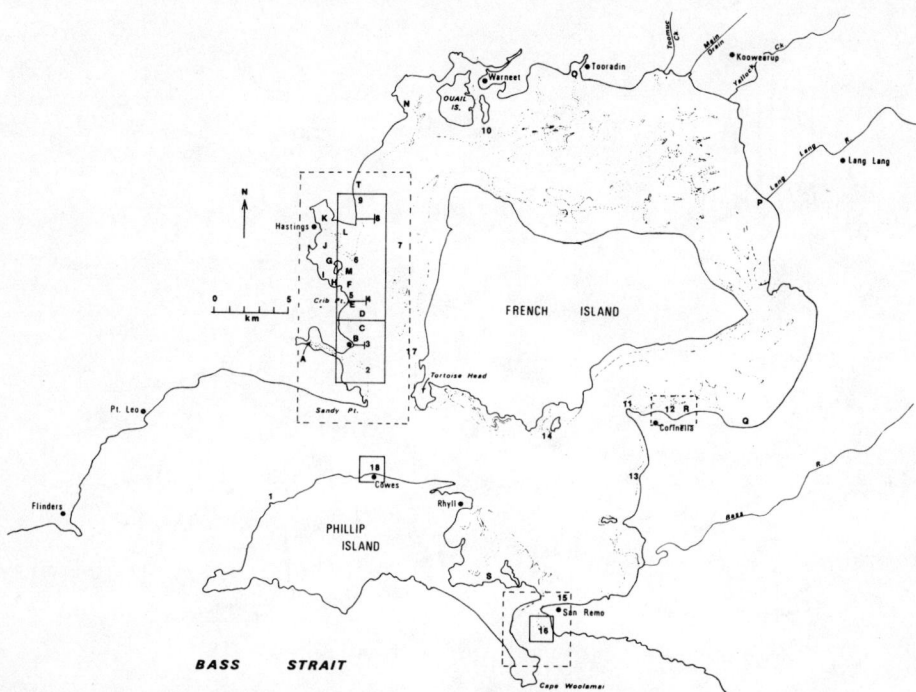

Fig. 45.1 Map of Westernport Bay (Victoria, Australia) showing mussel (numbers) and sediment (letters) sampling stations. Areas bounded by solid lines are subject to high chronic input of petroleum hydrocarbons (>450 ppm found in mussels). Areas bounded by dashed lines show approximately 100 to 200 ppm in mussels.

Water

Water so far has been sampled from three locations in Westernport Bay. One sample taken from a station North of Pt. Leo (Fig. 1), an unpolluted area, showed a total hydrocarbon concentration of 2.25 µg/l. Twenty l of seawater from this station was spiked with 200 µg of diesel oil. Recovery through the water sampling system was 103% when compared to direct solvent extraction of the stock solution. Most of the diesel oil was retained on the glass fiber filters as were the natural hydrocarbons (Table 3). Two samples were taken at station 3 on a day showing no visible pollution. Total concentrations were 1.15 µg/l on the ebb tide (water flowing from industrialized area) and 0.34 µg/l on the flood tide (water flowing from Bass Strait). The third site sampled was a boat dock near station K. This was the only sample which contained petroleum hydrocarbons in high concentrations (8.2 µg/l). The chromatograms showed both diesel and crude oils most of which passed through the glass filter and were retained on the resin column.

Discussion

Mussels

Hydrocarbon levels found in mussels can be used as indicators of the amount of petroleum suspended in the water column. Understanding of uptake, retention and discharge patterns of hydrocarbons in marine shellfish can be gathered from the literature. Lee $et\ al.$, (1972a) showed that $Mytilus$ incorporate hydrocarbons into body tissues with little, if any, biochemical modification. Stegeman and Teal (1973) demonstrated that uptake of petroleum hydrocarbons in shellfish is related to lipid content of the organisms and that the rate of uptake was linear from 0 to 500 µg/l. Uptake can be viewed as an equilibration process controlled by the relative solubilities of various classes of hydrocarbons in water compared to lipids. Uptake and discharge processes proceed simultaneously and the net amount retained by the organism is related to the saturation of its lipid stores and concentrations in the water. The equilibrium magnification factor calculated by analyses of seawater and animal lipids was about 2×10^5 for oysters up to a maximum equilibrium concentration of 22 mg/g lipid (Stegeman and Teal, 1973). Fossato and Canzonier (1976) reported similar magnification factors in $Mytilus$ experimentally dosed with diesel oil.

The maximum contamination level found in mussels from Westernport was 29.4 mg/g lipid (Table 1). Comparison with the literature indicates this is approximately the saturation

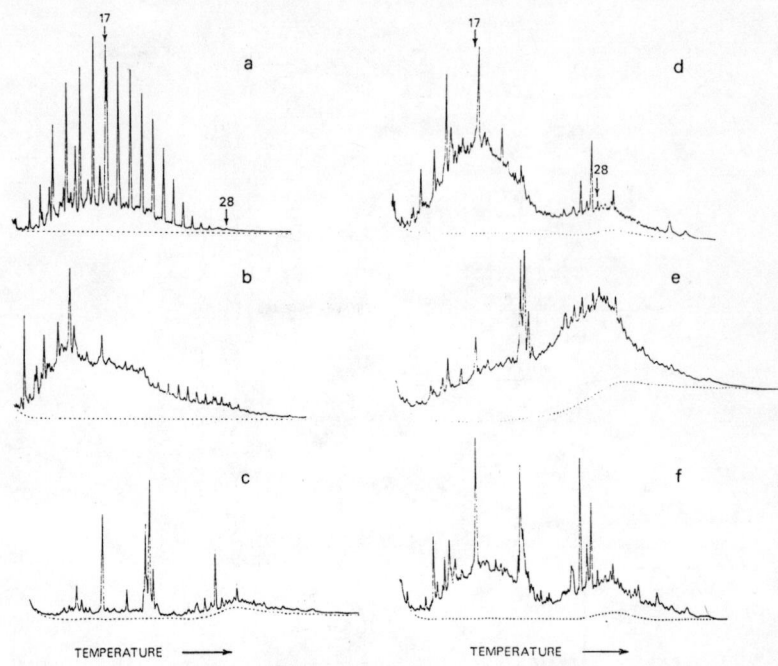

Fig. 45.2 Sample chromatograms of total hydrocarbon extracts of mussels collected in Westernport Bay showing various types of oil discharged into the system. Numbers are the positions of n-alkanes of the indicated carbon chain length. Dashed lines are column bleed signals a) fresh fuel oil (3220 ppm) from boat dock at Sta. 3; b) degraded crude oil (570 ppm) from refinery wharf Sta. 4; c) biogenics (no petroleum) from clean area Sta. 14; d) degraded diesel oil (100 ppm) from small boat dock Sta. 12; e) lube oil (126 ppm) from Sta 12; f) mixture of degraded diesel and lube oils (160 ppm) from Sta. 12.

level of hydrocarbons in living mussels. Unfortunately, few authors have reported results in terms of lipid weight. Values reported in terms of dry weight or wet weight of shellfish from oil spill sites or heavily chronically polluted areas indicate similar maximum concentrations (Zitko and Carson, 1970; Ehrhardt, 1972; Fossato and Siviero, 1974; Burns, 1975; etc.).

Assuming 2.94×10^4 μg/gm lipid is the maximum amount of oil a mussel can absorb and retain, then by applying the magnification factor of 2×10^5, a minimum water concentration of 147 μg/l is obtained. This is within the range of the solubility of oil in seawater conservatively estimated by Boylan and Tripp (1971) as 650 μg/l.

The result of analysis of an organism taken from the environment is therefore an estimate of the minimum amount to which the animal was exposed in the water column. If the concentration factor of 2×10^5 is applied to mussels collected from station 4 (near refinery outfall) where levels of petroleum hydrocarbons average 773 μg/g dry wt or 4.4 mg/g lipid a calculated water concentration of 22 μg/l is obtained. This calculated value is close to results obtained by analysis of seawater in other areas possibly subject to chronic oil pollution (Iliffe and Calder, 1974; Zsolnay, 1971; Barbier et al., 1973).

In areas such as boating sites pollution results from a series of small intermittent inputs (bilge pumpings, small spills, etc.). Mussels sampled at these sites reflect the intermittent nature of the input. For example, levels of petroleum in mussels from station 3 (a boat wharf about 3 km south of a refinery, Table 1, Fig. 1) vary from saturation values to residual levels left after the mussels depurated larger amounts into relatively clean tide waters. Residual levels of approximately 200 μg/g dry wt or 1.4 mg/gm lipid are common in these chronically polluted areas. Calculations predict approximately 7 μg/l average concentration of hydrocarbons in the water column. Similar residual values were reported by Farrington and Quinn (1973) in clams from Narragansett Bay and by Ehrhardt and Heinemann (1975) for *Mytilus* from Kiel Bight. Stegeman and Teal (1973) reported oysters retained 2 to 3 mg/g lipid of petroleum after depuration in the laboratory. Areas of Westernport Bay subject to less severe and less frequent contamination show residual petroleum concentrations of 20 to 50 μg/l dry wt (100 to 700 μg/g lipid) in mussels.

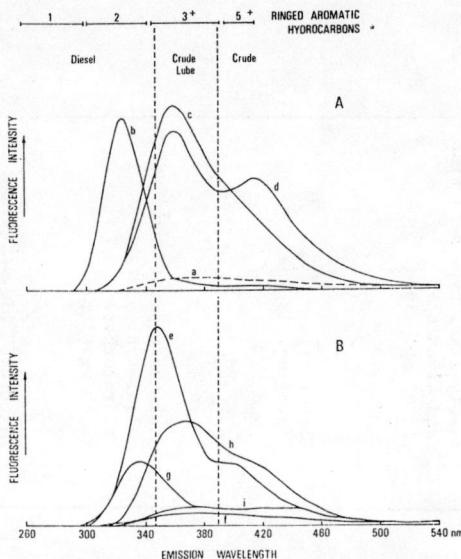

Fig. 45.3 Coupled excitation-emission fluorescence spectra of 20% benzene/hexane column fractions of : A. source oils and B. environmental samples (about 50 µg hydrocarbon/ml hexane). Samples scanned 260 to 540 nm emission with excitation monochrometers set 23 nm lower. Band pass width 5 nm; sensitivity 30. *Based on Gordon and Kiezer, 1974. a) huxane blank. b) diesel oil; c) lube oil; d) Gippsland crude oil; e) mussels from a refinery wharf; f) clean mussels; g) mussels from small boat wharf; h) sediments near refinery; i) clean sediments.

TABLE 2. Sediments

Total hydrocarbons in the C14 to C34 boiling range in surface sediments from Westernport Bay. (% unresolved was calculated by subtracting peak areas from total areas on gas chromatograms and is assumed to be petroleum.

STATION	DATE SAMPLED	HYDROCARBONS		PERCENT UNRESOLVED
		ug/g dry	mg/g lipid	
A	30. 7.74	36.2	109.7	50%
B	28.11.75	6.5	28.6	84%
C	28. 4.76	31.7	36.3	61%
D	28. 4.76	65.9	46.0	75%
E	28.11.75	6.5	12.6	79%
F	28. 4.76	31.4	49.7	80%
G	28. 4.76	47.7	28.9	72%
H	28.10.75	85.4	178.0	90%
H	11. 2.76	22.8	43.0	75%
I	28.10.75	5271.0	226.0	70%
I	11. 2.76	50.9	11.5	45%
J	28. 4.76	17.9	40.8	65%
K	28. 4.76	24.6	39.5	76%
L	28. 4.76	3.2	14.8	60%
M	28. 4.76	36.5	18.6	75%
N	27.11.75	1.8	8.7	22%
O	27.11.75	2.9	11.1	26%
P	1. 7.75	2.4	6.6	16%
P	27.11.75	2.3	8.9	54%
Q	27.11.75	2.6	10.0	52%
R	1. 7.75	9.4	44.0	83%
R	27.11.75	9.0	15.7	80%
R	11. 2.76	10.3	50.2	80%
S	7. 5.74	2.8	4.2	32%
S	27.11.75	3.6	7.4	54%
T	27.11.75	10.2	15.6	29%

TABLE 3. Water

Total hydrocarbons in the C14 to C34 boiling range in water samples from Westernport Bay.

	HYDROCARBONS RETAINED ON FILTERS	HYDROCARBONS TRAPPED ON RESIN COLUMNS	QUALITATIVE G.C. DESCRIPTION & LIPID WT.
1. Station north of Pt. Leo. Ebb tide. a. hexane fraction b. 10% benzene fraction c. 20% benzene fraction Total in 60 l.	54.9 ug 25.2 ug 54.9 ug ——— 135.0 ug	0 ug 0 ug 0 ug ——— 0 ug	mostly algal type paraffins and alkenes (1.40 mg. lipid)
2. 20 l seawater spiked with diesel a. hexane fraction b. 10% benzene fraction c. 20% benzene fraction Total in 60 l.	112.0 ug 40.8 ug 6.6 ug ——— 159.4 ug	0 ug 20 ug 0 ug ——— 20 ug	diesel oil (0.85 mg. lipid)
3. Station 3 flood tide a. hexane fraction b. 10% benzene fraction c. 20% benzene fraction Total in 60 l.	16.4 ug 3.9 ug 0.2 ug ——— 20.5 ug	0 ug 0 ug 0 ug ——— 0 ug	mixture of plant waxes and algal type biogenic hydrocarbons (1.40 mg. lipid)
4. Station 3 ebb tide. a. hexane fraction b. 10% benzene fraction c. 20% benzene fraction Total in 60 l.	40.5 ug 14.2 ug 14.4 ug ——— 69.1 ug	0 ug 0 ug 0 ug ——— 0 ug	mixture of biogenics and petroleum similar to diesel oil (1.97 mg. lipid)
5. Station K ebb tide. a. hexane fraction b. 10% benzene fraction c. 20% benzene fraction Total in 60 l.	37.9 ug 14.6 ug 9.9 ug ——— 62.4 ug	195 ug 233 ug 0 ug ——— 428 ug	mixture of diesel, crude, and some biogenics (3.38 mg. lipid)

Oil pollution in Westernport Bay varies considerably depending on sampling locations. As shown in Table 1, of the 18 sites sampled over the past two years, only three showed no detectable levels of petroleum hydrocarbons. These results may only be relevant for localized areas around sampling sites since many of these were in areas of boating activities, etc. The highest levels of petroleum contamination occur in the industrialized Western part of the Bay. Mussels in this region contain petroleum hydrocarbons of both crude oil and boating origin.

Sediments
Low levels of petroleum hydrocarbons in Westernport Bay sediments result from present rates of discharge. It is probable that an increase in chronic input would enlarge the area of contamination of both water and sediments. Large oil spills which would penetrate the anoxic mangrove sediments would be expected to have a long residence time, similar to that observed in salt marsh sediments after the West Falmouth oil spill (Burns, 1975).

Sanders et al., (1972) correlated biological data with chemical analyses of the subtidal sediments (Blumer and Sass, 1972) and showed that Ampeliscid amphipods provided a sensitive indicator of contamination levels which lethally affected the marine biota. Levels of sediment petroleum above 72 µg/g dry wt were shown to be lethally toxic to Ampeliscids. Using this value as a guide, levels of sediment petroleum hydrocarbons in most areas of Westernport Bay do not appear to be high enough to cause lethal toxicity to benthic organisms. Only two sites showed petroleum levels in excess of 72 µg/g dry wt. Both these sites had been subjected to small oil spills (Table 2). Sediments in areas of chronic input from refinery and boats (Stations A, C, D, F, G, M), contain as much as 44% to 92% of the petroleum levels assumed to be lethal to sensitive species. Thus, build up of petroleum in these areas may be high enough to cause sublethal effects including changes in the benthic communities. A preliminary benthic survey in Westernport suggested a decrease in macrofaunal diversity in the shallow areas near where our analyses of mussels showed residual petroleum contamination (Shapiro, 1974). Oil pollution in this region may be one of a number of stresses perturbing these animal populations.

Water
Adsorption on resin columns with a glass fiber filter in the line is an efficient means of sampling hydrocarbons in natural waters, especially when levels are too low to be detected by direct solvent extraction of seawater in a separatory funnel. Blank values are exceedingly low and do not contribute significantly to even very dilute samples.

Hydrocarbons in most water samples were retained on the glass fiber filers. This was also true of seawater spiked with diesel oil. Sample 5 (Table 3) had appreciable contamination from diesel and crude oils which passed through the glass filters. The results indicate some interesting partitioning between "dissolved" and "particulate" organic matter in the water column. More samples are required to resolve the problem. Workers who prefilter samples of seawater for hydrocarbon analyses with glass fiber filters may obtain erroneous results as most of the hydrocarbons may not pass through the prefilter.

Gas chromatograms showed patterns predicted by visual observation in the areas sampled. The seabed in shallow areas north of Pt. Leo where the first sample was taken, is colonized by dense beds of colonial algae. Hydrocarbons typical of algae (Youngblood and Blumer, 1973) appeared in water from this station and probably reflected algal detritus. Extensive areas north of station 3 are covered by mangroves and *Zostera* sea grass. Water on ebb tide at station 3 showed plant waxes in addition to a diesel oil pattern reflecting natural detritus as well as contamination (1.15 µg/l on the day we sampled). The hydrocarbon level in sample 5 (8.2 µg/l) was close to that predicted by the calculations for areas of chronic contamination.

Recommendations and Conclusions

In areas subject to chronic input of petroleum to the water column, analysis of mussel tissue provides a means of identifying input sources, types of hydrocarbons present in the system, and approximate average concentrations of hydrocarbons suspended in coastal waters. Values obtained by mussel analysis provide an integrated sample over more time than can be measured with an intermittent water sampling program. Di Salvo et al., (1975) and Fossato and Canzonier (1976) support this view. Results of mussel analysis are easier to interpret than water samples with low level petroleum contamination since mussels tend to show less interference from biogenic plant hydrocarbons. Thus, analysis of mussels provides a clearer picture of average water column condition. Mussels are easier and less costly to sample, however results must be interpreted with the biological mechanisms of uptake, retention and depuration in mind. Petroleum monitoring projects in coastal waters should emphasize analysis of an indicator species such as *Mytilus*.

Refinery and industrial outfalls appear to be the major sources of chronic contamination in Westernport Bay. Similar sources are probably important in many coastal ecosystems throughout the world. Every possible effort should be made to minimise petroleum input into marine waters. Outfall licences should be strictly enforced and should be amended to specify permissible levels on the basis of the most toxic fractions of petroleum released into the environment. Soluble aromatic hydrocarbons exhibit the most pronounced acute physiological effects on marine organisms (reviewed by Moore *et al.*, 1974) and should be carefully monitored in marine outfalls. Oil pollution associated with boating activities could be controlled by enforcing stringent regulations on oil discharge and providing pump out stations similar or superior to the facilities for tanker ballast water treatment already established in Westernport Bay.

The next stop in the Westernport hydrocarbon program is to provide data on whether levels of petroleum contamination are high enough to exhibit toxic effects on the marine fauna and flora. A monitoring project using mesh bags full of clean *Mytilus* suspended in problem areas for later analysis has already commenced. In addition a biochemical assay is being developed to establish whether monitoring mussels are being stressed by sublethal levels of petroleum contamination. The literature indicates that mussels are fairly resistant to chemical stress and that most other marine organisms are more sensitive (Moore *et al.*, 1974) Thus, if mussels show physiological stress from petroleum levels present in the environment this would indicate that the ecosystem as a whole is stressed.

Acknowlegements

We thank C. Haworth and A. Murrell for technical assistance in the analytical work, the divers and crew of the Marine Pollution Studies Group (Ministry for Conservation Victoria) research vessels for help in field sampling, and Dr. J.B. Robinson for advice on initial setup of analytical procedures. This research was carried out with the support of the Environmental Studies Program, Ministry for Conservation, Victoria, Australia. Contents of this report do not necessarily represent the official views of the Ministry. This is publication No. 120 in the Ministry for Conservation/Victoria Environmental Studies Series.

References

Anderson, J., Laboratory studies on the effects of oil on marine organisms : an overview, American Petroleum Institute Publication No. 4249. 70 pp. (1975)

Barbier, M., D. Joly, A. Saliot, & D. Tourres, Hydrocarbons from seawater Deep Sea Res. 20, 305 (1973).

Blumer, M. & J. Sass, The West Falmouth oil spill. II. Chemistry Woods Hole Oceanographic Institution Tech. Report 72-19. 123 pp. (1972).

Blumer, M., M. Ehrhardt, & J.H. Jones, The environmental fate of stranded crude oil, Deep Sea Res. 20, 239 (1973).

Boylan, D.B. & B.W. Tripp, Determination of hydrocarbons in seawater extracts of crude oil and crude oil fractions, Nature 230, 44 (1971).

Burns, K.A. & J.M. Teal, Hydrocarbons in the pelagic *Sargassum* Community Deep Sea Res. 20, 207 (1973).

Burns, K.A., Hydrocarbon metabolism in the intertidal fiddler crab, *Uca pugnax*, Mar. Biol. 36, 5 (1976).

Clark, R.C. & J.S. Finley, Uptake and loss of petroleum hydrocarbons by the mussel *Mytilus edulis*, in laboratory experiments, Fishery Bull. 73, 508 (1975).

DiSalvo, L., H.E. Guard, & L. Hunter, Tissue hydrocarbon burden of mussels as a potential monitor of environmental hydrocarbon insult, Environ. Sci. & Tech. 9, 247 (1975).

Ehrhardt, M., Petroleum hydrocarbons in oysters from Galveston Bay, Environ. Pollut. 3, 257 (1972).

Ehrhardt, M. & J. Heinemann, Hydrocarbons in blue mussels from the Kiel Bight, Environ. Pollut. 9, 263 (1975).

Farrington, J.W. & J.G. Quinn, Petroleum hydrocarbons in Narragansett Bay I. Survey of hydrocarbons in sediments and clams, Estuar. & Coast. Mar. Sci. 1, 71 (1973).

Farrington, J.W., J. Quinn, J.M. Teal, G.C. Medeiros, T.L. Wade, & K.A. Burns, Intercalibration of gas chromatographic analyses for hydrocarbons in tissues and extracts of marine organisms, Analyt. Chem. 48, 1711 (1975).

Farrington, J.W., J.M. Teal, & P.L. Parker, Petroleum Hydrocarbons in : Strategies for Marine Pollution Monitoring. ed. by E.D. Goldberg, Wiley, New York, (1976).

Fossato, V.U. & E. Siviero, Oil pollution monitoring in the Lagoon of Venice using the mussel, *Mytilus galloprovincialis* Mar. Biol. 25, 1 (1974).

Fossato, V.U. & W.J. Canzonier, Hydrocarbon uptake and loss by the mussel, *Mytilus edulis* Mar. Biol. 36, 243 (1976).

Gordon, D.C. & P.D. Keizer, Estimation of petroleum hydrocarbons in seawater by fluorescence spectroscopy : improved sampling and analytical methods, Fish. Res. Bd. Can. Tech. Report 481. 29 pp. (1974).

Grice, G.D., G.R. Harvey, V.T. Bowen, & R.H. Backus, The collection and preservation of open ocean marine organisms for pollutant analysis, Bull. Environ. Contam. Toxicol. 7, 125 (1972).

Hamelink, J.L., R.W. Waybrant, & R.C. Ball, A proposal : exchange equilibria control the degree chlorinated hydrocarbons are biologically magnified in lentic environments, Trans. Amer. Fish. Soc. 100, 207 (1971).

Harvey, G.R., Adsorption of chlorinated hydrocarbons from seawater by a cross-linked polymer, W.H.O.I. Techn. Report 72-86. 20 pp + appendix, (1972).

Iliffe, T.M. & J.A. Calder, Dissolved hydrocarbons in the eastern Gulf of Mexico Loop Current and the Caribbean Sea, Deep Sea Res. 21, 481 (1974).

Jackson, B.W., R.W. Judges, & J.L. Powell, Characterization of Australian crudes and condensates by gas chromatographic analysis, Environ. Sci. & Tech. 9, 656 (1975).

Lee, R.F., R. Sauerheber, & A.A. Benson, Petroleum hydrocarbons : uptake and discharge by the marine mussel *Mytilus edulis*, Science 177, 344 (1972).

Lee, R.F., R. Sauerheber, & G.H. Dobbs, Uptake, metabolism and discharge of polycyclic aromatic hydrocarbons by marine fish, Mar. Biol. 17, 201 (1972).

Moore, R.F., G.R. Chirlin, G.J. Puccia, & B.P. Schrader, Potential biological effects of hypothetical oil discharge on the Atlantic Coast and Gulf of Alaska, Rept. to the Council on Environ. Qual. MITSG 74-19. 121 pp. (1974).

Reinert, R.E., The accumulation of Dieldrin in an alga, daphnia, guppy food chain, Doctoral Dissertation University of Michigan. 76 pp. (1967).

Sanders, H.L., J.F. Grassle, & G.R. Hampson, The West Falmouth oil spill I. Biology W.H.O.I. Technical Report 72-20. 49 pp. (1972).

Shapiro, M.A., ed. Westernport Bay Environmental Study. 653 pp. available from : Environmental Studies Program, Ministry for Conservation, 240 Victoria Parade, E. Melbourne, Victoria (Australia), 3002. (1974).

Stegeman, J.J. & J.M. Teal, Accumulation, release, and retention of petroleum hydrocarbons by the oyster, *Crassostrea virginica* Mar. Biol. 22, 37 (1973).

Youngblood, W.W. & M. Blumer, Alkanes and alkenes in marine benthic algae, Mar. Biol. 21, 163 (1973).

Zitko, V. & W.V. Carson, The characterization of petroleum oils and their determination in the aquatic environment, Fish. Res. Bd. Can. Tech. Report No. 217, 29 pp. (1970).

Zsolnay, A., Preliminary study of the dissolved hydrocarbons and hydrocarbons on particulate material in the Gotland Deep and the Baltic, Kieler Meeresforschungen 27, 129 (1971).

PANEL DISCUSSION ON RESEARCH NEEDS

Friday, November 12 - 3:45 PM
Spanish Ballroom

Panel Members

Douglas A. Wolfe, Chairman
NOAA Outer Continental Shelf Environmental Assessment Program
Boulder, Colorado

Jack Anderson
Battelle Northwest Marine
  Research Laboratory
Sequim, Washington

Donald K. Button
University of Alaska
Institute of Marine Sciences
Fairbanks, Alaska

Donald C. Malins
Environmental Conservation Division
National Marine Fisheries Service
Northwest and Alaska Fisheries Center
Seattle, Washington

Edward W. Mertens
American Petroleum Institute and
  Chevron Research Company
Richmond, California

Gary O'Neil
Surveillance and Analysis Division,
  NW Region
Environmental Protection Agency
Seattle, Washington

Discussion

Wolfe: I want to call this final session of the Conference to order. For two and three-quarters days now, we've been hearing about the science that has been going on in this field of marine fate and effects of hydrocarbons over the last few years, and especially over the past several months. I hope that some of the papers you've heard have promoted awareness of the current state of the art and the current problems in the field and that these papers will stimulate us to new and better approaches to future research in this area.

This final session of the symposium was organized to provide some provocative discussion about where we go from here. That is, what are the principal problems in the areas of fate and effects of petroleum hydrocarbons which should be occupying scientific attention over the course of the next few years. And I guess the converse of that question should also be considered, i.e., what are some of the areas which probably have peaked out and should be deemphasized in the future. I want to thank all of you hardcore symposium goers who have stayed over for this session, and I hope you will participate in the discussion on those questions. The panel that you see before you was asked to express their individual opinions on research priorities primarily to serve as a basis of discussion. I hope that their thoughts will in fact provoke or stimulate some of you in the audience to express your own opinions as well. I've asked each of these gentlemen to take about five minutes to express some of their thoughts on research priorities, emphasizing their own areas of interest or perhaps their organization's area of interest. I'd like to request that each discussant from the floor use the microphone and identify himself before addressing his comment or question. The session is being recorded and we will include the tape script in the final proceedings of this symposium. With that I'd like to turn the discussion to Dr. Malins who will carry on with some of this thoughts. Don?

Malins: Thank you Dr. Wolfe. I have prepared to take somewhat the protagonist's position which Dr. Wolfe has suggested because of the intent to stimulate the audience to thought. I'm perfectly happy in that position. I have a number of ideas which of course run around in my mind as being involved in this matter of fate and effects. A lot of them are really concerns with areas which I think we are, for whatever reason, avoiding. I want to go through some of these, and I don't promise you they are in any particular order, but they are thoughts which come to mind from time to time. I think we're moving generally from concerns of acute toxicity and uptake and depuration studies into concern with long-term effects. And I think when we get into the area of long term effects, the picture of course gets very complicated indeed. It requires experts who would be working hopefully in a multi-disciplinary way to solve the problems which we face. I think we in the Environmental Conservation Division like to take somewhat of a clinical approach. In other words we approach the matter of whether petroleum is altering the life processes of organisms from the point of view of biochemistry in terms of enzymes and formation of metabolic products, possible incidence of disease, changes in immune responses and things like that.

Specifically I'd like to tell you about my views on the question of what is petroleum. I'm concerned that we have taken a very, very limited view all along as to what is petroleum in these fate and effect studies. It seems to me that petroleum in the practical sense, but perhaps not in the intellectual sense, has been treated as though consisting of only those hydrocarbons which we can analyze by gas chromatography and confirm from time to time by mass spectrometry. Now let's see what that amounts to. This really amounts to the alkanes and some of the alicyclics and takes us up to approximately 5 or perhaps 6 benzenoid rings. Yet if we introduce a functional group into those molecules so they don't gas chromatograph too conveniently, they seem to cease to be petroleum contamination in the organism because we don't look for it. It's true that we pick up the occasional phenol but beyond that point we miss just about everything. And that's a shame because many of the marine organisms which we're dealing with rapidly degrade these structures to other compounds. I think we must start looking at metabolic products. We know from terrestrial animal studies that metabolic products are considered to be some of the most noxious of the aromatic structures that can invade tissues and chemical systems. So I think that's a very important consideration for the future. I'm concerned as to what the animal is in fact receiving in the real world. What does petroleum, in fact, become under conditions of further oxidation and the conditions of microbial oxidations eloquantly described by Dave Gibson at this meeting? What, in fact, are these animals getting and how toxic are these products of degradation in the environment to various organisms? I'm also very concerned that there's a tendency to work with microbes with aromatics without considering that there may be tremendous changes in their biological effects when these molecules, existing in parts per billion, parts per million, are perhaps binding, forming covalent complexes with a host of macromolecules in the marine environment. We know from our own studies that there are tremendous differences with larvae in terms of metabolism of these structures. I'm very concerned about the fact that there are tremendous possibilities for synergism where other chemicals enter the picture and affect the degradation of the aromatic hydrocarbons. I think we've got to learn more about this, the effects of PCBs and metals and a whole host of other chemicals which we can imagine.

As far as the enzyme studies which you've heard about here are concerned, I think they should continue, and I think we should try to get some evaluation of the capability various organisms have for degrading aromatic hydrocarbons. There's obviously great diversity from Mytilus which seems to do nothing to a trout which degrades hydrocarbon at various levels, i.e., the various enzyme conversions are accomplished by the latter organism but not by the mollusk. I do urge that we press on with this matter of metabolites with the capabilities which exist now; we have strong capabilities now, in fact, to look at metabolite profiles in fish. We have the mass spectrometry, the GLC, and high pressure liquid chromatography, and the ability to make derivatives. These profiles can be accomplished.

I'm also concerned with the small residues. Although a great deal of hydrocarbons can be depurated from marine organisms, we know that quite a significant amount can remain in tissues. I'd like to know how much of that is associated with key biochemical systems such as DNA and with certain proteins and membranes. With regard to the ultrastructure I think this is very good work such as the electronmicroscopy and I think it should continue but it must have the backing of biochemical investigations and other investigations to determine how, in fact, some of these anomolies are appearing. We must attempt to confirm by other methods other sorts of anomolies which exist. I think it would be very nice to be able to take a look at some of the electron micrographs under conditions which animals are going through natural stresses, such as molting and through maturation. A number of the processes which involve mutagenic changes have very, very long latent periods and whatever we can do to hurry up these processes will give us some idea as to what may happen. I'd very much like to see some more work done in the rarely touched area of electrophysiology which would be an attempt to see how perhaps the petroleum hydrocarbons interfere with chemical cues. In marine organisms chemical stimulae are tremendously important in behavior. I'd like to see behaviorial studies continued and I'd like to see some good electrophysiology continued.

With regard to species I think this is a very difficult question. I'm not sure that one can prescribe a procedure for determining which species should be looked at but I think a rather broad spectrum approach should be taken. I personally feel that larval forms and immature forms may be particularly susceptible to some of the petroleum hydrocarbons and I'd very much like to see more work done in those areas. Studies which are being conducted in the laboratory and looking at biochemical, physiological and other parameters should, in fact, at the earliest possible time be taken to field to be tested to see whether, in fact, what we're learning in the laboratory has application in a field situation. I think conceptual models are very important. I think that good experimentation, which sometimes one does not always see, regretfully, should be introduced into our scheme of events. I think we've got to start getting time-dependent studies which are based on a multidisciplinary mode. I think this will give us one of the better opportunities for seeing potential damage if in fact problems should exist. I think I'm going to end up by saying that one of the most neglected areas is the food web. I think it's fair to say from Dr. Teal's review of the subject we still need very much to know, for example, how metabolites are transferred through the food web. I'd also very much like to know how persistent are the metabolic products which seem to be formed in quite large amounts in some species. Will these depurate or do they continue to remain in the tissues long after the hydrocarbons have largely left the organism.

Clayton McAuliffe (Chevron Oil Field Research Co.): I would like to make a few comments concerning some of the metabolic processes and products transformed through food chains and other things that Dr. Malins has proposed as areas for study. I fail to see where one can differentiate these sorts of processes for petroleum from all the other myriad organic materials that also undergo metabolic processes. I think you end up in a situation where you have very little opportunity to successfully conclude such studies in a definitive manner. Further petroleum has been a common substance as well as biogenic hydrocarbons throughout geological time, therefore, I fail to see why one would attempt to specifically isolate petroleum as a source of concern.

Malins: I disagree I'm afraid with your position that this cannot be done. We have plans without our laboratory where we hope to start some of these studies using radio tracers. In fact, that's how it's done and that together with sophisticated instrumental techniques which we have are going to allow us to get some ideas about this. You can use double-labeled systems, you can work with a binary system and follow the radioactivity from organism to another. I think you have to know something about the organism and if, in fact, it is tending to degrade these aromatics. I'm not going to get into it now, but I'd be very happy to demonstrate how this can be done on a blow-by-blow basis.

McAuliffe: But I think I raised the point why be more concerned about petroleum than you would other metabolic processes from all sorts of materials that undergo various conversions and likewise that are in the food web of marine organisms and others.

Malins: That's like saying why be concerned with cancer when you have heart disease, it seems to me.

Virginia Stout (National Marine Fisheries Service): You mentioned something about the big envelope of unknown compounds in petroleum. I was a little concerned to see people mainly emphasizing water soluble compounds. I come from the field of chlorinated hydrocarbons and I really know very little about petroleum hydrocarbons, I came here to learn. But there is evidence that chlorinated hydrocarbons, although they are extremely insoluble in water, many of them in the range of only parts per trillion in sea water, for instance, do cross membranes, both respiratory and gastrointestinal membranes. They do accumulate in organisms throughout the food chain and they depurate very slowly in many cases. I think that this big envelope of unknown and difficult-to-handle compounds very probably are causing many effects that we are not looking at. And in many cases flow-through systems for highly insoluble compounds have been developed for chlorinated hydrocarbons and presumably these methods could be used in petroleum studies as well.

Malins: Yes, I quite concur with those general comments and I am afraid that due to time I didn't mention the fact that I think that the sediments are very important to the animals which essentially cannot escape a petroleum impregnated sediment. In other words sediments may be an area which we should have primary concern with. We know very little, it seems to me, about those aromatics, let alone the metabolites which don't conveniently come to us by gas chromatography. In other words those substances which are not conveniently volatilized at temperatures used in gas chromatography tend to escape unnoticed. There are ways of handling them where you don't have to worry so much about volatilization. It may very well be that you can't resolve everything but by a method such as high pressure liquid chromatography and other techniques it would be possible to isolate certain fractions which would have certain different characteristics which I think could be tested independently of each other. In terms of toxicity I very much would like to see that happen.

Wolfe: I would like to go on from this interesting discussion to allow Dr. Mertens to express some of his thoughts.

Mertens: Thank you, Doug. As chairman of the American Petroleum Institute's Committee on the Effects on Fate and Effects of Oil on the Environment, I've been especially interested in this conference. This API Committee has been in existence for some 6 or 7 years and we've been funded at something like about 3/4 of a million dollars a year in a rather comprehensive integrated program. We're trying to cover the whole ballgame right from the beginning as far as fate is concerned right down to the ultimates as near as we can identify it at this point on biological effects.

I would agree with Dr. Malins on many of his points and I don't really disagree too much with any of the others. The ones that I particularly am in agreement with are matters of acute toxicity. I think we have fairly well consulted that part of the overall problem. I don't think additional acute toxicity studies are going to be all that worthwhile. Carcinogenicity of course is, as I see it, about the ultimate problem of them all. And for Dr. Malins' comment of correlating laboratory and field studies I couldn't agree more. I think we need to do more in the field and try to relate back to the laboratory. We should conduct both efforts simultaneously.

My impressions from the sessions that we've had, however, are not so much what are the research problems that we should be working on but what are the problems of research. We need to get realism into our experimental conditions and our exposure conditions. It doesn't really mean too much if in our laboratory work we're using fresh oil or if we're exposing at concentrations much higher than exist in the field or for durations of time that are much longer than one would expect in the field. I think we need to get such realism into our conditions so that we will have then greater relevance of results. Now many of the papers that have been given in the last couple days have been very excellent papers if they are looked at from the standpoint of mechanistic studies. But there are too many of these papers where you simply cannot take the results and say this is then what we can expect from the field, and this is where I would agree very wholeheartedly with Dr. Malins' comments. Now, it's very difficult to try to devise laboratory conditions that reflect what the field conditions are.

This morning, Dr. McAuliffe reported on some results of experimental oil spills. He reported that, within an hour or so all the benzene was gone from the water column and within 6 hours everything else was gone from the water column. Soon after that I was listening to some papers where they were exposing benzene for 96 hours. Well, if it's gone in one hour what are we doing with the other 95 hours of exposure? This is what I'm talking about when I say we have to emphasize the realism of results. The same applies, for example to the use of dispersants. You can't use a dispersant and then maintain a constant volume of water any longer. You have to allow for dilution as a function of time, otherwise there's no point in using a dispersant.

As far as the major problems that I see that we need to do work on, I would emphasize again the need to do field studies and I think we should take more advantage of the natural laboratories that we have, the natural seep areas. People working on the effects of different types of petroleum fractions or even the whole petroleum itself on different physiological characteristics of fish or shell fish, could very well at this point go to a natural seep area or to an offshore platform where there are a lot of fish that make these platforms their home. They could sample these fish and shellfish to see whether they see in these field organisms the same thing that they have observed in the laboratory.

In API·in our research work we have what we call a five-year plan. I've been taking many notes on what I've been hearing the last couple days because we are in the midst of revising our five-year plan for about the third year now. We revise it annually. We want to give prime importance to the study of carcinogenicity and here again we're starting out in the field. We're not starting out in the laboratory. That's the way we're approaching the work. Now, API isn't against laboratory work. As a matter of fact, a lot of the papers that were presented here the last couple of days were labroatory studies conducted under API sponsorship and indeed we will continue to do laboratory work.

I think one of the major areas that has not been touched enough is the matter of uptake of hydrocarbons by sediments and then the role of the sediments as a reservoir for subsequent contamination and exposure to marine organisms. This we have to do a lot more work on and this is an area where API will be zeroing in on a lot more than we have in the past. We think we ought to be taking a look at something like population dynamics, for example. And looking at the effects of dispersants. We need to do more laboratory work on this.

Finally, my one last item I would say is, and there's quite a bit of material available on this subject, the matter of tainting. A lot is known about tainting but so far as I know no one has ever tried to relate tainting to terms of an oil spill conditions. This we need to do fairly quickly. It should not be too difficult but it is an area that I believe has been overlooked just a little too long.

<u>Maurice Stansby</u> (National Marine Fisheries Service): With respect to what Dr. Mertens was saying and what we've been saying throughout the meeting about the need for both field studies and laboratory studies, I think we should go quite a distance in trying to take the results of the field studies to modify the types of laboratory approaches that we use. I think we've been all too rigid in the past, for example in the case of bioassays. It's customary to use a constant concentration. I think that just as in gas chromatography where we use temperature programming, we ought to be able to take the results such as Dr. McAuliffe has described to us and set up concentration programming to simulate what goes on in the natural environment. I'm sure there are many other ways, as we get more and more information from these practical field tests, that we can make the laboratory tests a lot more meaningful.

Mertens: I certainly would agree but I do want to emphasize that I'm not criticizing the laboratory work that's been done and has been reported but I do consider this to be very, very good mechanistic work that has been reported here at the symposium.

Lou DeSalvo (Naval Bioscience Lab, Berkely): I've been concerned with the fact that a lot of attention's been focused on spills and the fact that the spill's there and then it goes away - in fact, it probably does. I don't argue with these results but we have places where we have a constant spill going on, such as San Francisco Bay. There's a constant input, an integrated input coming down from the rivers, coming down from municipal sewage effluent, street runoff, and refineries. We all know the standard suite of places where we have outfalls. In San Francisco Bay, we have a system that is well mixed throughout. In our work we found that the sediments in San Francisco Bay are heavily loaded with various types of hydrocarbons. The animals that are exposed to this and the fisheries related to these sediments in the bay are affected by what I would consider to be an integrated, fairly steady-state system. Based on the urbanization, it's probably been increasing for a long time. I think that if we look back historically we can see that, for instance, the oyster industry peaked out in 1899 with 3 million pounds of meat which was produced in the Bay and it went down to about 0 in 1920. There were 13 salmon canneries in San Francisco Bay and the salmon went kaput. Dungeness crab started to decline in 1957. Dungeness crab migrate in and out of the Bay, and the life history of the dungeness crab is built around the young crabs moving through the sediments of the Bay. I've just got recent word that Striped Bass fishery is declining around San Francisco. All this is occurring in a system that I feel is receiving an integrated constant spill and analogous to a constant steady-state spill. I think research emphasis should be placed on chronic long-term inputs of hydrocarbons.

Mertens: I'd like to make a few comments on that. In the San Francisco Bay situation, yes, there has been quite a bit of oil getting into the sediments over a good number of decades at this point. But also a lot of other materials have been going in. So, it's going to be rather difficult to separate out what effect the oil in the sediments has had on the marine life there as compared to these other materials. And this is here again a good reason for going to another area where there's been a chronic exposure to oil over say centuries or so on, for example, a seep area around Coal Oil Point in Santa Barbara. That area down there does have its influx of municipal discharge as well as San Francisco Bay but it isn't nearly to the degree that we have in San Francisco Bay. There isn't nearly the manufacturing down there to modify the area. So if you take a place by Santa Barbara and compare it to a study in San Francisco Bay, I think you could begin to see how much of a factor the oil and sediments in San Francisco Bay have.

Allen Michael (University of Massachusetts): I agree somewhat with the previous speaker and I also can see the problem is that places like New York Harbor and San Francisco have a lot of other contaminants which pose a serious problem. I happen to think the chronic pollution is the issue we should face. The problems with seeps, of course, is that it tends to be a crude oil or a weathered oil which comes out of these areas. I think laboratory experiments and even field oil spills have demonstrated that it's really the refined products that tend to have greater effect. These more refined products are more likely to wash into systems like New York Harbor from land runoff. There are ways of attacking this, such as a field experimental approach. For example, I currently have an experiment going in the New York Harbor system where we have taken relatively clean sediment, brought it to the surface and innoculated it with different types of oil and then put it back down on the bottom. We are looking at the effects *in situ* of oil alone because we are well aware of the fact that any changes we see in the benthos of the New York Bight is probably the result of about fifty different pollutants, not one. Following on from there, I think we could attack these sorts of problems with field experimental approaches in such places where there's chronic pollution.

What I would like to see is some effort to develop models of cycling through the system. There's no question we're going to be using petroleum hydrocarbons for many years to come, and in spite of the fact that there are high levels of hydrocarbons in the sediment in the New York Bight, no doubt there is considerable cycling going on. If we want to get some predictabilities as to how pollutants may build up in the system, I think that a very worthwhile area of research would be doing some work on aquatic inputs: how pollutants get into the sediment, how the animals churn the sediment over and pump the material back out, because there's no doubt that there's a great deal of cycling going on.

Wolfe: Are there other comments? I'd like to move on and hear Dr. Button's remarks.

Button: I think I'd like to suggest a middle ground between those who advocate laboratory studies and those who have been talking about field studies. I think that middle ground really ought to be theoretical work. I think it's time to move, at least to some degree, from cataloging of the *status quo* of hydrocarbons and hydrocarbon distributions, to one of the mechanism of hydrocarbons movement, i.e., the mechanism of hydrocarbon action and the theory of these movements and these actions on a molecular basis. You can take any number of cases you wish---solubility, stability of emulsions, the mechanism of removal through sedimentation that's been discussed a number of times, the mechanism of bioaccumulation or of transport into biological tissues. We need to examine the kinetics of the turnover, how concentrations affect the velocities of hydrocarbon movement into these various sinks, and the thermodynamics that control these stead-state levels. We know that these steady-states happen, that they're perturbed by various influences of spills and what-not, but we really can't currently predict the concentrations at which these steady-states will arise or the composition of those steady states. Now going along hand-in-hand with this, of course, has to be development of analytical chemistry. I've been deeply impressed in this meeting with those developments over the past 2 or 3 years.

Wolfe: Are there comments on that?

It seems to me that the question of kinetics is related also to the question of which compounds in petroleum we should be most concerned about in terms of biological effects. It seems to me that Clayton McAuliffe's comment earlier about the chronic introduction of petroleum into the ocean is very pertinent here. The compounds we should be addressing in effects studies are probably those compounds for which the rates or avenues of introduction will be most drastically changed relative to the capacity of the environment to accommodate to that change. Such compounds, which are likely to be trace constituents of petroleum, may go unmetabolized in biological systems because the rates of introduction and accumulation have never necessitated it. I think that question might also be addressed by Dr. Malins in connection with which compounds require attention as far as metabolic byproducts are concerned.

Malins: Yes, I agree that we need an understanding of these processes. As far as the compounds are concerned, I think that we probably want to continue on at least with a little bit of work really which has gone on with isotopically labeled compounds. I think that the napthalenes and the alkylated napthalenes would be an area of considerable interest. As you know, Jack Anderson pointed out at this meeting that these structures apparently tend to accumulate rather preferentially in a number of tissues. I think we could work with the napthalenes and the alkyl-substituted napthalenes, that's the methyl napthalenes and possibly with benzopyrene. The only problem from a laboratory point of view is the carcinogenicity of benzopyrene, which causes some problems, I think benzanthracene, might be another compound we might want to look at. There are a number of compounds which in fact can be purchased and this would make a good beginning but I would hope at the same time that while such studies are in progress we are making some effort to find out which ones in the marine environment should be possibly given preferential study.

Mertens: May I make a comment. Don Button called for development of a model, I believe that's the most simplistic way I can describe it, and I certainly would agree with this. This has got to be the ultimate goal. I just simple wish to relate the experience we've had in API. We've had a project for developing a model on fate of oil, without including biological effects. To do so would be far more complicated. But just the different factors that have to do with the fate of oil, for example, evaporation, dissolution, dispersion, sedimentation, photooxidation, and so on, make a complex problem. We've been at this for some three or four years now at the University of Southern California with Ron Kolpack as the principal investigator. He's identified something like 15 different processes with about 45 different parameters. This gets to be a rather complex model as you can see. Needless to say, we haven't attained our goal yet. We think we have parts of that model that work reasonably well but we've got a lot more work to do and by the time we can eventually plug in all the biological factors into it, we're looking down the road quite a while. But that doesn't mean we should stop looking towards our goal.

Keith Hay (American Petroleum Institute): A lot of the money behind much of the research that is being done here is coming from the Bureau of Land Management based upon the development of the Outer Continental Shelf and effect of oil operations on the marine biota in that area. I don't know of too much impact that's been demonstrated from offshore development or even oil spills on fishery populations in the open marine environment. We've seen a number of studies concerned with fish; but compared to natural perturbations under open sea conditions, the impact of OCS development and even oil spills on fishery populations, I think is a very minor consideration, very frankly. I think we're seeing the real impact of oil on marine animals in confined areas like San Francisco Bay where most of the oil that gets into the marine environment comes through municipal outlets. This situation has nothing to do with those OCS operations, incidentally. I think that more emphasis is going to have to be put on chronic conditions in cases like that where organisms and pollutants where these may be are confined in estuarine areas, where there is a continuous input, and influence from refineries. I was shocked to see effects being tested by intermuscular injection of petroleum in trout, for example. What relationship that has to OCS development or any real world situation I just can't imagine.

Malins: This sort of approach is definitely worthwhile to the OCS programs I can assure you, because the question at hand was to determine what products, in fact, formed. Intraperitoneal injection is an excellent way of doing it by avoiding the gut and eliminating a number of complicating factors to see what you, in fact, get. And these sorts of studies showed in these animals for the first time the spectrum of metabolic products which the enzymes are capable of producing. It also showed in a feeding animal that these compounds can be deposited in tissues right throughout the body. Now this seems to me to be a very important beginning. Now, it is true, that it is not the beginning and the end of the matter but it certainly is a good start in my judgment so I'd strongly take exception to the sort of casual way in which you refer to such a study.

A. Prakash (Environment Canada, Ottawa): I'm speaking to you not as an expert in petroleum hydrocarbons but as a marine scientist involved in the development of policy regulations for the protection of coastal environment. For that I have to depend on scientific research information and scientific criteria developed by you people. I have enjoyed the last two days' deliberations and have learned a lot. My concern at the moment is a few comments which were made in Dr. Malins' presentation. He rightly pointed out that we're moving from a focus on acute toxicity to long-term effects of oil spills. We all know that oil imposes a certain amount of stress which is reflected in changes in physiological mechanisms, changes in behavioral patterns, induction of cytopathological, and pathological changes. This is all very fine but I was just wondering if those people who are involved in this type of study should not go a step beyond these observations to interpret what these changes mean in terms of survival, especially of commercial important species in the coastal zone. Also, what do spills mean in terms of changes in the community structure, since we know that animals and plants are very intimately interdependent on each other? These are some of the things which I think those who are working in the mechanistic respects of oil should address themselves to.

Malins: I think when you say the people who are involved in mechanistic studies should be doing this, perhaps you're asking too much of this particular group of individuals. If you try to do a good job, say on a biochemical study, one would hope that you would be able to interface as well as possible with a number of people doing other things. So I would agree with you in terms of the concerns that are in fact important, but I'm not sure that you can place the burden upon specialists who are in certain areas such as in biochemistry. It may be just too much. Problems of chronic effects are extremely complicated in many cases, and this is often not realized. I remember someone who came to our facility who had a very high position in Washington and we mentioned some of the work of Dr. McCain on tumors and he asked me "Well, when are you going to get this solved? Are you going to get it solved in the next three months? Is it going to last a year?" So you see, I had to tell him as politely as possible that the matter of tumors in fish is not any more simple because fish are low on the phylogenic tree than it is for man. It's a very complicated matter.

Mertens: I'd like to add just a little bit to Dr. Malins' comments in response to the question whether a biochemist should be able to solve a problem in a rather short time. I'm simply going to state something here that is quite self-evident to everybody, that this is a tremendously complicated problem. In solving it we have to have quite a diverse number of disciplines represented and involved in working out the problem. For example, just take a simple thing like what type of exposure to use in laboratory testing. A good biologist can do the work but the biologist needs to know what is the probable level, and what are probable level, and what are probable fractions, and what is the probable time of exposure. That kind of advice can't come from the biologist, of course, it needs to come from a very good chemist working hand in hand with the biologist. And that's just a very simple example.

Wolfe: We really must move on now to hear from Jack Anderson.

Anderson: I think the first thing I want to say is to reiterate the necessity of bringing together all of the various types of physiological studies, perhaps biochemical studies and even some field investigations. We should attempt to bring these to some common ground so that we know basically how they interrelate and try to get this confused assortment of data down to some reasonable level that we can begin to predict and understand. My next plea is that in future studies monitoring of tissue levels should be done with as many hydrocarbons as possible, regardless of exposure duration, and regardless of the oil used. Of course, this of necessity means that biologists must get together with people with good analytical expertise, work on the problem together, try to make some sense of their own data and hopefully try to relate their data to those produced in other laboratories. With that it's my hope that one could begin to look at the time of exposure, the dose given, the hydrocarbons present in the exposure medium and the effects elicited in the organism and come to some common ground. I think I've seen at least some hope of that in some of our data.

I think that we should begin with very detailed analyses of the tissues of whatever organism was used. We should start at that point and begin to watch the flux of those particular compounds from the water or from the sediment through the organism and find out how long that material lasts in the tissues. I think there are enough good analyses being conducted now where we can begin to see a fairly broad range of hydrocarbons in tissues of organisms. We've seen some of those at this symposium. I would hope that we'd also spend more time on the hydrocarbons that stay in the tissues longer. That's been our excuse for spending time with napthalenes. There are others like phenanthrenes and some of the four- and five-ring compounds that seem also to remain for some time. I would hope that we'd spend more time on those and less time on compounds that we see in sediments or that we see in water which may or may not pay off from the standpoint of biological investigations.

I think the point of metabolic products is a good one. I just don't feel that we're quite ready for that yet. Perhaps we're ready for a beginning but my feeling is we don't have a good enough handle on the original compounds yet. If it can be done at the same time, then that's fine.

We've talked about why we should look at long-term versus short-term tests and my point there is, what are we trying to simulate? Are we trying to simulate a spill in deep water where we have proved that the oil is not going to be very long or should we simulate the chronic exposure system such as mentioned for harbors where there is a constant input? At least we should try to design our work around a particular matural situation and keep that in mind. Surely we shouldn't design a constant flow system with concentrations of 1 ppm or 10 ppm or greater because I don't think there are very many instances where we would see that in the real world. I do believe that the long-term exposure is going to tell us more and it'll also give us a better handle on the hydrocarbons presumably.

As to which parameters should be measured, in my view, judging from the results that we've seen in biological studies I would still promote growth and reproduction studies from larval stages of invertebrates and fish. I don't think we've really seen much on behavior studies in terms of petroleum hydrocarbons. I would expect to see these as fairly sensitive indicators of effects although we really don't have enough information yet.

I certainly want to underline sediment work. We certainly don't see anything in the literature on bioavailability and we need to be concerned as the sediment hydrocarbon levels are generally two orders of magnitude or greater higher than we see in the water column. In polluted areas people have tried again and again to find the hydrocarbons in the water column and it's just too big a problem, so we're going to have to know how much of that material in the sediment is really available to the organism. With respect to which compounds should we worry about, my answer would be those that get into the organism; rather than spending all the time describing hydrocarbons in that sediment.

Another argument we often come up against is should we do laboratory, or field experiments? We've heard about a few field experiments and those seem to have paid off quite well in that there is at least a touch of the real world there and natural rates of degradation, release of hydrocarbons are going to occur even if it's a confined area. It seems to me that we should try to work toward that kind of controlled field experiment. I think the combination of studies in marsh areas, intertidal zones and big bags in the sea might give us a lot more information and more meaningful information. Before you design those kinds of experiments, however, you need good laboratory data. Probably as a result of those field experiments you're going to have to go back to the laboratory and try to answer some more questions that have come up as a result of your findings. That kind of moving back and forth from field to laboratory experiment I think is going to be the fruitful way to go. Admittedly from the field biologists we know that there are so many fluctuations in the parameters in the field that it takes years to really get a feel for the population in a given area. It doesn't seem reasonable to spend a great deal of time trying to characterize an area when you don't know if there's ever going to be spill there. You might have an extremely good handle on an area and spend a great deal of time doing it and really end up with very nice baseline data but until you see a spill in that area probably can't say very much about it. I'm not saying we shouldn't do baseline studies but I think we know how hard they are to do and the number of years required to get good information. I think the field experiment would be a little better because we add the oil under controlled conditions and we can see what happens.

Finally, I'd like to address the differences between the Alaskan OCS and studies elsewhere. The only thing that seems to make conclusions somewhat different in the Alaskan work as compared to others is the fact that volatiles aren't as volatile in cold water and at cold air temperatures. So I would think that cold water studies conducted elsewhere would directly relate to effects in the Alaskan OCS.

<u>H. Hofelich</u> (Seattle Wild Bird Clinic): One thing I noticed in this symposium is that a great deal of the studies have been concentrated toward the microenvironment, parts per million, parts per billion, things like that. Well, I'd just like to interject that in the marine ecosystem you also have a lot of animals that suffer only once in an acute oil spill. It only happens once and they die from it and that's what happens with a lot of our seabirds.

When they hit an oil spill they never come out of it, they never get a change to experience the chronic oil spills. These aspects might be very difficult to do any real experiments on. The point is that there is another side of the ecosystem and another side of the food web that reacts very quickly to acute changes associated with oil spills. Also, Washington state does have rather cold water. The solvents like benzene would probably stick around a lot longer here in this particular type of ecosystem. If benzene stays there for a long time and the birds come in contact with it, they may exhibit lung hemorrhage, and damage to the kidneys and the liver. These birds can suffer from long-term concentrations of these really volatile hydrocarbons.

Wolfe: I concur that these effects on birds are still very poorly understood, even though they are probably the most visible socially and politically of any effects.

Malins: I might point out that two individuals who are very well informed in these areas may very well be in the audience, Dr. Holmes and Dr. Cronshaw from California.

Usha Varanasi (National Marine Fisheries Service): Dr. Atlas presented a paper on oil biodegradation in the Arctic environment and he said that not only is the rate of degradation reduced but also that both n-alkanes and branched alkanes are oxidized at about the same rate. That would mean that in Arctic regions the oil remains more like a fresh oil than a weathered oil. If we are going to look at the effects under Arctic conditions, we perhaps have to use fresh oil instead of weathered oil because you see that may have some different consequences.

Anderson: I agree. I think the important thing to look at there would probably be sediment work again, i.e., oiled sediments. We know very little about the alterations of hydrocarbons in sediments over long time periods and what the effects of those might be on the organisms whether it's the effect on the organism or the accumulation in the tissue of the organism. I think there is where the temperature variable might be very important.

Wolfe: Thank you. I'd like to cut this short and invite a different view, that of an information user. Dr. O'Neil, Environmental Protection Agency.

O'Neil: Since we've been hearing about very detailed oil research for the last two or three days, it occurred to me maybe a different view might be appropriate at this point in the program. I think right along with all of the research that goes on there is a corrollary need as to how do we get that information into the decision-making processes that are going on. These processes tend to be mission oriented than basic oriented. From my perspective I'm in technical support in a regulatory agency and that's somewhere between the research group and the program people, the decision-making people. Time and time again we see research work which is probably very good and very correct; but for a whole variety of reasons, it's not at the right place at the right time to influence what may be very far-reaching and long-range decisions. These reasons may be as simple as the format that the report is put in or failure to recognize the timing or understanding the process that goes on. Many people in the research community say that's a failing on the part of the management side of these agencies, i.e., we did it but somehow they don't use it. That's a valid criticism but I think there's an equally valid criticism which can be pointed at the research community.

Sometimes researchers don't go very far in trying to help the manager and help the resource director or the decision maker to use that information. I think there are a couple of steps perhaps that all of us in the technical end of the business, i.e., researchers and technical support people, ought to be thinking about in order to get the most of our dollar. I think in the oil area or in the toxic substances area, we really need to get a better understanding of just what is the process by which major decisions in this area are made. What influences those decisions? We can have major research programs and it may turn out that the decisions are made on some piece of technical information perhaps. It may not be the technical information where most of the research money is going, however, because the managers don't really know what to do with that information. I think in many agencies, and ours is probably no exception, in many resource management types of situations the individual in charge can tell you he would like more information in this area, or in that area. I would guess it's probably equally true that when you get right down to the specific; if you could present him with that specific information, he's not exactly sure how he's going to use it. He can think in terms of general categories of information. He hasn't really thought through, or gone through the process of how he's going to use the specifics in that area. I think this is where the research community could do a great service to the other side, in helping these people verbalize and think through how that would be factored into their decision-making process. I'm firmly convinced in many cases that the mass of information that descends on some of these people when they have to make a decision, is just beyond assimilation because preconditioning, preplanning, has not been built into the process. We really need to get the most out of the limited amount of dollars that are available, and there obviously has got to be a balance between basic research and applied research. The research community's efforts on oil, because of it's visibility and political sensitivity, are no exception.

There'll probably be more research money available longer if this education process of some of the people in the decision-making role occurs rather than simply handing them reports that for whatever reason are never put into the form where the individual that really has the clout or that really makes the decision can use it. I really feel if all of us did our homework perhaps on that side as well as the detailed research plan, or the detailed analytical measurements, then the effects of our research and our work would be far more reaching. In the long run we would end up with better use of our resource dollars, less misdirected research, and probably much better decisions in the long run.

Clayton McAuliffe: I'd like to amplify a little bit on that. We have to consider what we're doing these exercises for and whether, once we have identified problems what we can do about them. I think we have examples in our current program, where we're spending tremendously large amounts of funds for studies in the outer continental shelf which accounts for only approximately 1% of the hydrocarbon contribution to the marine environment. We've also heard allusions to the bays which receive larger quantities of petroleum hydrocarbons, perhaps 40% of the total contribution, to a much smaller body of water, whereas the offshore areas are large bodies of water. We should attempt to identify the most adverse conditions such as birds in the nearshore area where oil strands, i.e., the bays and the estuaries, and then address our effort to correcting those areas, reducing the input of hydrocarbons if they are indeed causing problems---to the extent that you can do it economically. We should not try to reduce concentrations below the normal contribution from seepage, erosion of source rocks containing petroleum and the contribution of biogenic sources (some of which are similar to petroleum hydrocarbons. We must be realistic about what we can accomplish. If we can do something, fine. If you cannot do anything no matter what you find with reference to the effects of petroleum, if you can't do anything about it, you have to live with it.

O'Neil: Well, I guess I couldn't agree with you more. I have the feeling quite frequently that where the breakdown occurs is that the depth of research and degree of detail goes beyond what the normal decision maker in that agency, given the time that he has and the pressures that he's under, will be able to use. Quite frequently they might well be able to make a strong regulatory decision based on less information properly presented and properly interpreted to them.

Garvin Bucaria (Alaska OCS Office, Bureau of Land Management): I would heartily agree with Dr. O'Neil's comments about how we get this information to the decision makers. I find that environmental impact statements leave something to be desired. I find very useful such documents as the one by Evans and Rice (1974. Effects of oil on marine ecosystems: a review for administrators and policy makers. U. S. Natl. Mar. Fish. Serv. Fish. Bull. 72:625-638). I think perhaps we could use additional such documents periodically to update the general public and decision makers relative to possible concerns. These summary type of documents offer a possible means of coming to a consensus among people who have been working in a particular area maybe even across disciplinary lines. There's the other problem of ranking areas in terms of Dr. McAuliffe's statements. Our baseline is a rather relative thing. The 1% figure he cited was in terms of the entire United States. Relative to Alaska, however, any spill in the last few years was more than occurred before and it's a cumulative thing. The offshore may not be identified specifically as contributing a great deal of hazard, yet we know so little about it and we find it's a very difficult area to sample. The offshore is tied to the inshore, oil development is tied to the amount of economic activities that take place in any given area, therefore, everything is tied together. When we're talking about Alaska, we're talking about areas that are extremely productive relative to other areas, other seaports, other fisheries. We're talking, for instance, in the Kodiak area, about a fishery which is dependent on immediately nearshore waters. It does not include the tremendous vast fisheries that are being harvested by the foreign effort. When compared to other areas in the lower 48, we're talking about fisheries species which often times travel great distances and so we're involving values, or at least products which come from a large geographic area. So I would say that in terms of some of these chronic effects, acute effects, in places like the Bering Sea, the Western Gulf of Alaska, (off Kodiak, that is), we have some immediate concerns. We're presently dealing with an area that has had very little perturbation and the chances are that things will occur in the future. Some of us are extremely concerned about this, just the thought of a hazard, even if it is very low. If there are chronic potentials, how long will it take for these to show and then by the time they show can we do anything about it? This in my view indicates that we should be conservative relative to protecting the resources and this may cost money.

Wolfe: I think that in the interests of the hour we'll stop the discussion now. I want to thank all the panel members and the audience for staying till such a late hour and expressing so lucidly their ideas on this complex problem. Before we adjourn I'd like to make a few announcements. I'd like very much to thank the Planning and Editorial Committee who set up the symposium and are working on the manuscripts which have been submitted. I'd like to thank the Local Arrangements Committee and particularly the projectionists, Tom and Beth Arpke and Suanne Smith, as well as all the session chairmen who helped make this symposium a success and I'd also like to thank you, the audience, for joining us here and participating in this symposium. I guess we'll see you next at the Oil Spill Conference in New Orleans or at the symposium on "Recovery Potential of Oil Spills" next October in New Bedford. Meeting's adjourned.

LIST OF SYMPOSIUM PAPERS PRESENTED AND AUTHORS

Invited Papers

Harrald, J.R., Diane Boyd, and C.C. Bates, Oil Spills in the Alaskan Coastal Zone - The Statistical Picture.

Shaw, D.G. Hydrocarbons in the Water Column.

Kinney, P.J. Oil Slick Movement in Alaskan Waters.

McAuliffe, Clayton D. Oil Slick Weathering.

Gibson, David T. Biodegradation of Aromatic Petroleum Hydrocarbons.

Malins, D.C. Bioconversions and Metabolism of Petroleum Hydrocarbons.

Lee, Richard F. Accumulation and Turnover of Hydrocarbons in Marine Organisms.

Teal, J. Food Web Transport of Petroleum Hydrocarbons.

Rice, S.D. Comparative Toxicities of Petroleum Hydrocarbons in Marine Organisms.

Anderson, J. Sublethal Effects of Petroleum Hydrocarbons.

Hawkes, Joyce, Morphological abnormalities Produced by Hydrocarbon Exposure.

Michael, Allan P. Ecological Effects of Petroleum in Marine Systems.

Biological Effects

Mills, E.R., and S.M. Ray, Marine phytoplankton and petroleum: Influence on growth rates, chlorophyll levels and recovery from toxic effects.

Alexander, V. Preliminary results of studies on the toxicity and effects of petroleum hydrocarbons on marine phytoplankton in Alaskan coastal waters.

Steele, R.L. Effects of certain petroleum products on reproduction and growth of zygotes and juvenile states of the alga, Fucus endentatus De la Pyl.

Korn, S., D.A. Moles, and S.D. Rice, Effects of temperature on acute toxicity of benzene, naphthalene, and the water-soluble fraction of Cook Inlet crude oil to Pink Shrimp.

Hall, L.W., A.L. Buikema, and J. Cairns, The effects of Number 2 fuel oil and a simulated refinery effluent on the Grass Shrimp and the Pinfish.

Hodgins, H.O., W.O. Gronlund, J.L. Mighel, J.W. Hawkes, and P.A. Robisch, Effect of crude oil on trout reproduction.

DeVries, A.L. The effect of naphthalene on survival and protein synthesis of Bering Sea fishes.

Kleerekoper, H. The locomotor and orientation response of two species of fish and the sea turtle to the water soluble fraction of Louisiana crude (API ref. oil no. 2).

Kooyman, G.L. The effects of oil contamination on the conductance of marine mammal pelts, especially fur seals, and its effects on diving behavior.

Albers, P.H. Effects of external applications of crude oil on hatchery of Mallard eggs.

Szaro, R.C. and P.H. Albers, Effects of external applications of oil on the eggs of the common eider (Somateria mollisima).

Carr, R.S. and D. J. Reish, The effect of petrochemicals on the survival and life history of five species of Polychaetous annelids.

Fries, C.R. and M.R. Tripp, Cytological damage in Mercenaria mercenaria.

Laughlin, R.B. and J.M. Neff, The effects of chronic exposure of no. 2 fuel oil on hatching, growth, development rate and respiration of Limulus polyphemus.

Percy, J.A. Effects of dispersed crude oils upon the respiratory metabolism of an arctic marine amphipod Onisimus affinis.

Busdosh M. and R.M. Atlas, Toxicity of petroleum to arctic amphipods.

Tatem, H.E. Toxicity and physiological effects of petroleum hydrocarbons on estuarine grass shrimp Palaemonetes pugio.

Caldwell, R.S., E.M. Calderone, and M.H. Mallon, Effects of seawater-soluble fraction of Alaska crude oil and selected chemical components on the larval stages of the Dungeness crab, Cancer magister, Dana.

Smith, M.A. and M.B. Bonnett, Effects of crude oil exposure on King Crab (Paralithodes camtschatica) gill morphology.

Mecklenburg, T.A., S.D. Rice, and J.F. Karinen, Effects of Cook Inlet crude oil water-soluble fraction on survival and molting of King Crab (Paralithodes camschatica) and Coonstripe Shrimp (Pandulus hypsinotus) larvae.

Forns, J.M. Methods development and initial evaluations of crude oil flow-through bioassays with first through fourth-stage larval lobsters Homarus Americanus.

Taylor, T.L. and J.F. Karinen, Response of the clam Macoma balthica (L), exposed to Prudhoe Bay crude oil as unmixed oil, water-soluble fraction, and sediment-absorbed fraction in the laboratory.

Thomas, M.L.H. Long-term biological effects of Bunker C oil in the intertidal zone.

Naidu, A.S. and H.M. Feder, Effects of Prudhoe Bay crude oil on a tidal-flat ecosystem in Port Valdez, Alaska.

McCain, B.B., S.R. Wellings, C.E. Alpers, M.S. Myers, and W.P. Gronlund, Baseline data on the health of fishes from Alaska's outer continental shelf.

Spies, R.B. and P. Davis, Ecological studies around natural oil seeps in the Santa Barbara Channel.

Straughan, D. Biological survey of intertidal areas in the Straits of Magellan in January 1975, five months after the Metula oil spill.

Pequegnat, W.E. Petroleum in deep benthic ecosystems of the Gulf of Mexico and Caribbean Sea.

Bioaccumulation and Metabolism

Atlas, R.M. Studies on Petroleum Biodegradation in the Arctic.

Arhelger, S.D., B.R. Robertson, and D.K. Button, Arctic Hydrocarbon Biodegradation.

Ho, C.L. and T. Karim, Impact of Surface Adsorbed Petroleum Hydrocarbons on Organisms.

Anderson, J.W. and L.J. Moore, Bioavailability of Sediment-Sorbed Naphthalenes to the Sipunculid Worm, Phascolosoma agassizii.

Harris, R.P. Uptake and Metabolism of Hydrocarbons by Copepods.

Fucik, K.W., J.M. Neff, and M.L. Byington, The Uptake of Petroleum Hydrocarbons by the Clam, Rangia cuneata, Under Varying Conditions of Temperature and Salinity.

Stainken, D.M. The Chemical Accumulation and Depuration of No. 2 Fuel Oil by the Soft Shell Clam, Mya arenaria L.

Gruger, E.H., and M.M. Wekell, Effects of Chlorinated Biphenyls on the Induction of Aryl Hydrocarbon Hydroxylase by Petroleum Hydrocarbons.

Bieri, R.H. and V. Stamoudis, The Fate of Petroleum Hydrocarbons from a No. 2 Fuel Oil Spill in a Seminatural Estuarine Environment.

Meyer, P.A. Effect of Drilling Operations on the Hydrocarbon Content of Crustaceans.

Kiceniuk, J.W. Effect of Oil Dispersants and Crude Oil-Dispersant Mixtures of Heart Rate of Resting Fish (Tautogalobrus adspersus).

Measurement, Movement and Distribution of Hydrocarbons

Wise, S.A., S.N. Chesler, H.S. Hertz, and W.E. May, Interlaboratory calibration for the analysis of petroleum levels in sediment.

Reed, W.E. and I.R. Kaplan, Criteria for identification of sources of hydrocarbon pollutants.

Templeton, W.L. and R.M. Bean, Stable carbon isotope ratios in biota and sediments from Lake Maracaibo, Venezuela.

Smith, Craig L. Determination of the leeway of oil slicks.

McAuliffe, C.D. Evaporation and solution of C1-C10 hydrocarbon from crude oils on the sea surface.

Brooks, James M., Bernie B. Bernard and William M. Sackett, Input of low-molecular weight hydrocarbons from petroleum operations into the Gulf of Mexico.

Farrington, John W. Hydrocarbons in water, organisms, and surface sediments of the western North Atlantic.

Cheatham, D.L., S.J. Way, J.W. Short, and S.D. Rice, Effects of temperature, volatility, and biodegradation on persistence of aromatic hydrocarbons in seawater.

MacLeod, W.D., D.W. Brown, R.G. Jenkins, and L.S. Ramos, Intertidal hydrocarbon levels at two sites on the Strait of Juan de Fuca.

Bean, Roger W. and J.W. Blaylock, Characterization of volatile hydrocarbons in flowing seawater suspensions of Number 2 fuel oil.

Lytle Julia S. and Thomas F. Lytle, Sediment hydrocarbons as environmental indicators in the NE Gulf of Mexico.

Huang, C.P. and H.A. Elliott, Stability of emulsified crude oils as affected by suspended particles.

Mix, M.C., R.T. Riley, K.I. Kings, S.R. Trenholm, and R.L. Schaffer, Distribution and abundance of environmental chemical carcinogens from petroleum products in economically-important bivalve molluscs from Oregon bays and estuarines.

Calder, John A. Seasonal variations of hydrocarbons in the water column of the MAFLA Lease area.

Burns, Kathryn A. and Jonathan L. Smith, Distribution of petroleum hydrocarbons in Westernport Bay, Australia.

SUBJECT INDEX

A

Accommodation 19-20, 26-28, 79, 332, 338
Adenosine tri-phosphate, 270,272
AHH, see Aryl hydrocarbon hydroxylase
Alaska 1-7, 270-274, 463
Alcaligenes eutrophus 41
Algae
  effects of hydrocarbons 130, 138, 240-1, 256
Alkanes
  dissolved 434
  particulate 437-8
  sediments 405, 407-10
Alkylnaphthalene, see Naphthalene
American Petroleum Institute (API) 456-7
Amphipods
  effects of dispersions 192-200
  respiration 194-7
Analysis of hydrocarbons 82, 90, 334-5, 348-9, 388-9, 398-9, 433, 455, 457, 460
Analytical Intercalibration 345-350
Anas platyrhynchos, 158
Arctic ocean
  microflora 261-6, 270-4
Aromatics (see also Hydrocarbons)
  accumulation 313-321, 338, 340
  in clams 313-21
  in crude oil 217
  dissolved 331, 434
  evaporation 363
  fractional toxicities 84-6, 90, 219
  metabolism 36-45, 47-49
  metabolites 55, 455-6
  oxidation by microorganisms 38-44
  in oysters 338, 340
  particulate 438-9
  in seawater 21-24, 341, 434
  in sediments 342-3, 406
  in slicks 366
  solubility 12
  toxicity 86, 218-9
Arrow 239
Aryl hydrocarbon hydroxylase 49-50, 52, 323-8
  analysis 325-7
  induction 52, 323, 327-8
  occurrence 49-50
  pH optimum 326
  and temperature 325

B

Bacteria
  hydrocarbon metabolism 25-6, 36-45
Beaufort Sea 261-3
Beijerinckia sp., 42-3
Benthos, effects of oil 130-1
Benzene
  in brines 375, 380
  and growth 216
  and larval development 215-6
  metabolism 40
  in seawater 21
  toxicity 86, 213-4
Benzo(a) anthracene
  metabolism 48
  in oysters 66
  turnover 66

Benzo(a) pyrene, 73, 459
  metabolism 39, 42-3, 50
  in shellfish 421-9
  substrate for AHH 325
Benzpyrene hydroxylase, 50
Bioccumulation of HC
  crustacea 61, 202-3
  fish 65, 115-6
  molluscs 60-5, 306-8, 313-21, 338
  polychaetes 40
  zooplankton 37
Bioassay 25, 78, 81, 83, 457
  algal sensivity 138
  methodology 81, 84
  polychaete mortality 168-73
  statistics 84
Bioavailability 276-84
Biodegradation of petroleum 78, 135, 261-8, 270-4, 462
  nutrient requirements 261, 265, 270
  effect of temperature 26
Biphenyl chlorinated (PCB)
  effects on AHH 323-31
  synergisms with HC 323, 327-30
Birds
  fuel oil and eggs 158-63, 164-7
  effects of spills 19, 30, 131, 461-2
Bivalves (see also Clams, Effects, Mussels)
  as biological indicators 67
  cytology 174-81
  effects of phenol 174-81
Butane 378-9

C

Calanus helgolandicus (see also Copepod) 286
Callorhinus ursinus 152, 154
Cancer magister, see Crab, dungeness
Capitella capitata 133, 172
Carcinogen 421, 429, 456-7
  Casco Bay (Maine) 133
Chedabucto Bay (Nova Scotia) 20, 129, 131, 135, 238-45
Chlorinated biphenyls, see Biphenyl
Chromatography (see also Gas chromatography)
  column separations 334,388-9
  gas 335, 348-9, 388-9, 398-9, 433
  high pressure liquid 455
  and mass spectrometry 332, 335, 348-9, 388, 399
Cirriformia spirabranchia, 171
Clams (see also Bivalves)
  accumulation of hydrocarbons 306-8, 313-21, 342
  benzo(a)pyrene in 1, 7-8
  burrowing activity 230-34
  cytological effects 174-81
  depuration of hydrocarbons 306-8, 319-20
  effects of oil on 174-81, 229-36, 242
  filtration rates 308
  naphthalene in 306-8
Colloidal hydrocarbons 12-14, 26-8
  electrophoresis of 414-15
Cook Inlet crude oil (see also Crude oil) 221
  aromatics in 217
  effects 215-16
  toxicity 213, 219

Copepods 133
 and naphthalene 286-304
Corexit 195
Crab
 benzene toxicity 213-4
 crude oil toxicity 88, 90, 210-20, 221-28
 dungeness 86, 210-20
 effects of hydrocarbons on 182-90, 210-20, 221-8
 horseshoe 182-90
 king 88, 221-28
 larval rearing 222
 growth 186-7, 215-6, 223-6
 molting 215-6, 223-6
 naphthalene toxicity 213-4
 respiration 188-9
 survival 88, 90, 185, 223, 225
Crassostrea virginica
 hydrocarbon turnover 336-8
Crude oil (see also Cook Inlet, Prudhoe Bay crude oil)
 characteristics 193, 353, 414
 composition 38, 370
 dispersions 19-20, 27-8, 30, 193, 195
 effects 88-90, 98-9, 105, 107-8, 129-35, 138-41, 169-71, 192-200, 247, 256-8
 emulsions in seawater 19-20, 26-8, 413-20
 and growth 105, 107-8, 138-41
 pour point 193
 and reproduction 107-8, 170-71
 and respiration 98-99, 194-7
 in sediments 132, 252-3, 255
 specific gravity 193
 uptake of naphthalene from 307
 viscosity 193
 wind drift of slicks 351-62
Crustacea (see Amphipod, Copepod, Crab)
 hydrocarbon accumulation 61
Ctenodrillus serratus
 reproduction 171
Cytochrome P450 39, 48-9
Cytology
 hydrocarbon effects 115-28
 phenol effects 174-81

D

Depuration (of hydrocarbons), 25, 63-6, 74, 102, 202-3, 282, 289, 295-6, 300, 302, 306-8, 319-20, 332-43
Dihydrodiol formation 41
Dispersants 26-8, 30
Dispersions 79
 crude oil in seawater 19-20, 27-8, 193, 195
 in cold water 28
Dissolution
 of oil from slicks 19-22, 30, 363-71
^{14}C-Dodecane 270, 272

E

Effects of oil
 on algae 130, 138, 240-41, 256
 on amphipods 192-200
 on AHH 323, 327-8
 on behavior 103-4
 on Benthos 130-31
 on Birds 19, 30, 131, 158-63, 164-7
 on Bivalves 87, 89-90, 102-3, 174-81, 229-36, 242
 on burrowing activity 230-34
 in cold water 135
 on copepods 286-94
 on crabs 86-88, 90, 104-6, 182-90, 210-20, 221-28
 on cytology 115-28
 on enzymes 323, 327-8, 454-5
 on fish 87, 89-90, 98-9, 103-4, 115-26, 143-9, 323-31
 on Fucus 138-41, 242
 on growth 104-8, 138, 158-63, 186-7, 215-16, 223-6
 on hatching 147, 158-63, 164-7, 202, 204-5
 on intertidal zone 130-31
 on mammals 131, 151-6
 on marsh plants 130-31, 243, 247, 257-8
 on microfauna 133
 on osmoregulation 101-2
 on plankton 61, 130, 286-304
 on polychates 86, 88-90, 98, 105-6
 on reproduction 104-7, 138, 171
 on respiration 98-101, 188-9
 on shrimp 86-91, 96-7, 99-101, 104-5, 201-6, 221-8
 on survival 86-91, 96-8, 146-7, 158-63, 164-7, 168-73, 185, 213-4, 218-9, 221-8
Effect of Salinity
 on naphthalene uptake 305, 307, 309
Effects of temperature
 on evaporation 184
 on hydrocarbon metabolism 25, 53
 on hydrocarbon retention 293
 on AHH activity 325
Electron microscopy 115-28, 175-9, 455
 clam tissues 176-9
 fish tissues 115-28
 scanning 115
 transmission 115, 176-9
Electrophysiology 455
Emulsions
 crude oil-seawater 19-20, 26-8, 79, 413-20
 and ionic strength 419
 particle size 417
 stability 413-20
 and suspended particles 20, 413-20
Endoplasmic reticulum 119.126
Enhydra lutris 152
Environmental Protection Agency 462-3
Enzymes (see also Aryl hydrocarbon hydroxylase)
 analysis 325-7
 epoxide hydrase 48, 50-1
 β-glucuronidase 55
 glutathione-S-transferase 52,54
 hydrocarbon-metabolizing 36-46, 47-59
 induction 52, 323, 327-8
 pH optima 326
 temperature effects 325
Epoxide hydrase 48, 50-1
Ethane
 in seawater 377
Ethene
 depth distribution 377
 in seawater 377
Eurytemora affinis (see also Copepods) 286
Evaporation of hydrocarbons 22-3, 78, 169, 363
 effects of temperature 79-80, 184

F

Fish
  electron microscopy of 115-28
  histology 115-28
  hydrocarbon accumulation in 65
  hydrocarbon toxicity 86-7, 89-90
  sublethal effects 98-9, 103-4, 115-26
  tainting 60, 131
Fletcher ice island 271, 274
Fluoranthrene 394
Food chains 71-6
Food and water as sources of hydrocarbons 294-5
Fucus
  effects of petroleum 138-41, 242
Fuel, jet
  effects on algal growth 138-41
Fuel oil, No. 2 96-7
  accommodation in seawater 332, 338
  aromatic hydrocarbons in 340
  bioaccumulation 202-3, 313-21, 332-43
  characteristics of 353
  depuration 202-3, 313-21, 332-43
  effects of 88, 90, 98-103, 105-8, 138-41, 151, 154, 159-61, 165-6, 169-71, 185-8, 204-5
  emulsions in water 314
  in sediments 342
  solubility in seawater 402
  suspensions in seawater 397-402
  wind drift of slicks 354-7, 359

G

Gas chromatography (see Chromatography)
Gas chromatography - mass spectrometry 332, 335, 348-9, 388, 399
Gas venting
  hydrocarbon composition 378
  β-Glucuronidase 55
Glutathione-S-transferase 52, 54
Grooming (of fur) 156
Growth
  effects of hydrocarbons 138, 158-63, 186-7, 202, 204-5, 216, 223-6
Gulf of Mexico
  dissolved hydrocarbons 20, 381, 433-6
  hydrocarbon inputs to 383
  light hydrocarbons 381
  oil fields 382
  particulate hydrocarbons 436-8
  sediment hydrocarbons 406

H

Hatching
  effects of fuel oil 158-63, 164-7, 202, 204-5
Heterotrophic microorganisms 25-6, 262, 264
Histology of clams 174-81
  crude oil effects 115-28
  eye lens 125
  gill 117-8
  liver 121-2, 124
  mucous glands 120
  phenol effects 174-81
  skin 120
Hydrocarbons
  analytical methods 332, 334-5, 345-50, 388-9, 398-9, 433
  bioaccumulation 73-4, 313-21, 336-9, 341-2
  in brines 375, 38
  in Calanus 71
  in clams 342
  colloidal 12-14
  in copepods 71-2, 74
  depuration 63-5, 74, 290-4, 297-9, 313-21
  dissolution 19-22, 30, 363-71
  dissolved 433-6, 450-1
  effects on AHH 323-31
  evaporation of 22-3, 79-80, 363-71
  in food chains 71-7
  in gases 378-9
  low molecular weight 24, 363, 373-84
  metabolism 36-46, 47-59, 261-8, 455-6
  micelle formation 13
  monitoring 67
  in mussels 437-8
  in oysters 336-9
  particulate 15-16, 436-8
  photo-oxidation 19, 26
  relative toxicities 84-6
  in sediments 132, 278-80, 347, 390-5, 404-12, 449, 451, 456-7, 460, 462
  selective accumulation 72
  in sipunculids 279, 282
  solubility 8, 12
  sublethal effects 95-111
  synergisms with PCBs 323, 327-30
  turnover rates 110
  uptake kinetics 341-2, 458-9
  uptake from food vs. water 73-4
  in worms 279, 282

I

Indicator species 67, 229, 442, 451-2
Intertidal effects of oil 130-31
Intercalibration 345-50

K

Kuwait crude oil 247, 258
  in sediments 252-3, 255

L

Lanugo 154
LC50 78, 86, 88, 90, 95-8, 109, 169-72, 221, 223
Leeway of oil slicks 351-62
  and slick volume 352
  and waves 352
  and wind velocity 352, 354-7
Leptonychotes weddelli 152, 154
Limulus polyphemus
  growth 186-7
  effects of fuel oil 182-90
  respiration 188-9
  survival 185

M

Macoma balthica

burrowing activity 230-34
  effects of crude oil 229-36
MAFLA 404, 432-40
Mammals 131, 151-6
Mass spectrometry 332, 335, 348-9, 388, 399
Meiofauna 133
Mercenaria mercenaria
  cytology 174
  effects of phenol 174-81
  histology 176-9
  hydrocarbon accumulation 342
Metabolism of hydrocarbons 36-46, 47-59,
    261-9, 455
Metals
  effects in fish histology 123, 126
  distribution 305
Methane 376-7
Metula 135, 247-8, 250
Microorganisms
  population estimates 270-71
Micelle formation 13
Mollusks (see Bivalves, Clams)
Molting (crustacea) 226
  effects of crude oil 221, 223-5
Mousse 20, 247, 253, 258
Mussels 133, 247, 249, 256-7
  benzo(a)pyrene in 421, 427-8
  as biological indicators 67, 442, 447-8,
    451-2
  hydrocarbons in 257, 442, 447-8
Mutagenesis 455
Mya arenaria (see also Clams) 133, 242
  fuel oil turnover 313-21
Mytilus (see Mussels)

N

Napthalene 172, 323-4, 459
  assimilation 289, 295-6
  bioaccumulation 56, 66, 203, 276-84,
    306-8, 313-21
  bioavailability 276-84, 294-5
  biological half-life 313, 320
  complexation 56
  in copepods 74, 286-304
  depuration 66, 102, 282, 289, 295-6,
    300, 302, 306-8, 313-21
  effects 86, 215-6
  effect of temperature on evaporation
    80, 184
  metabolism 43, 55, 74-5
  in oysters 66, 102
  persistence in tissues 320
  in seawater exposure medium 310
  sediment-sorption 278-80
  in shrimp 102
  sublethal effects of 96-7
  toxicity 86, 213-4
  in worms 279, 282
Naphthalene-1-C^{14} 286-304
  food vs water as source 294-5
Naphthalene Uptake
  effect of temperature 305, 307, 309
  effect of salinity 305, 307, 309
New York Bight 132-3, 458
Nitrogen
  assimilation/retention 298
  requirement for biodegradation of
    petroleum 261, 266

O

Odobenus rosmarus 152, 154
Oil (see Petroleum, Fuel oil, Hydrocarbons)
Oil field brines
  discharge locations 382
  hydrocarbon composition 375, 380
Oil in ice 7-8
Oil seeps 457-8
Oil slicks (see Oil spills)
Oil Spills
  in Alaska 1-8
  numbers of 4-7
  sources of 2. 4. 6
  in United States 2, 4, 7
  volumes of 4, 5-7
Oil Spills
  Chedabucto 129, 131, 135, 238-46
  dissolution 19-22, 30, 363-72
  ecological effects of 129-35
  evaporation 22-3, 363-72
  experimental 21-4, 332-3, 363-71
  leeway of 351-62
  Metula 135, 247-8, 250
  spray injection 19
  spreading 19
  tanker Arrow 238-9
  Torrey Canyon 129, 131-2
  trajectory 351-62
  West Falmouth 129, 131, 133-5
  wind drift factor 358
Oil-water suspensions 26-8, 81, 84-5, 397-402
Oncorhynchus (see Salmon)
Opportunistic species
Onisimus affinis (see Amphypod)
Ophryotrocha puerilis
  reproduction 171
**Oysters (see also Bivalves, Clams,
  Crassotrea)**
  benzo(a)pyrene in 421, 427-8

P

Palaemonetes pugio 96-7
  effects of fuel oil 88, 201-6
Pandalus hypsinotus
  effects of crude oil 88, 90, 221-8
  molting 223-6
  rearing of larvae 222
  survival 223-5
Paralithodes camtschatica
  effects of crude oil 88, 221-8
  molting 223-6
  rearing of larvae 222
  survival 223, 225
Parophrys vetulus 115
Particulate hydrocarbons 15-16
PCB (see Biphenyl, chlorinated)
Penaeus sp. 88, 97
Petroleum (see also Hydrocarbons)
  bioaccumulation 115-6
  biodegradation 261-8
  biodegradation potential 270-74
  cycles in environment 36, 458-9
  effects 78-91, 158-63, 164-7, 240-43
  inputs to ocean 1-2
  monitoring 67, 442, 451-2
  origin 2, 36

persistence in environment 79, 131, 240
relative toxicities 84-5
weathering 131, 135, 368-71
<u>Phascolosoma agassizii</u> (see <u>Sipunculida</u>) 276-285
Phenanthrene 86, 323-4, 461
 in sediments 394
Phenol 174-81
Phocid seals 152
Phosphorus
 assimilation in copepods 298
 requirement for biodegradation 261, 266
Photo-oxidation 19, 26
Photosynthesis 130
Plankton, effects of oil 61, 130, 286-304
<u>Platichthys stellatus</u> 115
Pollution Incident Reporting System (PIRS) 1-7
Polychaeta 86, 88-90, 133
 hydrocarbon accumulation 64, 277
 metabolism of HC 64
Port Valdez (Alaska) 271, 274
Pristane 432, 436-7
 uptake 73-4
 in water 74
Pristane/Phytane ratio 71, 347
Propane
 in coastal water 377
<u>Protothaca staminea</u> 305-11
Prudhoe Bay crude oil 277
 biodegradation 261, 264-6
 effects 116-23, 143-9, 151-6, 229-37
 oiled sediment preparation 231
<u>Pseudomonas putida</u> 40-41, 43
Psychrophilic microorganisms 262
Psychrotrophic microorganisms 262
Puget Sound 385
Pyrene in sediments 394

R

<u>Rangia cuneata</u> 305-12
 filtration rates 308
Regulatory agencies 462-3
Reproduction
 effects of fuel oil 141, 171
Respiration
 effects of fuel oil 188, 194-7, 202-4
Retention of hydrocarbons
 feeding level and 294
 lipid content and 290-92, 298
 sex and 297
 starving and 293, 299
 temperature and 82, 293

S

Salinity
 effect on naphthalene uptake 305, 307, 309
Salinity shock
 and fuel oil effects 186, 189
<u>Salmo gardneri</u> (see Trout, rainbow) 115
Salmon
 AHH in 325-8
 coho 115
 effects of PCBs and hydrocarbons 323-31
Sea lion 151, 154
Sea otter 151-2, 154

Sediments
 composition 406
 hydrocarbons in 132, 252-3, 255, 280, 342, 347, 391-5, 406-10
 naphthalenes in 280
 paraffins in 280, 392-3
 petroleum interactions 231, 456-7, 460, 462
 unsaturated hydrocarbons in 391, 394-5
Sedimentation of oil 20
Shrimp 96-7
 effects of crude oil 86-91, 221-28
 effects of fuel oil 88, 90, 201-6
 larval rearing 222
 molting 223-6
Sipunculida
 naphthalene accumulation 276-284
Sole 115
<u>Somateria mollissima</u> 164
<u>Spartina alterniflora</u>
 effects of petroleum 243
Squalene
 in seawater 433-40
Strait of Juan de Fuca 385, 387
Straits of Magellan 247-8, 257
 intertidal species abundance/composition 254-5
Sublethal effects (see Effects)
 compared to lethal levels 109, 454-5

T

Tainting 60, 131
Tanker <u>Arrow</u> 238
Tanker <u>Metula</u> 247-8, 250
Tar 15
 as food for organisms 16, 75
Temperature
 and AHH activity 325
 and biodegradation 25
 and hydrocarbon retention 293
 interactions with effects 82, 186, 188
 and napthalene loss from solution 79-80, 184
 and napthalene uptake 305, 307, 309
Thermoregulation 132, 151-6
Thermal conductance of oiled fur 151-6
<u>Torrey Canyon</u> 129, 131-2
Toxicity of HC (see also Effects)
 Adults/larval sensitivity 87-9, 91, 107-8
 to various species 88-91
Trajectory of slicks 351-2
Trout, rainbow 115
 effects of crude oil 121-5, 143-9
 hatching success 147
 histology 121-5, 145-6
 maturation 145
 survival of eggs 146-7
Turnover (see also Bioaccumulation, Depuration) 110

U

United States Coast Guard (PIRS) 1-2

W

Walrus 151, 154
Weathering of petroleum 19, 131, 135, 368-71

West Falmouth spill 129, 131, 133-5
Westernport Bay (Australia) 442-53
Wind drift factor 358

Z

Zooplankton (see also Copepods)
  hydrocarbon accumulation by 61, 286-304
  effects of oil 130

AUTHOR INDEX

A

Ackman, R.G. 76
Adamson, R.H. 330
Ahokas, J.T. 330
Aizenshtat, Q. 411
Albers, P.H. 158, 164, 167
Alexander, V. 491
Allen, A.A. 30
Aller, C. 440
Alpers, C.E. 466
Ambike, S.H. 44
American Petroleum Institute 30, 207, 343, 395
Anderson, G.E. 244
Anderson, J.W. 30, 67, 91, 95, 111, 127, 172, 190, 199, 207, 219, 227, 236, 276, 284, 303, 311, 321, 343, 371, 403, 452, 454
Anderson, R.D. 67, 111, 343
Andrews, L.J. 16
Arcos, J.C. 57
Argus, M.F. 57
Arhelger, S.D. 270
Atema, J. 111
Atlas, R.M. 30, 227, 261, 268, 274, 465
Avigan, J. 76
Avolizi, R.J. 111, 199, 244

B

Baggi, G. 44
Bainbridge, R. 303
Baker, E.G. 16
Baker, J.M. 16, 30, 244, 259
Barbier, M. 303, 440, 452
Barnes, H. 303
Barnhart, R.A. 330
Barr, A.J. 311
Barry, M. 430
Bates, C.C. 1, 7
Baumberger, J.P. 190
Bean, R.M. 91, 397, 403, 466
Becher, P. 420
Becker, P. 30
Bellrose, F.G. 167
Bend, J.R. 57
Benville, P.E. 91
Berdugo, V. 286
Bernal, J.D. 16
Bernard, B.B. 373, 384
Berridge, S.A. 259, 360, 420
Beynon, L.R. 30
Bieri, R.H. 332, 344
Birkhead, T.R. 162
Blackman, R.A.A. 67
Blanton, W.G. 127
Blaylock, J.W. 276, 321, 397
Blumer, M. 16, 30, 44, 67, 76, 135, 180, 321, 344, 384, 411, 420, 452
Boehm, P.D. 16, 30, 68, 321, 403
Boesch, D.F. 135
Bohon, R.L. 17
Boney, A.D. 430

Bonnett, M.B. 466
Bookhout, C.G. 219
Borneff, J. 44
Boyd, B.D. 1, 7
Boylan, D.B. 219, 321, 344, 403, 452
Bradley, M.P.T. 395
Bray, E.E. 395
Bray, J.R. 259
Brenniman, G. 91
Brockson, R.W. 111, 199
Brodersen, C.C. 92, 227
Brooks, J.M. 373, 384
Brookes, P. 57
Brown, D.W. 41
Brown, R.A. 17, 31, 207, 227, 303, 411, 440
Bryden, M.M. 156
Buchanan, D.V. 219, 227
Buck, J.D. 274
Buckley, S.E. 384
Buhler, D.R. 330
Buikema, A.L. 465
Burdick, G.E. 149
Burns, K.A. 31, 57, 68, 76, 208, 303, 442, 452
Burwood, R. 57
Busdosh, M. 465
Butler, J.N. 17, 31
Butler, M.J.A. 244
Button, D.K. 17, 31, 270, 274, 454

C

Cairns, J. 465
Caldarone, E.M. 210
Calder, J.A. 411, 432, 441
Caldwell, P.J. 162
Caldwell, R.S. 92, 111, 210
Canevari, G.P. 31
Canonica, di.L. 44
Cardwell, R.C. 127
Carlson, C. 180
Carlson, G.P. 68
Carr, R.S. 168
Castellini, M.A. 151
Catelani, D. 44
Catterall, F.A. 44
Chan, G.L. 31, 135
Cheatham, D.L. 92, 227, 236, 466
Cheng, T.C. 180
Chesler, S.N. 345, 350
Chia, F. 92
Christensen, D.J. 430
Cimberg, R. 31
Clark, R.C. 68, 311, 321, 395, 411, 441, 452
Clarke, M.R. 76
Coan, E.V. 236
Colla, C. 44
Collier, T.K. 57
Conney, A.H. 330
Conover, R.J. 17, 31, 68, 76, 127, 303, 420
Coon, J.M. 57
Corner, E.D.S. 31, 57, 68, 76, 208, 286, 303

Couch, J. 127, 430
Cowell, E.G. 141
Cox, B.A. 68, 111, 208, 285, 311
Craddock, J.E. 76
Cundell, A.M. 31, 430
Cunningham, P.A. 311
Currier, H.B. 219

D

Dagley, S. 44
Daly, J.W. 45
Darrow, D.C. 76
Davis, C.C. 111
Davis, J.B. 31, 45
Davis, P. 466
Davis, R.W. 151
Dean, R.A. 420
Dean, Raymond D. 31
Defrank, J.J. 45
De Frenne, E. 45
De Pierre, J.W. 330
De Vries, A.L. 466
De waide, J.H. 57, 330
Diaz-Piferrer, M. 135
DiSalvo, L.H. 31, 68, 311, 321, 430, 452
Dixit, D. 112
Doebler, H.J. 360
Dooley, J.E. 45
Dorn, E. 45
Doss, M.P. 31
Doudoroff, P. 92, 227
Dow, R.L. 135, 236, 321
Drent, R.H. 156
Dunbar, M.J. 135
Duncan, J. 274
Dunn, B.P. 68, 430
Dunning, A. 199, 244
Duppel, W. 45

E

Eganhouse, R.P. 17, 31
Eglinton, G. 344, 411
Ehrhardt, M. 68, 321, 344, 430, 452
Eisenberg, D. 17
Eisler, R. 141
Elliott, H.A. 413
Elworthy, P.H. 17
Emery, K.O. 259
Emery, R.M. 92, 227
Epifanio, C.E. 112, 190, 227
Evans, D.R. 92, 208, 430

F

Fallah, M.H. 32
Farley, C.A. 430
Farrington, J.W. 136, 311, 344, 395, 411, 430, 452, 453, 466
Feder, H.M. 466
Ferris, J.P. 45
Finnerty, K.R. 76

Finney, D.J. 92, 227, 236
Fishbein, L. 330
Fitzgerald, D.E. 32
Fitzgerald, L.M. 360
Floodgate, G.D. 344
Foley, D.A. 180
Forns, J.M. 466
Forrester, W.D. 321
Fossato, V.U. 32, 68, 69, 285, 311, 453
Foster, M. 136
Fouts, J.R. 330
Frank, D.J. 384
Frank, H.S. 17
Frankenfeld, J.W. 303, 321
Franks, F. 17
Frazier, J.M. 311
Freegarde, M. 303
Friberg, S. 420
Fries, C.R. 174
Frisch, J. 156
Fritch, F.E. 141
Fucik, K.W. 190, 305, 311

G

Gabbiani, G. 127
Galtsoff, P. 181, 244, 321
Gardner, G.R. 127
Garrett, W.D. 360
Gatellier, C.R. 32
Gearing, P.J. 412
Gelboin, H.V. 330
Gelpi, E. 412
George, J.D. 172
Gibson, D.T. 32, 36, 45, 275
Giger, W. 45, 412, 430
Gilfillan, E.S. 32, 112, 136, 190, 199, 244, 245
Glaeser, J.L. 32
Goethe, F. 136
Goldberg, E.D. 17, 69
Gordon, D.C. 17, 112, 136, 227, 236, 321, 441, 453
Gordon, J.E. 17
Gorman, M.L. 167
Grassle, J.F. 136
Greenwood, J.J.D. 167
Grice, G.D. 453
Grimes, D.J. 76
Gronlund, W.D. 143, 149, 466
Gross, A.O. 162, 167
Grover, P.L. 57
Gruenfeld, M. 92, 227, 322, 420
Gruger, E.H., Jr. 57, 323, 330, 331
Grund, A. 275
Guillard, R.R.L. 303
Guinasso, N.L. 384
Gump, B.H. 345
Gunn, B.G. 384
Gunnerson, C.G. 259
Gutsell, J.S. 331

H

Haedrich, R.L. 76
Hagerman, L. 227
Haider, G. 127
Hall, D.G. 17
Hall, L.W. 465
Hammel, H.T. 156
Hamelink, J.L. 344, 453
Hann, R.W. 259
Hanson, H.C. 162
Hanson, H.P. 32
Hardy, R. 69, 76
Harger, J.R.E. 69
Hargrave, B.T. 112, 199, 245
Harrald, J.R. 1
Harris, M.J. 17
Harris, R.P. 286
Harrison, W. 32, 371
Hart, J.S. 156
Hartung, R. 162, 167
Harvey, G.R. 453
Hawkes, J.W. 112, 115, 127, 143, 149
Hedges, J.I. 412
Heinle, D.R. 303
Hela, I. 360
Hermann, R.B. 17
Hertz, H.G. 17
Hertz, H.S. 344, 345
Hiatt, R.W. 236
Higgenson, J. 430
Hites, R.A. 344
Hnatiuk, J. 32
Ho, C.L. 466
Hodgins, H.O. 57, 127, 143
Hogg, R. 420
Högn, T. 45
Holcomb, R.W. 331, 384
Holden, A.V. 77
Honjo, S. 77
Hood, D.W. 17
Hopkins, S.H. 311
Horn, M.H. 17, 77
Horowitz, A. 268
Huang, C.P. 413, 420
Huberman, E. 45
Hublon, W.F. 331
Huggett, R.J. 311

I

Iliffe, T.M. 441, 453
Irving, L. 156
Ishio, S. 430
Iverson, R.L. 441

J

Jackson, B.W. 453
Jacobson, S.M. 112
James, M.O. 57
James, R. 361
Jamison, V.W. 45
Janicki, R.H. 112

Jeffrey, A.M. 45
Jeffrey, G.A. 17
Jeffrey, L.M. 304
Jeffrey, P.G. 32
Jeffries, H.P. 77, 136, 181
Jenkins, R.G. 385
Jerina, D.M. 46
Joensen, A.H. 167
Johanson, E.E. 32, 371
Johnson, R.W. 412
Jones, E. 430
Jørgensen, C.B. 69

K

Kallio, R.E. 269
Kanter, R. 322
Kaplan, I.R. 466
Karim, T. 466
Karinen, J.F. 78, 92, 221, 227, 229, 236
Karrick, N.L. 57
Kash, D.E. 7
Kator, H. 92
Katz, L.M. 92, 112, 219, 227, 331
Kauss, P.B. 32, 92, 136
Kavanau, J.L. 17
Keevil, B.E. 199
Keith, L.N. 127
Kenyon, K.E. 361
Kenyon, K.S. 156
Keulegan, G.H. 361
Kiceniuk, J.W. 496
Kiesser, S.L. 276
Kilvington, C.C. 286
King, K.I. 421
Kinne, O. 112
Kinney, P.J. 32, 275, 465
Kinsey, D.W. 32
Kinter, W.B. 112
Kleerekoper, H. 465
Klein, A.E. 32
Knackmuss, H.J. 46
Kolpack, R.L. 32, 136, 361
Koons, C.B. 32
Kooyman, G.L. 151
Kopischke, E.D. 163
Korn, S. 69, 92, 465
Kraybill, H.F. 431
Kreider, R.E. 32, 361, 371
Kühnhold, W.W. 92, 112

L

Lance, G.N. 259
Langley, R. 149
La Roche, G. 92
Laughlin, R.B., Jr. 182, 190
Ledet, E.J. 33
Lee, R.F. 33, 58, 60, 69, 77, 112, 172, 181, 208, 275, 304, 322, 344, 412, 431, 453
Leitritz, E. 149
Levy, E.M. 304, 441
Lewis, E.G. 228

Lewis, E.L.   7
Lidman, U.   58, 331
Lie, U.   311
Lien, T.R.   17
Linden, O.   112, 208
Litchfield, J.T.   172
Litterest, C.L.   58
Lockwood, A.P.M.   228
Long, F.A.   18
Lowry, O.H.   331
Ludwick, J.C.   412
Lyklema, J.   18
Lysyj, I.   92
Lytle, J.S.   404, 412
Lytle, T.F.   404, 412

M

MacArthur, I.K.H.   361
Macek, K.J.   77, 149
Mackay, G.D.M.   33
Mackay, G.R.   127
Mackie, P.R.   69, 77, 412, 431
Mackin, J.G.   181
MacLeod, W.D.   236, 385, 396
Malaveille, C.   58
Malins, D.C.   47, 58, 396, 454
Mallon, M.H.   210
Mandelbaum, H.   361
Mangum, C.P.   190
Marshall, S.M.   304
May, W.E.   345, 350, 403
McAuliffe, C.D.   18, 19, 33, 93, 220, 344, 363, 371, 384, 403
McCain, B.B.   466
McCully, M.E.   141
McEwan, E.H.   156
McFarland, W.N.   112
McIntyre, A.D.   77
McLachlan, J.   142
McLean, A.Y.   33
McManus, J.J.   190
Mead, J.W.   136
Mecklenburg, T.A.   221, 228
Mellen, I.   127
Mertens, E.W.   454
Meyer, P.A.   466
Michael, A.D.   129, 136
Miget, R.J.   412
Mighell, J.L.   143
Mills, E.R.   465
Milne, H.   167
Mironov, O.G.   93, 112, 149
Mix, M.C.   421, 431
Moles, D.A.   465
Moore, B.J.   384
Moore, L.J.   276
Moore, R.F.   453
Moore, S.F.   93, 113, 163, 173, 208, 220, 245
Morris, B.F.   33
Morrow, J.   149, 190
Moss, B.L.   142
Moulder, D.S.   245
Mueller, J.A.   136

Mullin, M.M.   304
Munday, J.C.   361
Murray, J.   77
Murray, S.P.   33
Myers, J.J.   361
Myers, M.S.   466
Myers, P.   412
Myren, R.T.   236

N

Naidu, A.S.   466
Natrella, M.G.   236
Nebert, S.W.   331
Neff, J.M.   33, 69, 93, 113, 182, 191, 208, 228, 236, 285, 305, 312
Nelson-Smith, A.   33, 136, 220, 245
Neumann, G.   361
Neushul, M., Jr.   136
Newcombe, C.L.   322
Newell, R.C.   113, 199
Newman, J.W.   412
Newman, M.W.   127, 431
Nicholson, N.L.   136
Norstrom, F.J.   77
North, W.J.   136
Notten, W.R.F.   58
Nuzzi, R.   136

O

Oesch, F.   46
Ogata, M.   69
O'Hara, S.C.M.   286
Olla, B.   113
Omori, T.   46
O'Neil, G.   454
Orloci, L.   259
Oro, J.   412
Orton, J.H.   173
Oshida, P.S.   173
Ottway, S.   93
Owens, E.G.   245

P

Palacas, J.G.   412
Parizek, J.   127
Parker, P.L.   412
Pasteels, J.J.   69
Patterson, M.M.   259
Pauling, L.   18
Payne, J.F.   331
Payne, R.F.   58
Peake, E.   18, 93
Pederson, M.G.   58, 331
Pequegnat, W.E.   466
Percy, J.A.   33, 93, 113, 192, 200, 236, 245
Percy, K.L.   431
Perry, J.J.   33
Pfitzenmeyer, H.T.   322
Phillips, O.M.   361

Philpot, R.M. 331
Pierce, S.K. 113
Pilpel, N. 420
Pohl, R.J. 58
Pollock, E.G. 142
Pople, J.A. 18
Potts, W.T.W. 113
Price, L.C. 18
Pringle, B.H. 312
Prosser, C.L. 77, 113, 322
Pulich, W.M. 142

R

Ralston, A.W. 18
Ramos, L.S. 385
Ray, S.M. 465
Raymond, R.L. 46
Reeburgh, W.S. 18
Reed, A. 167
Reed, W.E. 466
Regnier, Z.R. 33
Reichenbach-Klinke, H. 127
Reimer, A.A. 113
Reiner, A.M. 46
Reinert, R.E. 77, 453
Reish, D.J. 69, 168
Renzoni, A. 93, 149
Reynolds, J.A. 18
Rice, S.D. 33, 69, 78, 93, 113, 150, 221, 228, 236, 371, 465, 466
Riley, R.T. 421
Risebrough, R.W. 331
Rittinghaus, H. 163
Robertson, B. 34, 270, 275
Robertson, J.D. 191
Robisch, P.A. 143, 323
Roesijadi, G. 113
Rogers, M.A. 412a
Rohrback, B.G. 396
Romanoff, A.L. 163
Rossi, S.S. 34, 70, 93, 113, 173, 208, 285
Roubal, W.T. 58, 70, 114, 128
Rudd, W.M. 275
Ruivo, M. 18
Russel, G. 142
Rutzler, K. 136

S

Sabo, D.J. 128
Sackett, W.M. 373
Sahyun, M.R.V. 58
Sanborn, H.R. 58, 70, 94
Sanders, H.L. 70, 453
Sangalang, G.B. 128
Scarpelli, G. 331
Scarratt, D.J. 137, 322
Schaffer, R.L. 421
Schatzberg, P. 361
Scheffer, V.B. 156
Schell, D.M. 275
Scholander, P.F. 157
Schwartz, J.R. 275

Schwartzberg, H.G. 361
Schwiger, G. 128
Seal, H.L. 259
Sever, J.R. 412a
Shapiro, M.A. 453
Shaw, D.G. 8, 237, 312, 431
Shaw, D.J. 18
Shelton, T.B. 34
Shipton, J. 70
Short, J.W. 78, 94, 228, 466
Siegel, S. 259
Sims, P. 58
Singhal, R.L. 128
Sivadier, H.O. 361, 372
Slowey, F. 312
Smith, C.L. 34, 351, 361, 372
Smith, J.E. 34, 137, 260, 361
Smith, J.L. 442
Smith, M.A. 465
Smith, R.W. 260
Smith, T.G. 157
Snedecor, G.W. 237, 361
Sorlini, C. 46
Soto, C. 304
Spies, R.B. 466
Sport, M.C. 384
Sprague, J.G. 114
Spurr, A.R. 128, 150
Stainken, D.M. 313, 322, 344
Stalling, D.L. 331
Stamoudis, V.C. 332
Stanier, R.Y. 46
Stanton, R.H. 331
Steed, D.L. 114
Steele, R.L. 138
Stegeman, J.J. 70, 77, 181, 208, 237, 304, 312, 322, 344, 431, 453
Stenhagen, E. 344
Stephenson, W. 260
Stokes, G.G. 361
Stowe, H.D. 128
Strachan, A. 34
Straughan, D. 34, 137, 142, 173, 247, 260
Stroop, D.V. 361
Struhsaker, J.W. 94, 128, 150, 200, 220
Sutton, C. 18, 34
Sverdrup, H.U. 361
Swaisland, A.J. 58
Swanson, V.E. 412a
Swedmark, M. 237, 420
Swinnerton, J.W. 18, 384, 403
Szaro, R.C. 164

T

Tafanelli, R. 128
Tanford, C. 18
Taniuchi, H. 46
Tarzwell, C.M. 420
Tatem, H.E. 114, 201, 208
Taylor, T.L. 229
Teal, J.M. 70, 71, 77, 128
Teeson, D. 361
Templeton, W.L. 466
Thomas, M.L.H. 238, 245, 322

Thomas, P.E.   46
Thompson, G.   228
Thompson, R.J.   70
Thorade, H.   361
Thuer, M.   420
Tomczak, G.   361, 362
Traxler, R.W.   34
Treccani, V.   46
Trenholm, S.R.   421
Tripp, M.R.   174, 181

V

Vale, G.H.   137
Vanderhorst, J.R.   34, 94, 208, 403
Vandermeulen, J.H.   137, 304
Van Dorn, W.G.   362
Varanasi, U.   58, 128
Vaughan, B.E.   94, 322, 403
Verwey, E.J.W.   420
Vines, R.G.   362
Vishnevetskii, F.E.   128
Voyer, R.A.   128

W

Wade, T.L.   18
Wakeham, S.G.   412a
Walker, J.D.   34, 269
Walker, N.   46
Walker, S.H.   412a
Walne, P.R.   322
Waluga, D.   128
Ward, M.E.   312
Warner, J.S.   114, 150, 208, 260, 285, 344
Warner, R.E.   200
Watable, T.   59
Way, S.J.   466
Wedemeyer, G.   331
Wegner, E.H.   46
Weinstein, I.B.   46, 59
Wekell, M.M.   323
Wellings, S.R.   59, 466
Wells, P.G.   94, 114, 208, 220, 228
Westfall, A.   137
White, A.   59
Whittle, K.J.   70
Wiebe, P.H.   77
Wieser, W.   312
Wilcoxon, R.   173
Williams, A.B.   114
Williams, R.T.   331
Williams, W.T.   260
Wilson, R.D.   34, 114, 191
Winters, K.   34
Wise, S.A.   345
Witmer, F.E.   18
Wohlschlag, D.E.   114
Wolfe, D.A.   454
Woodruff, D.L.   276
Wormald, A.P.   34, 137
Wu, J.   362

Y

Yamada, K.   46
Yang, C.S.   59
Yoshida, T.   59
Youngblood, W.W.   412a, 453

Z

Zajic, J.E.   35
Zar, J.J.   322
Zarella, W.M.   384
Zeller, M.V.   322
Ziffer, H.   46
Zitko, V.   70, 322, 431, 453
Zobell, C.E.   35, 344, 412a, 420
Zöllner, N.   304
Zsolnay, A.   304, 441, 453

DATE DUE

HETERICK MEMORIAL LIBRARY
574.52636 F252   onuu
/ Fate and effects of petroleum hydrocarb
3 5111 00070 8465